THE

Nucleon–Nucleon Interaction and the Nuclear Many-Body Problem

Selected Papers of Gerald E Brown and T T S Kuo

THE

Nucleon–Nucleon Interaction and the Nuclear Many-Body Problem

Selected Papers of Gerald E Brown and T T S Kuo

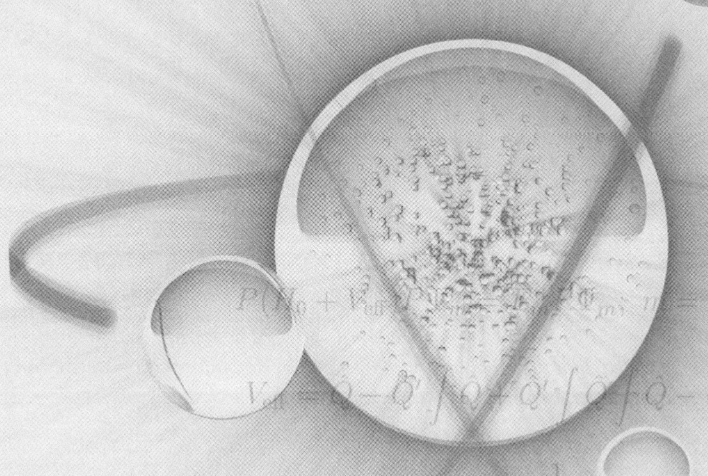

Editors

Gerald E Brown & Thomas T S Kuo
Stony Brook University, USA

Jeremy W Holt
Technische Universität München, Germany

Sabine Lee
University of Birmingham, UK

$$\frac{m^*}{m} \sim \frac{\langle \bar{q}q \rangle^*}{\langle \bar{q}q \rangle}$$

$$\sqrt{\frac{g_A}{g_A^*}}\frac{m_N^*}{m_N} = \frac{m_\sigma^*}{m_\sigma} = \frac{m_\rho^*}{m_\rho} = \frac{m_\omega^*}{m_\omega} = \frac{f_\pi^*}{f_\pi} = \Phi(n)$$

$$\frac{m^*}{m} \simeq 1 - 0.2\frac{n}{n_0}$$

World Scientific

NEW JERSEY • LONDON • SINGAPORE • BEIJING • SHANGHAI • HONG KONG • TAIPEI • CHENNAI

Published by

World Scientific Publishing Co. Pte. Ltd.
5 Toh Tuck Link, Singapore 596224
USA office: 27 Warren Street, Suite 401-402, Hackensack, NJ 07601
UK office: 57 Shelton Street, Covent Garden, London WC2H 9HE

British Library Cataloguing-in-Publication Data
A catalogue record for this book is available from the British Library.

THE NUCLEON-NUCLEON INTERACTION AND THE NUCLEAR MANY-BODY PROBLEM

ISBN-13 978-981-4289-28-3
ISBN-10 981-4289-28-0

Printed in Singapore by B & Jo Enterprise Pte Ltd

Preface

Why Now is a Good Time to Write About the Nucleon–Nucleon Interaction and the Nuclear Many Body Problem

Why do two old nuclear physicists, with the help of a junior colleague and a historian, now write about the nucleon–nucleon interaction to which they have devoted such a large portion of their research lives previously?

The immediate explanation is straightforward. The main problems at the level of meson exchange physics have been solved. We now have an effective nucleon–nucleon interaction V_{low-k}, pioneered in a renormalization group formalism by several of us at Stony Brook and our colleagues at Naples, which is nearly universally accepted as the unique low-momentum interaction that includes all experimental information to date.

Why does this make reconstructing the history of our understanding of the nucleon–nucleon interaction necessary or useful? There are several good reasons for engaging in a historical appreciation of the progression of research and the developments leading to our current knowledge in this subject area.

First, our understanding is based on a multi-step development in which a variety of different scientific insights and a wide range of physical and mathematical methodologies fed into each other. This is best appreciated by looking at the different "steps along the way", starting with the pioneering work of Brueckner and collaborators, which was just as necessary and important as the insightful, masterly improvements to Brueckner's approach by Hans Bethe and his students. The main achievement in the work of Brueckner and Bethe et al. was the "taming" of the hard core of the nucleon–nucleon potential, which has since been understood to result from the exchange of the ω-meson, a "heavy" photon. The off-shell effects which bedevilled Bethe's work that ended up in the 1963 Reference Spectrum Method were treated relatively accurately by introducing an energy gap between initial bound states and intermediate states. Kuo and Brown showed that this would be accurately handled by taking the intermediate states to be free; i.e. by just using Fourier components, as is now done in the effective theory resulting from the renormalization group formalism.

Well, one can say to the young people that this is "much ado about nothing". In fact, long ago, when Gerald E. Brown was Professor at Princeton, Murph Goldberger (turning on its head Winston Churchill's famous quote about the R.A.F. during the Battle of Britain) claimed in reference to the nuclear interaction that "never have so many contributed so little to so few." Admittedly, at the time it was hard going.

If we had a unique set of interactions, one for each angular momentum, spin and isospin channel, it could be argued that it would be justified to stop there. However, since Brueckner came on the scene, Bethe reorganized the theory, Kuo and Brown wrote their paper that prepared the effective theory by using the Scott–Moszkowski separation method, and chiral invariance hit the scene. Chiral invariance does not do anything for Yukawa's pion exchange, because the pion gets most of its mass from somewhere outside of the low-energy system, maybe by coupling to the Higgs boson. But the masses of the other mesons drop with increasing density, like

$$m_\rho^* \cong m_\rho(1 - 0.2n/n_0) \qquad \text{"Brown–Rho scaling"}$$

where n is the density and n_0 is nuclear matter saturation density. The change in masses of the scalar-σ and vector-ω mesons pretty much cancel each other in effects — the scalar exchange giving attraction and the vector repulsion. However, in the tensor force, the ρ-exchange "beats" against the pion exchange, the former cancelling more and more of the latter as the density increases. This decrease with density of the tensor force interaction has important effects:

(1) It is responsible for saturation in the nuclear many-body system.
(2) It converts an around hour-long carbon-14 lifetime from a superallowed transition in the Wigner $SU(4)$ for p-shell nuclei into an archeologically long 5,700-year transition.

"Brown–Rho scaling" is also important for neutron stars and may play an important role in turning them into black holes and for "cosmological natural selection". It must be admitted that the same effects could be given by three-body forces, but Brown–Rho scaling has a deep connection with chiral symmetry restoration. We shall review these facets in detail.

Undoubtedly, much more is to come, but we believe that now is a good time to summarize the interesting history of the nucleon–nucleon interaction.

CONTENTS

CHAPTER I
Introduction

1.1. Nuclear Matter: A Brief Overview

In the early 20th century physics moved from a macroscopic level down to atoms and molecules, and with this the nature of the discipline changed dramatically. Deterministic classical physics was replaced by a more probabilistic and often unintuitive or even counterintuitive quantum approach. As physical investigation moved to yet smaller entities, from atoms to their nuclei, the question arose whether the quantum mechanical principles that had been developed for the study of atoms would be applicable to the new and fast-developing field of nuclear physics. It seemed that many general principles of non-relativistic quantum theory could be applied at the nuclear level, but most of the calculations of nuclear properties were of a phenomenological, or at least semi-phenomenological kind. Instead of an ab initio approach, starting from first principles, scientists used a mixture of theoretical considerations and empirical data to develop models in a quest for an understanding of the nuclear realm.

In the years following the discovery of the neutron by James Chadwick in 1932, Dmitri Ivanenko and Werner Heisenberg independently proposed that the atomic nucleus consisted of protons and neutrons, now commonly referred to as nucleons.[1] This idea was soon commonly accepted, and scientists turned to a closer investigation of nuclear structure.

A striking feature of the binding energies of nuclei is the way in which, after the removal of electromagnetic effects, they follow a simple formula

$$\alpha A + \beta A^{2/3} + \gamma (N-Z)^2/A\,, \tag{1}$$

where A is the number of nucleons and the symmetry term at the end is small. From the absence of a term proportional to A^2, the number of pairs of nucleons, one can see that nuclear forces saturate. Various types of scattering experiments suggest that nuclei are roughly spherical and appear to have essentially the same density. The data are summarized in the expression called the Fermi model:

$$R = r_0 A^{1/3}\,, \tag{2}$$

[1] W. Heisenberg, 'Über den Bau der Atomkerne', *Z. Phys.* **77**, 1 (1932), *Z. Phys.* **78**, 156 (1932); *Z. Phys.* **80**, 587 (1933); D. Ivanenko, 'The neutron hypothesis', *Nature* **129**, 798 (1932); D. Ivanenko, 'Constitutive parts of atomic nuclei' *Comptes Rendus de l'Académie des Sciences de l'URSS* **1**, 50 (1933).

where R is the nuclear radius and r_0 is a constant fit to experiment. This indicates that nuclei have approximately constant density which is not dependent on their size and binding energies but dominated by volume and surface terms. Therefore it is possible to interpret α and r_o as constants relating to an infinite volume of nuclear matter.[2]

Among the first attempts to understand nuclear structure was the so-called liquid drop model. It was first conceived by George Gamow in 1929.[3] In the wake of Chadwick's discovery of the neutron, Heisenberg's seminal theory of the nucleus, and Ettore Majorana's work on nuclear exchange forces, Carl-Friedrich von Weizsäcker in his influential study of nuclear masses extended Gamow's model significantly in 1935.

The fact that the density and the binding energy per nucleon were approximately the same for all (stable) nuclei led to the comparison of the nucleus with a liquid drop, which similarly has a constant density, independent of the number of molecules. Using this analogy, Weizsäcker in 1935 developed his semi-empirical formula for the mass of a nucleus as a function of A (total number of nucleons) and Z (number of protons),[4] which became more widely used when Hans Bethe, in the first of his famous *Reviews of Modern Physics* articles reworked and simplified it.[5] The liquid drop model describes the nucleus as a classical fluid made up of neutrons and protons. It is envisaged to have a well-defined surface and the short-range forces, one attractive force holding the nucleons together and one repulsive force stopping the nucleons collapsing in. Systematic measurements of the binding energies of atomic nuclei, however, showed systematic deviations of observed values from those estimated using the liquid drop model.

Certain nuclei with special values for proton and neutron numbers (the so-called 'magic numbers') proved to be much more tightly bound than the liquid drop model predicted, and this suggested the existence of a shell structure within the nucleus. In 1949, Maria Goeppert Mayer[6] and Johannes Jensen[7] independently proposed an average potential which could reproduce the nuclear magic numbers. Thereby the shell structure in nuclei was theoretically established and it became a firm foundation of nuclear structure theory.[8]

The shell model treats the nucleons individually as opposed to treating the nucleus as a whole. The long-range repulsive Coulomb force and the strong short-range attractive nuclear force acting between nucleons are replaced by an average force. According to this model, the motion of each nucleon is governed by the average attractive force of all the other

[2]This matter is, however, immensely heavy. Hans Bethe, with Gerry Brown, estimated that a teaspoon full weighed as much as all the buildings in Manhattan.

[3]G. Gamow, 'Discussion on the structure of atomic nuclei', *Proc. Roy. Soc.* **A123**, 386 (1929), G. Gamow, 'Mass defect curve and nuclear constitution', *Proc. Roy. Soc.* **A126**, 632 (1929).

[4]C. F. von Weizsäcker, 'Zur Theorie der Kernmassen', *Z. Phys.* **96**, 431 (1935).

[5]H. A. Bethe and R. F. Bacher, 'Nuclear physics. A. Stationary states of nuclei', *Rev. Mod. Phys.* **8**, 165 (1937).

[6]M. Goeppert-Mayer and R. G. Sachs, 'On closed shells in nuclei', *Phys. Rev.* **235** (1948); M. Goeppert-Mayer, 'On closed shells in nuclei, II.', *Phys. Rev.* **75**, 1969 (1949).

[7]O. Haxel, J. H. D. Jensen and H. E. Suess, 'On the "Magic Numbers" in Nuclear Structure', *Phys. Rev.* **75**, 1766 (1949).

[8]See also M. Goeppert Mayer and J. H. D. Jensen, *Elementary Theory of Nuclear Shell Structure*, John Wiley & Sons, New York, 1955.

nucleons. The resulting discrete energy levels form 'shells', just as the orbits of electrons in atoms do. As nucleons are added to the nucleus, they drop into the lowest-energy shells permitted by the Pauli Principle, which requires each nucleon to have a unique set of quantum numbers to describe its motion.

A third nuclear model is the so-called optical model for nuclear reactions, used particularly for determining elastic scattering, total cross sections and transmission coefficients. The optical model is a quantum mechanical approach to the problem of scattering and absorption of particles impinging on a nucleus. Scattering of neutrons from nuclei is described by considering a plane wave in the potential of the nucleus, which comprises a real part and an imaginary part. This model is called the optical model since it resembles the theory of how light scatters when it enters a new medium.

Around the time of the development of the liquid drop model, Hideki Yukawa considered the question of the stability of the nucleus against break-up into its constituent nucleons, which could not be explained by the electromagnetic field of the protons or by gravitational forces which were negligible. Inspired by Heisenberg's 1932 nuclear theory and Enrico Fermi's 1934 theory of beta radioactivity, he attempted to find a unified picture of strong and weak interactions, and he suggested that a new field, a "meson field", could explain this phenomenon. In analogy with the photon in an electromagnetic field, he proposed a new kind of quantum, a field particle with finite mass that mediated the exchange forces within the nucleus.[9] The meson's finite mass was to ensure that the forces due to the meson field should have a finite range. From the size of the nucleus, which gave the range of the new interaction, Yukawa concluded that the mass of these conjectured particles (mesons) was about 200 electron masses. It was not until 1947, when Cecil Powell and his collaborators with the help of new experimental techniques, proved that there existed two distinct particles, both present in cosmic radiation, of which one was the previously unidentified, short-lived 'π-meson' (now called pion). It soon became clear that its properties were well-accounted for by Yukawa's theory.

It was hoped that Powell's discovery of the Yukawa meson would have been followed by the derivation of a potential governing the nucleon-nucleon interaction. The first attempt at calculating these interactions had been made by one of Heisenberg's students, Hans Euler, already in 1937. He calculated the properties of nuclear matter in second-order perturbation theory[10] working on the assumption that nucleons interacted via a two-body potential of Gaussian shape. He believed the force to be smooth at short distance and the comparative weakness of the force ensured that the contributions beyond the second order were small. But in order to achieve saturation, it was necessary for strong repulsion to occur in odd states, and this proved incompatible with later scattering experiments.

In Euler's calculation the second-order term from iterating the tensor force, although operating only in triplet states was very large and we will see later in this chapter that it basically gave the difference between the triplet attraction, which binds the deuteron

[9]H. Yukawa, 'On the Interaction of Elementary Particles', *Proc. Physico-Math. Soc. Japan* **17**, 48 (1935); See also H. Yukawa, 'Models and Methods in the Meson Theory', *Rev. Mod. Phys.* **21**, 474 (1949).

[10]H. Euler, 'Über die Art der Wechselwirkung in den schweren Atomkernen', *Z. Phys.* **105**, 553 (1937).

and the singlet attraction in which there is no tensor force and no bound state. Therefore investigators were worried that higher-order effects would also be large. However, the first-order term in the tensor force includes tensor-like correlations in the wave function, and the additional tensor interaction takes the wave function essentially to a (large) constant times the initial state. In other words, the coefficients pile up coherently for the second-order term, which acts as a triplet scalar interaction, and higher order terms are small. Gerry Brown, Gottfried Schappert and Chun Wa Wong[11] showed by Monte Carlo simulation that there are no coherent combinations in higher order which would contribute appreciably; thus, as far as the tensor force is concerned, the Euler second-order direct term is all that is left. It comes from the second-order pion and rho meson exchanges, as we shall develop, with opposite sign.

The properties of nuclear matter are known only for the saturation density $n_0 = 0.16$ nucleons/fm^3. However, the nuclear equation of state has been used in countless calculations of dense matter such as that in neutron stars, etc. None of these have, in a fundamental sense, described nuclear matter saturation at the one density n_0 in which we know its properties, although some calculations with three-body forces enforce essentially the same saturation as obtained by Brown–Rho scaling which we shall discuss in Chapter IV. We shall also discuss there that Brown–Rho scaling is essential in transforming the ^{14}C beta decay from a short half-life superallowed beta decay of some hours into an archeologically long $\tau_{1/2} \simeq 5370$ yr transition that makes carbon dating so effective.

The singular nature of the nuclear potential at short distances, i.e. the very strong repulsion at short distances, the so-called hard core, was described by Robert Jastrow in 1951.[12] The extreme case of an infinitely hard core would mean that all the potential matrix elements would be infinite in the uncorrelated wave functions. A possible solution of this problem was to build short-range correlations that would prevent the nucleons from being too close to each other.

A special many-body methodology needed to be developed to deal with the problem of the strongly repulsive core. Some of the features of the subsequent development of this many-body theory were already present in an earlier work by Eugene Feenberg on the structure of perturbation theory[13] and in the studies by Kenneth Watson of multiple scattering of a particle in a many-body medium.[14] Watson recognized the simplification resulting from the use of selective summations leading to vertex operators and modified particle propagators, although he never considered the full many-body problem.

Keith Brueckner used the algebra developed by Watson for studying the multi-scattering of a fast particle through an atomic nucleus. He modified it for his study of a particle in a

[11] G. E. Brown, G. T. Schappert, and C. W. Wong, 'Binding energy of nuclear matter', *Nucl. Phys.* **56**, 191 (1964).

[12] R. Jastrow, 'On the nucleon-nucleon interaction', *Phys. Rev.* **81**, 165 (1951).

[13] E. Feenberg, 'A note on perturbation theory', *Phys. Rev.* **74**, 206 (1948); E. Feenberg, 'Theory of scattering processes', *Phys. Rev.* **74**, 664 (1948).

[14] K. M. Watson, 'Multiple scattering and the many-body problem: applications to photomeson production in complex nuclei', *Phys. Rev.* **89**, 575 (1953); and N. C. Francis and K. M. Watson, 'The elastic scattering of particles by atomic nuclei', *Phys. Rev.* **92**, 291 (1953).

bound state and it was applied to the problem of nuclear saturation by Brueckner, Levinson and Mahmoud.[15] Their method depended on a treatment of the coherent particle motion which was exact in the limit of very many scatterers, and treated the incoherent motion as a perturbation. In this case the many-body potential energy could be expressed in terms of the low-energy scattering amplitudes. They applied the method to the two-body potentials given by pseudoscalar meson theory when the effects of nucleon pair formation were assumed to be small. In this approximation the many-body forces of the theory were negligible.

In a second paper, Brueckner extended the method developed for the treatment of the problem of nuclear saturation to the case of tensor forces.[16] The general result obtained expresses the many-body potential energy as a function of the triplet and singlet scattering phase shifts. One consequence was that the tensor force, which averaged to zero if the Born approximation was used to evaluate the scattering, gave a very sizable contribution to the potential energy.

In a third paper completing the series, Brueckner discussed the details of the structure of the nucleus.[17] He examined the characteristics of particle motion in the nuclear medium and he discussed the origin of the strong dependence of the potential energy on the nucleon momentum. Furthermore, he provided an equivalent formulation in which a uniform and constant potential was assumed but the nucleon moved with markedly reduced mass. The determination of the potential was shown to lead to a self-consistency equation which was to some extent similar to that appearing in the Hartree method of self-consistent fields.

Following these three papers on two-body forces and nuclear saturation, Brueckner and Levinson turned to the mathematical basis of their methods[18] and Richard Eden and Norman Francis considered how Brueckner's method related to the general theory of nuclear models. They described a framework for a unified theory of nuclear structure in which the wave functions for different nuclear models were obtained by transformations on the actual nuclear wave function. This formulation provided a basis for explaining the success of weak-coupling models of the nucleus and showed that these were not in conflict with the assumption that nucleons had very strong mutual interactions. Brueckner, Eden and Francis then moved on to relate their methods to correlations in the nucleus and their effect on high-energy nuclear reactions,[19] the nuclear shell model[20] and to the optical model.[21]

---

[15]K. A. Brueckner, C. A. Levinson and H. M. Mahmoud, 'Two-body forces and nuclear saturation. I. Central forces', *Phys. Rev.* **95**, 217 (1954).

[16]K. A. Brueckner, 'Nuclear saturation and two-body forces. II. Tensor forces', *Phys. Rev.* **96**, 508 (1954).

[17]K. A. Brueckner, 'Two-body forces and nuclear saturation. III. Details of the Structure of the Nucleus', *Phys. Rev.* **97**, 1352 (1955).

[18]K. A. Brueckner and C. A. Levinson, 'Approximate reduction of the many-body problem for strongly interacting particles to a problem of self-consistent fields', *Phys. Rev.* **97**, 1344 (1955).

[19]K. A. Brueckner, R. J. Eden, and N. C. Francis, 'High-energy reactions and the evidence for correlations in the nuclear ground-state wave function', *Phys. Rev.* **98**, 1445 (1955).

[20]K. A. Brueckner, R. J. Eden and N. C. Francis, 'Nuclear energy level fine structure and configuration mixing', *Phys. Rev.* **99**, 76 (1955).

[21]K. A. Brueckner, R. J. Eden and N. C. Francis, 'Theory of neutron reactions with nuclei at low energy', *Phys. Rev.* **100**, 891 (1955).

Keith Brueckner had recognized that the strong short-range interactions, such as the infinite hard core, would scatter nucleons to momenta well above the Fermi sea. Therefore the exclusion principle would have very little effect and could be treated as a small correction. In analogy to Watson's T-matrix

$$T = V + V\frac{1}{E - H_0}V + V\frac{1}{E - H_0}V\frac{1}{E - H_0}V + \cdots$$

with V the two-body potential, E the unperturbed energy and H_0 the unperturbed Hamiltonian, containing only the kinetic energy, he defined an operator G from the potential V. This G-Matrix obeys the well-known equation

$$G = V + V\frac{Q}{E - H_0}G\,,$$

where Q is the Pauli exclusion operator ensuring the intermediate states being above the Fermi surface. He showed that one could carry out a perturbation theory calculation in a Fermi gas treating G as the effective potential.

Brueckner's pioneering approach of solving the two-body scattering problem in the nuclear medium consisted of rearranging perturbation theory in such a way that the contribution to the total energy at each order was proportional to the number of particles. Energies per particle were manifestly finite. But its formulation was ambiguous, and it was not readily accepted by nuclear physicists, largely as a result of the very formal nature of the central proof of the theory. The theoretical structure needed a more concrete machinery to make it work. This was provided by Jeffrey Goldstone and Hans Bethe at Cambridge and Cornell.

Brueckner has rightly been attributed with 'taming' the short-range interactions between two nucleons; Hans Bethe, the master of organization and communication, set out 'to tame' the body of theory that Watson, Brueckner and their colleagues had created by recreating and reformulating Brueckner's work. As he stated in the introduction to his first important paper on the Nuclear Many-Body Problem,[22] while the success of the Brueckner model had been beyond question for many years, a theoretical basis for it had been lacking. He pointed out "it is well established that the forces between two nucleons are of short range, and of very great strength, and possess exchange character and probably repulsive cores. It has been very difficult to see how such forces could lead to any over-all potential and thus to well-defined states for the individual nucleons." Further explaining this point he emphasized that Brueckner had developed a powerful mathematical method for calculating the nuclear energy levels using a self-consistent field method, even though the forces are of short range. But the definitions on which the various concepts had been based had remained unsatisfactory.

Bethe's approach, based on a diagrammatic expansion of perturbation theory in a series ordered by the number of interacting particles, facilitated the understanding of Brueckner's work significantly. Bethe gave a self-contained and largely new description of Brueckner's method for studying the nucleus as a system of strongly interacting particles with the aim

[22]H. A. Bethe, 'Nuclear many-body problem', *Phys. Rev.* **103**, 1353 (1956).

of developing a method that was applicable to a nucleus of finite size while at the same time eliminating any ambiguities of interpretation and approximations required for computation. Thus Bethe, using the work of Brueckner and collaborators, produced an orderly formalism in which the evaluation of the two-body operator G would form the basis for calculating the shell model potential.

One of Bethe's students at Cambridge, Jeffrey Goldstone, by means of perturbation theoretical methods, established the so-called linked cluster expansion.[23] Using Feynman graphs to enumerate the terms of the perturbation series, and describing the states in a way that was equivalent to treating the independent-particle Fermi sea as a 'vacuum state', he proved the 'linked cluster' theorem for the non-degenerate case. About a decade later, Morita,[24] Brandow,[25] Johnson and Baranger,[26] Kuo, Lee and Ratcliff (KLR)[27] and others extended the linked cluster theorem to the degenerate case. The linked cluster theorems for the degenerate case are usually referred to as the folded-diagram methods. The KLR formalism has been widely applied to calculations of finite nuclei using realistic nucleon-nucleon interactions.[28]

As in other scientific contexts, analytic solutions to specific problems were a source of additional insight for Hans Bethe. Therefore, with Jeffrey Goldstone he went on to investigate the evaluation of G for the extreme infinite-height hard core potential. Bethe and Goldstone[29] defined a spatial wave function for two nucleons and derived the Bethe–Goldstone integro-differential equation for this function. The calculation of the Brueckner G-matrix is rather complicated. As we shall discuss in Chap. III, a simple and ingenious separation method for the calculation of G has been developed by Moszkowski and Scott.[30] In this method, the G-matrix in the $^1S_0$ and $^3S_1$ channels can be well approximated by a long-range potential $V_{\rm long}$ obtained by cancelling the repulsive core against part of the attractive well up to a separation distance $d \simeq 1.2$ fm. For $r < d$, $V_{\rm long} = 0$ and for $r > d$, $V_{\rm long} = V_{NN}$. We shall also discuss there that $V_{\rm long}$ is qualitatively similar to the low-momentum interaction V_{low-k}.

Starting from realistic NN potentials, the Brueckner–Bethe–Goldstone theory, which to first order is referred to as the Brueckner-Hartree-Fock (BHF) method, has been extensively applied to symmetric nuclear matter. However, the binding energy per particle BE/A and saturation density n_0 given by such calculations are all off the empirical value or $BE/A \simeq 16$

[23] J. Goldstone, 'Derivation of the Brueckner many-body theory', *Proc. Roy. Soc.* **A239**, 267 (1957).

[24] T. Morita, 'Perturbation theory for degenerate problems of many-fermion systems', *Prog. Theor. Phys.* **29**, 351 (1963).

[25] B. H. Brandow, 'Linked-cluster expansions for the nuclear many-body problem', *Rev. Mod. Phys.* **39** 771 (1967).

[26] M. B. Johnson and M. Baranger, 'Folded diagrams', Ann. Phys. (N.Y.) **62**, 172 (1971).

[27] T. T. S. Kuo, S. Y. Lee and K. F. Ratcliff, 'A folded-diagram expansion of the model-space effective Hamiltonian', *Nucl. Phys.* **A176**, 65 (1971).

[28] See Chap. III.

[29] H. A. Bethe and J. Goldstone, 'Effect of a repulsive core in the theory of complex nuclei', *Proc. Roy. Soc.* **A238**, 551 (1957).

[30] S. A. Moszkowski and B. L. Scott, 'Nuclear forces and the properties of nuclear matter', *Ann. Phys.* **11**, 65 (1960).

MeV and $n_0 = 0.16$ fm^{-3}. In fact they all lie on a so-called Coester band,[31] none reproducing the empirical BE/A and n_0 values simultaneously. Much effort has been devoted to improve the situation. In BHF, one includes only the lowest-order G-matrix diagram. As we shall discuss in Chap. III, a ring-diagram extension of the BHF method has been developed.[32] In this method the particle-particle hole-hole ring diagrams are summed to all orders. The ring-diagram results are an improvement over the BHF ones, but the obtained BE/A and n_0 are still significantly larger than the empirical values. There are indications that the inclusion of the Brown–Rho scaling, to be discussed in Chapter IV, may play an important role in reproducing the nuclear matter saturation properties.[33]

1.2. Development of Kuo–Brown Effective Interactions

"One of the authors, Gerry Brown, arrived at Princeton in early September, 1964. The next morning, as he came to the Palmer Physics Laboratory, Eugene Wigner, who just preceded him, opened the door for him. (It was a real contest to get ahead of Eugene and open the door for him which very few succeeded in doing.) Eugene asked Gerry, as we went into the building, what he planned to work on. "I plan to work out the nucleon–nucleon interaction in nuclei." Eugene said that it would take someone cleverer than him, to which Gerry replied that they probably disagreed what it meant to "work out". Gerry wanted to achieve a working knowledge, sufficiently good to be able to work out problems in nuclear physics. What follows in this book is the story of what Gerry and Tom Kuo, with considerable help from their students, collaborators and friends, put together."

Gerald E. Brown

Around the time of Gerry's arrival at Princeton in 1964, there were two main streams of activity. One was the Brueckner theory, which we have outlined in the preceeding section. Gerry called this a "realistic" approach because it dealt with the strongly repulsive hard core potential, which was later understood as having come from the strongly repulsive potential from ω-meson exchange, the ω being a massive photon. The other was the application of the Brueckner theory to finite nuclei. Hans Bethe had spent nearly two decades of his life organizing the Brueckner theory so that it would be amenable to calculations in finite systems such as nuclei. Brueckner theory had begun from many-body scattering theory where the effect of a vertical core was simply to keep the wave function excluded from the region occupied by the core.

The major work on the nucleus as a many body system, on the structure of the nucleus and on the spectra, nature of excited levels, etc. was led by Aage Bohr and Ben Mottelson

[31] F. Coester, S. Cohen, B. D. Day and C. M. Vincent, 'Variation in nuclear-matter binding energies with phase-shift-equivalent two-body potentials', *Phys. Rev. C* **1**, 769 (1970).

[32] H. Q. Song, S. D. Yang and T. T. S. Kuo, 'Infinite order summation of particle-particle ring diagrams in a model-space approach for nuclear matter', *Nucl. Phys.* **A462**, 491 (1987).

[33] L. W. Siu, J. W. Holt, T. T. S. Kuo and G. E. Brown, 'Low-momentum NN interactions and all-order summation of ring diagrams in symmetric nuclear matter', *Phys. Rev.* **C79**, 0540004 (2009).

in Copenhagen in the 1960's, the time of interest here.[34] We shall not treat their work in detail but will show how two-body interactions following from meson exchange give roughly equivalent descriptions. The various processes involving the mesonic interactions with each other and the way in which mesons fit into symmetry schemes are of great interest in themselves. Furthermore, the meson-focused approach helps to elucidate how nuclear physics relates to many modern developments in particle physics and other subdisciplines of physics.

Many years were spent in using the G-matrix to tame the infinitely repulsive interactions used in nuclear matter problems. It was natural to try it in finite systems, and some of this effort has been outlined above. A very important effort applied to finite nuclei was carried out by Dawson, Talmi and Walecka[35] who solved the Bethe–Goldstone equation not only for the ground state but also for excited states and then looked at the spectra of ^{18}O.

Igal Talmi[36] systematized the properties of subsystems of particles in partly filled shells of the shell model; e.g. the calcium isotopes where n particles are in the $f_{7/2}$ orbit. We would call these j^n where $j = 7/2$. He had an extremely simple formula for the energy of the partly filled shell

$$E(j^n, g.s.) = Cn + \frac{1}{2}n(n-1)\alpha + \left[\frac{1}{2}n\right]\beta\,, \tag{3}$$

where $[\frac{1}{2}n]$ is the step function which is equal to $\frac{1}{2}n$ if n is even and $\frac{1}{2}(n-1)$, if n is odd. The coefficients α and β are given by

$$\alpha = \frac{2(j+1)\overline{V}_2 - V_0}{2j+1}, \qquad \beta = \frac{2(j+1)}{2j+1}(V_0 - \overline{V}_2)\,, \tag{4}$$

$$\overline{V}_2 = \frac{\sum\limits_{J=\text{even}} (2J+1)V_J}{\sum\limits_{J=\text{even}} (2J+1)}\,, \tag{5}$$

$$V_J = (j^2 J|G|j^2 J)\,. \tag{6}$$

Talmi realized that any G-matrix model like that of Dawson, Talmi and Walecka mentioned earlier would be such that the force between the nucleons in the j^n configuration would be attractive such that α would be negative; i.e., that in the quadratic term there would be attraction between the nucleons. However, in the shell model nuclei he had looked at α was positive. This argument disabled essentially all of the models using the G-matrix as interaction to date, and spurred Kuo and Brown to go on to the next higher order.

[34]The work is summarized in Aage Bohr and Ben R. Mottelson, *Nuclear Structure*, (Vol. I Single Particle Motion; Vol. II Nuclear Deformations), W. A. Benjamin, New York, 1969.

[35]J. F. Dawson, I. Talmi and J. D. Walecka, 'Calculation of the level spectrum of O^{18} from the free two-nucleon potential', *Ann. Phys.* **18**, 339 (1962).

[36]I. Talmi, 'Effective interactions and coupling schemes in nuclei', *Rev. Mod. Phys.* **34**, 704 (1962).

We shall describe the Kuo–Brown calculations in Chapter II. Most of nuclear structure calculations in the 1960s had been carried out in Copenhagen under the guidance of Aage Bohr and Ben Mottelson[37] using only two types of effective interactions: (i) the pairing force and (ii) the quadrupole–quadrupole interactions

$$V_{ij} = -k \sum_{i.j} r_i^2 r_j^2 Y_2^m(\theta_i, \phi_i) Y_2^{-m}(\theta_j, \phi_j)(-1)^m$$

which were introduced by Elliott.[38]

Polarization in quantum mechanics is achieved in the nucleus by lifting a particle from a filled nuclear state to an unfilled one. Doing this with an operator having no angular dependence $P_0(\cos\theta)$ involves no change of shape, just an increase in size. With a $P_1(\cos\theta)$ it involves a small translation, which doesn't change shape, and comes out as a spurious mode. Thus a quadrupole excitation with $P_2(\cos\theta)$ gives the most favorable shape, and one expects two neighboring nucleons to interact importantly via this mode of core polarization. In Copenhagen a tremendous amount of work had been done treating nuclei between two closed shells with the quadrupole–quadrupole interaction.

George Bertsch[39] carried out the first microscopic calculation of the core polarization process and found that it is very important for the effective nucleon-nucleon interaction in nuclei. The calculation employed the Kallio–Kolltveit potential[40], which we viewed at the time as a realistic nucleon–nucleon interaction (in spite of it not having a spin-orbit or tensor force, it did give the correct scattering length and effective range). It is a hard-core, spin dependent (spin-singlet 's' and triplet 't') interaction of the form

$$V_i(r) = \begin{cases} \infty & \text{for } r \leq 0.4 \text{ fm} \\ -A_i e^{-\alpha_i(r-0.4\text{fm})} & \text{for } r > 0.4 \text{ fm} \end{cases} \quad \text{for } i = s, t, \tag{7}$$

where

$$A_s = 330.8 \text{ MeV}, \quad \alpha_s = 2.4021 \text{ fm}^{-1}$$
$$A_t = 475.0 \text{ MeV}, \quad \alpha_t = 2.5214 \text{ fm}^{-1}.$$

The core-polarization contribution was evaluated with second-order perturbation theory, as given by the core-polarization diagram of Fig. 1, where each vertex is a G-matrix interaction derived from the above potential. The Moszkowski–Scott separation method, which we shall discuss in Chapter III, was employed for calculating the G matrix.

[37] A. Bohr and B. R. Mottelson, *Nuclear Structure*, (Vol. I Single Particle Motion; Vol. II Nuclear Deformations), New York: W. A. Benjamin, 1969.

[38] J. P. Elliott, 'Collective motion in the nuclear shell model. I. Classification schemes for states of mixed configurations', *Proc. Roy. Soc.* **A245**, 128 (1958); J. P. Elliott, 'Collective motion in the nuclear shell model. II. The introduction of intrinsic wave functions", *Proc. Roy. Soc.* **A245**, 562 (1958).

[39] G. F. Bertsch, 'Role of core polarization in two-body interaction', *Nucl. Phys.* **74**, 234 (1965).

[40] A. Kallio and K. Kolltveit, 'An application of the separation method in shell-model calculation', *Nucl. Phys.* **53**, 87 (1964).

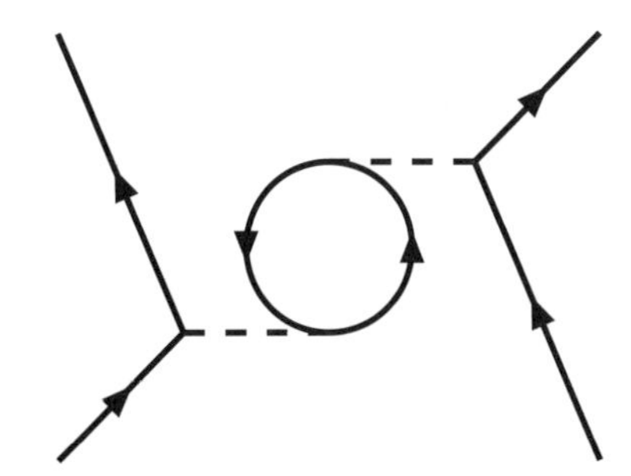

Fig. 1. The core polarization process calculated by Bertsch.

A more extensive calculation for the core polarization process was later performed by Kuo and Brown.[41] In their calculation, they employed a more realistic nucleon-nucleon interaction, the Hamada–Johnston potential[42] which was the 'best' nucleon-nucleon interactions at that time. The effects from the core-polarization process were found to be very important and desirable in both the Bertsch and Kuo–Brown calculations. Of particular interest is that the sign problem concerning the Talmi coefficient α, which was mentioned a little earlier, was largely resolved: its sign was negative when the effective interaction was given by the G interaction alone, but it became positive when G and core polarization were both included. In fact, as we shall discuss in Chapter III, the Kuo–Brown interactions turned out to be remarkably successful for a wide range of nuclei in both the (sd) and (pf) shells.[43] The rather successful results of the above initial calculations were 'very encouraging', and have led to many questions for further study.

The concept of Model Space is very useful in the theory of the nuclear shell-model. In treating nuclear many-body problems, one usually reduces the "full" many-body problem to a much smaller and more manageable one, often referred to as the model-space problem. Many-body problems are very difficult to solve, as we are all aware of, when A, the number of particles in the system, is large. In fact a many-body problem with A=3 is already a very hard problem, and in the real world, A is usually much larger. For example, A=18 when the nucleus ^{18}O is treated as a 18-nucleon problem. The wave function Ψ for the A=18 system is clearly very complicated, but to describe the low-energy properties of this nucleus, it may not be necessary to consider Ψ in its full glory; it may be sufficient to include just some low-energy parts of them. This is actually the approach commonly used in the shell-model description of ^{18}O, where this nucleus is treated as composed of an inert ^{16}O core with two

[41]G. E. Brown and T. T. S. Kuo, 'Structure of finite nuclei and the free nucleon-nucleon interaction–General discussion of the effective force', *Nucl. Phys.* **A92**, 481 (1967); T. T. S. Kuo and G. E. Brown, 'Structure of finite nuclei and the free nucleon-nucleon interaction–An application to ^{18}O and ^{18}F', *Nucl. Phys.* **85**, 40 (1966).

[42]T. Hamada and I. D. Johnston, 'A potential model representation of two-nucleon data below 315 MeV', *Nucl. Phys.* **34**, 382 (1962).

[43]T. T. S. Kuo and G. E. Brown, 'Reaction matrix elements for the 0f-1p shell nuclei', *Nucl. Phys.* **A114** (1968) 241.

valence neutrons in the 0d1s shell, corresponding to using a model space of

$$P = \sum_{2p'} |2p'0h\rangle\langle 2p'0h| \,, \tag{8}$$

where $2p'$ indicates the restriction that the two neutrons are confined in the $0d1s$ shell. The secular equation used in the shell model is of the form

$$PH_{\text{eff}}P\Psi_m = E_m P\Psi_m; \quad m = 1, d \tag{9}$$

with $H_{\text{eff}} = H_0 + V_{\text{eff}}$, where V_{eff} is the effective interaction and H_0 denotes the single-particle Hamiltonian. The dimension of the model space is labeled d.

The above model-space many-body problem is much simpler than the original full space one. But 'there is no free lunch'; in this model space approach there is the difficult task of deriving the model space effective interaction V_{eff}. Before the Kuo–Brown interactions, V_{eff} was generally determined empirically by fitting certain experimental data. That the Kuo–Brown interactions have worked well is an indication that V_{eff} can be microscopically derived from the free nucleon-nucleon interactions.

In Chap. II, we shall describe a folded-diagram theory for deriving the model-space effective interaction starting from the free NN interaction. The Kuo–Brown interactions have indicated that V_{eff} is mainly given by the sum of G and the second-order core polarization diagram often referred to as G_{3p1h}. But they are only a low-order approximation for the effective interaction V_{eff} between a pair of nucleons inside a nucleus. Using this folded diagram framework, we shall study the contribution to V_{eff} from higher-order diagrams. We shall discuss in Chap. III that the second-order results for V_{eff} and the corresponding all-order Kirson–Babu–Brown results are in fact quite similar. There are a number of different realistic nucleon-nucleon potential models, creating the uncertainty about which of them should be employed in nuclear structure calculations. This question will also be studied in Chapter III, and we shall show how one can derive a nearly universal low-momentum interaction V_{low-k} from the various NN potential models.

1.3. Towards a Unique Low-Momentum Nucleon–Nucleon Interaction

A fundamental problem in nuclear physics has long been the nucleon-nucleon (NN) potential V_{NN}. There have been a number of successful models for V_{NN}, such as the CD-Bonn,[44] Argonne,[45] Nijmegen[46] and Idaho[47] potentials. A common feature of these potentials is that they all can reproduce the empirical deuteron properties and low-energy phase shifts

[44]R. Machleidt, 'High-precision, charge-dependent Bonn nucleon-nucleon potential', *Phys. Rev. C* **63** (2001) 024001.

[45]R. B. Wiringa, V. G. J. Stoks and R. Schiavilla, 'Accurate nucleon-nucleon potential with charge-independence breaking', *Phys. Rev. C* **51**, 38 (1995).

[46]V. G. J. Stoks, R. A. M. Klomp, C. P. F. Terheggen and J. J. de Swart, 'Construction of high-quality NN potential models', *Phys. Rev. C* **49**, 2950 (1994).

[47]D. R. Entem, R. Machleidt, 'Accurate charge-dependent nucleon-nucleon potential at fourth order of chiral perturbation theory', *Phys. Rev. C* **68**, 041001 (2003).

very accurately. But they are, however, quite different from each other (see Chapter III). Certainly we would like to have a "unique" NN potential. (The gravitational potential $V(r) = -Gm_1m_2/r_{12}$ is unique as is the Coulomb potential $V(r) = kq_1q_2/r_{12}$.) In principle, the nonrelativistic NN potential should also be so. If there does exist one unique NN potential, then we will be confronted with the difficult question in deciding which, if any, of the existing potential models is the correct one. A partial answer to this question may be the low-momentum nucleon-nucleon potential V_{low-k} which was first developed, about 10 years ago, at Stony Brook (USA) and Napoli (Italy).[48–53]

An initial purpose for developing the V_{low-k} interaction was to have an energy-independent effective interaction which is more convenient for nuclear many-body calculations than the well-known Brueckner G-matrix. Because of its strong short-range repulsions, the bare interaction V_{NN} is not suitable for being directly used in perturbative calculations; it first needs to be 'tamed' into a smooth potential. The G-matrix is such a tamed interaction, and has been widely used for many many years. But G is not convenient to use, mainly because of its off-shell energy dependence. For example, to evaluate the matrix element $\langle k_1k_2|G(\omega)|k_3k_4\rangle$ for a certain vertex in a diagram we need to know the energy variable ω. But knowing the external indices (k_1, k_2, k_3, k_4) alone is in general not enough to determine ω; we need to have also information about what other particles in the diagram are doing. This makes the calculation using the G-matrix rather complicated, particularly for high-order diagrams. The V_{low-k} interaction to be described later does not have this off-shell energy dependence of the G-matrix.

The V_{low-k} project was started in about 1997 when Kuo, Coraggio and Bogner were visiting Arturo Polls and Angela Ramos in Barcelona. The Brueckner G-matrix had been the pillar in nuclear many body problems for 50 years or so. Maybe it was time to try something different! We thought about getting an energy-independent low-momentum interaction by way of a folded-diagram method, but were not at all sure how it would work. After much trial and error, we finally adopted the following renormalization (RG) method, namely deriving the V_{low-k} interaction by integrating out the high-momentum $(k > \Lambda)$ components of V_{NN}, Λ being a decimation scale. Certain low-energy physics, such as phase shifts, should be preserved by the integrating-out process. Thus, we have used the following

[48]S. K. Bogner, T. T. S. Kuo and L. Coraggio, 'Low momentum nucleon-nucleon potentials with half-on-shell T-matrix equivalence', *Nucl. Phys.* **A684**, 432c (2001).

[49]T. T. S. Kuo, S. K. Bogner, and L. Coraggio, 'A new theory of shell model effective interactions', *Nucl. Phys.* **A704**, 107c (2002).

[50]S. K. Bogner, T. T. S. Kuo, L. Coraggio, A. Covello and N. Itaco, 'Low-momentum nucleon-nucleon potential and shell model effective interaction', *Phys. Rev. C* **65**, 051301(R) (2002).

[51]T. T. S. Kuo, S. K. Bogner, L. Coraggio, A. Covello, and N. Itaco, 'Realistic low-momentum nucleon-nucleon potential', in *Challenges of Nuclear Structure* (Proceedings of the 7th International Seminar on Nuclear Structure; Maiori, Italy, May 27-31, 2001), ed. by A. Covello, p. 129, World Scientific Pub. Co. (2002).

[52]A. Schwenk, G. E. Brown, and B. Friman, 'Low-momentum nucleon-nucleon interaction and Fermi liquid theory', *Nucl. Phys.* **A703**, 191 (2002).

[53]S. K. Bogner, T. T. S. Kuo, and A. Schwenk, 'Model-independent low momentum nucleon interaction from phase shift equivalence', *Phys. Rep.* **386**, 1 (2003).

T-matrix equivalence formalism. We start from the half-on-shell T-matrix

$$T(k',k,k^2) = V_{NN}(k',k) + \mathcal{P}\int_0^\infty q^2 dq \frac{V_{NN}(k',q)T(q,k,k^2)}{k^2-q^2}\,, \tag{10}$$

where $\mathcal{P}$ denotes the principal value integration. We then define an effective low-momentum T-matrix by

$$T_{low-k}(p',p,p^2) = V_{low-k}(p',p) + \mathcal{P}\int_0^\Lambda q^2 dq \frac{V_{low-k}(p',q)T_{low-k}(q,p,p^2)}{p^2-q^2}\,, \tag{11}$$

noting that the integration limit is Λ. To preserve phase shifts, we require the following half-on-shell T-matrix equivalence:

$$T(p',p,p^2) = T_{low-k}(p',p,p^2);\ p',p \leq \Lambda\,, \tag{12}$$

which ensures that the low energy ($E_{lab} \leq 2\Lambda^2\hbar^2/m$) phase shifts of V_{NN} are preserved by V_{low-k}. The above equations define the effective low-momentum interaction V_{low-k}.

How to solve the above coupled equations was in fact a rather difficult task. It is a special inverse scattering problem, determining the V_{low-k} interaction backwards from the half-on-shell T-matrix. Also V_{low-k} has a special feature that it is nonvanishing only within the momentum model space $k < \Lambda$. As discussed in Chapter III, a folded-diagram solution for the above equations can be obtained. It is of the form

$$V_{low-k}(p',p) = \langle p'|\hat{Q} - \hat{Q}\int\hat{Q} + \hat{Q}\int\hat{Q}\int\hat{Q} - \cdots|p\rangle\,, \tag{13}$$

with the $\hat{Q}$-box given by

$$\hat{Q}(k',k) = V_{NN}(k'k) + \left\langle k'\left|V_{NN}\frac{P_\Lambda}{k^2-H_0}V_{NN}\right|k\right\rangle\,, \tag{14}$$

where each intergral sign represents a 'fold', H_0 is the kinetic energy operator and P_Λ denotes that the intermediate states must have momenta greater than Λ.

The above solution for V_{low-k} can be rewritten as $V_{low-k} = V_{NN}(1+\Omega(0,-\infty))$ where Ω is the wave operator which satisfies $\Omega P\Psi = Q\Psi$ where Ψ is the full-space eigenstate and P is the model space projection operator ($Q = 1 - P$). The wave operator can be conveniently calculated using the Lee–Suzuki iteration method.[54] Clearly V_{low-k} as given above is not Hermitian. There are a number of ways to transform it into a Hermitian interaction, as discussed by Holt et al.[55]

A remarkable feature of V_{low-k} is its near uniqueness. As discussed in Chap. III, there are a number of high-precision models for the V_{NN} potential. Although they all fit low-energy NN phase shifts and deuteron properties very accurately, these potentials themselves are, however, quite different from each other. But after integrating out the high-momentum

[54] S. Y. Lee and K. Suzuki, 'The effective interaction of two nucleons in the s-d shell', *Phys. Lett.* **B91**, 173 (1980); K. Suzuki and S. Y. Lee, 'Convergent theory for effective interaction in nuclei', *Prog. Theor. Phys.* **64**, 2091 (1980).

[55] J. D. Holt, T. T. S. Kuo and G. E. Brown, 'Family of hermitian low-momentum nucleon-nucleon interactions', *Phys. Rev. C* **69**, 034329 (2004).

($k > \Lambda$, with $\Lambda \sim 2$ fm^{-1}) components of these potentials, the resulting low-momentum interactions are nearly identical to each other. The NN potentials are constrained by scattering data up to $E_{lab} \sim 350$ MeV, corrresponding to the above Λ value. Thus, we only know the low-momentum NN potentials up to this Λ. How to determine the NN potential at higher momenta is by and large still an open question and is model dependent.

1.4. Brown–Rho Scaling and Density-Dependent Nuclear Interactions

In the last few decades of the twentieth century, our understanding of hadron physics has undergone a profound transformation. This has resulted from the development of a fundamental relativistic quantum field theory of strong interactions, going by the name of quantum chromodynamics, or QCD. In this theory protons, neutrons, mesons, and all other hadrons are described as color-neutral bound-states of quarks held together through a force mediated by the exchange of gluons. Our confidence in QCD as *the* theory of strong interactions comes from the precise agreement between theory and experiment at high energies where the remarkable property of asymptotic freedom[56] tells us that QCD is weakly-coupled and therefore solvable with perturbative techniques. Moreover, QCD is in principle capable of describing essentially all of traditional nuclear physics, including the masses and other properties of protons and neutrons, as well as the force binding them together in nuclei. However, at the low energy scales characteristic of nuclear physics ($E < 1$ GeV), QCD is notoriously difficult to solve because the scale-dependent coupling constant has grown to the point that one can no longer employ perturbation theory. Merging traditional nuclear physics with the underlying fundamental theory of QCD is therefore a significant challenge and inspires many of the modern developments in nuclear theory.

As we've already discussed, for problems in low-energy nuclear physics, QCD cannot be solved with the tools usually employed for relativistic quantum field theories. An alternative method first suggested by Kenneth Wilson[57] is to discretize spacetime and let powerful supercomputers solve the resulting equations that describe how an interacting system of quarks and gluons evolves. This "lattice QCD" approach has been able to yield accurate hadron masses,[58] a qualitatively correct description of the nuclear force in relative S-waves,[59] and many other hadronic properties. However, it appears that much effort is still required before lattice QCD will be capable of producing realistic nuclear forces comparable in accuracy to models based on meson exchange.

An alternative to lattice gauge theory is to construct a low-energy effective theory of interacting hadrons by exploiting the known symmetry structure of QCD, an idea whose

[56]D. J. Gross and F. Wilczek, 'Asymptotically free gauge theories. I', *Phys. Rev. D* **8**, 3633 (1973); D. J. Gross and F. Wilczek, 'Ultraviolet behavior of non-abelian gauge theories', *Phys. Rev. Lett.* **30**, 1343 (1973); H. D. Politzer, 'Reliable perturbative results for strong interactions?', *Phys. Rev. Lett.* **30**, 1346 (1973).
[57]K. G. Wilson, 'Confinement of quarks', *Phys. Rev. D* **10**, 2445 (1974).
[58]S. Dürr et al., 'Ab initio determination of light hadron masses', *Science* **322**, 1224 (2008).
[59]N. Ishii, S. Aoki, and T. Hatsuda, 'Nuclear force from lattice QCD', *Phys. Rev. Lett.* **99**, 022001 (2007).

roots lie in the seminal work of Weinberg.[60] Although the QCD Lagrangian is invariant under chiral symmetry transformations in the limit of massless bare quark, the QCD vacuum breaks this symmetry. This "spontaneous breaking" of chiral symmetry gives rise to a set of light pseudo-Goldstone bosons (e.g., the pions) and constrains their dynamics. Together with the nucleons, the Goldstone bosons comprise the low-energy degrees of freedom of the effective theory. The nuclear force is then obtained by calculating pion-exchange processes in a well-defined power counting scheme that organizes them according to importance. Besides pion-exchange, a set of short-range contact interactions that model the exchange of heavier mesons are fit to experiment. The resulting theory is called chiral effective field theory and has been remarkably successful in describing properties of the Goldstone bosons as well as $\pi\pi$ and πN scattering. In systems with two or more nucleons, chiral effective field theory is considerably more difficult, owing to the slower convergence in the chiral expansion. Nevertheless, all terms contributing to the NN interaction at fourth-order in the chiral expansion (next-to-next-to-next to leading order or N^{3}LO) have been calculated, and by fitting the 24 low-energy constants at this order one can very well reproduce NN scattering phase shifts and deuteron properties.[61]

Given the near universality of low-momentum interactions evolved to a scale of $\Lambda \simeq 2.0$ fm^{-1}, it may seem that there is little practical difference between traditional one-boson-exchange interactions and those derived from chiral effective field theory. However, it is well known that two-nucleon forces alone are insufficient to describe many properties of dense nuclear systems. Nowadays the nuclear many-body problem can be solved almost exactly for few-nucleon systems, and one finds that unevolved realistic two-nucleon forces systematically underbind ^{3}H, ^{3}He, and ^{4}He and significantly underpredict the nucleon-deuteron differential cross section at intermediate energies and backward angles. Moreover, as we've previously discussed, two-nucleon forces appear unable to predict simultaneously the saturation energy and density of symmetric nuclear matter. Given the small uncertainties associated with solving the above nuclear many-body problems, one can reasonably conclude that something is missing in the description of the nuclear force. Chiral effective field theory addresses this problem by introducing many-body forces with parameters fit to properties of light nuclei. Within a one-boson-exchange model, similar effects can be achieved through in-medium meson masses. Three-nucleon forces have the advantage that the uncertainties are currently better controlled, but density-dependent meson masses have the potential to directly connect with chiral symmetry breaking/restoration as we now discuss.

At low temperatures chiral symmetry is spontaneously broken by the QCD vacuum, but at large densities and/or temperatures the symmetry can be restored. The order parameter for the chiral phase transition is the chiral condensate $\langle \bar{q} q \rangle$, which gives the amplitude for finding a virtual quark-antiquark pair in the vacuum. A number of models suggest that this condensate is responsible for generating much of a hadron's total mass. Indeed, since the bare masses of the two lightest quarks are on the order of 5 MeV while the masses of

[60] S. Weinberg, 'Phenomenological Lagrangians', *Physica A* **96**, 327 (1979).

[61] D. R. Entem and R. Machleidt, 'Accurate charge-dependent nucleon-nucleon potential at fourth order of chiral perturbation theory', *Phys. Rev. C* **68**, 041001 (2003).

hadrons composed of them are typically 1 GeV, most hadronic mass (and therefore, most of the observable mass in the universe) is generated dynamically.

There are a number of models that connect the order parameter for chiral symmetry restoration with dynamical mass generation. In the Nambu-Jona-Lasinio (NJL) model,[62] which models QCD as a system of quarks interacting through zero-range contact interactions, bare quarks with small masses evolve into constituent quarks with masses on the order of ~ 300 MeV in vacuum. This constituent quark mass is directly proportional to the scalar quark condensate:

$$\frac{m^*}{m} \sim \frac{\langle \bar{q}q \rangle^*}{\langle \bar{q}q \rangle} \,. \tag{15}$$

The above scaling law is known as "Nambu scaling" and is characterized by hadron masses that scale linearly with the scalar quark condensate. This scaling law also comes out in QCD sum rule calculations.[63] Within the NJL model, the temperature and density dependence of the chiral condensate can be calculated, with the startling result that even at normal nuclear matter density found at the center of heavy nuclei, the chiral condensate can decrease by approximately 20–30%. This leads to the intriguing possibility that evidence for chiral symmetry restoration may be found even in normal nuclei.

The pion mass is generated by a different mechanism, since in the chiral limit (massless bare quarks) pions would be true Goldstone bosons and therefore massless. The pion in fact gets its mass from the explicit breaking of chiral symmetry due to the nonzero quark mass term in the QCD Lagrangian, and therefore ultimately by interacting with the Higgs particle which gives a bare quark its mass. One can therefore reasonably assume that the pion mass does not depend as sensitively on dynamical symmetry breaking as the non-Goldstone bosons.

In Brown–Rho scaling (BRS), which was derived in an attempt to address the question of scale invariance of chiral effective Lagrangians, one obtains that hadronic masses scale according to[64]

$$\sqrt{\frac{g_A}{g_A^*}}\frac{m_N^*}{m_N} = \frac{m_\sigma^*}{m_\sigma} = \frac{m_\rho^*}{m_\rho} = \frac{m_\omega^*}{m_\omega} = \frac{f_\pi^*}{f_\pi} = \Phi(n) \,, \tag{16}$$

where g_A is the axial coupling constant, f_π is the pion decay constant (an alternative order parameter for chiral symmetry breaking/restoration), Φ is a function of the nuclear density n, and all starred quantities represent in-medium (nonzero temperature and/or density) values as opposed to vacuum values. As pointed out by Lutz et al.,[65] the pion decay

[62]Y. Nambu and G. Jona-Lasinio, 'Dynamical model of elementary particles based on an analogy with superconductivity. I', *Phys. Rev.* **122**, 345 (1961).

[63]T. Hatsuda and S. H. Lee, 'QCD sum rules for vector mesons in the nuclear medium', *Phys. Rev. C* **46**, R34 (1992).

[64]G. E. Brown and M. Rho, 'Scaling effective Lagrangians in a dense medium', *Phys. Rev. Lett.* **66**, 2720 (1991).

[65]M. Lutz, S. Klimt and W. Weise, 'Meson properties at finite temperature and baryon density', *Nucl. Phys.* **A542**, 521 (1992).

constant can be connected to the scalar quark condensate through the Gell-Mann, Oakes, Renner relation[66]

$$f_\pi^2 m_\pi^2 = -(m_u + m_d)\langle \bar{q}q \rangle \,, \tag{17}$$

where m_u and m_d are the bare up and down quark masses and $\langle \bar{q}q \rangle$ is the scalar quark condensate for the up quark. Assuming that the pion mass is protected by chiral invariance, this relation would produce

$$\left(\frac{f_\pi^*}{f_\pi} \right)^2 = \frac{\langle \bar{q}q \rangle^*}{\langle \bar{q}q \rangle} \,. \tag{18}$$

The resulting dependence of hadron masses on the square root of $\langle \bar{q}q \rangle$ in Brown–Rho scaling at low densities is different from the linear scaling obtained in the NJL model and QCD sum rules. At higher densities, Koch and Brown[67] showed that the entropy from reduced-mass hadrons fit the entropy from lattice gauge simulations if one had Nambu scaling (15). Thus, the exact connection between the scalar quark condensate and the in-medium pion decay constant remains an open question.

The density dependence of the quark condensate is given (at low densities) by[68]

$$\frac{\langle \bar{q}q \rangle^*}{\langle \bar{q}q \rangle} = 1 - \frac{\sigma_{\pi N}}{f_\pi^2 m_\pi^2} n + \cdots \,, \tag{19}$$

where $\sigma_{\pi N}$ is the pion-nucleon sigma term that describes how the nucleon mass depends on the masses of bare quarks. From πN scattering experiments one can infer $\sigma_{\pi N} \simeq 45$ MeV. Thus, at nuclear matter density one would obtain from equations (16), (18), and (19) that $\Phi(n_0) \simeq 0.8$. One therefore often takes a linear drop with density of the masses:

$$\frac{m^*}{m} \simeq 1 - 0.2 \frac{n}{n_0} \,. \tag{20}$$

Moreover, to the extent that the phenomenological form factors associated with nucleon-meson interaction vertices are governed by the nucleon radius, they too will decrease with density due to the increase in the nucleon size.[69]

In Chapter IV we shall explore the consequences of Brown–Rho scaling for nuclear structure. Since the masses of light mesons are expected to decrease as the nuclear density increases, the nuclear force in-medium would be different from that in free space. Such a density-dependent nuclear interaction could address the well known deficiencies in free-space NN interactions fit to NN scattering phase shifts. The exchange of the σ, ρ, and ω mesons are all important components of the nuclear force. The σ and ω act opposite to one

[66] M. Gell-Mann, R. J. Oakes and B. Renner, 'Behavior of current divergences under $SU_3 \times SU_3$', *Phys. Rev.* **175**, 2195 (1968).

[67] V. Koch and G. E. Brown, 'Model of the thermodynamics of the chiral restoration transition', *Nucl. Phys.* **A560**, 345 (1993).

[68] E. G. Drnkarev and E. M. Levin, 'The QCD sum rules and nuclear matter', *Nucl. Phys* **A511**, 679 (1990); **A516**, 715 (1990); T. D. Cohen, R. J. Furnstahl and D. K. Griegel, 'From QCD sum rules to relativistic nuclear physics', *Phys. Rev. Lett.* **67**, 961 (1991).

[69] M. Rho,'Axial currents in nuclei and the skyrmion size', *Phys. Rev. Lett.* **54**, 767 (1985).

another and to a large extent cancel in central forces, even as their masses decrease. The two most important contributions to the tensor force come from π and ρ-meson exchange, which act opposite to each other:

$$V_\rho^T(r) = -\frac{f_\rho^2}{4\pi} m_\rho \vec{\tau}_1 \cdot \vec{\tau}_2 \, S_{12} \, f_3(m_\rho r) \,, \tag{21}$$

$$V_\pi^T(r) = \frac{f_\pi^2}{4\pi} m_\pi \vec{\tau}_1 \cdot \vec{\tau}_2 \, S_{12} \, f_3(m_\pi r) \,, \tag{22}$$

where

$$f_3(mr) = \left(\frac{1}{(mr)^3} + \frac{1}{(mr)^2} + \frac{1}{3mr} \right) e^{-mr} \tag{23}$$

and S_{12} is the tensor operator $S_{12} = 3(\vec{\sigma}_1 \cdot \vec{r} \vec{\sigma}_2 \cdot \vec{r})/r^2 - (\vec{\sigma}_1 \cdot \vec{\sigma}_2)$. In Brown–Rho scaling the ρ meson is expected to decrease in mass at finite density while the pion mass remains nearly unchanged due to chiral invariance. Therefore, one unambiguous prediction of Brown–Rho scaling is the decreasing of the tensor force in a nuclear medium. As we shall discuss later, this decrease in the tensor force plays an important role for nuclear saturation as well as for the Gamow–Teller transition strength of ^{14}C, which is responsible for the archaeologically long lifetime of ^{14}C.

CHAPTER II
Nuclear Many-Body Problems and the Brueckner G-Matrix

In this chapter we summarize our efforts to construct from the free-space nucleon–nucleon potential an effective interaction in nuclei, based on the Brueckner G-matrix approach. At the time (many years ago), the G-matrix was the preferred method for dealing with the problematic strong short-distance repulsion in the free nucleon–nucleon interaction. We introduce each reprint with a short summary and some personal recollections.

[G.E. Brown and T.T.S. Kuo (1967); T.T.S. Kuo and G.E. Brown (1966)]: The works reported in these two papers were carried out by the authors at Princeton around 1965. The nuclear theory group coordinated by Gerry Brown at that time was indeed very active and 'crowded' (Ben Bayman, Jon Blomqvist, Leonardo Castelejo, Bill Friedman, Tony Green, T.T.S. Kuo, Alex Lande, Erlend Ostergarrd, Harvey Picker, Igal Talmi, Chun Wa Wong, Larry Zamick, ... and graduate students G. Bertsch, M.Y. Chen, Bill Gerace, H. Mavromatis, J. Noble, I. Sharon, F. Wong, ...). We were all having a very good time there and then. The nuclear theory "bull session" was held every Thursday night from 8:00 till 11:00 pm or later. Afterwards people would walk to the computing center for punching cards. Our colleagues (Akito Arima, Aldo Covello, G. Satoris, George Temmer, ...) from Rutgers, which is not close to Princeton, also came to the bull session. People from Princeton also went to the Rutgers weekly seminar. Looking back, we were working really quite hard those days.

Up to now, 2010, the nuclear shell model is still the most successful model for nuclear structure. This model assumes a closed core with few effective valence nucleons confined in a small oscillator space (model space) and interacting with an effective interaction V_{eff}. Prior to the above papers, V_{eff} was by and large determined by fitting its parameters to certain experimental data. The main result of the above works was that V_{eff} can be microscopically derived from the free NN interaction. This effective interaction, often known as the Kuo–Brown interaction, has been remarkably successful. There have been many attempts to improve the Kuo–Brown interaction, but it has stood firm after more than 40 years since its first introduction.

[T.T.S. Kuo, S.Y. Lee and K.F. Ratcliff (1971); E.M. Krenciglowa and T.T.S. Kuo (1974)]: These two papers derived a linked-diagram expansion for determining the shell-model effective interaction $V_{\rm eff}$ from the free NN interactions. It provides a framework for reducing, or renormalizing, the full space A-body problem with Hamiltonian $H = (H_0 + V_{NN})$ to a model-space one with fewer degrees of freedom and a smaller number of active particles. For example, this framework allows us to reduce the 18-body ^{18}O nucleus to an effective 2-body problem composed of two neutrons confined in a small valence space of the $0d1s$ shell. The effective Hamiltonian for the reduced many-body problem is $PH_{\rm eff}P = P(H_0 + V_{\rm eff})P$. The effective interaction $V_{\rm eff}$ is given as a folded diagram expansion, which reduces to the well-known Goldstone linked diagram expansion when P is a one-dimensional model space.[1] An iteration method, the so-called Krenciglowa-Kuo method, was formulated to sum the folded diagrams to all orders. Clearly $H_{\rm eff}$ can only reproduce a subset of the physics contained in H. $H_{\rm eff}$ given by this iteration reproduces the eigenvalues correrspondiing to the states with maximum P-space overlaps.

[E.M. Krenciglowa, C.L. Kung, T.T.S. Kuo and E. Osnes (1976)]: For a long time a major problem for the calculation of the Brueckner G-matrix for nuclear shell model calculations was the treatment of its Pauli exclusion operator which involves both plane-wave and oscillator intermediate states. It was S.F. Tsai[2] who found an exact and convenient method for treating this operator. This paper further studied this method and applied it to a wide range of calculations.

[M.R. Anastasio, T.T.S. Kuo, T. Engeland and E. Osnes (1976)]: This paper investigated the convergence behavior of the folded-diagram effective interaction expansion by way of an exactly solvable model, a three-level SU(3) Lipkin model. The effect of the third-order diagrams was found to be negligible compared with that from the second-order ones when the self-consistent Hartree–Fock mean field was employed. This paper indicated that improved convergence could be attained by the use of the Hartree–Fock mean field. One of the authors (TTSK) is very grateful to the other (Gerry Brown) for introducing him to Prof. Eivind Osnes of Oslo. In the early' 60s, Eivind was working with Gerry in Copenhagen. TTSK first met Eivind in 1970 when he came to Stony Brook. This started our long collaboration of more than 20 years. TTSK has visited Oslo numerous times, and even learned some 'elementary' cross-country skiing there. Several Stony Brook students (Anastasio, De Guzman, Krenciglowa, Kung, Tam) have all visited Oslo for one or two semesters. We have collaborated on many projects, as discussed in a comprehensive review by Hjorth-Jensen, Kuo and Osnes.[3] (This review is fairly long (146 pages) and is not included in the present volume.) The Norwegian hospitality from Eivind, Torgier and Morten are heartily acknowledged.

[1] T.T.S. Kuo and E. Osnes, *Folded-Diagram Theory of the Effective Interaction in Nuclei, Atoms and Molecules*, (Springer-Verlag, Berlin, 1990).

[2] S.F. Tsai and T.T.S. Kuo, 'A new treatment for the projection operator in nuclear G-matrix equation', *Phys. Lett.* **B39**, 427 (1972).

[3] M. Hjorth-Jensen, T.T.S. Kuo and E. Osnes, 'Realistic Effective Interactions for Nuclear Systems', *Phys. Rep.* **261**, 126 (1995).

[T.T.S. Kuo, F. Osterfeld and S.Y. Lee (1980)]: The optical model potential for nucleon-nucleus scatttering is usually energy dependent (i.e. dependent on the scattering energy $E_{\rm lab}$). Using a folded-diagram method, this paper explored the possiblity of having an energy-independent optical model potential. Applications of this approach were later made.[4]

[T.T.S. Kuo, J. Shurpin, K.C. Tam, E. Osnes and P.J. Ellis (1981)]: The Goldstone diagrams in microscopic nuclear structure calculations are angular-momentum coupled, and the evaluation of such diagrams involves the summation of many Clebsch–Gordan, Racah and 9j symbols. It is a chore to perform this summation even for third order diagrams. For higher-order diagrams, it becomes prohibitively more difficult. This paper worked out a set of simple diagram rules, enabling one to write down the formulas for evaluating any general angular momentum coupled diagrams essentially by inspection.

[H.M. Sommermann, H. Müther, K.C. Tam, T.T.S. Kuo and A. Faessler (1981)]: The core polarization effective interaction in the Kuo–Brown calculation was carried out using 2nd-order perturbation theory with the intermediate states restricted to those of $N\hbar\omega$ ($N = 2$) excitation energy. How about the contribution from the $N > 2$ excitations? Vary–Sauer–Wong[5] carried out an extensive calculation for the core polarization, including oscillator intermediate states up to $\sim 25\hbar\omega$, and the converged values were significantly different from the $2\hbar\omega$ ones. In this paper, this intermediate state problem was reinvestigated. Basically we used the oscillator intermediate states for low-energy excitations while orthogonalized plane-wave states were used for excitation energies beyond. In this way, the final core polarization effects turned out to be very close to the $2\hbar\omega$ ones.

[J. Shurpin, T.T.S. Kuo and D. Strottman (1983)]: This paper also addressed the core polarization convergence problem. In a well-known paper by Barrett and Kirson,[6] the contribution of the 3rd-order core polarization diagrams was found to be generally comparable to that from the 2nd-order diagrams, raising concern about the convergence of the perturbation expansion. In this paper the core polarization was calculated using RPA-phonon exchanges and with the folded diagrams summed to all orders. The main result was that the core polarization so obtained was remarkably close to that from the 2nd-order perturbation. As to be discussed in Chap. III, the core polarization effect was reinvestigated using the Kirson-Babu-Brown induced interaction approach where certain classes of planar diagrams were summed to all orders. It is puzzling that the end result was again close to that from the 2nd-order perturbation theory.

[4] S.Y. Lee, F. Osterfeld, K.C. Tam and T.T.S. Kuo, 'General properties of energy-independent nuclear optical potentials', *Phys. Rev. C* **24**, 329 (1981).

[5] J.P. Vary, P.U. Sauer and C.W. Wong, 'Convergence rate of intermediate-state summations in the effective shell-model interactions', *Phys. Rev. C* **7**, 1776 (1973).

[6] B.R. Barrett and M.W. Kirson, 'Higher-order terms and the apparent non-convergence of the perturbation expansion for the effective interaction in finite nuclei', *Nucl. Phys.* **A148**, 145 (1970).

[A. Polls, H. Müther, A. Faessler, T.T.S. Kuo and E. Osnes (1983)]: The effective interaction given by the folded-diagram framework is of the form $V_{\text{eff}} = V(1b) + V(2b) + V(3b) + ...$, where $V(nb)$ represents the $n - body$ interaction. For a system with A nucleons, there are n-body effective interactions with $n \leq A$. This paper reported the first 3-body folded-diagram effective interactions for the $A = 19$ nuclei.

[T.T.S. Kuo, Z.Y. Ma and R. Vinh Mau (1986)]: In this paper the oscillator model space approach for finite nuclei was extended to nuclear matter calculations, leading to a so-called model-space Brueckner–Hartree–Fock (MBHF) method for nuclear matter. A momentum model space defined by $k \leq \Lambda$ with $\Lambda \sim 3$ fm^{-1} was employed. In the usual BHF nuclear matter theory, the single-particle spectrum has a sizable discontinuity at k_F, the Fermi surface. An advantage of the MBHF method is that its single particle spectrum is continuous at k_F.

[H.Q. Song, S.D. Yang and T.T.S. Kuo (1987)]: Both the BHF and MBHF methods for nuclear matter are a lowest-order G-matrix theory. Based on the MBHF framework, this paper developed an RPA method by which the particle-particle hole-hole diagrams of nuclear matter were summed to all orders, providing a rather elaborate nuclear matter framework. In this way, long-range correlations were included by way of particle-hole excitations around the Fermi surface. As is well known, both BHF and MBHF are unable to satisfactorily describe nuclear matter saturation properties. The binding energy per particle and saturation density of nuclear matter given by these methods were typically both too large compared with empirical values. The ring diagram method of this paper gave slightly better saturation results, but they were still far from satisfactory. Nuclear matter saturation remains to be a challenge to nuclear matter theory. As to be discussed in Chap. IV, to attain nuclear matter saturation the inclusion of medium corrections from Brown–Rho scaling or three-body interactions may be necessary.

1.C

Nuclear Physics **A92** (1967) 481—494; © *North-Holland Publishing Co., Amsterdam*

STRUCTURE OF FINITE NUCLEI AND THE FREE NUCLEON-NUCLEON INTERACTION

General discussion of the effective force

G. E. BROWN and T. T. S. KUO

Palmer Physical Laboratory, Princeton University, Princeton, New Jersey †

Received 30 September 1966

Abstract: It is shown that the main features of the observed effective forces in nuclei can be calculated directly from the free nucleon-nucleon interaction. Although the unrenormalized interaction acts as a short-range force, the renormalized one contains P_2, P_4 and other multipole forces. The calculated strength of the P_2 force agrees well with that obtained empirically.

1. Introduction

One of the long-standing problems in nuclear physics has been the connection between the free nucleon-nucleon force and the effective force between nucleons in nuclei. The first of such calculations [1]) attacked the problem by approximate solution of the Bethe-Goldstone equation [2]). This handled the hard-core effects conveniently but was somewhat clumsy in the treatment of many-body effects. The latter turn out to be of primary importance only for the triplet-even, nucleon-nucleon interaction (because of the strong tensor force) however, which did not enter into the work of ref. [1]). We believe the many-body effects to be conveniently and reasonably accurately handled along the lines given in our work of ref. [3]) and shall follow the development outlined there.

The hard core in modern nucleon-nucleon interactions, which was once thought to cause great difficulties, can be handled almost trivially using the Moszkowski-Scott separation method [4]) as applied to finite systems by Kallio and Kolltveit [5]); this is discussed in ref. [3]).

Inaccuracies remain in the evaluation of two-body matrix elements. The major one is concerned with the use of harmonic-oscillator wave functions in all of our work. It is estimated that this results in matrix elements roughly 5–15 % too large. Further, dispersion corrections discussed in ref. [3]) have not been introduced. These are the part of the effective matrix element most sensitive to the free nucleon-nucleon force; they decrease markedly in the transition from hard-core to soft-core force [6]). We

† This work was supported by the U.S. Atomic Energy Commission and the Higgins Scientific Trust Fund. This report made use of the Princeton Computer Facilities supported in part by the National Science Foundation Grant NSF-GP 579.

plan to include them later, when one knows more in detail about the force, but do not believe that they will be large for forces with reasonable types of central repulsion (not vertical hard cores). It is felt that our "bare" two-body matrix elements are accurate to within ≈10–15 %; we explain below bare and renormalized.

Two valence particles in the nucleus can interact either directly or through the core by means of virtual interactions of core particles. We call the direct interaction the bare one; inclusion of core excitations leads to the renormalized interaction. The calculation of these latter effects has been carried out thus far only in perturbation theory, and we cannot claim the same accuracy for the remormalized matrix elements as for the bare ones.

However, a number of essential features arise clearly. Whereas the bare force acts mainly as a short-range force when used between low-lying, shell-model states (because of its state dependence, it cuts off rapidly as the energies of the states increase), the renormalization leads to increased pairing matrix elements [7]) and gives a strong P_2 force, as well as weaker forces of other multipolarities. The strength of the P_2 force has been calculated directly from the free nucleon-nucleon interaction, as we shall outline. The pairing and P_2 parts of the force are the main features of the effective force [8]).

In addition to the P_2 force, an important P_4 component is present, which we also calculated, as well as many weaker pieces of the renormalized interaction.

We shall also discuss certain sum rules on the various parts of the renormalized interaction and shall calculate the effective interaction between particles in different shells.

The procedure for calculating the various matrix elements is precisely as outlined in ref. [3]), and we shall not repeat a description of it here. All matrix elements have been calculated starting from the Hamada-Johnston force [9]).

2. Effects from renormalization and the P_2 force

When treated in perturbation theory, the renormalization via virtual core excitation is given by the processes of fig. 1. We shall discuss mainly the process of fig. 1(a) here,

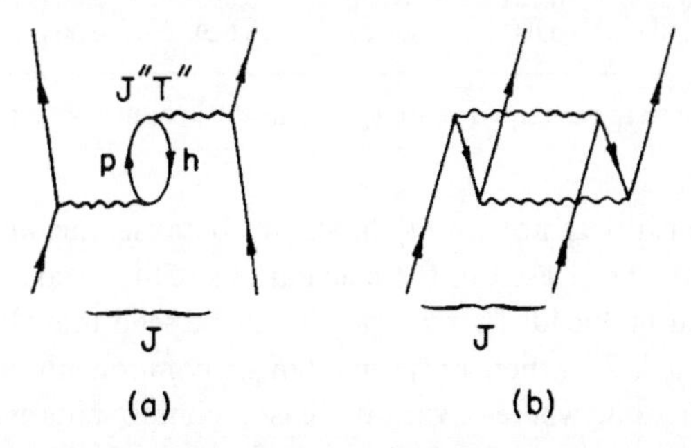

Fig. 1. Renormalization of the force through core excitation. In a) the particle and hole in the intermediate state are coupled to J'', T'' which are later summed over.

since that fig. 1(b) has been found to lead primarily to an increase in the pairing matrix elements.

In the process of fig. 1(a), the particle and hole generally differ in energy by $\sim 2\ \hbar\omega$, where $\hbar\omega$ is the distance between major shells, although in some cases [10]) where the core includes only part of a major shell, the energy differences can be less, and this leads to marked effects as noted in ref. [10]).

We show in table 1 the contributions of various J'' and T'' to the matrix elements of the effective interaction in the $f_{\frac{7}{2}}$ shell. Each of the figures in this table summarizes contributions from all particle-hole excitations of excitation energy $2\hbar\omega$ which could be coupled to a given J'', T''. Here $\hbar\omega$ was chosen as 10.5 MeV.

If we consider only the direct term of fig. 1(a), then we see that contributions from a given J'', T'' behave as a irreducible tensor force† of degree J'', T''. This is because each vertex in fig. 1(a) must be a scalar, since integration over angles is carried out independently for each vertex, and if the particle and hole are coupled to J'', T'', then the corresponding tensor component of the effective interaction is picked out. In particular, the $J'' = 2$, $T'' = 0$ contributions are just as those from a P_2 force, as we shall see.

TABLE 1

Contribution in MeV to the effective interaction $(f_{\frac{7}{2}}{}^2 J|G_{\text{eff}}|f_{\frac{7}{2}}{}^2 J)$, $T=1$, from all particle-hole bubbles corresponding to excitation energy $2\hbar\omega$ and coupled to J''

J \ J''	0	1	2	3	4	5	6	7
0	−0.026 −0.007	0.031 0.031	−0.544 −0.124	0.039 0.049	−0.275 −0.087	0.023 0.032	−0.103 −0.040	0.052 0.012
2	−0.026 −0.007	0.025 0.025	−0.254 −0.058	0.002 0.002	0.092 0.029	−0.012 −0.017	0.034 0.013	0.024 0.005
4	−0.026 −0.007	0.011 0.011	0.181 0.041	−0.017 −0.022	−0.031 −0.010	0.010 0.014	0.034 0.013	0.004 0.001
6	−0.026 −0.007	−0.010 −0.010	0.181 0.041	0.020 0.025	0.092 0.029	0.003 0.004	0.003 0.001	0.000 0.000

The upper of the two figures in each square corresponds to $T'' = 0$; the lower to $T'' = 1$.

The above relationship is not exact, however, because the antisymmetrized G-matrices of ref. [3]) have been used in the calculations. This means that the exchange term (fig. 2), has been included. In fig. 2(a), it can be seen that, if particle and hole are here coupled to J'', T'', then only that tensor component of the force enters into the upper interaction, whereas various tensor components will enter into the

† Here note that J'' is built out of the orbital angular momentum L'' and spin S''.

lower interaction. One can also see that various tensor components enter into the process (fig. 2b). However, this does not change our arguments in a practical way, as we shall show.

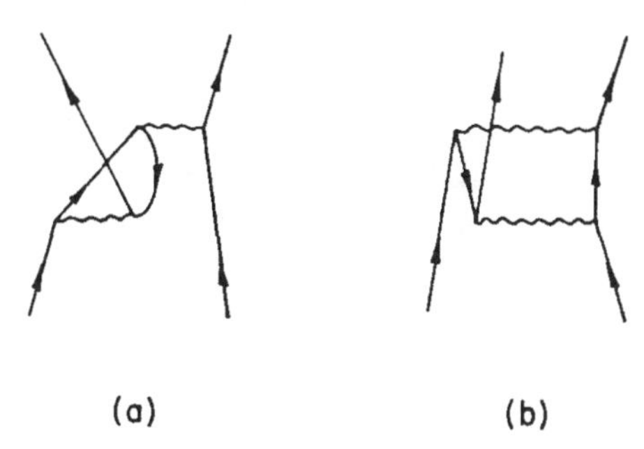

Fig. 2. Exchange terms corresponding to the direct term of fig. 1a.

We write the multipole interaction we wish to compare with as

$$H_\lambda = -X_\lambda \frac{\nu^\lambda}{(2n+l+\frac{3}{2})^\lambda} r_1^\lambda r_2^\lambda P_\lambda(\cos\theta_{12}), \tag{1}$$

where

$$\nu = m\omega/\hbar, \tag{1.1}$$

with $\hbar\omega$ the oscillator energy spacing. The above is for $T'' = 0$; for $T'' = 1$ we would need a $\tau_1 \cdot \tau_2$ in the interactions. Following similar argument, the above is also for $S'' = 0$. Thus H_λ corresponds to $J'' = \lambda$ and $T'' = 0$. We show in table 2 those values of X_2 and X_4, which would give the same matrix element as the corresponding $J'' = 2$ and 4, $T'' = 0$, renormalization. In the table, the values of X_2 used by Kisslinger and Sorensen [11]), which in our convention equal to $125/A$, are also shown. It is seen that the value calculated directly from the free nucleon-nucleon force is roughly the same. Furthermore, the calculated strength of the P_4 force is given. The latter fluctuates much more.

TABLE 2

Strengths of P_2 and P_4 forces which could give the same $(j^2J|G_{\text{eff}}|j^2J)$ as the corresponding bubble renormalization

Nucleus	j	X_2	$125/A$	X_4
^{18}O	0d$_{\frac{5}{2}}$	4.24	6.95	2.38
^{42}Ca	0f$_{\frac{7}{2}}$	2.81	2.98	2.08
^{58}Ni	1p$_{\frac{3}{2}}$	1.96	2.15	0
	0f$_{\frac{5}{2}}$	3.71		3.73
^{92}Zr	1d$_{\frac{5}{2}}$	1.31	1.36	0.78

If instead of matrix elements involving only one j, more j are involved, then the difference in radial dependences of the irreducible tensor force and the radial parts entering into the bubble term cause differences in the matrix elements.

In order to show the effect of different radial functions, we plot in fig. 3 various matrix elements in the Ni isotopes calculated with the Kisslinger-Sorensen P_2 force, and the contribution from the $J'' = 2$ ($T'' = 0, 1$) bubble terms. The correspondence is reasonably close. It should, of course, be noted that we are not trying to reproduce the Kisslinger-Sorensen matrix elements in detail; our matrix elements are much closer to the empirical ones necessary to give a detailed fit to spectra (see, for example, Lawson, McFarlane and Kuo [12]). However, in so far as the pairing and quadrupole forces are the main features of the empirical effective force, we should reproduce them.

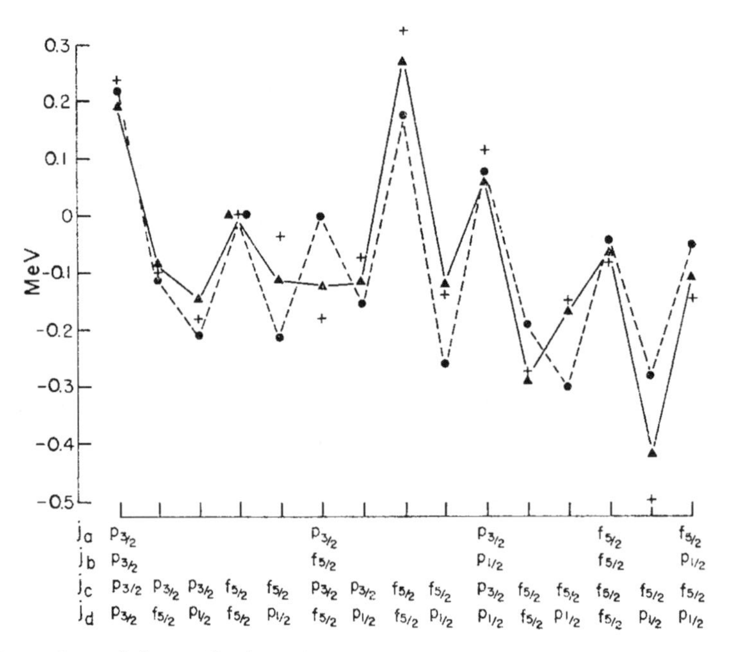

Fig. 3. Comparison of the matrix elements

$$(j_a j_b J = 2)|V|j_c j_d J = 2)_{T=1}$$

for the P_2 force and the $J'' = 2$, $T'' = 0$ bubble in the nickel region. Those for the P_2 force of Kisslinger and Sorensen are denoted by ●; those for the bubble renormalization are denoted by ▲. The $J'' = 2$ bubbles including both $T'' = 0$ and 1 components are shown as +.

In table 1 it is seen that there are many other contributions from other J'', T'' which, although smaller than the $J'' = 2$ and 4, $T'' = 0$ ones, are still appreciable. We now discuss certain sum rules.

Assuming seniority to be a good quantum number, Talmi [13]) has given an extremely simple formula for the binding energies of ground states in the j^n configuration.

$$E(j^n, \text{g.s.}) = Cn + \tfrac{1}{2}n(n-1)\alpha + [\tfrac{1}{2}n]\beta, \tag{2}$$

where $[\frac{1}{2}n]$ is the step function which is equal to $\frac{1}{2}n$ if n is even and $\frac{1}{2}(n-1)$, if n is odd. The coefficients α and β are given by

$$\alpha = \frac{2(j+1)\bar{V}_2 - V_0}{2j+1}, \qquad \beta = \frac{2(j+1)}{2j+1}(V_0 - \bar{V}_2), \tag{2.1}$$

where

$$\bar{V}_2 = \sum_{\substack{J>0 \\ \text{even}}} (2J+1)V_J / \sum_{\substack{J>0 \\ \text{even}}} (2J+1), \tag{2.2}$$

$$V_J = (j^2 J|G|j^2 J) \tag{2.3}$$

in our notation.

Talmi [13]) notes that the sign of α is found to be such that the quadratic term in eq. (2) is repulsive (negative, in his convention; positive in ours). Values of α from the bare interaction are of the wrong sign, but the renormalization (fig. 1a) changes the sign of α, as we shall show.

Given a sum of irreducible tensor forces

$$V = \sum_{\lambda} V_\lambda, \tag{3}$$

Arima, Nomura and Kawarda [14]) have shown that

$$E(j^n, \text{g.s.}) = [\tfrac{1}{2}n] \left\{ (j^2 J = 0|V_{\text{odd}}|j^2 J = 0) + \frac{2j+1}{2j-1} (j^2 J = 0|\underset{\lambda>0}{V_{\text{even}}}|j^1 J = 0) \right\}$$

$$+ \tfrac{1}{2}n(n-1) \left\{ (j^2 J = 0|V_0|j^2 J = 0) - \frac{2}{2j-1} (j^2 J = 0|\underset{\lambda>0}{V_{\text{even}}}|j^2 J = 0) \right\}, \tag{4}$$

where odd and even refer to the odd- and even-tensor components. For multipole forces (1) with X_λ positive all matrix elements in (4) are negative; i.e., correspond to attractive interactions. As for the contributions to $E(j^n, \text{g.s.})$ from the renormalization effects (fig. 1a), V_λ corresponds to the $J'' = \lambda$ bubble term.

It is first of all seen that the odd-tensor components contribute only to the pairing interaction, i.e., only to β in eq. (2). This follows directly from the fact that they must be compounded out of spins and even multipoles of the force; odd multipoles are rules out by parity conservation for the j^n configuration. Such spin-dependent forces are diagonal in seniority and contribute only to the "pair "term β.

From eq. (4) one sees that the even-tensor components, other than the zeroth one, make the quadratic term repulsive. We now consider the contribution to the quadratic term from the multipole force H_λ. We shall take X_λ to be positive [see eq. (1)]. Of course the P_λ in H_λ can be either positive or negative, depending on the angle, and the weighting is such that the contributions from negative P_λ predominate and give a net repulsion. This is largely an exclusion-principle effect; because of their identity, the particles are kept apart, lowering the probability of them being at zero angle with respect to each other, where all H_λ are attractive. This results in a net repulsion.

The repulsive contributions to the quadratic term come as small corrections to the strong pairing term, so that the interaction between particles is basically an attractive one, with small repulsive corrections.

In table 3, we show the values of α and β calculated with the bare and with the renormalized interactions. In sect. 5, we shall return to further discussion of α and β, and the contribution of effective three-body forces to them.

TABLE 3

Values of α and β

Nucleus	Bare interaction		Renormalized interaction		Empirical [a]	
	α	β	α	β	α	β
^{42}Ca (considered as $(f_{\frac{7}{2}}{}^2)^J$)	−0.21	−0.663	0.15	−1.96	0.23	−3.33

[a]) Values quoted from ref. [13]).

Although the P_2 and P_4 forces arising in the renormalization are the largest multipole forces, it is a distinct advantage to have calculated values for the other multipole forces. The $J'' = 1$, $T'' = 1$ interaction, for example, enters sensitively into configurational mixing of the type which would violate l-forbiddenness in M1 transitions. Knowing the relevant matrix elements, one could compute this effect. The fact that we find this interaction to be small in all cases calculated thus far is consistent with the rather good degree to which l-forbiddenness is obeyed.

3. The pairing interaction

The bare interaction acts as a short-range force. At first sight this seems strange, because with the use of the separation method, the inner part of the two-nucleon potential up to the cut-off distance of the order of 1 fm is removed. That the force is, indeed, short range can be seen from looking at the KK potentials [5]) which, although much simpler than the one employed here, is essentially the same in most features. In any such force with a hard core, the attractive potential must be much deeper and shorter in range than a well-behaved potential (without hard core) of the same effective range and scattered length. Thus, the KK potential drops off from maximum value at $\approx$ 1 fm with a range of $\approx$ 0.4 fm, whereas Yukawa or Gaussian potentials commonly used have ranges of $\approx$ 1.3–1.7 fm. Of course, the effective potential coming from the Hamada-Johnston potential is more complicated than the KK one, but has the same behaviour in this respect.

The fact that the bare interaction acts as a short-range one can be checked in several ways. Firstly, one sees that the spacings of the E_J, where

$$E_J = (j^2J|G|j^2J) \tag{5}$$

are somwhat similar to those for a δ-function force.

Furthermore, many spectra calculated with the bare interaction were nearly the same as those calculated with a δ-force.

If the bare interaction is of short range in character, then the "bubble" renormalization should arise mainly from the neutron-proton $T = 0$ interaction †. The argument is as follows. If the nucleus consisted of only one kind of particles, then a δ-force can equally well be expressed as

$$\delta(\boldsymbol{x}_1 - \boldsymbol{x}_2) = -\tfrac{1}{3}(\boldsymbol{\sigma}_1 \cdot \boldsymbol{\sigma}_2)\delta(\boldsymbol{x}_1 - \boldsymbol{x}_2), \tag{5.1}$$

since a pair of identical particles can only be in a singlet state when coincident in space. This force is manifestly of a pairing type, the interaction of a pair of nucleons in a singlet state with any other nucleon being zero. We might then expect a generalized seniority description, in which the low-lying states are formed by pairing as many nucleons as possible, to be good.

Now, treating interactions of only valence particles and including the bubbles in perturbation theory is a way of obtaining an approximate solution to the problem of many nucleons in many shells. According to the above, not much P_2 type force should arise from bubbles containing the same kind of particles as the valence particles; in other words, the $T = 1$ interaction should not produce much P_2 force. Otherwise, the generalized seniority picture would be upset, since it is known that a P_2 force breaks up pairing. This argument is rough, and may not be convincing, but it is true that the renormalizations from the $T = 0$ interaction are much stronger than from the $T = 1$ one.

Bubble renormalizations were broken up into separate parts coming from $T = 0$ and $T = 1$ interactions, and the latter were found to contribute much less than the former in the ratio of roughly $\frac{1}{3}$ or $\frac{1}{4}$ or less, although this ratio varied widely. For example, in the s, d-shells the $T = 0$ and 1 contributions to $(\mathrm{d}^2_{\frac{5}{2}}J|G_{\mathrm{eff}}|\mathrm{d}^2_{\frac{5}{2}}J)$ from process fig. 1a are

T \ J	0	2	4	
0	−1.10	0.011	0.48	[MeV]
1	0.04	0.055	0.05	

In any case, one can say that the renormalization results mainly from the neutron-proton force.

If the bare interaction is short-range in character, then it clearly has appreciable matrix elements of the pairing type

$$(j_a^2 J = 0|G|j_c^2 J = 0).$$

We show such matrix elements in fig. 4 for the Ni region, together with renormalized ones and those that would follow from the Kisslinger-Sorensen pairing force. As

† We follow here an argument of A. Arima.

noted by Bertsch [7]), the renormalization increases the pairing matrix elements in most cases.

One should note, however, that the bare interaction does not act like a zero-range force when leading to high-lying configurations. The interaction is state dependent, and the separation distance moves steadily outward with increasing excitation, so that the interaction cuts off and finally becomes repulsive. It is clear from our calculations, however, that the pairing matrix elements are still strongly attractive for the last filled states of the nucleus.

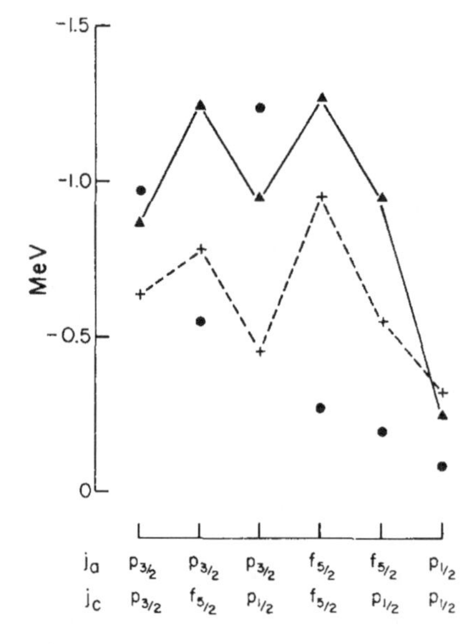

Fig. 4. Comparison of the pairing matrix elements in the nickel region. Those for the pairing force of Kisslinger and Sorensen are denoted by $+$; they are calculated with

$$(j_a j_b J|V|j_c j_d J)_{T=1} = -\tfrac{1}{2}G\delta_{J0}\delta j_a j_b \delta j_c j_d[(2j_a+1)(2j_c+1)]^{\frac{1}{2}}$$

with $G = 0.317$ MeV. The values for the bare and renormalized Hamada-Johnston force are denoted by ● and ▲, respectively.

4. Effective interaction between particles in different shells

One of the surprising features of the effective interaction is the repulsion between like particles in different shells (or, in fact, in different subshells in most cases) pointed out by Talmi and Unna [15]).

In fig. 5, we show calculated matrix elements for a particle in the s-d shell interacting with a particle in the p-shell. The plotted matrix elements are

$$V'_J = (jj'J|G_{\text{eff}}|jj'J)_{T=1},$$
$$V_J = \tfrac{1}{2}\{(jj'J|G_{\text{eff}}|jj'J)_{T=0}+(jj'J|G_{\text{eff}}|jj'J)_{T=1}\}. \tag{6}$$

The $V_{J'}$ are, in fact, just the neutron-neutron interaction, and the V_J the neutron-proton interaction. The latter is seen from the figure to be attractive.

Two calculations of renormalized $V_{J'}$ and V_J have been made; one calculated with a ^{12}C core and the other with ^{16}O core. As shown by fig. 5, the empirical values of Talmi and Unna are somewhere in between for most cases.

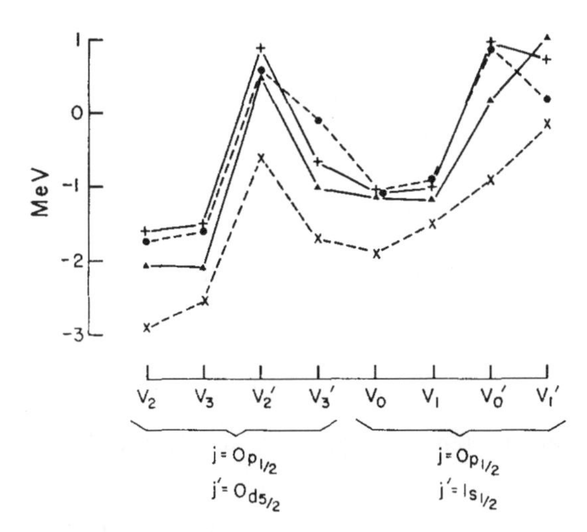

Fig. 5. Comparison of the matrix elements V_J and $V_{J'}$. ● – empirical values of Unna-Talmi, × – bare HJ (Hamada-Johnston) force, ▲ – renormalized HJ force calculated with ^{12}C core, + – renormalized HJ force calculated with ^{16}O core.

In neither the empirical nor theoretical work were spurious states removed. In two cases – those of V_0' and V_1' – major parts of the renormalization came from $J'' = 1$, $T'' = 0$ bubbles and these would probably be changed considerably by removal of spurious states. It is not clear to us how the empirical matrix elements would be modified by a correct treatment of spurious states.

5. Renormalization through collective core excitations

Our calculations of renormalization have been carried out only in perturbation theory. A more complete theory would allow for any number of bubbles as shown in fig. 6. This is equivalent to coupling in collective (and other) excitations of the core, calculated in the random-phase approximation; that is, to using as intermediate states

$$|i\rangle = Q_n^\dagger a_j^\dagger a_j^\dagger |0\rangle, \tag{7}$$

where the $Q_n^\dagger$ are creation operators for the normal modes of the core [16].

Our arguments for using perturbation theory considered in noting that some of the collective excitations went up in energy and some of them down from the unperturbed particle-hole energies, so that these changes should average out to some extent.

Furthermore, in most cases none of the collective excitations can come low in energy, since they involve as components particle-hole excitations of unperturbed energy $2\hbar\omega$; i.e. they are even-parity vibrations of closed shells.

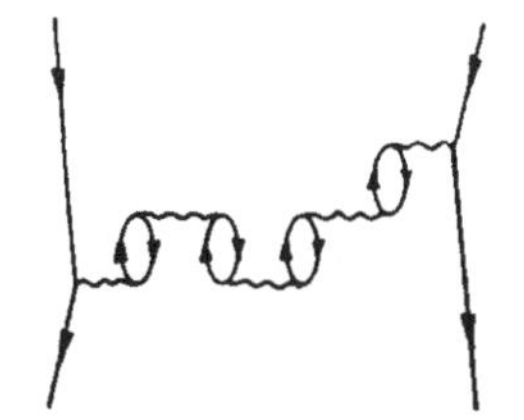

Fig. 6. Renormalization through exchange of collective excitations.

However, it would clearly be more satisfactory to carry out the calculation by first obtaining the normal modes of the core, and such calculations are underway.

6. Other renormalizations

Matrix elements are also renormalized by the processes of fig. 7.

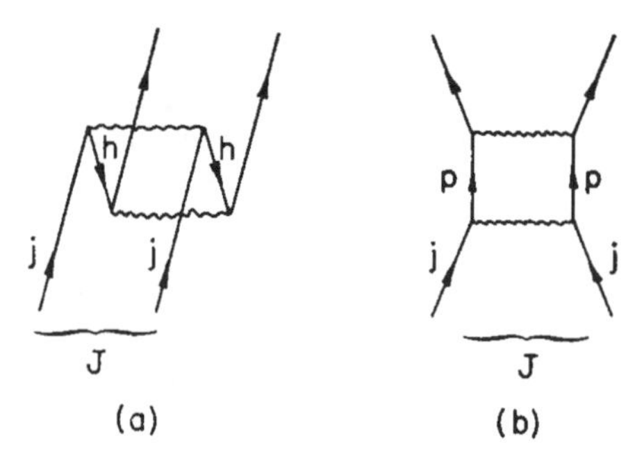

Fig. 7. Other renormalizations of the basic matrix element $(j^2J|G|j^2J)$. In a) h denotes holes. In b) p denotes particles up at least one shell higher than those labelled by j.

It can be argued that the process of fig. 7(a) should not be treated by perturbation theory, but the intermediate four-particle, two-hole states should be included explicitly, since, in the oxygen and calcium regions, at least one of these states comes down to a very low energy [17]). In heavier nuclei they are not so important, and it may be that this process can be adequately included in perturbation theory.

The process of fig. 7(b) is more difficult to discuss along the simple lines adopted here, and a more extensive formalism is needed [18]) to treat it adequately. The point is that – in some approximation – a complete set of intermediate two-particle states was used in the initial calculation of the G-matrix, and use of the renormalization (fig. 7b), involves some double counting. We argue, however, that low intermediate states were essentially dropped in our G-matrix calculation, because of the use of plane wave intermediate states, and that we should therefore add them back in explicitly.

In any case, contributions from processes of this type are small and are appreciable only for the pairing-type matrix elements

$$(j^2 J = 0|G_{\text{eff}}|j^2 J = 0).$$

7. Effective three-body forces

In addition to renormalizations of the two-body force, configuration admixture will also produce effective three-body forces, which will, however act by and large like n-dependent, two-body forces, where n is the number of valence particles. To be specific, let us consider the case of n neutrons in the $0f_{\frac{7}{2}}$ shell. The three-body forces will then be of the type shown in fig. 8. Considering the interaction to be between the two outside particles, the middle $0f_{\frac{7}{2}}$ particle clearly provides a renormalization. But this is dependent on the number of n $0f_{\frac{7}{2}}$-particles in the shell, and consequently can be discussed either as an n-dependent, two-body force or as a three-body one.

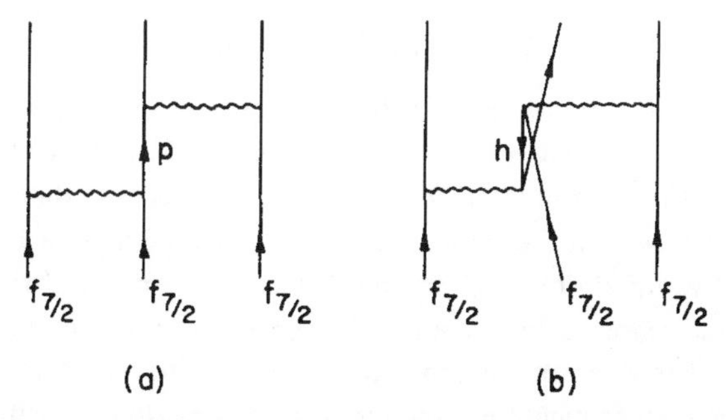

Fig. 8. Effective three-body interaction. In a) the particle p is thought of here as being a $2p_{\frac{3}{2}}$, $2p_{\frac{1}{2}}$, $0h_{\frac{9}{2}}$, $0h_{\frac{11}{2}}$ or $1f_{\frac{7}{2}}$ state. (The three initial particles are in the $0f_{\frac{7}{2}}$ shell.) In b), the hole is thought of as being in the $0p_{\frac{3}{2}}$ or $0p_{\frac{1}{2}}$ shell.

In the earlier renormalization, e.g., fig. 1(a), the particle p was allowed to be in either of the states occupied by the two interacting particles, and the process of fig. 8(b) contains terms which remove this possibility, in line with the exclusion principle.

It can be seen that, depending on the multipole of the interaction, the processes of fig. 8 contain multipole interactions which increase roughly linearly with n. We expect these to be small, especially since the two processes of figs. 8(a) and (b) give contributions of opposite sign.

One term, where p in fig. 8(a) refers to a particle in the $1f_{\frac{7}{2}}$ shell and the monopole part of the interaction is used, is of particular interest. This term represents a small expansion in the $0f_{\frac{7}{2}}$ orbit of the middle particle, its wave function can be represented as

$$\psi_{\frac{7}{2}} = |0f_{\frac{7}{2}}) + \delta|1f_{\frac{7}{2}}).$$

Since one knows empirically [19]) the rate of expansion as neutrons are added in the Ca isotopes, it is simple to estimate the effects of such an expansion †. Assuming the increase in volume to go as $0.4n/40$, roughly as observed, then the pairing matrix element will decrease as $(1\text{–}0.4\, n/40)$, so that an n^2 term of the size

$$-\beta\tfrac{1}{2}n(0.4)\tfrac{1}{40}n = -0.1\beta\tfrac{1}{2}n^2$$

results for even isotopes. This gives an additional repulsive contribution of ≈ 0.03 to α, since $\beta \approx -3.3$ (ref. [13])).

Even with inclusion of three-body forces, our calculated values of α in table 3 are smaller than the empirical ones. However, inclusion of low-lying deformed states will reduce the discrepancy. Since the depressions of the ground states decreases with increasing n (ref. [20])), viewed from the standpoint of effective forces in the $f_{\frac{7}{2}}$ shell, this would correspond to a repulsion which increases with n. It is clear that we should not expect a fit without introduction of these deformed states.

Calculation of the three-body forces discussed here is proceeding; we do not expect them to be large; although they may have an appreciable effect on α.

8. The particle-hole interaction

A number of calculations involving the particle-hole interaction have now been carried out for the Brueckner-Gammel-Thaler potential [21]) and the Hamada-Johnston potential [22]) using the bare interaction. This is correct, because then the particle-hole matrix elements are used in the RPA, the "bubble" diagrams are summed to all orders. Consequently, one should use the unrenormalized G-matrix for the particle-hole interaction and the renormalized one for the particle-particle interaction.

We note here that this corresponds to the use in the Landau theory of Fermi liquids [23]) of f as the particle-hole interaction, the transport coefficients A corresponding to the particle-particle interaction.

9. Discussion

We have shown that the main features of observed effective forces in nuclei can be obtained from a straightforward connection with a realistic free nucleon-nucleon interaction.

The main approximations in the calculations were the use of harmonic-oscillator wave functions and the neglect of dispersion corrections.

Possible extensions and improvements involving the inclusion of collective excitations in the renormalization and calculation of effective three-body forces were discussed.

In earlier papers [3,24]), we have shown that the calculated matrix elements agree surprisingly well with empirical ones derived directly from the experimental spectra

† This was pointed out to the authors by A. K. Kerman.

in various regions of the periodic table. Although this agreement is good, the theoretical matrix elements do not produce the observed spectra well in detail [12]), the errors shown in the empirical matrix elements being highly correlated. Thus, the predictive power of the theoretical matrix elements calculated with the estimated accuracy of ≈ 10–$15\,\%$, may not be high, at least as regards relative positions of close-lying levels. It seems to be too ambitious at the present to calculate the matrix elements more accurately, especially since uncertainties in the free nucleon-nucleon force may cause similar uncertainties of almost this magnitude.

We would like to thank G. Bertsch, A. M. Green, C. W. Wong and L. Zamick for many discussions concerning this work. We are also indebted to A. Arima and A. K. Kerman for specific suggestions.

References

1) J. F. Dawson, I. Talmi and J. D. Walecka, Ann. of Phys. **18** (1962) 339
2) H. A. Bethe and J. Goldstone, Proc. Roy. Soc. **A238** (1957) 551
3) T. T. S. Kuo and G. E. Brown, Nuclear Physics **85** (1966) 40
4) B. L. Scott and S. A. Moszkowski, Ann. of Phys. **14** (1961) 107
5) A. Kallio and K. Kolltveit, Nuclear Physics **53** (1964) 87
6) H. A. Bethe, Phys. Rev. **138** (1965) B804;
   C. F. Wong, Nuclear Physics **56** (1964) 213
7) G. F. Bertsch, Nuclear Physics **74** (1965) 234
8) Endless private and public communications by A. Bohr and B. R. Mottelson
9) T. Hamada and I. D. Johnston, Nuclear Physics **34** (1962) 383
10) L. Zamick, Phys. Lett. **20** (1966) 176
11) L. S. Kisslinger and R. A. Sorensen, Mat. Fys. Medd. Dan. Vid. Selsk. **32**, No. 9 (1960);
    Revs. Mod. Phys. **35** (1963) 853
12) R. D. Lawson, M. H. Macfarlane and T. T. S. Kuo, Phys. Lett. **22** (1966) 168
13) I. Talmi, Revs. Mod. Phys. **34** (1963) 704
14) A. Arima, M. Nomura and H. Kawarda, Phys. Lett. **19** (1965) 400
15) I. Talmi and I. Unna, Ann. Rev. Nucl. Sci., **10** (1960) 353
16) G. E. Brown, Unified theory of nuclear models (North-Holland Publ. Co., Amsterdam 1964) chapt. V
17) T. Engeland, Nuclear Physics **72** (1965) 68;
    G. E. Brown, Congrés International de Physique Nucléaire, Vol. I, Paris (1964) p. 129
18) C. Bloch and J. Horowitz, Nuclear Physics **8** (1958) 91;
    B. H. Brandow, Revs. Mod. Phys., to be published
19) R. Hofstadter *et al.*, Phys. Rev. Lett. **15** (1965) 758
20) W. Gerace and A. M. Green, to be published
21) A. M. Green, A. Kallio and K. Kolltveit, Phys. Lett. **14** (1965) 142; Nuclear Physics, to be published
22) H. A. Mavromatis, W. Markiewicz and A. M. Green, to be published
23) L. D. Landau, JETP (Sov. Phys.) **3** (1957) 920
24) T. T. S. Kuo, Nuclear Physics **A90** (1967) 199

1.C

Nuclear Physics **85** (1966) 40—86; © *North-Holland Publishing Co., Amsterdam*

STRUCTURE OF FINITE NUCLEI AND THE FREE NUCLEON-NUCLEON INTERACTION

An Application to ^{18}O and ^{18}F

T. T. S. KUO and G. E. BROWN

Palmer Physical Laboratory, Princeton University, Princeton, New Jersey †

Received 4 March 1966

Abstract: The intention of this work is to investigate the applicability of the free nucleon-nucleon potential determined by the scattering data in the shell-model description of finite nuclei. The potential is chosen to be that of Hamada and Johnston. We have chosen ^{18}O and ^{18}F as our first numerical calculations. A major part of the work reported here concerns the evaluation of the shell-model reaction matrix elements. They are evaluated using the separation method for the singlet-even and triplet-even states and the reference spectrum method for the singlet-odd and triplet-odd states. The second-order Born term for the triplet-even tensor force is found to be very important. It can be calculated conveniently and with good accuracy using the closure approximation with an energy denomiator of $\approx$220 MeV. Assuming the single-particle wave functions to be those of a harmonic oscillator well, the single-particle energy levels of $0d_{\frac{5}{2}}$, $1s_{\frac{1}{2}}$ and $0d_{\frac{3}{2}}$ for ^{17}O are calculated by letting the valence nucleon interact with the ^{16}O core. The results are very encouraging. The $0d_{\frac{5}{2}}$–$0d_{\frac{3}{2}}$ splitting is found to come mostly from the triplet-odd $l \cdot s$ force through the reaction matrix Hartree-Fock process. The spectra of ^{18}O and ^{18}F are satisfactorily reproduced by diagonalizing the model interaction $G\Omega_{3p1h}$ in the sd shell, except for the states which are presumed to be largely deformed; G is the reaction matrix and Ω_{3p1h} is the wave operator which takes care of the corrections arising from the one-particle-one-hole excitation of the ^{16}O core. It is found that the conventional shell-model effective interactions are well reproduced by the model interaction deduced from the free nucleon-nucleon potential.

1. Introduction

The question of residual interaction used in the nuclear shell-model has long been a very ambiguous and chaotic problem. The conventional approaches have been the following: One assumes them to take some simple and reasonable forms, like Gaussian or Yukawa. Their strengths, ranges and mixture constants are then taken as parameters and adjusted until an optimum fit to the experimental data is achieved [1]). Another approach is to consider directly the shell-model matrix elements themselves as adjustable parameters, without specifying the algebraic forms of the interactions [2,3]). Although both approaches have been rather successful in correlating many experimental facts, they are however unsatisfactory because of the following aspects: First,

† This work was supported by the U.S. Atomic Energy Commission and the Higgins Scientific Trust Fund. This report made use of the Princeton Computer Facilities supported in part by the National Science Foundation Grant NSF-GP 579.

†† There have been a large number of such calculations. A discussion of typical examples can be found in ref. [1]).

the origin of these interactions is rather obscure. Moreover, the interactions so determined are usually highly uncertain and can often be erroneous. They depend upon which nuclei one is considering, and in particular which configurations are included and which experimental levels one tries to fit by adjusting the parameters of the interactions. Since there is no way to know beforehand which configurations should be included and which experimental levels should be fitted while the interactions themselves are considered as adjustable parameters to be determined by these choices, it is obvious that such determinations are sometimes futile. This feature is very undesirable. On the other hand, if the interactions can be deduced independently, the structure of the observed levels can be studied unambiguously by comparing them with the results calculated using these interactions and a chosen set of configurations.

There is a third approach in which the shell-model matrix elements are obtained directly from the experimental spectra of particularly simple nuclei where the states are considered to be composed of two particles, each in a state of definite angular momentum j, coupled to a resultant J. The applicability of it is rather limited, namely it applies only to those nuclei with few valence nucleons. For instance, the $(f_{\frac{7}{2}})^n$ nuclei [4]). Moreover, as in the two preceding approaches, the origin of these matrix elements is equally obscure. The residual interactions determined by the above procedures are generally known as the "effective" interactions.

In this paper, we intend to clarify this problem by deducing the effective interactions from free two-nucleon interactions, about which our knowledge has increased considerably in recent years [5]). Up to the present, the best two-nucleon potentials which fit the two-nucleon scattering data up to about 350 MeV and the deuteron properties are probably those of Hamada and Johnston [6]) (hereafter referred to as HJ) and Lassila *et al.* [7]) (hereafter referred to as the Yale potential). Both potentials have infinite repulsive cores and approach the one-pion-exchange-potential at large distances. These potentials have been used in nuclear matter calculations and yielded reasonably satisfactory results [8-10]) (although neither gives quite enough binding energy). This suggests that the nucleon-nucleon interaction inside a nucleus can be obtained from the free two-nucleon interaction.

The presence of hard cores certainly causes difficulties in calculation. One early method of handling hard core potentials is that of energy shift of Bauer and Moshinsky (ref. [11])). This method is, however, not justifiable because of the neglect of many-body effects when compared with the more recent method of the reaction matrix theory [12]). Basically, this method has been used in several shell-model calculations (refs. [13,14])), where, however, some corrections for many-body effects has been made. In these cases, many-body effects are relatively unimportant since the nucleon-nucleon interaction is relative ^{3}S states does not enter. In cases where it does, many-body effects are large as we shall show. In this paper, we shall follow closely the method of the reaction matrix theory. The details of it will be presented in sect. 2. Kallio, Kolltveit and others have recently reported works along this line [15-17]). They used a simplified purely central interaction which acts in only relative S-states. A convenience of using this interaction is its extreme simplicity in calculating the reaction matrix elements. This simplicity disappears when the more realistic HJ or Yale potential is used. In this paper, we shall use the HJ potential in our calculations. The Yale potential is practically identical to the HJ for our purposes.

Once the nucleon-nucleon interaction is chosen, the effective interaction to be used in any shell-model calculation will be fixed, at least in principle. The single-particle wave functions and energies should be those determined by the self-consistent calculation starting out from the interaction. The self-consistent calculation is certainly very difficult, primarily because of the hard core. But for potentials without hard cores, the self-consistent calculation is easier because it is basically the ordinary Hartree-Fock calculation. Calculations along this line have been carried out by Levinson *et al.* using a Gaussian interaction with the Rosenfeld mixture [18]) and by Baranger *et al.* [19]) using a velocity dependent potential [20]). Both results indicate that the calculated wave functions are similar to the oscillator wave functions. So, instead of solving the self-consistent problem, we shall assume the single-particle wave functions to be those of a harmonic oscillator. The consequences of this assumption can be checked as follows. We can use the oscillator wave functions together with the HJ potential in computing other shell-model quantities to see if consistent results are obtained. In particular, in sect. 3 we shall compute the single-particle energies (the spin-orbit splittings) of valence nucleons, using the above scheme.

In sect. 4, the nuclear, shell-model, effective interaction will be deduced from the free two-nucleon interaction. In shell-model calculations, usually only a very limited configuration space can be included. This limited space is usually known as the model space. The model space is generally too limited to give an adequate description of the nuclear states, particularly when realistic potentials are used. Therefore, the effective interaction to be used in the model space should be the free two-nucleon interaction plus corrections due to configurations outside of the model space. This effective interaction will be called the model interaction. Various aspects concerning it will be discussed.

We have chosen ^{18}O and ^{18}F for our first numerical calculations, because there are both a number of existing calculations for comparison and a wealth of experimental data. Values of representative reaction matrix elements for these nuclei will be given in sect. 2. In sect. 3, the single-particle energies of the valence nucleon of ^{17}O will be computed and compared with experimental data [21]) from the stripping reaction ^{16}O(d, p)^{17}O. The nuclei ^{18}O and ^{18}F may be pictured as basically consisting of two valence nucleons outside of the ^{16}O core. This core, as we shall see, is not inert and corrections due to its excitation must be taken into account. Such corrections have been made to some extent, in the empirical matrix elements of Federman and Talmi [2, 3]). In sect. 4, they will be compared with the matrix elements of the sd shell model interaction deduced from the Hamada-Johnston potential. The spectra of ^{18}O and ^{18}F are obtained by diagonalizing the model interaction in the sd shell. The results are also shown in sect. 4.

All computations involved in this paper were carried out in the Princeton IBM-7094 computer.

2. Shell-Model Reaction Matrix Elements

We shall first give the formulae for the shell-model matrix element of a non-singular potential V. In j-j coupling, it can be written as

$$(abJT|V|cdJT) = \frac{1}{\sqrt{(1+\delta_{ab})(1+\delta_{cd})}} \sum_{LL'S} X\begin{Bmatrix} l_a & \frac{1}{2} & j_a \\ l_b & \frac{1}{2} & j_b \\ L & S & J \end{Bmatrix} X\begin{Bmatrix} l_c & \frac{1}{2} & j_c \\ l_d & \frac{1}{2} & j_d \\ L' & S & J \end{Bmatrix} (-1)^{L+L'}$$
$$\times \sum_{nln'l'N\mathscr{L}} \langle nlN\mathscr{L}, L|n_a l_a n_b l_b, L\rangle\langle n'l'N\mathscr{L}, L'|n_c l_c n_d l_d, L'\rangle$$
$$\times(-1)^{l+l'}[1-(-1)^{l+S+T}] \sum_{\mathscr{J}} U(\mathscr{L}lJS; L\mathscr{J})U(\mathscr{L}l'JS; L'\mathscr{J})(nlST\mathscr{J}|V|n'l'ST\mathscr{J}). \tag{2.1}$$

The index a stands for the quantum numbers n_a, l_a, j_a of a spherical oscillator orbital (ref. [22]) and t_z, the z-component of the isospin. The ket $|abJT)$ represents a normalized and antisymmetrized state where a and b couple to total angular momentum J and isospin T. The X-coefficients are the transformation coefficients from j-j to L-S coupling [22, 23]. The bracket $\langle nlN\mathscr{L}, L|n_a l_a n_b l_b, L\rangle$ is the Brody-Moshinsky transformation bracket between the laboratory frame and the centre-of-mass and relative frame; nl and $N\mathscr{L}$ are respectively the relative and the centre-of-mass oscillator quantum numbers [24]). The U-coefficient is related to the Racah coefficient by

$$U(\mathscr{L}lJS; L\mathscr{J}) = \sqrt{(2L+1)(2\mathscr{J}+1)}W(\mathscr{L}lJS; L\mathscr{J}). \tag{2.2}$$

The integral $(nlST\mathscr{J}|V|n'l'ST\mathscr{J})$ will be called the reduced integral, where we note that $l(l')$ and S couple to $\mathscr{J}$.

Generally, the potential V is written as $\sum_i \mathscr{O}_i V_i$, where all the radial dependence is contained in V_i. For instance, the HJ potential which we shall use in this paper is given as

$$V = V_c + S_{12}V_T + (\boldsymbol{l}\cdot\boldsymbol{s})V_{LS} + L_{12}V_{LL}, \tag{2.3}$$

where the subscripts c, T, LS and LL denote respectively central, tensor, spin-orbit and quadratic spin-orbit terms. The explicit forms of the radial functions are given in appendix 1. The operator S_{12} is the ordinary tensor operator, and the quadratic spin-orbit operator L_{12} is defined by

$$L_{12} = [\delta_{l\mathscr{J}} + (\boldsymbol{\sigma}_1\cdot\boldsymbol{\sigma}_2)]\boldsymbol{l}^2 - (\boldsymbol{l}\cdot\boldsymbol{s})^2. \tag{2.4}$$

It is convenient to separate the reduced integral into the angular spin part and the radial part, namely

$$(nlST\mathscr{J}|V|n'l'ST\mathscr{J}) = \sum_i (lS\mathscr{J}|\mathscr{O}_i|l'S\mathscr{J})(nl|V_i(r)|n'l'). \tag{2.5}$$

The index T in the right-hand side has been absorbed in $V_i(r)$ as usual. The angular spin parts can be readily obtained for the HJ potential:

$$(lS\mathscr{J}|S_{12}|l'S\mathscr{J}) = \delta_{S1}(-)^{1-\mathscr{J}}[24(2l+1)(2l'+1)]^{\frac{1}{2}}W(lSl'S;\ \mathscr{J}2)(l0l'0|20), \quad (2.6)$$

where $(l0l'0|20)$ is the vector coupling coefficient. Note that the value of the above matrix element is $\sqrt{8}$ for $l = 0$, $l' = 2$ and $S = 1$.

$$(lS\mathscr{J}|\boldsymbol{l}\cdot\boldsymbol{s}|l'S\mathscr{J}) = \delta_{ll'}\delta_{S1}\tfrac{1}{2}[\mathscr{J}(\mathscr{J}+1)-l(l+1)-2], \quad (2.7)$$

$$(lS\mathscr{J}|L_{12}|l'S\mathscr{J}) = \delta_{ll'}(\delta_{l\mathscr{J}}+\delta_{S1}-3\delta_{S0})l(l+1)-(lS\mathscr{J}|\boldsymbol{l}\cdot\boldsymbol{s}|l'S\mathscr{J})^2. \quad (2.8)$$

So far everything has been straightforward, because we have pretended that $V_i(r)$ is a well-behaved function. But for the HJ potential, $V_i(r)$ is infinitely repulsive for $r < r_c$, the core radius. Then the reduced integral will be infinite, and the above formalism is invalid unless modified. According to the reaction matrix theory, an appropriate modification is to replace V by the Brueckner reaction matrix G for the scattering of two nucleons [12]). This point will be further discussed in sect. 4. The G-matrix is defined by the integral equation

$$G = V - V\frac{Q}{e}G, \quad (2.9)$$

where Q is the Pauli operator which excludes all the occupied states. The energy denominator e is defined as $E(i)+E(j)-E(l)-E(m)$, where $E(l)$ and $E(m)$ are the initial state self-consistent energies. The indices i and j label the intermediate states whose energies are given by $E(i) = T(i)+U(i)$ where $T(i)$ is the kinetic energy and $U(i)$ the potential energy. G also satisfies

$$G\phi = V\Omega\phi = V\psi, \quad (2.10)$$

where ϕ is the unperturbed wave function, Ω the wave operator and $\psi(=\Omega\phi)$ the correlated wave function which vanishes inside of the core radius. An immediate consequence is that the reduced integral (eq. (2.5)) is no longer singular for potentials with infinite repulsive cores when V is replaced by G. From eqs. (2.9) and (2.10), we have

$$\Omega = 1 - \frac{Q}{e}V\Omega, \quad (2.11)$$

$$\psi = \phi - \frac{Q}{e}V\psi. \quad (2.12)$$

The evaluation of the G-matrix elements is very complicated due to the presence of the propagator Q/e. First, the Pauli operator Q is non-local. This causes the matrix element $\langle a(1)b(2)|G|c(1)d(2)\rangle$ to be dependent not only on nucleons 1 and 2 but also on all the other nucleons in the nucleus. Second, the determination of the energy denominator e is so far still rather unsettled. Even so, the presence of e will destroy the simplicity of the reduced integral in eq. (2.1), namely it will depend on other quantum numbers like $N\mathscr{L}$ as well. We shall discuss these aspects later in this section. Hence, it is clear that in order to evaluate the G-matrix elements one must use some appropriate approximation method. Following Bethe, Brandow and Petschek [8], the basic idea of the approximation is the following:

Let us denote the propagator Q/e by P. Then eq. (2.9) can be written as $G = V-VPG$, and eq. (2.11) can be written as $\Omega = 1-PV\Omega$. We now define the approximate reaction matrix G_A and wave operator Ω_A by

$$G_A = V_A - V_A P_A G_A, \tag{2.13}$$

$$\Omega_A = 1 - P_A V_A \Omega_A, \tag{2.14}$$

where the approximate propagator and potential P_A and V_A are chosen such that the evaluation of G_A and Ω_A can be largely simplified. Then the exact G-matrix is equal to G_A plus correction terms, namely

$$G = G_A + \Omega_A^+(V-V_A)\Omega + G_A^+(P_A-P)G. \tag{2.15}$$

The correction terms will be treated by perturbation methods and may be made small and converge rapidly by a suitable choice of P_A and V_A. In the following, we shall apply this approach to our present problem.

2.1. THE SEPARATION METHOD

The separation method was first introduced by Moszkowski and Scott for evaluating the reaction matrix elements in nuclear matter [25,26]. The idea is to divide the potential V into two parts, the short-range part V_s and the long-range part V_L. Namely

$$V = V_s\theta(d-r) + V_L\theta(r-d), \tag{2.16}$$

where $\theta(x)$ is the step function which equals to one if $x > 0$ and zero otherwise. Roughly speaking, the separation distance d is chosen so that the attractive part of V_s balances the repulsive core. Then what remains is essentially V_L. Let us choose the approximate reaction matrix and wave operator as

$$G_s = V_s - V_s \frac{1}{e_A} G_s, \tag{2.17}$$

$$\Omega_s = 1 - \frac{1}{e_A} V_s \Omega_s. \tag{2.17a}$$

Unlike the case for nuclear matter where the unperturbed wave functions are plane waves, the unperturbed wave functions for finite nuclei are taken to be the oscillator wave functions. Hence it will be convenient to choose the approximate energy denominator as

$$e_A = H_0(r) - E_{nl} + H_0(R) - E_{N\mathscr{L}} = \frac{p^2}{2\mu} + \tfrac{1}{2}\mu\omega^2 r^2 - E_{nl} + \frac{P^2}{2M} + \tfrac{1}{2}M\omega^2 R^2 - E_{N\mathscr{L}}, \tag{2.18}$$

where $H_0(r)$ and $H_0(R)$ are the relative and centre-of-mass oscillator Hamiltonian whose eigenfunctions are correspondingly $\phi_{nl}(\boldsymbol{r})$ and $\phi_{N\mathscr{L}}(\boldsymbol{R})$ with eigenvalues E_{nl} and $E_{N\mathscr{L}}$. Note that the transformation from

$$\phi_{n_1 l_1}(\boldsymbol{r}_1)\phi_{n_2 l_2}(\boldsymbol{r}_2) \quad \text{to} \quad \phi_{N\mathscr{L}}(\boldsymbol{R})\phi_{nl}(\boldsymbol{r}),$$

where

$$\boldsymbol{r} = (\boldsymbol{r}_1 - \boldsymbol{r}_2)/\sqrt{2}, \qquad \boldsymbol{R} = (\boldsymbol{r}_1 + \boldsymbol{r}_2)/\sqrt{2},$$

can be easily obtained by means of the Moshinsky transformation [24])†. The quantities μ and M are the reduced and centre-of-mass mass ($\mu = M = m$, the nucleon mass). $E_{nl} = (2n+l+\frac{3}{2})\hbar\omega$ and $E_{N\mathscr{L}} = (2N+\mathscr{L}+\frac{3}{2})\hbar\omega$, $n, N = 0, 1, 2 \ldots$. By iterating eq. (2.15), with the approximate reaction matrix defined by eq. (2.17), the following expression is obtained:

$$G \approx G_s + V_L - V_L \frac{Q}{e} V_L - 2G_s \frac{Q}{e} V_L - G_s \frac{Q-1}{e} G_s - G_s \left(\frac{1}{e} - \frac{1}{e_A}\right) G_s. \tag{2.19}$$

Note that here we take G to be Hermitian. We shall now evaluate the right-hand side term by term.

The separation distance d will be chosen so that the diagonal matrix element of G_s vanishes. With our choice of e_A, we can write the correlated wave function as $\Omega_s[\phi_{N\mathscr{L}}(\boldsymbol{R})\phi_{nl}(\boldsymbol{r})] = \phi_{N\mathscr{L}}(\boldsymbol{R})\psi_A(\boldsymbol{r})$. The subscript A denotes approximate. Then from eqs. (2.17) and (2.12) and noting that $(H_0(R) - E_{N\mathscr{L}})\phi_{N\mathscr{L}} = 0$, $\psi_A(\boldsymbol{r})$ satisfies

$$\psi_A(\boldsymbol{r}) = \phi_{nl}(\boldsymbol{r}) - \frac{1}{H_0 - E_{nl}} V_s \psi_A(\boldsymbol{r}), \tag{2.20}$$

which defines the correlated wave function in the relative coordinate. Note that here we have suppressed the quantum numbers S, T and $\mathscr{J}(=l+S)$ for clarity; ψ_A has the same l-value as ϕ_{nl} unless V_s contains tensor force. In this section we shall just consider the SE (singlet-even) case which has no tensor force. The TE (triplet-even) case where tensor force dominates will be treated in subsect. 2.2. Let $R_A(r)$ and $R_{nl}(r)$ stand for the radial parts of $\psi_A(r)$ and $\phi_{nl}(r)$ and noting that $(H_0(r) - E_{nl})\phi_{nl} = 0$, the radial part of eq. (2.20) for the SE HJ potential becomes

$$\left[-\frac{d^2}{dr^2} + \frac{l(l+1)}{r^2} + v^2 r^2 + \frac{2\mu}{\hbar^2} V_c(r)\right] R_A(r) = 2v(2n+l+\tfrac{3}{2}) R_A(r), \tag{2.21}$$

† Note well that we use the definitions of Moshinsky [24]). The more conventional definitions are $\boldsymbol{r} = \boldsymbol{r}_1 - \boldsymbol{r}_2$, $\boldsymbol{R} = \frac{1}{2}(\boldsymbol{r}_1 + \boldsymbol{r}_2)$, $\mu = \frac{1}{2}m$, $M = 2m$, $E_{nl} = (2n+l-\frac{1}{2})\hbar\omega$ and $E_{N\mathscr{L}} = (2N+\mathscr{L}-\frac{1}{2})\hbar\omega$. $n, N = 1, 2, 3 \ldots$

where ν is the oscillator size parameter defined by $\nu = \mu\omega/\hbar$ and $V_c(r)$ the central part of eq. (2.3). Note that eq. (2.21) is a bound state problem but not an eigenvalue problem.

Using Green's theorem and noting that $G_s\phi_{nl} = V_s\psi_A$, one readily has

$$(\phi_{nl}|G_s|\phi_{nl}) = \frac{\hbar^2}{2\mu}\int_{r=d}(\phi_{nl}\nabla\psi_A - \psi_A\nabla\phi_{nl})\cdot \mathrm{d}\boldsymbol{S}. \tag{2.22}$$

TABLE 1

Singlet-even separation distances

l	n	14	$\hbar\omega$(MeV) 12	8.4	7.0
0	0	1.015	1.011	1.004	1.002 (fm)
	1	1.060	1.048	1.027	1.020
	2	1.131	1.099	1.056	1.042
	3	1.285	1.184	1.092	1.068
	4		1.437	1.142	1.101
	5			1.223	1.145
	6			1.431	1.210
2	0	0.816	0.815	0.814	0.814
	1	0.820	0.818	0.816	0.815
	2	0.823	0.821	0.818	0.817
	3	0.827	0.824	0.820	0.819
	4		0.827	0.822	0.820
	5			0.824	0.822
	6			0.826	0.824

Then to make $(\phi_{nl}|G_s|\phi_{nl})$ vanish, one needs

$$\frac{R'_A(r)}{R_A(r)} = \frac{R'_{nl}(r)}{R_{nl}(r)} \quad \text{at} \quad r = d. \tag{2.23}$$

The separation distance d is therefore determined by integrating eq. (2.21) outward from the core radius r_c where $R_A = 0$ to d where the boundary condition (2.23) is satisfied. In addition, we let ψ_A heal to ϕ_{nl} for $r \geqq d$ by letting $R_A(d) = \phi_{nl}(d)$. Values of d so determined for the SE HJ potential are shown in table 1.

Using these separation distances, the diagonal matrix elements $(\phi_{nl}|G_s|\phi_{nl})$ will vanish. Note that d varies with n and l. So the separation method for the off-diagonal matrix elements is not yet defined. One approximate method to overcome this difficulty is to use an average separation distance $d = \frac{1}{2}(d(nl)+d(n'l'))$ for $(\phi_{nl}|G|\phi_{n'l'})$. Of course $(\phi_{nl}|G|\phi_{n'l'})$ is by no means made to vanish by so doing. A simpler and possibly better way is just taking an average state independent separation distance. It has been in fact found that results computed with an average state independent separation distance are very similar to those computed with a state dependent separation distance. Similar results have been found by Kallio for the Kallio-Kolltveit force.[17]). We shall use the state-independent separation. An average state independent separation distance of $d = 1.05$ fm will be used for $\hbar\omega = 14$ MeV (see table 1).

Once the sparation into V_s and V_L is made, the contributions from G_s vanish in principle. Contributions to $(\phi_{nl}|G|\phi_{n'l'})$ from various other terms of eq. (2.19) can then be computed in a straightforward way. Results for $^1S_0$ waves and the SE HJ potential are shown in table 2. The contribution from V_L can be computed most easily because it is just the ordinary radial integral $(\phi_{nl}|V_L|\phi_{n'l'})$. The method for computing the Pauli and dispersion corrections $-G_s[(Q-1)/e]G_s$ and $-G_s[1/e-1/e_A]G_s$ will be discussed in subsect. 2.4. The cross term $-2G_s(Q/e)V_L$ vanishes (except in the case tensor forces are present which we discuss later) because the correlated wave function ψ_A has been chosen to heal to ϕ_{nl} for $r \geqq d$. This can be seen as follows: Define the

TABLE 2

$^1S_0$ reduced integrals with $\hbar\omega = 14$ MeV and $d = 1.05$ fm

N	$\mathscr{L}$	n	l	V_L	$-V_L\left(\frac{Q}{e}\right)V_L$	$-G_s\frac{Q-1}{e}G_s$	$-G_s\left(\frac{1}{e}-\frac{1}{e_A}\right)G_s$	V_L(KK)
0	4	0	0	−5.61	−0.35	0.17	0.63	−6.64(MeV)
1	2				−0.35	0.15		
2	0				−0.35	0.14		
0	2	1	0	−4.53	−0.37	0.22	1.09	−5.38
1	0				−0.39	0.21		
0	0	2	0	−3.34	−0.34	0.20	1.54	−3.89

V_L(KK) are computed using the Kallio-Kolltveit potential with the same $\hbar\omega$ but $d = 1.025$ fm.

defect wave function χ_A as

$$\chi_A = \phi_{nl} - \psi_A = (1-\Omega_s)\phi_{nl}, \tag{2.24}$$

where Ω_s is defined by eq. (2.20). It follows then from eq. (2.12) that

$$(\phi_{nl}|-2G_s\frac{Q}{e}V_L|\phi_{nl}) \approx -2(\chi_A|V_L|\phi_{nl}), \tag{2.25}$$

which vanishes since χ_A has been set to zero for $r \geqq d$. In the last column of table 2, integrals of V_L for the Kallio-Kolltveit force are given for comparison. It is seen that they are very similar to those of HJ.

As shown by the table, the Pauli corrections are small compared to V_L. This is expected. Since V_s is strong and short-ranged, it therefore mainly induces transitions to intermediate states of very high energy where occupation is practically zero. Hence Q for G_s can be well approximated by 1. That the second Born term $-V_L(Q/e)V_L$ is small can also be explained.

Since V_L is central, weak and long-ranged, the intermediate states induced by V_L should be mainly near the Fermi sea where states are mostly occupied. Thus the contribution is depressed by the Pauli operator Q. Based on the same argument, higher-order corrections terms like $V_L(Q/e)V_L(Q/e)V_L$ are expected to be negligible.

2.2. TENSOR FORCE

In this section we shall discuss how to use the separation method to handle the HJ potential where the tensor force dominates. The closure approximation which has been reported recently [27]) will be discussed in more detail. As before, the potential is separated into V_s and V_L. The approximate reaction matrix G_s will be defined in the same way as the previous section, namely by eqs. (2.17) and (2.18). We shall first show how to determine the separation distance d so that $(\phi_{nl}|G_s|\phi_{nl})$ vanishes. We shall be primarily concerned with the $^3S_1$ state, since it is the most important and states of $l \geqq 2$ hardly feel the hard core.

With the presence of the tensor force, the correlated wave function ψ_A for the uncorrelated $^3S_1$ state will be a mixture of S- and D-waves. This may be expressed symbolically as

$$\psi_A = \Omega_s\phi_{n0} = u+w, \tag{2.26}$$

where u and w denote respectively the S- and D-wave and ϕ_{n0} the $^3S_1$ relative oscillator wave function. The wave operator Ω_s has been defined in the preceding section. Following the same procedure in obtaining eq. (2.21), the radial parts of u and w should satisfy the following coupled equations:

$$\left\{\frac{d^2}{dr^2}+(2n+\tfrac{3}{2})2\nu-\nu^2r^2-\frac{2\mu}{\hbar^2}V_c(r)\right\}u(r) = \sqrt{8}\,\frac{2\mu}{\hbar^2}V_T(r)w(r),$$

$$\left\{\frac{d^2}{dr^2}+(2n+\tfrac{3}{2})2\nu-\nu^2r^2-\frac{6}{r^2}+\frac{2\mu}{\hbar^2}[-V_c(r)+2V_T(r)\right.$$
$$\left.+3V_{LS}(r)+3V_{LL}(r)]\right\}w(r) = \sqrt{8}\,\frac{2\mu}{\hbar}V_T(r)u(r). \tag{2.27}$$

Here V_c, V_T, V_{LS} and V_{LL} are the central, tensor, spin-orbit and quadratic spin-orbit HJ radial potentials as shown by eq. (2.3).

It can be easily verified that in order to make $(\phi_{n0}|G_s|\phi_{n0}) = 0$ for the TE case, we need

$$\frac{u'(r)}{u(r)} = \frac{R'_{n0}(r)}{R_{n0}(r)} \quad \text{at} \quad r = d. \tag{2.28}$$

To see this, one just follows the same procedure in obtaining eqs. (2.22) and (2.23) and notes that w is orthogonal to ϕ_{n0}. Before we can determine the separation distance d, it is necessary to know the boundary condition of $w(r)$ at $r = d$. Defining the defect wave function χ_A as $\phi_{n0}-\psi_A$, it is seen that w is $-\chi_A$ in the D-channel because originally there is no D-wave in ϕ_{n0}. Then according to the reference-spectrum method [8]) which shall be discussed in the next section, the radial part of the defect wave function in the D-channel should be proportional to

$$\mathscr{H}_2(\gamma r) = \gamma r h_2^{(1)}(i\gamma r),$$

for $r > d$ where $V_s = 0$; $h_2^{(1)}$ is the exponentially decaying Hankel function [28]) and γ is the decay constant. The meaning as well as the determination of it will be discussed in subsect. 2.3. Hence, the exterior boundary condition of $w(r)$ is

$$\frac{w'(r)}{w(r)} = \frac{\mathscr{H}_2'(\gamma r)}{\mathscr{H}_2(\gamma r)} \quad \text{at} \quad r = d. \tag{2.29}$$

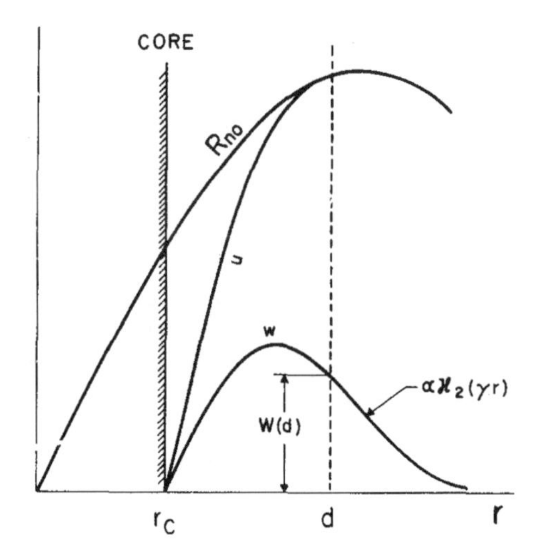

Fig. 1. Determination of the $^3S_1$ separation distance.

The separation distance d can now be determined by integrating eq. (2.27) outward from the core radius r_c till eqs. (2.28) and (2.29) are satisfied.

The numerical integration of eq. (2.27) is more complicated than the uncoupled case of eq. (2.21), because the ratio of $u'(r)$ to $w'(r)$ at r_c is not known. We only know that $u(r)$ and $w(r)$ must be zero at r_c. The standard technique for handling this problem is to take two sets of solutions, starting out from r_c with different ratios of $u'(r)$ to $w'(r)$. Denoting them by u_1, w_1 and u_2, w_2, ψ_A is then given by au_1+bu_2 and aw_1+bw_2. Constants a, b and the separation distance d are determined at the same time by integrating eq. (2.27) outward from the core till d where the following conditions are satisfied:

$$\begin{aligned} au_1+bu_2 &= R_{n0}, \\ au_1'+bu_2' &= R_{n0}', \\ \frac{aw_1'+bw_2'}{aw_1+bw_2} &= \frac{\mathscr{H}_2'(\gamma r)}{\mathscr{H}_2(\gamma r)}. \end{aligned} \tag{2.30}$$

Note that we choose u to heal to R_{n0} for $r \geqq d$. Fig. 1 summarizes how the separation distance is determined. Results of d and $w(d)$ for the TE HJ potential with $\hbar\omega = 14$ MeV and various values of γ^2 are shown in table 3. As we shall see in subsect. 2.3, the range of γ^2 used in table 3 has covered essentially the whole physical region; in fact γ^2 is state-dependent and model-dependent (see subsect. 2.3). Fortunately, one sees from table 3 that d and $w(d)$ are very insensitive to γ^2. This is partly due to the slow variation of $\mathscr{H}'_2(\gamma r)/\mathscr{H}_2(\gamma r)$ in the region of interest. Parallel to what we did for the SE case (subsect. 2.1), we decide to use a state independent d and $w(d)$. For $\hbar\omega = 14$ MeV (table 3), we shall use $d = 1.07$ fm and $w(d) = 0.10$.

TABLE 3

$^3S_1$ separation distances for $\hbar\omega = 14$ MeV

γ^2	2		4		6		8	
	d(fm)	$W(d)$	d	$W(d)$	d	$W(d)$	d	$W(d)$
$n = 0$	1.04	0.094	1.04	0.090	1.05	0.088	1.05	0.085
1	1.06	0.109	1.07	0.105	1.07	0.102	01.7	0.099
2	1.09	0.117	1.09	0.112	1.10	0.108	1.10	0.104
3	1.12	0.120	1.13	0.114	1.14	0.109	1.14	0.105

Once the separation into V_s and V_L is made, the contribution from G_s in the expansion of $(\phi_{n0}|G|\phi_{n0})$ according to eq. (2.19) vanishes in principle. The contribution from V_L can be computed in a straightforward way. One must notice however $(\phi_{n0}|V_{TL}|\phi_{n0}) = 0$, where V_{TL} is the long range part of the tensor force. Hence to compute the contribution from the tensor force which dominates in the TE HJ potential, one has to compute $-V_{TL}(Q/e)V_{TL}$ which is rather complicated. We shall now describe how to evaluate the second Born term $-V_L(Q/e)V_L$ in general.

The main difficulty is the treatment of Q/e. As usual, one writes $e = E(i)+E(j) -E(l)-E(m)$, where $E(l)$ and $E(m)$ are the self-consistent energies of the initial states. Since the initial states for our present problem are bound states in a nucleus, $E(l)$ and $E(m)$ can be taken as the binding energy of a nucleon in states l and m. Thus $E(l)$ and $E(m)$ are clearly state-dependent, namely depend on how strong states l and m are bound, $E(i)$ and $E(j)$ are the energies of the intermediate states allowed by the Pauli operator Q. It is well known that they should be evaluated off the energy shell. According to the recent investigations by Bethe and Moszkowski [29]), it is clear that taking intermediate states simply as free particles may well be good approximation (see ref. [1]), chapt. 12). We shall thus use plane wave intermediate states and consequently $E(i)$ and $E(j)$ are simply the kinetic energies.

Using plane-wave intermediates states suggests that it would be suitable to use the angle averaged Pauli operator [30]) for nuclear matter. That is, the Pauli operator

$Q(kKk_F)$ is taken to be

$$\begin{aligned} Q(kKk_F) &= 0 && \text{if } k^2+\tfrac{1}{4}K^2 < k_F^2 \\ &= 1 && \text{if } k-\tfrac{1}{2}K > k_F \\ &= \frac{k^2+\frac{1}{4}K^2-k_F^2}{kK} && \text{otherwise.} \end{aligned} \quad (2.31)$$

Here $\boldsymbol{k}$ and $\boldsymbol{K}$ are respectively the relative and centre-of-mass momentum and k_F the Fermi momentum. The angle-average approximation has been shown to be accurate in nuclear matter calculations [30]). We then have

$$(N\mathscr{L}nlS\mathscr{J}|-V_L\frac{Q}{e}V_L|N\mathscr{L}nlS\mathscr{J}) = -\int f(k)\mathrm{d}k, \quad (2.32)$$

where we recall that $N\mathscr{L}$ and nl denote respectively the centre-of-mass and relative oscillator wave function, and $\mathscr{J} = l+S$. Here $f(k)$ is given by

$$f(k) = k^2\int K^2\mathrm{d}k\mathrm{d}\Omega_K\mathrm{d}\Omega_k(\boldsymbol{kK}|V_L|N\mathscr{L}nlS\mathscr{J})^2\frac{Q(kKk_F)}{e(kK,\, lm)}, \quad (2.33)$$

with

$$e(kK,\, lm) = \frac{k^2\hbar^2}{2\mu}+\frac{K^2\hbar^2}{2M}-E(l)-E(m). \quad (2.34)$$

The integration over spatial variables is understood. By expanding the plane waves into spherical Bessel Functions, $f(k)$ can be written explicitly as

$$f(k) = \frac{8}{\pi^2}k^2\int K^2\mathrm{d}K\,\frac{Q(kKk_F)}{e(kK,\, lm)}R_{N\mathscr{L}}\left(\frac{K}{\sqrt{2}}\right)^2$$
$$\times\sum_i\sum_{l'}\langle lS\mathscr{J}|\mathcal{O}_i|l'S\mathscr{J}\rangle^2\left[\int R_{nl}(r)j_{l'}(\sqrt{2}kr)V_L^i(r)r^2\mathrm{d}r\right]^2, \quad (2.35)$$

where $R_{N\mathscr{L}}(K/\sqrt{2})$ is the oscillator wave function in the momentum representation, the oscillator size parameter for it being simply the inverse of that for $R_{N\mathscr{L}}(R)$. Here V_L has been expressed as $\sum_i Q_i V_{Li}(r)$ similar to eq. (2.3). The angular spin integrals $(lS\mathscr{J}|\mathcal{O}_i|l'S\mathscr{J})$ are given by eqs. (2.6)–(2.8); $j_{l'}(\sqrt{2}kr)$ is the spherical Bessel function. The summation over l' is just one term ($l = l'$) except for tensor force where l and l' may differ by 2. The treatment of the Pauli operator can certainly be improved upon, as is being investigated by Wong [31]).

The contribution from $-V_L(Q/e)V_L$ is then evaluated by numerically integrating eqs. (2.32) and (2.35). Before doing so, the value of k_F and range of integration for K must be chosen. In the centre region of the nucleus k_F should be very close to the nuclear matter value (1.36 fm^{-1}). However k_F decreases rapidly near the nuclear surface. For the present, we shall just take k_F as 1.3 fm^{-1}. This can certainly be improved upon. A better choice would be $k_F(\rho)$, where ρ is the local density. Such an improvement is being planned. We shall take the range of integration for K to be 0 to $2k_F$. The initial state energies $E(l)$ and $E(m)$ are state-dependent. For instance, if both l and m are the $0d_{\frac{5}{2}}$ state in ^{18}O, one should use $E(l) = E(m) \approx -4.14$ MeV which is the last neutron separation energy observed for ^{17}O. Let us first use $E(l) = E(m) = -10$ MeV to obtain some numerical results for $-V_L(Q/e)V_L$. Results for the TE HJ potential and $\hbar\omega = 14$ MeV are shown in table 4, where V_{TL} and V_{cL}

TABLE 4

$^3S_1$ second-order Born terms and the determination of e_{eff}

N	$\mathscr{L}$	n	l	$-V_{TL}\frac{Q}{e}V_{TL}$	e_{eff}	$-V_{cL}\frac{Q}{e}V_{cL}$
0	4	0	0	−4.65	221	−0.08 (MeV)
1	2			−4.54	226	−0.08
2	0			−4.52	227	−0.08
0	2	1	0	−4.37	215	−0.08
1	0			−4.42	213	−0.08
0	0	2	0	−3.21	227	−0.07

$k_F = 1.3$ fm^{-1}, $d = 1.07$ fm and $\hbar\omega = 14$ MeV are used.

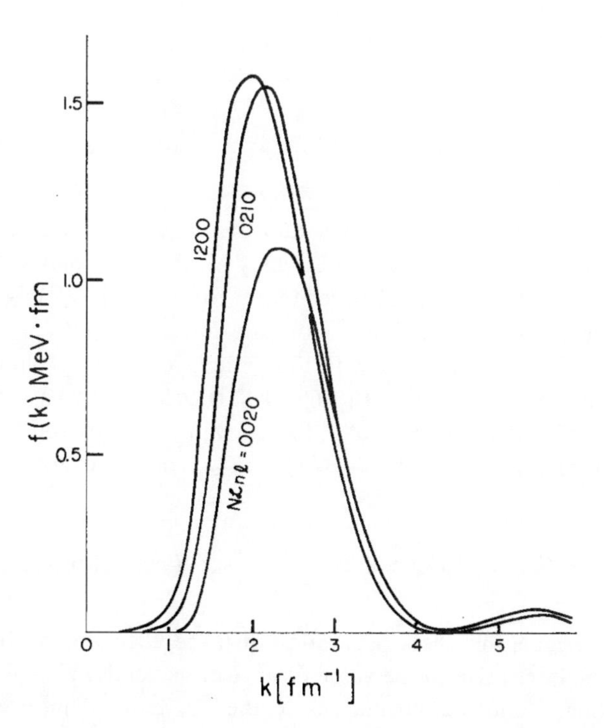

Fig. 2. Plot of $f(k)$ for $V_{Tl}(Q/e)V_{Tl}$ with $\hbar\omega = 14$ MeV, $k_F = 1.3$ fm^{-1} and $d = 1.07$ fm.

denote respectively the long-range tensor and central potential. One sees that $-V_{\mathrm{cL}}(Q/e)V_{\mathrm{cL}}$ is negligible, but $-V_{\mathrm{TL}}(Q/e)V_{\mathrm{TL}}$ is very sizable.

It would not be practical to calculate every reduced integral for $-V_{\mathrm{TL}}(Q/e)V_{\mathrm{TL}}$. Instead, it would be desirable to have a simple but accurate approximation. It is found that the integrand $f(k)$ of eq. (2.32) peaks sharply around $k \approx 2.2\ \mathrm{fm}^{-1}$ as shown by fig. 2. This is mainly due to the presence of the Pauli operator and the property of the Fourier transform of $V_{\mathrm{TL}}(r)$. This indicates clearly that the contribution to $-V_{\mathrm{TL}}(Q/e)V_{\mathrm{TL}}$ predominantly comes from intermediate states of energy around 200 MeV. It follows that one would expect using closure with an effective energy denominator to be a good approximation; namely

$$\left\langle -V_{\mathrm{TL}}\frac{Q}{e}V_{\mathrm{TL}}\right\rangle \rightarrow -\frac{1}{e_{\mathrm{eff}}}\langle V_{\mathrm{TL}}^2\rangle. \tag{2.36}$$

This is indeed found to be true. As shown by table 4, $e_{\mathrm{eff}} \approx 220$ MeV would practically reproduce the original numbers for all cases. That e_{eff} is so large indicates that the values of $E(l)$ and $E(m)$ are not very important in evaluating $-V_{\mathrm{TL}}(Q/e)V_{\mathrm{TL}}$. For example, a change of $E(l)+E(m)$ by 20 MeV would change it roughly by 10 %. However, the values of $E(l)+E(m)$ would be important when l and m are far below the Fermi surface. It is found that e_{eff} is only weakly dependent on k_{F}. Hence it would be adequate to simply use the closure approximation when not too high an accuracy is demanded. In this paper, we shall use the closure approximation with $e_{\mathrm{eff}} = 220$ MeV for the calculation of the reaction matrix elements. Eq. (2.36) can be rewritten as

$$-V_{\mathrm{TL}}\frac{Q}{e}V_{\mathrm{TL}} \rightarrow -\frac{8V_{\mathrm{TL}}(r)^2}{e_{\mathrm{eff}}}+\frac{2V_{\mathrm{TL}}(r)^2S_{12}}{e_{\mathrm{eff}}}, \tag{2.37}$$

where $V_{\mathrm{TL}}(r)$ is the radial part alone. The above is valid for triplet states, because in obtaining it we used the relation $S_{12}^2 = 8-2S_{12}$ which is true only for triplet states. As shown by eq. (2.37), the second-order tensor force $-V_{\mathrm{TL}}(Q/e)V_{\mathrm{TL}}$ now behaves like a first-order effective force. The major part of it, $8V_{\mathrm{TL}}(r)^2/e_{\mathrm{eff}}$, acts in fact as the central force. This is consistent with the findings of Feingold and Irwin [32, 33].

The magnitudes of $(N\mathscr{L}nlS\mathscr{J}|-V_{\mathrm{TL}}(Q/e)V_{\mathrm{TL}}|N'\mathscr{L}'n'l'S\mathscr{J})$ have been found to be negligibly small when $N\mathscr{L} \neq N'\mathscr{L}'$. Typical values are around 0.1 MeV.

Unlike the SE case, the cross term $-2G_{\mathrm{s}}(Q/e)V_{\mathrm{TL}}$ is no longer zero for the TE case. It is given by

$$(\phi_{n0}|-2G_{\mathrm{s}}\frac{Q}{e}V_{\mathrm{TL}}|\phi_{n0}) \approx 2\sqrt{8}\,\frac{w(d)}{\mathscr{H}_2(\gamma d)}(\mathscr{H}_2(\gamma r)|V_{\mathrm{TL}}(r)|\phi_{n0}(r)). \tag{2.38}$$

Just like the determination of the separation distance, the value of this integral has been found to be insensitive to the value of γ^2. Consequently, $\gamma^2 = 4$ has been used in the calculation. Various contributions to the $^3\mathrm{S}_1$ reduced integrals for the HJ potential are summarized in table 5, for the case of $\hbar\omega = 14$ MeV. The corresponding reduced integrals for the long-range part of the Kallio-Kolltveit potential are also shown in the table, under the heading of $V_L(KK)$. It is interesting to see the close numerical similarity between them and the sums of V_L, $-V_L(Q/e)V_L$ and $-2G_s(Q/e)V_L$ for the HJ potential. The Pauli and dispersion correction terms are also shown in the table. Their evaluation will be discussed in subsect. 2.4. Higher-order correction terms like $V_L(Q/e)V_L(Q/e)V_L$ are presumed to be small because they have similar structure as the third-order ring diagrams whose contributions are small as has been discussed by Dahlblom *et al.* for the case of nuclear matter [34]).

TABLE 5

$^3S_1$ reduced integrals with $\hbar\omega = 14$ MeV, $d = 1.07$ fm and $e_{eff} = 220$ MeV

	$n\,l$	0 0	1 0	2 0
I	V_{cL}	−3.13	−2.29	−1.61(MeV)
	V_{TL}	0	0	0
II	$-\dfrac{8V_{TL}(r)^2}{e_{eff}}$	−4.66	−4.28	−3.31
III	$-2G_s\dfrac{Q}{e}V_{TL}$	−1.94	−1.86	−1.61
		$\underline{N\mathscr{L}}$	$\underline{N\mathscr{L}}$	$\underline{N\mathscr{L}}$
	$-G_s\dfrac{(Q-1)}{e}G_s$	0.21(04)	0.25(02)	0.22(20)
		0.19(12)	0.26(10)	
		0.18(20)		
	$-G_s\left(\dfrac{1}{e}-\dfrac{1}{e_A}\right)G_s$	0.97	1.66	2.31
	I+II+III	−9.73	−8.44	−6.53
	V_L(KK)	−9.73	−8.56	−6.67

V_L(KK) are computed using the Kallio-Kolltveit potential with the same $\hbar\omega$ but $d = 0.925$ fm.

Recall that in table 2 we gave the contributions from $-V_L(Q/e)V_L$ for the SE HJ potential. They were evaluated using eqs. (2.32)–(2.34) with $k_F = 1.3$ fm^{-1}, $E(l)+E(m) = -20$ MeV and $d = 1.05$ fm.

2.3. REFERENCE SPECTRUM METHOD

When the potential outside of the repulsive hard core is never attractive, the separation method we just described is clearly no longer applicable because the essence of the separation method is that the attractive part of V_s balances in some sense the repulsive core. For the TO (triplet-odd) and SO (singlet-odd) HJ potential, the potentials are repulsive throughout for most cases. So the separation method is not applicable for these states. We shall describe in this section how to handle them using basically the reference spectrum method of Bethe, Brandow and Petschek [8]).

This method follows the same basic principle as the separation method. Namely, one first constructs an approximate reaction matrix which can be handled conveniently and is to a large extent a good approximation to the true reaction matrix. Correction terms are then treated by perturbation methods. The approximate reaction matrix used in the reference spectrum method will be denoted by G_R. It is defined by

$$G_R = V - V\frac{1}{e_R}G_R. \tag{2.39}$$

The choice of the reference spectrum energy denominator is as follows. The true energy denominator is given by $e = E(i)+E(j)-E(l)-E(m)$. In nuclear matter calculations, it is possible to treat $E(l)+E(m)$ exactly, where l and m are below the Fermi surface. The treatment of $E(i)$ and $E(j)$ where i and j are above the Fermi surface is however very difficult as we mentioned in subsect. 2.2. In the reference spectrum method for nuclear matter, the particle spectrum $E(i)$ is approximated by

$$E(i) = A + \hbar^2 k_i^2/2m^*M,$$

where the parameters m^*, the effective mass and A are calculated by a self-consistent process [8,9]). However, as we have mentioned in subsect. 2.2, the particle spectrum may well be approximated by free particles according to the recent works of Bethe and Moszkowski [29]). This removes the problem of determining A and m^*. We therefore choose e_R as

$$e_R = \frac{p_i^2}{2M} + \frac{p_j^2}{2M} - E(l) - E(m). \tag{2.40}$$

We shall in fact consider e_R as the *true* energy denominator e. Then G_R differs from G, the exact reaction matrix, only in the Pauli operator Q.

We shall now evaluate $(\phi_{N\mathscr{L}}\phi_{nl}|G_R|\phi_{N\mathscr{L}}\phi_{nl})$. Let the correlated wave function for $\phi_{N\mathscr{L}}\phi_{nl}$ be denoted by $\psi_R = \Omega_R(\phi_{N\mathscr{L}}\phi_{nl})$, where $\Omega_R = 1-(1/e_R)V\Omega_R$. Transforming to the R and r coordinates, e_R can be rewritten as

$$e_R = [H_0(R) - E_{N\mathscr{L}}] + \frac{p^2}{2\mu} + \tfrac{1}{2}E_{N\mathscr{L}} - E(l) - E(m), \tag{2.41}$$

where $H_0(R)$ is the centre of mass oscillator Hamiltonian, p is the relative momentum and $E_N = (2N+\mathscr{L}+\frac{3}{2})\hbar\omega$. In obtaining eq. (2.41), we have replaced $\frac{1}{2}M\omega^2R^2$ by the average value $\frac{1}{2}M\omega^2\langle R^2\rangle = \frac{1}{2}E_{N\mathscr{L}}$. Approximating ψ_R by $\phi_{N\mathscr{L}}(R)\psi_R(r)$, we have

$$\psi_R(r) = \phi_{nl}(r) - \frac{1}{p^2/2\mu + \gamma^2\hbar^2/2\mu}V(r)\psi_R(r), \tag{2.42}$$

with

$$\frac{\hbar^2}{2\mu}\gamma^2 = \tfrac{1}{2}E_{N\mathscr{L}} - E(l) - E(m). \tag{2.43}$$

It should be noted that the subscript R denotes the reference spectrum method rather than the centre-of-mass coordinate. The quantity γ^2 is clearly state-dependent. and moreover that it depends in fact on R implies that $\psi_R = \phi_{N\mathscr{L}}(R)\psi_R(\boldsymbol{r})$ is an approximation. As will be justified later, we shall consider γ^2 as a parameter with its value guided by eq. (2.43). The value of γ^2 will be chosen somewhat larger than indicated by the equation for the purpose of making $\psi_R(r)$ heal to $\phi_{nl}(r)$ faster. This will make G_R a better approximation to G because it approximately takes into account the Pauli operator Q which has been replaced by 1 in the reference-spectrum method. It may be mentioned that γ^2 is always positive. Typical values for $(\hbar^2/2\mu)\gamma^2$ range from ≈ 40 MeV to ≈ 150 MeV.

The defect wave function shall be denoted, as usual, by $\chi = \phi_{nl}(\boldsymbol{r}) - \psi_R(\boldsymbol{r})$. Note that ψ_R vanishes at the core radius r_c. From eq. (2.42), the radial part of χ satisfies

$$\left[\frac{d^2}{dr^2} - \frac{l(l+1)}{r^2} - \gamma^2 - \frac{2\mu}{\hbar^2} V(r)\right] \chi(r) = -\frac{2\mu}{\hbar^2} V(r) R_{nl}(r), \tag{2.44}$$

where for the SO HJ potential

$$V(r) = V_c(r) - 2\mathscr{J}(\mathscr{J}+1) V_{LL}(r), \tag{2.45}$$

and for the TO HJ potential †

$$\begin{aligned} V(r) &= V_c(r) + 2V_T(r) - V_{ls}(r) + [2\mathscr{J}(\mathscr{J}+1) - 1] V_{LL}(r), & \mathscr{J} &= l, \\ &= V_c(r) - \frac{2(\mathscr{J}+2)}{2\mathscr{J}+1} V_T(r) - (\mathscr{J}+2) V_{ls}(r) - (\mathscr{J}+2) V_{LL}(r), & \mathscr{J} &= l-1, \\ &= V_c(r) - \frac{2(\mathscr{J}-1)}{2\mathscr{J}+1} V_T(r) + (\mathscr{J}-1) V_{ls}(r) + (\mathscr{J}-1) V_{LL}(r), & \mathscr{J} &= l+1. \end{aligned} \tag{2.45a}$$

Recall that the meaning of the various potentials V has been given by eq. (2.3). The defect wave function can then be obtained by solving numerically †† eqs. (2.44) and (2.45), with the boundary conditions $\chi(r_c) = R_{nl}(r_c)$ and $\chi(\infty) = 0$. Typical solutions are shown in figs. 3 and 4. For the SO potential, χ has the same l-value as R_{nl}. For the TO potential where there is tensor force, different l-values may couple. However χ for ${}^3P_0$ and ${}^3P_1$ remain to be pure P-wave because of the conservation of $\mathscr{J}$. The defect wave function χ for ${}^3P_2$ will be a mixture of P- and F-waves. This coupling should be weak because the F-wave hardly feels the potential which is short-ranged. It therefore has been ignored.

The defect wave function χ will be obtained by solving eq. (2.44) for P-waves only. For higher partial waves, the following approximation is made:

$$\begin{aligned} \chi(r) &= R_{nl}(r), & r &< r_c, \\ &= R_{nl}(r_c)\left[2 - \frac{r}{r_c}\right], & r_c &< r < 2r_c, \\ &= 0, & r &> 2r_c. \end{aligned} \tag{2.46}$$

This should be adequate.

† We would like to thank Dr. J. Blomquist for pointing out an error in the original manuscript.

†† In numerically solving eq. (2.44), χ is integrated from both the core radius and a distant exterior point. By combining with a proper amount of the homogeneous solutions of eq. (2.44) obtained in a similar way, the two solutions of χ are matched at some interior point to give the solution for χ.

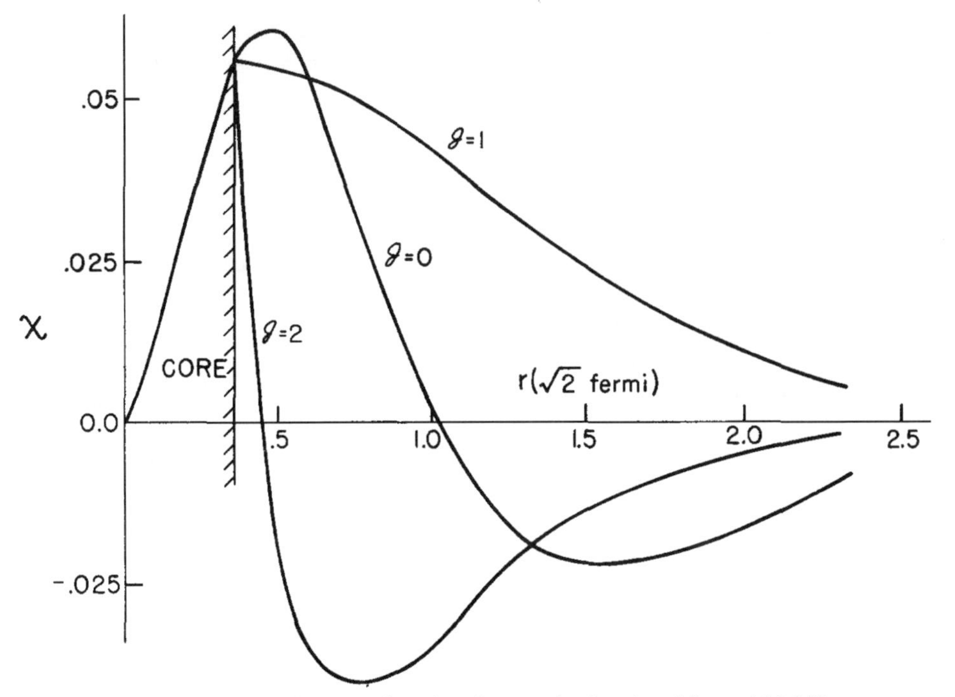

Fig. 3. $^3P_{0,1,2}$ defect wave functions for $n = 1$, $\gamma^2 = 4$ and $\hbar\omega = 14$ MeV.

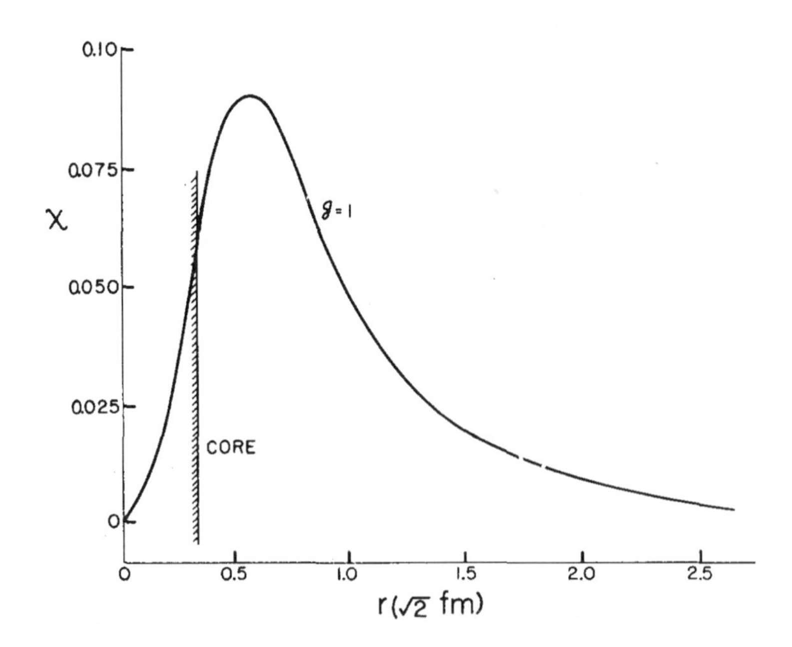

Fig. 4. $^1P_1$ defect wave function for $n = 1$, $\gamma^2 = 4$ and $\hbar\omega = 14$ MeV.

Once the defect wave function χ is known, the integral $(\phi|G_R|\phi)$ can be computed by straightforward integration since $(\phi|G_R|\phi) = (\phi|V|\psi)$ and $\psi = \phi-\chi$. It should be noted that ψ' is discontinuous at the core radius r_c. We thus break the integral into two parts: $\int_0^{r_c+}$ and $\int_{r_c+}^{\infty}$. Writing explicitly, we have

$$(\phi_{nl}|G_R|\phi_{nl}) \approx \left(E_{nl}+\frac{\hbar^2}{2\mu}\gamma^2\right)\tfrac{1}{3}r_c R_{nl}(r_c)^2 + \frac{\hbar^2}{2\mu}[R_{nl}(r)\psi'_{nl}(r)]_{r=r_c}+(\phi_{nl}|V|\psi_{nl})_{r>r_c}, \qquad (2.47)$$

TABLE 6

$^1P_1$ reduced integrals using the reference-spectrum method for $\hbar\omega = 14$ MeV

n	l	$\mathscr{J}$	$\gamma^2 = 2$	$\gamma^2 = 4$	$\gamma^2 = 6$	$\gamma^2 = 8$	$\gamma = 4$ $-G_R \frac{Q-1}{e} G_R$	N	$\mathscr{L}$
0	1	1	1.81	1.91	1.99	2.06	0.13 (MeV)	0	3
							0.12	1	1
1	1	1	2.85	3.07	3.24	3.40	0.21	0	1

TABLE 7

$^3P_{0,1,2}$ reduced integrals using the reference-spectrum method for $\hbar\omega = 14$ MeV

n	l	$\mathscr{J}$	$\gamma^2 = 2$	$\gamma^2 = 4$	$\gamma^2 = 6$	$\gamma^2 = 8$	$\gamma^3 = 4$ $-G_R \frac{Q-1}{e} G_R$	N	$\mathscr{L}$
0	1	0	−1.88	−1.82	−1.78	−1.75	0.06 (MeV)	0	3
							0.06	1	1
		1	1.92	1.99	2.04	2.08	0.13	0	3
							0.12	1	1
		2	−0.92	−0.89	−0.86	−0.85	0.04	0	3
							0.04	1	1
1	1	0	−0.95	−0.87	−0.79	−0.72	0.03	0	1
		1	2.73	2.87	2.97	3.05	0.18	0	1
		2	−1.68	−1.62	−1.57	−1.53	0.06	0	1

where $\psi'_{nl}(r)$ is $(d/dr)\psi_{nl}(r)$; $\psi_{nl}(r)$ is only the radial part of the correlated wave function corresponding to ϕ_{nl}. The first term on the right hand side shall be referred to as the volume term. It is sensitive to γ^2 and E_{nl}. The second term is the discontinuity term, and it is usually small. The third term is an ordinary integral where range of integration is from r_c to ∞. For the off-diagonal matrix elements the approximation of $(\phi_{nl}|G_R|\phi_{n'l'}) = \frac{1}{2}[(\phi_{nl}|G_R|\phi_{n'l'})+(\phi_{n'l'}|G_R|\phi_{nl})]$ will be used.

Sample values of $(\phi_{nl}|G_R|\phi_{nl})$ for the relative P-waves are given in table 6 for the

SO HJ potential, and in table 7 for the TO HJ potential. Both are computed for a wide range of γ^2 using $\hbar\omega = 14$ MeV. It is significant to see that $(\phi_{nl}|G_R|\phi_{nl})$ varies very slowly with the value of γ^2. It follows that γ^2 which is state-dependent as shown by eq. (2.43) may be taken as state-independent. This is a tremendous simplification. It is made possible by the following: First, γ^2 is small compared to V and $l(l+1)/r^2$ for small r. This makes $\int_{r_c}^{\infty}\phi V\psi dr$ insensitive to γ^2 because V is short-ranged and for small r, ψ is insentitive to γ^2. Second, the amplitude of the P-wave is very small at r_c. This makes the contribution from the volume term relatively unimportant. Recall that the volume term of eq. (2.47) is strongly dependent on γ^2. We shall simply use $\gamma^2 = 4$ for future calculation with the SO and TO HJ potential. It should be a reliable approximation.

The Pauli correction terms are also given in tables 6 and 7. Their evaluation will be discussed in subsect. 2.4.

Recall that in subsect. 2.2 while determining the separation distance for the TE HJ potential, we let the defect wave function in the D-channel for $r > d$ be proportional to $\mathscr{H}_2(\gamma r) = \gamma r h_2^{(1)}(i\gamma r)$. The reason for this can now be seen easily. Substituting V_s for V in eq. (2.44) and noting that $V_s = 0$ for $r > d$ by definition, one obtains at once that $\chi(r) \propto \mathscr{H}_2(\gamma r)$ for $r > d$ in the D-channel.

We shall now compare the reference spectrum method with the separation method. As we have just seen, it is really rather simple to use the reference spectrum method in handling the SO and TO HJ potential. One would then naturally inquire how would it work for the SE and TE HJ potential, particularly for the S-waves. For the S-waves, the reference spectrum method will not work so nicely as for the P-waves for the following reasons: First, $(\phi_{nl}|G_R|\phi_{nl})$ will be sensitive to γ^2. This largely complicates the calculation because the state-dependence of γ^2 should now be followed carefully. The simplicity of using a state-independent γ^2 no longer exists. Second, the approximation of replacing the Pauli operator Q by 1 in G_R is made less satisfactory for the S-waves than the P-waves. As a consequence, the Pauli correction terms to G_R for the S-waves are usually large, comparable in size with $(\phi_{nl}|G_R|\phi_{nl})$. On the other hand, the separation method is more satisfactory in handling the Pauli operator, because the correlated wave function can be made to heal to ϕ_{nl} for $r \geqq d$ and setting $Q = 1$ for V_s is acceptable. As we shall see in subsect. 2.4, the Pauli correction terms are directly measured by the defect wave function.

Hence for the odd-parity potentials, we have no choice but to use the reference-spectrum method and it is convenient and satisfactory here. For the even-parity potentials and particularly the S-waves, both the separation method and the reference-spectrum method are equally complicated. This is expected, because it is the S-wave which is distorted the most by the hard core and most difficult to handle. No really simple way in handling it has been established. Hence it is somwehat a matter of taste to choose one of the two methods. Calculations have in fact been done in applying the reference-spectrum method to the SE HJ potential. It is found that using $\gamma^2 \approx 4$ in the reference-spectrum method yields roughly the same $^1S_0$ reduced integrals as the separation method. This is partly why the have decided to use $\gamma^2 = 4$ in the reference-spectrum method.

It appears clear that the separation method provides a better treatment for the Pauli operator Q, while the reference-spectrum method is supposed to be more satisfactory in handling the energy denominator e providing e_R is truly a good approximation. Thus to improve the separation method, one might use e_R instead of e_A of eq. (2.18) in defining G_s of eq. (2.17). This is known as the modified separation method [8]). This method was not adopted because it would introduce certain dispersion corrections into G_s.

2.4. PROPAGATOR CORRECTION TERMS

We shall first evaluate the Pauli corrections terms. Recalling eq. (2.19), they are given by $-G_s(Q-1/e)G_s$ for the separation method. The following approximation is made:

$$(\phi| - G_s \frac{Q-1}{e} G_s|\phi) \approx -(\chi|e \frac{Q-1}{e} e|\chi). \tag{2.48}$$

The approximation is that we take $\chi = (1/e)G_s\phi$ while in fact χ is computed by $(1/e_A)G_s\phi$. The evaluation of $(\chi|e[(Q-1)/e]e|\chi)$ is then identical to that of $(N\mathscr{L}nl\mathscr{J}| -V_L(Q/e)V_L|N\mathscr{L}nl\mathscr{J})$ described in subsect. 2.2 (eqs. (2.32)–(2.35)), except replacing Q by $Q-1$, V_L by e and $\phi_{nl}(r)$ by $\chi(r)$. The calculation is simpler because $|\boldsymbol{kK})$ is the eigenstate of e with eigenvalue $k^2\hbar^2/2\mu + K^2\hbar^2/2M - E(l) - E(m)$. Note that in eq. (2.48) we have suppressed the quantum numbers N, $\mathscr{L}$, S and $\mathscr{J}$. Written in full, ϕ should be $\phi_{N\mathscr{L}}(R)\phi_{nl}(r)$ and χ should be $\phi_{N\mathscr{L}}(R)\chi(r)$; l and S couple to $\mathscr{J}$.

For the separation method, the defect wave function χ is obtained numerically while determining the separation distance. Once χ is known, the Pauli correction term can be computed according to the method we just described. Typical results are given in table 2 for the SE HJ potential and in table 5 for the TE HJ potential. They are computed using $E(l)+E(m) = -20$ MeV and $k_F = 1.3$ fm^{-1}. Note that for the $^3S_1$ case, χ has components in both the S- and D-channel. Accordingly, the Pauli correction term for it is a sum of two terms. Recall that in the calculation of the reduced integrals, we used a state-independent separation distance. Certainly no approximation of this kind was used in evaluating the Pauli correction terms.

The Pauli correction terms for the reference-spectrum method are given by $(\phi| - G_R[(Q-1)/e_R]G_R|\phi)$. They can be computed in a very similar way:

$$(\phi| - G_R \frac{Q-1}{e_R} G_R|\phi) = -(\chi|e_R \frac{Q-1}{e_R} e_R|\chi), \tag{2.49}$$

where $\chi = (1/e_R)G_R\phi$ is obtained by solving eq. (2.44). After χ is known, everything is identical to the case we just described. Typical results are given in table 6 for the SO HJ potential and table 7 for the TO HJ potential. They are computed using $E(l)+E(m) = -20$ MeV and $k_F = 1.3$ fm^{-1}. As shown by the tables, they are small in both the separation method for the S-waves and the reference-spectrum method for the P-waves. This indicates that these methods have been suitably chosen in the treatment of the Pauli operator Q, as we have discussed in subsect. 2.3. Note that all the Pauli correction terms are positive.

The other propagator correction term is the dispersion term $-G_s(1/e - 1/e_A)G_s$ in the separation method. There is no dispersion term in the reference-spectrum method because e_R is presumed to be equivalent to the exact e. The following approximation is used:

$$(\phi| - G_s \left(\frac{1}{e} - \frac{1}{e_A}\right) G_s|\phi) \approx -(\chi|(e_A - e)|\chi) \approx \overline{(e_A - e)}(\chi \mid \chi). \tag{2.50}$$

The relation $\chi = (1/e_A)G_s\phi$ has been used in obtaining the above equation; $\overline{e_A - e}$ denotes the average value of $e_A - e$ for χ which is highly localized. Recalling that e_A is defined as $H_0(R) + H_0(r) - E_{N\mathscr{L}} - E_{nl}$ and taking e as $p^2/2\mu + P^2/2M - E(l) - E(m)$, eq. (2.50) becomes

$$(\phi| - G_s \left(\frac{1}{e} - \frac{1}{e_A}\right) G_s|\phi) \approx [\tfrac{1}{2}E_{N\mathscr{L}} + E_{nl} - E(l) - E(m)](\chi|\chi). \tag{2.51}$$

Note that ϕ is certainly specified by $N\mathscr{L}nl$. In obtaining eq. (2.51), we have taken the average value of $-\frac{1}{2}M\omega^2R^2$ as $-\frac{1}{2}E_{N\mathscr{L}}$ and ignored the average value of $-\frac{1}{2}\mu\omega^2r^2$ because $\chi(r)$ is very short-ranged. The dispersion term will be slightly smaller if the average value of $-\frac{1}{2}\mu\omega^2r^2$ is not ignored, $\chi(r)$ is obtained numerically while determining the separation distance. Then the dispersion term can be evaluated easily.

Results shown in table 2 for the $^1S_0$ waves and table 3 for the $^3S_1$ waves are calculated according to eq. (2.51). It is seen that they are all positive and increase rapidly with n. The rapid increase is expected because both E_{nl} and $(\chi|\chi)$ increase rapidly with n. Because our $\chi(r)$ has no dependence on $N\mathscr{L}$, the dispersion term, unlike the Pauli term, does not depend on N and $\mathscr{L}$ individually. Instead, it depends on $E_{N\mathscr{L}}$ as a whole as shown by eq. (2.51). An important point to be noted is that the dispersion term is proportional to $(\chi|\chi)$. Since $\chi = \phi - \psi$ and vanishes inside of the hard core, the dispersion term therefore depends very sensitiviely on the size of the hard core.

We shall, however, not include the dispersion correction in our matrix elements, as discussed later.

2.5. SUMMARY

In this section, we summarize our methods for evaluating the shell model reaction matrix elements. The basic formula is eq. (2.1) with V replaced by the reaction matrix G. The main task is then to evaluate the reduced integral $(nlST\mathscr{J}|G|n'l'ST\mathscr{J})$. We shall ignore all partial waves with l greater than 5 in computation, because their contributions are negligible.

For the SE HJ potential, we used the separation method. The reduced integral will be calculated by

$$(nlST|G|n'l'ST) \approx \delta_{ll'}(nlST\mathscr{J}|V_L|n'l'ST\mathscr{J}), \tag{2.52}$$

where the separation distance d defining V_L, the long-range part of the potential, is taken to be state-independent. It is an average of the ${}^1S_0$ separation distances. Recalling the expansion of eq. (2.19), eq. (2.52) amounts to leaving out the following terms, namely $-V_L(Q/e)V_L$, $-G_s[(Q-1)/e)]G_s$ and $-G_s[(1/e-1/e_A)]G_s$ noting that the cross term $-2G_s(Q/e)V_L$ vanishes. The first two terms are ignored because they are small and tend to cancel each other as shown by table 2. Although the dispersion term $-G_s(1/e-1/e_A)G_s$ is large as shown by the table, it is left out for the following reasons:

As we have just discussed (eq. (2.50)), the dispersion term is given by $-\overline{(e_A-e)}(\chi|\chi)$. The quantity $(\chi|\chi)$ is extremely sensitive to the size of the hard core. It will decrease rapidly as the core diminishes. The HJ potential gives too little binding energy per nucleon in nuclear matter [10]), which indicates that its core radius may be too large. Furthermore, the infinite hard core is anyhow highly phenomenological. More realistically, we should have a soft repulsive core resulting presumably from the exchange of heavier mesons. Such a potential is being investigated by Wong [35]). When either a soft-core potential is used or the hard-core radius is reduced, the dispersion term is small. As has been found by Wong [36]), in going from a vertical core to a (more realistic) Yukawa repulsion in fitting the same nucleon-nucleon data, the quantity $(\chi|\chi)$ drops by a factor of approximately 2. Moreover, the dispersion term is sensitive to off-the-energy shell effects. The potential we are using is deduced from the two-body scattering data which provide only on-the-energy shell information. Therefore, before one is more sure about the nucleon-nucleon potential and how to handle off-the-energy effects, it would be advisable to simply leave out the dispersion corrections, which should end up being much smaller than estimated here for a vertical core.

For the TE HJ potential, we again use the separation method. The reduced integral is calculated by

$$(nlST\mathscr{J}|G|n'l'ST\mathscr{J}) \approx (nlST\mathscr{J}|V_L - V_{TL}\frac{Q}{e}V_{TL} - G_s\frac{Q}{e}V_{TL} - V_{TL}\frac{Q}{e}G_s|n'l'ST\mathscr{J}), \tag{2.53}$$

with $-V_{TL}(Q/e)V_{TL}$ given by $-8V_{TL}(r)^2/e_{eff}+2V_T(r)^2S_{12}/e_{eff}$, the closure approximation of eq. (2.37). The contributions from the cross terms are evaluated according to the approximation

$$-V_{TL}\frac{Q}{e}G_s|l) = -\delta_{l0}V_{TL}|\chi_D), \tag{2.54}$$

where $|\chi_D)$ is given by $[-w(d)/\mathscr{H}_2(\gamma d)]\mathscr{H}_2(\gamma r)$ as discussed in subsect. 2.2. Both the separation distance d and $w(d)$ are taken to be state-independent. They are obtained from averaging the $^3S_1$ separation distances and $w(d)$'s. Thus terms which we shall omit are the Pauli terms, dispersion terms and $-V_L(Q/e)V_L$ for V_L other than the tensor forces. These terms are all found to be negligible, as shown partly by table 5, except the dispersion corrections. The reasons for leaving them out are the same as we just discussed for the SE case.

For the SO and TO HJ potential, we employ the reference-spectrum method. We adopt

$$(nlST\mathcal{J}|G|nlST\mathcal{J}) \approx \tfrac{1}{2}(nlST\mathcal{J}|V\Omega_R+\Omega_R V|nlST\mathcal{J}), \tag{2.55}$$

where $\Omega_R\phi$ yields the correlated wave function $\psi = \phi-\chi$ with χ given by eqs. (2.44)–(2.46). A state-independent γ^2 will be used. The terms we shall omit are the Pauli corrections. As shown by tables 6 and 7, they are in fact negligible.

TABLE 8

Decomposition of the $(0d_{\frac{5}{2}}{}^2JT|G|0d_{\frac{5}{2}}{}^2JT)$ matrix elements, calculated with $d = 1.05$ fm for the SE potential, $d = 1.07$ fm and $w(d) = 0.1$ for the TE potential, $\gamma^2 = 4$, $e_{\text{eff}} = 220$ MeV and $\hbar\omega = 14$ MeV. These specifications will be used from now on unless otherwise specified

T		1			0	
J	0	2	4	1	3	5
V_c	−2.50	−0.95	−0.49	−0.43	−0.49	−1.33 (MeV)
V_T	0.59	0.01	0.10	0.41	0.42	0.20
V_{LS}	0 33	−0.39	−0.14	−0.01	0.02	0.01
V_{LL}	−0.04	−0.02	−0.01	1.02	0.26	0.01
$-2G_s\frac{Q}{e}V_{TL}$				−0.57	−0.35	−0.73
$-8V_{TL}{}^2/e_{\text{eff}}$				−1.10	−0.82	−1.80
$2V_{TL}{}^2S_{12}/e_{\text{eff}}$				0.08	0.03	−0.01
$S = 0$	−2.44	−0.84	−0.45	1.39	0.43	0
$S = 1$	0.97	−0.18	0.07	−1.93	−1.36	−3.64
total	−1.47	−1.02	−0.39	−0.54	−0.93	−3.64
K−K	−2.85	−0.77	−0.50	−2.49	−1.61	−3.65

The shell-model reaction matrix elements are then computed following the above techniques. Typical results are shown in table 8 for the HJ potential. As shown, they are decomposed so that the contributions from various parts can be visualized. It should be noticed, however, that the decomposition is not exact. Recall that in the reference-spectrum method for the odd-parity potentials, the radial reduced integrals are divided into two parts, i.e.

$$\int_0^{r_c{}^+}, \quad \int_{r_c{}^+}^{\infty}.$$

The decomposition is practical only for the latter. Therefore, for the odd-parity potentials, the terms V_c, V_T, V_{LS} and V_{LL} in table 8 are just for $\int_{r_c{}^+}^{\infty}$. Contributions from $\int_0^{r_c{}^+}$ are included as a whole in the respective numbers shown under the heading of

$S = 0$ (spin singlet), $S = 1$ (spin triplet) and total. The corresponding matrix elements for the Kallio-Kolltveit force (KK) are also given in the table for comparison. It is interesting to note the close similarity between the KK and the even-parity HJ matrix elements.

3. Single-Particle Energies

As we mentioned earlier, the single-particle wave functions and energies should be determined by the self-consistent calculation using the nucleon-nucleon interaction we have chosen, namely the HJ potential. This is, however, very difficult and we shall not attempt it. Instead, we assume the self-consistent wave functions to be those of an oscillator, namely ϕ_{nlj}. We shall consider ^{17}O which has a valence nucleon outside the ^{16}O core. Assuming that the single-pargicle wave functions of the 17 nucleons to be given by those of an oscillator, the single-particle energy of the valence nucleon can be easily evaluated by letting it interact with the nucleons in the ^{16}O core, via the two-nucleon interactions we have chosen. These energies can then be compared with the observed values, which are usually furnished by the ^{16}O(d, p)^{17}O experiments [21].

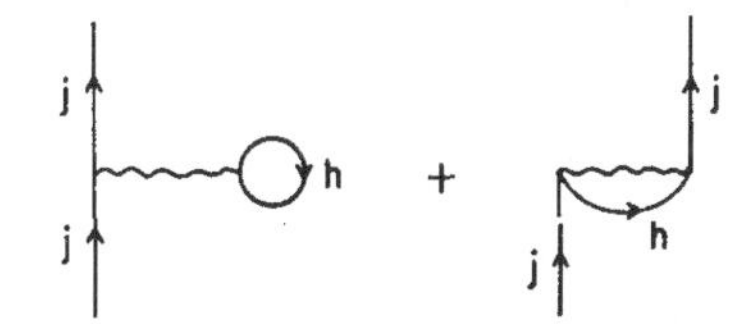

Fig. 5. First-order diagrams for the evaluation of the single-particle energies.

It may be noted that the oscillator wave function defined by a potential well of infinite depth can certainly be improved upon. A better choice would be those of a finite potential well, e.g. the Wood-Saxon wave functions. They are in fact pretty much the same except in the tail region. The former are chosen because it is much more convenient to transform them to the relative and centre-of-mass coordinates than the latter †. The size parameter $\hbar\omega$ of the oscillator well is determined so that the mean square radius of the nucleus is reproduced. This leads to $\hbar\omega = 41A^{-\frac{1}{3}} = 15.6$ MeV for $A = 18$, if the nuclear radius [31]) is taken to be 1.2 $A^{\frac{1}{3}}$ fm. However, according to the electron scattering data, the value of $\hbar\omega$ should be $\approx$ 14 MeV for the oxygen isotopes [38]). Calculations will be carried our for both of these values.

We shall use the linked cluster perturbation formula [39,40]) to evaluate the *potential* energy of the valence nucleon of ^{17}O due to its interaction with the ^{16}O core. We shall average over the isospin quantum numbers. Thus the single-particle energies we shall determine are charge-independent. The first-order contributions are the G-matrix Hartree-Fock terms shown by fig. 5, where the wavy line denotes the G-interaction.

† The transformation of the Wood-Saxon wave function to the centre-of-mass and relative coordinates cannot be carried out simply. It is usually done by first expanding the Wood-Saxon wave functions into the oscillator wave functions which can then be transformed conveniently using the Moshinsky method.

The contribution from this class of diagrams is given by

$$V_{\mathrm{HF}} = \frac{1}{2(2j+1)} \sum_{J,\,T,\,h} (2T+1)(2J+1)(jhJT|G|jhJT), \tag{3.1}$$

where $(jhJT|G|jhJT)$ is the antisymmetrized and normalized particle-particle reaction matrix element whose evaluation has been discussed in sect. 2. This notation will be

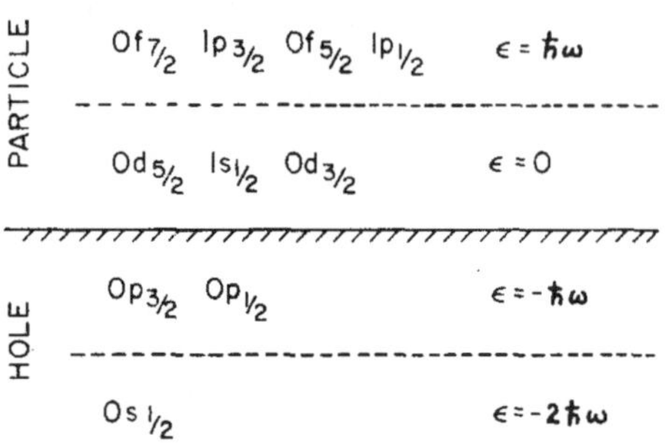

PARTICLE

$0f_{7/2}$ $1p_{3/2}$ $0f_{5/2}$ $1p_{1/2}$ $\epsilon = \hbar\omega$

$0d_{5/2}$ $1s_{1/2}$ $0d_{3/2}$ $\epsilon = 0$

HOLE

$0p_{3/2}$ $0p_{1/2}$ $\epsilon = -\hbar\omega$

$0s_{1/2}$ $\epsilon = -2\hbar\omega$

Fig. 6. Configurations included in the calculation of the single-particle energies of ^{17}O. The indicated ε values are used to determine the energy denominators in the perturbation theory. The same configurations are used for the ^{18}O and ^{18}F spectra except using $\varepsilon(0d_{\frac{5}{2}}) = 0$, $\varepsilon(1s_{\frac{1}{2}}) = 0.87$ and $\varepsilon(0d_{\frac{3}{2}}) = 5.08$, all in MeV.

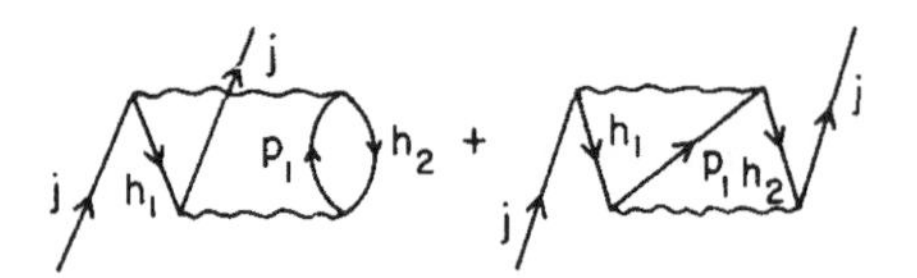

Fig. 7. Diagrams which give the V_{3p2h} corrections to the single-particle energies.

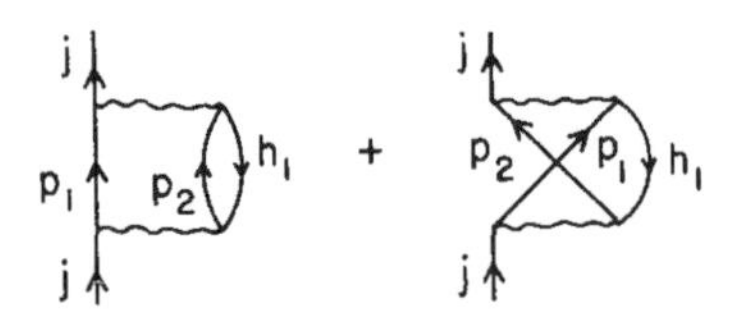

Fig. 8. Diagrams which give the V_{2p1h} corrections to the single-particle energies.

used from now on. The index j refers to the orbitals $0d_{\frac{5}{2}}$, $1s_{\frac{1}{2}}$ and $0d_{\frac{3}{2}}$. The index h refers to the hole orbitals as shown by fig. 6 where we consider $\hbar\omega = 14$ MeV.

Since the ^{16}O core is certainly not absolutely closed, we must take into account the correction terms due to the p-h (particle-hole) excitation of the core. One kind of them come from the processes shown by fig. 7 where two particles are lifted from the core. The corresponding contributions are given by

$$V_{\mathrm{3p2h}} = \frac{1}{2(2j+1)} \sum_{J,\,T;\,p_1,\,h_1>h_2} (2T+1)(2J+1)(1+\delta_{jp_1}) \frac{(jp_1JT|G|h_1h_2JT)^2}{\varepsilon_j+\varepsilon_{p_1}-\varepsilon_{h_1}-\varepsilon_{h_2}}, \tag{3.2}$$

where the configurations for h_1, h_2 and p_1 to be summed over are shown in fig.6, and $h_1 > h_2$ means that permutations between h_1 and h_2 are excluded in the sum. Note that while summing over h_1, h_2 and p_1, the combined parity of them must be the same as j. The energy denominator $\varepsilon_j+\varepsilon_{p_1}-\varepsilon_{h_1}-\varepsilon_{h_2}$ is taken to be a multiple of $\hbar\omega$ (= 14 MeV) as shown by fig. 6.

Another kind of correction comes from the processes shown by fig. 8 where the particle states p_1, p_2 and the hole states h_1 to be summed over are given by fig. 6. It must be noted that these contributions seem to have been already included in V_{HF} of eq. (3.1). This is, however, not quite true. Recall that in our evaluation of the G-

TABLE 9

Single-particle potential energies of ^{17}O

		$0d_{\frac{5}{2}}$	$1s_{\frac{1}{2}}$	$0d_{\frac{3}{2}}$
	total	−25.51	−25.21	−20.48 (MeV)
	$T=0$	−14.08	−14.91	−14.17
V_{HF}	$T=1$	−11.43	−10.30	− 6.31
	$l\cdot s$	− 3.95	− 2.59	2.20
	total	1.86	2.32	3.10
	$T=0$	1.49	1.98	2.67
$V_{3\mathrm{p}1\mathrm{h}}$	$T=1$	0.37	0.34	0.43
	$l\cdot s$	0.05	0.10	0.04
	total	− 3.49	− 4.14	− 4.25
	$T=0$	− 2.68	− 3.04	− 3.33
$V_{2\mathrm{p}1\mathrm{h}}$	$T=1$	− 0.80	− 1.10	− 0.92
	$l\cdot s$	− 0.23	− 0.41	− 0.20
total		−27.14	−27.03	−21.63
relative		0	0.11	5.51
exp		0	0.87	5.08
relative	$(l\cdot s)$	0	1.23	6.17

matrix elements, we used the plane wave intermediate states and a rather large value of γ^2 (≈ 80 MeV). This indicates that the intermediate states included in the G-matrix are predominantly those of rather high energy, which should have small overlap with states p_1 and p_2. Hence the contributions from the above diagrams are to large extent left out in V_{HF}, although there is certainly some overlap. The contributions from this class of diagrams are given by

$$V_{2\mathrm{p}1\mathrm{h}} = \frac{-1}{2(2j+1)} \sum_{J,\,T,\,h_1,\,p_1>p_2} (2J+1)(2T+1)\frac{(jh_1JT|G|p_1p_2JT)^2}{\varepsilon_{p_1}+\varepsilon_{p_2}-\varepsilon_{h_1}-\varepsilon_j}, \tag{3.3}$$

where the energy denominator will be taken as a multiple of $\hbar\omega$ as before.

Results of the potential energies V_{HF}, $V_{3\mathrm{p}2\mathrm{h}}$ and $V_{2\mathrm{p}1\mathrm{h}}$ using the HJ potential and $\hbar\omega$ = 14 MeV are shown in table 9. Here $T=0$, $T=1$ and $\boldsymbol{l}\cdot\boldsymbol{s}$ denote respectively the contributions from the isospin singlet, triplet and the $\boldsymbol{l}\cdot\boldsymbol{s}$ potentials. It is seen that the largest contribution comes from the G-matrix Hartree-Fock diagrams with

V_{3p2h} and V_{2p1h} both an order of magnitude smaller. This indicates that higher-order terms may not be important. It is amazing to see that the relative spacings between the single-particle levels agree so well with the observed ones. Furthermore, as shown by the table, the doublet splitting comes overwhelmingly from V_{HF} via the two-body $T = 1$ $\boldsymbol{l} \cdot \boldsymbol{s}$ force. It may be noted that the $\boldsymbol{l} \cdot \boldsymbol{s}$ force is non-vanishing only for the states of spin triplet ($S = 1$) and $l \neq 0$. It is then clear that it is the TO (triplet-odd) $\boldsymbol{l} \cdot \boldsymbol{s}$ force which is largely responsible for the doublet splitting. The result differs strongly from many other approaches where the doublet splittings resulted predominantly from the second-order phenomenological tensor force [41]). Our result, however, is generally consistent with the findings of Brueckner *et al.* [42]), where the Gammel-Thaler potential was used.

It should be pointed out that our result for the separation between 0d and 1s orbitals will be altered when better wave functions are used. As is well known, one may add a correction term $Dl^2 (D < 0)$ to the oscillator potential to account for approximately the difference between the oscillator potential well and the more realistic potential well of finite depth. This indicates that the calculated 0d levels will be further pushed down slightly relative to the $1s_{\frac{1}{2}}$ level when better wave functions are employed.

We shall now evaluate the absolute binding energy of the valence nucleon of ^{17}O. As shown by table 9, the total potential energy of $0d_{\frac{5}{2}}$ is -27.14 MeV. Recall that this is obtained by using the oscillator wave functions together with the HJ potential. To be consistent, the kinetic energy of the $0d_{\frac{5}{2}}$ nucleon is therefore taken to be $\frac{1}{2}(2n+l+\frac{3}{2})\hbar\omega = 24.5$ MeV because $n = 0$, $l = 2$ and $\hbar\omega$ has been chosen to be 14 MeV. Hence the net binding energy for the $0d_{\frac{5}{2}}$ nucleon is 2.64 MeV. The observed value is 4.15 MeV obtained by taking the binding energy difference of ^{17}O and ^{16}O (ref. [43])). It is encouraging to see that the calculated binding energy is of the right order of magnitude. There are certainly many other processes which we have left out in our calculation. The inclusion of them should somewhat increase the binding energy but possibly with small effect on relative spacings. We have already seen this in table 9, because the sum of V_{3p2h} and V_{2p1h} give essentially an additional binding of 1 MeV to all three levels with their relative spacings practically unaffected.

The above calculation has been repeated for $\hbar\omega = 15.6$ MeV. It is found that the total potential energy $(V_{HF} + V_{3p2h} + V_{2p1h})$ for $0d_{\frac{5}{2}}$ varies almost in direct proportion to the value of $\hbar\omega$. This is interesting, because the average potential energy of a nucleon inside an oscillator well is proportional to $\hbar\omega$. The relative spacings are about the same as the $\hbar\omega = 14$ MeV case. This result shows that the calculated single-particle energies are not sensitive to the value of $\hbar\omega$ as long as it stays within a reasonable range.

Similar calculations have been done for ^{41}Ca, and equally satisfactory results are obtained [44]). Therefore, that we assume the self-consistent wave function for the HJ potentials to be those of the oscillator well does appear to be largely valid.

Since our calculation for the single-particle energies has yielded reasonable results, it would be interesting to evaluate the total binding energy of the nucleus using the same set ($\hbar\omega$ = 14 MeV) of matrix elements. This can be done easily for ^{16}O. The first-order potential energy of ^{16}O is given by summing up the diagrams shown by fig. 9. Writing explicitly, the contribution is

$$V_{2h} = \sum_{h_1 \geqq h_2} \sum_{JT} (2J+1)(2T+1)(h_1 h_2 JT|G|h_1 h_2 JT), \tag{3.4}$$

where the hole states h_1 and h_2 to be summed over are given by fig. 6. Recall that $(h_1h_2JT|G|h_1h_2JT)$ is the antisymmetrized and normalized reaction matrix element. Using the HJ matrix elements for $\hbar\omega$ = 14 MeV, it is found that V_{2h} = -340 MeV.

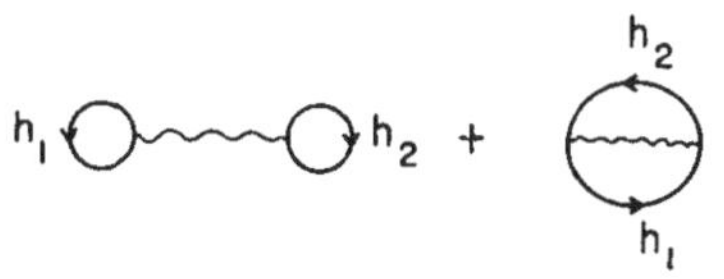

h_1 h_2 + h_2 h_1

Figure 9

Fig. 9. Diagrams including in evaluating the binding energy of ^{16}O.

Taking the kinetic energy of each nucleon as $\frac{1}{2}E_{nl}$, where $E_{nl} = (2n+l+\frac{3}{2})\hbar\omega$, the total kinetic energy is 252 MeV. The sum of the two gives the total binding energy of ^{16}O as 88 MeV, compared to the observed [43]) value of 127.6 MeV.

Clearly there are higher-order terms we have not calculated. In particular, the ring diagrams. But since they have been found to be small in nuclear matter [29]), they might also be unimportant for the present problem. Note that we have not included the Coulomb energy yet which is approximately 14 MeV. So, we do obtain some binding for ^{16}O, but it is far from enough. This is similar to the result of nuclear matter calculations using the HJ potential [10]).

4. Effective Interaction and Core Polarization

We shall now calculate the spectra of ^{18}O and ^{18}F. The methods we shall develop are, however, applicable in general.

For low-lying states, these nuclei may be considered basically as an inert ^{16}O core plus two valence nucleons confined in the sd shell. Then to first approximation, the spectra can be obtained by diagonalizing the two-nucleon secular matrix in the sd shell. This is a rather simple problem and makes feasible the conventional approaches of considering the matrix elements themselves as adjustable parameters. Their values are determined by requiring an optimum fit to the observed data. The matrix elements so determined may be called the effective interactions. A most recent determination of such effective interactions has been given by Federman-Talmi [2,3]) (referred to as FT hereafter) for ^{18}O. We shall investigate whether their values can be deduced from the free nucleon-nucleon potential which is chosen to be the HJ potential in the present work.

Let us first compare their values with our G-matrix elements as shown in table 10. The latter are calculated with the same specifications as given in the caption of table 8. Note that in both their work and what we shall do, the single-particle energies are taken from experiments as $\varepsilon(0d_{\frac{5}{2}}) = 0$, $\varepsilon(1s_{\frac{1}{2}}) = 0.87$ and $\varepsilon(0d_{\frac{3}{2}}) = 5.08$, all in MeV (ref. [21]). As shown by the table, no impressive agreement is seen. In particular, the $J = 4$ and $J = 3$ matrix elements of Federman-Talmi are repulsive while for any reasonable potential they would be attractive. It is thus obvious that the calculated

TABLE 10
Comparison of the $T = 1$ matrix elements

j_a	j_b	j_c	j_c	J	Federman Talmi	G	HJ $-G\frac{Q_{3p1h}}{2\hbar\omega}G$	Sum
$0d_{\frac{5}{2}}$	$0d_{\frac{5}{2}}$	$0d_{\frac{5}{2}}$	$0d_{\frac{5}{2}}$	0	−3.24	−1.47	−1.06	−2.53(MeV)
				2	−1.59	−1.02	0.08	−0.94
				4	0.03	−0.39	0.53	0.14
$0s_{\frac{1}{2}}$	$0s_{\frac{1}{2}}$	$0s_{\frac{1}{2}}$	$0s_{\frac{1}{2}}$	0	−1.97	−2.28	0.07	−2.21
$0d_{\frac{5}{2}}$	$0s_{\frac{1}{2}}$	$0d_{\frac{5}{2}}$	$0s_{\frac{1}{2}}$	2	−0.76	−1.04	−0.05	−1.09
				3	0.72	−0.30	0.54	+0.24
$0d_{\frac{5}{2}}$	$0d_{\frac{5}{2}}$	$0s_{\frac{1}{2}}$	$0s_{\frac{1}{2}}$	0	−0.77	−0.76	−0.33	−1.10
$0d_{\frac{5}{2}}$	$0d_{\frac{5}{2}}$	$0d_{\frac{5}{2}}$	$0s_{\frac{1}{2}}$	2	−0.48	−0.52	−0.29	−0.81
$0d_{\frac{5}{2}}$	$0d_{\frac{5}{2}}$	$0d_{\frac{3}{2}}$	$0d_{\frac{3}{2}}$	0		−3.18	−0.93	−4.11
				2		−0.58	−0.36	−0.94

spectra will be by no means satisfactory if we barely diagonalize the G-matrix in the sd shell. The comparison we just made, is however, not appropriate. As we shall see, it is the *model* interaction rather than the G-interaction to which the effective interaction of FT should be compared.

4.1. THE MODEL INTERACTION

Because of the strong short-ranged nucleon-nucleon interaction such as the HJ potential, to confine the valence nucleons in the sd shell with the ^{16}O core closed is simply too limited to give an adequate representation for the nuclear states. Even after elimination of the hard-core and short-range attraction in obtaining the reaction matrix, the latter is still sufficiently strong so that it is still too limited to confine the valence nucleons in the sd shell with the ^{16}O core closed. This limited space may be referred to as the model space. To have a better representation, a straightforward way is to include more configurations in constructing the secular matrix. This, however, is not practical because the difficulty in calculation increases very rapidly with the number of configurations included. Thus we shall use the perturbation method to account for the configurations left out in the model space. We shall basically follow the approach of Eden and Francis [45])[†].

Suppose the eigenvalue problem to be solved is $(H_0+V)\psi = E\psi$. The unperturbed wave functions ϕ_i are defined by $H_0\phi_i = \varepsilon_i\phi_i$. When the perturbing potential is weak, we may approximately expand ψ as a finite sum of ϕ_i. Then to solve $(H_0+V)\psi = E\psi$ amounts to diagonalizing a secular matrix of finite size. But when V induces strong short-range correlations like the HJ potential, ψ must be expanded as an infinite sum of ϕ_i, i.e. $\psi = \sum_{i=1}^{\infty} a_i\phi_i$. This will lead to an infinite secular matrix which is rather meaningless. We now define the model space to be spanned by ϕ_i, $i = 1, m$. The eigenfunction in the model space is denoted by $\psi_M = \sum_{i=1}^{m} a_i\phi_i$. The wave functions ψ and ψ_M are formally related by

$$\psi = \Omega_M\psi_M, \tag{4.1}$$

with

$$\Omega_M = 1 - \frac{Q_M}{H_0 - E} V\Omega_M, \tag{4.2}$$

where $Q_M = \sum_{i=m+1}^{\infty} |\phi_i)(\phi_i|$. By making use of the orthogonality property among ϕ_i's, the infinite dimensional problem $(H_0+V)\psi = E\psi$ is formally reduced to a finite dimensional problem

$$(H_0 + V\Omega_M)\psi_M = E\psi_M \tag{4.3}$$

in the model space of ϕ_i, $i = 1, 2, \dots m$. The configurations ϕ_i, $i = m+1, \dots \infty$ are to be formally taken care of by the wave operator Ω_M. The operator $V\Omega_M$ shall be called the model interaction. It is clear now that it is the model interaction $V\Omega_M$ rather than V which we shall diagonalize when we choose to work in the finite dimensional model space.

It now can be seen why we used the G-matrix instead of V in sect. 2. If in eq. (4.2) we restrict Q_M to project on the two nucleon states outside of the sd shell, then $V\Omega_M$ is essentially the G-matrix we have defined. It should be pointed out that the intermediate states we used in defining the G-matrix must have some overlap with the sd shell. The associated double-counting is presumed to be small because the G-matrix intermediate states are mostly of high energy. So we may rewrite eq. (4.3) as

$$(H_0 + G\Omega'_M)\psi_M = E\psi_M, \tag{4.4}$$

where the wave operator Ω'_M is supposed to take care of the configurations which are significant but not included in both the model space and that in defining the G-matrix. For instance, Ω'_M may take care of the 3p1h (three-particle-one-hole) configurations where a particle is lifted out from the ^{16}O core. We illustrate these points in fig. 10. As shown, $a, b \dots f$ refer to the shell-model, single-particle states in the sd shell; α and β are single-particle states outside of the sd shell, and usually they have much higher energy. The processes like those shown by (i) are included in the G-matrix, where the dotted line stands for a V-interaction. The processes like (ii) are included

[†] See Bloch and Horowitz [45]) and Brandow [40]) for a more complete and rigorous formulation.

in the diagonalization of G in the sd shell, where a wavy line stands for the G-interaction. The operator Ω'_{M} of eq. (4.4) is then, for example, to take care of processes as shown by (iii), where p and h are respectively the shell-model particle and hole states.

The interaction $V\Omega_{\mathrm{M}}$ or $G\Omega'_{\mathrm{M}}$ will be referred to as the model interaction. And it is the model interaction which we should diagonalize in the model space and compare with the results of FT. Namely, the ordinary effective interaction should in fact correspond to the model interaction.

4.2. METHOD OF CALCULATION

In this section, we describe how to calculate $G\Omega'_{\mathrm{M}}$, where Ω'_{M} arises from corrections due to the low-lying 3p1h configurations. That is, we shall evaluate $G\Omega_{3\mathrm{p1h}}$.

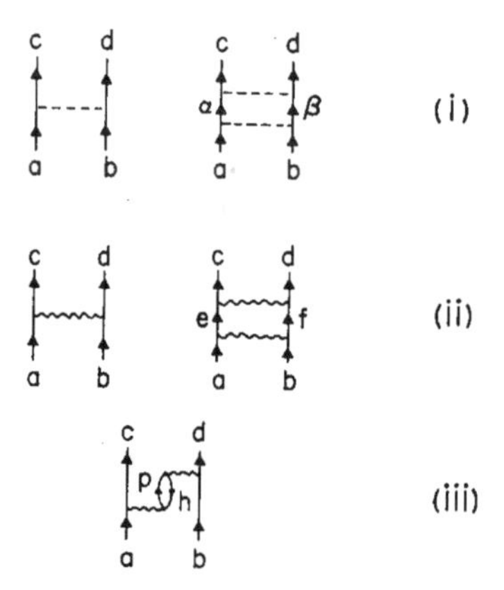

Fig. 10. Diagrams involving the valence nucleons.

This has been investigated by Bertsch [46]) for the Kallio-Kolltveit force by directly summing up diagrams as shown by fig. 10 (iii). We shall evaluate these corrections using the second quantized formalism which appears to be more straightforward. Of course there are corrections other than those of 3p1h. They will be discussed in subsect. 4.3.

The following notation for angular momentum coupling will be used: The normalized and antisymmetrized state $|abJM, TT_z)$ of two nucleons is constructed by

$$A^+(abJM, TT_z)|0) = \frac{1}{\sqrt{1+\delta_{ab}}}[c_a^+ c_b^+]^{JM}_{TT_z}|0)$$

$$= \frac{1}{\sqrt{1+\delta_{ab}}}\sum_{m_a, t_{za}}(j_a m_a j_b m_b|JM)(\tfrac{1}{2}t_{za}\tfrac{1}{2}t_{zb}|TT_z)c^+_{j_a m_a t_{za}}c^+_{j_b m_b t_{zb}}|0), \quad (4.5)$$

where $|0)$ stands for the ^{16}O core. As usual the operator $c^+(c)$ is the fermion creation (destruction) operator and $(j_a m_a j_b m_b|JM)$ and $(\tfrac{1}{2}t_{z_a}\tfrac{1}{2}t_{z_b}|TT_z)$ the vector coupling coefficients. Similarly the corresponding particle-hole state is constructed by

$$A^+(a\underline{b}JM,\,TT_z)|0) = [c_a^+ c_b]^{JM}_{TT_z}|0)$$
$$= \sum_{m_a t_{za}} (j_a m_a j_b - m_b|JM)(-)^{j_b - m_b}(\tfrac{1}{2}t_{za}\tfrac{1}{2} - t_{zb}|TT_z)(-)^{\frac{1}{2}-t_{zb}} \tag{4.6}$$
$$\times c^+_{j_a m_a t_{za}} c_{j_b m_b t_{zb}}|0).$$

The G-interaction in the second quantized form can be written as

$$H_i = \sum_{\substack{a \geq b,\, c \geq d \\ JM,\, TT_z}} (abJT|G|cdJT)A^\dagger(abJM,\,TT_z)A(cdJM,\,TT_z), \tag{4.7}$$

where A is the Hermitian conjugate of $A^\dagger$. It can be easily seen that $(abJM, TT_z|H_i|cdJM, TT_z)$ is just the matrix element $(abJT|G|cdJT)$ which is independent of M and T_z.

The following matrix elements are then derived:

$$L(phJ''T''a,\,b)$$
$$= (0|c_{j_b m_b t_{zb}} H_i [A^\dagger(p\underline{h}J''M'',\,T''T_z'')c_a^+]^{j_b m_b}_{\frac{1}{2}t_{zb}}|0)$$
$$= (-)^{j_p + j_h + J'' + T'' + 1} \sum_{J'''T'''} \left[\frac{2J'''+1}{2j_b+1}\,\frac{2T'''+1}{2}\right]^{\frac{1}{2}} U(j_a j_p j_b j_h;\,J'''J'')$$
$$\times U(\tfrac{1}{2}\tfrac{1}{2}\tfrac{1}{2}\tfrac{1}{2};\,T'''T'')(bhJ'''T'''|G|paJ'''T'''), \tag{4.8}$$

$$M(abJT||cdJ'T',\,phJ''T'';\,JT)$$
$$= (0|A(abJM,\,TT_z)H_i[A^+(cdJ'M',\,T'T_z')A^+(p\underline{h}J''M'',\,T''T_z'')]^{JM}_{TT_z}|0)$$
$$= \frac{(-1)^{J''+T''+1+j_b+j_a}}{[(1+\delta_{ab})(1+\delta_{cd})]^{\frac{1}{2}}}\, U(\tfrac{1}{2}\tfrac{1}{2}TT'';\,T'\tfrac{1}{2})$$
$$\times \{\delta_{ac}U(j_c j_d JJ'';\,J'j_b)L(phJ''T''d,\,b)$$
$$+\delta_{ad}U(j_d j_c JJ'';\,J'j_b)L(phJ''T''c,\,b)(-)^{J'+T'+1}$$
$$+\delta_{bc}U(j_c j_d JJ'';\,J'j_a)L(phJ''T''d,a)(-)^{J+T+1}$$
$$+\delta_{bd}U(j_d j_c JJ'';\,J'j_a)L(phJ''T''c,\,a)(-)^{J+T+J'+T'}\}. \tag{4.9}$$

Here the U-coefficients are related to the Racah coefficients as shown by eq. (2.2). In deriving these formulae, we have simply expanded the L- and M-matrix elements in terms of basic particle-particle G-matrix elements, using the standard angular momentum recoupling techniques and the well-known Wick theorem in contracting the operators. Recall that a, b, c, d and p denote the particle states and h denotes the hole states. We consider the p-h excitation of the core as a quasi-boson, the core polarization phonon. Therefore we consider p to be distinct from a, b, c and d. As a consequence, we do not have terms arising from contracting a or b to p, and we do not worry about the exclusion principle between c or d and p. Graphically, the matrix elements L and M correspond to the vertices shown by fig. 11 where the rectangular box stands for H_i as given by eq. (4.7), all possible contractions included except those just mentioned. The brackets indicate the schemes of coupling.

Recall that in eq. (4.2) which defines the wave operator Ω_M, there is the difficulty that the perturbed energy E appears in the denominator. This difficulty may be removed by using the linked cluster perturbations formalism [39,40]. For our present problem, however, the following approximation should be adequate. We are dealing with Ω_{3p1h}. For parity conservation, only the positive parity p-h excitation of the core enters in Ω_{3p1h}. Thus the average excitation energy of the low-lying 3p1h con-

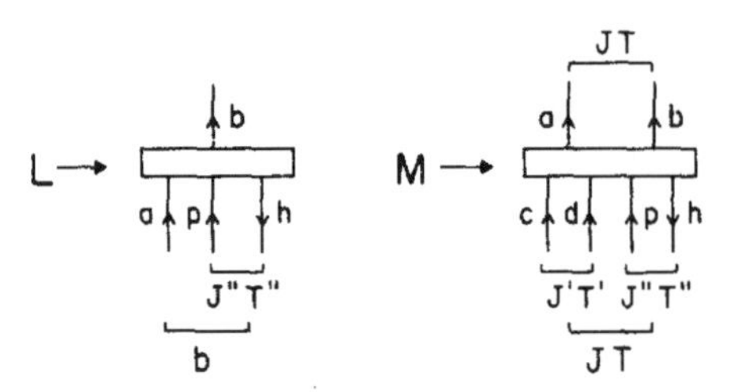

Fig. 11. Vertices corresponding to the matrix elements L and M.

figurations is $\approx 2\hbar\omega$. Hence the energy denominator for them may be replaced by $2\hbar\omega$ and we have, recalling eq. (4.2),

$$G\Omega_{3p1h} \approx G - G\frac{Q_{3p1h}}{2\hbar\omega}G, \tag{4.10}$$

where we have approximated $G\Omega_{3p1h}$ by G in the second-order terms. $Q_{3p1h} = |3p1h)(3p1h|$ and we sum over all 3p1h states of excitation energies $\approx 2\hbar\omega$.

Symbolically, the matrix element $(abJT|-G(Q_{3p1h}/2\hbar\omega)G|cdJT)$ is evaluated by

$$-G\frac{Q_{3p1h}}{2\hbar\omega}G = \text{(a)} - \text{(b)} \tag{4.11}$$

(a) (b)

where as before the rectangular box stands for H_i. Since all contractions will be included in evaluating diagram (a), it will consist of both connected diagrams as shown by fig. 10 (iii) and disconnected diagrams as diagram (b) shown above. The disconnected diagrams are the single-particle, self-energy diagrams. Since we shall use the experimentally observed single-particle energies, these disconnected diagrams are presumably included already as we have discussed in sect. 3. To avoid double counting, they should be subtracted out as we did in the above equation. Recalling eqs. (4.8) and (4.9) and noting that diagram (a) is essentially the product of two M-matrix elements and diagram (b) is essentially the product of two L-matrix elements, we write eq. (4.11) explicitly as

$$\langle abJT|-G\frac{Q_{3\text{p1h}}}{2\hbar\omega}G|cdJT\rangle$$

$$= -\frac{1}{2\hbar\omega}\sum_{phJ''T''}\left\{\sum_{efJ'T'}M(abJT||efJ'T', phJ''T'', JT)M(cdJT||efJ'T', phJ''T'', JT)\right.$$

$$\left.-\frac{\delta_{ab,cd}}{1+\delta_{ab}}\sum_{g=a,b}[L(phJ''T''g, a)^2+L(phJ''T''g, b)^2]\right\}. \tag{4.12}$$

The values of ef to be summed over are ac, ad, bc and bd when a, b, c and d are distinct. They become ac and ad if $a = b$; ac and bc if $c = d$; aa, bb and ab if $a \neq b$ but $ab = cd$; and merely aa if $a = b = c = d$. These values are summed over because all others give either zero or entirely disconnected graphs. It may be noted that for the off-diagonal cases ($ab \neq cd$), there are no disconnected graphs because of the conservation of angular momentum.

When $a = b = c = d$, eq. (4.12) simplifies to

$$\langle aaJT|-G\frac{Q_{3\text{p1h}}}{2\hbar\omega}G|aaJT\rangle$$

$$= -\frac{1}{\hbar\omega}\sum_{phJ''T''}L(phJ''T''a, a)^2(-)^{J''+T''}U(j_aJ''Jj_a; j_aj_a)$$

$$\times U(\tfrac{1}{2}T''T\tfrac{1}{2}; \tfrac{1}{2}\tfrac{1}{2}), \tag{4.13}$$

which is identical to the expression obtained by Bertsch [46] †. There are many aspects concerning the evaluation of $-G(Q_{3\text{p1h}}/2\hbar\omega)G$ which should be discussed. Before doing so, we would like first to show some numerical results.

Using the configurations shown in fig. 6, eq. (4.12) is evaluated for a, b, c and d in the sd shell. Here $\hbar\omega = 14$ MeV is used both in the energy denominator and in the calculation of the matrix elements. For parity conservation, we may have either the hole in the 0p shell and particle in the 1p-0f shell, or the hole in the 0s shell and particle in the 1s-0d shell. Results are shown in table 10. It is very important to see that some of them are positive. In particular, that the values of $-G(Q_{3\text{p1h}}/2\hbar\omega)G$ are positive for the $J = 3$ and 4 cases is especially desirable, because the Federman-Talmi values are repulsive while the G-matrix elements alone are attractive. That the diagonal matrix elements of $-G(Q_{3\text{p1h}}/2\hbar\omega)G$ may have alternate signs can be easily understood. By looking at either eq. (4.11) or eq. (4.12), the first term is just the square of the M-matrix elements. So, it is always negative. But the second term which is the square of the L matrix elements has opposite sign. Hence the total sign of the two depends on the relative magnitudes of them. It should be noted that $-G(Q_{3\text{p1h}}/2\hbar\omega)G$ is different from the ordinary second-order perturbation energy correction which is always negative.

† This was pointed out to the authors by Dr. L. Zamick.

As shown by the table, the sum of G and $-G(Q_{3p1h}/2\hbar\omega)G$ agree much better with the values of FT than G alone. The strengths of the former are, however, generally smaller than those of FT. This can be attributed in part to that the $0d_{\frac{3}{2}}$ configuration was not included in the work of FT. As illustrated by the last two rows of table 10, the coupling between $0d_{\frac{3}{2}}$ and the rest of the sd shell has been found to be strong for the HJ potential. It is now seen that the effective interactions of FT may well be deduced from the model interaction $G\Omega_{3p1h}$. This point will become even clearer later, when the spectra of ^{18}O and ^{18}F are obtained by diagonalizing $G\Omega_{3p1h}$ (subsect. 4.3).

We now go back to discuss the evaluation of $-G(Q_{3p1h}/2\hbar\omega)G$. We first discuss the exclusion principle. Recall that in evaluating both the L- and M-matrix elements,

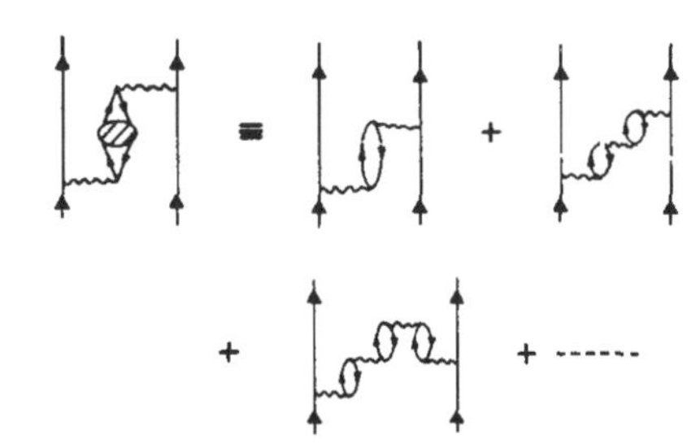

Fig. 12. Interaction of the valence nucleons via the core excitation phonon.

we consider the particle p excited from the core to be distinct from the valence nucleons. In so doing, there must be terms which violate the exclusion principle. This, however, should not be serious. First, the violation occurs only when a hole is excited in the 0s shell. The number of such cases are relatively small. Second, there will be cancellations among the terms which violate the exclusion principle. This is due to the following: In evaluating $-G(Q_{3p1h}/2\hbar\omega)G$ as shown by eq. (4.11), the exclusion principle is ignored in evaluating diagram (a). But it is also ignored in diagram (b) as shown by eq. (4.12).

We are dealing with the corrections arising from the p-h excitation of the ^{16}O core. So far we have just included one bubble in our calculation as shown by fig. 10 (iii). In fact the p-h excitation of the core consists of a series of such bubbles, usually known as a phonon. Thus instead of renormalizing the interaction between the valence nucleons by one bubble, it should be renormalized by a phonon. This is shown in fig. 12 where the hatched circle denotes the p-h excitation phonon of the core. It is well-known that some of these phonons may be highly collective. Thus their excitation energies may be far less than $2\hbar\omega$. For them, it seems that we have underestimated their effects because we used $2\hbar\omega$ as the energy denominator. This underestimation is however not as serious as it appears. First, there are just a few of such low-lying phonons and some of them are spurious and should be discarded. Second, we must sum over all the phonons and many of them must have excitation energy much larger than $2\hbar\omega$. Hence, on the average, the use of perturbation method as we did is acceptable. For simplitiсy, we choose to use the perturbation method, although it would be worthwhile to look into this problem more carefully.

Another point which should be noticed is the spurious states. As is well-known, some components of the p-h excitation of the ^{16}O core merely correspond to the centre-of-mass motion of the core and therefore are spurious [1]). Hence we should prohibit them to enter into our calculation. But since only the $2\hbar\omega$ p-h excitations of the core enter into our calculation, the contamination of the spurious components in

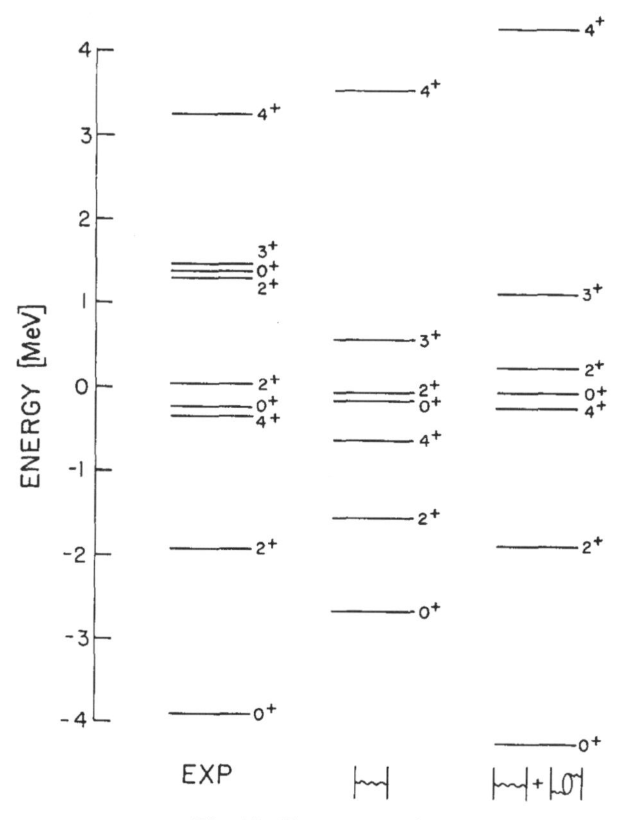

Fig. 13. The spectra of ^{18}O.

what we did is rather slight. The reason is that the $2\hbar\omega$ spurious state consists of predominantly two-particle-two-hole excitations of the core, as shown recently by Giraud [47]). It would be necessary, however, to handle this problem more carefully if the whole problem is to be treated more carefully instead of using the perturbation method as we have chosen.

4.3. SPECTRA OF ^{18}O AND ^{18}F

The spectra of ^{18}O and ^{18}F are obtained by diagonalizing $G\Omega_{3p1h}$ given by eqs. (4.10) and (4.12) in the s-d shell two-nucleon subspace. Results are shown in figs. 13

and 14. The spectra shown in the second columns are obtained by diagonalizing the G-matrix alone. Those in the third column are for $G\Omega_{3p1h}$. The experimental spectrum for ^{18}O is taken from Hewka *et al.* [48]). That for ^{18}F is taken from Polletti and Warburton [49]). The absolute ground state energies shown in the figures are determined as follows. Using the binding energy (BE) data furnished by Koenig *et al.* [43]), the ground state energies of the valence nucleons are taken to be

$$-\text{BE}(^{18}\text{O})-\text{BE}(^{16}\text{O})+2\text{BE}(^{17}\text{O}) = -3.90 \text{ MeV}$$

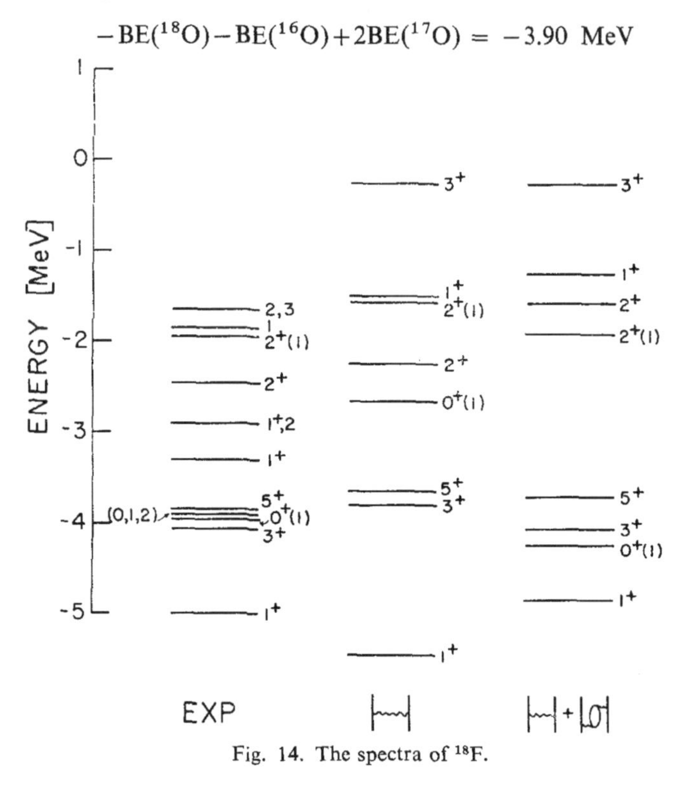

Fig. 14. The spectra of ^{18}F.

for ^{18}O and

$$-\text{BE}(^{18}\text{F})-\text{BE}(^{16}\text{O})+\text{BE}(^{17}\text{O})+\text{BE}(^{17}\text{F}) = -5.00 \text{ MeV}$$

for ^{18}F.

For the ^{18}O spectrum, it is clearly seen that the results for $G\Omega_{3p1h}$ agree much better with the observed values than G alone. The bubble diagram $[-G(Q_{3p1h}/2\hbar\omega)G]$ spreads apart the energy levels: it lowers the lower states and raises the higher ones. The states in the middle are rather unaffected by the bubble diagram. This feature is worth noting. All the ^{18}O states have isospin $T = 1$. For the ^{18}F spectrum, again the results with the bubble diagram are better than those without. Here all states have isospin $T = 0$ except few $T = 1$ states which are the same as ^{18}O and are specified by the symbol (1). The calculated eigenvalues and eigenfunctions with the bubble diagram are given in table 11.

The agreement between the calculated and experimental energy levels is rather outstanding, except for a few states whose structure should be basically different from the states which can be generated from the configurations we have provided. Therefore they are not supposed to be reproduced here. In order to understand the nature of the calculated states, it is necessary to look into the wave functions very carefully. The nature of the wave functions we show in table 11 must be emphasized. They are the model wave functions ψ_M. The physical wave function is $\psi = \Omega_M \psi_M$ as defined by eq. (4.1). So the physical wave functions are in fact very complicated. In addition to their two-nucleon component is the sd shell, they have both the 3p1h components and the two-nucleon components outside of the sd shell which are introduced by the G-matrix.

TABLE 11

Eigenvalues and eigenfunctions of ^{18}O and ^{18}F

	J^π	Energy (MeV) exp	calc	$d_{\frac{5}{2}}^2$	$d_{\frac{5}{2}}s_{\frac{1}{2}}$	$d_{\frac{5}{2}}d_{\frac{3}{2}}$	$s_{\frac{1}{2}}^2$	$s_{\frac{1}{2}}d_{\frac{3}{2}}$	$d_{\frac{3}{2}}^2$
	0^+	−3.90	−4.22	0.901			0.324		0.287
	2^+	−1.92	−1.89	0.782	0.579	0.090		0.190	0.099
$T = 1$	4^+	−0.35	−0.27	0.954		0.300			
	0^+	−0.27	−0.06	−0.322			0.945		−0.055
	2^+	0.02	0.23	−0.607	0.788	−0.008		0.105	−0.007
	3^+	1.47	1.11		1.000	0.025			
	4^+	[3.22]	4.26	−0.300		0.954			
	1^+	−5.0	−4.83	0.571		−0.629	0.507	−0.143	−0.040
$T = 0$	3^+	−4.06	−4.04	0.527	0.817	−0.234			0.016
	5^+	−3.88	−3.69	1.0					
	1^+	−1.87	−1.23	0.554		−0.163	−0.803	0.110	−0.103
	2^+	−1.65	−1.59		0.753	0.536		0.382	

It is well-known that the nucleus ^{18}O is abundant in deformed (non-spherical) components [50, 51]). The observed B(E2) transition rates from the second 0^+ to the first 2^+ is far too large (30 e^2 fm^4) to be reproduced by the two-particle model. To account for this, these two states are usually taken as largely composed of deformed components - 4p2h (four-particle-two-hole) states [50, 51]). What we have found in this work, however, seems to disagree with this. The energy values of these two states are rather satisfactorily reproduced with their wave functions being largely two-particle states with 3p1h corrections. On the contrary, we would like to assign the third 0^+ and third 2^+ as states composed of predominantly 4p2h components. To settle this, it is necessary to look into the wave functions we have computed here and evaluate the transition rates. Such an investigation is in the progress. The following outcome is inevitable: The wave functions computed here will not be collective enough to reproduce the observed B(E2) values. Therefore, it will be necessary to bring in the deformed components. That the mixing of an appreciable amount of deformed components may have very little effect on the relative energies to most states has been recently indicated by the calculation of Federman and Talmi using empirical matrix elements [2,3]).

For ^{18}F, there is much less information available both experimentally and theoretically. Our calculation shows that quite a few states can be accounted for by the present picture. There are many states which we do not find at all in the calculation. They are presumed to be states composed of more complicated configurations than we have provided. Very likely, they are largely deformed. Using Zamick's formula and parameters [52]), one finds the 4p-2h deformed state with $T_p = T_h = T = 0$ to lie at 2.67 MeV excitation. Thus, the group of states at ≈ -3 MeV in fig. 14 may well come largely from this parentage. It will be very rewarding to reinvestigate this nucleus including the deformed components and see if the other states can be reproduced.

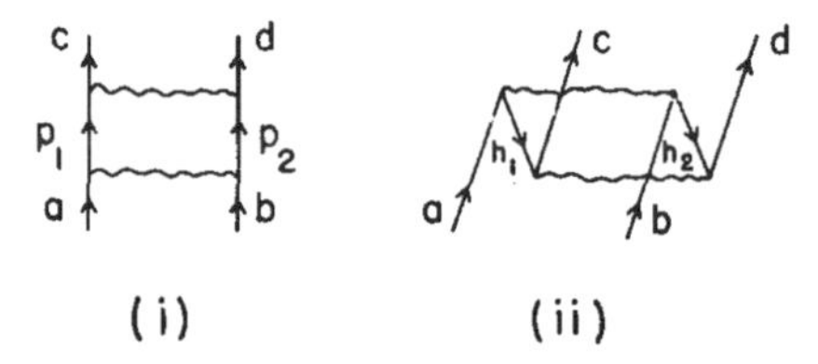

Fig. 15. Diagrams showing other corrections to the s-d shell model interaction.

As shown by table 11, there is very strong configuration admixture in ^{18}F. For instance, the 1^+ ground state is dominated by the $0d_{\frac{3}{2}}0d_{\frac{5}{2}}$ component although the $0d_{\frac{3}{2}}$ level is 5.08 MeV above the $0d_{\frac{5}{2}}$ level. This strong admixture is caused by the strong tensor force in the HJ potential.

We now discuss corrections other than the 3p1h configurations. So far we have just dealt with $G\Omega_{3p1h}$, where Ω_{3p1h} takes care of the corrections from the 3p1h configurations which are approximately $2\hbar\omega$ apart in energy. There are other corrections which we have left out in our model space but are also $2\hbar\omega$ apart or less. These corrections must also be investigated. They are the processes shown by fig. 15, where a, b, c and d are states in the s-d shell. In diagram (i), p_i and p_2 are particles in the 0f-1p shell. In diagram (ii), h_1 and h_2 are holes in the 0p shell. The contribution of diagram (i) can be easily evaluated; it is

$$(abJT|-G\frac{Q_{2p}}{2\hbar\omega}G|cdJT) = -\frac{1}{2\hbar\omega}\sum_{p_1>p_2}(abJT|G|p_1p_2JT)(p_1p_2JT|G|cdJT). \quad (4.14)$$

Here $Q_{2p} = |2p)(2p|$ where $|2p)$ denotes the state of two nucleons in the 0f-1p shell. In table 12, we tabulate the matrix elements of G, $-G(Q_{3p1h}/2\hbar\omega)G$ and $-G(Q_{2p}/2\hbar\omega)G$ for comparison. In diagonalizing the secular matrix, the effects of

TABLE 12

The s-d shell reaction matrix elements and correction terms

T = 1, J = 0

	$0d_{5/2}^2$	$1s_{1/2}^2$	$0d_{3/2}^2$
$0d_{5/2}^2$	−1.47	−0.76	−3.18 (MeV)
	−1.06	−0.33	−0.93
	−0.28	−0.11	−0.10
$1s_{1/2}^2$		−2.28	−0.62
		0.07	−0.22
		−0.07	−0.09
$0d_{3/2}^2$			−0.17
			−0.37
			−0.25

T = 1, J = 2

	$d_{5/2}^2$	$d_{5/2}s_{1/2}$	$d_{5/2}d_{3/2}$	$s_{1/2}d_{3/2}$	$d_{3/2}^2$
$d_{5/2}^2$	−1.02	−0.52	−0.40	−0.51	−0.58 (MeV)
	0.08	−0.29	0.03	−0.28	−0.36
	−0.04	−0.03	−0.01	−0.03	−0.03
$d_{5/2}s_{1/2}$		−1.04	−0.15	−1.34	−0.71
		−0.05	0.00	−0.02	0.08
		−0.06	−0.03	−0.03	−0.02
$d_{5/2}d_{3/2}$			−0.35	−0.63	−0.77
			0.30	−0.14	−0.27
			−0.04	−0.01	−0.01
$s_{1/2}d_{3/2}$				−0.50	−0.01
				0.43	−0.14
				−0.06	−0.03
$d_{3/2}^2$					−0.27
					0.53
					−0.02

T = 1, J = 4

	$d_{5/2}^2$	$d_{5/2}d_{3/2}$
$d_{5/2}^2$	−0.39	−0.94(MeV)
	0.53	−0.35
	−0.02	−0.02
$d_{5/2}d_{3/2}$		−1.80
		0.57
		−0.04

T = 1, J = 3

	$d_{5/2}s_{1/2}$	$d_{5/2}d_{3/2}$
$d_{5/2}s_{1/2}$	−0.30	−0.06 (MeV)
	0.54	−0.05
	−0.00	−0.00
$d_{5/2}d_{3/2}$		−0.42
		−0.61
		−0.00

TABLE 12 (continued)

	$d_{5/2}^2$	$d_{5/2}d_{3/2}$	$s_{1/2}^2$	$s_{1/2}d_{3/2}$	$d_{3/2}^2$
$d_{5/2}^2$	−0.53 −0.21 −0.56	3.34 −0.11 0.68	−0.45 −0.16 −0.32	−0.04 −0.31 0.09	2.19(MeV) −0.44 0.15
$d_{5/2}d_{3/2}$		−5.78 0.37 −1.37	2.16 −0.42 0.63	−1.86 0.20 −0.34	−0.26 0.45 −0.29
$s_{1/2}^2$			−3.94 0.40 −0.50	0.26 0.31 0.06	−0.46 0.22 0.10
$s_{1/2}d_{3/2}$				−3.28 0.14 −0.27	0.92 −0.18 0.05
$d_{3/2}^2$					−0.13 0.05 −0.24

$T = 0,\ J = 1$

	$d_{5/2}^2$	$d_{5/2}s_{1/2}$	$d_{5/2}d_{3/2}$	$d_{3/2}^2$
$d_{5/2}^2$	−0.93 0.07 −0.16	−1.41 −0.08 −0.20	1.67 0.28 0.14	0.30(MeV) 0.25 −0.04
$d_{5/2}s_{1/2}$		−3.36 −0.19 −0.35	1.17 −0.16 0.16	0.09 −0.05 0.02
$d_{5/2}d_{3/2}$			−1.33 0.30 −0.28	1.87 0.28 0.11
$d_{3/2}^2$				−2.49 0.13 −0.16

$T = 0,\ J = 3$

	$d_{5/2}^2$
$d_{5/2}^2$	−3.64(MeV) −0.02 −0.20

$T = 0,\ J = 5$

In each block, the numbers of the first, second and third row are respectively values of G, $-G(Q_{3p1h}/2\hbar\omega)G$ and $-G(Q_{2p}/2\hbar\omega)G$.

$-G(Q_{2p}/2\hbar\omega)G$ have been found to concentrate on depressing the ground state, by about 0.5 MeV. Since it is rather unclear to what extent the contribution of $-G(Q_{2p}/2\hbar\omega)G$ is false because there is some overlap between it and that of G itself, we feel that itshould be better to leave out these corrections rather than include them.

The diagram (ii) of fig. 15 is more important. The intermediate states of it are the 4p2h states which are supposed to be largely deformed and may lie very low in energy [50, 51]. For example, those low-lying states which are observed but not at all reproduced by our calculation as shown by figs. 13 and 14 may well be of the 4p2h nature. Therefore it is obvious that the contribution of diagram (ii) can no longer be evaluated by the perturbation method which we have been using so far. To handle this properly, the deformed 4p2h components must be treated on the same footing as the 2p components in the sd shell. This has been done to some extent using the effective interactions [2, 3]. More extensive investigations using the HJ potential as described here are underway. Hopefully, this may lead to a satisfactory account of the deformed states which have not been attempted at in the present work.

5. Conclusions

In this work, we first showed how to evaluate the shell model reaction matrix elements for the HJ potential. Using these matrix elements, the single-particle energies of ^{17}O and the spectra of ^{18}O and ^{18}F were calculated. It is really encouraging to see the results are so satisfactory. It is even more impressive if one notes that there is no parameter which we have adjusted at our disposal. All parameters entering this work are fixed by the two-body scattering data and the experimental single-particle spectrum of ^{17}O. Our results seem to assert that the spectroscopic properties of finite nuclei may well be reproduced to a large extent, by the structure of a group of nucleons interacting via the free nucleon-nucleon potential which is chosen to be the HJ potential in this work.

The matrix elements generated in this work are being applied to various other nuclei as well. They all seem to give additional supports to the above assertion. Preliminary results for ^{42}Ca and ^{42}Sc (ref. [44]) and the 0p shell nuclei [53] are both highly satisfactory. The results for the tin isotopes show that the pairing effects are well reproduced [54]. An extensive application to the calculation of the nuclear magnetic moments over a wide range of nuclei has led to very impressive results [55]. So, it does appear that the long persisting problem of the "effective" interactions in the shell-model spectroscopy is being clarified. The so-called effective interactions should in fact correspond to the model interactions deduced from the free nucleon-nucleon potentials as we have discussed. This also explains why the effective interactions have been highly model-dependent. Moreover, it may be noted that they would be rather meaningless if the model space was improperly chosen.

One may inquire as to how our results will be affected if different potential models which also fit the same two-body scattering data are used. They would give essentially the same results. For instance, the Yale potential is known to be numerically very similar to that of HJ. There are other potential models which fit the same two-body scattering data but do not have vertical repulsive cores, e.g. the non-local potentials of Tabakin [56]. Even for this case where the analytical forms of the potential are so different, the matrix elements of them are found [54]) to be very similar to the HJ. The above result is in fact not surprising. As we have seen, our reaction matrix elements largely come from the first-order, long-range potentials, i.e. the potentials on-the-energy shell which will be nearly the same for different potentials as long as they fit the same two-body scattering data. So, the results of this work do not depend significantly on the potential models, as long as they fit the same two-body data.

An important aspect in evaluating the reaction matrix elements is the state dependence. It is particularly important for the S-waves. We have not treated this very carefully. We used the average state-independent separation distances for the SE and TE potentials. This would have necessarily somewhat overestimated the S-state attractions for the partial waves of high n-values. In evaluating the TE, second-order, tensor terms of $-V_{TL}(Q/e)V_{TL}$, the effective energy denominator e_{eff} was also chosen to be state-independent. We took it to be $e_{eff} = 220$ MeV which was determined for $\varepsilon(l)+\varepsilon(m) = -20$ MeV. This somewhat underestimates the contribution for the pair of nucleons which are bound by less than 20 MeV, but overestimates for the pair which are much more strongly bound. For example, $\varepsilon(l)+\varepsilon(m)$ should be ≈ -60 MeV for two 0s nucleons which may lead to $e_{eff} \approx 260$ MeV. The dispersion terms which we have decided to leave out are also strongly state-dependent. The SO and TO contributions are, however, almost state-independent; as we have seen that they are rather unsensitive to the value of γ^2. So, our results for them should be more certain. The above treatments where the state dependence is averaged out seem to suffice for our present work. But the state dependence should be treated very carefully when calculating other nuclear properties like the saturation properties, the binding energies, the isotope shifts, etc. The binding energies can probably be calculated fairly well by using an average state-independent treatment.

We have somewhat overestimated the attraction of the HJ potential, because we have left out the propagator correction terms. But even so, the binding energies we calculated are still too small. This may be due to the presence of the vertical hard core whose radius is possibly too large. So it seems that an improved potential model should have a somewhat smaller hard core or, even better, a soft core, but with its exterior components rather unchanged. Such a potential would certainly give more binding. Since the energy spectra we have calculated here depend largely on the long-range parts of the potential, we expect that our results may not be affected much when such an improved potential model is used.

The last remark we would like to make concerns the deformed states. An investigation using the matrix elements so generated but with the deformed configurations properly included will be very interesting as well as important in further verifying the approaches of the present work. Such an investigation is in progress, and not surprisingly, preliminary results of it have been very encouraging as well [57]).

The authors would like to thank Drs. C. W. Wong and L. Zamick for helpful conversations. One of them (T.K.) is in particular indebted to Dr. C. W. Wong for many illuminating discussions.

Appendix 1

The radial dependences of the Hamada-Johnston potential (eq. (2.3)) are given by

$$V_c = 0.08(\tfrac{1}{3}\mu)(\tau_1 \cdot \tau_2)(\sigma_1 \cdot \sigma_2)Y(x)[1+a_c Y(x)+b_c Y^2(x)],$$
$$V_T = 0.08(\tfrac{1}{3}\mu)(\tau_1 \cdot \tau_2)(\sigma_1 \cdot \sigma_2)Z(x)[1+a_t Y(x)+b_t Y^2(x)],$$
$$V_{LS} = \mu G_{LS} Y^2(x)[1+b_{LS} Y(x)],$$
$$V_{LL} = \mu G_{LL} x^{-2} Z(x)[1+a_{LL} Y(x)+b_{LL} Y^2(x)],$$

where $Y(x) = e^{-x}/x$, $Z(x) = (1+3/x+3/x^2)Y(x)$, μ is the pion mass (139.4 MeV) and x is measured in $\hbar/\mu c = 1.415$ fm. The hard core radius is 0.343 $\hbar/\mu c$ in all states. The values of the parameters shown above are given in table A1.

TABLE A.1
Parameters of the Hamada-Johnston potential

State	a_c	b_c	a_t	b_t	G_{LS}	b_{LS}	G_{LL}	a_{LL}	b_{LL}
singlet-even	8.7	10.6					−0.000891	0.2	0.2
triplet-odd	−9.07	3.48	−1.29	0.55	0.1961	−7.12	0.000891	−7.26	6.92
triplet-even	6.0	− 1.0	−0.5	0.2	0.0743	0.1	0.00267	1.8	0.4
singlet-odd	−8.0	12.0					−0.00267	2.0	6.0

References

1) G. E. Brown, Unified theory of nuclear models and nucleon-nucleon forces, 2nd ed. (North-Holland Publ. Co., Amsterdam, to be published)
2) P. Federman and I. Talmi, Phys. Lett. **19** (1965) 490
3) I. Talmi, Revs. Mod. Phys. **34** (1962) 704
4) J. D. McCullen, B. F. Bayman and L. Zamick, Phys. Rev. **134** (1964) B515
5) M. J. Moravcsik, The two-nucleon interaction (Clarendon Press, Oxford, 1963);
R. Wilson, The nucleon-nucleon interaction (Interscience Publishers, New York, 1963)
6) T. Hamada and I. D. Johnston, Nuclear Physics **34** (1962) 382
7) Lassila *et al.*, Phys. Rev. **126** (1962) 881
8) H. A. Bethe, B. H. Brandow and A. G. Petschek, Phys. Rev. **129** (1963) 225
9) M. Razavy, Phys. Rev. **130** (1963) 1091
10) N. Azziz, Nuclear Physics **85** (1966) 15
11) M. Bauer and M. Moshinsky, Nuclear Physics **4** (1957) 615
12) G. E. Brown, Lectures on theory of nuclear matter, NORDITA, Copenhagen (1964) unpublished
13) J. F. Dawson, I. Talmi and J. D. Walecka, Ann. of Phys. **18** (1962) 339
14) J. F. Dawson and J. D. Walecka, Ann. of Phys. **22** (1963) 133
15) A. Kallio and K. Kolltveit, Nuclear Physics **53** (1964) 87;
T. Engeland and A. Kallio, Nuclear Physics **59** (1964) 211
16) R. K. Bhaduri and E. L. Tomusiak, Proc. Phys. Soc. (London) **86** (1965) 451
17) A. Kallio, Phys. Lett. **18** (1965) 51
18) L. Kelson and C. A. Levinson, Phys. Rev. **134** (1964) B269
19) K. T. R. Davis, S. J. Krieger and M. Baranger, to be published
20) A. M. Green, Nuclear Physics **33** (1962) 218
21) B. L. Cohen, R. H. Fulmer, A. L. McCarthy and P. Mukherjee, Revs. Mod. Phys. **35** (1963) 332
22) A. de-Shalit and I. Talmi, Nuclear shell theory (Academic Press, New York, 1963)

23) J. M. Kennedy and M. J. Cliff, CRT Report 609, Chalk River, Ontario, (1955)
24) T. A. Brody and M. Moshinsky, Tables of transformation brackets (Monografias **del Institute** de Fisica, Mexico, 1960)
25) B. L. Scott and S. A. Moszkowski, Ann. of Phys. **14** (1961) 107
26) B. L. Scott and S. A. Moszkowski, Nuclear Physics **29** (1962) 665
27) T. T. S. Kuo and G. E. Brown, Phys. Lett. **18** (1965) 54
28) L. I. Schiff, Quantum mechanics (McGraw-Hill Co., New York, 1955) p. 79
29) H. Bethe, Phys. Rev. **138** (1965) B804;
S. A. Moszkowski, Phys. Rev. **140** (1965) B283
30) G. E. Brown, G. T. Shappert and C. W. Wong, Nuclear Physics **56** (1964) 191
31) C. W. Wong, private communication
32) A. M. Feingold, Phys. Rev. **101** (1956) 258; **105** (1957) 944
33) E. R. Irwin, Ph.D. Thesis, Cornell University (1963)
34) T. Dahlblom, K. G. Fogel, B. Quist and A. Törn, Nuclear Physics **56** (1964) 177
35) C. W. Wong, private communication
36) C. W. Wong, Nuclear Physics **56** (1964) 213
37) S. A. Moszkowski, Handbuch der Physik, Bd. 39, (Springer-Verlag, Berlin, 1957) p. 411
38) L. R. B. Elton, Nuclear sizes (Oxford University Press, 1961)
39) J. Goldstone, Proc. Roy. Soc. **A239** (1957) 267;
40) B. H. Brandow, Revs. Mod. Phys., to be published
41) P. Goldhammer, Revs. Mod. Phys. **35** (1963) 40
42) K. A. Brueckner, A. M. Lockett and M. Rotenberg, Phys. Rev. **121** (1961) 255
43) L. A. König, J. H. E. Mattauch and A. H. Wapstra, Nuclear Physics **31** (1962) 18
44) T. T. S. Kuo, to be published
45) R. J. Eden and N. C. Francis, Phys. Rev. **97** (1955) 1367;
C. Bloch and J. Horowitz, Nuclear Physics **8** (1958) 91
46) G. F. Bertsch, Nuclear Physics **74** (1965) 234
47) B. Giraud, Nuclear Physics **71** (1965) 373
48) P. Hewka, R. Middleton and J. Wiza, Phys. Lett. **10** (1964) 93
49) A. R. Polletti and E. K. Warburton, Phys. Rev. **137** (1965) B595
50) T. Engeland, Nuclear Physics **72** (1965) 68
51) G. E. Brown, Cong. Int. Physique Nucléaire, Vol. 1 (Paris, 1964) p. 129;
G. E. Brown and A. M. Green, Nuclear Physics **75** (1966) 401
52) L. Zamick, Phys. Lett. **19** (1965) 580
53) E. C. Halbert, Y. E. Kim and T. T. S. Kuo, Phys. Lett. **20** (1966) 657
54) T. T. S. Kuo, E. Baranger and M. Baranger, Nuclear Physics **81** (1966) 241
55) H. Mavromatis, L. Zamick and G. E. Brown, Nuclear Physics **80** (1966) 545
56) F. Tabakin, Ann. of Phys. **30** (1964) 51
57) G. E. Brown and A. M. Green, Nuclear Physics **85** (1966) 86

1.C

Nuclear Physics **A176** (1971) 65—88; © *North-Holland Publishing Co., Amsterdam*

A FOLDED-DIAGRAM EXPANSION OF THE MODEL-SPACE EFFECTIVE HAMILTONIAN

T. T. S. KUO and S. Y. LEE

Department of Physics,

State University of New York, Stony Brook, New York 11790 †

and

K. F. RATCLIFF

Department of Physics,

State University of New York, Albany, New York 12203

Received 12 July 1971

Abstract: A linked-diagram expansion for the model-space effective Hamiltonian is derived. This expansion which contains the folded diagrams is independent of the energy eigenvalue. Our result is generally in agreement with the expansion obtained earlier by Brandow, Morita and Oberlechner *et al.*, but our method of derivation is different from theirs and is probably simpler. The properties of the model-space wave functions and their connection with the true eigenfunctions are investigated. In particular, a linked-diagram expansion is obtained for the evaluation of the expectation value of a general operator with respect to the true eigenfunction. This expansion also contains the folded diagrams and is independent of the energy eigenvalues.

1. Introduction

An important problem in nuclear physics and many other branches of physics as well has been the derivation of the model-space effective Hamiltonian, namely the modified Hamiltonian which when diagonalized within a chosen model space will yield eigenvalues which are also eigenvalues of the original Hamiltonian when diagonalized in the entire Hilbert space. If such a modified Hamiltonian can be constructed, the solution of our physical problem will then be largely simplified due to the small dimensionality of the chosen model space. To diagonalize the original Hamiltonian in the entire Hilbert space usually implies the diagonalization of a matrix of very enormous dimensionality. A typical application of this idea is the shell model for nuclear structure, where the model space is chosen to consist of an inert core plus a few valence nucleons restricted to a few harmonic-oscillator orbits. Then if we know how to construct the effective Hamiltonian for this model space, the diagonalization of it within this model space will thus yield eigenvalues which are also eigenvalues of the original nuclear Hamiltonian for the entire Hilbert space. Much work has in fact been done in recent years with regard to the determination of the shell-model nuclear effective Hamiltonian.

† Work supported by the US Atomic Energy Commission.

Generally speaking, the model-space effective Hamiltonians H_{eff} can be classified into two types; those which are dependent on the energy eigenvalue E and those which are not dependent on E. The E-dependent H_{eff} has been formulated by Feshbach [1]) and Bloch and Horowitz [2]). It can also be derived [3]) conveniently from the Green's function method. Although the structure of the E-dependent H_{eff} is usually not very complicated, its dependence on E will, however, cause difficulty in the solution of the model-space eigenvalue problem. The eigenvalue problem must then be solved self-consistently when H_{eff} itself is dependent on the eigenvalue E. In fact we have found in a previous paper [3]) that for nuclear structure calculations using the Reid nucleon-nucleon potential and the Green function formalism the E-dependence of H_{eff} is rather weak, and consequently the requirement of self-consistency can be achieved without much difficulty. But in general the E-dependence of H_{eff} is not guaranteed to be weak. Hence it is generally more desirable if we can devise a model-space effective Hamiltonian which is not dependent on E.

The purpose of this paper is mainly to derive and discuss the E-independent model-space effective Hamiltonian. The problem of how to construct such a Hamiltonian has been investigated by several authors [4-7]). Motivated in particular by the pioneering work of Brandow [5]) and Baranger and Johnson [7]), we have re-investigated this problem. In this paper, we give a different and probably simpler derivation for the E-independent model-space effective Hamiltonian. A desirable feature of our approach seems to be that one can easily visualize the connection between the model-space wave functions and the true eigenfunctions, the removal of the unlinked diagrams from the effective Hamiltonian, and the removal of the unlinked diagrams from the expectation value of a general operator with respect to the true eigenfunction. In addition, the physical meaning of the unlinked diagrams can be easily understood.

In the following sections, we first define and explain the folded diagrams in sect. 2. The folded diagrams will be of essential importance in the construction of the effective Hamiltonian and the evaluation of the expectation value of a general operator with respect to the true eigenfunction. In sect. 3, we outline the proof of the decomposition theorem which deals with the factorization of diagrams arising from the operation of the time-evolution operator on the model-space basis vector. In sect. 4, we establish a one-to-one correspondence between the model-space wave functions and the true eigenfunctions of the original Hamiltonian. Based on this and the decomposition theorem, the model-space effective Hamiltonian and the associated secular equation are then derived. In sect. 5, we discuss how to evaluate the expectation value of a general operator with respect to the true eigenfunctions. It is shown that the expectation value of the operator with respect to the ground state of the core can be separated out as an additive constant. In this paper, we do not carry out any numerical calculations. It is in fact not clear as to how one should group the various terms in the effective Hamiltonian to enhance the convergence of the series expansion of H_{eff}. Discussions in this regard and comparisons of the present formalism with the Green's function formalism are given in sect. 6.

2. Folded diagrams

To facilitate our discussions, let us first define some terminologies, namely the model space, the active line and the passive line. This can be conveniently accomplished by considering an example. Consider that we are calculating the low-lying

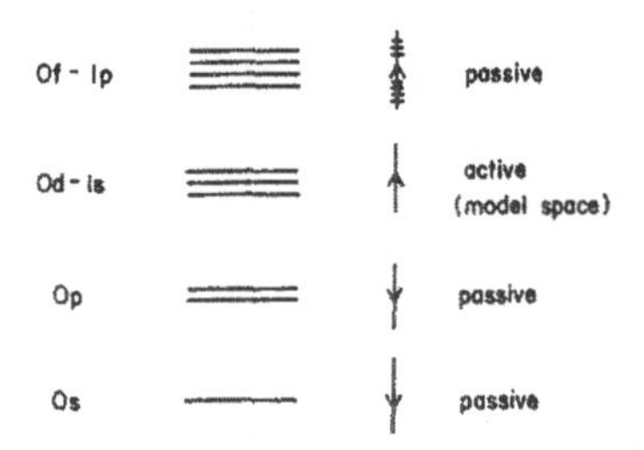

Fig. 1. A typical choice of the model space suitable for the calculation of the low-lying spectrum of ^{18}O.

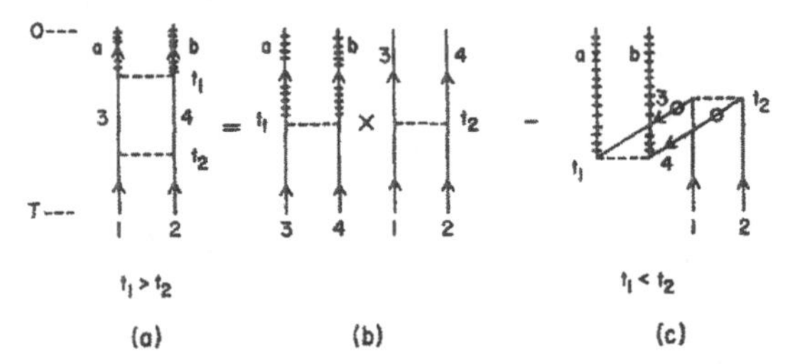

Fig. 2. A diagrammatic identity which defines the folded diagram. Diagram (c) is the folded diagram.

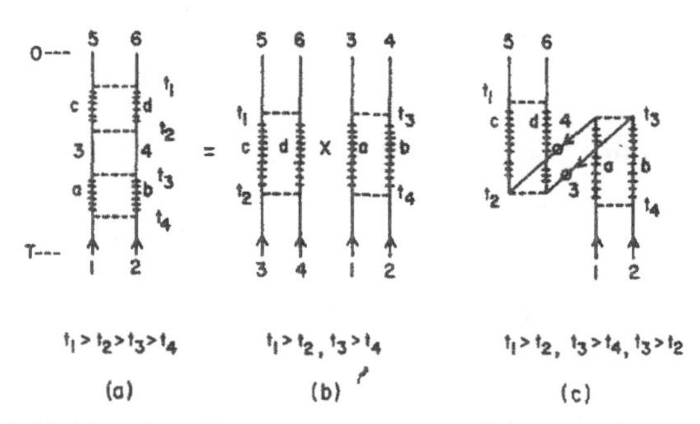

Fig. 3. A diagrammatic identity which illustrates the generalized time-ordering of the folded diagram.

spectrum of the nucleus ^{18}O. For this calculation, we may choose the model space as composed of the harmonic oscillator orbits in the 0d-1s shell. We shall refer to the particle lines in the 0d-1s shell as the *active* lines, and the hole lines and particle lines outside the 0d-1s shell as the *passive* lines. This is shown in fig. 1. From now on, we shall frequently use this example for illustration.

Consider now the time-ordered Goldstone diagram (a) of fig. 2. This is a typical diagram which may occur in the calculation of the energies of ^{18}O. Here 1, 2, 3 and 4 are active particle lines while a and b are passive particle lines. The dotted vertex line represents a V-interaction. The folded diagram then arises from the factorization of the diagram (a) of fig. 2. As shown by the figure, we factorize diagram (a) into the product of two independent diagrams as shown by diagram (b) of the figure. The time sequence for diagram (a) is $0 \geqq t_1 \geqq t_2 \geqq T$, but the time sequence for diagram (b) is $0 \geqq t_1 \geqq T$ and $0 \geqq t_2 \geqq T$ with no constraint on the relative ordering of t_1 and t_2. Thus (b) is not equal to (a) unless we subtract from it the time-incorrect part, namely that with $t_1 \leqq t_2$. The time-incorrect part is the folded diagram shown by diagram (c) of the figure. We note here that the lines 3 and 4 of diagram (c) are not hole lines, but they are the folded active particle lines. From now on we shall draw a little circle to denote the folded lines, as shown by the figure.

To illustrate the calculation of the folded diagrams, we consider now another example. As shown by fig. 3, the diagram (a) of the figure is equivalent to the product of the two diagrams shown by (b) minus the folded diagram (c) of the same figure. Here again the folded diagram is to take care of the time-incorrect contributions arising from the factorization. The contributions from the various diagrams of fig. 3 are:

$$\text{(a)} = (-i)^4 \int_T^0 \mathrm{d}t_1 \int_T^{t_1} \mathrm{d}t_2 \int_T^{t_2} \mathrm{d}t_3 \int_T^{t_3} \mathrm{d}t_4 \exp\left[-it_1(\varepsilon_c+\varepsilon_d-\varepsilon_5-\varepsilon_6)\right.$$
$$\left. - it_2(\varepsilon_3+\varepsilon_4-\varepsilon_c-\varepsilon_d) - it_3(\varepsilon_a+\varepsilon_b-\varepsilon_3-\varepsilon_4) - it_4(\varepsilon_1+\varepsilon_2-\varepsilon_a-\varepsilon_b)\right], \tag{1}$$

$$\text{(b)} = (-i)^2 \int_T^0 \mathrm{d}t_1 \int_T^{t_1} \mathrm{d}t_2 \exp\left[-it_1(\varepsilon_c+\varepsilon_d-\varepsilon_5-\varepsilon_6) - it_2(\varepsilon_3+\varepsilon_4-\varepsilon_c-\varepsilon_d)\right]$$
$$\times (-i)^2 \int_T^0 \mathrm{d}t_3 \int_T^{t_3} \mathrm{d}t_4 \exp\left[-it_3(\varepsilon_a+\varepsilon_b-\varepsilon_3-\varepsilon_4) - it_4(\varepsilon_1+\varepsilon_2-\varepsilon_a-\varepsilon_b)\right], \tag{2}$$

$$\text{(c)} = (-i)^4 \int_T^0 \mathrm{d}t_1 \int_T^{t_1} \mathrm{d}t_2 \int_{t_2}^{0} \mathrm{d}t_3 \int_T^{t_3} \mathrm{d}t_4 \exp\left[-it_1(\varepsilon_c+\varepsilon_d-\varepsilon_5-\varepsilon_6)\right.$$
$$\left. - it_2(\varepsilon_3+\varepsilon_4-\varepsilon_c-\varepsilon_d) - it_3(\varepsilon_a+\varepsilon_b-\varepsilon_3-\varepsilon_4) - it_4(\varepsilon_1+\varepsilon_2-\varepsilon_a-\varepsilon_b)\right]. \tag{3}$$

We note that in the above equations we have left out the common factors

$$\langle 56|V|\mathrm{cd}\rangle\langle \mathrm{cd}|V|34\rangle\langle 34|V|\mathrm{ab}\rangle\langle \mathrm{ab}|V|12\rangle.$$

In actual calculations, these factors should of course be included in all of the above three equations. Since the time T in the above equations will be taken to be $-\infty$, we shall assume as usual that the interaction is switched on adiabatically, namely the interaction has a time dependence of

$$V(t) = V \mathrm{e}^{-\eta|t|}, \tag{4}$$

with η being positive and approaching to zero. This time dependence will make the integrals in the above equations definite when T approaches to $-\infty$. The ε in the above equations are just the single-particle energies of the free propagators. The i in the above equations represent in fact $i/\hbar$.

Eq. (3) gives a clear definition for the folded diagram. As mentioned above, the purpose of the folded diagram is just to take care of the time-incorrect contribution arising from the factorization. It is clearly seen in fig. 3 and eqs. (1), (2) and (3) that (a) is simply equal to (b)–(c). We may calculate the folded diagrams in two equivalent ways. One is to evaluate the integral of eq. (3) directly. The other is to use the usual rules for evaluating the time-ordered Goldstone diagrams. For the latter we must, however, include all the Goldstone diagrams corresponding to the generalized time

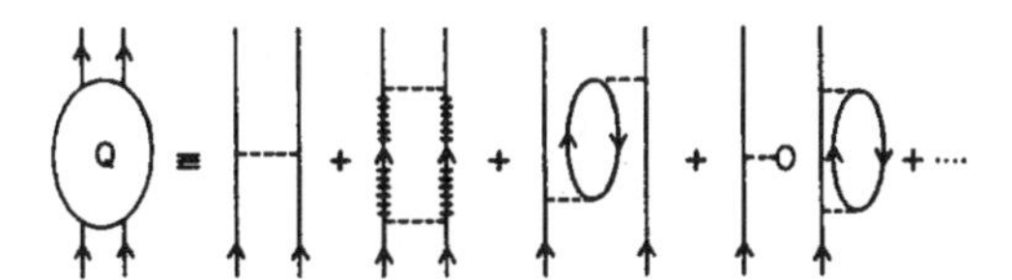

Fig. 4. Typical irreducible diagrams which are contained in a Q-box.

ordering. This can be seen by studying diagram (c) of fig. 3. If this diagram were the usual time-ordered Goldstone diagram, it would have a time sequence of $t_1 > t_3 > t_2 > t_4$. But there are other time-incorrect contributions in addition to this, namely those corresponding to $t_3 > t_1 > t_2 > t_4$, $t_3 > t_1 > t_4 > t_2$, $t_3 > t_4 > t_1 > t_2$ and $t_1 > t_3 > t_4 > t_2$. We note that the time constraints common to all of them are just $t_1 > t_2$, $t_3 > t_4$ and $t_3 > t_2$. Thus if we choose to use the rules for the Goldstone diagrams to evaluate the folded diagram (c) of fig. 3, we need to evaluate all of the five Goldstone diagrams corresponding to the preceding time sequences. This is usually known as the generalized time ordering. The sum of the five previously mentioned Goldstone diagrams can be easily shown to be equivalent to the integral given by eq. (3). In using the usual Goldstone rules to evaluate the folded diagrams, we need to make one modification of the rules; in the determination of the energy denominators we should treat a folded particle line as a hole line and conversely a folded hole line as a particle line.

Let us now define the Q-box which denotes a collection of diagrams of some identical structures. As shown by fig. 4, a Q-box is a collection of irreducible diagrams each of which contains at least one vertex interaction and is linked to at least one external active line. Here by irreducible we mean that the propagators between two successive vertex interactions must contain at least one passive line which we defined at the beginning of this section. Hence we see that no intermediate state in a Q-box may consist solely of active lines. In addition, the "incoming" lines of each Q-box must consist entirely of active lines. In other words, the external lines which enter the Q-box from below are entirely active lines. Thus there are two types of Q-box depending on the nature of the "outgoing" lines of the Q-box. We may refer to those Q-boxes whose "outgoing" lines consist entirely of the active lines as the closed Q-boxes, while those with their "outgoing" lines consisting of at least one passive line as the open Q-boxes. We can now generalize the factorization operation to a sequence of Q-boxes where except for the last Q-box in the sequence all of the Q-boxes are of the closed type. For example, we show in fig. 5 that a sequence of two Q-boxes where the last one is of the open type can be factorized into two independent Q-boxes. The last diagram represents all the folded diagrams which correspond to time-incorrect contributions arising from the factorization. The diagrams we showed in figs. 2 and 3 are typical examples arising from the factorization of two Q-boxes.

Our approach for the folded diagrams is similar to and to a large extent motivated by the work of Baranger and Johnson [7]) where they investigated the folded diagrams of Morita [4]) and Brandow [5]) using a time-dependent point of view. There is, however, a basic difference between their approach and ours. To illustrate this, let us consider the example shown in fig. 3. The diagram (a) of the figure is composed of two irreducible vertex parts of finite times, namely the top part is from time t_2 to t_1 and the lower part is from time t_4 to t_3. In the Baranger-Johnson approach, one would "shrink" the finite-time (or time-delayed) irreducible vertex parts of diagram (a) of fig. 3 to those of zero time (or instantaneous) by folding the external lines of each irreducible vertex part to a common time. This operation will bring in the time-incorrect contributions and to correct for them one needs therefore to include the

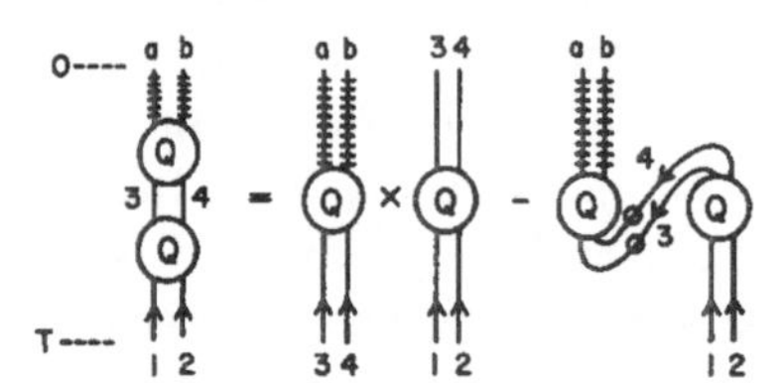

Fig. 5. Factorization of a sequence of two Q-boxes and the associated folded diagrams.

folded diagrams. In our approach, the factorization operation in fact extends each of the two finite-time irreducible vertex parts of diagram (a) of fig. 3 to that of infinite time, and thus have two irreducible vertex parts having the same time boundaries 0 and T with T approaching $-\infty$ as shown by diagram (b) of the figure. This operation also brings in the time-incorrect contributions and to correct for them the folded diagrams must be included. Thus the difference is that in the approach of Baranger and Johnson they shrink the finite-time irreducible vertex parts to those of zero time while we extend them to those of infinite time. As we shall see later both have the same merit in removing the E-dependence of H_{eff}.

It should be pointed out that the Q-box we have just defined is just a collection of diagrams of some common structure as mentioned before. Although each individual irreducible diagram which belongs to a given Q-box has definite time boundaries, the Q-box as a whole has, however, no definite time boundaries. This is simply because the various irreducible diagrams belonging to the Q-box do not have identical time boundaries. Thus it should be noted that the diagrammatic equation of fig. 5 does not imply any relative time-ordering of the Q-boxes as a whole, but rather it represents the collection of all the individual diagrams which have the same general structure shown by the figure.

3. Decomposition theorem

Let $|0j\rangle$ denote a model-space basis vector. For example it may represent a state composed of two valence particles and a closed ^{16}O core, namely

$$|0j\rangle = [a_1^+ a_2^+]_j|0\rangle, \tag{5}$$

where $|0\rangle$ represents the unperturbed ^{16}O core state, the subscripts 1 and 2 denote the single-particle states (above the closed shells) and j represents the other quantum numbers necessary in specifying the state. The single-particle states are defined by

$$H_0 \phi_\mu = \varepsilon_\mu \phi_\mu, \tag{6}$$

where ε_μ and ϕ_μ are respectively the single-particle energy and wave function. The unperturbed Hamiltonian H_0 is related to H, the total Hamiltonian, by

$$\begin{aligned} H &= H_0 + H_1, \\ H_0 &= T + U, \\ H_1 &= V - U, \end{aligned} \tag{7}$$

where T is the kinetic energy, V is the nucleon-nucleon interaction potential and U is a one-body potential which we can choose at will. For nuclear structure calculations, V and U are commonly chosen as a phenomenological nucleon-nucleon potential and a harmonic oscillator potential, respectively.

Let us now operate on $|0j\rangle$ with the time-development operator $U(0, -\infty)$. In the interaction representation, the time-development operator $U(t, t')$ can be written [8]) as

$$U(t, t') = 1 + \sum_{n=1}^{\infty} I_n, \tag{8}$$

with

$$I_n = (-i)^n \int_{t'}^{t} dt_1 \int_{t'}^{t_1} dt_2 \ldots \int_{t'}^{t_{n-1}} dt_n H_1(t_1) H_1(t_2) \ldots H_1(t_n). \tag{9}$$

Thus when operating on $|0j\rangle$ with $U(0, -\infty)$ we will obtain an infinite series each term of which can be represented by a time-ordered diagram. Since H_1 is given by $V-U$, the diagrams will generally contain vertices of both V and U. For example, in addition to the V-diagrams shown in fig. 4, the Q-box will contain diagrams with U-vertices as well. Some examples are shown in fig. 6. The factorization procedure we described in the previous section is clearly also applicable to diagrams with U-vertices.

In the evaluation of the wave function $U(0, -\infty)|0i\rangle$ there are certain terms which can be factorized out and treated as a multiplicative factor. Consider the example shown in fig. 7. The three diagrams in the l.h.s. of the figure all belong to $U(0, -\infty)$

$|0i\rangle$. They arise from the operation of I_3 of eq. (9) on $|0i\rangle$. They have identical labels for all of the particle and hole lines and differ from each other only in the relative time ordering of the interaction vertices, as shown by the figure. Similar to what we did in eqs. (1), (2) and (3) we can easily show that the sum of the three diagrams in the l.h.s.

Fig. 6. Typical diagrams which belong to the Q-box and contain the U-vertex. Here - - - × represents a $-U$ interaction.

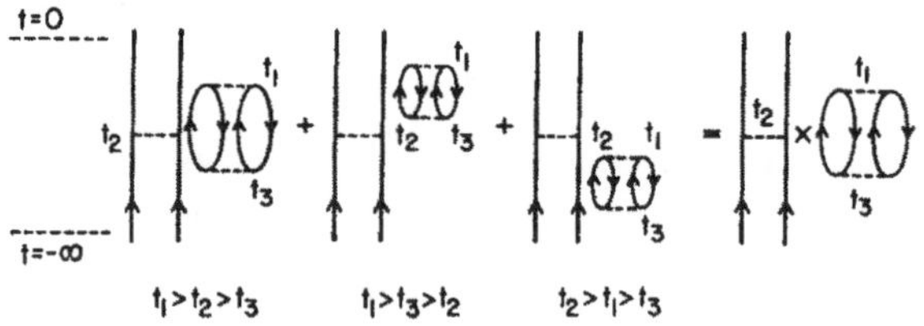

Fig. 7. Factorization of unlinked diagrams.

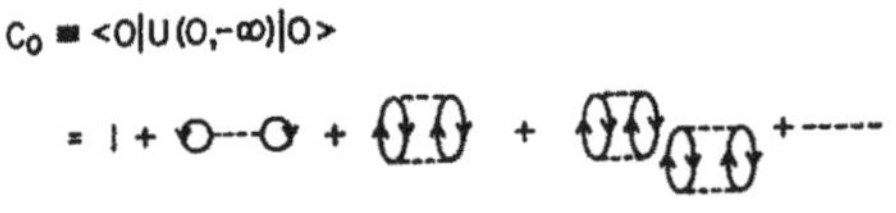

Fig. 8. Diagrams contained in C_0.

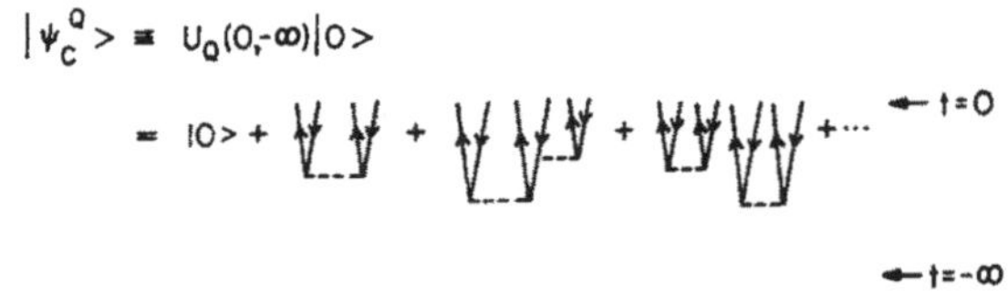

Fig. 9. Diagrams contained in the wave function $|\psi_c{}^Q\rangle$. $|0\rangle$ represents the unperturbed closed-shell core state.

is equal to the product of two *independent* diagrams as shown by the r.h.s. of the figure. The time integration for these two independent diagrams is

$$\int_{-\infty}^{0} dt_2 \times \int_{-\infty}^{0} dt_1 \int_{-\infty}^{t_1} dt_3 .$$

Thus we see that by collecting all the unlinked diagrams corresponding to all possible distinct time orderings, we can factorize out these unlinked diagrams, calculate them independent of the other diagrams and treat them as a multiplicative factor.

The above argument can be generalized to show that the diagrams contained in C_0 of fig. 8 can be factorized out as a whole. Similarly the unlinked wave function diagrams contained in $|\psi_c^Q\rangle$ of fig. 9 can also be factorized out. Thus symbolically we can write

$$U(0, -\infty)|0i\rangle = \{\text{diagrams linked to } i\} \times \{\text{diagrams not linked to } i\}, \quad (10)$$

where the diagrams not linked to i are just $C_0 \times |\psi_c^Q\rangle$ given respectively by figs. 8 and 9. In fig. 9, we introduce the operator $U_Q(0, -\infty)$ which generates all the wave-function diagrams when operating on $|0\rangle$. We note that the above factorization of the unlinked

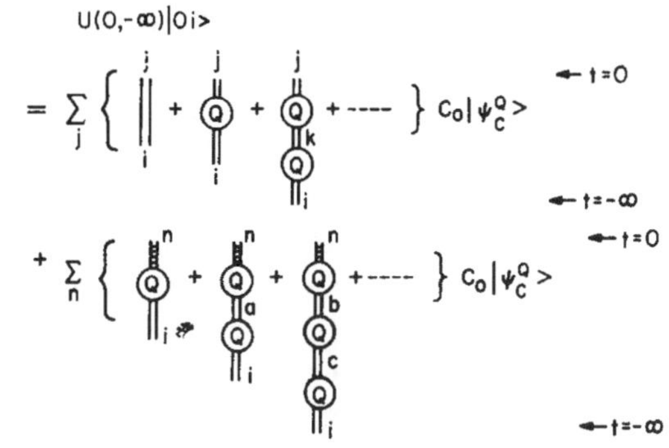

Fig. 10. Diagrams contained in $U(0, -\infty)|0i\rangle$. Note that the intermediate indices k, a, b, c, ... represent summations over all allowed states in the model space. The sum of all the diagrams in the upper parenthesis equals to $\langle j|U_L(0, -\infty)|i\rangle$.

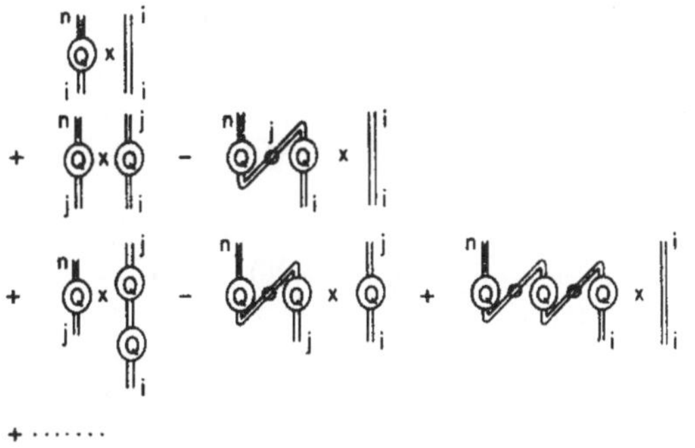

Fig. 11. Factorization of the diagrams in the second parenthesis of fig. 10.

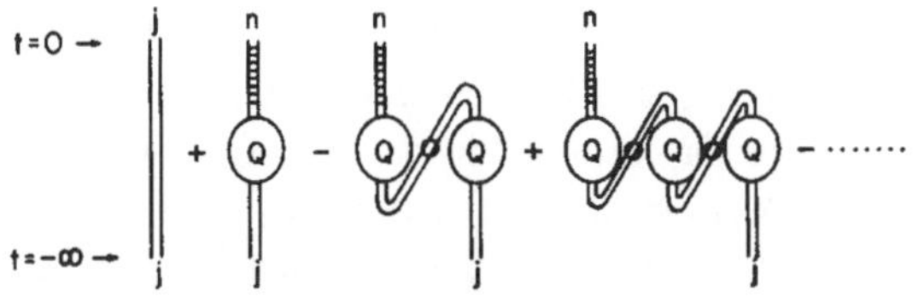

Fig. 12. Diagrams contained in the wave function $U_{QL}(0, -\infty)|j\rangle$. Note that the "railed" propagator $|n\rangle$ contains at least one passive line and the "folded" propagators between two successive Q-boxes consist entirely of active lines.

diagrams is basically very similar to the theorem of Bethe, Brandow and Petchek [9]) for the evaluation of the hole line self-energy insertions in the theory of nuclear matter.

The diagrammatic identity shown in fig. 10 for $U(0, -\infty)|0i\rangle$ should now be obvious. The diagrams which are linked to i must belong to one of the following two categories: those with the propagator at $t = 0$ being j and those with the propagator at $t = 0$ being n where j represents a propagator composed entirely of active lines and n represents a propagator which contains at least one passive line. At time $t = -\infty$, the propagator for these diagrams should of course be i for all of them. In fig. 10, these two kinds of linked diagrams are placed respectively in the upper and lower parenthesis. We note that we sum over all allowed j- and n-states. If we recall the definition of the Q-boxes (see sect. 2 and fig. 4), it is clear that the diagrams which are linked to i can be grouped into those with zero Q-box (i.e., no interaction), one Q-box, two Q-box and so on as shown by the figure. The first term in the upper parenthesis is just the Kronecker delta function δ_{ij}. The corresponding term in the lower parenthesis is identically zero because $|n\rangle$ is always orthogonal to $|i\rangle$.

We shall now show that in fig. 10 the diagrams in the lower parenthesis and those in the upper parenthesis have a common factor. Using the factorization procedure shown in fig. 5, the diagrams in the lower parenthesis of fig. 10 can be rewritten as those shown in fig. 11. The first row of fig. 11 is just equivalent to the first term of the lower parenthesis of fig. 10, where the free propagator from i to i merely represents the constant one. The terms in the second row are equivalent to the second term of the parenthesis. The terms in the third row are equivalent to the third term of the parenthesis, and note that here we have applied the factorization procedure shown by fig. 5 twice. By continuing the above process and collecting terms column-wise, we see clearly each column of fig. 11 contains a factor which is equal to the sum of the diagrams contained in the upper parenthesis of fig. 10. Thus we can rewrite the expression shown by fig. 10 as

$$U(0, -\infty)|0i\rangle = \sum_j U_Q(0, -\infty)|0j\rangle\langle 0j|U(0, -\infty)|0i\rangle, \tag{11}$$

with

$$U_Q(0, -\infty)|0j\rangle = |\psi_c^Q\rangle \times U_{QL}(0, -\infty)|j\rangle, \tag{12}$$

$$\langle 0j|U(0, -\infty)|0j\rangle = \langle j|U_L(0, -\infty)|i\rangle \times C_0, \tag{13}$$

where the wave functions contained in $U_{QL}(0, -\infty)|j\rangle$ are shown in fig. 12 and the matrix elements which contribute to $\langle j|U_L(0, -\infty)|i\rangle$ are shown in fig. 10. The constant C_0 and the wave function $|\psi_c^Q\rangle$ have been defined respectively in fig. 8 and fig. 9. In general, the state vector $|0i\rangle$ represents the wave function of a core denoted by $|0\rangle$ and some active particles (and/or holes) denoted by $|i\rangle$. For example, in the case of ^{18}O $|0\rangle$ denotes the unperturbed ^{16}O core and $|i\rangle$ denotes the two-valence particles in the 1s–0d shell.

Eqs. (11) to (13) will be referred to as the decomposition theorem. It states that the results arising from operating on $|0i\rangle$ with $U(0, -\infty)$ can be written as a product of two parts; one part consists entirely of wave functions $U_Q(0, -\infty)|0j\rangle$ and the other part consists entirely of numbers, namely the matrix element $\langle 0j|U(0, -\infty)|0i\rangle$. It also states that the "core" wave function $|\psi_c^Q\rangle$ and the "core" matrix element C_0 both enter $U(0, -\infty)|0i\rangle$ as multiplicative factors as shown by the above equations. The subscript L denotes that the diagrams must be linked to at least one external active line. As we shall discuss in sect. 4, $|\psi_c^Q\rangle$ will be proportional to the true ground state wave function of the core (see eq. (33)).

Since $|n\rangle$ is always orthogonal to $|i\rangle$, it can be seen easily from eq. (12) and fig. 12 that

$$\langle 0i|U_Q(0, -\infty)|0j\rangle = \delta_{ij}, \tag{14}$$

where δ_{ij} is the Kronecker delta function.

4. The model-space secular equation

By making use of the decomposition theorem, we shall derive in this section the model-space secular equation

$$H_{\rm eff} P\psi_\alpha = E_\alpha P\psi_\alpha, \tag{15}$$

where E_α and ψ_α are the true eigenvalue and eigenfunction defined by

$$H\psi_\alpha = E_\alpha \psi_\alpha, \tag{16}$$

and H is the total Hamiltonian given by eq. (7). P is the projection operator

$$P = \sum_{i=1}^{d} |0i\rangle\langle 0i|, \tag{17}$$

where $|0i\rangle$ represents the model-space basis vector with $\langle 0i|0i\rangle = 1$ and d is the dimensionality of the model space. Thus $P\psi_\alpha$ is the projection of the true wave function onto the model space. $H_{\rm eff}$ is the model-space effective Hamiltonian which we shall derive in this section. As we shall see shortly, $H_{\rm eff}$ is independent of E but will contain the folded diagrams. We shall also show that it will not contain the unlinked diagrams.

Let us first establish a one-to-one correspondence between the model-space parent state ϕ_α and the true eigenfunction ψ_α for $\alpha = 1, 2, \ldots, d$. The construction of ϕ_α is explained in the following. Suppose we choose a trial parent state ϕ_1 as

$$|\phi_1\rangle = \frac{1}{\sqrt{d}} \sum_{i=1}^{d} |0i\rangle, \tag{18}$$

and operate on it by the time-development operator $U(0, -\infty)$ with $U(t, t')$ given by

$$U(t, t') = e^{-iH(t-t')}. \tag{19}$$

Following Thouless [10]), we consider that $U(t, t')$ can be analytically continued into the complex time-plane and thus allow t' to have a small imaginary part. Thus we have by eq. (19)

$$\frac{U(0,-\infty)|\phi_1\rangle}{\langle\phi_1|U(0,-\infty)|\phi_1\rangle} = \lim_{\varepsilon\to 0^+}\lim_{t'\to-\infty}\frac{e^{-iH[0-t'(1-i\varepsilon)]}|\phi_1\rangle}{\langle\phi_1|e^{-iH[0-t'(1-i\varepsilon)]}|\phi_1\rangle},$$

where ε is a very small positive number such that $t'\varepsilon\to-\infty$ when $t'\to-\infty$ and $t'\varepsilon\to 0$ when t' is finite. By inserting a complete set of eigenstates of H between $|\phi_1\rangle$ and the time-evolution operator for both the numerator and denominator of the r.h.s. of the above equation we have

$$\frac{U(0,-\infty)|\phi_1\rangle}{\langle\phi_1|U(0,-\infty)|\phi_1\rangle} = \lim_{\varepsilon\to 0^+}\lim_{t'\to-\infty}\frac{\sum_\alpha e^{iE_\alpha t'}e^{E_\alpha\varepsilon t'}|\psi_\alpha\rangle\langle\psi_\alpha|\phi_1\rangle}{\sum_\beta e^{iE_\beta t'}e^{E_\beta\varepsilon t'}\langle\phi_1|\psi_\beta\rangle\langle\psi_\beta|\phi_1\rangle}$$

$$= \frac{|\psi_1\rangle}{\langle\phi_1|\psi_1\rangle}$$

$$\equiv |\tilde{\psi}_1\rangle, \tag{20}$$

where $|\psi_1\rangle$ is the lowest eigenstate of H with $\langle\psi_1|\phi_1\rangle\neq 0$. Here the lowest eigenstate refers to the eigenstate with its eigenvalue E_1 being the most negative (or the least positive if all the eigenvalues are positive). The above result is because of the presence of the real exponential damping factor in the above equation which suppresses all the other non-vanishing terms much more rapidly than the term containing $|\psi_1\rangle$, and therefore only the $|\psi_1\rangle$ term survives.

Since $\langle\phi_1|\psi_1\rangle$ is just a number and is non-zero, the state $|\tilde{\psi}_1\rangle$ is clearly also an eigenstate of H. We note that $|\tilde{\psi}_1\rangle$ is not normalized to unity, it satisfies instead

$$\langle\phi_1|\tilde{\psi}_1\rangle = 1. \tag{21}$$

Note that for $|\psi_1\rangle$ we have $\langle\psi_1|\psi_1\rangle = 1$.

Since we now know in principle $|\psi_1\rangle$, we can construct another trial parent state $|\phi_2\rangle$ which satisfies

$$\langle\phi_2|P\psi_1\rangle = 0. \tag{22}$$

Here $|\phi_2\rangle$ is a state vector of unit length contained entirely in the model space. There is clearly more than one way to choose $|\phi_2\rangle$ to satisfy eq. (22). Since the basis vectors outside the model space are orthogonal to those contained in the model space, it follows from eq. (22) that

$$\langle\phi_2|\psi_1\rangle = 0. \tag{23}$$

To explain this result more explicitly, we can write

$$|\psi_1\rangle = P|\psi_1\rangle + Q|\psi_1\rangle, \tag{24}$$

with Q defined by

$$P+Q = 1, \qquad PQ = QP = 0. \tag{25}$$

Eq. (23) then clearly follows from the fact that $Q|\phi_2\rangle = 0$.

Because of $\langle\phi_2|\psi_1\rangle = 0$ by eq. (23), the procedure which leads to eq. (20) will now yield

$$\frac{U(0 \ -\infty)|\phi_2\rangle}{\langle\phi_2|U(0,-\infty)|\phi_2\rangle} = \frac{|\psi_2\rangle}{\langle\phi_2|\psi_2\rangle} \equiv |\tilde{\psi}_2\rangle, \tag{26}$$

where $|\psi_2\rangle$ is the lowest eigenstate of H with its eigenvalue higher than E_1 and $\langle\phi_2|\psi_2\rangle \neq 0$. Since $\langle\phi_2|\psi_2\rangle$ is just a non-zero number, $|\tilde{\psi}_2\rangle$ is clearly also an eigenstate of H. It follows readily from eq. (26) that $\langle\phi_2|\tilde{\psi}_2\rangle = 1$. The above procedure can be easily continued, and hence we obtain the general result

$$|\tilde{\psi}_\alpha\rangle = \frac{U(0,-\infty)|\phi_\alpha\rangle}{\langle\phi_\alpha|U(0,-\infty)|\phi_\alpha\rangle} = \frac{|\psi_\alpha\rangle}{\langle\phi_\alpha|\psi_\alpha\rangle}, \tag{27}$$

with

$$\langle\phi_\alpha|\phi_\alpha\rangle = 1, \tag{28}$$

$$\langle\phi_\alpha|\psi_\alpha\rangle \neq 0, \tag{29}$$

$$\langle\phi_\alpha|\psi_1\rangle = \langle\phi_\alpha|\psi_2\rangle = \ldots = \langle\phi_\alpha|\psi_{\alpha-1}\rangle = 0, \tag{30}$$

for $\alpha = 1, 2, \ldots, d$. Thus starting from the parent state ϕ_α we can construct the eigenstates $|\tilde{\psi}_\alpha\rangle$ of H for $\alpha = 1, 2, \ldots, d$ according to eq. (27). $|\tilde{\psi}_\alpha\rangle$ is not normalized to unity, but rather it is normalized by $\langle\phi_\alpha|\tilde{\psi}_\alpha\rangle = 1$. It should be noted that these d eigenstates are not necessarily the lowest d eigenstates of H, but they are the lowest d eigenstates which satisfy eq. (29).

We would like to point out that the above correspondence between $|\tilde{\psi}_\alpha\rangle$ and $|\phi_\alpha\rangle$ as shown by eqs. (27) to (30) can be established only under the condition that the projections of $|\psi_\alpha\rangle$, $\alpha = 1, 2, \ldots, d$, onto the model space are linearly independent. We assume that this condition is true. Then we can use the Schmidt procedure to orthonormalize the parent states ϕ_α and write the general form of them as

$$|\phi_\alpha\rangle = \sum_{i=1}^{d} C_i^\alpha|0i\rangle, \qquad \alpha = 1, 2, \ldots, d, \tag{31}$$

with

$$\sum_{i=1}^{d} C_i^{\alpha *} C_i^\beta = \delta_{\alpha\beta}. \tag{32}$$

Thus we have shown that we can construct d eigenstates from d parent states according to eq. (27). Our purpose is, however, not to use eq. (27) to calculate the eigenfunction $|\tilde{\psi}_\alpha\rangle$, but to use it to derive the model-space effective Hamiltonian. As we shall discuss later, it would not be convenient to calculate the eigenfunction $|\tilde{\psi}_\alpha\rangle$ directly from eq. (27).

As a by-product, the following result can be readily shown [10]) in exactly the same manner as we derived eq. (20). Namely if $|0\rangle$ denotes the unperturbed core state such as the unperturbed core state of ^{16}O, then the state $|\tilde{\psi}_c\rangle$ given by

$$|\tilde{\psi}_c\rangle = \frac{U(0, -\infty)|0\rangle}{\langle 0|U(0, -\infty)|0\rangle} = |\psi_c^Q\rangle \tag{33}$$

is the true ground state of the core Hamiltonian provided that the overlap of the true ground state with $|0\rangle$ is not zero. Recall that $|\psi_c^Q\rangle$ was defined in fig. 9. It is readily seen that $|\tilde{\psi}_c\rangle$ satisfies $\langle 0|\tilde{\psi}_c\rangle = 1$. In the case of ^{16}O, $|\tilde{\psi}_c\rangle$ is just the true ground state of ^{16}O. Thus we see that in eqs. (11) to (13) of the previous section it is the true ground state of the core which enters as a "multiplicative" factor. It is well known that eq. (33) can also be derived by using the adiabatic condition of Gell-Mann and Low [11]).

Since $|\tilde{\psi}_\alpha\rangle$ is an eigenstate of H by eq. (27), we have

$$H\frac{U(0, -\infty)|\phi_\alpha\rangle}{\langle\phi_\alpha|U(0, -\infty)|\phi_\alpha\rangle} = E_\alpha\frac{U(0, -\infty)|\phi_\alpha\rangle}{\langle\phi_\alpha|U(0, -\infty)|\phi_\alpha\rangle}. \tag{34}$$

Here it should be noted that it is not appropriate to remove the denominator $\langle\phi_\alpha|U(0, -\infty)|\phi_\alpha\rangle$ from both sides of the above equation. The reason is that neither $U(0, -\infty)|\phi_\alpha\rangle$ nor $\langle\phi_\alpha|U(0, -\infty)|\phi_\alpha\rangle$ is finite by itself; only the ratio of them is a well-defined quantity. Then by eq. (31) and eq. (11), the decomposition theorem, we can rewrite the above equation as

$$H\frac{\sum_{ij} U_Q(0, -\infty)|0j\rangle\langle 0j|U(0, -\infty)|0i\rangle C_i^\alpha}{\sum_{km}\langle 0k|U(0, -\infty)|0m\rangle C_k^{\alpha *}C_m^\alpha} = E_\alpha\frac{\sum_{ij} U_Q(0, -\infty)|0j\rangle\langle 0j|U(0, -\infty)|0i\rangle C_i^\alpha}{\sum_{km}\langle 0k|U(0, -\infty)|0m\rangle C_k^{\alpha *}C_m^\alpha}. \tag{35}$$

To simplify this equation, we define

$$b_j^\alpha = \frac{\sum_i \langle 0j|U(0, -\infty)|0i\rangle C_i^\alpha}{\sum_{km}\langle 0k|U(0, -\infty)|0m\rangle C_k^{\alpha *}C_m^\alpha}. \tag{36.1}$$

By using eq. (13) we see that both the denominator and the numerator of this equation contain the term C_0, the vacuum fluctuation diagrams shown by fig. 8. After cancelling out this term, the above equation is rewritten as

$$b_j^\alpha = \frac{\sum_i \langle j|U_L(0, -\infty)|i\rangle C_i^\alpha}{\sum_{km}\langle k|U_L(0, -\infty)|m\rangle C_k^{\alpha *}C_m^\alpha}. \tag{36.2}$$

Then by multiplying eq. (35) from the left by $\langle 0k|$ and noting that $\langle 0k|U_Q(0, -\infty)|0j\rangle = \delta_{kj}$ by eq. (14), eq. (35) becomes

$$\sum_{j=1}^{d}\langle 0k|HU_Q(0, -\infty)|0j\rangle b_j^\alpha = E_\alpha b_k^\alpha. \tag{37}$$

This is the model-space secular equation which we have set out to derive. The U_Q operator was defined by eq. (12) and figs. 12 and 9. Eq. (37) is exactly of the form of eq. (15) with H_{eff} given by $HU_Q(0, -\infty)$. To see this, we need to investigate the physical meaning of b_k^α. By eqs. (27), (31), (11) and (36) we readily have

$$|\tilde{\psi}_\alpha\rangle = \sum_j U_Q(0, -\infty)|0j\rangle b_j^\alpha. \tag{38}$$

Then by eq. (14) we have from the above equation

$$\langle 0j|\tilde{\psi}_\alpha\rangle = b_j^\alpha. \tag{39}$$

Hence b_j^α is just the projection of the eigenstate $|\tilde{\psi}_\alpha\rangle$ on the model-space basis vector $|0j\rangle$, and thus eq. (37) is indeed of the form of eq. (15). Eq. (37) has been obtained by Morita [4]) and Oberlechner *et al.* [6]) using different methods of derivation and with different definitions for the coefficients b_j^α.

It should be pointed out that eigenstate $|\tilde{\psi}_\alpha\rangle$ is *not* normalized to unity. In addition, the vector $[b_1^\alpha, b_2^\alpha \ldots b_d^\alpha]$ is usually not of unit length and not orthogonal to vectors with different α. Namely we have in general

$$\sum_{i=1}^{d} b_i^{\alpha *} b_i^\beta \neq \delta_{\alpha\beta}. \tag{40}$$

The above result is indeed expected, since $HU_Q(0, -\infty)$ is not Hermitian and hence the eigenvectors with different E_α obtained from eq. (37) are generally not orthogonal to each other. Furthermore, since it is the true eigenfunction $|\psi_\alpha\rangle$ which should be normalized to unity, we should normalize neither $|\tilde{\psi}_\alpha\rangle$ nor the projection of $|\tilde{\psi}_\alpha\rangle$ onto the model space to unity.

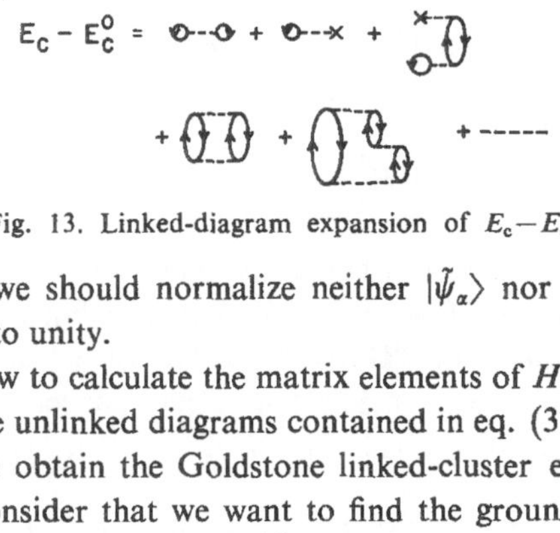

Fig. 13. Linked-diagram expansion of $E_c - E_c^0$.

In order to show how to calculate the matrix elements of $HU_Q(0, -\infty)$ in eq. (37) and how to remove the unlinked diagrams contained in eq. (37), let us first, as a by-product, show how to obtain the Goldstone linked-cluster expansion for the non-degenerate system. Consider that we want to find the ground state energy of ^{16}O. Denoting the unperturbed core state of ^{16}O by $|0\rangle$, we know by eq. (33) that $|\tilde{\psi}_c\rangle$ is the true ground state of ^{16}O and hence we have

$$H \frac{U(0, -\infty)|0\rangle}{\langle 0|U(0, -\infty)|0\rangle} = E_c \frac{U(0, -\infty)|0\rangle}{\langle 0|U(0, -\infty)|0\rangle}, \tag{41}$$

where E_c denotes the true ground state energy of ^{16}O. Applying the decomposition theorem eq. (11) to the above equation and noting that we have now only one unperturbed state in the model space, eq. (41) can be rewritten as

$$HU_Q(0, -\infty)|0\rangle = E_c U_Q(0, -\infty)|0\rangle. \tag{42}$$

Since $H = H_0+H_1$ and $H_0|0\rangle = E_c^0|0\rangle$ where E_c^0 is the unperturbed energy of ^{16}O, we have from eq. (42) that

$$\langle 0|H_1 U_Q(0, -\infty)|0\rangle = E_c - E_c^0. \tag{43}$$

Here we note that the diagrams contained in the wave function $U_Q(0, -\infty)|0\rangle$ have been given by fig. 9, and as shown there are both linked and unlinked diagrams. But to have non-vanishing contributions to $\langle 0|H_1 U_Q(0, -\infty)|0\rangle$, it is obvious that only those terms which can be "closed" by one H_1 interaction at time $t = 0$ will survive. Hence the energy shift $E_c - E_c^0$ contains linked diagrams only, as shown by fig. 13. This result is just the Goldstone linked cluster expansion.

Let us now remove the core energy E_c from eq. (37). We first write H as H_0+H_1 in eq. (37). Then by figs. 9 and 13 and eqs. (12) and (14) it is easily seen that the matrix element

$$\langle 0k|H_1 U_Q(0, -\infty)|0j\rangle,$$

contains a term $\delta_{kj}(E_c - E_c^0)$ where $E_c - E_c^0$ are just the linked diagrams given by fig. 13. This term is obtained by attaching H_1 entirely to $|\psi_c^Q\rangle$ of eq. (12) and fig. 9. The remaining terms of the above matrix element are then all linked to the external active lines, originating either from attaching the last interaction H_1 entirely to $U_{QL}(0, -\infty)|j\rangle$ or linking up $U_{QL}(0, -\infty)|j\rangle$ with $|\psi_c^Q\rangle$ via the last interaction H_1. Denoting these terms by

$$\langle 0k|[H_1 U_Q(0, -\infty)]_L|0j\rangle,$$

eq. (37) can now be rewritten as

$$\langle k|H_0|k\rangle b_k^\alpha + \sum_j \langle 0k|[H_1 U_Q(0, -\infty)]_L|0j\rangle b_j^\alpha = (E_\alpha - E_c)b_k^\alpha. \tag{44}$$

Eq. (44) is the principal result of the present paper. The eigenvalue of it is $E_\alpha - E_c$ rather than E_α. For example, if we are calculating the energies of ^{18}O, the eigenvalue of eq. (44) will be the energy of ^{18}O relative to the true ground state energy of ^{16}O. The state vector $|k\rangle$ represents merely the valence-particle wave functions in the model space, such as $|d_{\frac{5}{2}}^2\rangle$ in the calculation of ^{18}O. The effective Hamiltonian is now given by

$$H_{\text{eff}} = H_0 + [H_1 U_Q(0, -\infty)]_L \equiv H_0 + H_1^{\text{eff}}, \tag{45}$$

where subscript L denotes that the diagrams which contribute to H_{eff} must be linked to at least one of the external active lines. The diagrams which are unlinked to any of the external active lines are contained in the core energy E_c and have been removed, as shown by eq. (44). Diagrammatically, the diagrams contained in $\langle 0k|[H_1 U_Q(0, -\infty)]_L|0j\rangle$ are shown in fig. 14. Here we note that we have two kinds of Q-boxes, Q and Q′. Both Q and Q′ are a collection of irreducible vertex diagrams which are linked to at least one external active line, and by irreducible we mean that the propagator between two successive H_1 interactions contain at least one passive line. The difference between Q and Q′ is that the lowest-order term of the Q-box is a term with two H_1 interactions while the lowest-order term of the Q-box is a term with one H_1 interaction. This can be seen from fig. 12. By adding one last

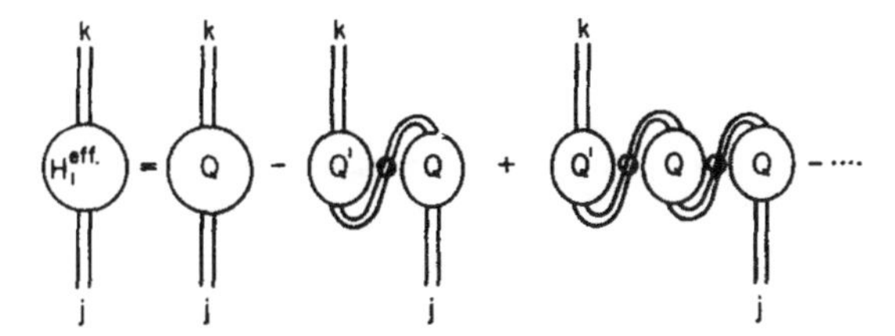

Fig. 14. Diagrammatic representation of $\langle k|H_1^{\text{eff}}|j\rangle$ where $H_1^{\text{eff}} = [H_1 U_Q(0, -\infty)]_L$.

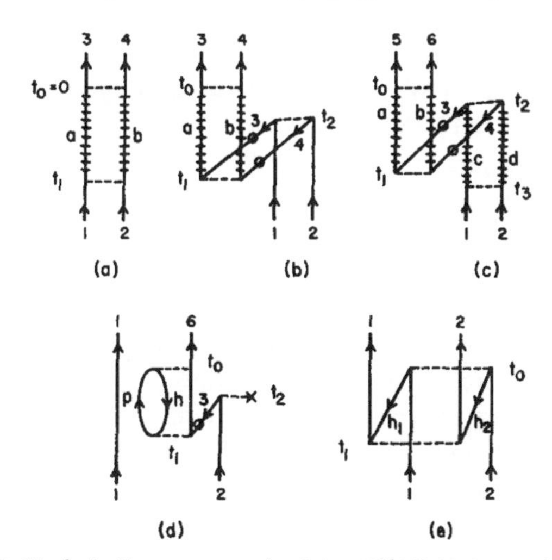

Fig. 15. Typical diagrams contained in $\langle 0k|[H_1 U_Q(0, -\infty)]_L|0j\rangle$.

H_1 interaction H_1 at $t_0 = 0$, all the last Q-boxes of the figure will have at least two H_1 interactions and thus become the Q′-boxes. But there is a one H_1 interaction term arising from adding a H_1 vertex to the free propagator. Thus adding a H_1 interaction to the first two terms of fig. 12 give rise to the one-Q-box term of fig. 14. Some typical diagrams contained in fig. 14 are shown in fig. 15. Because of the nature of the Q′-box, the lowest-order folded diagram contain *three* H_1 interactions. As an example, the diagram (b) of fig. 15 is a lowest-order folded diagram. In fig. 15, diagrams (a) and (e) originate from the first term of fig. 14, and diagrams (b), (c) and (d) originate from the second term of the same figure.

Because of the time argument in $[H_1 U_Q(0, -\infty)]_L$, it should be pointed out that for all of the diagrams contained in fig. 14 the last interaction is always at $t_0 = 0$ and the time t_0 is clearly not to be integrated. All the other interactions occur at times earlier than t_0. To illustrate this point, the contribution from diagram (c) of fig. 15 is given by

$$(-1)^1(-i)^3\langle 56|V|\text{ab}\rangle\langle \text{ab}|V|34\rangle\langle 34|V|\text{cd}\rangle\langle \text{cd}|V|12\rangle$$
$$\times\int_{-\infty}^{0}\text{d}t_1\int_{t_1}^{0}\text{d}t_2\int_{-\infty}^{t_2}\text{d}t_3\exp\left[-it_1(\varepsilon_3+\varepsilon_4-\varepsilon_\text{a}-\varepsilon_\text{b})-it_2(\varepsilon_\text{c}+\varepsilon_\text{d}-\varepsilon_3-\varepsilon_4)\right.$$
$$\left.-it_3(\varepsilon_1+\varepsilon_2-\varepsilon_\text{c}-\varepsilon_\text{d})\right], \tag{46}$$

where we note that the time $t_0 = 0$ is of course not integrated over and the times t_1, t_2 and t_3 are *all* less than 0. The $(-1)^1$ comes from the minus sign in front of the second term of fig. 14. In general, we should have a factor $(-1)^f$ where f is equal to the "number" of foldings. For example, the diagrams (b), (c) and (d) are all once-folded diagrams and hence all have $f = 1$. The diagrams corresponding to the third term of fig. 14 are all twice-folded and hence will have $f = 2$.

The evaluation of the diagrams which constitute H_1^{eff} of fig. 14 is indeed straightforward. As shown by eq. (46), the rules for evaluating these diagrams with and without folded lines can be easily obtained. We should like to point out that the contributions of all these diagrams are independent of either E_α or E_c of eq. (44). They only depend on the single-particle energies ε, as can be seen from eq. (46). Thus our H_1^{eff} is independent of the energy eigenvalue.

Fig. 14 shows only the general structure of H_1^{eff}. Starting from it, some simplifications can be derived. For example, some diagrams contained in the one-Q-box term will be cancelled exactly by the corresponding diagrams in the Q′-Q-box term of the figure. Similarly there are cancellations among the corresponding diagrams arising from the Q′-Q-box and the Q′-Q-Q-box terms, and so on. There are unlinked diagrams contained in H_1^{eff} where by unlinked we mean that the diagram is composed of more than one unlinked part and each part itself is linked to at least one external active line. It can be shown that most of those unlinked diagrams of H_1^{eff} will cancel out among themselves. The last diagram in fig. 4 is such an example. This diagram is contained in both the one-Q-box term and the Q′-Q-box term of fig. 14 but with opposite signs. Thus they cancel out. The details of this cancellation together with some numerical calculations will be discussed in a separate paper.

The effective interaction H_1^{eff} given by fig. 14 and that of Brandow [5]) are generally consistent with each other. Brandow's result was derived using the energy representation. Namely the starting point is the E-dependent effective Hamiltonian and then by making a series expansion of the energy denominator the E-dependence is removed. As we have shown, our result is obtained starting from the time-dependent representation of the folded diagrams. It is likely that these two different approaches are in fact exactly equivalent to each other.

Since we have removed the unlinked diagrams in the secular equation as shown by eq. (44) and some of these unlinked diagrams have violated the Pauli exclusion principle, the corresponding diagrams of H_1^{eff} which also violate the Pauli exclusion principle must also be included. An example of such a diagram is the diagram (e) of fig. 15.

Thus we have completed the derivation of the model-space secular equation with the model-space effective Hamiltonian being independent of the energy eigenvalue but containing the folded diagrams. The effective interaction H_1^{eff} is composed of all the irreducible and linked vertex diagrams shown in fig. 14. It is easily seen that all the diagrams of H_1^{eff} satisfy the property that the propagator between two successive H_1 interactions must contain at least one passive line with the folded line counted as an active line.

We shall not perform any calculation of H_1^{eff} in this paper. Some discussion about it will be given in sect. 6.

5. Linked-diagram expansion for expectation values

In sect. 4 we have derived a method for determining the eigenvalues of H by solving the model-space secular equation. The wave functions so obtained are just the projection of the eigenfunctions of H onto the model space. In this section we shall derive a method for the calculation of the expectation value.

$$\langle\theta\rangle_\alpha = \frac{\langle\tilde{\psi}_\alpha|\theta|\tilde{\psi}_\alpha\rangle}{\langle\tilde{\psi}_\alpha|\tilde{\psi}_\alpha\rangle}, \tag{47}$$

where θ is a general operator and $\tilde{\psi}_\alpha$ is the eigenfunction of H defined by eq. (27). The obvious difficulty in evaluating $\langle\theta\rangle_\alpha$ is that we do not know $\tilde{\psi}_\alpha$ explicitly; we only know the projection of $\tilde{\psi}_\alpha$ onto the model space. Although $\tilde{\psi}_\alpha$ is related to the coefficients b_j^α by eq. (38), the evaluation of $\langle\theta_\alpha\rangle$ by directly substituting eq. (38) into eq. (47) is still quite complicated. The approach which we shall take is as follows.

We first construct an auxiliary Hamiltonian

$$H_\lambda = H+\lambda\theta = (T+V)+\lambda\theta, \tag{48}$$

where H is the original Hamiltonian and λ is a small parameter. The eigenvalue and eigenfunction of H_λ are given by

$$H_\lambda\tilde{\psi}_\alpha^\lambda = E_\alpha^\lambda\tilde{\psi}_\alpha^\lambda. \tag{49}$$

By multiplying eq. (16) and eq. (49) respectively by ψ_α^λ and ψ_α and then taking the difference of the two products, it is readily obtained that

$$\frac{\langle\tilde{\psi}_\alpha|\theta|\tilde{\psi}_\alpha^\lambda\rangle}{\langle\tilde{\psi}_\alpha|\tilde{\psi}_\alpha^\lambda\rangle} = \frac{E_\alpha^\lambda - E_\alpha}{\lambda}. \tag{50}$$

It then follows that

$$\langle\theta\rangle_\alpha = \lim_{\lambda\to 0}\frac{E_\alpha^\lambda - E_\alpha}{\lambda}. \tag{51}$$

By eq. (44) we have

$$(E-E_0-E_c) = \frac{\sum_{i,j} b_i^{\alpha *} b_j^{\alpha} \langle 0i|[H_1\, U_Q(0,\,-\infty)]_L|0j\rangle}{\sum_k |b_k^{\alpha}|^2}, \tag{52}$$

where E_c is the ground state energy of the core and E_0 is the unperturbed energy of the valence particles, namely $H_0|k\rangle = E_0|k\rangle$. Here the single-particle energies for the model-space basis vectors are taken to be degenerate. Thus we see that the energy shift $(E_\alpha - E_0 - E_c)$ can be expanded into a series of linked diagrams as shown by eq. (52). The quantity $(E_\alpha^\lambda - E_0 - E_c^\lambda)$ can be similarly expanded into linked diagrams according to eq. (52) except that all the H_1 interactions, including those contained in $U_Q(0,\,-\infty)$ and those used in the determination of b_j^α, must be replaced by $H_1 + \lambda\theta$. After expanding E_α^λ and E_α into linked diagrams according to eq. (52), it is clearly seen from eq. (51) that the only diagrams which will contribute to $\langle\theta\rangle_\alpha$ are those with only one θ-vertex. This is because the diagrams with no θ-vertex of E_α^λ and E_α will cancel with each other and the diagrams with more than one θ-vertex will vanish since they approach zero with higher powers of λ. Thus eq. (51) becomes

$$\langle\theta\rangle_\alpha = \langle\theta\rangle_c + \langle\theta\rangle_{\alpha L}, \tag{53}$$

where $\langle\theta\rangle_c$ is just the expectation value of θ for the ground state wave function of the core, namely

$$\langle\theta\rangle_c = \lim_{\lambda\to 0} \frac{E_c^\lambda - E_c}{\lambda} = \frac{\langle\tilde{\psi}_c|\theta|\tilde{\psi}_c\rangle}{\langle\tilde{\psi}_c|\tilde{\psi}_c\rangle}, \tag{54}$$

and $\langle\theta\rangle_{\alpha L}$ is given by

$$\langle\theta\rangle_{\alpha L} = \frac{\sum_{i,j} b_i^{\alpha *} b_j^{\alpha} \langle i|\theta_L^{\text{eff}}|j\rangle}{\sum_k |b_k^{\alpha}|^2}, \tag{55}$$

with

$$\langle i|\theta_L^{\text{eff}}|j\rangle = \left\{\begin{matrix}\text{the set of all diagrams that are linear in } \theta \text{ and arise from} \\ \text{replacing } H_1 \text{ by } H_1+\theta \text{ in } \langle 0i|[H_1\, U_Q(0,\,-\infty)]_L|0j\rangle.\end{matrix}\right\}. \tag{56}$$

Note that in the above equation the H_1 contained in $U_Q(0,\,-\infty)$ are also replaced by $H_1+\theta$ and the diagrams contained in $\langle 0i|[H_1\, U_Q(0,\,-\infty)]_L|0j\rangle$ have been given by fig. 14. Thus the diagrams which are contained in $\langle\theta\rangle_{\alpha L}$ are those which are linked to at least one external active line and have only one θ-vertex. They contain the folded diagrams and are independent of the energy E_α and E_c. Some typical diagrams of $\langle i|\theta_L^{\text{eff}}|j\rangle$ are shown in fig. 16. Note that the structure of them is identical to those of the effective interaction $[H_1\, U_Q(0,\,-\infty)]_L$ except for the presence of one θ-vertex.

By eq. (43) and fig. 13, it is easily seen from eq. (54) that $\langle\theta\rangle_c$ is composed of just the same set of linked diagrams shown by fig. 13 except that each of them must have one θ-vertex. We show in fig. 17 some typical diagrams which belong to $\langle\theta\rangle_c$. This linked-diagram expansion of $\langle\theta\rangle_c$ is a well-known result, and has been derived before using a different method [10]).

Thus we have derived a linked-diagram expansion for the evaluation of the expectation value of a general operator θ with respect to eigenfunction of H, as shown by eqs. (53) to (56). Similar to the appearance of E_c in eq. (44), the core expectation value $\langle\theta\rangle_c$ enters eq. (53) as an additive factor. This is a useful result, since we can just calculate the linked-diagram part $\langle\theta\rangle_{\alpha L}$ and compare it with the difference of the

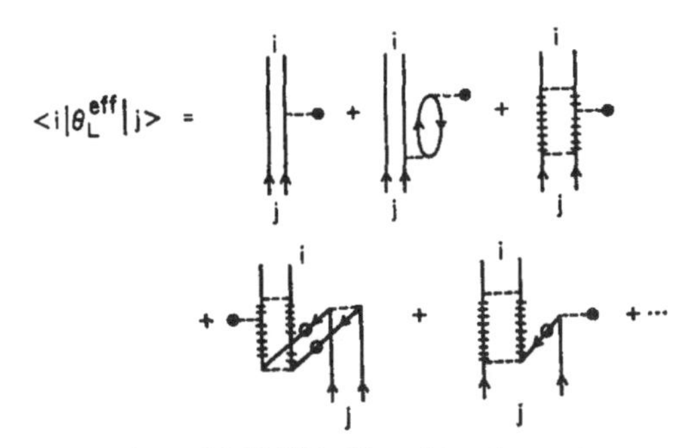

Fig. 16. Linked-diagram expansion of $\langle i|\theta_L^{eff}|j\rangle$. Here θ is taken to be a one-body operator for the sake of simplicity in drawing diagrams, and is represented by the symbol - - -●. Note that each diagram has only one θ-vertex, and in every diagram the last vertex is always at time $t_0 = 0$ and all the other vertices are at times earlier than t_0.

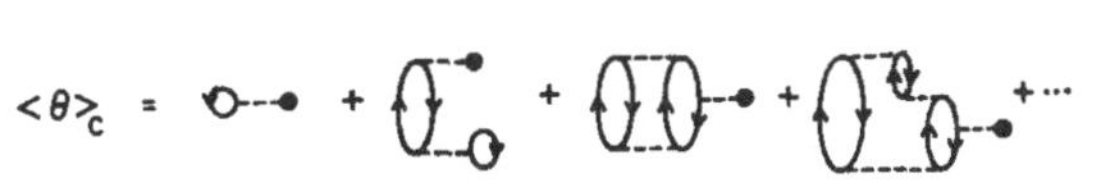

$\langle\theta\rangle_c$ = + + + + ...

Fig. 17. Linked-diagram expansion of $\langle\theta\rangle_c$. See the captions of fig. 16 for explanations.

experimental values of $\langle\theta\rangle_\alpha$ and $\langle\theta\rangle_c$. For example, in the usual shell-model calculation of ^{18}O we normally calculate only the $\langle\theta\rangle_{\alpha L}$ part. Then eq. (53) states that the expectation value of θ for the state ψ_α of ^{18}O is equal to $\langle\theta\rangle_{\alpha L}$ plus the expection value of θ for the ground state of ^{16}O. In many cases, $\langle\theta\rangle_c$ will be identically zero. For example, if θ is the E2 operator and the ground state wave function of the core has $J^\pi = 0^+$, then clearly $\langle\theta\rangle_c = 0$. There are also cases that $\langle\theta\rangle_c$ is not identically zero, such as in the calculation of the mean-square radius of a nucleus.

6. Discussion

As shown by eq. (44), we have derived a model-space secular equation where the eigenvalue is $E_\alpha - E_c$ with E_c being the core energy, the eigenvector is the projection of the complete eigenfunction onto the model space and the effective interaction is given by $[H_1 U_Q(0, -\infty)]_L$. The diagrams which constitute $[H_1 U_Q(0, -\infty)]_L$ are clearly specified in fig. 14. Since the effective interaction consists of an infinite series of diagrams, a crucial question is how to group and sum the various diagrams so that its convergence may be reasonably fast. In nuclear structure calculations where the nucleon-nucleon interaction is presumed to be very strong, we feel that we should first sum up all the iterative V-interactions within each Q and Q′ box to form the Brueckner reaction matrix G [ref. [3])]. Then all the Q and Q′ boxes are expressed in terms of the G-interactions.

After we have expressed $H_1^{\text{eff}} = [H_1 U_Q(0, -\infty)]_L$ in terms of the G-matrices, it is however not clear which will be the most efficient way to group and calculate the various G-matrix diagrams of H_1^{eff}. The following scheme seems to be promising. If the diagrams with higher order of G-interactions in the Q and Q′ boxes of fig. 14 are not important relative to the lower-order ones, then we can neglect these higher-order diagrams and the structure of the Q and Q′ boxes become fairly simple. When the structure of Q and Q′ is simple, it will not be a complicated job to sum up the folded diagrams to all orders. This is because that the Q and Q′ series shown in fig. 4 is in fact of the form of a geometric series, and can be summed up to all orders by solving a matrix equation provided that the structure of the Q and Q′ boxes is not too complicated. We have not yet investigated the above scheme in detail, partly because we feel that for further investigation it will be necessary to carry out some numerical calculations for H_1^{eff} using the folded diagram formalism. Such calculations are definitely needed in order to study the relative importance of various diagrams which constitute the Q and Q′ boxes. As far as we know, no such calculations have been made.

The effective interaction H_1^{eff} as given by eq. (45) and fig. 14 is not Hermitian. This, however, should be expected. The reason is that it is the complete eigenfunctions of H which should be orthogonal to each other if their energies are not degenerate, but their projections onto the model space are not necessarily orthogonal to each other which is consistent with the fact that H_1^{eff} is not Hermitian. If H_1^{eff} is Hermitian, then the projections of these eigenfunctions onto the model space, are also orthogonal to each other. In fact we can make a similarity transformation to make H_1^{eff} Hermitian, if the eigenvalues of eq. (44) are all real as they should be. This is a similarity transformation of the model space of dimensionality d so that the projections of d non-degenerate eigenfunctions onto the model space are orthogonal to each other, within the model space. From a calculational point of view, it appears to be more convenient to use the non-Hermitian form of H_1^{eff} as it is given in this paper.

We now discuss the choice of the model space. As shown in sect. 4, the eigenvalues given by the model-space secular equation eq. (44) are not necessarily the lowest eigenvalues of the original Hamiltonian H. Rather, they are the eigenvalues of H whose corresponding eigenfunctions have non-zero projections onto the model space which we have chosen. Thus to describe a given physical problem, it is very important to choose the appropriate corresponding model space. In general, we feel that the best choice of model space should be that with the largest overlap with the true eigenfunctions of the physical problem under consideration. A basic difficulty in making such a choice is that we simply do not know the true eigenfunctions of the physical system before we solve the corresponding eigenvalue problem. There does not seem to be any systematic method to avoid this difficulty. Thus in practice to judge whether we have chosen an appropriate model space we can only rely on the comparison of the calculated results with experiments.

Let us now compare the folded-diagram H_1^{eff} of the present paper with the effective interaction derived [3]) from the Green's function formalism. Let us denote the latter as $\mathscr{H}_1^{\text{eff}}(E)$. A clear difference between the two is that $\mathscr{H}_1^{\text{eff}}(E)$ is dependent on the energy eigenvalue E and does *not* contain the folded diagrams, while H_1^{eff} is independent of E but contains the folded diagrams. In terms of the diagrams shown in fig. 14, $\mathscr{H}_1^{\text{eff}}(E)$ will contain just the first one-Q-box term of the figure with a simple modification of the form of the energy denominator. Namely the energy denominator for $\mathscr{H}_1^{\text{eff}}(E)$ contains E while that for H_1^{eff} does not contain E. For example, the contribution of diagram (e) of fig. 15 is

$$\frac{\langle 12|V|\mathrm{h}_1\mathrm{h}_2\rangle^2}{\varepsilon_1+\varepsilon_2-(\varepsilon_1+\varepsilon_2-\varepsilon_{\mathrm{h}_1}-\varepsilon_{\mathrm{h}_2})}$$

for H_1^{eff}, but is

$$\frac{\langle 12|V|\mathrm{h}_1\mathrm{h}_2\rangle^2}{E-E_{\mathrm{c}}-(\varepsilon_1+\varepsilon_2-\varepsilon_{\mathrm{h}_1}-\varepsilon_{\mathrm{h}_2})}$$

for $\mathscr{H}_1^{\text{eff}}(E)$, namely the effective interaction in the Green's function formalism. Recall that E_{c} is the core energy. Thus although it is remarkable that in the present formalism we can remove the E-dependence from the effective interaction, we are also paying a high price for that. Namely we now have to calculate many more additional folded diagrams in constructing H_1^{eff}, as compared with the construction of $\mathscr{H}_1^{\text{eff}}(E)$. It seems to be difficult to decide which of the two ways for constructing the effective Hamiltonian is a more desirable one. It will probably depend on the nature of the physical problems being investigated. In a previous paper [3]), we have performed some calculations for $\mathscr{H}_1^{\text{eff}}(E)$ and found that its dependence on E is very smooth and weak for nuclear structure calculations using the Reid nucleon-nucleon potential. Since the folded diagrams are mainly to take care of the E-dependence, the weak and smooth E-dependence of $\mathscr{H}_1^{\text{eff}}(E)$ suggests that the folded-diagram series of fig. 14 may converge rapidly, for the same type of calculations. In comparing the two methods for constructing the effective interaction, it should be pointed out that for the E-dependent case the model-space eigenvectors are also dependent on E. This may cause considerable difficulty and inconvenience in the evaluation of the expectation value and transition matrix element of some physical operator where we need the model-space eigenvectors.

As shown in sect. 5, we have derived a linked-diagram expansion for the evaluation of the expectation value of a general operator with respect to the complete eigenfunctions. Similar to the case for the expansion of the effective interactions, it will also be of great interest to investigate the significance of the various terms in the series expansion of the expectation value and the convergence of the series. Investigations along this direction remain to be done. A natural extension of our treatment for the expectation value is the evaluation of the transition matrix element

$$\langle\theta\rangle_{\alpha\beta}=\frac{\langle\tilde{\Psi}_\alpha|\theta|\tilde{\Psi}_\beta\rangle}{[\langle\tilde{\Psi}_\alpha|\tilde{\Psi}_\alpha\rangle\langle\tilde{\Psi}_\beta|\tilde{\Psi}_\beta\rangle]^{\frac{1}{2}}}, \tag{57}$$

where θ represents a general physical operator and the ψ represent the complete eigenfunctions. To have the proper dependence on the number of particles for $\langle\theta\rangle_{\alpha\beta}$, we expect that only the linked diagrams in eq. (57) will contribute to $\langle\theta\rangle_{\alpha\beta}$. The derivation for such a linked-diagram expansion for $\langle\theta\rangle_{\alpha\beta}$ is found, however, to be much more complicated than that for the expectation value. Further investigations on the evaluation of $\langle\theta\rangle_{\alpha\beta}$ are in progress.

The authors wish to thank Professor G. E. Brown for discussions and a critical reading of the manuscript. One of the authors (K.F.R.) wishes to thank the Research Foundation of the State University of New York for a fellowship in support of this work.

References

1) H. Feshbach, Ann. of Phys. **19** (1962) 287
2) C. Bloch and J. Horowitz, Nucl. Phys. **8** (1958) 91
3) T. T. S. Kuo, Reaction matrix theory for nuclear structure, in Proc. Fourth Int. Symp. on light-medium mass nuclei, October 1970, to be published by the University of Kansas Press
4) T. Morita, Prog. Theor. Phys. **29** (1963) 351
5) B. H. Brandow, Rev. Mod. Phys. **39** (1967) 771; Lectures in theoretical physics, vol. 11, Boulder 1968, ed. K. T. Mahanthappa (Gordon and Breach, New York, 1969); Ann. of Phys. **57** (1970) 214
6) G. Oberlechner *et al.*, Nuovo Cim. **68B** (1970) 23
7) M. Johnson and M. Baranger, Ann. of Phys. **62** (1971) 172
8) See, e.g., P. Nozières, The theory of interacting fermi systems (Benjamin, New York, 1965)
9) H. A. Bethe, B. H. Brandow and A. G. Petschek, Phys. Rev. **129** (1963) 225
10) D. J. Thouless, The quantum mechanics of many-body systems (Academic Press, New York and London, 1961) ch. IV
11) M. Gell-Mann and F. Low, Phys. Rev. **84** (1951) 150

1.C

Nuclear Physics **A235** (1974) 171—189;

CONVERGENCE OF EFFECTIVE HAMILTONIAN EXPANSION AND PARTIAL SUMMATIONS OF FOLDED DIAGRAMS

E. M. KRENCIGLOWA and T. T. S. KUO

Department of Physics, State University of New York at Stony Brook, Stony Brook, New York 11790 †

Received 22 April 1974
(Revised 29 July 1974)

Abstract: Starting from a time-dependent formulation of the energy-independent effective hamiltonian, a sequence of partial summations for the folded-diagram series is defined and a connection between the energy-independent and energy-dependent effective hamiltonians is shown. The partial summations are shown to be convergent in the presence of intruder states which are weakly coupled to the model space.

1. Introduction

The question of convergence of the expansion for the effective hamiltonian in practical nuclear applications has been raised by Schucan and Weidenmüller [1,2]), in view of known low-lying intruder states and the formal analytic properties of the effective hamiltonian that reproduces the lowest eigenvalues of the full hamiltonian. The question of convergence was also raised by Barrett and Kirson [3]) who examined the third order, and part of the fourth order, in the G-matrix terms of the effective interaction. However, it is well known that approximating the effective hamiltonian by the bare G-matrix plus core polarization, as originally done by Kuo and Brown [4]), in fact works quite well. This represents somewhat of a perplexing situation. The Q-box formulation of the folded-diagram series [5,6]) constructs the effective hamiltonian in essentially two steps: one, the Q-box itself, when expanded in powers of the interaction, essentially represents an expansion of the Q-space eigenvalues about some unperturbed value(s) and two, the folded diagrams essentially represent an expansion of some of the eigenvalues of the full hamiltonian about some unperturbed value(s). Hence, a resolution of possible divergences into two possible sources. Approximation schemes to the Q-box have been suggested by Kirson [7]) and Krenciglowa, Kuo, Osnes and Giraud [8]). It is the second step at which this presentation is principally directed. We will show that, by means of a sequence of partial summations, the folded-diagram series may be summed exactly.

In sect. 2, the notation and problem are defined. Sect. 3 represents a collection of previous results that may not be well known but are necessary for the development

† Work supported by the U. S. Atomic Energy Commission under contract AT (11-1)-3001.

of the partial summations for the effective hamiltonian in sect. 4. Also, in sect. 4, a formal connection between the starting point for the derivation of the folded-diagram series as given by Brandow [9]) and the time-dependent derivation of Kuo, Lee and Ratcliff [5]) is made; the folded-diagram series in the time dependent formulation is explicitly summed exactly, by means of a sequence of partial summations, into a form expressed in terms of energy-dependent interactions of Bloch and Horowitz [10]). In sect. 5, we apply the partial summations and folded-diagram series to two simple model hamiltonians. We find that the partial summations can converge in the presence of intruder states and that convergent partial summations yield eigenvalues of the full hamiltonian which correspond to eigenvectors with the largest overlap with the P-space. We also show that the problem of intruder states can be consistently dealt with within the framework of the folded-diagram series and the partial summations by means of some appropriate P-space modifications.

2. Notation

The notation we use will be almost identical to the notation adopted by Kuo, Lee and Ratcliff [5]) and Ratcliff, Lee and Kuo [6]). In studying truncations of a Hilbert space, the notions of a projection operator P onto the model space and its complement Q are introduced. We denote the model space of *active* state labels by $A, B, C, \ldots$ and *passive* state labels by $\alpha, \beta, \gamma, \ldots$. Then the P- and Q-space projection operators are defined by

$$P = \sum_A |A\rangle\langle A|, \qquad Q = \sum_\alpha |\alpha\rangle\langle\alpha|, \qquad \langle A|\alpha\rangle = 0,$$

and

$$P+Q = 1$$

defines the Hilbert space we are dealing with.

The nuclear hamiltonian has been traditionally divided into two pieces

$$H = H_0+H_1,$$
$$H_0 = T+U,$$
$$H_1 = V-U,$$

and the study of a truncation of the full hamiltonian usually requires that the model space be degenerate with respect to H_0. Thus, traditionally H_0 is chosen first and this choice is then used to define the P- and Q-spaces. Then, the many-particle states $|A\rangle$ and $|\alpha\rangle$ are eigenstates of H_0 and we denote the corresponding eigenvalues as ε_A and ε_α, respectively,

$$H_0|A\rangle = \varepsilon_A|A\rangle,$$
$$H_0|\alpha\rangle = \varepsilon_\alpha|\alpha\rangle, \; \varepsilon_A \neq \varepsilon_\alpha.$$

In diagrams, active *states* are denoted by lines and passive *states* by railed lines. Dots denote H_1 interactions.

3. Previous results

In this section, we collect results from three sources (Kuo *et al.*, Ratcliff *et al.* and Lee [11]), two of which are unpublished. The formal structure of the energy-independent effective hamiltonian (H_{eff}) is systematically analysed by Ratcliff *et al.*

The starting point for their derivation of H_{eff} is the time-development operator ($U(0, -\infty)$) in the interaction representation. The key result is the *decomposition theorem* and the central ideas developed are the notions of a Q-box ($\hat{Q}$) and a generalized folding operation. The Q-box is defined as a collection of one or more H_1 interactions and the fermion lines between these interactions which together form an irreducible, linked part of a valence diagram. Irreducible, here, means that the intermediate states between two successive H_1 interactions must be passive. The Q-box was given the formal definition

$$\langle B|\hat{Q}(\varepsilon_A)|A\rangle = \langle B|H_1\,|A\rangle + \langle B|H_1 \frac{Q_{\text{L}}}{\varepsilon_A - H_0} \hat{Q}(\varepsilon_A)|A\rangle, \tag{1}$$

where the subscript L on the projection operator Q_{L} specifies that the intermediate states must not only lie outside the model space but must also result in linked-valence diagrams.

The energy-independent effective hamiltonian is defined as

$$H_{\text{eff}} = H_0 + H_1\, U_Q(0, -\infty). \tag{2}$$

If the P-space is d-dimensional, the eigenvalues of $\langle A|H_{\text{eff}}|B\rangle$ coincide with d eigenvalues † of H and the eigenvectors of $\langle A|H_{\text{eff}}|B\rangle$ coincide with the projections of the corresponding eigenvectors of H onto the model space. The operator $U_Q(0, -\infty)$ stems from a factorization of $U(0, -\infty)$. This factorization is known as the decomposition theorem and is stated as

$$U(0, -\infty)|A\rangle = \sum_B U_Q(0, -\infty)|B\rangle\langle B|U(0, -\infty)|A\rangle.$$

The vector $U_Q(0, -\infty)|B\rangle$ is defined by a series whose terms are grouped or classified according to the number of Q-boxes or, equivalently, the number of generalized folds. The vector $U_Q(0, -\infty)|B\rangle$ is defined [5, 6] in fig. 1 where $|\psi_c^Q\rangle$ represents core excitations and the railed semi-circle is essentially the Q-box without the last H_1 interaction at

† In the derivation of H_{eff} by Kuo *et al.*, it is assumed that the time-development operator can be analytically continued into the complex time-plane. This leads to the statement that the lowest-d eigenvectors of H having non-zero overlap with so called "parent states" can be constructed by appropriate application of the time-development operator and that H_{eff} reproduces the eigenvalues corresponding to these eigenvectors. However, we find, numerically, that the H_{eff} obtained by partial summations (subsect. 4.2 and sect. 5) yields eigenvalues of H which correspond to eigenvectors with the largest overlap with the P-space.

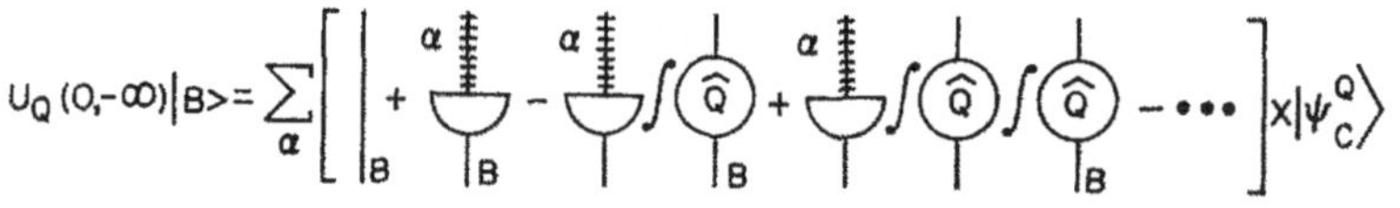

Fig. 1. Diagrammatic structure of the vector $U_Q(0, -\infty)|B\rangle$.

$t = 0$. The symbol, $\int$, denotes a generalized fold. The notion of a generalized fold [5, 6]) is illustrated here by means of example. Consider two irreducible collections of energy diagrams, C_1 and C_2, not necessarily the same. The generalized fold is defined in fig. 2 where the times t_1', t_1 are the times of the first and last H_1 interactions of C_1 and t_2 is the time of the last H_1 interaction of C_2. The irreducible collections C_1 and C_2 also carry internal time labels indicating the times of the other H_1 interactions. The only restriction on the time labels of C_2 in diagram (c) of fig. 2 is $t_2 < t_1$; all possible ordering of the internal time labels of C_2 relative to the internal time labels of C_1 are included in diagram (c) of fig. 2 subject to the constraint, $t_2 < t_1$. The right-hand side of fig. 2 provides a means of evaluating a generalized folded diagram.

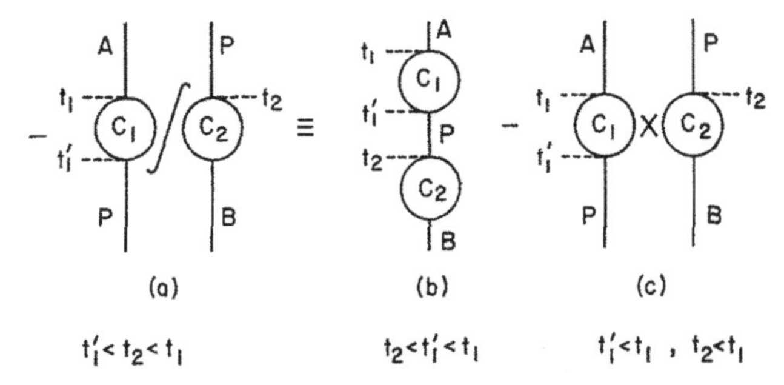

Fig. 2. Definition of a once generalized fold for two irreducible collections of energy diagrams, C_1 and C_2. The time constraints for the various terms are indicated and the state label P indicates a summation over all states in the P-space.

Thus, H_{eff} is formally defined by eq. (2) and was analyzed, in detail, for 0-, 1- and 2-valence nucleons by Ratcliff *et al.* For the case of 2-valence nucleons, the final result was a clean separation of the 0-, 1- and 2-body contributions to H_{eff}, the 0-body and 1-body terms being, respectively, the exact core energy and exact (experimental) one-body energies. Of more interest to this work is the structural simplicity of this formulation. It was noted that a formulation of H_{eff} in terms of Q-boxes represents, at least, a clean computational procedure.

The schematic structure of H_{eff} is given in fig. 3 and we denote this as

$$H_{\text{eff}} = \Delta E_c + F_0 + F_1 + F_2 + F_3 + \ldots, \tag{3}$$

where F_n denotes the n-generalized folded contribution to H_{eff} and ΔE_c is the exact core energy minus the unperturbed core energy. The n-generalized folded contribu-

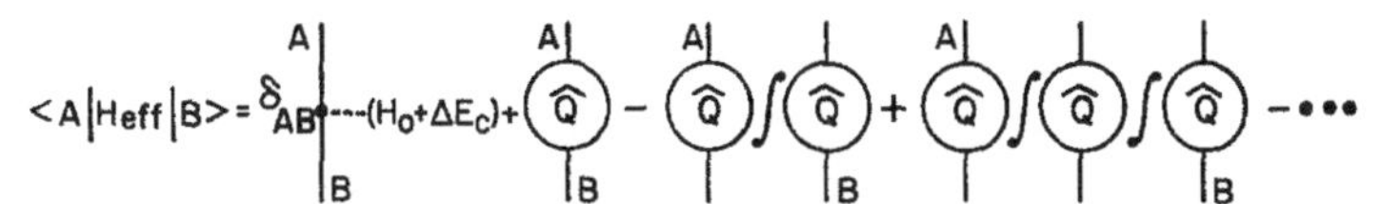

Fig. 3. Diagrammatic representation of $\langle A|H_{\text{eff}}|B\rangle$.

tion can be expressed in terms of the Q-box and its derivatives as follows:

$$F_0 = H_0+\hat{Q},$$
$$F_1 = \hat{Q}_1\hat{Q},$$
$$F_2 = \hat{Q}_2\hat{Q}\hat{Q}+\hat{Q}_1\hat{Q}_1\hat{Q},$$
$$F_3 = \hat{Q}_3\hat{Q}\hat{Q}\hat{Q}+\hat{Q}_2\hat{Q}_1\hat{Q}\hat{Q}+\hat{Q}_2\hat{Q}\hat{Q}_1\hat{Q}+\hat{Q}_1\hat{Q}_2\hat{Q}\hat{Q}+\hat{Q}_1\hat{Q}_1\hat{Q}_1\hat{Q},$$

where

$$\hat{Q}_m \equiv \frac{1}{m!}\frac{\mathrm{d}^m}{\mathrm{d}\varepsilon_P^m}\hat{Q}.$$

Here, we have explicitly assumed that the P-space is degenerate with respect to H_0 and defined $\varepsilon_A = \varepsilon_B = \ldots = \varepsilon_P$. The number of terms in F_n increases rapidly with n; F_4 has fourteen terms. We state a set of rules (11) for obtaining the terms of F_n, $n \geqq 1$:

(i) The total number of Q-boxes, differentiated or not, is $n+1$.

(ii) The first Q-box must be at least once differentiated.

(iii) The last Q-box must not be differentiated.

(iv) Each individual Q-box can be differentiated up to n times subject to the constraints; (a) The sum of the powers of the energy derivatives is n. (b) If a Q-box is differentiated k times, there must be at least k undifferentiated Q-boxes to its right, not necessarily in succession.

(v) With each mth order energy derivative, associate a factor of $(m!)^{-1}$.

Thus a general term of F_n appears as

$$\hat{Q}_{m_1}\hat{Q}_{m_2}\hat{Q}_{m_3}\ldots\hat{Q}_{m_n}\hat{Q}\ ,$$

and F_n is the sum of all terms with the above form. Each term is characterized by the combination $(m_1, m_2, m_3, \ldots m_n)$, and all combinations subject to the above rules are allowed.

This factorization of the terms of F_n into a sum of products of Q-boxes and its derivatives is accomplished by the introduction and formal definition of a generalized folding operation. This factorization is not necessarily possible for individual diagrams that contribute to F_n. The once-folded class of diagrams in fig. 4 carries the time constraints, $t_1' < t_2 < t_1$, where t_1' is the time of the first H_1 interaction of the first Q-box, t_2 is the time of the last H_1 interaction of the second Q-box and t_1 is the time of the last H_1 interaction (this is the time associated with the H_1 in eq. (2)). To

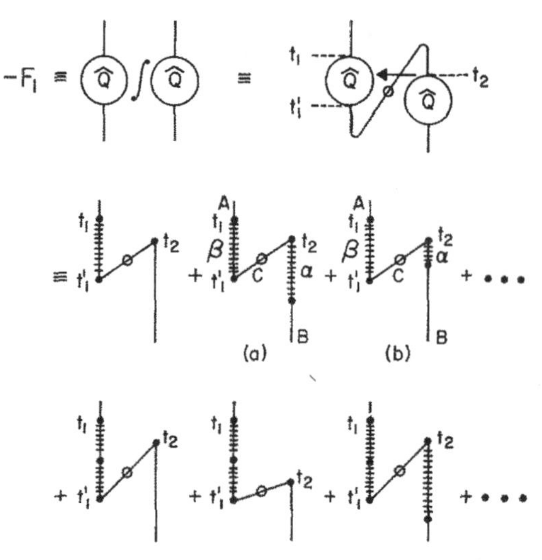

Fig. 4. Diagrams contained in F_1. The time constraints are $t'_1 < t_2 < t_1$. The sum of diagrams (a) and (b) exhibits the character of the overall factorization of $F_1 = \hat{Q}_1\hat{Q}$.

illustrate the character of the factorization of F_1, we turn our attention to the diagrams labelled (a) and (b) of fig. 4. Diagrams (a) or (b), individually, do not exhibit any factorization. However, the sum (a)+(b) factorizes. This is explicitly shown in two ways. First, the sum of diagrams (a) and (b) constitute a generalized fold and the definition of a generalized fold [5, 6]) yields immediately

$$-[(a)+(b)] = \sum_{C,\alpha,\beta} \langle A|H_1|\beta\rangle\langle\beta|H_1|C\rangle\langle C|H_1|\alpha\rangle\langle\alpha|H_1|B\rangle$$

$$\times[((\varepsilon_B-\varepsilon_\beta)(\varepsilon_B-\varepsilon_C)(\varepsilon_B-\varepsilon_\alpha))^{-1}-((\varepsilon_C-\varepsilon_\beta)(\varepsilon_B-\varepsilon_C)(\varepsilon_B-\varepsilon_\alpha))^{-1}]$$

$$= \sum_C \langle A|\frac{1}{\varepsilon_B-\varepsilon_C}\left[\sum_\beta\frac{H_1|\beta\rangle\langle\beta|H_1}{\varepsilon_B-\varepsilon_\beta}-\sum_\beta\frac{H_1|\beta\rangle\langle\beta|H_1}{\varepsilon_C-\varepsilon_\beta}\right]|C\rangle\langle C|\sum_\alpha\frac{H_1|\alpha\rangle\langle\alpha|H_1}{\varepsilon_B-\varepsilon_\alpha}|B\rangle$$

$$= \sum_C \langle A|\frac{1}{\varepsilon_B-\varepsilon_C}[q(\varepsilon_B)-q(\varepsilon_C)]|C\rangle\langle C|q(\varepsilon_B)|B\rangle$$

$$= \langle A|\left[\frac{\mathrm{d}}{\mathrm{d}\varepsilon}q(\varepsilon)\right]_{\varepsilon=\varepsilon_P} q(\varepsilon_P)|B\rangle,$$

where q, analogous to the Q-box for the purposes of this example, is defined as

$$q(\varepsilon) = \sum_\delta \frac{H_1|\delta\rangle\langle\delta|H_1}{\varepsilon-\varepsilon_\delta}.$$

We assume that the model space is degenerate and interpret

$$\frac{1}{\varepsilon_B-\varepsilon_C}[q(\varepsilon_B)-q(\varepsilon_C)]$$

as a derivative. Second, if we use the usual diagrammatic rules, then

$$-[(a)+(b)] = -\sum_{C,\alpha,\beta} \langle A|H_1|\beta\rangle\langle\beta|H_1|C\rangle\langle C|H_1|\alpha\rangle\langle\alpha|H_1|B\rangle$$

$$\times [((\varepsilon_B-\varepsilon_\beta)(\varepsilon_B+\varepsilon_C-\varepsilon_\alpha-\varepsilon_\beta)(\varepsilon_B-\varepsilon_\alpha))^{-1}+((\varepsilon_B-\varepsilon_\beta)(\varepsilon_B+\varepsilon_C-\varepsilon_\alpha-\varepsilon_\beta)(\varepsilon_C-\varepsilon_\beta))^{-1}]$$

$$= -\sum_{C,\alpha,\beta} \frac{\langle A|H_1|\beta\rangle\langle\beta|H_1|C\rangle\langle C|H_1|\alpha\rangle\langle\alpha|H_1|B\rangle}{(\varepsilon_B-\varepsilon_\beta)(\varepsilon_C-\varepsilon_\beta)(\varepsilon_B-\varepsilon_\alpha)}$$

$$= \sum_C \langle A| \left[\frac{\mathrm{d}}{\mathrm{d}\varepsilon}\left[\sum_\beta \frac{H_1|\beta\rangle\langle\beta|H_1}{\varepsilon-\varepsilon_\beta}\right]_{\varepsilon=\varepsilon_P}\right] |C\rangle\langle C|\sum_\alpha \frac{H_1|\alpha\rangle\langle\alpha|H_1}{\varepsilon_P-\varepsilon_\alpha}|B\rangle$$

$$= \langle A| \left[\frac{\mathrm{d}}{\mathrm{d}\varepsilon}q(\varepsilon)\right]_{\varepsilon=\varepsilon_P} q(\varepsilon_P)|B\rangle.$$

Here we note, again, that $\varepsilon_B = \varepsilon_C = \ldots = \varepsilon_P$.

4. Partial summation of the folded-diagram series

4.1. VALENCE PART

In this section, we examine the structure of the valence part of the effective interaction. We write the valence part of the effective interaction as

$$H_0+v_\mathrm{v} = F_0+F_1+F_2+\ldots,$$

where F_n, $n \geqq 1$, is the n-generalized folded contribution composed of sums of terms that are products of $n+1$ Q-boxes, differentiated or not, as described previously. It is convenient to indicate the time constraints, hence terms, of F_n diagrammatically. In fig. 5, we indicate the diagrammatic structure of the terms of F_1 and F_2. All time references indicated in fig. 5 are either the time of the first or last H_1 interaction of a given Q-box. The time constraint arrow fixes the time associated with the tail of the arrow to lie within the time boundaries of the Q-box referred to by the head of the arrow.

We define a partial summation, S_1, of those terms in the effective valence interaction (v_v) which have the time references to the first Q-box only. Thus, we have

$$S_1 = \hat{Q}+\sum_{l=1}^{\infty}\hat{Q}_l(\hat{Q})^l,$$

where the Q-box and its derivatives are evaluated at ε_P,

$$\hat{Q} = \hat{Q}(\varepsilon_P).$$

The diagrammatic structure of S_1 is given in fig. 6. Now, S_1 does not contain those terms in v_v which have derivatives on Q-boxes other than the first Q-box. Define S_2 as a partial summation of the S_1

$$S_2 = \hat{Q}+\sum_{l=1}\hat{Q}_l(S_1)^l.$$

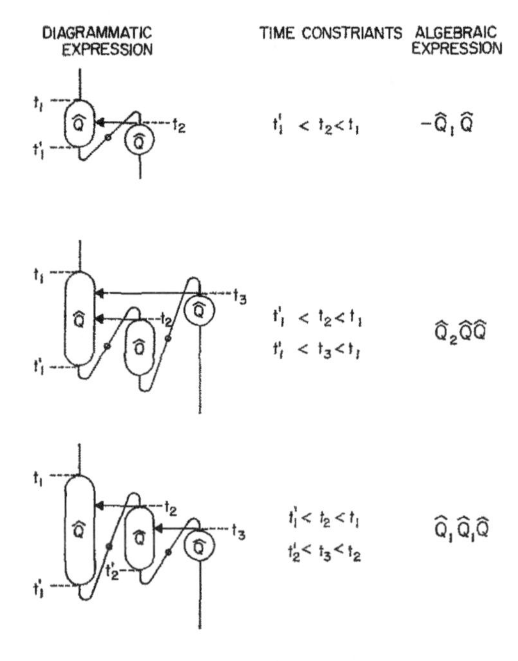

Fig. 5. Examples of generalized time constraints. Diagrammatic representation of generalized time constraints and corresponding algebraic expression are listed.

The diagrammatic structure of S_2 is given in fig. 6. Here, the time associated with the last interaction of any S_1 is constrained to lie within the time boundaries of the first Q-box only. It is clear that the time constraints for each term of S_2 are well-defined and distinct and thus S_2 can be expressed solely in terms of Q-boxes with distinct

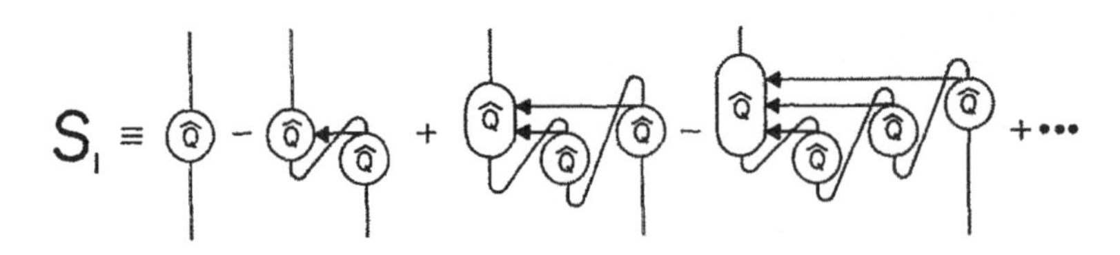

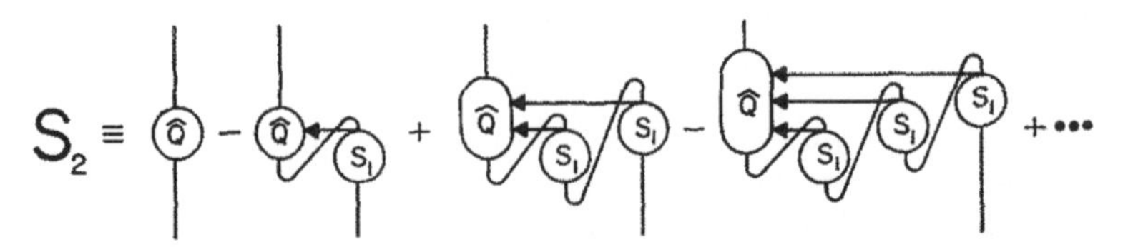

Fig. 6. Diagrammatic definition of the partial sums S_1 and S_2. Note that all time constraint arrows reference the first Q-box only and S_2 is defined in terms of S_1.

time constraints. That is, all the diagrams contained in S_2 are also contained in v_V. Thus, if we defined

$$S_{n+1} = \hat{Q} + \sum_{l=1}^{\infty} \hat{Q}_l (S_n)^l \,, \tag{4}$$

then all diagrams contained in S_{n+1} are also contained in v_V. This suggests that if

$$\lim_{n\to\infty} S_n \equiv S_\infty$$

exists, then

$$S_\infty \equiv v_V = \hat{Q} + \sum_{l=1}^{\infty} \hat{Q}_l v_V^l .$$

The suggestion is based, first, on the "counting of terms" argument presented above and, second, on the observation that the final form, S_∞, is of the form of Brandow's result [9]).

$$\mathscr{W}_\infty = \sum_{l=0}^{\infty} \mathscr{V}^{(l)} (-\mathscr{W}_\infty)^l .$$

We stress that, here, we have been dealing with the valence part of the effective interaction and that the Q-box is defined with a projection operator (Q_L) in eq. (1). Finally, S_n is not analogous to Brandow's $\mathscr{W}_n$ since $\mathscr{W}_n$, for finite n, depends on the exact energy.

4.2. EFFECTIVE HAMILTONIAN

In this section, we show that when the folded diagram series for H_{eff} is convergent, it can be summed up to all orders and is equivalent to the Bloch-Horowitz, energy-dependent effective hamiltonian. We view the expression for S_∞ as a suggestion for the following assertions. Here we take the particle vacuum as the "core". Then, the Q-box becomes

$$\hat{Q}(\varepsilon) = H_1 + H_1 \frac{Q}{\varepsilon - H_0} \hat{Q}(\varepsilon), \tag{5.1}$$

where Q is just the Q-space projection operator. Then

$$\hat{Q}(\varepsilon) = H_1 + H_1 \frac{1}{\varepsilon - QHQ} H_1 \tag{5.2}$$

is a solution for $\hat{Q}(\varepsilon)$. If ε is an eigenvalue of H, then $\hat{Q}(\varepsilon)$ is just the energy-dependent interaction and eq. (5) is just the Bloch-Horowitz equation. The assertion is that if

$$\lim_{n\to\infty} S_n \equiv S_\infty$$

exists, then

$$H_{\text{eff}} = H_0 + S_\infty \,, \tag{6}$$

where S_n is defined by eq. (4) and $\hat{Q}$ by eq. (5) with $\varepsilon = \varepsilon_P$. Eq. (6) has been given by Hoffmann *et al.* [12]) and quoted by Schucan and Weidenmüller [1]) with references to Brandow [9]) and Des Cloizeaux [13]). In the following, S_∞ is explicitly summed to a closed form.

Assume that the P-space is d dimensional and that $|\psi_i\rangle$ is an eigenvector of H_{eff} with corresponding eigenvalue E_i. We assume that the P-space is degenerate with respect to H_0, therefore $|\psi_i\rangle$ is an eigenvector of S_∞ with eigenvalue $\Delta E_i = E_i - \varepsilon_P$. We have

$$\hat{Q} = PH_1P + PH_1QA^{-1}QH_1P,$$

$$\hat{Q}_l = \frac{1}{l!}\frac{\mathrm{d}^l}{\mathrm{d}\varepsilon^l}\hat{Q}(\varepsilon)|_{\varepsilon=\varepsilon_P} = PH_1QA^{-1}(-A^{-1})^lQH_1P,$$

where $A = \varepsilon_P - QHQ$. We want to explicitly sum the right-hand side of

$$S_\infty = \hat{Q} + \sum_{l=1}^{\infty}\hat{Q}_l(S_\infty)^l$$

$$= PH_1P + \sum_{l=0}^{\infty} PH_1QA^{-1}(-A^{-1})^lQH_1P(S_\infty)^l. \tag{7}$$

But, we have $S_\infty|\psi_i\rangle = \Delta E_i|\psi_i\rangle$. Therefore, if we multiply eq. (7) from the right by $|\psi_i\rangle$, we obtain

$$S_\infty|\psi_i\rangle = [PH_1P + \sum_{l=0}^{\infty} PH_1QA^{-1}(-A^{-1})^lQH_1P(\Delta E_i)^l]|\psi_i\rangle$$

$$= [PH_1P + PH_1Q(\Delta E_i + A)^{-1}QH_1P]|\psi_i\rangle$$

$$= [PH_1P + PH_1Q(\varepsilon_P + \Delta E_i - QHQ)^{-1}QH_1P]|\psi_i\rangle.$$

Thus,

$$S_\infty|\psi_i\rangle = \hat{Q}(E_i)|\psi_i\rangle, \tag{8}$$

$$[H_0 + S_\infty]|\psi_i\rangle = E_i|\psi_i\rangle = [H_0 + \hat{Q}(E_i)]|\psi_i\rangle, \tag{9}$$

where the r.h.s. is identically the energy-dependent effective hamiltonian for the energy E_i operating on the corresponding eigenvector. Eq. (8) is true for any $|\psi_i\rangle$, $i = 1, 2, \ldots, d$. Let U be a matrix with $|\psi_i\rangle$ as its columns

$$U = (|\psi_1\rangle, |\psi_2\rangle, \ldots, |\psi_d\rangle).$$

Then, by eq. (8)

$$S_\infty U = (\hat{Q}(E_1)|\psi_1\rangle, \hat{Q}(E_2)|\psi_2\rangle, \ldots, \hat{Q}(E_d)|\psi_d\rangle)$$

or

$$S_\infty = (\hat{Q}(E_1)|\psi_1\rangle, \hat{Q}(E_2)|\psi_2\rangle, \ldots, \hat{Q}(E_d)|\psi_d\rangle)U^{-1} \tag{10.1}$$

$$= U\Delta EU^{-1}, \tag{10.2}$$

where each $\hat{Q}(E_i)|\psi_i\rangle$ is a column vector and ΔE is a diagonal matrix with elements $\Delta E_1, \Delta E_2, \ldots, \Delta E_d$. Thus, the final result is that $H_{\text{eff}} = H_0 + S_\infty$ and the eigenvalues and eigenvectors of H_{eff} have been shown, explicitly, to be solutions of the Bloch-Horowitz equation. That S_∞ has the form given is obvious, at least in retrospect.

The partial summations are defined recursively and with the above prescription, S_{n+1} can be summed into closed form immediately. Let $|\psi_i^{(n)}\rangle$ be an eigenvector of $H_0 + S_n$ with eigenvalue $E_i^{(n)}$ and $U^{(n)}$ be a matrix with $|\psi_i^{(n)}\rangle$ as its columns, then

$$S_{n+1} = (\hat{Q}(E_1^{(n)})|\psi_1^{(n)}\rangle, \hat{Q}(E_2^{(n)})|\psi_2^{(n)}\rangle, \ldots, \hat{Q}(E_d^{(n)})|\psi_d^{(n)}\rangle)(U^{(n)})^{-1}. \tag{11}$$

This procedure preserves the notions of an active state and passive state (or P- and Q-space) and it is of interest to note that in the derivation of the Q-box formulation of H_{eff} many general results are formulated in terms of active and passive states. With a state formulation of H_{eff}, we can choose the P-space states first and define H_0 so that the P-space is degenerate with respect to H_0. Specifically, assume that the full hamiltonian, H, in the full space, $P+Q$, is given (that is, the P-space has been chosen). Now, define

$$H_0 \equiv \varepsilon_P P + \sum \varepsilon_\alpha |\alpha\rangle\langle\alpha|,$$

$$H_1 \equiv H - H_0,$$

where ε_P, ε_α are constants. The model space is thus defined to be degenerate with respect to H_0. In this separation of $H = H_0 + H_1$, only the matrix elements of H_0 and H_1 between the many-particle *states* in the $P+Q$ space are defined. This is not a statement about the traditional introduction of a one-body potential U which is then used to determine the many-nucleon matrix elements $\langle A|T+U|B\rangle$ or $\langle A|H-U|B\rangle$, say. It is clear that with this definition of H_0 and H_1, ε_P is a parameter introduced to start the partial summations S_n and that $H_{\text{eff}} = H_0 + S_\infty$ is, in fact, independent of ε_P.

5. Model applications and discussion

In this section, we discuss truncations of the Hilbert space for two exactly solvable model hamiltonians, labelled H^{I} and H^{II}. The results will be based on the state formulation of the expansion for H_{eff} (i.e., the state formulation entails the notions of passive and active states without any specific reference to the many-body nature of the problem). H_{eff} is computed by partial summations and by the folded-diagram series and the results are compared. Our results will show that the problem of an intruder state can be dealt with consistently within the framework of the folded-diagram series and the partial summations.

The full hamiltonian is separated into H_0 and H_1 parts

$$H_0 = \varepsilon_P P + \sum_\alpha \varepsilon_\alpha |\alpha\rangle\langle\alpha|,$$

$$H_1 = H - H_0,$$

TABLE 1

Matrix elements of model hamiltonian $H^{\rm I}$ ($\langle i|H^{\rm I}|j\rangle = \langle j|H^{\rm I}|i\rangle$)

State labels	$|1\rangle$	$|2\rangle$	$|3\rangle$	$|4\rangle$	$|5\rangle$
$\langle 1|$	−10.74	−0.61	3.23	−0.35	1.75
$\langle 2|$		−13.54	1.74	0.57	−0.24
$\langle 3|$			−10.41	−1.66	0.19
$\langle 4|$				−8.14	0.74
$\langle 5|$					−0.08

Here the matrix elements of H_0 are −10, −10, −5, −5, 0 and the strength parameter x is equal to 1.

and we will be interested in truncations to the full hamiltonian

$$H(x) \equiv H_0+xH_1$$
$$= (H_0+\Delta P)+(xH_1-\Delta P),$$

as a function of x and Δ. The strength parameter x allows us to vary the interacting hamiltonian H_1 and the shift parameter Δ allows us to vary the separation between the Q-space and P-space unperturbed spectrum. The matrix elements of $H^{\rm I}$ are the Kuo-Brown bare G-matrix elements plus the single particle energies for $A = 18$, $T = 0$, $J^\pi = 1^+$ and the full Hilbert space, of dimension 5, is taken to be the s-d shell. The matrix elements of $H^{\rm I}$ are given in table 1. Truncations in the s-d shell have been studied by Barrett, Halbert and McGrory [14]) who explicitly construct $H_{\rm eff}$ by a technique analogous to eq. (10.2). The second model hamiltonian ($H^{\rm II}$) is from Hoffman *et al.* [15]) and is given in table 2. References to a particular hamiltonian, $H^{\rm I}$ or $H^{\rm II}$, will be exhibited with superscripts, I or II, respectively.

To start, the P-space is chosen as

$$P = |1\rangle\langle 1|+|2\rangle\langle 2|$$

for both hamiltonians. The Q-box is obtained by eq. (5.2) and is evaluated at the

TABLE 2

Matrix elements of model hamiltonian $H^{\rm II}$ ($\langle i|H^{\rm II}|j\rangle = \langle j|H^{\rm II}|i\rangle$)

State labels	$|1\rangle$	$|2\rangle$	$|3\rangle$	$|4\rangle$
$\langle 1|$	1	5	0	5
$\langle 2|$		26	5	0
$\langle 3|$			−2	1
$\langle 4|$				4

Here the matrix elements of H_0 are 1, 1, 3, 9 and the strength parameter x is equal to 1.

"unperturbed energy", $\varepsilon_P + \Delta$. The partial summations are evaluated according to eq. (11) and in this presentation, the maximum value for n in S_n is 1000. We found that if the partial summations converged to $H_{\rm eff}$, say, for a given x, H_0 and Δ they converge to the same $H_{\rm eff}$ for the same x and H_0 but different Δ. If the partial summations are divergent for a given x, H_0 and Δ, they remain divergent for different choices of Δ. However, divergent sequences of partial summations for $H_{\rm eff}$ sometimes yielded one convergent eigenvalue and corresponding projected eigenvector (recall, the P-space is 2-dimensional). This can be understood from the structure of $H_{\rm eff}$ in equation (10.1). When convergent, the partial sums yield an $H_{\rm eff}$ which reproduces the eigenvalue and corresponding projected eigenstates of the full hamiltonian which have the largest overlap with the P-space. Some of these points are illustrated in figs. 7 and 8, where the eigenvalues of various operators related to $H^{\rm II}(x)$ are plotted as a function of x, $0 \leq x \leq 0.3$. Here $\Delta = -11$ and H_0 is the same as in Hoffman *et al.* [15]) and is the diagonal matrix with matrix elements (1, 1, 3, 9). Only the eigenvalues of the first two partial sums are shown but partial sums of higher order were evaluated to establish convergence. In fig. 8, we see that different pairs of eigenvalues of $H^{\rm II}(x)$ are reproduced by the partial summations; the pairs of eigenvalues (E_1, E_2), (E_1, E_3) and (E_1, E_4) are reproduced as x increases. The eigenvectors corresponding to these

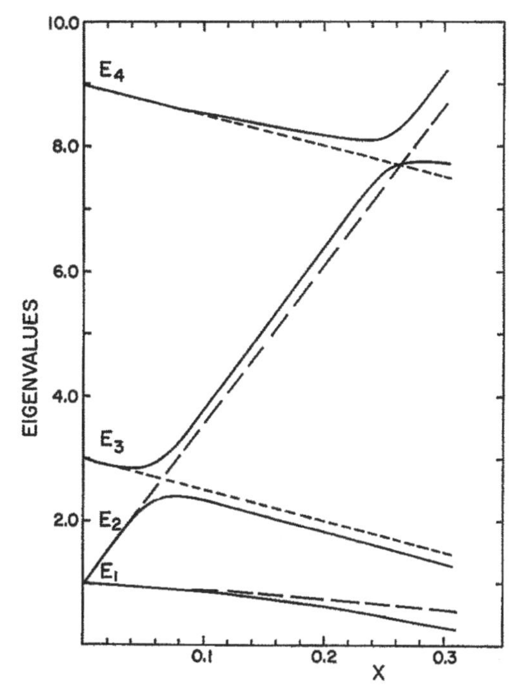

Fig. 7. Eigenvalues of operators relating to $H^{\rm II}$ as a function of x. The exact eigenvalues of $H^{\rm II}(x)$ are indicated by solid lines and labelled as E_1, E_2, E_3, and E_4. Short dashed lines indicate Q-space eigenvalues $(QH^{\rm II}(x)Q)$ and long dashed lines indicate eigenvalues of $H_0+\hat{Q}(x)$.

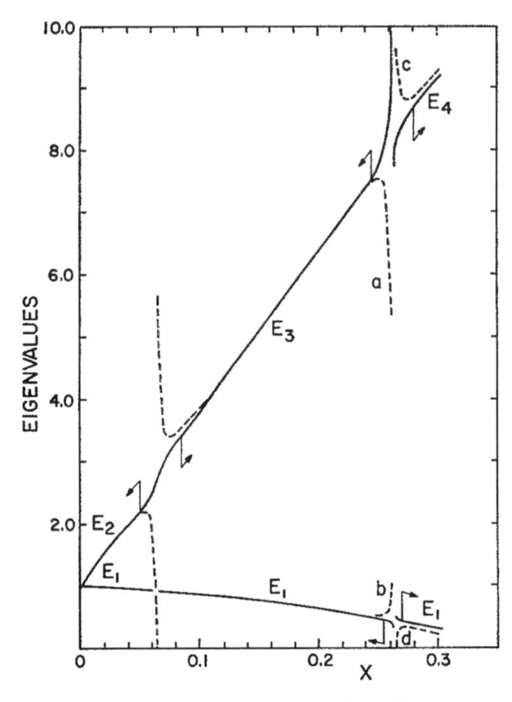

Fig. 8. Eigenvalues of the first two partial summations for $H^{\text{II}}_{\text{eff}}$ as a function of x. The dashed lines indicate the eigenvalues of $H_0+S_1(x)$ and solid lines indicate the eigenvalue of $H_0+S_2(x)$. The bent arrows and labels E_i, $i = 1, 2, 3, 4$ indicate regions for which the eigenvalues of $H_0+S_2(x)$ reproduce the exact eigenvalues E_i, $i = 1, 2, 3$ or 4, to the scale of the graph.

pairs were observed to have the largest overlap with the P-space. The total probability of an eigenvector ($|\Psi_i\rangle$) of $H^{\text{II}}(x)$, corresponding to the eigenvalues reproduced by the partial summations, to be in the P-space is

$$\langle\Psi_i|P|\Psi_i\rangle/\langle\Psi_i|\Psi_i\rangle \gtrsim 0.8,$$

except for neighborhoods of $\Delta x \approx 0.02$ about the singularities at $x \approx 0.06$ and 0.26. The details of the eigenvalues of H_0+S_1 in fig. 8 are not shown for $0.0628 < x < 0.0632$; the two eigenvalues of H_0+S_1 do not cross or coincide in this region of x. The region, $0.262 < x < 0.264$, is also not shown in fig. 8 and we describe the behavior of the eigenvalues of H_0+S_1 in this region as follows. In fig. 8, the two eigenvalues of H_0+S_1 for $x \lesssim 0.262$ are labelled (a) and (b). As x increases from 0.262, the eigenvalues corresponding to curves (a) and (b) of fig. 8 approach each other, coincide, become complex conjugates and then coincide again. As x increases further, one eigenvalue of H_0+S_1 emerges in fig. 8 as the curve labelled (d). The other eigenvalue goes more negative, experiences a discontinuity, becomes positive and emerges in fig. 8 as the curve labelled (c). This discontinuity arises because $S_1(x)$ has a pole when the upper eigenvalue of $H_0+\hat{Q}(x)$ coincides with the upper eigenvalue of $QH(x)Q$. The eigenvalues of H_0+S_1 at some points in the interval $0.262 \leqq x \leqq 0.264$ are given in table 4. The partial summations for H^{II} fail to converge for $0.9 \lesssim x < 1.3$ but one eigenvalue and eigenvector does converge to the largest eigenvalue and corresponding projected eigenvector of $H^{II}(x)$. For example, at $x = 1.0$, the second partial sum yields an eigenvalue of 27.8229 whereas the corresponding exact eigenvalue is 27.8232. The total probability that the eigenvector corresponding to the largest eigenvalue be in the P-space is greater than 0.95 whereas this probability is less than 0.5 for the remaining eigenvectors for $0.9 < x < 1.3$. The folded-diagram series for H^{II}_{eff}, eq. (3) with $\Delta E_c = 0$, was evaluated up to and including F_4 with $\Delta = 0$ and it fails to converge for $x \gtrsim 0.06$.

The partial sums for $H^{I}(x)$ fail to converge for $x \gtrsim 0.8$ and reproduce the lowest two eigenvalues of $H^{I}(x)$ for $x \lesssim 0.8$. H_0 was the diagonal matrix with matrix elements $(-10, -10, -5, -5, 0)$. The folded-diagram series, with $\Delta = 0$, evaluated through F_4 diverges for $x \gtrsim 0.3$. The region of convergence for the folded-diagram series may perhaps be extended by choosing a different Δ but an important observation here is the potentiality of a large difference in x between the regions of convergence for $H_{eff}(x)$ when evaluated by partial summations and when evaluated by perturbation (the folded-diagram series).

The nature of the singularities in $H_{eff}(x)$ have been studied carefully by Weidenmüller and collaborators [1, 2, 12]), and several schemes have been offered to deal with the singularities. The partial summations presented can handle some of the singularities as exhibited by fig. 8. If the partial summations fail to converge, then so will the folded-diagram series. Moreover, the divergence of the folded-diagram series may merely reflect a poor choice of ε_P, rather than the more profound entity, the intruder state, which is the usual label for the cause of the singularity. That the partial summations for H_{eff} are convergent, can be given the interpretation that the separation of the full space, into a particular P- and Q-spaces chosen, in fact, represents an approximate decoupling of the full hamiltonian, and conversely, if the partial sums diverge. From fig. 8, it is seen that the decoupling need not coincide with the lowest eigenvalues of the full hamiltonian. The question of convergence of the partial summations for H_{eff} then reduces to the rather simple and obvious statement that, for a given choice of P-space of dimension d, there exist d eigenvectors of the full hamiltonian, $|\Psi_i\rangle$, $i = 1, 2, \ldots, d$, that are predominantly in the P-space. That is,

$$|\Psi_i\rangle \sim P|\Psi_i\rangle,$$

or equivalently,

$$\langle\Psi_i|P|\Psi_i\rangle/\langle\Psi_i|\Psi_i\rangle \sim 1, \qquad i = 1, 2, \ldots, d, \tag{12}$$

is a reasonable approximation. This is simply a statement of the implicit assumption that we are in fact dealing with a perturbation on these eigenvectors and is neither new nor profound. For the model hamiltonians presented, a typical number for the quality of approximation (12) is

$$\langle\Psi_i|P|\Psi_i\rangle/\langle\Psi_i|\Psi_i\rangle \gtrsim 0.8$$

for convergence of the partial sums in a few iterations. In terms of fig. 8, condition (12) is not satisfied for $0.05 \lesssim x \lesssim 0.09$ and $0.24 \lesssim x \lesssim 0.28$. The precise reason for failure of the partial summations to converge is not clear to us. It is useful to consider the properties of $H(x) \equiv H_0+xH_1$. The partial sums converge within a few iterations if approximation (12) is good. Near crossing points such as $x \approx 0.06$ and 0.26 in fig. 8, the full eigenvectors corresponding to the crossing have comparable probability in both the P- and Q-spaces. The crossing point near $x \approx 0.06$, for example, can be interpreted as a mixing of an intruder state, corresponding to the lowest QHQ eigenvalue which originates from E_3 at $x = 0$ in fig. 7, into the model space. The mixing of intruder states into the model space is strong near a crossing point and the partial sums do not converge within a few iterations. However, the mixing of intruder states into the model space need not be strong only near a so called crossing point but can extend over a region of x, obviously, and we do find regions of x where S_{1000} had not converged. In as much as the partial summations represent exact summations of classes of folded-diagrams, then approximation (12) must also be reasonable if the folded-diagram series is used to evaluate H_{eff}.

Within the framework of the partial summations presented here and the Q-box perturbation analysis of refs. [5, 6]), the P-space can be chosen at our discretion to suit particular needs. The conventional choice for the P-space leads to a linked-valence, connected perturbation expansion in the Q-box and its derivatives [5, 6]) or in $H_1(9)$. That the conventional choice for the P-space represents a good choice in practical nuclear applications has been questioned by Weidenmüller and collaborators, in view of experimentally observed low-lying nuclear energy levels whose structure is believed to be predominantly outside the conventional P-space. These low-lying states are labelled as intruder states and are essentially a consequence of the overlapping of the PHP and QHQ spectra. Various schemes to deal with intruder states have been presented and evaluated by Weidenmüller and collaborators.

The partial summations of H_{eff} can handle intruder states, if, in fact, approximation (12) is reasonable for the particular P-space chosen. In the following, we assume that we are interested in the effective hamiltonian that reproduces the lowest eigenvalues. If the partial summation for H_{eff} fail to converge, then assumption (12) is probably not reasonable. This can be translated to say that there is an intruder state and this intruder state is strongly coupled to the P-space by H. The presence of an intruder state is a manifestation of the choice of P-space. The intruder state can be removed by making a choice for the P-space consistent with approximation (12) but this can be difficult, apriori. It is clear that in order to make a better choice for the P-space, some knowledge of the structure of the intruder state is necessary. A new P-space can be defined to incorporate the knowledge of the intruder state, in some manner. This sort of procedure was mentioned by Hoffmann *et al.* [12]) within the different context of modification of the single particle hamiltonian. We label the procedure as a P-space modification and contend that the consistency of the procedure can be verified. The P-space modification can be tested by performing the partial summations but the goal is to have the folded-diagram series converge as a test for the P-space modification. This goal is desirable because it will demand a high quality of the truncation [approximation (12)] but suffers from the drawback that non-convergence may merely reflect an inappropriate choice of ε_P. Also, given that the folded-diagram series appears to converge to a given order, $n+1$, in Q-box, convergence (or lack of convergence) to higher order in Q-box can be readily tested by evaluating equation (7) to finite order in l, $l \leq n$, since $\hat{Q}_l$, $l \leq n$, have already been evaluated.

In the following, we discuss, by means of example, an application of P-space modifications for the effective hamiltonian for H^{I} given in table 1. We choose not to change the dimension of the P-space in this example although we do have the freedom to increase or decrease the dimensionality of the P-space. The original P-space is as discussed previously

$$P = |1\rangle\langle 1| + |2\rangle\langle 2|.$$

It is clear, from inspection of table 1, that the state $|3\rangle$ is strongly coupled to the P-space. Recall, that for this P-space the partial sums fail to converge for $x \gtrsim 0.8$. At $x \approx 0.8$, a $PH^{\mathrm{I}}P$ and $QH^{\mathrm{I}}Q$ eigenvalue cross. The state corresponding to the lowest eigenvalue of $QH^{\mathrm{I}}Q$, is the intruder state here. It is not difficult to obtain the intruder state but we choose to illustrate the procedure by making the guess for the intruder state, $|I\rangle$,

$$|I\rangle = \sqrt{\tfrac{1}{2}}\{|3\rangle + |4\rangle\}.$$

Then, the $\{|3\rangle, |4\rangle\}$ subspace of the full hamiltonian, H^{I}, is transformed by the unitary transformation

$$\sqrt{\tfrac{1}{2}}\begin{bmatrix} 1 & 1 \\ 1 & -1 \end{bmatrix}.$$

That is, the full hamiltonian, H^{I}, is appropriately transformed so that the guess for the intruder state becomes a basis state for the representation of H^{I}. The hamiltonian matrix is then transformed by the unitary transformation that diagonalizes the 3-dimensional subspace $\{|1\rangle, |2\rangle, |I\rangle\}$ of H^{I}. The projection operator formed from the two eigenvectors corresponding to the lowest two eigenvalues of H^{I} in the subspace $\{|1\rangle, |2\rangle, |I\rangle\}$ is defined to be the new P-space. The new P-space is, thus, 2-dimensional and each P-space state is a linear combination of $|1\rangle, |2\rangle, |I\rangle$. The net transformation of the full hamiltonian in this procedure is, of course, unitary. For the old P-space, $\langle\Psi_i|P|\Psi_i\rangle/\langle\Psi_i|\Psi_i\rangle$ was 0.66 and 0.86 for the eigenvectors of the lowest two eigenvalues, whereas, for the new P-space this ratio was 0.91. The partial summations and folded-diagram series are computed as outlined previously. The results for a diagonal and an off-diagonal matrix element of H_{eff} for the new P-space are given in table 3. In table 3, the ε_P dependence of the partial summations and folded-diagram series is illustrated. The choices for ε_P are the eigenvalues of PHP. The partial summations converge for both values of ε_P. The folded-diagram series appears nicely convergent for ε_P equal to the smallest PHP eigenvalue. It is clear from table 3 that

TABLE 3

Convergence of a diagonal and off-diagonal matrix element of the effective hamiltonian for H^{I} after P-space modification

	$\varepsilon_P = -14.892$				$\varepsilon_P = -11.635$			
	diagonal matrix element		off-diagonal matrix element		diagonal matrix element		off-diagonal matrix element	
n	H_0+S_n	$\sum_{l=0}^{n} F_l$	H_0+S_n	$\sum_{l=0}^{n} F_l$	H_0+S_n	$\sum_{l=0}^{n} F_l$	H_0+S_n	$\sum_{l=0}^{n} F_l$
0	−15.546	−15.546	0.525	0.525	−16.187	−16.187	1.059	1.059
1	−15.428	−15.442	0.844	0.692	−15.348	−13.887	0.705	0.276
2	−15.455	−15.456	0.737	0.745	−15.466	−19.854	0.777	2.053
3	−15.448	−15.448	0.770	0.759	−15.447	−0.903	0.759	−3.426
4	−15.450	−15.450	0.760	0.762	−15.450	−68.141	0.764	15.828
exact	−15.449		0.762		−15.449		0.762	

The ε_P dependence of the partial summations (H_0+S_n) and folded-diagram series $(\sum F_l)$ is illustrated. The corresponding exact matrix elements are shown. Note that the folded-diagram series for $\varepsilon_P = -11.635$ is diverging.

choosing ε_P as the largest PHP eigenvalue is not a good choice for the folded-diagram series and illustrates that divergence of the folded-diagram series may merely reflect an inappropriate choice of ε_P. Choosing ε_P has been discussed in a slightly different context, by Schucan and Weidenmüller [1]). The partial summations and the folded-diagram series (for $\varepsilon_P = -14.892$) reproduce the lowest two eigenvalues and corresponding projected eigenvectors. The convergent results in table 3, themselves, verify the consistency of the P-space modification. The particular technique for a P-space modification presented was for illustration and not meant to be the only technique, clearly.

TABLE 4

Eigenvalues of $H_0+S_1(x)$ for H^{II} for some values of x in the interval $0.262 \leqq x \leqq 0.264$

x	Eigenvalues of $H_0+S_1(x)$	
0.262	0.999	5.349
0.26253760	2.534	2.555
0.26253761	$2.545\pm$	$4.775\times10^{-3}i$
0.263	$0.814\pm$	$2.986i$
0.26327519	$-4.340\pm$	$4.943\times10^{-2}i$
0.26327520	−4.368	−4.312
0.2633	−8.991	−2.362
0.2634	−43.401	−1.176
0.2635	88.799	−0.776
0.264	14.605	−0.133

Note the complex conjugate eigenvalues ($i = \sqrt{-1}$).

In concluding this presentation, we stress that discussion on the folded-diagram series has been solely for H_0 that is degenerate with respect to the P-space and that discussion has been directed at the state formulation as opposed to a linked-valence and connected formulation. In practical perturbation calculations, a linked-valence and connected formulation is used and H_{eff} is separated into a sum of effective hamiltonian operators of given particle rank. For example, for ^{18}O, H_{eff} is separated as

$$H_{\text{eff}} = H_{\text{eff}}(\text{0-body}) + H_{\text{eff}}(\text{1-body}) + H_{\text{eff}}(\text{2-body}), \tag{13}$$

when ^{16}O is taken as a closed core, and each term in eq. (13) can be calculated separately. In practice, H_{eff} (2-body) is calculated using degenerate perturbation theory and H_{eff} (1-body) is taken from experiment. However, H_{eff} (1-body) can be calculated and the unperturbed hamiltonian (H_0) used to compute H_{eff} (1-body) need not coincide with the unperturbed hamiltonian used to calculate H_{eff} (2-body). Thus, in practical applications, the procedure for calculating H_{eff} is not fully equivalent to degenerate perturbation theory and therefore the implication, on practical application, of our results for the degenerate folded-diagram series, particularly divergence in the presence of an intruder state, must be judged with this in mind.

We would like to thank Dr. E. Osnes for numerous interesting discussions in this area, and Dr. K. F. Ratcliff for very informative and enlightening discussions on the problem of intruder states.

Note added in proof: We kindly thank M. R. Anastasio and J. W. Hockert for pointing out that the partial summation, eq. (11), is more simply derived by noting that $S_{n+1}|\psi_i^{(n)}\rangle$ is a Taylor series.

References

1) T. H. Schucan and H. A. Weidenmüller, Ann. of Phys. **73** (1972) 108
2) T. H. Schucan and H. A. Weidenmüller, Ann. of Phys. **76** (1973) 483
3) B. R. Barrett and M. W. Kirson, Nucl. Phys. **A148** (1970) 145
4) T. T. S. Kuo and G. E. Brown, Nucl. Phys. **85** (1966) 40
5) T. T. S. Kuo, S. Y. Lee and K. F. Ratcliff, Nucl. Phys. **A176** (1971) 65
6) K. F. Ratcliff, S. Y. Lee and T. T. S. Kuo, unpublished
7) M. W. Kirson, Ann. of Phys. **66** (1971) 624
8) E. M. Krenciglowa, T. T. S. Kuo, E. Osnes and B. Giraud, Phys. Lett. **47B** (1973) 322
9) B. H. Brandow, Rev. Mod. Phys. **39** (1967) 771;
B. H. Brandow, Proc. Int. School of Physics "Enrico Fermi", Course 36, Varenna, 1965, ed. C. Bloch (Academic Press, New York, 1966)
10) C. Bloch and J. Horowitz, Nucl. Phys. **8** (1958) 91
11) S. Y. Lee, thesis, SUNY, Stony Brook 1972, unpublished
12) H. M. Hoffmann, S. Y. Lee, J. Richert, H. A. Weidenmüller and T. H. Schucan, preprint
13) J. Des Cloizeaux, Nucl. Phys. **20** (1960) 321
14) B. R. Barrett, E. C. Halbert and J. B. McGrory, in Symposium on correlations in nuclei, Balatonfüred, Hungary, September 3–8, 1973
15) H. M. Hoffmann, S. Y. Lee, J. Richert, H. A. Weidenmüller and T. H. Schucan, Phys. Lett. **45B** (1973) 421

ANNALS OF PHYSICS **101**, 154–194 (1976)

The Nuclear Reaction Matrix*

E. M. KRENCIGLOWA, C. L. KUNG, T. T. S. KUO,† AND E. OSNES

Institute of Physics, University of Oslo, Oslo 3, Norway; and
Department of Physics, State University of New York at
Stony Brook, Stony Brook, New York 11794

Received May 4, 1976

Different definitions of the reaction matrix G appropriate to the calculation of nuclear structure are reviewed and discussed. Qualitative physical arguments are presented in support of a two-step calculation of the G-matrix for finite nuclei. In the first step the high-energy excitations are included using orthogonalized plane-wave intermediate states, and in the second step the low-energy excitations are added in, using harmonic oscillator intermediate states. Accurate calculations of G-matrix elements for nuclear structure calculations in the $A \approx 18$ region are performed following this procedure and treating the Pauli exclusion operator Q_{2p} by the method of Tsai and Kuo. The treatment of Q_{2p}, the effect of the intermediate-state spectrum and the energy dependence of the reaction matrix are investigated in detail. The present matrix elements are compared with various matrix elements given in the literature. In particular, close agreement is obtained with the matrix elements calculated by Kuo and Brown using approximate methods.

1. INTRODUCTION

A fundamental concept in the nuclear-matter theory [1] pioneered by Brueckner, Bethe, and Goldstone is the introduction of the reaction matrix G defined as the sum of all the particle ladder-type interactions between a pair of nucleons in nuclear matter. This is necessitated by the presence of very (or infinitely) strong short-range repulsion contained in nearly every modern model for the nucleon–nucleon interaction. Nuclear matter is a hypothetical nuclear system; it is an infinite homogeneous many-body system consisting of neutrons and protons and is chosen mainly for its mathematical simplicity as compared with the real finite nuclei. Our main purpose in studying nuclear matter theory is of course not limited

* Work supported in part by the U.S.E.R.D.A. contract AT(11-1)-3001 and by the Norwegian Research Council for Science and Humanities.

† NORDITA guest professor at University of Oslo (1974/75).

to the understanding of nuclear matter itself. We would like to extend the methods developed in nuclear matter theory to the study of real nuclei.

The first step in extending the nuclear matter theory to real nuclei is perhaps the generalization of the nuclear-matter reaction matrix to the reaction matrix for real nuclei. Since the early work of Dawson, Talmi, and Walecka [2], there has indeed been a very large amount of work performed for the purpose of calculating the nuclear reaction matrix [3–15]. (Note that nuclear reaction matrix will be used from now on to denote the reaction matrix for finite nuclei in contrast to nuclear-matter reaction matrix which denotes the reaction matrix for nuclear matter.) Many of these works have been discussed in the review articles by Baranger [16], Barrett and Kirson [17], Kuo [18] and Becker [19]. In the calculation of the reaction matrix for finite nuclei, the treatment of the Pauli exclusion operator in the integral equation for the reaction matrix was usually the most difficult part. A method for the simplification of the treatment of the Pauli-exclusion operator was recently proposed by Tsai and Kuo [15]. But only preliminary applications of this method have been made [20, 21]. The main purpose of the present paper is to perform a detailed study of the Tsai–Kuo method by actually applying it to a rather extensive calculation of the reaction matrix elements appropriate to *s-d* shell nuclei. Since the Tsai–Kuo method provides a practically exact method for the calculation of nuclear reaction matrix elements with orthogonalized plane-wave intermediate states, it will be of interest to compare the matrix elements so calculated with those calculated with other methods. Such a comparison will be given in this paper.

Although nuclear reaction matrices have been used in nuclear structure calculations for a long time, a number of fundamental elements in the calculation and definition of these matrices are still rather ambiguous. Examples are the choice of single-particle spectrum for the intermediate states in the integral equation for the reaction-matrix, the determination of the energy variable of the reaction matrix, what Pauli-exclusion operator should be used and others. It is also the purpose of the present paper to discuss and investigate these ambiguities.

The organization of this paper is as follows. In Section 2, the definition of the reaction matrix for finite nuclei is derived and discussed starting from a linked-diagram many-body perturbation theory for the model-space effective interactions. Two types of reaction matrices will be discussed; one with harmonic-oscillator intermediate states and the other with orthogonalized plane-wave intermediate states. In Section 3, we discuss in some detail, the actual calculational procedures employed in this paper, together with a brief review of the Tsai–Kuo method. The results of our calculations are presented and discussed in Section 4, with emphasis on the convergence properties in connection with the truncation of the intermediate state projector in the reaction-matrix integral equation. Section 5 contains a summary of the findings of the present paper.

2. Reaction Matrix and Model-Space Effective Interactions

We consider that the nuclear many-body problem can be described by non-relativistic quantum mechanics and the nucleon–nucleon interaction can be represented by a two-body potential V. Then the nuclear many-body problem is described by the usual many-nucleon Schroedinger equation

$$\begin{gathered} H\psi_\lambda(1, 2 \cdots A) = E_\lambda\psi_\lambda(1, 2 \cdots A) \\ H = T + V \end{gathered} \tag{1}$$

where T is the kinetic energy operator and A is the number of nucleons in the nucleus. As is well known [18], a perturbation approach has to be employed in the solution of the above equation. This is because, for a general interaction potential V, such as the Reid nucleon–nucleon potential [22], it is just practically impossible to solve Eq. (1) exactly when A is large (such as $A \geqslant 3$). The commonly used procedure for the solution of the above many-nucleon equation is to convert it into a model-space Schroedinger equation

$$PH_{\text{eff}}P\psi_\mu = E_\mu P\psi_\mu, \qquad \mu = 1, 2 \cdots d \tag{2}$$

where the effective Hamiltonian H_{eff} is to be calculated using linked-diagram perturbation methods [18]. Here P is the projection operator (projector) for the model space whose dimensionality is d.

To illustrate the connection between the reaction matrix for finite nuclei and H_{eff}, we consider the following two typical cases. First we consider the calculation of the ground state energy of ^{16}O using a nondegenerate linked-diagram perturbation method. Then we will consider a calculation of the low-lying states of ^{18}O using a degenerate valence-linked perturbation method. For both cases, we must first choose a convenient single-particle basis on which the perturbation expansion of H_{eff} is formulated. This is done by introducing a one-body potential U and rewriting H as

$$\begin{aligned} &H = H_0 + H_1 \\ &H_0 = T + U, \qquad H_1 = V - U \\ &H_0 = \sum_{i=1}^{A} h_0(i), \qquad T = \sum_{i=1}^{A} t(i), \qquad U = \sum_{i=1}^{A} u(i). \end{aligned} \tag{3}$$

The single-particle wavefunction ϕ_n and energy ϵ_n are then the eigenfunction and eigenvalue of $h_0(= t + u)$, i.e.,

$$h_0\phi_n = \epsilon_n\phi_n. \tag{4}$$

For nuclear matter, it is convenient to simply choose plane waves as basis states. This is because nuclear matter is an infinite homogeneous medium and is therefore translationally invariant. For finite nuclei, it is convenient to use a single-particle basis which is of definite rotational symmetry. This is usually done by taking U as a spherical harmonic-oscillator potential, i.e.,

$$u(i) = \tfrac{1}{2}mr_i^2\Omega^2 \tag{5}$$

where m is the nucleon mass, $\mathbf{r}_i$ is the position vector for the ith nucleon and Ω is the oscillator angular velocity. Then ϕ_n and ϵ_n are just the familiar oscillator wave function and energy. It should be noted that U is not originally contained in $H(= T + V)$. We add U to T to obtain H_0. Thus we must correct for this in the remaining part of the Hamiltonian by writing $H_1 = H - H_0 = V - U$. This is an important point for defining the reaction matrix for a pair of nucleons in a real nucleus.

Consider now the calculation of the ground-state energy of ^{16}O. The basic theory for this calculation is the Goldstone linked-diagram expansion [18, 23]. This is a nondegenerate perturbation method where the model space is one-dimensional, i.e.,

$$P = |\Phi_0\rangle\langle\Phi_0| \tag{6}$$

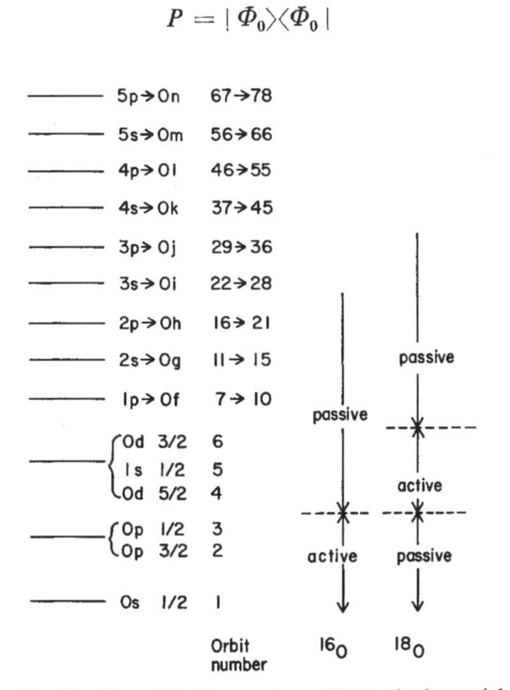

FIG. 1. Classification of the harmonic-oscillator single-particle orbits.

where Φ_0 is the Slater determinant corresponding to the unperturbed ^{16}O core (i.e., the oscillator orbits 1, 2, and 3 of Fig. 1 are all filled). Let W_0 denote the unperturbed energy of $\Phi_0(H_0\Phi_0 = W_0\Phi_0)$ and E_0 the ground-state energy of ^{16}O. Then it follows from the Goldstone theorem that the energy shift

$$\Delta E_0 \equiv E_0 - W_0 = \langle \Phi_0 \mid H_{\text{eff}} - H_0 \mid \Phi_0 \rangle \tag{7}$$

is given by the sum of all the linked diagrams, as shown in Fig. 2. Thus if we can evaluate this linked-diagram series, we will obtain ΔE_0. There are, however, a number of difficulties in evaluating this series. Suppose V is a hard-core nucleon-nucleon potential, then each individual diagram (which has at least one V vertex) of Fig. 2 is infinite. For example, the first term of the figure is

$$\sum_{\alpha,\beta=1}^{3} \langle \alpha\beta \mid V \mid \alpha\beta \rangle$$

which is infinite when V is a hard-core potential. This is because α and β are both harmonic-oscillator wave functions (see Eqs. (4) and (5)) and hence their relative wave function is in general nonvanishing inside the hard core of V. Note that the summation limits 1 and 3 imply that α and β are summed over the lowest three orbits of Fig. 1.

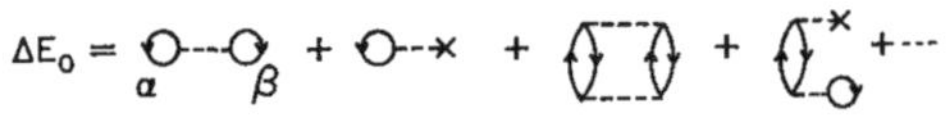

$\Delta E_0 =$ α β + + + + ···

FIG. 2. Linked diagram expansion for the ground state energy shift of ^{16}O. Here the dotted-line vertex represents a V interaction and the X vertex (---X) represents a U interaction of Eq. (3).

Thus to calculate this diagrammatic series for ΔE_0, it is necessary to first overcome the above difficulty that each single diagram in the expansion is infinite. To accomplish this, we need to use the reaction-matrix approach developed in nuclear matter theory. It was precisely this same type of difficulty which necessitated the introduction of the nuclear-matter reaction matrix when one calculated ΔE_0 for nuclear matter. It is most convenient to define the nuclear reaction matrix in terms of diagrams. It corresponds to a procedure where diagrams in the expansion of ΔE_0 are grouped according to some common structure and the diagrams in each group are then summed up to all orders of V. For example, corresponding to the first diagram of Fig. 2, we define a nuclear reaction matrix $\langle \alpha\beta \mid G \mid \alpha\beta \rangle$ in terms of the diagrams shown in Fig. 3. Using Goldstone diagram rules, Figure 3 is equivalent to the following integral equation

$$\langle \alpha\beta \mid G \mid \alpha\beta \rangle = \langle \alpha\beta \mid V \mid \alpha\beta \rangle + \sum_{i,j>3} \frac{\langle \alpha\beta \mid V \mid ij \rangle \langle ij \mid G \mid \alpha\beta \rangle}{\epsilon_\alpha + \epsilon_\beta - \epsilon_i - \epsilon_j} \tag{8}$$

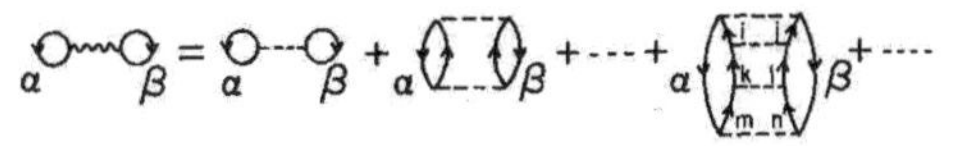

FIG. 3. Definition of a reaction matrix. A wavy-line vertex is used to denote a reaction matrix.

which defines the reaction-matrix operator G. It should be noted that the G-matrix so defined is dependent on our choice of U in Eq. (3). This is because the single-particle wave functions i, j, α, and β and the single-particle energies ϵ_α, ϵ_β, ϵ_i, and ϵ_j are all given by Eq. (4) which depends on our choice of U. The meaning of Fig. 3 may be made clearer if we cut open the hole lines α and β as shown in Fig. 4.

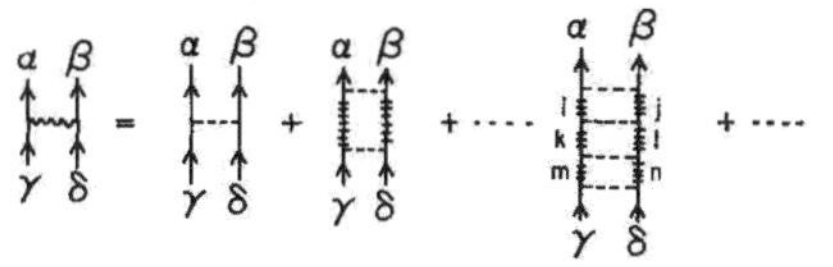

FIG. 4. A more transparent definition of the reaction matrix. Here a "railed" line denotes a passive single-particle state as defined in Fig. 1. For the diagonal case of $\alpha = \gamma$ and $\beta = \delta$, the present definition and that of Fig. 3 are equivalent.

In this way, we see that particles α and β (which are both within the Fermi sea, i.e., α and β are both less than or equal to orbit 3 of Fig. 1) are allowed to interact any number of times with the restriction that all intermediate single-particle states must be particle states (i.e., with orbit numbers higher than 3 of Fig. 1). It is because of this restriction that, in Eq. (8), we have the limitation that the summation indices i and j must both be greater than 3. For convenience, we will call the single-particle state with orbit number greater than 3 a *passive* single-particle state and the corresponding particle line a *passive* line, as indicated in Fig. 1 for the ^{16}O case.

Defining a two-particle projector Q_{2p}

$$Q_{2p} = \sum_{i,j>3} | ij\rangle\langle ij |, \tag{9}$$

we have from Eq. (8) an operator equation for G as

$$G = V + V \frac{Q_{2p}}{\epsilon_\alpha + \epsilon_\beta - H_0(2p)} G \tag{10}$$

where $H_0(2p)$ is the part of the unperturbed Hamiltonian which is operative only on the two-particle intermediate states of the G matrix. For the above case, we have $H_0(2p) \mid ij\rangle = (\epsilon_i + \epsilon_j) \mid ij\rangle$. Since $\mid ij\rangle$ is an eigenstate of $H_0(2p)$, we have $[Q, H_0(2p)] = 0$ and consequently

$$Q_{2p} \frac{1}{\epsilon_\alpha + \epsilon_\beta - H_0(2p)} = \frac{1}{\epsilon_\alpha + \epsilon_\beta - H_0(2p)} Q_{2p} = \frac{Q_{2p}}{\epsilon_\alpha + \epsilon_\beta - H_0(2p)}. \tag{11}$$

It is common to write Q_{2p} in a numerator as is done in Eq. (10). The energies ϵ_α and ϵ_β are dependent on external factors. For the above case, they depend on what hole lines α and β actually are. It is convenient to rewrite Eq. (10) for a general energy variable ω as

$$G(\omega) = V + V \frac{Q_{2p}}{\omega - H_0(2p)} G(\omega). \tag{12}$$

Usually, different G vertices will have different energy variables ω, depending on the actual location of the G vertex. For example, in Fig. 5 the energy variable ω for the middle G vertex (i.e., $\langle j\gamma \mid G(\omega) \mid k\gamma\rangle$) is $\epsilon_\alpha + \epsilon_\beta + \epsilon_\gamma - \epsilon_i$ and for both the other two G vertices ω is $\epsilon_\alpha + \epsilon_\beta$. Thus the middle G vertex is off-energy-shell by the amount $\epsilon_\gamma - \epsilon_i$ while the other two are on-energy shell. The projector Q_{2p} is usually called the Pauli exclusion operator.

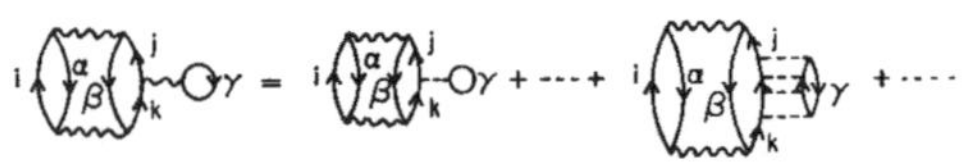

FIG. 5. A diagram containing an off-energy-shell G matrix.

The reaction matrix defined in Eq. (12) is rather similar to the nuclear matter reaction matrix; they differ only in using different Q_{2p} and $H_0(2p)$. For the nuclear matter G, Q_{2p} is defined in terms of plane-wave single-particle states and an $H_0(2p)$ which has plane wave eigenfunctions. It is useful to make a graph of

$$Q_{ab} \equiv \langle ab \mid Q_{2p} \mid ab\rangle \tag{13}$$

and

$$P_{ab} \equiv \langle ab \mid P_{2p} \mid ab\rangle \tag{14}$$

where

$$P_{2p} + Q_{2p} = I_{2p} = \sum_{\text{all}} \mid ij\rangle\langle ij \mid . \tag{15}$$

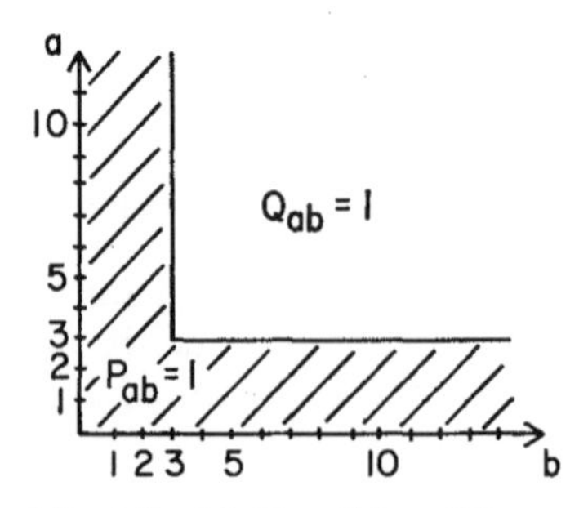

FIG. 6. A graph of Q_{ab} and P_{ab} defined in Eqs. (13) to (15), appropriate for a ^{16}O calculation. The orbital labels are explained in Fig. 1.

This will give us a clear picture of what intermediate two-particle states are allowed in Eq. (12). Such a graph is given in Fig. 6. This graph will be used later in discussing approximations in connection with treatments of Q_{2p}. As we will soon see, a large part of the difficulty in calculating $G(\omega)$ comes from the presence of the projector Q_{2p}.

Since $H_0(2p) = T(2p) + U(2p)$ and we have chosen U to be a one-body harmonic oscillator potential, the above G matrix will be referred to as the G-matrix with harmonic-oscillator intermediate states. In terms of this G-matrix, the energy shift ΔE_0 of Fig. 2 can be reexpressed as the diagrammatic series shown in Fig. 7.

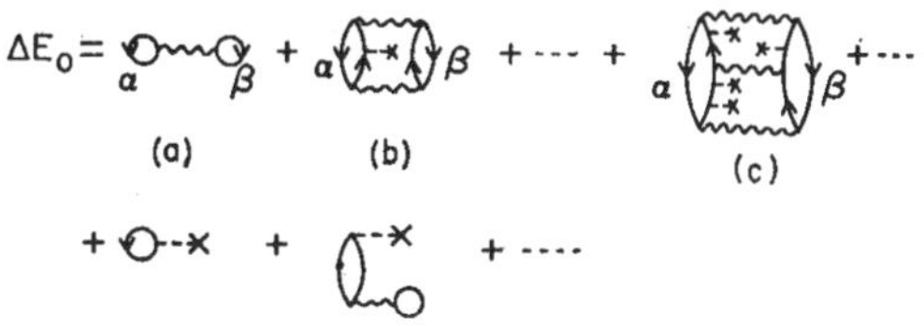

FIG. 7. Expansion of ΔE_0 in terms of the G-matrix vertex of Eq. (12). G-matrix vertices are denoted by wavy lines.

In so doing, one must be careful about the following points. First we must avoid double counting. For example diagram (i) of Fig. 8 should not be included in the expansion of ΔE_0. This is because this diagram is already included in diagram (a) of Fig. 7. The other point is that a G-matrix is composed of repeated interactions between two "particle" lines. Thus we should be careful about the meaning of hole–hole and particle–hole G-matrix vertices. For example, the meaning of a particle–hole G vertex is illustrated in diagram (ii) of Fig. 8 (middle vertex). It is helpful to use this diagram to illustrate once more the determination of the energy variable ω of the G-matrix. Here $\omega = \epsilon_\alpha + \epsilon_\beta + \epsilon_\gamma - \epsilon_k$ for the middle G-matrix, $\epsilon_\alpha + \epsilon_\beta$ for the top G-matrix and $\epsilon_\alpha + \epsilon_\beta$ for the lowest G-matrix.

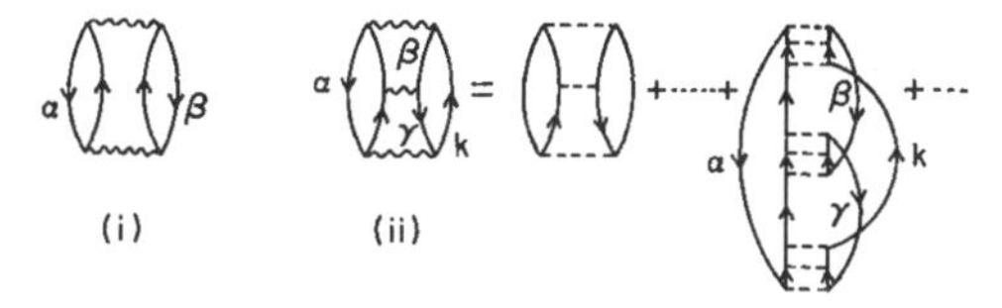

FIG. 8. Some precautions in drawing G-matrix diagrams; (i) is not allowed, but (ii) is allowed.

It is well known that although V is infinite (for a hard core potential) the G-matrix defined by Eq. (12) is in general finite except when ω happens to be a singularity of $G(\omega)$. Thus each term in the expansion of ΔE_0 in Fig. 7 is in general finite.

This makes a term-by-term evaluation of this series feasible and avoids the difficulty that each term in the expansion of Fig. 2 may be infinite.

Since we have $H = H_0 + H_1$ and $H_1 = V - U$ as shown in Eq. (3), the Goldstone expansion for ΔE_0 is a power series in terms of H_1. This is of course expected, because U is not originally present in H. We have added U to T to form H_0 and hence this added U must be taken out, making the interaction Hamiltonian become $V - U$. Thus, as shown in Fig. 7, we have diagrams with U vertices in the expansion of ΔE_0. Here in diagrams (b) and (c) the U vertices are all attached to the intermediate *particle* lines between successive G vertices. The diagrams (a), (b), (c) have similar general structure, and in fact we can define a new G matrix which includes all these diagrams. As shown in Fig. 9, we define a reaction matrix where

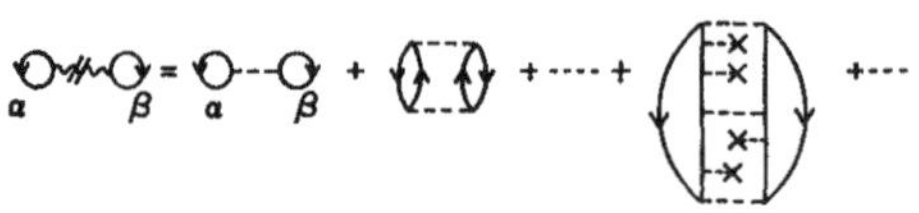

FIG. 9. The diagrammatic series defining G_T of Eq. (16).

U-insertions, both diagonal and off-diagonal, to the intermediate particle lines are summed up to all orders. This gives [16, 15] a new reaction matrix G_T which satisfies

$$G_T(\omega) = V + VQ_{2p} \frac{1}{\omega - Q_{2p}T(2p)\, Q_{2p}} Q_{2p}G_T(\omega). \tag{16}$$

Note that in Fig. 9 the last and the first vertex of each individual diagram must be a V vertex. In terms of G_T, diagrams (a), (b), and (c) of Fig. 7 are all contained in one G_T diagram, namely, diagram (a) with its wavy-line vertex replaced by a G_T vertex. Which reaction matrix, G or G_T, should we use in the calculation of ΔE_0? Before answering this question, it is convenient to first discuss G and G_T appropriate for an ^{18}O calculation.

Up to now we have studied reaction matrices designed for a calculation of the ground-state energy of ^{16}O, based on the Goldstone linked-diagram perturbation expansion. To calculate the low-lying states of ^{18}O using a realistic nucleon-nucleon interaction, a convenient starting point is the degenerate folded-diagram perturbation theory for the effective Hamiltonian H_{eff} (see Refs. [18, 24, 25] and references quoted therein). As usual, we may choose a model space P composed of a closed ^{16}O core (i.e., orbits 1, 2, and 3 of Fig. 1 are filled) and two valence neutrons in the s-d shell (i.e., orbits 4, 5, and 6 of Fig. 1). Then, as shown in Fig. 1, the active single-particle states (or lines as used in diagrams) are orbits 4, 5, and 6, the rest being all passive. After a P space has been chosen, the next step is to determine the effective Hamiltonian H_{eff} of Eq. (2). Writing H_{eff} as $H_0 + V_{eff}$ where V_{eff} is the model-space effective interaction, V_{eff} is given by [18, 24] the folded-diagram

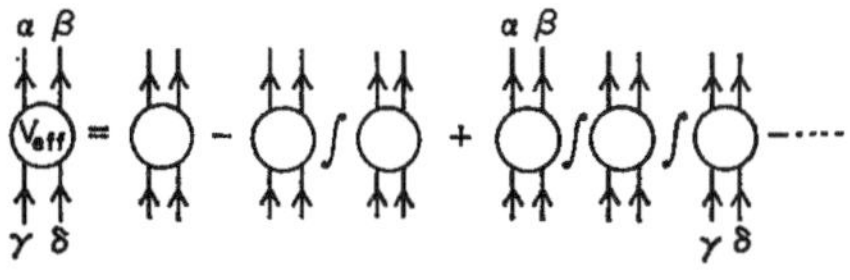

FIG. 10. A folded-diagram expansion of the effective interaction V_{eff}.

series shown in Fig. 10. Here each circle which is often referred to as a $\hat{Q}$-box represents a collection of all the irreducible valence interaction diagrams and the integral sign represents a generalized folding operation. Every diagram contained in a $\hat{Q}$-box must have at least one V or U vertex, all vertices linked to at least one valence line and at least one passive line between any two successive vertices. Further discussion is given in Section 4.3.

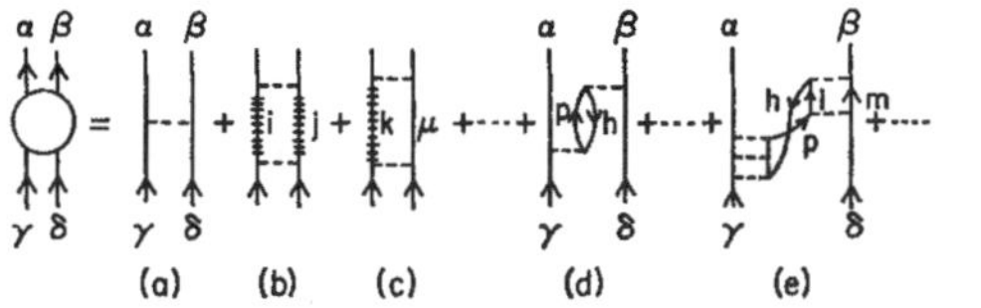

FIG. 11. Typical diagrams contained in a $\hat{Q}$-box. Here i, j, k, and h are passive lines, while μ, l, and m are active lines.

In Fig. 11, we give typical diagrams contained in a $\hat{Q}$ box. How do we calculate this diagrammatic series ? Suppose we have a hard-core nucleon–nucleon potential. Then we encounter the same type of difficulty as in the calculation of the diagrammatic series of Fig. 2 where we deal with the ground state energy shift ΔE_0 ; namely, each individual diagram of Fig. 11 is infinite. A reaction matrix is now needed to overcome this difficulty. We define a reaction matrix by the diagrammatic series shown in Fig. 4. Here α, β, γ, and δ are all within the *s-d* shell (i.e., orbits 4, 5, and 6 of Fig. 1). The restriction for the intermediate states depends on the choice of the model space, and is enforced by the projection operator Q_{2p} in Eq. (12). Note that the reaction matrix appropriate for the present ^{18}O calculation is also given by Eq. (12) except that Q_{2p} and ω are now different from those used for the ΔE_0 calculation of ^{16}O. There are different ways of choosing Q_{2p}. A common procedure of defining Q_{2p} is to specify its boundaries labelled by three numbers n_I, n_{II}, and n_{III}. This is explained in Fig. 12. For example, Q_{2p} of Fig. 6 is specified by $(n_I, n_{II}, n_{III}) = (3, 3, \infty)$.

If we choose to require both intermediate single-particle states in the G matrix being passive, then we have a Q_{2p} specified by $(n_I, n_{II}, n_3) = (6, 6, \infty)$. This is because, as seen in Fig. 1, our choice of the P space implies that the passive single-particle states start from orbit 7. Note that orbits 1, 2, and 3 are passive *hole*

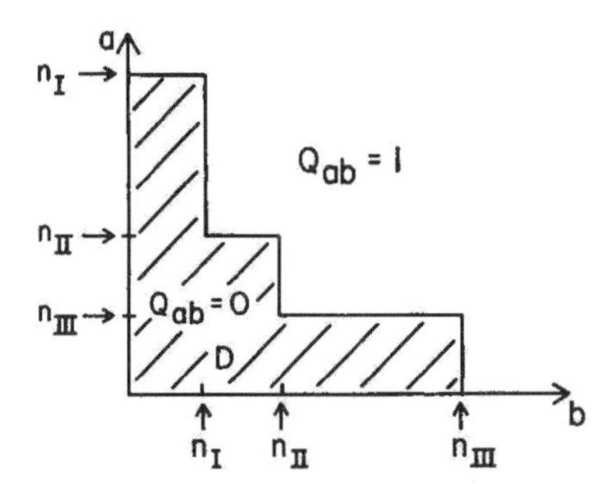

FIG. 12. A definition of the projection operator Q_{2p} by specifying its boundaries n_I, n_{II}, and n_{III} in terms of the orbit numbers described in Fig. 1. Here, $Q_{ab} = \langle ab \mid Q_{2p} \mid ab \rangle$. The region where $Q_{ab} = 0$ is denoted as D.

states. Using the above Q_{2p}, we are including diagrams (a) and (b) of Fig. 11 in the first-order G-matrix element $\langle \alpha\beta \mid G \mid \gamma\delta \rangle$. But diagram (c) is excluded from this matrix element. Since it is sufficient to have just one passive line in the intermediate states, diagram (c) is of course an allowed diagram of the $\hat{Q}$-box. Another way to choose Q_{2p} is to also include diagrams such as diagram (c) in the G matrix. This will require a choice of Q_{2p} with $(n_I, n_{II}, n_{III}) = (3, 6, \infty)$. This Q_{2p} clearly allows all intermediate two-particle states having only one passive line (i.e., orbit number greater than 6). There are other reasonable choices. For example if we want to use a somewhat larger model space, then we may use a Q_{2p} specified by $(3, 10, \infty)$. (See discussions in Section 4.4.)

Once a Q_{2p} is chosen, we must make sure that the G-matrix diagrams included in the $\hat{Q}$-box are consistent with the particular choice of Q_{2p}. Consider that we use a Q_{2p} specified by $(3, 6, \infty)$. In Fig. 13 we show several G-matrix diagrams relevant to $\hat{Q}$-box calculations. Here diagram (i) is clearly an allowed diagram in the Q-box. Since we use a $(3, 6, \infty)$ Q_{2p}, diagram (i) contains all the ladder-type diagrams indicated by diagrams (a), (b), and (c) of Fig. 11. Diagram (iii) is also an allowed diagram arising from summing diagrams of the structure shown by diagrams (d) and (e) of Fig. 11. Whether diagram (ii) is allowed or not depends on what orbits x and y are. If $(x, y) = (4, 4)$, then this diagram is not allowed. This is because 4 is an active orbit, as explained in Fig. 1. Hence this (4, 4) intermediate state contains no passive line and is disallowed because all intermediate states of any Q-box diagrams must have at least one passive line. If $(x, y) = (7, 7)$ (in this case we should draw x and y as railed lines) and we use a $(3, 6, \infty)$ Q_{2p}, then diagram (ii) is also disallowed because of double counting. Namely, diagram (ii) is already contained in diagram (i). Clearly for $(x, y) = (7,7)$ and a $(3, 10, \infty)$ Q_{2p}, diagram (ii) becomes an allowed diagram. Diagrams (iv) and (v) are both diagrams third

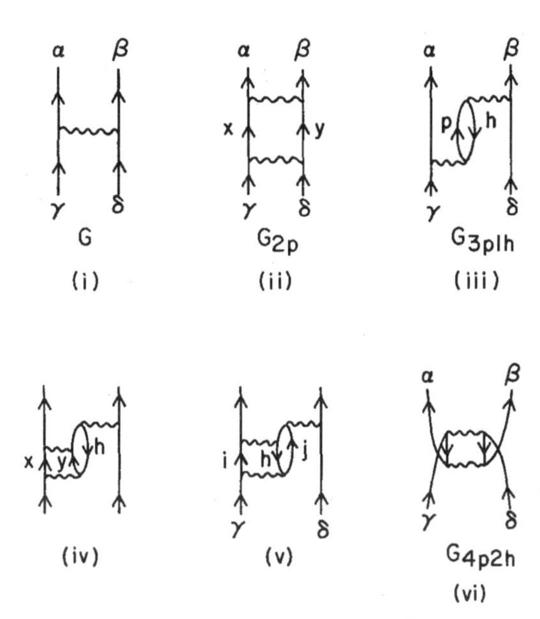

FIG. 13. G-matrix diagrams of a $\hat{Q}$-box.

order in G. Because these two diagrams tend to cancel with each other and because of possible double-counting errors involved in the calculation of diagram (iv), these two diagrams have received much attention in effective interaction calculations [26]. The following is worth noting. Since h is a passive line, diagram (iv) is an allowed diagram for x and y both being in the s-d shell (orbits 4, 5, and 6) when the G-matrix is calculated with a (3, 6, ∞) Q_{2p}. It becomes a disallowed diagram for the same (x, y) values if we use a (3, 3, ∞) Q_{2p}. This is because of double counting, namely for this situation diagram (iv) is already contained in diagram (iii). However, it should be pointed out that a (3, 3, ∞) Q_{2p} gives a bare G-matrix (i.e., diagram (i) of Fig. 13) which is not appropriate for ^{18}O. This is because this G-matrix for a (3, 3, ∞) Q_{2p} contains intermediate active states. Note that a G-matrix with a (3, 3, ∞) Q_{2p} may, however, be appropriate for diagram (iii) of Fig. 13 because of the presence of the passive line h.

We have mentioned that Q_{2p} and ω for the ^{18}O reaction matrix are generally different from those for the ^{16}O reaction matrix. Much attention has just been given to Q_{2p} for the ^{18}O reaction matrix.

Some examples will now be given to illustrate the determination of the energy variable ω for the ^{18}O reaction matrix. We will follow the folded-diagram formulation [18], and hence ω depends on the single-particle energies of the incoming (entering from below) single-particle lines. Diagram (i) of Fig. 13 has $\omega = \epsilon_\gamma + \epsilon_\delta$.

For the upper vertex $\langle h\beta \mid G(\omega) \mid p\delta\rangle$ of diagram (iii), we have $\omega = \epsilon_\gamma + \epsilon_\delta + \epsilon_h - \epsilon_\alpha$ as indicated clearly by diagram (e) of Fig. 11. As shown by diagrams (d) and (e) of Fig. 11, the G matrix for this vertex is given by

$$\begin{aligned}\langle h\beta \mid G \mid p\delta\rangle &= \langle h\beta \mid V \mid p\delta\rangle + \sum_{|l,m\rangle \in Q_{2p}} \frac{\langle h\beta \mid V \mid lm\rangle\langle lm \mid V \mid p\delta\rangle}{\epsilon_\gamma + \epsilon_\delta - (\epsilon_l + \epsilon_m + \epsilon_\alpha - \epsilon_h)} + \cdots \\ &= \langle h\beta \mid V \mid p\delta\rangle + \sum_{|l,m\rangle \in Q_{2p}} \frac{\langle h\beta \mid V \mid lm\rangle\langle lm \mid G \mid p\delta\rangle}{(\epsilon_\gamma + \epsilon_\delta - \epsilon_\alpha + \epsilon_h) - H_0(2p)}\end{aligned} \tag{17}$$

where $H_0(2p)$ operates only on the two-particle intermediate state $\mid lm\rangle$, with $H_0(2p) \mid lm\rangle = (\epsilon_l + \epsilon_m) \mid lm\rangle$. From Eq. (17), we see clearly that the energy variable ω is $(\epsilon_\gamma + \epsilon_\delta + \epsilon_h - \epsilon_\alpha)$. Note that in the above equation we have used a $(3, 6, \infty)$ Q_{2p} and hence have the summation indices l and/or m greater than 6 and both l and m greater than 3. To give one more example, the energy variable for the lowest G-matrix vertex $\langle ij \mid G(\omega) \mid \gamma h\rangle$ of diagram (v) of Fig. 13 is $\omega = \epsilon_\gamma + \epsilon_h$.

Because $H_1 = V - U$, there are clearly diagrams with U vertices in the Q box of Fig. 11. Just as we did for the reaction matrix appropriate for the ^{16}O calculation we can include all the U-vertex insertions to the intermediate states of the reaction matrix appropriate for ^{18}O calculations. This is done by adding all diagrams like (a) and (b) of Fig. 14 to the right-hand side of Fig. 4. The resulting reaction matrix is just the G_T given by Eq. (16) where we note that Q_{2p} and ω are now for the ^{18}O calculations. For instance, we may use a $(3, 6, \infty)$ Q_{2p}, as was just discussed.

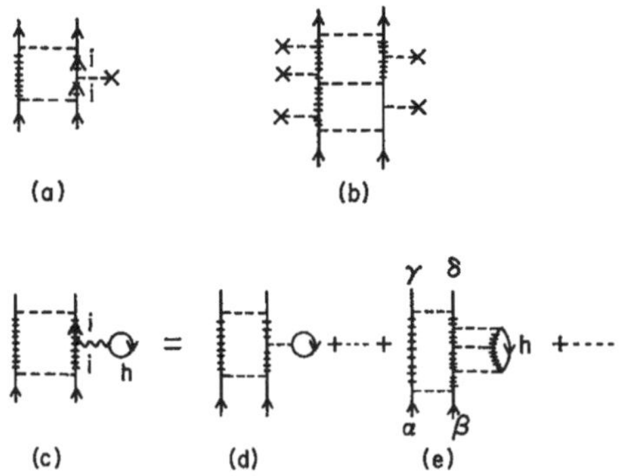

FIG. 14. Typical U and self-energy insertions to the G-matrix intermediate states.

We have now seen what diagrams are contained in the reaction matrices G of Eq. (12) and G_T of Eq. (16) for both the ^{16}O and ^{18}O cases. Let us now come back to the question which of these two reaction matrices should be used in actual calculations ? This question is closely related to the convergence properties of the effective interaction. If we could calculate reaction-matrix diagrams in the effective interaction to all orders, then it does not matter whether we use G or G_T, since they must give the same result. But in practice we can only calculate some low-order reaction-matrix diagrams of the effective-interaction expansion, hence the effective interaction calculated with G will in general be different from that calculated with G_T. Rigorously speaking, it is not known which of these two effective interactions is more reliable. There are, however, qualitative arguments that favor the choice of G_T. These arguments are the following.

The first argument concerns the dependence of the reaction matrix on the choice of U. Since our Hamiltonian $H = T + V$ does not have U to begin with, it is natural that we like to minimize the dependence of the reaction matrix on U which is artifically added by us. In this spirit, G of Eq. (12) is less desirable because Eq. (12) depends on the projector Q_{2p} and on the high lying spectrum of $H_0(2p)$ and hence $Q_{2p}\, UQ_{2p}$. In contrast, G_T of Eq. (16) depends only on the projector Q_{2p}. This may be understood from the following argument. Since U is originally not contained in H, we must substract U from V to correct for the addition of U to T. (Recall that $H = T + V = T + U + V - U$.) Thus the internucleon interaction becomes $H_1 = V - U$ instead of V. The U insertions to the intermediate states as shown by diagrams (a) and (b) of Fig. 14 arise from using H_1 as the intermediate-state two-nucleon interaction. Thus it is natural that such insertions serve to remove the U dependence of the reaction matrix, changing G to G_T.

The second argument which concerns the self-energy insertions follows from the study of three-body-clusters in nuclear matter [1]. As shown by diagram (c) of Fig. 14, there are self-energy insertions of the Brueckner–Hartree–Fock type to the intermediate particle lines. If we can choose a U such that all the U insertions and all the self-energy insertions cancel exactly, then we no longer have the U insertions. Consequently we would have the reaction matrix G defined by Eq. (12). However, it is unlikely that such a choice for U is possible in principle because the Brueckner–Hartree–Fock self-energy insertions on a particle line are off-energy-shell so that U-insertions can only give approximate cancellation of self-energy insertions. For low-lying single-particle states where the self-energy insertions are not far off the energy shell, this type of approximate cancellation is indeed possible as indicated by actual Brueckner–Hartree–Fock calculations [27]. But for high-lying single-particle states it is not at all clear that the U- and self-energy inserrtions can cancel with each other. Thus when i is a highly excited state, diagrams (a) and (c) of Fig. 14 are not expected to cancel with each other.

Another consideration is that we should sum the three-body-cluster diagrams to all orders, in analogy to the nuclear-matter calculations of three-body-clusters [1]. Diagram (c) is a three-body diagram in the sense that it represents repeated interactions among two valence particles and a core particle h. There are, however, other three-body diagrams of the same general structure as shown by the diagram of Fig. 15. The sum of all the three-body-cluster diagrams of this type was estimated to give negligible contribution to the binding energy of nuclear-matter [1]. To our knowledge, an accurate estimate of these diagrams for finite nuclei has not been made. Assuming that the nuclear-matter result remains generally valid for finite nuclei, we expect that the sum of diagram (c) of Fig. 14, the diagram shown in Fig. 15 and all the other similar three-body diagrams will be negligibly small. In other words, the self-energy-insertion diagram (c) is cancelled by other three-body diagrams. Hence the U-insertion diagrams survive, leading to the reaction matrix G_T.

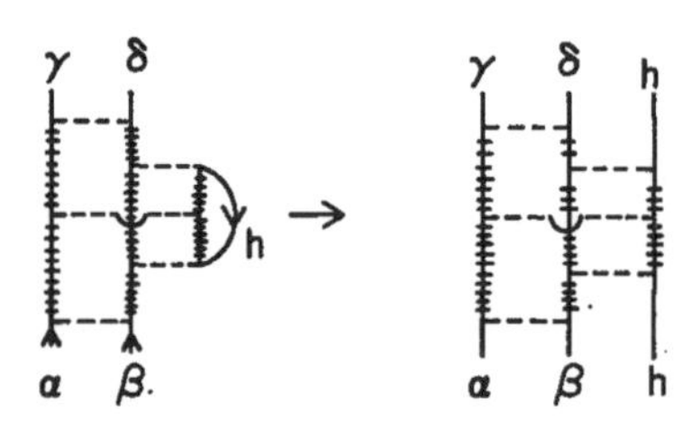

FIG. 15. A three-body cluster diagram. By opening up the hole line as shown, the left-hand side is clearly seen to be a three-body cluster diagram.

The above discussion suggests the following approach. We divide the intermediate states into two parts; high-lying and low-lying. For the former, we ignore all the self-energy insertions because of the above three-body-cluster consideration. Thus we should use G_T solely for the high-lying intermediate states. Low-lying intermediate states are treated differently. Here we consider that the U insertions and the self-energy insertions cancel with each other. Thus both insertions are ignored, leading to G of Eq. (12) with the V replaced by G_T and Q_{2p} restricted to low-lying intermediate particle states. This mixed approach will be referred to as a two-step approach and discussed in more detail in Section 4.4.

3. METHOD OF CALCULATION

In the preceding section we have derived reaction matrices for two interacting nucleons by summing up appropriate two-nucleon ladder diagrams to all orders. These reaction matrices are needed for doing microscopic nuclear structure calculations. We now discuss how to calculate these reaction matrices by solving their defining equations (12) and (16). We will refer to G as the reaction matrix with oscillator intermediate states and G_T as the reaction matrix with orthogonalized plane-wave intermediate states.

We use the notation $|\,abJT\rangle$ to represent a normalized and anti-symmetrized two-nucleon state where one nucleon is in the oscillator state $|\,n_a l_a j_a\rangle$ and the other in $|\,n_b l_b j_b\rangle$. The angular-momentum coupling for this state is as shown in Fig. 16, and T is the two-nucleon isospin quantum number. We can expand $|\,abJT\rangle$ into wave functions of the center-of-mass and relative coordinates;

$$|\,abJT\rangle = \sum \{\text{LAB-CMR}\} \times |\,NLnl, S\mathscr{J}TJ\rangle \tag{18}$$

where NL and nl are respectively the center-of-mass and relative harmonic-oscillator quantum numbers. Here we use the angular-momentum coupling scheme shown in Fig. 16 and $\mathscr{J}$ is defined by $\mathscr{J} = \mathbf{l} + \mathbf{S}$. The {LAB-CMR} transformation coefficients are well known (see, for example, Ref. [3]). Using the above transformation, we can express $\langle abJT\,|\,G\,|\,cdJT\rangle$ as

$$\langle abJT\,|\,G\,|\,cdJT\rangle = \sum \{\text{REDUC}\} \times \langle NLnl, S\mathscr{J}TJ\,|\,G\,|\,N'L'n'l', S'\mathscr{J}'TJ\rangle \tag{19}$$

and similarly for G_T. The {REDUC} transformation coefficients are again well known. In the present work we have used the formulas given in Ref. [3] in calculating the {REDUC} coefficients. The matrix elements $\langle N \cdots |\,G\,|\,N' \cdots\rangle$ and $\langle N \cdots |\,G_T\,|\,N' \cdots\rangle$ are usually called the reduced matrix elements of G and G_T. One is reminded that G and G_T are functions of ω as shown by Eqs. (12) and (16).

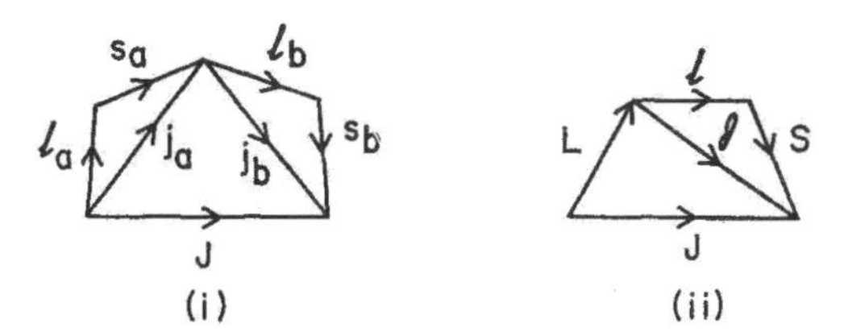

FIG. 16. Angular-momentum coupling for $|\,j_a j_b JT\rangle$ and $|\,NLnl, S\mathscr{J}JT\rangle$ shown respectively by (i) and (ii). In (ii) the spin quantum numbers obey $\mathbf{s}_a + \mathbf{s}_b = \mathbf{S}$.

In nuclear structure calculations it is the reaction matrix elements in the form of $\langle abJT\,|\,G\,|\,cdJT\rangle$ and $\langle abJT\,|\,G_T\,|\,cdJT\rangle$ which are most often used. In many earlier works [2–11], these matrix elements were calculated by first calculating the reduced matrix elements of G and G_T. Then the matrix elements in the $|\,abJT\rangle$ representation were obtained by means of Eq. (19). One major difficulty in this procedure arises from the treatment of the projection operator Q_{2p}. Consider the calculations of $\langle N \cdots | G | N' \cdots \rangle$ (we use the shorthand notation of $| N \cdots \rangle$ for $| NLnl, S\mathscr{J}TJ \rangle$). We first write Eq. (12) as

$$\langle N \cdots | G(\omega) | N' \cdots \rangle = \langle N \cdots | V | N' \cdots \rangle + \sum_{\substack{N'',N''' \\ N}} \frac{\begin{bmatrix} \langle N \cdots | V | N'' \cdots \rangle \langle N'' \cdots | Q_{2p} | N''' \cdots \rangle \\ \times \langle N''' \cdots | Q_{2p} | N'''' \cdots \rangle \langle N'''' \cdots | G | N' \cdots \rangle \end{bmatrix}}{\omega - \langle N''' \cdots | H_0(2p) | N''' \cdots \rangle} \tag{20}$$

by noting that $Q_{2p} Q_{2p} = Q_{2p}$. By defining the correlated wave function ψ by $V | \psi(N \cdots) \rangle = G | N \cdots \rangle$, we have from eq. (20)

$$| \psi(N \cdots) \rangle = | N \cdots \rangle + \sum_{\substack{N'',N''' \\ N''''}} \frac{\begin{bmatrix} | N'' \cdots \rangle \langle N'' \cdots | Q_{2p} | N''' \cdots \rangle \\ \times \langle N''' \cdots | Q_{2p} | N'''' \cdots \rangle \langle N'''' \cdots | V | \psi(N \cdots) \rangle \end{bmatrix}}{\omega - \langle N''' \cdots | H_0(2p) | N''' \cdots \rangle} \tag{21}$$

If we could now solve for the correlated wave function $| \psi(N \cdots) \rangle$, then the reduced matrix element would be obtained by evaluating the integral $\langle N \cdots | V | \psi(N' \cdots) \rangle$.[1] But in so doing, one usually must make a number of approximations. This is because Q_{2p} is *not* diagonal in the $| NLnl, S\mathscr{J}TJ \rangle$ representation; Q_{2p} is diagonal only in T and J, but not the other six quantum numbers. This implies the coupling of many different $(NLnl, S\mathscr{J}TJ)$ channels and makes an "exact" calculation of $\langle N \cdots | G | N' \cdots \rangle$ practically impossible. Typical approximations used in the literature [18, 19] involve setting Q_{2p} diagonal in $(N, L, S, \mathscr{J})$, truncating Q_{2p} so that it couples only a small number of neighboring $(NL \cdots)$ channels or replacing Q_{2p} by its angle average. Because of the presence of $Q_{2p} T(2p) Q_{2p}$ in Eq. (16), the calculation of $\langle N \cdots | G_T | N' \cdots \rangle$ is even more difficult.

Since Q_{2p} is diagonal in the $| abJT \rangle$ representation, one expects that most of the above-mentioned difficulties can be avoided if one calculates G and G_T directly on the $| abJT \rangle$ representation, without first calculating the reduced matrix elements of G and G_T. Barrett *et al.* [12] have proposed a method for calculating G (the reaction matrix with oscillator intermediate states) where Q_{2p} is treated directly in the $| abJT \rangle$ representation. They have applied this method in the $A = 18$ region and found it both convenient and accurate. For the calculation of G_T (the reaction matrix with orthogonalized plane-wave intermediate states) directly in the $| abJT \rangle$ representation, a convenient and accurate method has been proposed by Tsai and Kuo [15, 20]. The operator identity of Bethe, Brandow and Petschek [28] is essen-

[1] For a hard-core potential, $\langle N \cdots | V | N' \cdots \rangle$ of Eq. (20) is generally not defined. Thus, the G-matrix is usually calculated by evaluating $\langle N \cdots | V | \psi(N' \cdots) \rangle$ of Eq. (21).

tial for the calculation of G. Consider two reaction matrices, G_A and G_B. They are identical except that their propogators P_A and P_B are different:

$$\begin{aligned} G_A &= V + VP_A G_A \\ G_B &= V + VP_B G_B . \end{aligned} \tag{22}$$

Then the operator identity reads

$$G_B = G_A^+ + G_A^+(P_B - P_A)\, G_B . \tag{23}$$

The use of this formula is most readily seen in the calculations of G. In connection with Eq. (12), we define a free reaction matrix G_F by

$$G_F(\omega) = V + V \frac{1}{\omega - H_0(2p)} G_F(\omega). \tag{24}$$

Then by Eq. (23) we have $G(\omega)$ given simply as

$$G(\omega) = G_F(\omega) - G_F(\omega) \left(\frac{1 - Q_{2p}}{\omega - H_0(2p)} \right) G(\omega) \tag{25}$$

where we note that G_F is Hermitian. Since the complement of Q_{2p} is P_{2p} as given by Eq. (15), we rewrite Eq. (25) as

$$G(\omega) = G_F(\omega) - G_F(\omega) \frac{P_{2p}}{\omega - H_0(2p)} G(\omega). \tag{26}$$

Note that in Eqs. (25) and (26) we have both P_{2p} and Q_{2p} commute with $H_0(2p)$. The calculation of the $\langle abJT \mid G_F(\omega) \mid cdJT \rangle$ matrix elements is straightforward. One may calculate them by using a "reference-spectrum" method [3] where the two-body correlated wave function is obtained by numerical solution of the differential equation for the defect wave function. Note that this is simple, and exact, because there is no Pauli-exclusion operator for G_F. Another method [12] is to write G_F as

$$G_F(\omega) = (H_0(2p) - \omega) + (H_0(2p) - \omega) \frac{1}{\omega - H_0(2p) - V} (H_0(2p) - \omega) \tag{27}$$

where the propagator $(\omega - H_0(2p) - V)^{-1}$ is expanded as

$$\frac{1}{\omega - H_0(2p) - V} = \sum_{n=1}^{n_{\max}} \frac{\mid \chi_n \rangle \langle \chi_n \mid}{\omega - \lambda_n} . \tag{28}$$

Here $\mid \chi_n \rangle$ and λ_n are eigenfunctions and eigenvalues of $(H_0(2p) + V)$. Then $\langle abJT \mid G_F \mid cdJT \rangle$ is calculated conveniently using Eqs. (27) and (28). Although we should use $n_{\max} = \infty$ in principle, an approximation of finite-$n_{\max}$ truncation is always made [12]. Using either methods, we can calculate the matrix elements of G_F in the $|\, abJT\rangle$ representation conveniently.

From Eqs. (9) and (15), we have

$$P_{2p} = \sum_{e,f\in D} |\, efJT\rangle\langle efJT\,| \tag{29}$$

where D is the region defined in Fig. 12. As mentioned in Section 2, the boundary n_{III} should be infinite and therefore the domain of D is in fact infinite. This means that we sum over an infinite number of terms in Eq. (29), in general. With G_F known, our procedure for calculating G is to solve the matrix equation (26) which is rewritten as

$$\langle ab \,|\, G(\omega) \,|\, cd\rangle = \langle ab \,|\, G_F(\omega) \,|\, cd\rangle + \sum_{e,f\in D} \frac{\langle ab \,|\, G_F(\omega) \,|\, ef\rangle\langle ef \,|\, G(\omega) \,|\, cd\rangle}{\omega - (\epsilon_e + \epsilon_f)} \tag{30}$$

where the quantum numbers J and T are common to all state vectors and have been suppressed for brevity. Eq. (30) is a simple matrix equation to solve provided that D is finite. We do not know how to solve this equation if D is infinite. Thus a common and indispensible approximation is to make a finite-n_{III} approximation. For example we may use a (3, 6, 21) Q_{2p} where n_{III} is 21. The accuracy of the finite-n_{III} approximation can be investigated by solving Eq. (30) using various n_{III} values. It appears that the finite-n_{III} approximation is the only necessary approximation in the calculation of G, the reaction matrix with oscillator intermediate states. As discussed earlier, G depends on the choice of U. An example is that for a given ω one may make $G(\omega)$ accidentally singular if one uses a U such that ω is equivalent to one of the eigenvalues of $Q_{2p}(H_0 + V)\, Q_{2p}$. This is clearly not a desirable feature.

We now discuss the calculation of G_T in the $|\, abJT\rangle$ representation. We first define a free reaction matrix.

$$G_{TF}(\omega) = V + V \frac{1}{\omega - T(2p)} G_{TF}(\omega). \tag{31}$$

By Eqs. (23) and (16), we have

$$G_T(\omega) = G_{TF}(\omega) + G_{TF}(\omega) \left\{ Q_{2p} \frac{1}{\omega - Q_{2p}T(2p)\, Q_{2p}} Q_{2p} - \frac{1}{\omega - T(2p)} \right\} G_T(\omega). \tag{32}$$

Even when we know G_{TF} and use a finite n_{III}, it is still very difficult to solve for $G_T(\omega)$ from Eq. (32). This is mainly because of the presence of Q_{2p} in the denominator and $[Q_{2p}, T(2p)] \neq 0$. Equation (32) has been solved by Sauer with several approximations [10]. Tsai and Kuo [15] have devised a method which allows a convenient and practically exact method for calculating G_T of Eq. (16). The only necessary approximation in this method is the finite-n_{III} truncation. We outline the essence of their method in the following. They first proved a simple matrix identity[2]

$$Q_{2p} \frac{1}{Q_{2p} A Q_{2p}} Q_{2p} = \frac{1}{A} - \frac{1}{A} P_{2p} \frac{1}{P_{2p}(1/A) P_{2p}} P_{2p} \frac{1}{A} \tag{33}$$

where P_{2p} is the complement of Q_{2p} (see Eq. (15)) and A is a matrix contained in the $(P_{2p} + Q_{2p})$ space. Using the above identity and the fact that the solution of Eq. (16) can be written as

$$G_T(\omega) = V + VQ_{2p} \frac{1}{Q_{2p}(\omega - T(2p) - V) Q_{2p}} Q_{2p} V \tag{34}$$

we have readily

$$G_T(\omega) = G_{TF}(\omega) + \Delta G(\omega) \tag{35}$$

with

$$\begin{aligned} \Delta G(\omega) &= -V \frac{1}{A} P_{2p} \frac{1}{P_{2p}(1/A) P_{2p}} P_{2p} \frac{1}{A} V \\ A &\equiv \omega - T(2p) - V. \end{aligned} \tag{36}$$

An equivalent and probably more convenient expression for $\Delta G(\omega)$ is

$$\begin{aligned} \Delta G(\omega) &= -G_{TF}(\omega) \frac{1}{e} P_{2p} \frac{1}{P_{2p}[(1/e) + (1/e) G_{TF}(1/e)] P_{2p}} P_{2p} \frac{1}{e} G_{TF}(\omega) \\ e &\equiv \omega - T(2p). \end{aligned} \tag{37}$$

In the above, G_{TF} is the free reaction matrix defined in Eq. (31).

Can we calculate $G_T(\omega)$ as given by Eqs. (35) and (36) or by Eqs. (35) and (37) in a convenient way ? The answer is yes, provided that we make a finite-n_{III} approximation. The free reaction matrix G_{TF} can be conveniently calculated by either solving the defect-wave function differential equation [3] in r-space or solving the integral equation for G_{TF} using a momentum-space matrix-inversion method [29] which will be discussed in more detail later (see Section 4.1). With the aid of the momentum-space matrix-inversion method, $\Delta G(\omega)$ can be calculated conveniently using either Eq. (36) or Eq. (37). Tsai [20] has calculated $\Delta G(\omega)$ by solving Eq. (36). In the present work, we calculate $\Delta G(\omega)$ by solving Eq. (37). Our judgement is that Eq. (37) is more suitable for numerical computation. (See discussion in Section 4.1.)

[2] Becker [19] has pointed and that this identity could be obtained in a different way.

It is readily seen that Eq. (37) can be calculated by means of simple matrix operations. Defining

$$\begin{aligned} G_L &\equiv G_{TF}(1/e) \\ G_R &\equiv (1/e)\, G_{TF} \\ G_{LR} &\equiv (1/e) + (1/e)\, G_{TF}(1/e) \end{aligned} \tag{38}$$

we rewrite Eq. (37), in detail, as

$$\begin{aligned} \langle abJT \mid G_T(\omega) \mid cdJT\rangle &= \langle abJT \mid G_{TF}(\omega) \mid cdJT\rangle - \sum_{\substack{e,f,g,h \\ \epsilon P_{2p}}} \langle abJT \mid G_L \mid efJT\rangle \\ &\times \langle efJT \mid G_{LR}^{-1} \mid ghJT\rangle\langle ghJT \mid G_R \mid cdJT\rangle \end{aligned} \tag{39}$$

where we note that G_{LR}^{-1} is defined as the inverse of the P_{2p} part of G_{LR} in the P_{2p} space. To be specific, we have

$$G_{LR}^{-1} \equiv P_{2p} \frac{1}{P_{2p} G_{LR} P_{2p}} P_{2p}\,. \tag{40}$$

We see now why a finite-n_{III} approximation is necessary. If n_{III} is finite, the domain of P_{2p} is finite (see Fig. 12). Then $p_{2p}G_{LR}P_{2p}$ is a finite matrix and G_{LR}^{-1} can be calculated with ease. We do not know how to calculate G_{LR}^{-1} if n_{III} is infinite. A finite n_{III} will be used in this work. As shown in Section 4.2, we will investigate the reliability of this approximation by calculating ΔG using a sequence of n_{III} values.

Our calculation of G_T proceeds as follows. We first calculate $G_{TF}(\omega)$ using the momentum-space matrix-inversion methods [29]. This gives us $\langle KLkl, S\mathcal{J}TJ \times \mid G_{TF}(\omega) \mid KLk'l', S\mathcal{J}TJ\rangle$ where K and k are the center-of-mass and relative momentum, respectively. It is now clear why the momentum-space method is convenient. Since $\mid K, k\rangle$ is an eigenstate of $e \equiv \omega - T(2p)$, the matrix elements of G_L, G_R and G_{LR} are readily obtained:

$$\begin{aligned} \langle KLkl \mid G_L \mid KLk'l'\rangle &= \langle KLkl \mid G_{TF}(\omega) \mid KLk'l'\rangle \frac{1}{\omega - (K^2/4) - k'^2} \\ \langle KLkl \mid G_R \mid KLk'l'\rangle &= \frac{1}{\omega - (K^2/4) - k^2} \langle KLkl \mid G_{TF}(\omega) \mid KLk'l'\rangle \\ \langle KLkl \mid G_{LR} \mid KLk'l'\rangle &= \delta_{kk'}\delta_{ll'} \frac{1}{\omega - (K^2/4) - k^2} + \frac{1}{\omega - (K^2/4) - k^2} \\ &\times \langle KLkl \mid G_{TF}(\omega) \mid KLk'l'\rangle \frac{1}{\omega - (K^2/4) - k'^2}\,. \end{aligned} \tag{41}$$

Here note that the common quantum numbers S, $\mathcal{J}$, T, and J have been suppressed. Using the above momentum-space matrix elements, the reduced matrix elements introduced in Eq. (19) for G_{TF}, G_L, G_R and G_{LR} are obtained by a simple transformation. For example

$$\langle NLnl \mid G_{TF} \mid N'Ln'l' \rangle = (2/\pi) \int_0^\infty dK\, P_{NL}(K)\, P_{N'L}(K) \int_0^\infty dk\, dk'\, kk' \times P_{n'l'}(k')\, P_{nl}(k) \langle KLkl \mid G_{TF} \mid KLk\, l' \rangle \tag{42}$$

where the P's are the momentum-space harmonic oscillator wave functions given by

$$P_{nl}(k) = (2/\pi)^{1/2}\, k \int_0^\infty j_l(kr)\, R_{nl}(r)\, r\, dr = (-1)^n\, R_{nl}(k, \nu') \tag{43}$$

with $\nu' = \hbar/\mu\Omega$. Here μ is the two-nucleon reduced mass. The expression for $P_{NL}(K)$ is identical in form to Eq. (43) and $P_{NL}(K)$ is just $(-1)^N R_{NL}(K, \nu'')$ with $\nu'' = \hbar/M_c\Omega$ where M_c is the two-nucleon total mass. Here R_{nl} is the r-space oscillator wave function. In this work, the above momentum integrals are carried out numerically, using the Gauss integration method. In this way, we obtain all the

$$\langle N \cdots | G_{TF} | N' \cdots \rangle, \quad \langle N \cdots | G_L | N' \cdots \rangle, \quad \langle N \cdots | G_R | N' \cdots \rangle, \quad \text{and} \quad \langle N \cdots | G_{LR} | N' \cdots \rangle$$

reduced matrix elements. Substituting them into Eq. (19), we obtain the matrix elements of G_{TF}, G_L, G_R, and G_{LR} in the $\mid abJT \rangle$ representation. With these matrix elements, it is simple to calculate the $\langle abJT \mid G_T(\omega) \mid cdJT \rangle$ matrix elements using Eq. (39).

It is clear that the above procedure for the computation of G_T is indeed a simple one. Once the momentum-space matrix elements of $G_{TF}(\omega)$ are obtained, G_T is obtained by straightforward numerical integrations and matrix operations. As we mentioned earlier, the only approximation which is necessary in our method is the finite-n_{III} approximation.

A noteworthy feature of the present method is that G_T is equal to the simple sum of G_{TF} and ΔG as seen in Eq. (35). This states that G_T—the reaction matrix with orthogonalized plane-wave intermediate states—differs from the free reaction matrix G_{TF} by an additive term (i.e., ΔG). For different nuclei, we will have different Q_{2p}. For example, an appropriate Q_{2p} for ^{42}Ca may have $(n_{\text{I}}, n_{\text{II}}, n_{\text{III}}) = (6, 10, 28)$. Thus, ΔG is nucleus dependent while G_{TF} is not nucleus dependent. This suggests that the nucleon–nucleon effective interaction in a nucleus may be a sum of two parts; one is nucleus independent and the other is nucleus dependent. In fact a number of empirical effective interactions [18] do have this structure.

It should be pointed out that $G_T(\omega)$ may have singularities. We calculate $G_T(\omega)$ in terms of $G_{TF}(\omega)$. From Eq. (31), we see that $G_{TF}(\omega)$ has a singularity at ω equal to the deuteron ground state energy $E_D(-2.22$ MeV). It is not clear where are the singularities of $G_T(\omega)$. From Eq. (34), these singularities are formally located at energies where the determinant of $Q_{2p}(\omega - T(2p) - V)\, Q_{2p}$ vanishes. But since the domain of Q_{2p} is infinitely large, it is difficult in practice to locate these singularities numerically. We calculate G_T in terms of G_{TF}. Hence, for $\omega < E_D$, G_T is likely to be a non-singular quantity as seen in Eqs. (35) and (37). Fortunately ω is usually more than several MeV more negative than E_D for most nuclear structure calculations. For these energies, $G_T(\omega)$ should be well behaved.

4. Results

4.1 *Momentum-Space Calculations*

Our calculation of G_T employs the momentum-space matrix inversion method to evaluate the k-space matrix elements of G_{TF}. As seen in Eqs. (39), (40), and (41), the k-space matrix elementx of G_{TF} are needed for the calculation of G_T. And the calculation of G_L, G_R and G_{LR} is made simple if these matrix elements are known. Hence to calculate G_T accurately, it is essential to calculate G_{TF} accurately. In this section we investigate the accuracy of our G_{TF} calculation, together with some calculational details.

The momentum-space matrix inversion method has been used in a number of nuclear structure calculations [29–31]. Here, we shall only present those features of this method which are necessary for the discussion of our results. Consider an uncoupled partial wave l. Using the notation

$$G_{ll\alpha}(kk'K, \omega) \equiv \langle KLkl, S\mathcal{J}TJ \mid G_{TF}(\omega) \mid KLk'l, S\mathcal{J}TJ \rangle \tag{44}$$

where α represents the quantum numbers not explicitly shown, we have [29],

$$G_{ll\alpha}(kk'K, \omega) = V_{ll\alpha}(kk') - \frac{2}{\pi}\int_0^\infty q^2\, dq\, \frac{V_{ll\alpha}(kq)\, G_{ll\alpha}(qk'K, \omega)}{-\omega + (K^2\hbar^2/2M_c) + (q^2\hbar^2/2\mu)} \tag{45}$$

and

$$V_{ll\alpha}(kk') \equiv (2/\pi)\langle j_l(kr)\, Y(lS\mathcal{J}) \mid V \mid j_l(k'r)\, Y(lS\mathcal{J})\rangle.^3 \tag{46}$$

Here $Y(lS\mathcal{J})$ is the angular-momentum (relative) and spin wave functions with $\mathbf{l} + \mathbf{S} = \boldsymbol{\mathcal{J}}$. Note that $G_{ll\alpha}$ is diagonal in K and L.

[3] In this work we use the Reid soft-core potential throughout. The matrix elements $V_{ll'\alpha}(kk')$ are calculated by numerical integration. These matrix elements can also be calculated analytically using the analytical formulas for $V_{ll'\alpha}(kk')$ for the Reid soft-core potential given by Tsai (unpublished) or Haftel and Tabakin [31], for example. In fact we have used the analytical formulas of Tsai to check the accuracy of our numerical integrations.

Equation (45) is of the general form

$$X(k, k') = I(k, k') + \int_0^\infty K(k, q)\, X(q, k')\, dq \tag{47}$$

and we want to solve for $X(k, k')$. Defining x by

$$q = S_n + \tan((1 + x)\pi/4), \qquad -1 \leqslant x \leqslant 1, \tag{48}$$

where S_n is finite and positive, we rewrite Eq. (47) as

$$X(k, k') = I(k, k') + \sum_{n=1}^{N} \int_{S_{n-1}}^{S_n} K(k, q)\, X(q, k')\, dq + \int_{-1}^{1} F(k, x)\, X(x, k')\, dx \tag{49}$$

where $S_0 = 0$ and $S_n < S_{n+1}$. Each of the above integrals is converted into a finite sum using the Gauss integration formula [32], namely,

$$\int_{S_{n-1}}^{S_n} K(k, q)\, X(q, k')\, dq = \sum_{i=1}^{N_s(n)} K(k, q_i)\, X(q_i, k')\, w(q_i) \tag{50}$$

and

$$\int_{-1}^{1} F(k, x)\, X(x, k')\, dx = \sum_{j=1}^{N_t} F(k, x_j)\, X(x_j, k')\, w(x_j) \tag{51}$$

where q_i and $w(q_i)$ are the Gauss mesh points and weights, respectively, and so are x_j and $w(x_j)$. Furthermore, $N_s(n)$ and N_t are the number of Gauss mesh points used in each integration interval. The total number of these mesh points is

$$N_{\text{tot}} = \sum_{i=1}^{N} N_s(i) + N_t\,. \tag{52}$$

Denoting the k-space values of these mesh points as k_i, $i = 1, 2, \ldots N_{\text{tot}}$, Eq. (47) can be written as a matrix equation

$$X(k_i, k_j) = I(k_i, k_j) + \sum_{l=1}^{N_{\text{tot}}} K(k_i, k_l)\, X(k_l, k_j)\, w(k_l) \tag{53}$$

which can be readily solved to yield $X(K_i, k_j)$, i.e., the values of X on the mesh points. Note that for $l > (N_{\text{tot}} - N_t)$, we have made a mapping of q into x (see Eq. (49)). Hence in Eq. (53), the weight factor $w(k_l)$ for $l > (N_{\text{tot}} - N_t)$ contains all the necessary factors for transforming x back into k. Equation (35) is thus converted into a matrix equation of dimension N_{tot} and we can easily obtain $G_{ll\alpha}(k_ik_jK, \omega)$, i.e., the values of $G_{ll\alpha}$ on the k-space mesh points. For coupled partial waves such as the ${}^3S_1 - {}^3D_1$ channel, the integral equation for $G_{ll'\alpha}(kk'K, \omega)$

is converted into a matrix equation in a similar way. Note that the matrix equation to be solved now has a dimension $4N_{tot}^2$ if N_{tot} k-space Gauss mesh points are used. Thus $G_{ll'\alpha}(k_i k_j K, \omega)$ is also obtained readily by solving matrix equation of finite dimensions.

The underlying idea of the above k-space matrix-equation method is to consider $G_{ll'\alpha}(kk'K, \omega)$ as a smooth and well-behaved function of k and k'. Then it is sufficient to know the values of $G_{ll'\alpha}$ on the k-space mesh points and use interpolation methods to determine the values of $G_{ll'\alpha}$ for k-values between these mesh points. An obvious question to ask is how many Gauss points should one use and how should they be distributed. To answer this question, we have used an empirical approach [29–31] where the number and distribution of the k-space Gauss points are varied until stable final results are obtained.

We have made extensive calculations of the k-space matrix elements of G_{TF} using various numbers and distributions of the k-space Gauss points. In Table I, we show some representative results for the reduced matrix element $\langle NLnl \mid G_{TF} \mid NLnl\rangle$ of Eq. (42). Note that the k, k', and K integrations of Eq. (42) are also carried out using Gauss integration formulas and the same set of Gauss points used in the solution of Eq. (45). As shown by the table, satisfactory stability for the calculation of these reduced matrix elements has clearly been reached. The following observations are worth mentioning. Because we use the Reid soft-core potential which has a very strong short-range repulsion for $r \lesssim 0.5$ fm, the intermediate states in the vicinity of $k \approx 4\ \text{fm}^{-1}$ ($\sim \pi/(2r_c)$, $r_c \approx 0.5$ fm) are expected to be important. Hence, we should place enough Gauss points near $k \approx 4\ \text{fm}^{-1}$. Since the two-nucleon wave functions have very small k-space components beyond $k \approx 3\ \text{fm}^{-1}$

TABLE I

Reduced Matrix Elements $\langle NL\, nl \mid G_{TF}(\omega) \mid NL\, nl\rangle$ for the $^1S_0$ Channel Calculated with Different Sets of Gauss Points[a]

S_0	S_1	S_2	S_3	$N_s(1)$	$N_s(2)$	$N_s(3)$	N_t	N_{tot}	$n = 0$	$n = 2$
0	2.6	10.0	20.0	18	12	4	1	35	−9.08578	−2.88163 (MeV)
0	2.6	5.0	18.0	20	8	20	1	49	−9.08573	−2.88161
0	2.6	10.0	25.0	20	16	12	0	48	−9.08573	−2.88161
0	2.6	10.0	25.0	16	16	13	0	45	−9.08573	−2.88161
0	2.6	10.0	18.0	16	14	4	1	35	−9.08573	−2.88161

[a] Here $N = L = l = 0$, $\hbar\Omega = 14$ MeV and $\omega = -10$ MeV. The intervals of integrations are denoted by S_n (in fm^{-1}) and the number of Gauss points in each interval are specified by $N_s(n)$ and N_t as explained in Eqs. (49) to (51).

and there are rapid oscillations of $P_{nl}(k)$ of Eq. (43) for large n, we should place enough Gauss points in the region of $k \lesssim 3\,\text{fm}^{-1}$. These arrangements are clearly observed in Table I. We have found that the contribution from very high momentum components is not very important. This is shown by rows 3 and 5 of Table I where contributions from momentum components higher than $25\,\text{fm}^{-1}$ are ignored. Results so obtained are clearly in good agreement with those where higher momentum components are included. The majority of the calculations reported in this paper is done using the set of Gauss points shown in row 5 of Table I.

To summarize, we point out that the k-space matrix elements of G_{TF} can be calculated very accurately. As shown by Table I, we estimate the fractional error of G_{TF} to be less than 10^{-4}. As seen in Eq. (41), the matrix elements of G_L, G_R, and G_{LR} are related to those of G_{TF} in a simple way. Hence, they also can be calculated very accurately. We have used the k-space matrix inversion method [29] to compute phase shifts of simple scattering potentials and compared with those obtained by solving differential equations in the r-space. It is found that highly satisfactory results (i.e., a fractional error of less than 10^{-3}) can be obtained by using merely 10 Gauss points. This lends further support to the reliability of the k-space matrix inversion method.

If V is a hard-core potential such as the Hamada-Johnston potential [33], we can not obtain the G_{TF} matrix elements using the k-space matrix-inversion method outlined above. The reason is that $V_{ll'\alpha}(kq)$ of Eq. (46) becomes infinite for hard-core potentials. This, however, is not a real difficulty. For hard core potentials, we can obtain the k-space matrix elements of G_{TF} by solving the defect-wave function differential equation [3] where hard cores are taken care of by r-space boundary conditions for the correlated wavefunctions. Then ΔG expressed in terms of G_{TF} as given by Eq. (37) can still be readily calculated, when V is a hard-core potential. But if we use Eq. (36) where ΔG is expressed directly in terms of V, then we can not calculate ΔG for hard-core potentials. For this reason and the fact that $G_{TF}(k, k')$ is found to be more localized in k-space than $V(k, k')$ and thus is more suitable for numerical calculation, we feel that Eq. (37) is more suitable than Eq. (36) for ΔG calculations.

4.2 *Treatment of the Exclusion Operator* Q_{2p}

After obtaining the reduced matrix elements of G_{TF}, G_L, G_R, and G_{LR} using methods described in the preceeding section, the matrix elements of these operators in the $|\,abJT\rangle$ representation are readily obtained by way of Eq. (19) and similar relations for G_L, G_R, and G_{LR}. Substituting these matrix elements into Eq. (39), we can calculate the $\langle abJT\,|\,G_T(\omega)\,|\,cdJT\rangle$ matrix elements, provided we use a finite-n_{III} approximation (see Section 3 and Fig. 12). As discussed earlier this is the only necessary approximation in our $G_T(\omega)$ calculation.

The following numerical procedure is used in assessing the reliability of the finite-n_{III} approximation. For fixed n_I and n_{II}, we have calculated $G_T(\omega)$ with Eq. (39) using a rather wide range of values for n_{III}. In Table II we show typical dependence of the $T = 1$ $\langle abJT | G_T(\omega) | cdJT \rangle$ matrix elements on n_{III}. After $n_{III} = 36$, we see that these matrix elements change very little with further increase of n_{III}. Generally speaking, we have found that the $T = 1$ $G_T(\omega)$ matrix elements are satisfactorily stable with n_{III} variation when n_{III} is sufficiently large. Hence, for sufficiently large n_{III}, the finite-n_{III} approximation should be reliable. The largest value of n_{III} which we have used is 78 (see Fig. 1) as shown in Table II. Computational difficulty has prevented us from using larger n_{III} values.

TABLE II

Dependence of $\langle abJT | G_T(\omega) | cdJT \rangle$ on n_{III}[a]

T	J	a	b	c	d	$n_{III} = 6$	10	15	21	36	55	78
1	0	4	4	4	4	−1.576	−1.558	−1.549	−1.533	−1.532	−1.531	−1.530 (MeV)
				5	5	−0.803	−0.775	−0.790	−0.779	−0.778	−0.777	−0.777
				6	6	−3.279	−3.267	−3.260	−3.247	−3.245	−3.244	−3.244
		5	5	5	5	−2.324	−2.276	−2.249	−2.241	−2.224	−2.221	−2.221
				6	6	−0.656	−0.632	−0.645	−0.636	−0.635	−0.634	−0.634
		6	6	6	6	−0.237	−0.224	−0.218	−0.207	−0.207	−0.207	−0.206
0	1	4	4	4	4	−0.707	−0.561	−0.558	−0.550	−0.540	−0.533	−0.529
				5	5	−0.475	−0.405	−0.412	−0.408	−0.416	−0.427	−0.428
				6	6	2.208	2.049	2.049	2.047	2.039	2.032	2.026
		5	5	5	5	−3.397	−3.264	−3.209	−3.173	−3.088	−3.015	−3.008
				6	6	−0.440	−0.503	−0.522	−0.517	−0.519	−0.508	−0.504
		6	6	6	6	−0.426	−0.216	−0.208	−0.192	−0.178	−0.169	−0.160

[a] These matrix elements are calculated with $\omega = -10$ MeV, $\hbar\Omega = 14$ MeV and $(n_I, n_{II}) = (3, 6)$. The orbit labels of Fig. 1 are used for specifying a, b, c, d and (n_I, n_{II}, n_{III}).

The $T = 0$ $\langle abJT | G(\omega) | cdJT \rangle$ matrix elements show a stronger dependence on n_{III}, as illustrated in the lower half of Table II. Hence, for the $T = 0$ matrix elements we may need to use a larger n_{III} to ensure good reliability of the finite-n_{III} approximation than for the $T = 1$ matrix elements. Although there still appears to be some small oscillations of the $T = 0$ matrix elements as n_{III} is varied, even in the vicinity of $n_{III} \approx 55$, we do expect, however, these matrix elements to become stable as n_{III} is further increased. This is because of considerations of the approximate conservation of the center-of-mass momentum. Suppose we use a (3, 6, 21)

Q_{2p} to calculate $G_T(\omega)$. Then this Q_{2p} properly excludes diagrams like diagram (b) of Fig. 17 from $G_T(\omega)$. But this Q_{2p} fails to exclude diagrams like diagram (a) of Fig. 17. Hence, the error caused by using a (3, 6, 21) Q_{2p} is the inclusion in $G_T(\omega)$ of diagrams where *one* intermediate particle line is in a very high single-particle state while the other intermediate particle line remains in a low-lying single particle state, as illustrated by diagram (a) of Fig. 17. Low-lying single-particle states have momentum components mainly in the vicinity of k_F. Hence, lines 3 and 4 in diagram (a) all have momentum components mainly around k_F. This means, because of conservation of the center-of-mass momentum, the "railed" intermediate particle line cannot have a very large momentum. The above reasoning indicates that the finite-n_{III} approximation is reliable as long as n_{III} is as large as $\sim 3k_F$. There is of course ambiguity in the meaning of k_F for finite nuclei. But it is clear that the ladder diagrams of $G_T(\omega)$ where *only one* intermediate particle line is very high in energy while the other remaining in a low-lying state are not important. Hence, in principle, the finite-n_{III} approximation is expected to be reliable when a sufficiently large n_{III} is used.

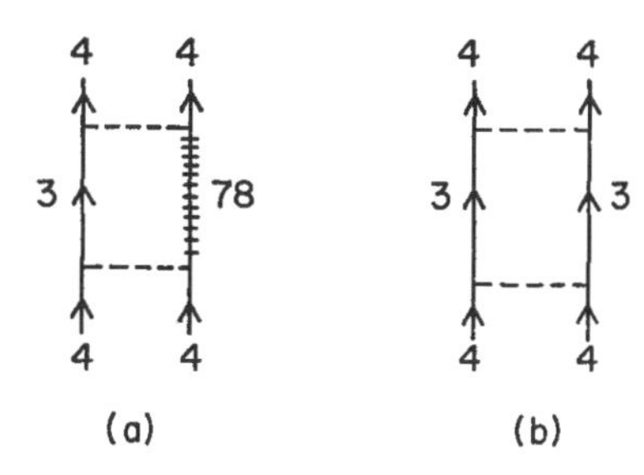

FIG. 17. Diagrams related to the finite-n_{III} approximation for $G_T(\omega)$. The orbit numbers of Fig. 1 are used in labeling the fermion lines.

Limitations of the applicability of our present method for the $G_T(\omega)$ calculation is now obvious. In solving for $G_T(\omega)$ from Eq. (39), we must use a finite-n_{III} approximation. As just discussed, we need to use a sufficiently large n_{III} to ensure the reliability of this approximation. The number of matrix elements of G_{TF}, G_L, G_R, and G_{LR} required for the calculation of Eq. (39) increases rapidly as n_{III} increases. Hence, in practice one simply can not afford to use a very large n_{III}. In fact we have used a (3, 10, 21) Q_{2p} to generate all the $G_T(\omega)$ matrix elements for orbits 1 to 10 (see Fig. 1) for five values for ω. (These results will be discussed in section 4.5.) The error associated with a use of $n_{\text{III}} = 21$ is estimated to be less than 5% for the $T = 0$ matrix elements and even smaller for the $T = 1$ matrix elements, as indicated by Table II. For "light" nuclei such as ^{18}O or ^{42}Ca, it is still feasible to use a n_{III} somewhat larger than the ones used in the present work (see Table II). Hence, for light nuclei the present method for the $G_T(\omega)$ calculation can be satisfactorily applied. But for heavy nuclei such as ^{208}Pb, there may be computational difficulty in applying the present method. This is because the number of matrix elements of G_{TF}, G_L, G_R, and G_{LR} required for the calculation of $G_T(\omega)$ Eq. (39) would be immensely large for heavy nuclei. For example, one may use a (15, 28, 78) Q_{2p} for Pb calculations, and for such a Q_{2p} we need a very large number of G_{TF}, G_L, G_R, and G_{LR} matrix elements. It is clear that we need a more convenient method for calculating $G_T(\omega)$ for heavy nuclei.

Some properties of the reduced matrix elements of $G_T(\omega)$ are worth noting. By transforming the bra and ket vectors $\langle abJT \mid$ and $\mid cdJT\rangle$ of Eq. (39) into center-of-mass and relative wave functions $\langle NLnl \mid$ and $\mid N'L'n'l'\rangle$, the reduced matrix elements $\langle NLnl \mid G_T(\omega) \mid N'L'n'l'\rangle$ have been calculated. Note that our method allows us to compute the NL nondiagonal reduced matrix elements of $G_T(\omega)$ exactly except for the finite-n_{III} approximation. Typical results are shown in Table III.

TABLE III

Dependence of Reduced Matrix Elements $\langle nlNL \mid G_T(\omega) \mid nlNL\rangle$ on n_{III} [a]

J	n	l	N	L	$G_T(\omega)$				
					(3, 6, 6)	(3, 6, 10)	(3, 6, 15)	(3, 6, 21)	$G_{TF}(\omega)$
1	0	0	0	0	−9.96	−9.33	−9.20	−9.03	−19.25 (MeV)
			0	2	−10.75	−10.04	−9.56	−9.21	−14.32
			1	0	−10.24	−10.04	−9.60	−9.45	−15.91
			1	2	−11.33	−10.26	−10.17	−9.92	−13.12
			2	0	−11.46	−10.34	−10.25	−10.03	−14.11
			2	2	−11.76	−11.45	−11.41	−11.18	−12.21
			3	0	−12.10	−11.69	−11.64	−11.37	−12.89
3	0	0	0	2	−10.71	−10.10	−9.56	−9.22	−14.32

[a] These matrix elements are calculated for $\omega = -10$ MeV, $\hbar\Omega = 14$ MeV and the $^3S_1 - {}^3S_1$ ($T = 0$) channel. The last column gives the corresponding reduced matrix elements of $G_{TF}(\omega)$. Numbers in parentheses are n_{I}, n_{II}, and n_{III}.

As shown, these reduced matrix elements have a clear dependence on center-of-mass quantum numbers N and L. Generally speaking, matrix elements become more attractive as (N, L) increases. This agrees with the familiar density dependence of the effective nucleon–nucleon interactions in nuclei. For small (N, L) the pair of interacting nucleons has a larger probability of being inside the nucleus than for large (N, L). Hence, the presence of other nucleons in the nucleus has a larger "blocking" effect on a pair of nucleons in a small (N, L) state than in a large (N, L) state. Our results indicates clearly that this blocking effect makes the reaction-matrix interaction *less* attractive. In fact, this blocking effect is rather large. Consider the first row of Table III. G_{TF} is the case where there is no blocking at all. Its reduced matrix element is -19.25 MeV. The corresponding matrix element for G_T with a (3, 6, 6) Q_{2p} becomes -9.96 Mev. Hence, we see that ΔG of Eq. (35) is in fact quite large (note $G_T = G_{TF} + \Delta G$). The blocking effect is represented by ΔG and becomes less important as (N, L) becomes large, as indicated by Table III.

From Table III we note that the change in G (i.e., $G_T - G_{TF}$) between different $(n_{\text{I}}, n_{\text{II}}, n_{\text{III}})$ values is much smaller than the difference between G_{TF} and the (3, 6, 6) G_T. For example the former is about 0.9 MeV between (3, 6, 6) and (3, 6, 21) while the latter is about 9 MeV, for the first row of the table. This is consistent with our earlier observation that intermediate states with only *one* particle in a very high single-particle state and the other particle in a low-lying occupied single-particle state are *not* important for $G_T(\omega)$ calculations. In other words, the wing regions ($a \gg n_{\text{II}}$ and $b \leqslant n_{\text{I}}$, and $b \gg n_{\text{II}}$ and $a \leqslant n_{\text{I}}$) of Fig. 12 are not important for G_T calculations, lending support to the finite-n_{III} approximation. From Table III we note also the J dependence of the reduced matrix elements of G_T. In general, ΔG becomes small when J becomes rather large. This is because for rather large J's the number of two-particle states contained in P_{2p} which can couple to such large J values is usually small, compared with that with smaller J values.

As mentioned earlier, $G_T(\omega)$ is *not* diagonal in (N, L). We have, however, found that the (N, L)-nondiagonal reduced matrix elements of $G_T(\omega)$ are generally small. The largest such nondiagonal matrix element we have found in the present work has a magnitude of approximately 0.5 MeV. Similar observations were noted earlier by Tsai [20]. He pointed out that the (N, L) nondiagonal matrix elements were typically less than 5%, in magnitude, of the neighboring (N, L) diagonal matrix elements. This is in agreement with what we have found.

In summary, we point out that the present method treats the exclusion operator Q_{2p} in $G_T(\omega)$ "exactly" within the context of a finite-n_{III} approximation. This is an improvement over earlier methods [4, 10] where in addition to this approximation one needs to make a number of other approximations such as the angle-average approximation and the (NL)-diagonal approximation for Q_{2p}. Tsai [20] has investigated the accuracy of these two approximation schemes for Q_{2p} and found them to be rather accurate, errors being typically less than 5%. In fact the approximation method of Sauer [10] is found [20] to be highly accurate. (A comparison of typical matrix elements is given in Table VI.)

Since the present method is already quite straightforward (when a finite-n_{III} approximation is made), clearly there is no need to make further approximations.

4.3 *Energy Dependenze of* $G_T(\omega)$

In the calculation of effective nucleon–nucleon interactions in nuclei we need not only the matrix elements of $G_T(\omega)$ but also their energy derivatives. The effective interaction V_{eff} as shown in Fig. 10 can be expressed as[4]

$$V_{\text{eff}} = \hat{Q} + \frac{d\hat{Q}}{d\omega} P \hat{Q} + \frac{1}{2!}\frac{d^2\hat{Q}}{d\omega^2} P \hat{Q} P \hat{Q} + \frac{d\hat{Q}}{d\omega} P \frac{d\hat{Q}}{d\omega} P \hat{Q} + \cdots \tag{54}$$

where $\hat{Q}$ represents the $\hat{Q}$-box as explained in Fig. 11 and P is the model space projector of Eq. (2). Note that it is only in the above form where folded diagrams are expressed as products of $\hat{Q}$-boxes and their energy derivatives that all the $\hat{Q}$'s in Eq. (54) are identical. If we write V_{eff} in terms of folded diagrams, we have [18, 24]

$$V_{\text{eff}} = \hat{Q} - \hat{Q}' \int \hat{Q} + \hat{Q}' \int \hat{Q} \int \hat{Q} - \cdots \tag{55}$$

where $\hat{Q}'$ differs from $\hat{Q}$ in that $\hat{Q}'$ does not have terms first order in H_1. Thus we do not have enough diagrams in any of the $\hat{Q}'$-boxes to form a reaction matrix. Diagrams in a $\hat{Q}'$ box can only form $(G_T(\omega) - V)$, i.e., a reaction matrix without the V term. But since V is ω independent, we have

$$\frac{d^n}{d\omega^n}(G_T(\omega) - V) = \frac{d^n}{d\omega^n} G_T(\omega), \qquad n = 1, 2, 3, \ldots . \tag{56}$$

Hence, in Eq. (54) all $\hat{Q}$-boxes can be expressed in terms of reaction matrix vertices. It is now clear that in the calculation of V_{eff} we need $G_T(\omega)$ as well as their energy derivatives. Thus it is essential to investigate the energy dependence of $G_T(\omega)$.

We have calculated $G_T(\omega)$ for five values of ω, i.e., $\omega = -5, -10, -20, -40$, and -85 MeV. For most nuclear structure calculations, the energy variable ω is typically within the above range. In Fig. 18, we show some typical energy dependence of $G_T(\omega)$. The following points are worth noting. It is seen that $G_T(\omega)$ varies quite smoothly with ω. This is very convenient in enabling us just to calculate $G_T(\omega)$ for a few energies and determine $G_T(\omega)$ for other energies by interpolation. In fact $G_T(\omega)$ varies with ω almost linearly and with a rather small "slope." The $T = 0$ matrix elements are found to have, in general, a stronger dependence on ω than the $T = 1$ matrix elements. Within the energy range shown in Fig. 18, the

[4] Here we utilize the folded-diagram expansion for the effective interaction, as formulated in [18]. For the present discussion, detailed knowledge of this expansion is not required; it will suffice to note that folded diagrams may be expressed in terms of energy derivatives of the various non-folded diagrams (contained in the $\hat{Q}$-box). Thus, it is of interest to know the energy derivatives of the reaction matrix elements and hence their energy dependence.

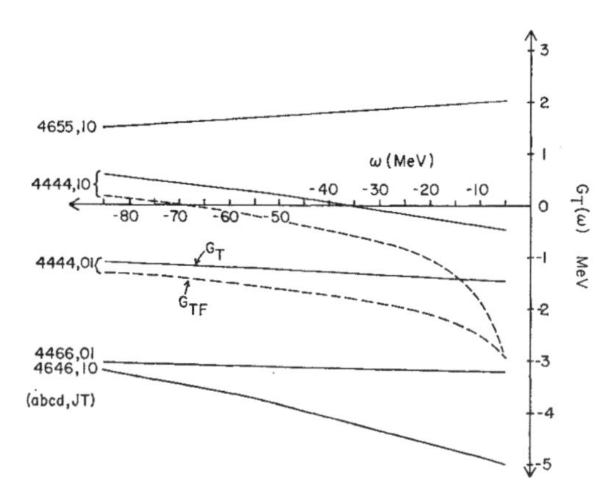

FIG. 18. Typical energy dependence of $\langle abJT | G_T(\omega) | cdJT \rangle$ calculated with $(n_{\rm I}, n_{\rm II}, n_{\rm III}) = (3, 10, 21)$ and $\hbar\Omega = 14$ MeV. The dotted-line curves are for $G_{TF}(\omega)$.

largest "slope" of the $T = 0$ curves is typically about -0.03 while it is about -0.01 for the $T = 1$ curves. Thus the dependence of $G_T(\omega)$ on ω is indeed rather weak. It is clear that for diagonal matrix elements we must have $dG_{TF}(\omega)/d\omega$ less than or equal to zero. As shown by Eq. (34), we should also have $dG_T(\omega)/d\omega \leqslant 0$ for diagonal matrix elements. This is seen from the lower four curves of Fig. 18. But for off-diagonal matrix elements, we may of course have $dG_T(\omega)/d\omega > 0$, as shown by the top curve of Fig. 18.

The energy dependence of $G_{TF}(\omega)$ is also illustrated in Fig. 18. Clearly $G_T(\omega)$ has a weaker dependence on ω then $G_{TF}(\omega)$. $G_{TF}(\omega)$ has a singularity at the deuteron binding energy of $\omega = -2.22$ MeV, as indicated by the $T = 0$ G_{TF} curve of the figure. As mentioned earlier, we do not know where the singularities of $G_T(\omega)$ are. The curves of Fig. 18 seem to indicate that the poles of $G_T(\omega)$ may lie at energies significantly higher than $\omega = -2.22$ MeV.

4.4 *Spectrum of Low-Lying Intermediate States*

We have pointed out in section 3 that there are two types of reaction matrix, $G_T(\omega)$ and $G(\omega)$ defined respectively by Eqs. (16) and (12). The former has orthogonalized plane-wave intermediate states while the latter has harmonic-oscillator intermediate states. Consider dividing Q_{2p} into two parts

$$Q_{2p} = Q_l + Q_h \tag{57}$$

and using different spectra for Q_l, the low-lying part of Q_{2p}, and Q_h. For example, we may use a Q_l consisting of all two-particle states constructed from orbits 7 to 10 (see Fig. 1). We define now a new reaction matrix $\bar{G}_T$ where for Q_h we use orthogonalized intermediate states and for Q_l we use harmonic-oscillator intermediate states. By the operator identity of Eq. (23) we have

$$\bar{G}_T(\omega) = G_T'(\omega) + G_T'(\omega)(Q_l/(\omega - H_0))\,\bar{G}_T(\omega) \tag{58}$$

with

$$G_T'(\omega) = V + VQ_h \frac{1}{\omega - Q_h T(2p)\, Q_h} Q_h G_T'(\omega) \tag{59}$$

where $[Q_l, H_0] = 0$. By studying the difference between $G_T(\omega)$ of Eq. (16) and $\bar{G}_T(\omega)$, we can assess the effect due to the choice of the spectrum for the low-lying intermediate states. It is clear from Eq. (58) that $\bar{G}_T$ is significantly dependent on the choice of H_0. For example, H_0 and a "shifted" harmonic-oscillator Hamiltonian $H_0 + C_0$ where C_0 is a constant can give very different results for $\bar{G}_T$. The idea behind $\bar{G}_T$ is that the low-lying single-particle states should be well represented by the harmonic-oscillator states. Namely, for low-lying single-particle states the average potential generated by the nucleons in the core nucleus is adequately represented by the harmonic-oscillator potential U. Thus, as discussed in Section 2, we should use harmonic-oscillator states for low-lying intermediate states. In so doing, there is the difficulty of where the boundary between Q_l and Q_h should be placed.

We have made calculations with the following choices. First we calculate $G_T(\omega)$ using a (3, 6, 21) Q_{2p}. Then we use a Q_h with $(n_{\rm I}, n_{\rm II}, n_{\rm III}) = (3, 10, 21)$ to calculate $G_T'(\omega)$. The difference between Q_{2p} and Q_h defines, of course, Q_l. Representative matrix elements of $G_T(\omega)$ and $G_T'(\omega)$ are given in Table IV. We note with emphasis that these two sets of matrix elements differ rather little from each

TABLE IV

Matrix Elements of G_T, G_T, G_{2p}, and G_{2p}' in the $|abJT\rangle$ Representation[a]

T	J	a	b	c	d	G_T (3, 6, 21)	G_T' (3, 10, 21)	G_{2p}	G_{2p}'
1	0	4	4	4	4	−1.533	−1.445	−0.262	−0.248
		4	4	5	5	−0.779	−0.738	−0.105	−0.094
		4	4	6	6	−3.247	−3.215	−0.074	−0.054
		5	5	5	5	−2.241	−2.215	−0.063	−0.056
		5	5	6	6	−0.636	−0.602	−0.086	−0.077
		6	6	6	6	−0.209	−0.133	−0.231	−0.226

[a] Calculated with $\hbar\Omega = 14$ MeV and $\omega = -10$ MeV. Note that $\bar{G}_T$ of Eq. (58) is approximately equal to $G_T' + G_{2p}'$.

other. This means that low-lying intermediate states are *not* so important in reaction matrix calculations, being consistent with the short-range nature of the nucleon–nucleon potential. For short-range nucleon–nucleon potentials with a "core" radius of approximately 0.5 fermi, important intermediate states are located around a relative energy of about 400 MeV. A reasonable choice for C_0 in an ^{18}O calculation is to have $\epsilon_i = -5$ MeV for $i = 4$, 5, and 6 (see Fig. 1). Hence, in Table IV we have used $\omega = -10$ MeV, corresponding to two valence particles in the 4, 5, and 6 orbits.

For two valence particles in orbits 4 to 6 and a Q_l consisting of two particles in orbits 7 to 10, $\omega - H_0$ of Eq. (58) is just $-2\hbar\Omega$. We use $\hbar\Omega = 14$ MeV which is considerably larger than the $G_T{}'$ matrix elements connecting orbits 4 to 6 to orbits 7 to 10. Hence, we can calculate $\bar{G}_T$ by perturbation method, i.e.,

$$\bar{G}_T(\omega) \approx G_T{}'(\omega) + G'_{2p}(\omega) \tag{60}$$

with

$$G'_{2p}(\omega) = (-1/2\hbar\Omega)\, G_T{}'(\omega)\, Q_l G_T{}'(\omega). \tag{61}$$

Note that G'_{2p} is just diagram (ii) of Fig. 13 where each reaction matrix vertex is a $G_T{}'(\omega)$ vertex of Eq. (59), and the intermediate states x and y are restricted to orbits 7 to 10. Typical results for G'_{2p} are shown in Table IV. Note that the $\bar{G}_T$ (i.e., $G_T{}' + G'_{2p}$) matrix elements are approximately 10 % larger in magnitude than G_T. Since we have used a (3, 6, 21) Q_{2p} for G_T and $Q_{2p} = Q_l + Q_h$ for G_T, the above difference is entirely due to the use of different spectra for Q_l.

There have been many discussions [26] about the G'_{2p} diagram. In the calculation of G'_{2p}, we should use $G_T{}'$ calculated with Q_h, as is done in Eq. (61). If we calculate $\bar{G}_T(\omega)$ by

$$\bar{G}_T(\omega) \approx G_T(\omega) + G_{2p}(\omega) \equiv \bar{G}_D(\omega) \tag{62}$$

with

$$G_{2p}(\omega) = -(1/2\hbar\Omega)\, G_T(\omega)\, Q_l G_T(\omega) \tag{63}$$

then we have a double-counting error in $\bar{G}_D(\omega)$. This is because the intermediate states belonging to Q_l are included twice, once in G_T which is calculated with a (3, 6, 21) Q_{2p}, and once in G_{2p}. This double-counting error may be defined as

$$\Delta \equiv (\langle\bar{G}_T\rangle - \langle\bar{G}_D\rangle)/\langle\bar{G}_T\rangle \tag{64}$$

where $\langle\bar{G}_T\rangle$ and $\langle\bar{G}_D\rangle$ represent the matrix elements of $\bar{G}_T$ and $\bar{G}_D$. As shown in Table IV, this double-counting error Δ is typically 5 %. Using different reaction matrices, Barrett and Kirson [26] have calculated G_{2p} and G'_{2p} and obtained values fairly close to those given in Table IV. They used, however, a different definition for Δ and hence obtained a much larger result for the so-called double-counting error. In fact, the G_{2p} calculated by Kuo and Brown [3] using reaction matrices calculated with a combination of the separation method and the reference spectrum method also has values fairly close to those given in Table IV. For example, for the (4444, 10) case their [3] value is -0.28 MeV compared with our present values of -0.262 for G_{2p} and -0.248 for G'_{2p}. It is clear that the calculation of G_{2p} (or G'_{2p}) does not depend sensitively on what bare reaction matrices we use. As long as we use the same spectrum for Q_l (which implies for the present case an energy denominator of $2\hbar\Omega$), G'_{2p} calculated with a (3, 10, 21) G_T' and G_{2p} calculated with a (3, 6, 21) G_T are very similar to each other.

In summary, we point out that $\bar{G}_T$ where a harmonic oscillator spectrum is used for Q_l can be fairly different from G_T where plane-wave intermediate states are used for both Q_l and Q_h. $\bar{G}_T$ is typically $\sim$10% larger in magnitude than G_T as shown in Table IV. Clearly $\bar{G}_T$ is dependent on where we place the boundary between Q_l and Q_h. Our choice of a (3, 10, 21) Q_h is probably a reasonable one. But it remains to be seen where is the most suitable boundary between Q_l and Q_h.

4.5 *Matrix Elements for Effective Interactions*

In this section we calculate some representative matrix elements for the effective interaction between two valence nucleons in the 1*s*-0*d* shell. As indicated by Eq. (54), matrix elements of V_{eff}, the effective interaction between two valence nucleons, can be expressed in terms of the $\hat{Q}$-boxes and their energy derivatives. The first step in the calculation of V_{eff} is the construction of the $\hat{Q}$-box. In fact, most earlier calculations of V_{eff} include only the first $\hat{Q}$-term of Eq. (54). An example is the calculation of Kuo and Brown [3]. To our knowledge, the folded diagrams (i.e., the terms after the first $\hat{Q}$-term in Eq. (54)) in V_{eff} have only been studied in a preliminary way [34, 35].

There is indeed a fairly large number of uncertainties [18] in the calculation of V_{eff}. But, on the other hand, the Kuo-Brown [3] matrix elements of V_{eff} have indeed been very successful [36] in reproducing experimental results, although a number of approximations were used in their calculations. Uncertainties in the V_{eff} calculation are mainly in the following areas: first, the accuracy of the G-matrix calculation and second, the effect of higher-order diagrams. Using the method reported in the present paper, we can now calculate the G-matrix very accurately. Hence, we feel that the first uncertainty is now by and large overcome. The second uncertainty is closely related to the basic question of convergence of the perturbation expansion of V_{eff}. Once a $\hat{Q}$-box is given, methods for summing up the $\hat{Q}$-box folded-diagram series have been studied by Krenciglowa and Kuo [37] and by Anastasio *et al.* [38]. Their results indicate that the following procedure for the calculation of V_{eff} appears to be both convenient and practical. First we construct the $\hat{Q}$-box which does *not* contain any folded diagrams. Then using their methods one calculates V_{eff} by summing up the $\hat{Q}$-box folded-diagram series. Using exactly solvable models, this procedure has been studied by Anastasio *et al.* [39, 40].

It is beyond the scope of the present paper to perform a detailed calculation of V_{eff} using the reaction matrices calculated with our present method. Such a calculation is in progress and we plan to publish our results in a forthcoming paper. In the present paper, we will just calculate several familiar diagrams which belong to the $\hat{Q}$-box term of Eq. (54). We have calculated $\hat{Q}$-box diagrams G, G_{2p}, G_{3p1h} and G_{4p2h} of Fig. 13, in close analogy to the Kuo-Brown [3] calculation. These are the four diagrams frequently considered for effective interaction calculations [3, 34]. A weak point in the Kuo-Brown [3] calculation is that the projector Q_{2p} was treated in an approximate way. Hence, it is not precisely clear what intermediate states were contained in the reaction matrix and what intermediate states should be taken care of by shell model diagonalizations. This ambiguity can now be definitey avoided because our present method for the G-matrix calculation treats Q_{2p} accurately. Based on our discussions in section 4.4, we have used a (3, 10, 21) Q_{2p} in calculating G_T. Then the G_{2p} diagram is given by G'_{2p} of section 4.4 and Table IV. Attention should be given to the diagram G_{3p1h} which represents the interaction between a pair of nucleons via core polarizations and is well known to be of essential importance in V_{eff}. Note that α, β, γ, and δ all belong to the $1s$-$0d$ shell (i.e., orbits 4 to 6). Aside from factors of angular-momentum coupling, diagram G_{3p1h} of Fig. 13 is given by

$$\langle\alpha\beta \mid G_{3p1h} \mid \gamma\delta\rangle = \sum_{p,h} \frac{\langle h\beta \mid G_T(\omega_1) \mid p\delta\rangle\langle\alpha p \mid G_T(\omega_2) \mid \gamma h\rangle}{\omega - (\epsilon_\alpha + \epsilon_\delta + \epsilon_p - \epsilon_h)} \tag{65}$$

where

$$\begin{aligned} \omega &= \epsilon_\gamma + \epsilon_\delta \\ \omega_1 &= (\epsilon_\gamma + \epsilon_\delta) - (\epsilon_\alpha - \epsilon_h) \\ \omega_2 &= (\epsilon_\gamma + \epsilon_\delta) - (\epsilon_\delta - \epsilon_h). \end{aligned} \tag{66}$$

Using harmonic-oscillator single-particle states and including only $2\hbar\Omega$ core excitations, Eq. (65) becomes

$$\langle\alpha\beta \mid G_{3p1h} \mid \gamma\delta\rangle = -(1/2\hbar\Omega) \sum_{p,h} \langle h\beta \mid G_T(\omega + \Delta) \mid p\delta\rangle\langle\alpha p \mid G_T(\omega + \Delta) \mid \alpha h\rangle \tag{67}$$

where Δ can be either $-\hbar\Omega$ or $-2\hbar\Omega$ depending on the location of h. We have used $\hbar\Omega = 14$ MeV. The value of ω depends on our choice of U in $H_0 = T + U$. If we could solve the nuclear many-body problem exactly, our result is clearly independent of our choice of U. But since we can only treat the nuclear many-body problem in an approximate way by perturbation, our result for V_{eff} will be dependent on our choice of U. It is not yet clear what is a good choice for U, although there are indications [40] that a U of the Hartree-Fock type may be helpful in facilitating the convergence of the effective interaction perturbation expansion. Since Brueckner-Hartree-Fock calculations [27] with realistic nucleon–nucleon interactions have yielded low-lying single-particle energies in close agreement with experiments, it is reasonable to choose $\omega = -10$ Mev for γ and δ in orbits 4 to 6. Since the G_T matrix does not depend on ω strongly as shown by Fig. 18 we have set $\Delta = 0$ in Eq. (67) to facilitate comparison with the Kuo–Brown matrix elements which are energy independent. Thus we have calculated diagrams G, G_{2p}, G_{3p1h}, and G_{4p2h} of Fig. 13 with $G_T(\omega)$ calculated with a fixed value of $\omega = -10$ Mev and a fixed energy denominator of $2\hbar\Omega$. The formulas we used for these diagrams are then identical to those used in [3].

In Table V we give representative values of these diagrams calculated with the above mentioned G_T matrix elements. The corresponding values calculated by Kuo and Brown [3] are also given for comparison. Since the Kuo and Brown matrix

TABLE V

Comparison of Matrix Elements[a]

T	J	a	b	c	d	G	G_{3p1h}	G_{2p}	G_{4p2h}
1	0	4	4	4	4	−1.445	−0.928	−0.248	−0.332 (MeV)
						−1.47	−1.06	−0.28	−0.254
						−1.54			
		4	4	5	5	−0.738	−0.362	−0.094	−0.040
						−0.76	−0.33	−0.11	−0.023
						−0.77			
		4	4	6	6	−3.215	−0.751	−0.054	−0.197
						−3.18	−0.93	−0.10	−0.146
						−3.33			
		5	5	5	5	−2.215	+0.106	−0.056	−0.006
						−2.28	+0.07	−0.07	−0.002
						−2.26			
		5	5	6	6	−0.602	−0.225	−0.077	−0.033
						−0.62	−0.22	−0.09	−0.019
						−0.64			
		6	6	6	6	−0.133	−0.355	−0.226	−0.252
						−0.17	−0.37	−0.25	−0.204
						−0.23			

[a] The matrix elements in the first row are calculated with the present method using a (3, 10, 21) Q_{2p}, $\hbar\Omega = 14$ MeV and $\omega = -10$ MeV. The matrix elements of Kuo and Brown [3] and Negele [41] are given in the second and third rows, respectively.

elements have been very successful in reproducing experimental nuclear properties [3, 36], it is interesting as well as encouraging that these two sets of matrix elements are in such close agreement, especially for the bare G term. It should be noted that our calculation of G_{3p1h} and G_{4p2h} are preliminary in the sense that we have neglected the ω dependence of $G_T(\omega)$ in the calculation of these diagrams. Judging from the rather weak ω-dependence of $G_T(\omega)$ shown in Fig. 18, our values of G_{3p1h} and G_{4p2h} are not expected to change very much when the ω-dependence of $G_T(\omega)$ is properly taken care of. Using a local density approximation, Negele [41] has calculated the bare G matrix elements appropriate for $A = 18$ calculations. His values are also given in Table V. It is of great interest to see that all the three sets of matrix elements are so close to each other.

As mentioned earlier, G, G_{3p1h}, G_{2p}, and G_{4p2h} are just low-order G-matrix diagrams of the $\hat{Q}$-box. To obtain V_{eff}, we need the $\hat{Q}$-box and its energy derivatives as indicated by Eq. (54). Many uncertainties [18] remain in the evaluation of V_{eff}. But we have at least made the first step in making accurate calculations of the above low-order $\hat{Q}$-box diagrams. More complete calculations of V_{eff} are in progress and will be reported in a future paper.

Effort has been made in ascertaining the numerical reliability of our calculated matrix elements. We have made two completely independent computer programs for the computation of these matrix elements. Very good agreement is obtained between results given by these two programs. We have also compared our results with those given by a third (independent) computer program constructed by Tsai [20]. As mentioned earlier, Tsai calculated $G_T(\omega)$ using Eqs. (35) and (36) while we use Eqs. (35) and (37) for its calculation. Occasional small differences ($\sim$1–2 %)

TABLE VI

Matrix Elements of G_T in MeV for the $^1S_0$ and $^3S_1$ Channels[a]

	n l N L J	$\langle nl\,NL \mid G_T(\omega) \mid nl\,NL\rangle$ Present	Tsai	Sauer
$^1S_0$	0 0 2 0 0	−6.81	−6.79	−6.76
	1 0 1 0 0	−4.57	−4.54	−4.55
	2 0 0 0 0	−2.06	−2.03	−2.05
$^3S_1$	0 0 2 0 1	−10.03	−10.00	−9.25
	1 0 1 0 1	−6.88	−6.83	−6.73
	2 0 0 0 1	−3.32	−3.28	−3.32

[a] Here J is the total angular momentum and all matrix elements are calculated with $\omega = -10$ MeV and $\hbar\Omega = 14$ MeV. The present G-matrix elements are evaluated according to Eqs. (35) and (37). The results of Tsai and Sauer are taken from [20] and [10], respectively. A (3, 6, 21) Q_{2p} is used in all three calculations.

between matrix elements of ours and of Tsai are found. A likely reason is that we have used a larger number of Gauss points in the present calculation. In Table VI we give a comparison of typical matrix elements calculated by the present method with those calculated by Tsai [20] and by Sauer [10].

5. Conclusion

Nuclear reaction matrix elements appropriate for nuclear structure studies in the $A \approx 18$ region have been calculated using the Tsai-Kuo [15] method. This method is found to be both convenient and accurate for calculations of nuclear reaction matrix elements with orthogonalized plane-wave intermediate states. In our calculations only one approximation is made, namely, the finite-$n_{\rm III}$ approximation in the treatment of the exclusion operator Q_{2p}. This is in fact the only necessary approximation. In contrast, a number of additional approximations for the treatment of Q_{2p} were necessary in previous methods for reaction matrix calculations. Since the present method is already so straightforward, there is indeed no need to make further approximations than the one of finite-$n_{\rm III}$ truncation. For the $T = 1$ matrix elements, the finite-$n_{\rm III}$ approximation is found to be highly satisfactory. It is found to be less satisfactory for the $T = 0$ matrix elements, however. Further investigation of the accuracy of the $T = 0$ finite-$n_{\rm III}$ approximation is needed. This may be done by doing $T = 0$ calculations using larger $n_{\rm III}$ values than those reported in the present work. It will be very interesting in devising a new method where the finite-$n_{\rm III}$ approximation is not necessary. We have studied the possibility of devising such a method and have not been successful.

It is of interest that close agreement is observed between our present reaction matrix elements and those calculated by Kuo and Brown [3] using approximate methods. The present calculation is in line with the Kuo-Brown calculation in the sense that both use free-particle (plane-wave) intermediate states for the reaction matrix. The present calculation treats the exclusion operator in a more accurate way. The observed close agreement between these two sets of matrix elements lends support to the approximation methods used in the Kuo–Brown calculation. Close agreement between the present matrix elements and those of Sauer [10] is also observed. It appears that now we have reached the point where bare reaction matrix elements with orthogonalized plane-wave intermediate states can be calculated conveniently and with high accuracy. Using these matrix elements, one can proceed in evaluating the effective interaction in the $A = 18$ region. Note that the present matrix elements are different from those [12] where a harmonic oscillator spectrum is used for the intermediate states of the reaction matrix.

Further investigations along the following directions may be interesting and worthwhile:

(1) In the present paper, we have only calculated the bare reaction matrix and a few low-order $\hat{Q}$-box diagrams. In order to compare with experimental results or empirical matrix elements, one may use the present reaction matrix elements to calculate matrix elements of V_{eff} as indicated by Eq. (54) or Eq. (55). Such a calculation is in progress.

(2) For heavier nuclei (such as ^{208}Pb) the calculation of reaction matrices using the present method will involve a large amount of numerical work. This is because of the large dimension of P_{2p} for heavy nuclei even when a finite-n_{III} approximation is employed. Hence, it is desirable to find a simpler treatment of Q_{2p} for heavy nuclei.

(3) The present method for nuclear reaction matrix calculation can also be used to perform accurate calculations of nuclear defect wave functions induced by the short-range nucleon–nucleon correlations. In particular, the present method enables us to study the effect of the exclusion operator Q_{2p} on these correlations. A preliminary calculation along this line has been done by Weng *et al.* [42]. A more complete investigation remains to be done.

(4) It is of great interest to perform G-matrix calculations with the exclusion operator Q_{2p} constructed in a more realistic way. In the present work, Q_{2p} is constructed in terms of harmonic oscillator wave functions. A more realistic way is to construct Q_{2p} (or P_{2p}) in terms of appropriate Wood–Saxon or Brueckner–Hartree–Fock wave functions. As indicated by Eqs. (35) and (37), the projection operator P_{2p} enters into our G-matrix calculation in a very convenient way. Hence, it appears that the present G-matrix calculation method is particularly convenient for G-matrix calculations with more realistic Q_{2p} projection operators.

Acknowledgments

The authors (especially *TTSK*) would like to thank Professor G. E. Brown, Dr. S. F. Tsai and Dr. K. F. Ratcliff for numerous helpful discussions during the long course of this work. Three of them (EMK, CLK, and TTSK) would like to thank Dr. T. Engeland for his assistance and hospitality during their stay in Oslo.

References

1. H. A. Bethe, *Ann. Rev. Nucl. Sci.* **21** (1971), 93.
2. J. F. Dawson, I. Talmi, and J. D. Walecka, *Ann. Phys. (N.Y.)* **18** (1962), 339.
3. T. T. S. Kuo and G. E. Brown, *Nucl. Phys.* **85** (1966), 40.
4. C. W. Wong, *Nucl. Phys.* **A91** (1967), 399.
5. C. M. Shakin, Y. R. Waghmare, M. Tomaselli, and M. H. Hull, *Phys. Rev.* **161** (1967), 1006.
6. R. L. Becker, A. D. MacKeller, and B. M. Morris, *Phys. Rev.* **174** (1968), 1264.

7. H. S. KÖHLER AND R. J. MCCARTHY, *Nucl. Phys.* **86** (1966), 611; R. J. MCCARTHY, *Nucl. Phys.* **A130** (1969), 305.
8. A. KALLIO AND B. D. DAY, *Nucl. Phys.* **A124** (1967), 177.
9. S. T. BUTLER, R. G. L. HEWITT, B. H. J. MCKELLAR, I. R. NICHOLLS, AND J. S. TRUELOVE, *Phys. Rev.* **186** (1969), 963.
10. P. U. SAUER, *Nucl. Phys.* **A150** (1970), 467.
11. D. GRILLOT AND H. MCMANUS, *Nucl. Phys.* **A113** (1968), 161.
12. B. R. BARRETT, R. G. L. HEWITT, AND R. J. MCCARTHY, *Phys. Rev.* **C3** (1971), 1137.
13. R. K. TRIPATHI AND P. GOLDHAMMER, *Phys. Rev.* **C6** (1972), 101.
14. J. R. DEMOS AND M. K. BANERJEE, *Phys. Rev.* **C5** (1972), 75.
15. S. F. TSAI AND T. T. S. KUO, *Phys. Lett.* **39B** (1972), 427.
16. M. BARANGER, *in* "Proc. Int. Sch. of Physics Enrico Fermi," Course XL, (M. Jean, Ed.), p. 511. Academic Press, New York, 1969.
17. B. R. BARRETT AND M. W. KIRSON, *in* "Advances in Nuclear Physics" (M. Baranger and E. Vogt, Eds.), Vol. 6, p. 219. Plenum Press, New York, 1973.
18. T. T. S. KUO, *Ann. Rev. Nucl. Sci.* **24** (1974), 101.
19. R. L. BECKER, *in* "Proc. Int. Top. Conf. on Effective Interactions and Operators in Nuclei," Tucson 1975, (B. R. Barrett, Ed.), Vol. 2, p. 96. Springer–Verlag, Berlin, 1975.
20. S. F. TSAI, Ph. D. thesis, State University of New York at Stony Brook, 1973, unpublished.
21. E. M. KRENCIGLOWA, C. L. KUNG, T. T. S. KUO, AND E. OSNES, *in* "Proc. Int. Top. Conf. on Effective Interactions and operators in Nuclei," Tucson 1975, (B. R. Barrett, Ed.), Vol. 1, p. 21. University of Arizona, Tucson, 1975.
22. R. V. REID, *Ann. Phys. (N.Y.)* **50** (1968), 411.
23. J. GOLDSTONE, *Proc. Roy. Soc.* **A239** (1957), 267.
24. T. T. S. KUO, S. Y. LEE, AND K. F. RATCLIFF, *Nucl. Phys.* **A176** (1971), 65.
25. B. H. BRANDOW, *Rev. Mod. Phys.* **39** (1967), 771.
26. B. R. BARRETT AND M. K. KIRSON, *Phys. Lett.* **55B** (1975), 129.
27. K. T. R. DAVIES, M. BARANGER, R. M. TARBURTON, AND T. T. S. KUO, *Phys. Rev.* **177** (1969), 1519.
28. H. A. BETHE, B. H. BRANDOW, AND A. G. PETSCHEK, *Phys. Rev.* **129** (1963), 225.
29. G. E. BROWN, A. D. JACKSON, AND T. T. S. KUO, *Nucl. Phys.* **A133** (1969), 481.
30. F. COESTER, S. COHEN, B. DAY, AND C. M. VINCENT, *Phys. Rev.* **C1** (1970), 769.
31. M. I. HAFTEL AND F. TABAKIN, *Nucl. Phys.* **A158** (1970), 1.
32. M. ABRAMOWITZ AND I. A. STEGUN, Eds. "Handbook of Mathematical Functions," Dover, New York, 1965.
33. T. HAMADA AND I. D. JOHNSTON, *Nucl. Phys.* **34** (1962), 382.
34. B. R. BARRETT AND M. W. KIRSON, *Nucl. Phys.* **A148** (1970), 145.
35. E. M. KRENCIGLOWA, T. T. S. KUO, E. OSNES, AND B. GIRAUD, *Phys. Lett.* **47B** (1973), 322.
36. E. C. HALBERT, J. B. MCGRORY, B. H. WILDENTHAL, AND S. P. PANDYA, *in* "Advances in Nuclear Physics," (M. Baranger and E. Vogt, Eds.), Vol. 4, p. 315. Plenum Press, New York, 1971.
37. E. M. KRENCIGLOWA AND T. T. S. KUO, *Nucl. Phys.* **A235** (1974), 171.
38. M. R. ANASTASIO, J. W. HOCKERT, AND T. T. S. KUO, *Phys. Lett.* **53B** (1974), 221.
39. M. R. ANASTASIO, T. T. S. KUO, AND J. B. MCGRORY, *Phys. Lett.* **57B** (1975), 1.
40. M. R. ANASTASIO, T. T. S. KUO, T. ENGELAND, AND E. OSNES, Hartree-Fock Self-Consistent Potential and Convergence of Linked-Valence Effective Interaction Expansion, *Nucl. Phys.* (1976), in press.
41. J. W. NEGELE, *Phys. Rev.* **C1** (1970), 1260.
42. W. T. WENG, T. T. S. KUO, AND K. F. RATCLIFF, *Phys. Lett.* **52B** (1974), 1.

1.C

Nuclear Physics A271 (1976) 109—132; © *North-Holland Publishing Co., Amsterdam*

HARTREE-FOCK SELF-CONSISTENT POTENTIAL AND LINKED-VALENCE EXPANSIONS OF THE EFFECTIVE INTERACTION †

M. R. ANASTASIO and T. T. S. KUO ††

Institute of Physics, University of Oslo, Norway

and

Department of Physics, State University of New York at Stony Brook, New York 11794, USA

and

T. ENGELAND and E. OSNES

Institute of Physics, University of Oslo, Norway

Received 6 October 1975

(Revised 20 April 1976)

Abstract: The convergence behavior of the linked-valence expansion of H_{eff} is studied within the context of an exactly solvable three-level model. All diagrams, through third order in the interaction, that contribute to H_{eff} are summed. Here H_{eff} is computed using the unperturbed and the Hartree-Fock (HF) single-particle potential. Significant differences between the convergence behavior of the perturbation expansion of H_{eff} in these two cases are observed. HF is found to be a practical and powerful procedure for constructing a good model space for effective-interaction calculations and consequently facilitates the convergence of the expansion. These differences are still evident even if the HF and the unperturbed bases are very similar.

1. Introduction

Ever since the work of Kuo and Brown [1]) showed that approximating the effective interaction by the bare G-matrix plus the core polarization gave good agreement with experimental results, much effort has been devoted to the question of the convergence of H_{eff}. Barrett and Kirson [2]) included terms through third order in the G-matrix in a calculation of H_{eff} and suggested that the expansion in powers of the G-matrix was not converging.

More recently Schucan and Weidenmüller [3, 4]) have studied the formal properties of the effective interaction as a power series in the interaction and cast doubt on the convergence of H_{eff} in cases of practical interest because of the presence of low-lying intruder states. This has stimulated further investigation of the convergence of H_{eff} expansions in simple models [5-12]) where the exact solutions are known. In at least two cases, the AHK algorithm [7]) and the Padé approximants [8-10, 12]), the H_{eff}

† Work supported in part by the US Energy Research and Development Administration under contract AT(11-1)-3001.

†† NORDITA guest professor at the University of Oslo (1974/75).

expansion is not a power series so that the formal properties of Schucan and Weidenmüller no longer apply.

The work in these last two cases has been done using a state formulation, while in realistic calculations the effective interaction is generally computed in a linked-valence formulation [16]. There are important differences between these two formulations that could affect the convergence properties of an expansion of H_{eff} like the AHK algorithm or the Padé approximants. For example, in a state formulation the absolute energy of the nucleus ^{18}O would be calculated by matrix techniques, implicitly including both linked and unlinked diagrams. But in a linked-valence formulation the energy differences between neighboring nuclei such as $E(^{18}O) - E_{g.s.}(^{16}O)$ or $E(^{18}O)+E_{g.s.}(^{16}O)-2E(^{17}O)$, would be calculated including only the linked-valence diagrams. Here $E(A)$ is the energy of nucleus A and the subscript g.s. indicates the ground state.

The aim of the present work is to study the convergence properties of the linked-valence expansion of H_{eff} within the context of an exactly solvable model. This model is a generalization to three levels, in a manner similar to Meshkov [13], of the well known two-level model of Lipkin *et al.* [14]. The model is described in sect. 2 along with some of the details of the methods used to obtain the exact solutions.

The linked-valence folded-diagram series of Kuo, Lee and Ratcliff [15, 16] and the AHK algorithm [7] are briefly described in sect. 3. Both the AHK algorithm, generalized to the linked-valence formulation, and the folded-diagram series grouped in powers of the interaction are used to compute H_{eff}. The diagrams that are necessary for these two methods are discussed in the appendix along with some of the details of their calculation.

An important question that is related to the computation of H_{eff} is the choice of the single-particle potential. In refs. [1, 2], H_{eff} is computed using the set of diagrams that implies the use of the Hartree-Fock (HF) potential (hereafter referred to as the HF set of diagrams), when in fact the harmonic oscillator potential is used. The HF single-particle potential is described in sect. 3. Subsequent work [17-19] indicates that the additional diagrams missing from refs. [1, 2] are also important. This question of the choice of the single-particle potential and how it may affect the convergence properties of expansions of H_{eff} are investigated here.

The results of the calculation of H_{eff} are discussed in sect. 4 for two different choices of the parameters in the model Hamiltonian. For each choice, H_{eff} is calculated using the HF set of diagrams and (a) the unperturbed single-particle potential, (b) the HF single-particle potential, and (c) the unperturbed basis with the HF energies. These results indicate that the convergence of the expansions of H_{eff} in low orders is not adversely affected by the presence of intruder states in the spectrum. However, the agreement with the exact results is crucially dependent on the choice of the single-particle potential.

Excellent agreement with the exact results is obtained when the HF choice is made, as in case (b). The results indicate that the HF procedure can be a powerful tool for constructing a good model space for effective interaction calculations and consequently facilitates the convergence of the linked-valence effective-interaction expansion. In contrast to this, the results show that poor agreement with the exact results and a misleading spectrum can result when the unperturbed potential is used with the HF set of diagrams, as in case (a).

2. Description of the model

The modified Lipkin two-level model presented in ref. [6]) is extended here to a three-level model in a manner similar to the extension of the original Lipkin model [14]) to a three-level model by Meshkov [13]). This extended model consists of three levels, each with a degeneracy L, and N spinless fermions distributed among the three levels. Each state of the system can be described by two quantum numbers: let the Greek letters α, β, γ and δ denote the quantum number necessary to distinguish between the different levels and let the Latin letters p and q specify which of the degenerate states the particle occupies within a level. The three levels are labeled with $\alpha = z$ for the lowest level, $\alpha = y$ for the intermediate level and $\alpha = x$ for the highest level.

The particles interact through the Hamiltonian

$$H = H_0 + H_{\rm I} = H_0 + H_1 + H_2 + H_3, \tag{2.1}$$

where

$$H_0 = \sum_{p\alpha} \varepsilon_\alpha a^\dagger_{p\alpha} a_{p\alpha}, \tag{2.2}$$

and

$$H_1 = V \sum_{\substack{pq \\ \alpha\beta\gamma\delta \\ \alpha\neq\gamma,\delta;\ \beta\neq\gamma,\delta}} a^\dagger_{p\alpha} a^\dagger_{q\beta} a_{q\delta} a_{p\gamma},$$

$$H_2 = W \sum_{\substack{pq \\ \alpha\beta,\ \alpha\neq\beta}} a^\dagger_{p\alpha} a^\dagger_{q\beta} a_{q\alpha} a_{p\beta},$$

$$H_3 = U \sum_{\substack{pq \\ \alpha\beta\gamma,\ \beta\neq\gamma}} a^\dagger_{p\alpha} a^\dagger_{q\beta} a_{q\gamma} a_{p\alpha}. \tag{2.3}$$

Here the sums over p and q run from 1 to L and the sums over α, β, γ and δ run over x, y and z. A diagrammatic representation of each of the terms in $H_{\rm I}$ is shown in fig. 1.

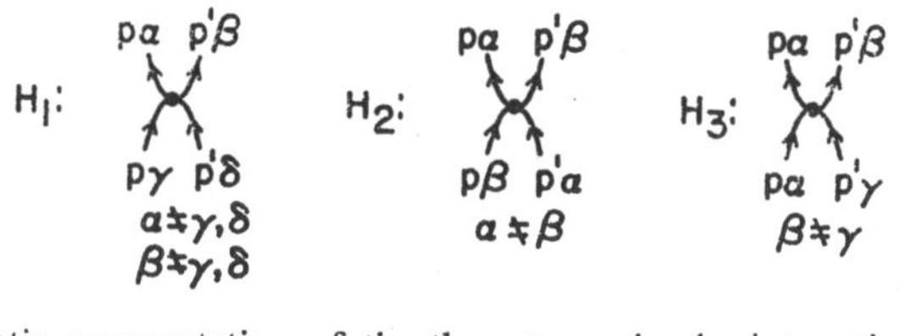

Fig. 1. A diagrammatic representation of the three terms in the interaction Hamiltonian $H_{\rm I}$.

Because of the simple form of the model, its solutions can be classified according to irreducible representations of the group SU(3). To see this the operators $A_{\alpha\beta}$ are defined as [the approach of Hecht [20]) is followed below]

$$A_{\alpha\beta} = \sum_p a^\dagger_{p\alpha} a_{p\beta}, \qquad \alpha, \beta = x, y, z. \tag{2.4}$$

From the commutation relations of the $a^\dagger$ and a operators it is easy to show that

$$[A_{\alpha\beta}, A_{\gamma\delta}] = A_{\alpha\delta}\delta_{\beta\gamma} - A_{\gamma\beta}\delta_{\alpha\delta}. \tag{2.5}$$

Thus the operators $A_{\alpha\beta}$ are the nine infinitesimal generators of the group U(3). Equivalently the set of operators given by

$$\begin{aligned} \hat{N} &= A_{zz}+A_{yy}+A_{xx}, \qquad Q_0 = 2A_{zz}-A_{yy}-A_{xx}, \\ \Lambda_0 &= \tfrac{1}{2}(A_{yy}-A_{xx}), \qquad A_{\alpha\beta}, \alpha \neq \beta, \end{aligned} \tag{2.6}$$

can be considered. Since the number of particles is fixed in the model, the number operator $\hat{N}$ can be excluded leaving the eight generators of SU(3).

The operators $A_{\alpha\beta}$, $\alpha \neq \beta$ are step operators, i.e., they correspond to moving a particle up (or down) one level. With these operators the notion of a maximum weight state $|\Psi_H\rangle$ can be defined such that

$$A_{zy}|\Psi_H\rangle = A_{zx}|\Psi_H\rangle = A_{yx}|\Psi_H\rangle = 0. \tag{2.7}$$

This is equivalent to maximizing first the number of particles in the z-level and then maximizing the number of particles in the y-level. For each maximum weight state there is a unique correspondence with an irreducible representation $(\lambda\mu)$ of the SU(3) group with $\lambda = n_z - n_y$ and $\mu = n_y - n_x$. Here n_α is the number of particles in the α-level. Each maximum weight state is equivalent to a Young tableau with three rows and at most L columns.

The model Hamiltonian can be rewritten in terms of the SU(3) generators as

$$\begin{aligned} H_0 &= \sum_\alpha \varepsilon_\alpha A_{\alpha\alpha}, \\ H_1 &= V[(A_{zy}^2+A_{zx}^2+A_{yz}^2+A_{yx}^2+A_{xz}^2+A_{xy}^2+2(A_{zy}A_{zx}+A_{yx}A_{zx}+A_{xy}A_{zy} \\ &\qquad +A_{yz}A_{yx}+A_{xz}A_{yz}+A_{xz}A_{xy})], \\ H_2 &= W(C_2-\tfrac{1}{6}Q_0^2-2\Lambda_0^2-2\hat{N}), \\ H_3 &= U(\hat{N}-1)(A_{zy}+A_{zx}+A_{yz}+A_{yx}+A_{xz}+A_{xy}), \end{aligned} \tag{2.8}$$

where

$$C_2 = \sum_{\alpha\neq\beta} A_{\alpha\beta}A_{\beta\alpha}+\tfrac{1}{6}Q_0^2+2\Lambda_0^2 \tag{2.9}$$

is the Casimir operator of the SU(3) group.

Since the Hamiltonian is a function of the generators of the group it commutes with the Casimir operator of the group and the eigenstates of the Hamiltonian can simultaneously be eigenstates of the Casimir operator. Since specifying the pair of numbers $(\lambda\mu)$ is equivalent to specifying the eigenvalues of the Casimir operator [21], the Hamiltonian will not connect states corresponding to different irreducible representations.

In order to solve the exact problem for a given irreducible representation, the basis states used are classified according to the SU(3) $\supset$ SU(2) chain. A complete set of operators $\{Q_0, \Lambda_0, \Lambda^2\}$ is defined with Q_0 and Λ_0 given by eq. (2.6) and Λ^2, the Casimir operator for the subgroup SU(2), given by

$$\Lambda^2 = \tfrac{1}{2}(A_{yx}A_{xy}+A_{xy}A_{yx})+\Lambda_0^2. \tag{2.10}$$

The basis states $|(\lambda\mu)\varepsilon\Lambda\nu\rangle$ are then defined by

$$\begin{aligned} Q_0|(\lambda\mu)\varepsilon\Lambda\nu\rangle &= \varepsilon|(\lambda\mu)\varepsilon\Lambda\nu\rangle, \\ \Lambda^2|(\lambda\mu)\varepsilon\Lambda\nu\rangle &= \Lambda(\Lambda+1)|(\lambda\mu)\varepsilon\Lambda\nu\rangle, \\ \Lambda_0|(\lambda\mu)\varepsilon\Lambda\nu\rangle &= \tfrac{1}{2}\nu|(\lambda\mu)\varepsilon\Lambda\nu\rangle. \end{aligned} \tag{2.11}$$

The possible ε, Λ and ν values for a given $(\lambda\mu)$ representation are

$$\begin{aligned} \varepsilon &= 2\lambda+\mu-3(r+s), \\ \Lambda &= \tfrac{1}{2}(\mu+r-s), \qquad -\Lambda \leqq \tfrac{1}{2}\nu \leqq \Lambda, \end{aligned} \tag{2.12}$$

with $r = 0, 1, 2, \ldots, \lambda$ and $s = 0, 1, 2, \ldots, \mu$.

To construct the Hamiltonian in this basis all that is now required are the matrix elements of the operators $A_{\alpha\beta}$ and $A_{\alpha\beta}A_{\gamma\delta}$. These matrix elements can be computed from the relations given by Hecht [20], i.e.

$$\begin{aligned} A_{yx}|(\lambda\mu)\varepsilon\Lambda\nu\rangle &= [(\Lambda-\tfrac{1}{2}\nu)(\Lambda+\tfrac{1}{2}\nu+1)]^{\frac{1}{2}}|(\lambda\mu)\varepsilon\Lambda(\nu+2)\rangle, \\ A_{xy}|(\lambda\mu)\varepsilon\Lambda\nu\rangle &= [(\Lambda+\tfrac{1}{2}\nu)(\Lambda-\tfrac{1}{2}\nu+1)]^{\frac{1}{2}}|(\lambda\mu)\varepsilon\Lambda(\nu-2)\rangle, \\ A_{yz}|(\lambda\mu)\varepsilon\Lambda\nu\rangle &= f[(\lambda\mu)\varepsilon\Lambda\nu]|(\lambda\nu)(\varepsilon-3)(\Lambda+\tfrac{1}{2})(\nu+1)\rangle \\ &\quad +f[(\lambda\mu)\varepsilon, -(\Lambda+1)\nu]|(\lambda\mu)(\varepsilon-3)(\Lambda-\tfrac{1}{2})(\nu+1)\rangle, \\ A_{xz}|(\lambda\mu)\varepsilon\Lambda\nu\rangle &= f[(\lambda\mu)\varepsilon\Lambda, -\nu]|(\lambda\mu)(\varepsilon-3)(\Lambda+\tfrac{1}{2})(\nu-1)\rangle \\ &\quad -f[(\lambda\mu)\varepsilon, -(\Lambda+1), -\nu]|(\lambda\mu)(\varepsilon-3)(\Lambda-\tfrac{1}{2})(\nu-1)\rangle, \\ A_{zy}|(\lambda\mu)\varepsilon\Lambda\nu\rangle &= f[(\lambda\mu)(\varepsilon+3), -(\Lambda+\tfrac{3}{2})(\nu-1)]|(\lambda\mu)(\varepsilon+3)(\Lambda+\tfrac{1}{2})(\nu-1)\rangle \\ &\quad +f[(\lambda\mu)(\varepsilon+3)(\Lambda-\tfrac{1}{2})(\nu-1)]|(\lambda\mu)(\varepsilon+3)(\Lambda-\tfrac{1}{2})(\nu-1)\rangle, \\ A_{zx}|(\lambda\mu)\varepsilon\Lambda\nu\rangle &= -f[(\lambda\mu)(\varepsilon+3), -(\Lambda+\tfrac{3}{2}), -(\nu-1)]|(\lambda\mu)(\varepsilon+3)(\Lambda+\tfrac{1}{2})(\nu+1)\rangle \\ &\quad +f[(\lambda\mu)(\varepsilon+3)(\Lambda-\tfrac{1}{2}), -(\nu+1)]|(\lambda\mu)(\varepsilon+3)(\Lambda-\tfrac{1}{2})(\nu+1)\rangle, \end{aligned} \tag{2.13}$$

where

$$f[(\lambda\mu)\varepsilon\Lambda\nu] = \left\{ \frac{(\Lambda+\frac{1}{2}\nu-1)[\Lambda+1+\frac{1}{3}(\lambda-\mu-\frac{1}{2}\varepsilon)][\Lambda+2+\frac{1}{3}(\lambda+2\mu-\frac{1}{2}\varepsilon)][\frac{1}{3}(2\lambda+\mu-\frac{1}{3}\varepsilon)-\Lambda]}{(2\Lambda+1)(2\Lambda+2)} \right\}^{\frac{1}{2}}. \tag{2.14}$$

For the rest of this work the level degeneracy will be fixed to be $L = 4$. The exact results are computed for the representation that contains the states that evolve from the unperturbed ground state of the system. This corresponds to the (40) and (22) representations for the $N = 4$ and $N = 6$ systems respectively. When the effective interaction is computed for the $N = 6$ system the model space is three dimensional. The three basis states used to span the model space are $|1\rangle$ = two particles in the y-level plus a filled z-level, $|2\rangle$ = two particles in the x-level plus a filled z-level and $|3\rangle$ = a symmetric linear combination of one particle in each of the y- and x-levels plus a filled z-level. With the notation used above these states are $|1\rangle = |(22)612\rangle$, $|2\rangle = |(22)61,-2\rangle$ and $|3\rangle = |(22)610\rangle$. The x, y and z levels are defined by the single-particle potential that is used.

3. Formalism

In this section the main ideas of the formalism used in this work are outlined. First, two methods of computing the effective interaction are described and then the problem of the choice of the single-particle potential is discussed with particular emphasis on the HF method.

3.1. EFFECTIVE INTERACTION

Two methods of constructing a linked-valence effective interaction are presented here. Both of them use concepts that can be introduced by first discussing the $\hat{Q}$-box formulation of the folded-diagram series of Kuo, Lee and Ratcliff [15, 16].

First the $\hat{Q}$-box is defined as the sum of all irreducible diagrams with at least one H_I vertex, every one of which must be linked to at least one external valence line in the model space, i.e. an active line. Here irreducible is defined to mean that the intermediate state between any two successive vertices must contain at least one line outside the model space, i.e. a passive line. Some of the low-order diagrams that make up the $\hat{Q}$-box are shown in fig. 2.

It should be noted that in the valence case a distinction must be made between linked diagrams and connected diagrams that does not need to be made for a closed shell case. As defined above a linked diagram must have all vertices linked (that is, attached either directly or indirectly) to at least one valence line. This implies that diagrams with vacuum fluctuations are not linked. If in addition to being linked the diagram contains only one linked piece, where there is at least one vertex (consequently a non-interacting valence line does not affect the definitions) and all vertices are linked together, then the diagram is also connected. The $\hat{Q}$-box contains both connected and disconnected diagrams.

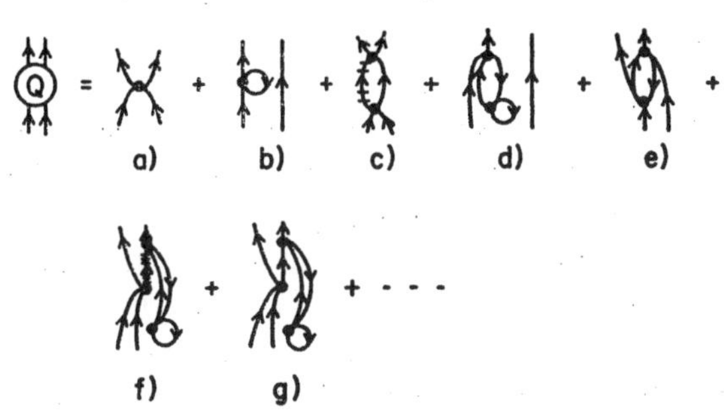

Fig. 2. Some low-order diagrams included in the $\hat{Q}$-box. Hugenholtz diagrams are used and a railed line represents a passive particle line. The rest of the symbols are explained in the appendix.

Two other sums of diagrams that are related to the $\hat{Q}$-box must also be defined. The $\hat{Q}'$ box differs from the $\hat{Q}$-box in that all diagrams in the $\hat{Q}$-box must have at least two H_{I} vertices and that immediately prior to the last H_{I} vertex in the diagram at least one valence line must be in a passive state. Consequently, diagrams (a), (b), (d) and (g) of fig. 2 and higher order diagrams of the same type are excluded from the $\hat{Q}'$ box. The $\hat{Q}_1$ box contains all the diagrams in the $\hat{Q}'$ box. In addition, the diagram with one interaction between valence lines and all the diagrams where there is only one vertex attached to a valence line, all other valence lines being non-interacting, and this vertex is connected to a core excitation. Consequently, diagram (g) of fig. 2 and all others of the same type are excluded from the $\hat{Q}_1$ box. When the Hartree-Fock single-particle potential is used and no diagrams higher than third order are considered the $\hat{Q}$-box, the $\hat{Q}'$ box [excluding diagram (a)] and the $\hat{Q}_1$ box are all identical.

With these definitions, the linked-valence effective interaction is given by

$$H_{\text{eff}}|P\Psi\rangle = (H_0^{\text{v}}+\mathcal{V}_{\text{eff}})|P\Psi\rangle = (E-E_{\text{c}})|P\Psi\rangle, \tag{3.1}$$

with

$$\mathcal{V}_{\text{eff}} = \hat{Q}_1 - \hat{Q}'\int\hat{Q} + \hat{Q}'\int\hat{Q}\int\hat{Q} - \dots . \tag{3.2}$$

Here E_{c} is the exact ground-state energy of the core system, H_0^{v} is the single-particle Hamiltonian which operates only on the valence particles and the symbol $\int$ represents a generalized folding operation. For example, $\hat{Q}'\int\hat{Q}\int\hat{Q}$ is just the sum of all generalized time ordered twice folded diagrams subject to the constraint that the time of the top H_{I} vertex in the $\hat{Q}$-box following the folding operation must be later than the time of the lowest H_{I} vertex in the $\hat{Q}$-box (or $\hat{Q}'$ box) preceding the folding operation.

The terms in eq. (3.2) are grouped according to the number of folding operations. However, the terms in each $\hat{Q}$-box can be expanded out and all non-folded and folded diagrams with the same number of H_I interactions can be collected together. This grouping of diagrams is equivalent to Brandow's expansion of the effective interaction [22]). This form of $\mathcal{V}_{\text{eff}}$ has been used in calculations for real systems, e.g. see ref. [2]), and will be compared with the results from a second method for computing $\mathcal{V}_{\text{eff}}$, the AHK algorithm which is discussed below.

It should be noted that, unlike the state formulation where the $\hat{Q}$-box can usually be obtained exactly [see, for example, ref. [12])], exact calculations of the $\hat{Q}$-box in the linked-valence formulation are generally not feasible. This results from the restriction that only linked diagrams can be included making it difficult to algebraically sum all the diagrams in a $\hat{Q}$-box. Consequently, low-order perturbation theory (or some technique based on perturbation theory such as the Padé approximants) must be used to compute the $\hat{Q}$-box in the linked-valence formulation. More details concerning the precise low-order diagrams of the $\hat{Q}$-box and the $\mathcal{V}_{\text{eff}}$ grouped in the manner described above that are used in this work are given in the appendix. For practical reasons, diagrams higher than third order will not be included.

The AHK algorithm proposed by Anastasio, Hockert and Kuo [7]) is the second method for constructing $\mathcal{V}_{\text{eff}}$ that will be used in this work. It is based on the method of partial summations of Krenciglowa and Kuo [5, 24]) which solves a Bloch-Horowitz type of equation for the linked-valence effective interaction by linear iteration. The AHK algorithm uses an improved iteration procedure that is defined recursively, once the $\hat{Q}$-box is given, by

$$(S_{n+1})_{ij} = (S_n + \Delta_n^1)_{ij}. \tag{3.3}$$

If for some n this sequence of the S_n has converged then $H_0^{\text{v}} + S_{n+1} = H_0^{\text{v}} + S_n \equiv H_{\text{eff}}$. Here Δ_n^1 is defined by

$$\sum_k (D(j) - I)_{ik} (\Delta_n^1)_{kj} = (S_n - \hat{Q}(E_j^n))_{ij}, \tag{3.4}$$

$$D(j)_{ik} = \left(\frac{\mathrm{d}\hat{Q}(E_j^n)}{\mathrm{d}E}\right)_{ik} \delta_{kj} + \left(\frac{\hat{Q}(E_j^n) - \hat{Q}(E_k^n)}{E_j^n - E_k^n}\right)_{ik} (1 - \delta_{kj}), \tag{3.5}$$

where I is the identity matrix, E_j^n and $|\Phi_j^n\rangle$ are the eigenvalues and eigenvectors of $H_0^{\text{v}} + S_n$ and all operators are evaluted in the biorthogonal basis of $|\Phi_j^n\rangle$ and $\langle \bar{\Phi}_j^n|$. In this work the iteration procedure is initialized by taking S_0 = diagram 1-1 of fig. 9.

3.2. HF SINGLE-PARTICLE POTENTIAL

Any choice of the division of H into an H_0 and H_I is arbitrary and can be modified by the introduction of a one-body potential U such that

$$H = H_0' + H_I', \tag{3.6}$$

where

$$H_0' = H_0 + U, \qquad H_I' = H_I - U. \tag{3.7}$$

The "best" choice for U has been a subject of much discussion over the years. One possible choice is $U = 0$, which would correspond to an H_0' equal to the unperturbed potential for the model problem discussed in sect. 2. This choice has the advantage that all calculations in this basis are straightforward (for the present three-level model) but it has the disadvantage that the number of diagrams that contribute to the effective interaction becomes quite unmanageable in this basis.

Another possible choice is to take U to be the HF self-consistent potential for the core system. It will become evident below that this choice greatly reduces the number of diagrams that must be summed for H_{eff}, since a large number of diagrams are made to cancel each other. These diagrams would have to be computed if $U = 0$ were used. Of course this advantage must be balanced against the increased complexity of the basis. However, there is also a more physical reason for the HF choice.

The HF potential yields a single-particle basis that minimizes the ground-state energy of the core system subject to the constraint that the core system is represented by a Slater determinant composed of the lowest filled single-particle levels. This choice should also provide a good single-particle basis for the low-lying states of the valence system as long as the valence particles do not induce any large changes in the self-consistent field of the core. Consequently, the choice of the HF potential may provide improved convergence properties of the effective interaction expansion over the results obtained with the unperturbed potential since recent work [5, 6, 12]) has exhibited the intuitive result that the convergence properties are crucially dependent on the quality of the model space used. With this in mind the effect of the HF choice for U on the convergence properties of H_{eff} will be studied.

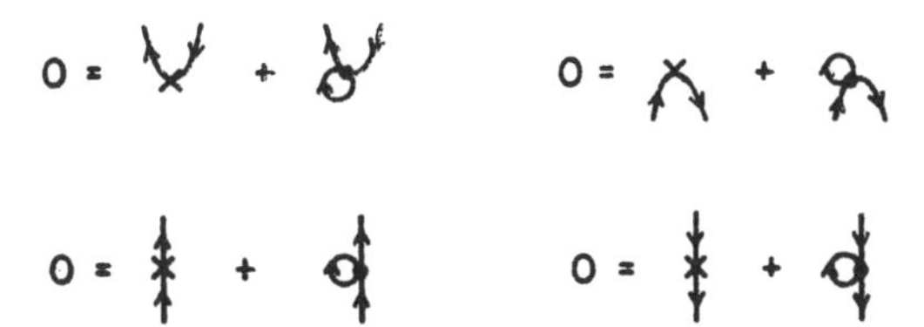

Fig. 3. A diagrammatic representation of the equations that must be solved self-consistently to construct the HF potential. Here a cross represents a $-U$ insertion and the rest of the symbols are explained in the appendix.

The equations that must be solved self-consistently to obtain the HF potential U^{HF} are straightforward. They amount to finding a basis such that the diagrams in fig. 3 sum to zero. The self-consistent HF single-particle wave functions $\phi_{p\alpha}^{\text{HF}}$ are expanded in the unperturbed single-particle basis $\eta_{p\alpha}$ as

$$|\phi_{p\alpha}^{\text{HF}}\rangle = \sum_{\alpha'} c_{\alpha'\alpha}|\eta_{p\alpha'}\rangle, \tag{3.8}$$

where

$$|\eta_{p\alpha'}\rangle = a_{p\alpha'}^{\dagger}|0\rangle. \tag{3.9}$$

The HF equations expressed in the unperturbed single-particle basis are then

$$\sum_{\alpha'} c_{\alpha'\alpha}\langle\eta_{p\beta}|(H_0+U^{\mathrm{HF}})|\eta_{p\alpha'}\rangle = \varepsilon_\alpha^{\mathrm{HF}} c_{\beta\alpha}, \tag{3.10}$$

where

$$\langle\eta_{p\beta}|U^{\mathrm{HF}}|\eta_{p\alpha'}\rangle = \sum_{q\gamma\delta} c_{\gamma z}^* c_{\delta z}\langle\eta_{p\beta}\eta_{q\gamma}|H_{\mathrm{I}}|\eta_{p\alpha'}\eta_{q\delta}\rangle, \tag{3.11}$$

and antisymmetrized matrix elements of H_{I} are used. Because the interaction H_{I} given in eq. (2.3) can only change the level quantum number α, both U^{HF} and $c_{\beta\alpha}$ are independent of p, the quantum number that specifies the state within a level. This quantum number has been suppressed from the labels on the c-coefficients and will also be suppressed on the single-particle wave functions throughout the rest of this work.

4. Results and discussions

4.1. "REALISTIC" FORCE

Here the parameters in the model Hamiltonian are chosen so that the exact results contain several of the features of a realistic system. For example, it is desired that the two lowest exact states possess wave functions dominated by model space components while the third lowest exact state possesses a wave function dominated by components outside the model space. This third state then serves as an intruder state which often exists in a realistic problem. In this way, some of the general properties of the approximations used to compute the effective interaction for a realistic system can be studied.

The force parameters used are (in arbitrary units)

$$U = 0.4, \quad V = 0.4, \quad W = -0.5, \qquad \varepsilon_z = 0, \quad \varepsilon_y = 10, \quad \varepsilon_x = 12.5. \tag{4.1}$$

These force parameters are weak since they are small when compared with the separations between the single-particle energies. With this unperturbed single-particle spectrum the model space is not degenerate. Although the formalism outlined in sect. 3 is still applicable, caution is required since energy denominators of individual diagrams in the linked-valence expansion for H_{eff} may vanish. However this problem will not arise in the examples studied here.

Since the expansions for H_{eff} outlined in sect. 3 yield the energies of the system with valence particles relative to the exact ground-state energy of the core, the exact low-lying spectrum for $E(N=6)-E_{\mathrm{g.s.}}(N=4)$ is shown in table 1 along with $E_{\mathrm{g.s.}}(N=4)$. From the values of the overlaps of the model space projections of the exact wave functions $\langle P\Psi_i|P\Psi_i\rangle$ it is clear that state 3 is an intruder state.

The results for the HF calculation in the $N=4$ system are shown in table 2. The HF single-particle wave functions $\phi_\alpha^{\mathrm{HF}}$ are not very different from the unperturbed wave functions η_α defined in eq. (3.9). The same is true for the single-particle energies $\varepsilon_\alpha^{\mathrm{HF}}$, although HF increases the separation between each of the levels somewhat. This new basis does not do such a good job of reproducing the ground-state energy of the core with $E_{\mathrm{g.s.}}(4) = -1.46$ and $E_{\mathrm{g.s.}}^{\mathrm{HF}}(4) = -0.83$, close to a 50% error.

TABLE 1

The exact results of the model calculation with $U = 0.4$, $V = 0.4$, $W = -0.5$, $\varepsilon_z = 0$, $\varepsilon_y = 10$ and $\varepsilon_x = 12.5$

State	$E(6)-E_{g.s.}(4)$	$\langle P\Psi\|P\Psi\rangle$	State	$E(6)-E_{g.s.}(4)$	$\langle P\Psi\|P\Psi\rangle$
1	19.252	0.889	5	29.588	0.055
2	22.591	0.832	6	30.835	0.088
3	27.479	0.181	7	34.841	0.078
4	28.246	0.597	8	36.246	0.077

Here $E(6)$ is an exact eigenvalue of the $N = 6$ system, $E_{g.s.}(4)$ is the exact ground-state energy of the $N = 4$ system, $E_{g.s.}(4) = -1.460$, and $|P\Psi\rangle$ is the model-space projection of an exact eigenvector of the $N = 6$ system.

TABLE 2

Results from the HF calculation for the $N = 4$ system, with $U = 0.4$, $V = 0.4$, $W = -0.5$, $\varepsilon_z = 0$, $\varepsilon_y = 10$ and $\varepsilon_x = 12.5$

α	$\langle\eta_z\|\phi_\alpha{}^{HF}\rangle$	$\langle\eta_y\|\phi_\alpha{}^{HF}\rangle$	$\langle\eta_x\|\phi_\alpha{}^{HF}\rangle$	$\varepsilon_\alpha{}^{HF}$
z	0.992	−0.102	−0.073	−0.588
y	−0.064	−0.912	0.406	10.068
x	0.108	0.398	0.911	13.689

Here $|\phi_\alpha{}^{HF}\rangle$ and $\varepsilon_\alpha{}^{HF}$ are HF single-particle wave functions and energies respectively and $|\eta_\alpha\rangle$ is an unperturbed single-particle wave function defined in eq. (3.9) of the text. Further, $E^{HF}_{g.s.}(4) = -0.834$.

The effective interaction is computed by first constructing the $\hat{Q}$-box in low-order perturbation theory. Only linked-valence diagrams are summed and no HF insertions are included. These diagrams are listed in fig. 9 and a more detailed discussion of them is given in the appendix. This approximate $\hat{Q}$-box is then used in the AHK algorithm, defined in eqs. (3.3)–(3.5), to give the effective interaction. This is done for three different combinations of the single-particle wave functions and energies: (a) both the unperturbed single-particle wave functions and energies are used, (b) both the HF single-particle wave functions and energies are used, and (c) the unperturbed single-particle wave functions and the HF single-particle energies are used.

Clearly for case (a) the HF insertions should be included on the low-order diagrams in the $\hat{Q}$-box. However, in calculations for a real system these insertions are generally not included to keep the number of diagrams from becoming too large. It is usually argued that the harmonic oscillator basis (the unperturbed basis in the case of this model) is sufficiently close to the HF basis so that these additional diagrams will not contribute much. As mentioned above, the unperturbed and the HF bases are very close so that this additional approximation can be tested by comparing the results in cases (a) and (b).

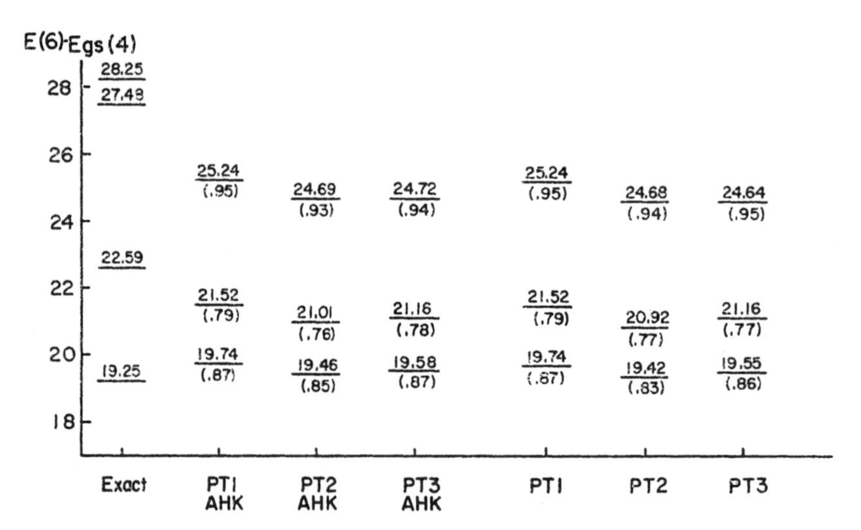

Fig. 4. The spectrum resulting from the effective-interaction calculation with the force parameters of eq. (4.1) and the unperturbed single-particle basis and energies. The numbers above the lines are the eigenvalues and the numbers in parentheses are the wave function overlaps χ_{43}, χ_{22} and χ_{11} for the upper, intermediate and lower states respectively. Here PT3-AHK indicates that third-order perturbation theory is used to approximate the $\hat{Q}$-box which is then used in the AHK algorithm to compute H_{eff}. PT3 indicates that the folded-diagram series, grouped in powers of H_I, is summed through third order in H_I to compute H_{eff}.

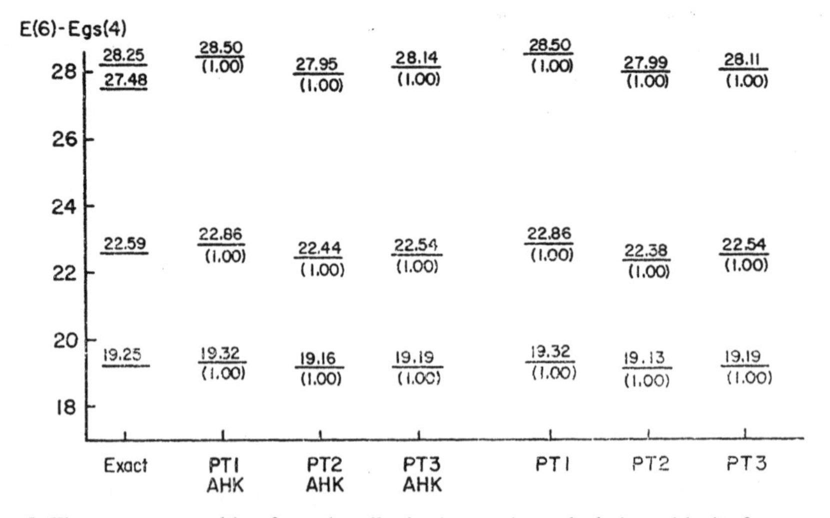

Fig. 5. The spectrum resulting from the effective-interaction calculation with the force parameters of eq. (4.1) and the HF single-particle basis and energies. The notation of fig. 4 is also used here.

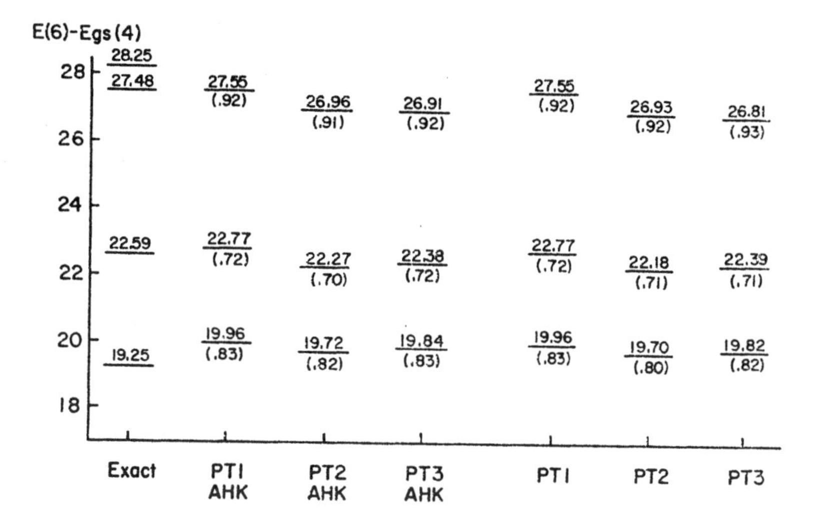

Fig. 6. The spectrum resulting from the effective-interaction calculation with the force parameters of eq. (4.1) and the unperturbed single-particle basis with the HF single-particle energies. The notation of fig. 4 is also used here.

Case (c) has been included in an attempt to isolate the origin of any differences between the results from cases (a) and (b). In particular, it is of interest to know if increasing the separation in energy of the single-particle levels by using the HF energies instead of the unperturbed energies will account for any of the differences between cases (a) and (b). It may be true that the use of the HF basis is even more important.

The results of the calculations for cases (a), (b) and (c) are shown on the left side of figs. 4, 5 and 6 respectively. For case (a), the eigenvalues appear to be converging fairly rapidly. However, by just considering the eigenvalues it is not clear to which of the exact states they are converging. In particular, the highest eigenvalue obtained from H_{eff} could correspond to the exact states 3 or 4. To make this identification it is necessary to look at the overlaps of the eigenvectors of H_{eff} with the projections of the exact wave functions Ψ_i onto the model space. The wave function overlap χ_{ij} is defined by

$$\chi_{ij} = \left| \frac{\langle P\Psi_i | \Phi_j \rangle}{\langle P\Psi_i | P\Psi_i \rangle^{\frac{1}{2}}} \right|, \tag{4.2}$$

where the $|\Phi_j\rangle$ are the normalized eigenvectors of H_{eff} and P is the operator that projects onto the model space. These numbers are shown in parentheses in the figures. From this it is clear that H_{eff} is converging to the exact states 1, 2 and 4 which are the ones with the largest $\langle P\Psi_i / P\Psi_i \rangle$ overlaps.

Although convergence, in low orders, is good, the spectrum produced is too compressed. The lowest two eigenvalues are reproduced fairly well but the corresponding

TABLE 3

The value of matrix elements of topologies calculated with different single-particle wave functions and energies for the force parameters given by eq. (4.1)

Topology	Case (a)		Case (b)		Case (c)	
	D_{11}	D_{23}	D_{11}	D_{23}	D_{11}	D_{23}
1-1	0.000	0.566	−0.640	−0.564	0.000	0.566
2-1	−0.454	−0.195	−0.197	0.086	−0.394	−0.264
2-2	0.363	0.256	0.074	−0.124	0.334	0.236
2-3	−0.134	−0.149	0.001	0.044	−0.120	−0.161
2-4	−0.032	−0.040	0.000	0.038	−0.030	−0.036
3-1	0.037	0.037	0.008	−0.022	0.029	0.035
3-2	−0.129	−0.098	−0.043	0.019	−0.099	−0.103
3-3	0.051	0.011	0.028	0.012	0.041	0.010
3-4	0.065	0.046	0.012	−0.020	0.055	0.039
3-5	0.064	0.047	0.011	−0.020	0.054	0.041
3-6	0.064	0.045	0.011	−0.019	0.054	0.038
3-7	−0.031	−0.039	0.005	0.014	−0.026	−0.034
3-8	−0.012	−0.009	−0.001	0.007	−0.010	−0.008
3-9	−0.012	−0.008	−0.001	0.006	−0.010	−0.007
3-11	0.086	0.074	0.017	−0.032	0.070	0.075
3-12	0.086	0.039	0.017	−0.015	0.070	0.031
3-13	0.018	0.039	−0.003	−0.017	0.015	0.039
3-14	0.018	0.017	−0.003	−0.006	0.015	0.013
3-15	−0.043	−0.041	−0.007	0.016	−0.035	−0.043
3-16	−0.043	−0.017	−0.007	0.005	−0.035	−0.013
3-17	0.011	0.013	0.001	−0.004	0.009	0.013
3-18	0.011	0.010	0.001	−0.002	0.009	0.008
3-28	0.015	0.110	−0.008	−0.112	0.011	0.158
3-29	−0.020	−0.060	−0.001	0.032	−0.015	−0.077
3-30	−0.068	−0.111	−0.007	0.092	−0.056	−0.119
3-31	0.015	0.003	0.002	−0.069	0.014	−0.010
3-32	0.015	0.060	0.002	−0.082	0.014	0.056
3-33	0.014	0.013	0.004	−0.007	0.011	0.015
3-34	0.014	0.011	0.004	−0.005	0.011	0.010
3-35	−0.007	−0.004	−0.002	0.000	−0.006	−0.003
3-36	−0.006	0.005	−0.003	−0.004	−0.005	0.004
3-37	−0.006	−0.020	0.000	0.016	−0.005	−0.017
3-38	−0.006	0.008	0.000	0.005	−0.005	0.003
3-39	−0.012	−0.014	0.000	0.011	−0.010	−0.011
3-40	−0.012	−0.014	0.000	0.012	−0.010	−0.011
3-41	0.002	0.000	0.000	0.000	0.002	0.000
3-42	0.002	0.003	0.000	−0.002	0.002	0.003
3-44	−0.011	−0.013	0.000	0.012	−0.009	−0.013
3-45	−0.011	−0.042	0.000	0.019	−0.009	−0.048
3-46	−0.003	−0.004	0.000	0.004	−0.002	−0.004
3-47	−0.003	−0.013	0.000	0.005	−0.002	−0.015
3-48	0.039	0.067	0.004	−0.051	0.032	0.066
3-49	0.039	0.046	0.004	−0.034	0.032	0.039
3-50	−0.031	−0.055	−0.001	0.042	−0.026	−0.050
3-51	−0.031	−0.043	−0.001	0.034	−0.026	−0.038
3-52	0.004	0.004	0.001	−0.002	0.003	0.004
3-53	0.004	0.004	0.001	−0.001	0.003	0.004

TABLE 3 (continued)

Topology	Case (a)		Case (b)		Case (c)	
	D_{11}	D_{23}	D_{11}	D_{23}	D_{11}	D_{23}
3-54	−0.013	−0.011	0.000	0.014	−0.011	−0.010
3-55	−0.013	−0.026	0.000	0.015	−0.011	−0.021
3-56	0.007	0.005	0.000	−0.006	0.006	0.004
3-57	0.007	0.012	0.000	−0.007	0.006	0.010
F-1	−0.010	−0.104	0.006	0.065	−0.010	−0.142
F-2	0.006	0.020	0.000	−0.016	0.005	0.017
F-3	0.013	−0.030	0.014	0.031	0.012	−0.036
F-4	−0.002	−0.003	0.000	0.002	−0.002	−0.003

Cases (a), (b) and (c) are explained in sect. 4 and the topologies are numbered according to the order used in fig. 9. Here D_{ij} represents a matrix element of one of these topologies where i and j are respectively the final and initial states. The different states are denoted by the numbers 1, 2 or 3 as defined in sect. 2.

wave function overlaps are not good. This contrasts with the results for the upper level where the wave function overlap is very good but the eigenvalue is too low with H_{eff} yielding a value of 24.7 versus the exact value of 28.2.

For case (b), where the HF single-particle potential is used, convergence in low orders, for both the eigenvalues and wave function overlaps χ of H_{eff}, is rapid. In fact, the first-order results are already quite good. This is shown in fig. 5. Of course, the model space projections of the exact wave functions are computed in the HF basis here. As indicated by the values of χ, the eigenvalues are converging to the exact states 1, 2 and 4 which also have the largest $\langle P\Psi_i|P\Psi_i\rangle$ overlaps in the HF basis. The agreement for these eigenvalues with the exact values is quite good as is the agreement for the overlap χ. This is a marked improvement over the results from case (a).

The results for case (c) are shown in fig. 6. Here the rate of convergence is very similar to case (a). But the use of the HF single-particle energies considerably improves the spectrum. Although the spectrum is still too compressed, the agreement for the upper level with the exact eigenvalue is much better, going from 24.7 in case (a) to 26.9 here, while the agreement for the lowest level is slightly worse.

Even though the use of the HF energies improves the spectrum when compared with case (a), the overall improvement in H_{eff} is not so large. This manifests itself in the values of χ which are even worse in case (c) than in case (a).

A second method, outlined in sect. 3, for computing H_{eff} is also considered. Here the folded-diagram series of eq. (3.2) grouped in powers of H_{I} is used. The diagrams that contribute to this expansion through third order in H_{I} are shown in fig. 9 and discussed in the appendix. As with the $\hat{Q}$-box, only linked-valence diagrams are summed and no HF insertions are included. This is equivalent to the expansion that has been used for H_{eff} in real systems, see e.g. ref. [2]). The results of this calculation for cases (a), (b) and (c) are shown on the right side of figs. 4, 5 and 6 respectively.

These results are very similar to those obtained with the approximate $\hat{Q}$-box and the AHK algorithm discussed above. For this set of force parameters there is very little numerical difference between the two methods. This indicates that the sum of folded diagrams, fourth order and higher, that are generated from the approximate $\hat{Q}$-box are not important here since it is essentially these diagrams that constitute the difference between the two methods.

The differences due to the different single-particle bases and energies that are used, seen in the spectrums discussed above, are also evident in the matrix elements of the diagrams shown in table 3. In going from case (a) to case (b) the matrix elements of a given diagram are reduced by a factor of about ten. When going from case (a) to case (c) the value of the matrix elements are still reduced, but only by about 15 %. This reduction in the size of the matrix elements of diagrams is even more dramatic in the strong force case as discussed in subsect. 4.2.

In summary, the eigenstates of H_{eff}, in all these cases, in low orders seem to be converging fairly rapidly in spite of the presence of the intruder state in the exact spectrum. The eigenstates converge to the exact states with the largest $\langle P\Psi_i|P\Psi_i\rangle$ overlaps just as they did for calculations of H_{eff} in a state formulation [12]. It should be noted that the linked-valence effective interaction $H_{\text{eff}}(x)$ used in the present work is formally, and exactly, equivalent to the difference $W_6(x)-W_4(x)$, where $W_6(x)$ and $W_4(x)$ are respectively the energy-independent effective interactions for the $N = 6$ and $N = 4$ systems. Here x is a strength parameter defined by $H(x) = H_0+xH_1$ and x is taken to be unity for the results quoted in this work. Thus when calculated as a power series in x, the radius of convergence for $H_{\text{eff}}(x)$ is determined by the intruder-state branch points of both $W_6(x)$ and $W_4(x)$ as in Schucan and Weidenmüller [3, 4]. However, since the AHK algorithm does not represent a power series expansion of $H_{\text{eff}}(x)$, its formal convergence properties are unknown. This could probably be a reason that the calculations here do not seem strongly influenced by the presence of intruder states in the exact spectrum. Certainly the results obtained here indicate that the results from low-order perturbation theory in the $\hat{Q}$-box can give a useful approximation to the exact results.

The spectrum that results in the different cases depends strongly on the single-particle basis and energies even though the HF basis is very close to the unperturbed one. This points out the importance of the HF insertions that are excluded from the diagrams summed in case (a). The results from case (c) indicate that increasing the separation between the unperturbed energies is not sufficient to account for the differences between cases (a) and (b). The small change in the basis when going from the unperturbed to the HF basis has considerable effect on the effective interaction that is generated.

4.2. VERY STRONG FORCE

The parameters in the model Hamiltonian are chosen here to exhibit the power of the HF method in providing a good model space. This choice is not meant to describe

TABLE 4

The exact results of the model calculation with the force parameters of eq. (4.3)

State	$E(6)-E_{g.s.}(4)$	$\langle P\Psi\|P\Psi\rangle$	State	$E(6)-E_{g.s.}(4)$	$\langle P\Psi\|P\Psi\rangle$
1	10.937	0.115	5	25.450	0.131
2	11.690	0.060	6	27.568	0.043
3	13.709	0.046	7	29.181	0.002
4	25.342	0.258	8	31.443	0.190

Here $E(6)$ is an exact eigenvalue of $N = 6$ system, $E_{g.s.}(4)$ is the exact ground-state energy of the $N = 4$ system, $E_{g.s.}(4) = -26.322$, and $|P\Psi\rangle$ is the model-space projection of an exact eigenvector of the $N = 6$ system.

TABLE 5

Results from the HF calculation for the $N = 4$ system with the force parameters of eq. (4.3)

α	$\langle\eta_z\|\phi_\alpha^{HF}\rangle$	$\langle\eta_y\|\phi_\alpha^{HF}\rangle$	$\langle\eta_x\|\phi_\alpha^{HF}\rangle$	ε_α^{HF}
z	−0.627	0.775	0.079	−14.498
y	−0.559	−0.518	0.648	5.966
x	−0.543	−0.362	−0.758	7.092

Here $|\phi_\alpha^{HF}\rangle$ and ε_α^{HF} are HF single-particle wave functions and energies respectively and $|\eta_\alpha\rangle$ is an unperturbed single-particle wave function defined in eq. (3.9) of the text. Further, $E_{g.s.}^{HF}(4) = -26.274$.

TABLE 6

The resulting eigenvalues and wave function overlaps from the effective interaction calculation with the force parameters of eq. (4.3) and the unperturbed single-particle basis and energies

	PT1 AHK	PT2 AHK	PT3 AHK	PT1	PT2	PT3
E_3	5.637	8.067	3.667	5.637	2.882	3.358
χ_{33}	(0.363)	(0.973)	(0.971)	(0.363)	(0.984)	(0.941)
E_2	4.000	3.477	0.678	4.000	0.379	−21.216
χ_{22}	(0.368)	(0.891)	(0.993)	(0.368)	(0.846)	(0.355)
E_1	1.863	2.947	−0.254	1.863	−3.381	−43.054
χ_{11}	(0.999)	(0.980)	(0.985)	(0.999)	(0.958)	(0.374)

Here PT3-AHK indicates that third-order perturbation theory is used to approximate the $\hat{Q}$-box which is used in the AHK algorithm to compute H_{eff}. PT3 indicates that the folded-diagram series, grouped in powers of H_1, is summed through third order in H_1 to compute H_{eff}. The χ_{ij} are defined in eq. (4.2) of the text. The corresponding exact results are given in table 4.

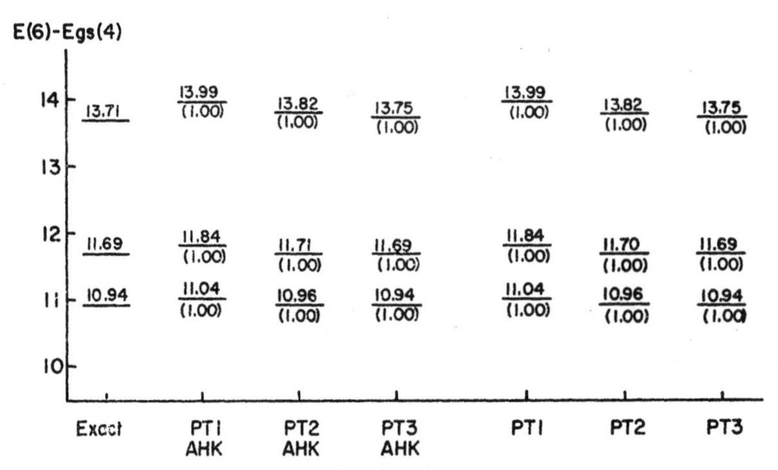

Fig. 7. The spectrum resulting from the effective-interaction calculation with the force parameters of eq. (4.3) and the HF single-particle basis and energies. The notation of fig. 4 is also used here with the exception that χ_{33} is given instead of χ_{43} for the wave function overlap associated with the upper level.

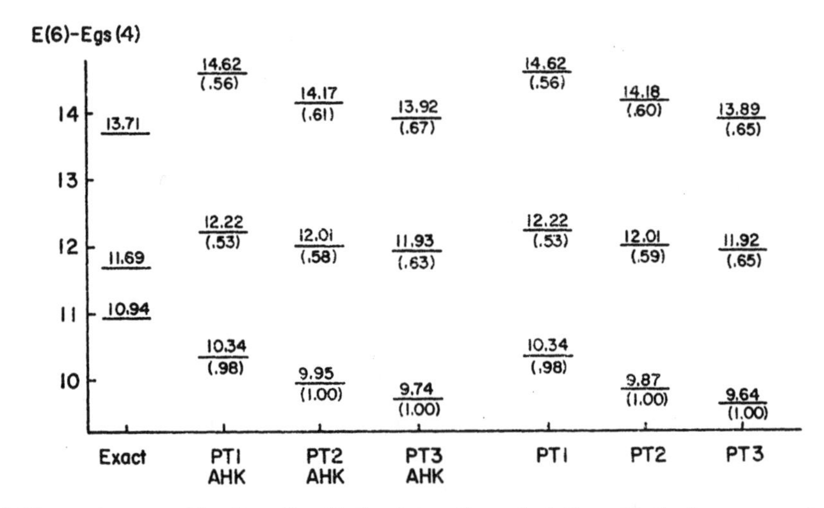

Fig. 8. The spectrum resulting from the effective-interaction calculation with the force parameters of eq. (4.3) and the unperturbed single-particle basis with the HF single-particle energies. The notation of fig. 7 is also used here.

a realistic system but to show how well HF can work in an extreme case. The force parameters used here are (in arbitrary units)

$$U = -0.5, \quad V = -0.5, \quad W = -0.1, \qquad \varepsilon_z = 0, \quad \varepsilon_y = 2, \quad \varepsilon_x = 2.5. \qquad (4.3)$$

These force parameters are clearly strong since they are comparable in size to the separation of the single-particle energies.

The exact low-lying spectrum for $E(6)-E_{g.s.}(4)$ is shown in table 4 along with $E_{g.s.}(4)$ and the model-space overlaps $\langle P\Psi_i|P\Psi_i\rangle$ in the unperturbed basis. Note that none of the low-lying states have a significant overlap in the model space. Based on previous calculations with the AHK algorithm it is expected that the expansion of H_{eff} will diverge when a model space defined with the unperturbed H_0 is used.

The results of the HF calculation in the $N = 4$ system are shown in table 5. Here the HF single-particle wave functions and energies are substantially different from the unperturbed values. Large admixtures of several of the unperturbed η_β components are present in each ϕ_α^{HF} and the separation between the different ε_α^{HF} are much larger than in the unperturbed case. Contrary to the "realistic" force used in subsect. 4.1, the HF single-particle basis does an excellent job of reproducing the ground-state energy of the $N = 4$ system with $E_{g.s.}(4) = -26.32$ and $E_{g.s.}^{HF}(4) = -26.27$.

When the effective interaction is computed for the three different cases many of the results, shown in table 6 and figs. 7 and 8, are qualitatively the same as those in subsect. 4.1. However, there are some important differences. As expected, the expansion of H_{eff} diverges in low orders for case (a), as shown in table 6, both when the AHK algortihm is used and when the folded-diagram series grouped in powers of H_I is used. However, the AHK algorithm yields eigenvalues that diverge more slowly indicating the importance of the higher order folded diagrams. In addition, the AHK algorithm yields wave function overlaps that are quite good whereas they are very poor when the folded-diagram series is used.

For the $N = 6$ system, just as for the $N = 4$ system, HF yields a better basis for H_{eff} calculations than it did for the "realistic" force of subsect. 4.1. This is evident in case (b) where the agreement of the eigenvalues and eigenvectors of H_{eff} with the exact results is outstanding. Here the eigenvalues converge to the lowest three states of the exact spectrum. Based on previous work in the state formulation, this indicates that they are the ones with the largest $\langle P\Psi_i|P\Psi_i\rangle$ overlaps in the HF basis.

The quality of the basis is also reflected in the matrix elements of the individual diagrams shown in table 7. The magnitude of the matrix elements of third order diagrams are reduced between three and six orders of magnitude in going from the unperturbed to the HF single-particle potential. This is quite a drastic change in size. In particular, the D_{23} matrix element of the diagram 3-41 changes from -0.210 to -0.000002 and the D_{11} matrix element of diagram 3-30 changes from -4.510 to -0.003. These results show how powerful HF can be in constructing a good model space for use in effective-interaction calculations.

Significant improvement in the resulting spectrum is also obtained when the HF energies are used with the unperturbed basis to compute H_{eff} as shown in fig. 8. The eigenvalues now appear to be converging in low orders and the upper two states reproduce the exact states 2 and 3 fairly well. However the overall spectrum is not as compressed as the exact one and the resulting wave function overlaps are poor indicating that further significant improvements are made when the correct HF single-particle basis is also used.

TABLE 7

The value of matrix elements of topologies calculated with different wave functions and energies for the force parameters given by eq. (4.3)

Topology	Case (a)		Case (b)		Case (c)	
	D_{11}	D_{23}	D_{11}	D_{23}	D_{11}	D_{23}
1-1	0.000	−0.707	−0.600	0.048	0.000	−0.707
2-1	−3.550	−1.520	−0.088	0.002	−0.430	−0.179
2-2	2.833	2.003	0.013	0.000	0.289	0.205
2-3	−1.050	−1.167	−0.017	−0.009	−0.117	−0.091
2-4	−0.250	−0.314	−0.002	0.000	−0.024	−0.034
3-1	−1.540	−2.970	−0.006	0.000	−0.020	−0.027
3-2	−8.270	−5.431	−0.014	0.000	−0.118	−0.040
3-3	−0.727	−6.686	0.002	0.000	−0.007	−0.069
3-4	3.361	2.377	0.001	0.000	0.035	0.025
3-5	3.428	2.309	0.001	0.000	0.035	0.025
3-6	3.428	2.424	0.001	0.000	0.035	0.025
3-7	−2.415	1.926	−0.004	0.000	−0.027	0.020
3-8	1.526	1.305	0.000	0.000	0.017	0.013
3-9	1.526	1.079	0.000	0.000	0.017	0.012
3-11	−4.176	−3.605	0.000	0.000	−0.047	−0.027
3-12	−4.176	−1.858	0.000	0.000	−0.047	−0.023
3-13	−0.879	−1.882	−0.001	0.000	−0.010	−0.014
3-14	−0.879	−0.846	−0.001	0.000	−0.010	−0.011
3-15	2.076	2.022	0.001	0.000	0.024	0.014
3-16	2.076	0.824	0.001	0.000	0.024	0.011
3-17	−0.521	−0.631	0.000	0.000	−0.007	−0.005
3-18	−0.521	−0.464	0.000	0.000	−0.007	−0.005
3-28	−1.184	−10.033	−0.001	0.001	−0.019	−0.059
3-29	−1.549	−0.130	−0.006	0.000	−0.022	−0.006
3-30	−4.510	−8.141	−0.003	−0.001	−0.052	−0.059
3-31	1.440	2.033	0.000	−0.001	0.014	0.027
3-32	1.440	4.375	0.000	−0.001	0.014	0.040
3-33	−0.407	0.267	−0.001	0.001	−0.005	−0.002
3-34	−0.407	−1.108	−0.001	0.000	−0.005	−0.010
3-35	−0.284	0.285	0.000	0.000	−0.003	0.004
3-36	−0.448	−0.741	0.001	0.000	−0.005	−0.008
3-37	0.300	0.963	0.000	0.000	0.003	0.010
3-38	0.300	1.374	0.000	0.000	0.003	0.015
3-39	−0.639	−0.697	0.000	0.000	−0.006	−0.008
3-40	−0.639	−0.681	0.000	0.000	−0.006	−0.008
3-41	−0.106	−0.210	0.000	0.000	−0.001	−0.002
3-42	−0.106	−0.168	0.000	0.000	−0.001	−0.002
3-44	0.521	0.631	0.000	0.000	0.007	0.005
3-45	0.521	2.051	0.000	0.000	0.007	0.016
3-46	0.144	0.203	0.000	0.000	0.002	0.002
3-47	0.144	0.655	0.000	0.000	0.002	0.005
3-48	−1.880	−3.294	−0.001	0.000	−0.021	−0.026
3-49	−1.880	−2.236	−0.001	0.000	−0.021	−0.024
3-50	1.504	2.666	0.001	0.000	0.016	0.024
3-51	1.504	2.111	0.001	0.000	0.016	0.022
3-52	−0.196	−0.203	0.000	0.000	−0.002	−0.002
3-53	−0.196	−0.194	0.000	0.000	−0.002	−0.002

TABLE 7 (continued)

Topology	Case (a)		Case (b)		Case (c)	
	D_{11}	D_{23}	D_{11}	D_{23}	D_{11}	D_{23}
3-54	0.654	0.553	0.000	0.000	0.007	0.005
3-55	0.654	1.245	0.000	0.000	0.007	0.014
3-56	−0.354	−0.250	0.000	0.000	−0.004	−0.002
3-57	−0.354	−0.609	0.000	0.000	−0.004	−0.007
F-1	0.508	5.079	0.002	0.000	0.005	0.030
F-2	−0.300	−0.963	0.000	0.000	−0.003	−0.010
F-3	0.650	1.461	−0.001	0.000	−0.006	0.012
F-4	0.106	0.168	0.000	0.000	0.001	0.002

The notation is the same as in table 3.

4.3. CONCLUSION

Clearly the model used here may be too simplistic, so that the linked-valence expansions of H_{eff} and the HF method are better approximations than they would be in a real system. However, these results do indicate that HF can be a useful procedure for constructing a good model space which may be a crucial factor in improving the convergence properties of the linked-valence expansions of the effective interaction. In addition the convergence of the expansions does not seem to be adversely affected by the presence of intruder states.

However, care must be taken if no HF insertions are included on the diagrams summed for H_{eff} when the HF self-consistent single-particle potential is not used. This approximation may yield poor agreement with the exact spectrum even if the single-particle basis used is very similar to the HF one.

We want to thank E. M. Krenciglowa and K. F. Ratcliff for many informative discussions and D. J. Rowe for helpful correspondence.

Appendix

All of the linked-valence topologies used in the construction of the $\hat{Q}$-box (and $\mathscr{V}_{\text{eff}}$) in this work are given in fig. 9. Each topology is a Hugenholtz diagram [23]), which is just a compact way of drawing a diagram with antisymmetrized vertices. Each topology represents from one to four diagrams, the one given in the figure plus all those that can be obtained from it by exchanging the external pairs of lines. This is done only if the exchange of the lines is not already implicitly included by the use of antisymmetrized vertices. In the figure, an upgoing arrow represents a particle line, a downgoing arrow represents a hole line and a dot represents an antisymmetrized matrix element of H_{I}.

If the terms of the folded-diagram series in eq. (3.2) are grouped in powers of H_{I}, then all the one-body and two-body topologies, both folded and non-folded, through third order in H_1 that are summed for $\mathscr{V}_{\rm eff}$ are shown in fig. 9. Here it is assumed that the HF basis is being used. In first order there is only one topology, 1-1, in second order there are five topologies, 2-1 through 2-5 and in third order there are sixty-two topologies, 3-1 through 3-57 and F-1 through F-5. However, because there are no possible particle states outside of the model space in the model, topologies 2-5, 3-19 through 3-27 and F-5 make no contribution to $\mathscr{V}_{\rm eff}$ or the $\hat{Q}$-box. In addition topologies 3-10 and 3-43 are identically zero because H_1 does not allow a two-hole to two-hole transition.

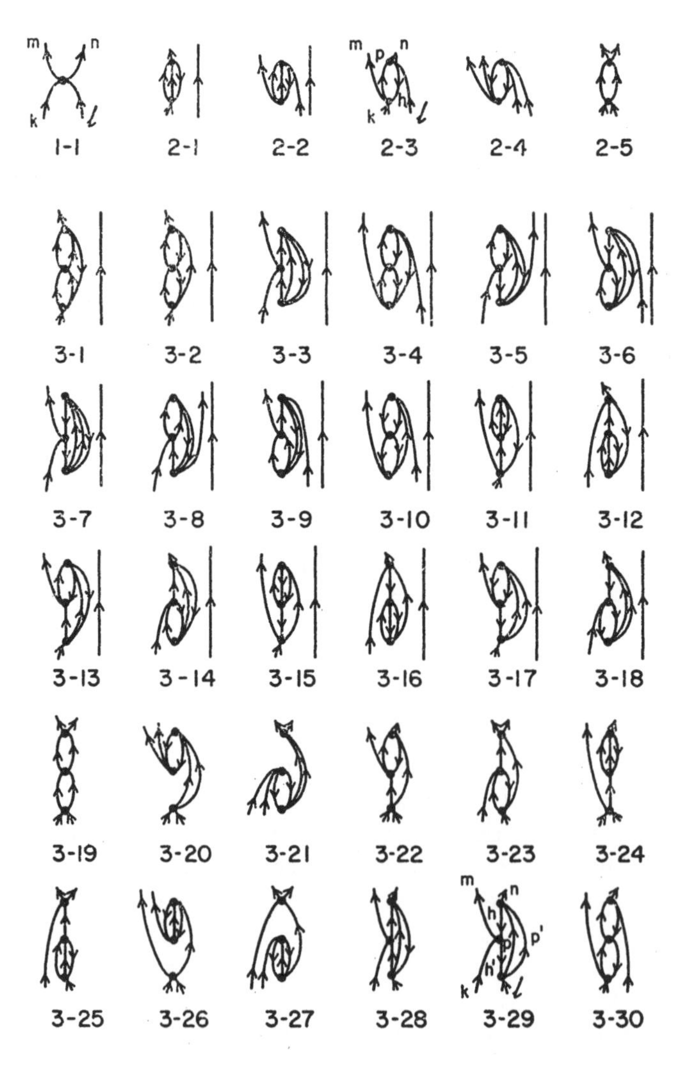

Fig. 9.

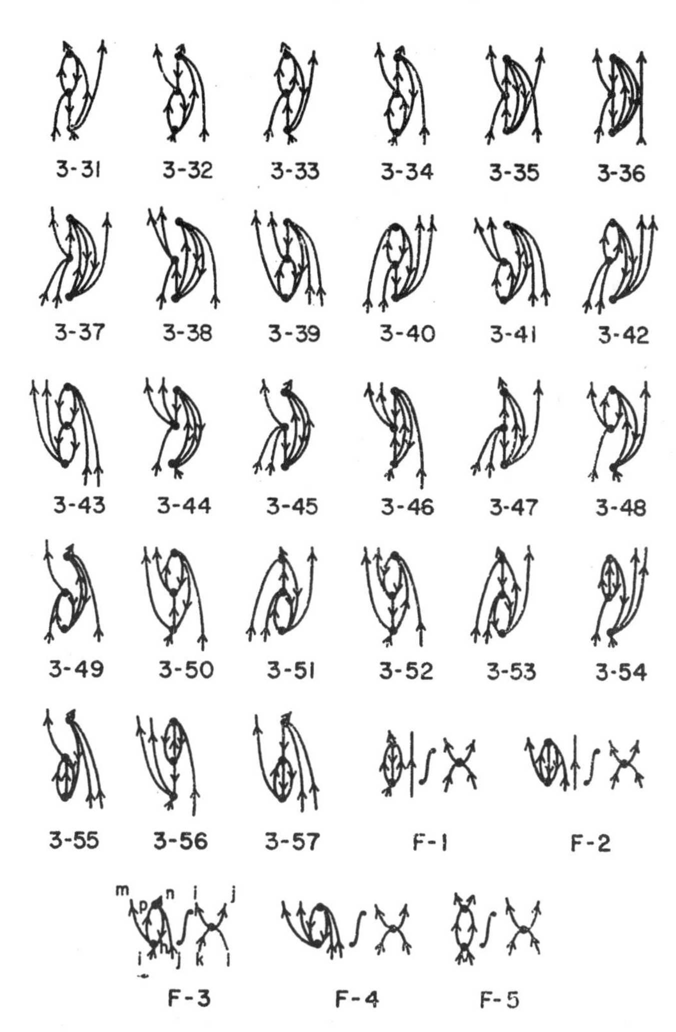

Fig. 9. The topologies used in the effective-interaction calculations. The notation and the evaluation of these topologies are explained in the appendix.

Since it is assumed that the HF basis is being used the topologies contained in the $\hat{Q}$-box, when it is expanded in perturbation theory through third order in H_1, are just the non-folded topologies of fig. 9. This is true because the $\hat{Q}_1$ box, which contains all the non-folded diagrams in the folded-diagram series, is identical to the $\hat{Q}$-box through third order in H_1 if the HF basis is used. It is only in fourth order that the $\hat{Q}$-box and $\hat{Q}_1$ box differ.

To illustrate the rules for evaluating Hugenholtz diagrams, the analytic expressions for a few representative diagrams of fig. 9 are given here:

$$1\text{-}1 = V_{mnkl}, \tag{A.1}$$

$$2\text{-}3 = +\sum_{ph} \frac{V_{hnpl}V_{mpkh}}{\omega+\varepsilon_h-\varepsilon_m-\varepsilon_l-\varepsilon_p}, \tag{A.2}$$

$$3\text{-}29 = -\tfrac{1}{2}\sum_{pp'hh'} \frac{V_{hnpp'}V_{mh'kh}V_{pp'h'l}}{(\omega+\varepsilon_h-\varepsilon_m-\varepsilon_p-\varepsilon_{p'})(\omega+\varepsilon_{h'}-\varepsilon_k-\varepsilon_p-\varepsilon_{p'})}, \tag{A.3}$$

$$F\text{-}3 = -\sum_{ijhp} \frac{V_{hnpj}V_{ijkl}V_{pmhi}}{(\omega+\varepsilon_h-\varepsilon_m-\varepsilon_p-\varepsilon_j)(\omega+\varepsilon_h+\varepsilon_i-\varepsilon_k-\varepsilon_l-\varepsilon_m-\varepsilon_p)}, \tag{A.4}$$

where ω is the starting energy and V_{mnkl} is an antisymmetrized two-body matrix element of H_{I}. The starting energy ω is just the sum of single-particle energies of the incoming external lines. For example, $\omega = \varepsilon_k+\varepsilon_l$ for topology 3-29.

No attempt has been made to calculate the linked-valence $\hat{Q}$-box topologies which are fourth order in H_{I}. The number of these topologies is very large and it becomes impractical to include them in real calculations. Typical numerical values of the diagrams included in the present work can be found in tables 3 and 7.

References

1) T. T. S. Kuo and G. E. Brown, Nucl. Phys. **85** (1966) 40
2) B. R. Barrett and M. W. Kirson, Nucl. Phys. **A148** (1970) 145
3) T. H. Schucan and H. A. Weidenmüller, Ann. of Phys. **73** (1972) 108
4) T. H. Schucan and H. A. Weidenmüller, Ann. of Phys. **76** (1973) 483
5) E. M. Krenciglowa and T. T. S. Kuo, Nucl. Phys. **A235** (1974) 171
6) M. R. Anastasio and T. T. S. Kuo, Nucl. Phys. **A238** (1975) 79
7) M. R. Anastasio, J. W. Hockert and T. T. S. Kuo, Phys. Lett. **53B** (1974) 221
8) H. M. Hofmann, S. Y. Lee, J. Richert, H. A. Weidenmüller and T. H. Schucan, Ann. of Phys. **85** (1974) 410
9) H. M. Hofmann, S. Y. Lee, J. Richert, H. A. Weidenmüller and T. H. Schucan, Phys. Lett. **45B** (1973) 421
10) H. M. Hofmann, J. Richert and T. H. Schucan, Z. Phys. **268** (1974) 293
11) C. M. Vincent and S. Pittel, Phys. Lett. **47B** (1973) 327
12) M. R. Anastasio, T. T. S. Kuo and J. B. McGrory, Phys. Lett. **57B** (1975) 1
13) N. Meshkov, Phys. Rev. **C3** (1971) 2214
14) H. J. Lipkin, N. Meshkov and A. J. Glick, Nucl. Phys. **62** (1965) 188
15) T. T. S. Kuo, S. Y. Lee and K. F. Ratcliff, Nucl. Phys. **A176** (1971) 65
16) T. T. S. Kuo, Ann. Rev. Nucl. Sci. **24** (1974) 101
17) P. J. Ellis and H. A. Mavromatis, Nucl. Phys. **A175** (1971) 309
18) P. J. Ellis and E. Osnes, Phys. Lett. **41B** (1972) 97
19) D. J. Rowe, Phys. Lett. **44B** (1973) 155
20) K. T. Hecht, Nucl. Phys. **62** (1965) 1
21) H. J. Lipkin, Lie groups for pedestrians (North-Holland, Amsterdam, 1965)
22) B. H. Brandow, Rev. Mod. Phys. **39** (1967) 771
23) N. M. Hugenholtz, Physica **23** (1957) 481
24) E. M. Krenciglowa, Ph.D. thesis, State University of New York at Stony Brook (1975)

Theory of Energy-Independent Nuclear Optical-Model Potentials

T. T. S. Kuo[(a)] and F. Osterfeld
Institut für Kernphysik, Kernforschungsanlage Jülich, D-5170 Jülich, Federal Republic of Germany

and

S. Y. Lee
Physics Department, State University of New York at Stony Brook, Stony Brook, New York 11794
(Received 11 January 1980)

With use of an equation-of-motion approach, a new microscopic theory of the nuclear optical-model potential which is explicitly energy independent has been derived. This theory suggests the possibility of replacing a wide range of conventional energy-dependent optical-model potentials by a single, energy-independent but nonlocal, optical-model potential.

PACS numbers: 24.10.Ht

Several authors[1,2] have already demonstrated quite generally that a generalized optical-model potential exists which produces the exact asymptotic wave function for elastic nucleon-nucleus scattering when used in solving the one-body Schrödinger equation. This potential is complex, nonlocal, and depends explicitly on the incident energy E of the full many-body wave function.[3-5] In the Green's-function formalism[2] (and also in Feshbach's formalism[1]) the optical-model wave function $\rho_E(\vec{k})$ (in the k representation) is obtained by solving the nonlocal Schrödinger equation

$$T\rho_E(\vec{k}) + \int d^3k' V_G(\vec{k},\vec{k}',E)\rho_E(\vec{k}') = E\rho_E(\vec{k}), \quad (1)$$

where T is the kinetic energy operator and $V_G(\vec{k},\vec{k}',E)$ is the mass operator of the one-particle Green's-function which depends on the independent variables $\vec{k}$, $\vec{k}'$, and E, where E is the exact eigenenergy of the many-body wave function.

The purpose of this paper is to present a new microscopic derivation of the optical potential which leads to a formally energy-independent potential; i.e., we shall show that the optical-model wave function can be obtained from the following nonlocal Schrödinger equation:

$$T\rho_E(\vec{k}) + \int d^3k' V^{\rm eff}(\vec{k},\vec{k}',\epsilon_{\vec{k}'})\rho_E(\vec{k}') = E\rho_E(\vec{k}). \quad (2)$$

VOLUME 45, NUMBER 10 PHYSICAL REVIEW LETTERS 8 SEPTEMBER 1980

The only but essential difference between Eqs. (1) and (2) is that $V^{\rm eff}(\vec{k},\vec{k}',\epsilon_{\vec{k}})$ depends only on the single-particle energies $\epsilon_{\vec{k}}$ of the unperturbed Hamiltonian T (or $H_0=T+\hat{V}$, where $\hat{V}$ is an arbitrary auxiliary potential) and *not* on the exact eigenenergy E. Since $\epsilon_{\vec{k}'}$ is a unique function of $\vec{k}'$, $V^{\rm eff}(\vec{k},\vec{k}',\epsilon_{\vec{k}'})$ has one variable less than $V_G(\vec{k},\vec{k}',E)$. We call the energy dependence of $V^{\rm eff}(\vec{k},\vec{k}',\epsilon_{\vec{k}'})$ *implicit* (with respect to $\epsilon_{\vec{k}'}$) in contrast to that of $V_G(\vec{k},\vec{k}',E)$ which is explicitly energy dependent.

The optical potential $V^{\rm eff}(\vec{k},\vec{k}')$ (we omit $\epsilon_{\vec{k}'}$ because it is not an independent variable) has the following two important properties:

(i) It is particularly suited for a microscopic calculation of the optical potential since the potential has to be calculated only *once* with respect to the basis of a solvable unperturbed Hamiltonian and *not* for various incident energies E.

(ii) A microscopically calculated optical potential which is only implicitly energy dependent may eventually uniquely replace a wide range of different potentials which are now in use in nuclear reaction theory.

In the following we shall prove that the potential $V^{\rm eff}(\vec{k},\vec{k}')$ exists and we shall also present an exact diagrammatic expansion of it. This expansion is based on the Rayleigh-Schrödinger type of perturbation theory and, as we shall show, contains folded diagrams. The method that we use for deriving the optical potential is based on the equation-of-motion technique.[6]

We start from the many-body Hamiltonian $H=T+V$, where T is the kinetic energy and V the realistic nucleon-nucleon interaction. We consider the $(A+1)$-body-system Schrödinger equation for stationary state,

$$H|\psi_{A+1}^{(+)}(E)\rangle\rangle = E|\psi_{A+1}^{(+)}(E)\rangle\rangle. \tag{3}$$

The function $\psi^{(+)}$ is the exact many-body wave function and obeys the standard asymptotic boundary conditions [superscript (+) indicates incident plane wave in the elastic channel, radially outgoing waves in reaction channels]. In the case of elastic scattering, we are only interested in that part of $|\psi_{A+1}^{(+)}(E)\rangle\rangle$ where A nucleons form the true ground state $|\psi_A^0\rangle$ of the A-body system and where one nucleon is in a scattering state. Therefore we project out from $|\psi_{A+1}^{(+)}(E)\rangle\rangle$ this particular part and define a model-space problem by

$$H_{\rm eff}P|\psi_{A+1}^{(+)}(E)\rangle\rangle = (E-E_A^0)P|\psi_{A+1}^{(+)}(E)\rangle\rangle, \tag{4}$$

where E_A^0 is the A-body ground-state energy. The projection operator P is defined by

$$P=\textstyle\sum_{\vec{k}}|\bar{\varphi}_{\vec{k}}\rangle\langle\varphi_{\vec{k}}|, \quad \langle\varphi_{\vec{k}}| = \langle\psi_A^0|a_{\vec{k}}, \tag{5}$$

where the operator $a_{\vec{k}}^+$ ($a_{\vec{k}}$) creates (destroys) a particle in a momentum state $\vec{k}$ and the wave functions $\langle\varphi_{\vec{k}}|$ fulfill the biorthogonal relation $\langle\varphi_{\vec{k}}|\bar{\varphi}_{\vec{k}'}\rangle = \delta(\vec{k}-\vec{k}')$. Now, the essential problem to be solved consists in the derivation of the effective Hamiltonian $H_{\rm eff}$, since knowing $H_{\rm eff}$ we obtain readily from Eq. (4) a one-body Schrödinger equation for the quantity

$$\rho_E(\vec{k}) = \langle\psi_A^0|a_{\vec{k}}|\psi_{A+1}^{(+)}(E)\rangle\rangle \tag{6}$$

which is just the optical-model wave function in the $\vec{k}$ representation.

For the derivation of $H_{\rm eff}$ we start from the equation of motion for the operator $a_{\vec{k}}$:

$$\begin{aligned}&\langle\psi_A^0|[a_{\vec{k}},H]|\psi_{A+1}^{+}(E)\rangle\rangle\\ &\quad= (E-E_A^0)\langle\psi_A^0|a_{\vec{k}}|\psi_{A+1}^{(+)}(E)\rangle\rangle.\end{aligned} \tag{7}$$

The commutator $[a_{\vec{k}},H]$ can be written as $\epsilon_{\vec{k}}a_{\vec{k}}+A_{\vec{k}}$, where $\epsilon_{\vec{k}}$ is the single-particle kinetic energy and $A_{\vec{k}}$ is defined by

$$A_{\vec{k}} = \sum_{\beta\gamma\delta} a_\beta^\dagger(\vec{k}\beta|V|\gamma\delta)a_\delta a_\gamma. \tag{8}$$

With use of the Gell-Mann–Low theorem,[7] the ground state ψ_A^0 can be obtained by applying the time evolution operator to the unperturbed ground state Φ_A^0. We shall prove also that the true scattering wave function $|\psi_{A+1}^{(+)}(E)\rangle\rangle$, satisfying Eq. (7), can be obtained in a similar procedure, i.e.,

$$\frac{|\psi_{A+1}^{(+)}(E)\rangle\rangle}{\langle\langle\Phi_{A+1}(E)|\psi_{A+1}^{(+)}(E)\rangle\rangle} = \frac{U(0,-\infty)|\Phi_{A+1}(E)\rangle\rangle}{\langle\langle\Phi_{A+1}(E)|U(0,-\infty)|\Phi_{A+1}(E)\rangle\rangle}, \tag{9}$$

with the parent state $\Phi_{A+1}(E)$ given by

$$|\Phi_{A+1}(E)\rangle\rangle = \mathbf{S}_{\vec{k}}\, c_{\vec{k}}(E)a_{\vec{k}}^\dagger|\Phi_A^0\rangle, \tag{10}$$

where a summation is carried over the discrete set together with an integration over the continuum set of states. The coefficients $c_{\vec{k}}(E)$ are to be determined. Substituting the above into Eq. (7), we ob-

VOLUME 45, NUMBER 10 PHYSICAL REVIEW LETTERS 8 SEPTEMBER 1980

tain

$$\langle \Phi_A{}^0 | U(\infty, 0) A_{\vec{k}} U(0, -\infty) | \Phi_{A+1}(E) \rangle / (D_A D_{A+1})$$
$$= (E - E_A{}^0 - \epsilon_{\vec{k}}) \langle \Phi_A{}^0 | U(\infty, 0) a_{\vec{k}} U(0, -\infty) | \Phi_{A+1}(E) \rangle / (D_A D_{A+1}), \qquad (11)$$

where D_A and D_{A+1} stand for $\langle \Phi_A{}^0 | U(\infty, 0) | \Phi_A{}^0 \rangle$ and $\langle \Phi_{A+1}(E) | U(0, -\infty) | \Phi_{A+1}(E) \rangle$, respectively.

We now analyze[8,9] the diagrammatic structure of this equation. The unlinked diagrams, denoted as S, of the right-hand-side and left-hand-side numerator are identical and can be factorized out as a whole, yielding

$$(N_L)_{\vec{k}} S / (D_A D_{A+1}) = (E - E_A{}^0 - \epsilon_{\vec{k}}) (N_R)_{\vec{k}} S / (D_A D_{A+1}) \qquad (12)$$

with

$$(N_L)_{\vec{k}} \equiv \langle \Phi_A{}^0 | U(\infty, 0) A_{\vec{k}} U(0, -\infty) | \Phi_{A+1}(E) \rangle_{\mathcal{L}} \qquad (12a)$$

$$(N_R)_{\vec{k}} \equiv \langle \Phi_A{}^0 | U(\infty, 0) a_{\vec{k}} U(0, -\infty) | \Phi_{A+1}(E) \rangle_{\mathcal{L}}, \qquad (12b)$$

where $\mathcal{L}$ denotes the linked diagrams defined as those with all their vertices linked to the external particle line. Hence obviously N_L and N_R have the chain structure shown in Fig. 1. The boxes, the so-called Q boxes,[8,9] represent the sum of all irreducible diagrams beginning and terminating with one single-particle line. As usual an irreducible diagram is defined as a linked diagram which cannot be separated into two connected pieces, each with at least one vertex, by cutting one particle line. Notice the similarity in structure between N_L and N_R of Fig. 1. The only difference between them is that N_L is terminated at time $t = 0$ by the operator $A_{\vec{k}}$ of Eq. (8) while N_R terminates at time $t = 0$ with the operator $a_{\vec{k}}$. Therefore the last box in N_L is a slashed Q box indicating that the operator $A_{\vec{k}}$ acts at time $t = 0$, and as a result N_L does not have the free propagator term contained in N_R. The chain structure of N_L and N_R suggests clearly that we can write N_L as a product of N_R and a quantity containing the slashed Q boxes. This can indeed be done by using a folded-diagram factorization procedure[8,9] leading to the basic result

$$(N_L)_{\vec{k}} = \int d^3k' V_{\vec{k}\vec{k}'}{}^{\text{eff}} (N_R)_{\vec{k}'}, \qquad (13)$$

where the effective interaction V^{eff} is given by the linked-diagram expansion shown in the bottom part of Fig. 1. Combining Eqs. (6), (7), (11), (12), and (13), we obtain as a final result the one-body Schrödinger equation for the optical-model wave function $\rho_E(\vec{k})$:

$$\epsilon_{\vec{k}} \rho_E(\vec{k}) + \int d^3k' V_{\vec{k}\vec{k}'}{}^{\text{eff}} \rho_E(\vec{k}') = (E - E_A{}^0) \rho_E(\vec{k}). \qquad (14)$$

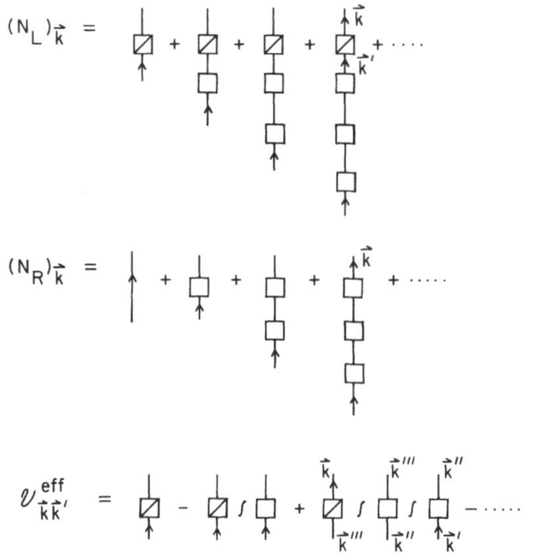

FIG. 1. Diagrammatic structures of N_L, N_R, and V^{eff}.

Clearly V^{eff} is the energy-independent optical-model potential of Eq. (2) we have been looking for. Equation (14) serves to determine the optical-model wave function $\rho_E(\vec{k})$. From Eqs. (6), (7), and (12) we see that $\rho_E(\vec{k})$ is just $(N_R)_{\vec{k}} S / (D_A D_{A+1})$, where $(N_R)_{\vec{k}}$ is related to the coefficients $c_{\vec{k}}(E)$ through Eqs. (12b) and (10). Thus Eq. (14), which determines $\rho_E(\vec{k})$, also formally determines $c_{\vec{k}}(E)$. This completes our proof that starting from the equation of motion (7) we derive the optical-model potential of Eq. (14) and Fig. 1. And this derivation is indeed quite straightforward.

A special feature of our optical-model potential is that it contains folded diagrams, as shown by its diagrammatic expansion in Fig. 1. For example, the third term is twice folded. The symbols $\int$ indicate folding operations.[8,9] The folded diagrams can be conveniently calculated and should not add much computational difficulty to the present theory. This is based on some bound-state folded-diagram calculations[10] with use of a realistic nucleon-nucleon potential. Some typical diagrams of our optical-model potential are given in Fig. 2. With use of standard methods,[8,9] the value for diagram α is obtained as

$$\sum \tfrac{1}{2}vv/(\epsilon_{\vec{k}'} - \epsilon_a - \epsilon_b + \epsilon_c + i\eta), \tag{15a}$$

and that of diagram β as

$$-\sum \tfrac{1}{2}vvv/[(\epsilon_{\vec{k}''} - \epsilon_a - \epsilon_c + \epsilon_b + i\eta)(\epsilon_{\vec{k}'} - \epsilon_a - \epsilon_c + \epsilon_b + i\eta)], \tag{15b}$$

where the summations run over all intermediate indices and for brevity we have omitted the indices associated with the antisymmetric v vertices. Note that diagram β is a folded diagram. In the above, we take the limit of $\eta \to 0+$. This comes from the boundary condition of $\psi_{A+1}^{(+)}(E)$ that, at time $t=0$, it contains a component where we have one particle in an outgoing spherical wave. By making a diagrammatic expansion of $\psi_{A+1}^{(+)}(E)$ by way of Eq. (9), we see readily that this boundary condition is assured by the above limiting procedure. These $i\eta$ factors will make our optical-model potential generally complex. Several observations are now in order:

(i) The central result of this paper is the derivation of an optical-model potential which is explicitly energy independent. As shown by Eqs. (15a) and (15b), individual diagrams of V^{eff} depend energetically only on the single-particle energies ϵ's defined by the unperturbed Hamiltonian. This is in contrast to the optical-model potential $V_G(\omega_E)$ of the Green's-function formalism[2,5,11,12] which is explicitly dependent on the exact energy $\omega_E = E - E_A{}^0$.

For the energy-dependent potentials we expect threshold effects since with increasing projectile energy new inelastic or reaction channels open up energetically suddenly. At least these threshold effects appear in the imaginary part of the potential. Below the inelastic threshold the energy-dependent potential is Hermitian (no imaginary part). Now the question arises whether there is a contradiction between the energy-dependent and energy-independent optical potentials. The answer is, no! Indeed, by using a partial summation method[8,13] for summing up the folded-diagram series one can obtain an exact mathematical relation between the energy-independent V^{eff} and the Green's-function optical potential, namely

$$(T + V^{\text{eff}}) = \int d\omega_E |\rho_E\rangle\langle\tilde{\rho}_E|[T + V_G(\omega_E)]|\rho_E\rangle\langle\tilde{\rho}_E|, \tag{16}$$

where the optical-model wave function ρ_E obeys the biorthogonal relation $\langle\tilde{\rho}_E|\rho_{E'}\rangle = \delta(E - E')$. From Eq. (16) one can see that V^{eff} is a particular energy-averaged potential obtained by averaging over $V_G(\omega_E)$ with respect to energy. Hence if we know ρ_E and $V_G(\omega_E)$ for all energies, we can construct, in principle, an energy-independent V^{eff} as shown by Eq. (16). In reality, this is of course not practical. What we have done in this paper is to have attained a systematic method for calculating V^{eff}, as indicated by Fig. 1, without prior knowledge of ρ_E and $V_G(\omega_E)$.

If we act with the optical-model operator in Eq. (14) onto a wave function $|\rho_{E_0}\rangle$ with E_0 below threshold, then, of course, only the Hermitian part of V^{eff} will be active. In this sense V^{eff} also includes threshold effects implicitly. This one can also easily understand from a physical point of view. The optical-model wave function $\rho_E(\vec{k})$ of Eq. (6) is expected to be a smooth function of k, peaked around $k_0 \simeq (2m\omega_E/\hbar^2)^{1/2}$. Then from Eq. (14) we see that $\rho_E(\vec{k})$ depends primarily on the portion of $V_{\vec{k}\vec{k}'}{}^{\text{eff}}$ with $k \simeq k' \simeq k_0$. Hence for $\rho_E(\vec{k})$ the important contribution to V^{eff} comes from the diagrams with starting energies $\epsilon_{\vec{k}}$ [see Eqs. (15a) and (15b)] in the vicinity of $\epsilon_{\vec{k}'} = \omega_E$. If ω_E is below inelastic threshold then the incoming nucleon is essentially sensitive to only those matrix elements $V_{\vec{k},\vec{k}'}$ which are real and therefore Hermitian. To give a further support of our theory, let us mention the work of Johnson[14] in which he showed that by the inclusion of folded diagrams one can obtain an energy-independent nucleon-nucleon potential.

(ii) In order to support our theory by numerical findings we refer to the analyses of neutron scattering data which Perey and Buck[15] have performed using different sets of nonlocal optical potentials. They found by variation of the nonlocal-

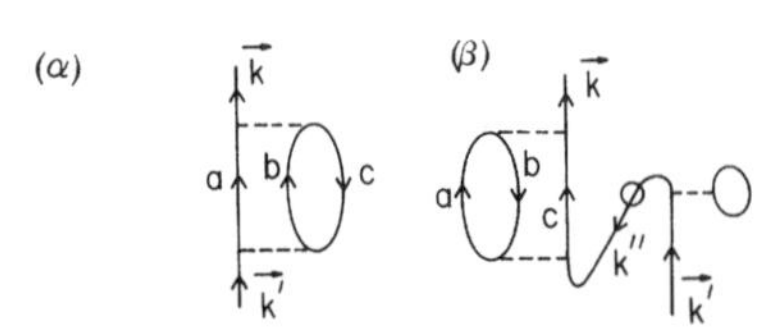

FIG. 2. Typical diagrams contained in the energy-independent optical-model potential $V_{\vec{k}\vec{k}'}{}^{\text{eff}}$.

ity a nonlocal potential which could fit neutron elastic scattering data in the energy range from 0.4 to 24 MeV and where the optical-model parameters were energy independent. This is, of course, a remarkable result since we know that the absorption, for instance, is changing quite dramatically by going from 0.4 to 24 MeV incident projectile energy. We believe that these findings of Perey and Buck strongly support our result that there exists a nonlocal potential, which is not explicitly energy-dependent, which describes elastic nucleon scattering in a wide energy range. The theory presented in this paper may serve as a convenient tool in deriving such a potential. Actual calculation of an energy-independent optical-model potential as outlined in this paper is in progress.

We are grateful to Professor A. Faessler, Professor A. Weiguny, Professor M. Stingl, and Professor G. Vagradov for many helpful discussions. One of us (T.T.S.K.) wishes to thank Professor A. Faessler for his hospitality and assistance during his stay at Jülich. We thank Professor G. E. Brown for a critical reading of the manuscript. One of us (T.T.S.K.) is an Alexander von Humboldt Foundation awardee. This work was partially supported by the U. S. Department of Energy under Contract No. EY-76-S-02-3001.

[a]Permanent address: State University of New York at Stony Brook, Stony Brook, N.Y. 11794.

[1]H. Feshbach, Ann. Rev. Nucl. Sci. 8, 49 (1958).

[2]J. S. Bell and E. J. Squires, Phys. Rev. Lett. 3, 96 (1959); J. S. Bell, in *Lectures on the Many-Body Problem*, edited by E. R. Caianello (Academic, New York, 1962).

[3]H. Feshbach, Ann. Phys. (N.Y.) 5, 357 (1968), and 19, 287 (1962).

[4]N. Austern, *Direct Nuclear Reaction Theory* (Wiley, New York, 1970).

[5]F. Villars, in *Fundamentals in Nuclear Theory*, edited by A. de Shalit and C. Villi (International Atomic Energy Agency, Vienna, 1967).

[6]D. J. Rowe, Rev. Mod. Phys. 40, 153 (1968).

[7]M. Gell-Mann and F. Low, Phys. Rev. 84, 150 (1951).

[8]T. T. S. Kuo, S. Y. Lee, and K. F. Ratcliff, Nucl. Phys. A176, 65 (1971).

[9]T. T. S. Kuo and E. M. Krenciglowa, Nucl. Phys. A342, 454 (1980). This reference reports a similar equation-of-motion approach for the bound-state problems.

[10]J. Shurpin, D. Strottman, T. T. S. Kuo, M. Conze, and P. Manakos, Phys. Lett. 69B, 395 (1977).

[11]J. P. Jeukenne, A. Lejeune, and C. Mahaux, Phys. Rep. 25C, 83 (1976).

[12]N. Vinh Mau and A. Bouyssy, Nucl. Phys. A257, 189 (1976).

[13]E. M. Krenciglowa and T. T. S. Kuo, Nucl. Phys. A235, 171 (1974).

[14]M. Johnson, Ann. Phys. (N.Y.) 97, 400 (1976).

[15]F. G. Perey and B. Buck, Nucl. Phys. 32, 353 (1962).

ANNALS OF PHYSICS **132**, 237–276 (1981)

A Simple Method for Evaluating Goldstone Diagrams in an Angular Momentum Coupled Representation*

T. T. S. KUO,[†] J. SHURPIN, AND K. C. TAM

Department of Physics, State University of New York at Stony Brook, Stony Brook, New York 11794

E. OSNES

Institute of Physics, University of Oslo, Oslo 3, Norway

AND

P. J. ELLIS[‡]

School of Physics and Astronomy, University of Minnesota, Minneapolis, Minnesota 55455

Received May 20, 1980

A simple and convenient method is derived for evaluating linked Goldstone diagrams in an angular momentum coupled representation. Our method is general, and can be used to evaluate any effective interaction and/or effective operator diagrams for both closed-shell nuclei (vacuum to vacuum linked diagrams) and open-shell nuclei (valence linked diagrams). The techniques of decomposing diagrams into ladder diagrams, cutting open internal lines and cutting off one-body insertions are introduced. These enable us to determine angular momentum factors associated with diagrams in the coupled representation directly, without the need for carrying out complicated angular momentum algebra. A summary of diagram rules is given.

1. INTRODUCTION

One of the fundamental problems in nuclear structure physics is to calculate the properties of finite nuclei starting from the free nucleon–nucleon interaction. Since only a few low-lying states are generally of physical interest, the first step is to formulate

* Work supported in part by DOE Contracts EY-76-S-02-3001 and AT(11-1)-1764 and by Norwegian Research Council of Science and Humanities.

[†] Nordita Guest Professor at University of Oslo (Fall Semester, 1978), Alexander von Humboldt Foundation Awardee at Institut für Kernphysik, Kernforschungsanlage Jülich (Spring Semester, 1979).

[‡] On quarter leave at State University of New York at Stony Brook (Spring Semester, 1978).

0003-4916/81/040237–40$05.00/0

the problem in terms of an effective interaction acting in a strongly truncated Hilbert space (a model space with a finite number of states).

To be more specific, consider the nuclear structure problem of ^{18}O—the prototype nucleus for effective interactions—which has two neutrons more than the closed-shell nucleus ^{16}O. Instead of considering the true 18-particle eigenvalue problem

$$H\Psi_n = E_n\Psi_n , \tag{1}$$

which of course cannot be solved exactly, we study the more tractable and probably more meaningful model problem consisting of two nucleons in the $1s0d$ shell outside an ^{16}O closed-shell core. Then, inside this model space we have to use an effective Hamiltonian H_{eff} to account for the neglected degrees of freedom. Thus, the ^{18}O model space eigenvalue problem can be written as

$$PH_{\text{eff}}P\Psi_m = (E_m - E_0{}^c)\, P\Psi_m , \qquad m = 1, 2,..., d. \tag{2}$$

Here, P is a projection operator

$$P = \sum_{i=1}^{d} | \Phi_i \rangle\langle \Phi_i |, \tag{3}$$

projecting onto the model space spanned by unperturbed shell-model states Φ_i consisting of an inert ^{16}O core and two valence nucleons in the $1s0d$ shell. Further, in Eq. (2) $E_0{}^c$ is the ^{16}O ground state energy. We may consider Eq. (2) as the defining equation for H_{eff}. Thus, H_{eff} is defined to reproduce exactly the binding energies of d states in ^{18}O relative to the binding energy of the ^{16}O ground state. The corresponding wave functions are then projections of the exact wave functions Ψ_m onto the model space.

For ^{18}O, and any other nucleus with two valence nucleons, the effective Hamiltonian can be divided into a one-body and a two-body part

$$\begin{aligned} H_{\text{eff}} &= H^{(1)}_{\text{eff}} + H^{(2)}_{\text{eff}} , \\ H^{(1)}_{\text{eff}} &= \sum_i \tilde{\epsilon}_i a_i{}^\dagger a_i , \\ H^{(2)}_{\text{eff}} &= \tfrac{1}{4} \sum_{ijkl} \langle ij \,|\, V_{\text{eff}} \,|\, kl\rangle\, a_i{}^\dagger a_j{}^\dagger a_l a_k . \end{aligned} \tag{4}$$

The operator $a_i{}^\dagger$ (a_i) creates (annihilates) a single fermion in the state i. These operators appear in diagrams as the external lines. In studies of ^{18}O, it is customary not to calculate $H^{(1)}_{\text{eff}}$. Rather, the $\tilde{\epsilon}_i$ are taken as the experimental energies of the lowest $J^\pi = 5/2^+$, $1/2^+$, and $3/2^+$ states in ^{17}O. Hence, our main effort is directed toward the calculation of $H^{(2)}_{\text{eff}}$. Nevertheless, in so doing we also need to evaluate one-body diagrams because they contribute to $H^{(2)}_{\text{eff}}$ by way of the folded diagrams which are part of V_{eff} (see below). However, our technique is quite general and is readily applied

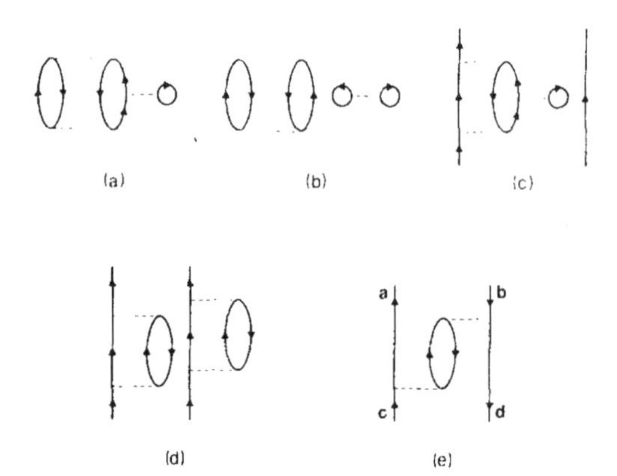

FIG. 1. Some linked and unlinked diagrams. Interaction vertices are represented by dashed lines.

to nuclei with more than two valence nucleons (i.e., Eq. (4) for H_{eff} contains terms up to $H_{\text{eff}}^{(n)}$ for n valence nucleons), as outlined in Subsection 3.7, below.

The effective interaction can be expressed as a perturbation series in the free nucleon–nucleon interaction, using either a time-dependent perturbation theory [1, 2] or a time-independent perturbation theory [3, 4]. This is formally exact, but, of course, actual calculations can only be approximate, given the many-body nature of the problem. Modern many-body perturbation theories are often expressed in a diagrammatic language. These methods have several advantages over the old-fashioned, purely algebraic methods. Firstly, many terms which cancel among each other are eliminated from the outset. Secondly, one has a systematic way of enumerating the various contributions, and is less likely to miss out terms. Thirdly, diagrams illustrate in a rather direct way the physical processes involved, allowing one to use physical intuition in selecting the important terms. Last, but not least, diagrams are spectacular and beautiful, and are even fun to work with.

As is well known, for the ground state of a closed-shell nucleus the energy shift (from the unperturbed energy) is given by the sum of all linked vacuum to vacuum diagrams. In Fig. 1, diagram (a) is of such a form, but not (b). For nuclei with valence particles, a convenient way to generate the effective interaction diagrams is the so-called Q-box formulation [2, 5, 6]. In this language the effective interaction can be written schematically as[1]

$$V_{\text{eff}} = \underset{\sim}{Q} - \underset{\sim}{Q}' \int \underset{\sim}{Q} + \underset{\sim}{Q}' \int \underset{\sim}{Q} \int \underset{\sim}{Q} - \underset{\sim}{Q}' \int \underset{\sim}{Q} \int \underset{\sim}{Q} \int \underset{\sim}{Q} + \cdots. \tag{5}$$

Here, $\hat{\underset{\sim}{Q}}$ (generally termed $\hat{\underset{\sim}{Q}}$-box) is an infinite sum of diagrams, typical members of which are diagrams (c), (d) and (e) in Fig. 1. They are valence linked and irreducible in the sense that all interactions must be linked to at least one valence line and all intermediate states must contain at least one passive line (i.e., a non-valence line, see,

[1] Here, it is understood that the purely one-body component of H_{eff} is removed, as described in Ref. [5].

for example, Ref. [1]). The $\hat{Q}$-box formulation can be used to generate all folded diagrams as indicated by the last three terms on the right-hand side of Eq. (5), where $\hat{Q}'$ is obtained from $\hat{Q}$ by removing all terms which are first order in the perturbing potential. When the folded diagrams are included, the disconnected valence diagrams such as diagram (d) of Fig. 1 will cancel among themselves. Similar techniques [7, 8] may be applied to evaluate the matrix elements of an operator.

The basic building blocks in any microscopic nuclear structure theory are thus these linked diagrams, and their evaluation is usually the first step in nuclear structure calculations. The purpose of the present paper is to devise a simple method to determine the angular momentum factors associated with such diagrams. In Section 2, we will first briefly review the ordinary diagram rules which are obtained from the Wick theorem and the anticommutation relations for fermions. These rules permit an easy evaluation of a diagram in the uncoupled representation or m-scheme, i.e. each fermion carries definite values for the z-components of the angular momentum and isospin. However, the nucleon–nucleon interaction is invariant under rotations in ordinary space and in isospin space and one should exploit this fact. Because of this, the fermion lines in diagrams should be coupled together to definite angular momentum J and isospin T. This leads to diagrams in the angular momentum coupled representation.

Hitherto, there seemed to be no simple systematic way of deriving the angular momentum factors associated with such diagrams, other than to start from the m-scheme, write out all the Clebsch–Gordan coefficients and perform the summations. This involves a great deal of angular momentum or Racah algebra and there is a significant possibility for error. As the example in the recent review of Ellis and Osnes [4] shows, the above procedure is complex even for the simple second-order core-polarization diagram. The third-order diagrams have been derived by Barrett and Kirson [9] using this approach, and in addition a few fourth-order ones were evaluated. However, the amount of Racah algebra necessary precludes detailed fourth-order studies. In fact to address this problem, Goode and Koltun [10] introduced the average interaction (averaged over J and T). Its calculation is then equivalent to that of the vacuum to vacuum diagrams and is thus computationally much simpler.

In Section 3 we will present a direct method for determining the angular momentum factors associated with linked Goldstone diagrams in the angular momentum coupled representation. Our method is general and can be used to evaluate effective interaction and/or effective operator diagrams of both closed-shell nuclei (vacuum to vacuum linked diagrams) and open-shell nuclei (valence linked diagrams). This allows one to write down the angular momentum factors of any diagram essentially by inspection. Obviously this can greatly simplify calculations and may allow one to tackle problems that would otherwise be intractable. Our method gives some attention to computer time considerations. In Section 4 we give a summary of our diagram rules and we evaluate a few representative diagrams to illustrate our technique.

2. Diagram Rules in the Uncoupled Representation

Using time-dependent perturbation theory [1, 2], the basic matrix elements which generate the model space effective interaction can be written as

$$M_{fi} = \langle C \mid A_f V(0) U(0, -\infty) A_i^\dagger \mid C \rangle_L , \tag{6}$$

where $\mid C\rangle$ is the closed-shell unperturbed ground state and L implies that we include only linked diagrams (valence linked diagrams in the case of valence nuclei). The time evolution operator $U(0, -\infty)$ is given by

$$U(0, -\infty) = \lim_{\substack{t' \to -\infty(1-i\epsilon) \\ \epsilon \to 0+}} \sum_{n=0}^{\infty} \left(\frac{-i}{\hbar}\right)^n \int_{t'}^{0} dt_1 \int_{t'}^{t_1} dt_2 \cdots \int_{t'}^{t_{n-1}} dt_n \, T[V(t_1) \cdots V(t_n)], \tag{7}$$

where T denotes the time-ordered product and all the perturbing operators $V(t)$ are in the interaction representation. For the ground state of closed-shell nuclei, $A_i^\dagger = A_f = 1$ and M_{fi} is just the sum of all the linked vacuum to vacuum diagrams such as diagram (a) of Fig. 1. For the case with valence particles and holes, $A_i^\dagger$ and A_f may have the form

$$\begin{aligned} A_i^\dagger &= a_1^\dagger a_3^\dagger a_2 , \\ A_f &= a_4^\dagger a_7 a_5 = (a_5^\dagger a_7^\dagger a_4)^\dagger . \end{aligned} \tag{8}$$

It may be noted that the ordering of the external lines is in accordance with the standard ordering chosen for the fermion operators in $A_i^\dagger$ and $A_f^\dagger$, as illustrated in Fig. 2. This will be the convention used in this work. With valence lines, the matrix element M_{fi} is just the sum of all the valence-linked (including folded) diagrams [1, 2] such as (c) and (e) of Fig. 1.

As each linked diagram is just a term in the perturbation expansion of the matrix element M_{fi}, as indicated by Eqs. (6) and (7), the starting point for the derivation of the diagram rules is clearly these two equations. In the uncoupled representation, the diagram rules can be readily derived from these two equations, using the Wick theorem and the symmetry properties of the operator $V(t)$. As they are well known (see, e.g., Refs. [3, 11, 12]), we will just state these rules in the following.

(i) From the time integrations of Eq. (7) we obtain an energy denominator factor

$$[\omega_i - (\Sigma\epsilon_p - \Sigma\epsilon_h)]^{-1} \tag{9}$$

between successive vertices. Here, ω_i is the starting energy, i.e., the unperturbed

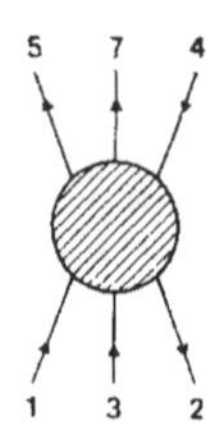

Fig. 2. Illustration of the ordering of external fermion lines discussed in the text.

energy at the time $t = -\infty$.[2] For example, $\omega_i = 0$ in Fig. 1a and $\omega_i = \epsilon_c - \epsilon_d$ in Fig. 1e. The ϵ's are the unperturbed single particle energies and we sum over all intermediate particle (p) and hole (h) indices between vertices, ignoring the exclusion principle. Note that our diagrams are time ordered (i.e., Goldstone diagrams) as indicated by the time constraints of Eq. (7).

(ii) For each vertex there is an antisymmetrized matrix element[3]

$$\langle ab \mid V \mid cd\rangle = \langle ab \mid V \mid cd\rangle - \langle ab \mid V \mid dc\rangle, \tag{10}$$

where a and b "leave" the vertex from the left and right side and c and d "enter" the vertex from the left and right side, respectively. With this definition, the two-body interaction V may be written as

$$V = \tfrac{1}{4} \sum_{ijkl} \langle ij \mid V \mid kl\rangle\, a_i^\dagger a_j^\dagger a_l a_k\,. \tag{11}$$

In the uncoupled representation, the unperturbed wave functions used above are, for example, harmonic oscillator wave functions in the m-scheme. Thus $\mid d\rangle = \mid (nljm, m_t)_d\rangle$.

We also need to deal with a one-body operator U, which may be a scalar operator in the perturbing potential or a tensor operator corresponding to an observable quantity. For each vertex U, there is a matrix element $\langle a \mid U \mid b\rangle$, where a leaves the vertex and b enters it. In the uncoupled representation, the operator U may be written as

$$U = \sum_{ij} \langle i \mid U \mid j\rangle\, a_i^\dagger a_j\,. \tag{12}$$

(iii) By applying the Wick theorem to Eqs. (6) and (7) we obtain for each diagram an overall factor of

$$(-1)^{n_h + n_l + n_c + n_{\mathrm{exh}}}/2^{n_{\mathrm{ep}}}, \tag{13}$$

where n_h is the number of hole lines, n_l is the number of closed loops, n_c is the number of crossings of different external lines as they trace through the diagram, n_{exh} is the number of external hole lines which continuously trace through the diagram, including those which enter as hole lines but emerge as particle lines, and n_{ep} is the number of

[2] The starting energies for effective operator diagrams [7, 8] are slightly different. Because of the presence of two time evolution operators $U(\infty, 0)$ and $U(0, -\infty)$, the starting energy may be the unperturbed single particle energies at either $t = -\infty$ or $t = +\infty$, depending on the location of the time intervals. The operator acts at the time $t = 0$.

[3] The interaction V is in principle the free nucleon–nucleon interaction. However, since V generally contains a strong repulsive core, it is in actual calculations replaced by the more well-behaved nuclear reaction matrix G, which already includes the short-range correlations induced by V (see, e.g., Ref. [2]). As far as diagram rules are concerned, we may just replace the V vertices by the respective G vertices. This will be discussed further in Subsection 3.2 in connection with the particle–particle ladder diagrams.

pairs of lines which start at the same interaction, end at the same interaction and go in the same direction (so-called equivalent pairs).

Note that for external particle-hole lines, we should draw the particle lines to the left of the hole lines, as indicated in Eq. (8) and Fig. 2. The above rules are obtained with this convention. We will further comment on this point near the end of this section. We will now give some examples to illustrate the determination of the above factors. (These examples will also be used in Sections 3 and 4 to illustrate our rules for determining the angular momentum factors.)

The above rules are indeed simple and would enable one immediately to write down expressions for any linked Goldstone diagram in the uncoupled representation. To illustrate, we have for diagram (e) of Fig. 1

$$(1e) = \frac{(-1)^5}{2^0} \sum_{ph} \frac{\langle hd \mid V \mid pb\rangle\langle ap \mid V \mid ch\rangle}{(\epsilon_c - \epsilon_d) - (\epsilon_a + \epsilon_p - \epsilon_h - \epsilon_d)}, \tag{14}$$

where the overall factor results from

$$n_h = 3,\ n_l = 1,\ n_c = 0,\ n_{\text{exh}} = 1,\ n_{\text{ep}} = 0.$$

For the diagrams in Fig. 3, the values of the quantities n_h, n_l, etc., determining the overall factor (13) are given in Table I.

In (d), we seem to have a crossing of external lines, but this crossing is between the same external line. Thus $n_c = 0$. In (f) we have a crossing between two external lines. There are cases where $A_i^\dagger$ and A_f of Eq. (6) may not have the same number of fermion operators, as shown by diagram (h). This diagram arises in a model space calculation including one-particle and two-particle one-hole basis states. For such cases, it is indeed essential to follow the convention that in $A_i^\dagger$ and $A_f^\dagger$ we arrange the particle creation operators to the left of the hole creation operators. This is because of the following consideration. If $A_i^\dagger$ and $A_f^\dagger$ have the same number of fermion operators, then a similar rearrangement of these operators in both $A_i^\dagger$ and $A_f^\dagger$ will not alter

TABLE I

Values of Quantities Determining the Overall Factor for the Diagrams in Fig. 3

Diagram	n_h	n_l	n_c	n_{exh}	n_{ep}
(a)	4	3	0	0	1
(b)	4	3	0	0	2
(c)	3	1	0	0	2
(d)	3	2	0	0	1
(e)	4	1	0	2	1
(f)	2	1	1	0	0
(g)	2	2	0	0	0
(h)	1	0	0	1	1
(i)	1	1	0	0	0
(j)	0	0	0	0	1

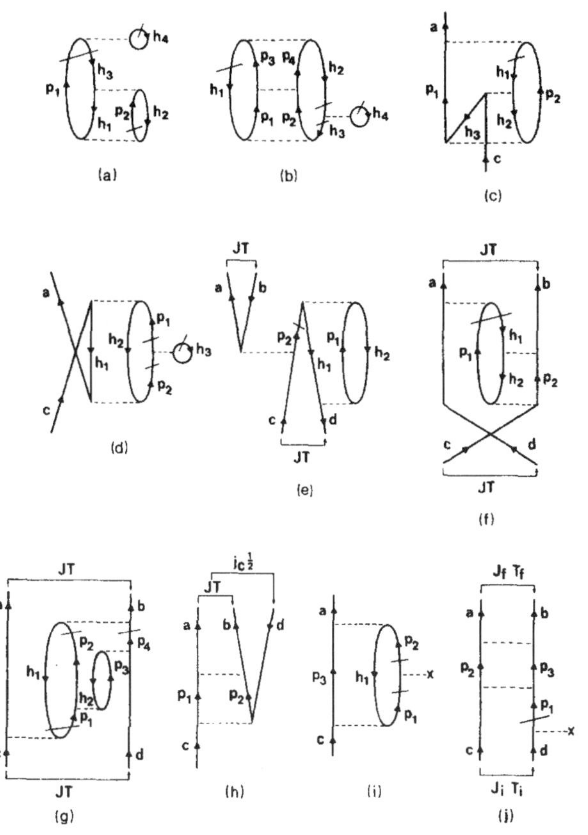

FIG. 3. Examples of effective interaction and operator diagrams.

the value of the diagram. Thus we have the freedom of, say, arranging all the hole creation operators in front of the particle creation operators in $A_i^\dagger$ and $A_f^\dagger$ and our diagram rules still apply. It is clear that we do not have such freedom if $A_i^\dagger$ and $A_f^\dagger$ do not have the same number of fermion operators.

Finally let us mention the familiar rule that from the set of exchange and topologically equivalent diagrams only one diagram is to be retained in the calculations. Such equivalent diagrams are easily identified by contracting the dashed-line vertex to a dot vertex (Hugenholtz notation). To illustrate, all the five diagrams of Fig. 4 are equivalent (i.e., they have the same value and correspond to the same physical process). Only one of them will be retained in actual calculations.

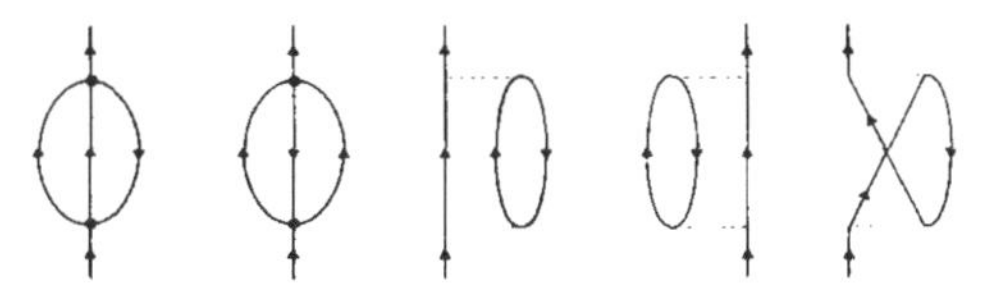

FIG. 4. Examples of equivalent diagrams.

3. Diagram Rules for the Coupled Representation

To determine the value of a given diagram, the rules given in Section 2 are first applied to obtain the energy denominators, factors, phases and the indices in the bra and ket vectors of the matrix elements involved. Our task here is to set out rules for determining the angular momentum couplings of these matrix elements, and associated factors, when the initial and final states of the diagram are coupled to a definite angular momentum and, if desired, isospin. In Subsection 3.1 we define various coupled matrix elements. In Subsection 3.2 we discuss the case of ladder diagrams (particle–particle, particle–hole or hole–hole ladders) since their angular momentum structure is particularly simple. Our strategy in the following subsections is then to express a given diagram in terms of ladder diagrams insofar as it is possible. In Subsection 3.3 we show that certain diagrams which are not in an obvious ladder form in the original coupling scheme, can be put into ladder form by angular momentum recoupling. As is explained, in order to perform the necessary recoupling transformations, it is convenient first to couple the diagram to a scalar. In Subsection 3.4 we show how other diagrams can be decomposed into products of ladder diagrams by cutting open internal lines. This technique can further be used to cut off one-body insertions, as discussed in Subsection 3.5. Then, in Subsection 3.6 we discuss the evaluation of diagrams for one-body tensor operators of rank greater than zero. Such diagrams are needed, for instance, if we wish to obtain matrix elements of the electromagnetic $E2$ operator so as to calculate the effective charge. Generalizations of our method to three-body cluster diagrams and other more complicated cases are discussed in Subsection 3.7.

3.1. *Cross-Coupled Matrix Elements*

In order to take full advantage of the power and flexibility of Racah algebra, it is worth realizing that in a two-particle matrix element the single-particle angular momenta can be coupled in several different ways.

First, we note that a single-particle ket vector $|jm\rangle$ transforms under rotation as a spherical tensor T_m^j of rank j, component m, while the corresponding bra vector $\langle jm|$ transforms as $(-1)^{j-m}T^j_{-m}$. Thus, we may define a coupled (unnormalized) two-particle ket vector as

$$|j_aj_bJM\rangle = \sum_{m_am_b} C^{j_a\ j_b\ J}_{m_am_bM}\,|(j_am_a)(j_bm_b)\rangle$$

$$\equiv \{|j_aj_b\rangle\}^J_M, \tag{15}$$

which transforms as a spherical tensor T_M^J. In Eq. (15) C is a standard Clebsch–Gordan coefficient which is defined according to the Condon–Shortley phase convention. The corresponding bra vector is

$$
\begin{aligned}
\langle j_a j_b JM | &= \sum_{m_a m_b} \langle (j_a m_a)(j_b m_b) | \, C^{j_a \, j_b \, J}_{m_a m_b M} \\
&= \sum_{m_a m_b} \langle (j_a m_a)(j_b m_b) | \, (-1)^{j_a - m_a} (-1)^{j_b - m_b} (-1)^{J-M} \, C^{j_a \;\; j_b \;\; J}_{-m_a \, -m_b \, -M} \\
&= (-1)^{J-M} \{ \langle j_a j_b | \}^{J}_{-M} ,
\end{aligned}
\tag{16}
$$

confirming that it transforms as a spherical tensor $(-1)^{J-M} T^{J}_{-M}$. Since the interaction operator V transforms as a scalar under rotation, the standard coupled (unnormalized) two-particle matrix element of V is given by

$$
\begin{aligned}
&\langle j_a j_b JM \mid V \mid j_c j_d JM \rangle \\
&= \sum_{m_a m_b m_c m_d} C^{j_a \, j_b \, J}_{m_a m_b M} \, C^{j_c \, j_d \, J}_{m_c m_d M} \langle (j_a m_a)(j_b m_b) \mid V \mid (j_c m_c)(j_d m_d) \rangle \\
&= (-1)^{J-M} \sum_{m_a m_b m_c m_d} (-1)^{j_a - m_a} (-1)^{j_b - m_b} \, C^{j_a \;\; j_b \;\; J}_{-m_a \, -m_b \, -M} \, C^{j_c \, j_d \, J}_{m_c m_d M} \\
&\quad \times \langle (j_a m_a)(j_b m_b) \mid V \mid (j_c m_c)(j_d m_d) \rangle .
\end{aligned}
\tag{17}
$$

It follows from the Wigner–Eckart theorem that the matrix element is independent of M. The uncoupled matrix element on the right-hand side of Eq. (17) is antisymmetrized as stated in Section 2.

In the above we have suppressed the principal quantum numbers n and orbital angular momentum quantum numbers l which are not of relevance here. Further, it is not necessary to consider explicitly the isospin quantum numbers, since the isospin factors follow by replacing j and m by $\frac{1}{2}$ and m_t and by replacing J and M by T and M_T. (This is readily seen near the end of this subsection and in the examples of Section 4 where isospin factors are explicitly given.) Let us also replace j_a by a and m_a by α, etc., so that we simplify Eq. (17) to

$$
\begin{aligned}
\langle abJM \mid V \mid cdJM \rangle &= \sum_{\alpha\beta\gamma\delta} C^{abJ}_{\alpha\beta M} \, C^{cdJ}_{\gamma\delta M} \langle (a\alpha)(b\beta) \mid V \mid (c\gamma)(d\delta) \rangle \\
&\equiv \langle \overset{JM}{\overrightarrow{ab}} \mid V \mid \overset{JM}{\overrightarrow{cd}} \rangle .
\end{aligned}
\tag{18}
$$

Using well-known properties of Clebsch–Gordan coefficients we can reverse the direction of the coupling of the single-particle angular momenta on one or both sides of the matrix element, obtaining

$$
\langle \overset{JM}{\overrightarrow{ab}} \mid V \mid \overset{JM}{\overleftarrow{cd}} \rangle = (-1)^{c+d-J} \langle \overset{JM}{\overrightarrow{ab}} \mid V \mid \overset{JM}{\overrightarrow{cd}} \rangle
\tag{19}
$$

and similar relations. Note that the ordering $abcd$ is unchanged, only the direction of the coupling is altered.

As indicated at the beginning of this subsection, it will be convenient to consider alternative coupling schemes for the two-particle matrix elements. Thus, we shall define "cross-coupled" matrix elements [13, 14] in which single-particle angular momenta from the bra and ket vectors are coupled, i.e., the coupling runs across the interaction operator. Two frequently encountered cross-coupled matrix elements are

$$\overset{JM}{\underset{JM}{\langle ab \mid V \mid cd \rangle}} = (-1)^{J-M} \sum_{\alpha\beta\gamma\delta} (-1)^{a-\alpha} C^{c\ a\ J}_{\gamma\,-\alpha\,-M} (-1)^{b-\beta} C^{d\ b\ J}_{\delta\,-\beta\,M} \langle (a\alpha)(b\beta) \mid V \mid (c\gamma)(d\delta) \rangle, \tag{20}$$

$$\overset{JM}{\underset{JM}{\langle ab \mid V \mid cd \rangle}} = (-1)^{J-M} \sum_{\alpha\beta\gamma\delta} (-1)^{a-\alpha} C^{d\ a\ J}_{\delta\,-\alpha\,-M} (-1)^{b-\beta} C^{c\ b\ J}_{\gamma\,-\beta\,M} \langle (a\alpha)(b\beta) \mid V \mid (c\gamma)(d\delta) \rangle, \tag{21}$$

which are defined in close analogy to Eq. (17). We emphasize that in each of Eqs. (20) and (21) the left-hand side should only be regarded as a short-hand notation for the expression given on the right-hand side. In both equations the phase factor $(-1)^{J-M}$ is included to ensure that the matrix elements are scalar and independent of M. We have chosen to associate the phase factor $(-1)^{J-M}$ and the value $-M$ of the total magnetic quantum number with the coupling involving the single-particle state a. This is of course arbitrary, since the matrix element is independent of M. Below we shall explicitly couple the total J's in Eqs. (20) and (21) to angular momentum zero, in which case the magnetic quantum numbers M and $-M$ do not show up at all.

This simple idea of cross-coupling is illustrated in Fig. 5. Diagram (a) corresponds to the ordinary coupled matrix element of Eq. (18). The cross-coupled matrix element of Eq. (20) corresponds to diagram (b), while the cross-coupled matrix element of Eq. (21) leads to diagram (c).

In the same way as for the standard coupled matrix elements we can reverse the direction of the coupling of the single-particle angular momenta in the cross-coupled matrix elements, obtaining

$$\overset{JM}{\underset{JM}{\langle ab \mid V \mid cd \rangle}} = (-1)^{b+d-J} \overset{JM}{\underset{JM}{\langle ab \mid V \mid cd \rangle}} \tag{22}$$

and similar relations. Again we note that the ordering $abcd$ is unchanged, only the

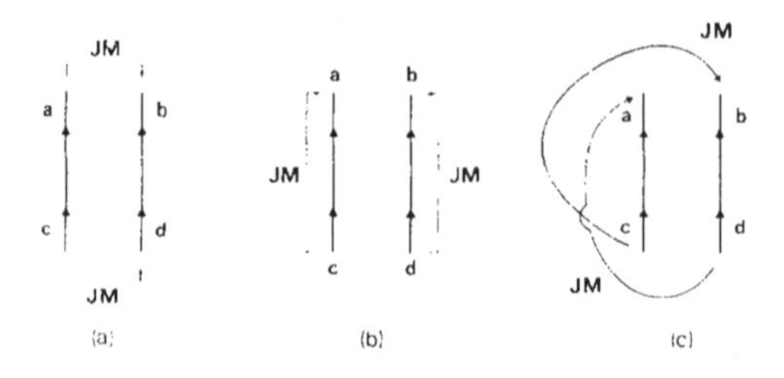

FIG. 5. Examples showing different ways of coupling matrix elements. If isospin couplings are also included, we just replace JM by $JMTM_T$.

direction of the coupling is reversed. In fact, we have as a general rule

$$\cdots \overbrace{a \cdots b}^{JM} \cdots = (-1)^{a+b-J} \cdots \overbrace{a \cdots b}^{JM} \cdots, \tag{23}$$

provided of course that the ordering of state labels remains unchanged.

The cross-coupled matrix elements (20) and (21) can be related to the standard coupled matrix element (17) or (18) and vice versa by means of angular momentum recoupling transformations. To that end it is useful explicitly to couple the matrix elements to scalars (as indicated below by the superscript 0). For the standard coupled matrix element (18) we have

$$\begin{aligned}
\langle \overset{JM}{\urcorner}ab \mid V \mid \overset{JM}{\urcorner}cd\rangle &= \hat{J}^{-2} \sum_M \langle \overset{JM}{\urcorner}ab \mid V \mid \overset{JM}{\urcorner}cd\rangle \\
&= \hat{J}^{-1} \sum_M (-1)^{J-M} C^{J\ J\ 0}_{M-M\ 0} \langle \overset{JM}{\urcorner}ab \mid V \mid \overset{JM}{\urcorner}cd\rangle \\
&= \hat{J}^{-1} \langle \overset{J}{\urcorner}ab \mid V \mid \overset{J}{\urcorner}cd\rangle^0,
\end{aligned} \tag{24}$$

where $\hat{J} \equiv \sqrt{2J+1}$. In the first step of Eq. (24) we used the fact that the matrix element is independent of M, while the second step was obtained using the well-known relation

$$C^{J\ J\ 0}_{M-M\ 0} = (-1)^{J-M} \hat{J}^{-1}. \tag{25}$$

We shall often refer to the scalar coupled matrix element on the right-hand side of Eq. (24) as a reduced matrix element. It is implied in Eq. (24) that the two-particle angular momenta are coupled in the order J_{in} (ket) to J_{out} (bra). Reversing this order gives rise to a phase $(-1)^{2J}$, which is of no consequence, since J is an integer.

Similarly, we can couple the total J's in the cross-coupled matrix elements (20) and (21) to zero resultant angular momentum, obtaining

$$\begin{aligned}
\langle \underbrace{\overbrace{ab \mid V \mid cd}^{JM}}_{JM} \rangle &= \hat{J}^{-2} \sum_M \langle \underbrace{\overbrace{ab \mid V \mid cd}^{JM}}_{JM} \rangle \\
&= \hat{J}^{-1} \sum_M (-1)^{J-M} C^{J\ J\ 0}_{M-M\ 0} \langle \underbrace{\overbrace{ab \mid V \mid cd}^{JM}}_{JM} \rangle \\
&= \hat{J}^{-1} \langle \underbrace{\overbrace{ab \mid V \mid cd}^{J}}_{J} \rangle^0,
\end{aligned} \tag{26}$$

$$\overset{JM}{\underset{JM}{\langle ab \mid V \mid cd\rangle}} = \hat{J}^{-1} \overset{J}{\underset{J}{\langle ab \mid V \mid cd\rangle^0}}. \tag{27}$$

In Eq. (26) we have explicitly coupled the J's in the order J_{ab} to J_{ca}, although the order of the coupling may be reversed without changing the sign of the matrix element since J is an integer.

Now, using well-known recoupling transformations of Racah algebra, it is straightforward to relate reduced cross-coupled matrix elements to reduced standard coupled matrix elements. For example, we obtain for the reduced cross-coupled matrix elements on the right-hand sides of Eqs. (26) and (27)

$$\begin{aligned} \overset{J}{\underset{J}{\langle ab \mid V \mid cd\rangle^0}} &= \sum_{J'} X\begin{pmatrix} d & b & J \\ c & a & J \\ J' & J' & 0 \end{pmatrix} \overset{J' \quad J'}{\langle ab \mid V \mid cd\rangle^0} \\ &= \sum_{J'} X\begin{pmatrix} c & a & J \\ d & b & J \\ J' & J' & 0 \end{pmatrix} \overset{J' \quad J'}{\langle ab \mid V \mid cd\rangle^0}, \end{aligned} \tag{28}$$

$$\begin{aligned} \overset{J}{\underset{J}{\langle ab \mid V \mid cd\rangle^0}} &= \sum_{J'} X\begin{pmatrix} c & b & J \\ d & a & J \\ J' & J' & 0 \end{pmatrix} \overset{J' \quad J'}{\langle ab \mid V \mid cd\rangle^0} \\ &= \sum_{J'} X\begin{pmatrix} c & b & J \\ d & a & J \\ J' & J' & 0 \end{pmatrix} \overset{J' \quad J'}{\langle ab \mid V \mid cd\rangle^0} (-1)^{a+b-J'} \\ &= \sum_{J'} X\begin{pmatrix} d & a & J \\ c & b & J \\ J' & J' & 0 \end{pmatrix} \overset{J' \quad J'}{\langle ab \mid V \mid cd\rangle^0}. \end{aligned} \tag{29}$$

Here, the X-coefficient is defined in terms of the Wigner 9-j symbol as

$$X\begin{pmatrix} r & s & t \\ u & v & w \\ x & y & z \end{pmatrix} = \hat{t}\hat{w}\hat{x}\hat{y} \begin{Bmatrix} r & s & t \\ u & v & w \\ x & y & z \end{Bmatrix}, \tag{30}$$

where we recall that $\hat{t} = \sqrt{2t+1}$.

Now, X-coefficients with zero arguments can be simplified. Some examples are

$$X\begin{pmatrix} r & s & t \\ u & v & t \\ x & x & 0 \end{pmatrix} = \hat{x}\hat{t}(-1)^{s+t+u+x} \begin{Bmatrix} r & s & t \\ v & u & x \end{Bmatrix}, \tag{31}$$

$$X\begin{pmatrix} r & r & 0 \\ u & u & 0 \\ x & x & 0 \end{pmatrix} = \hat{x}\hat{r}^{-1}\hat{u}^{-1}. \tag{32}$$

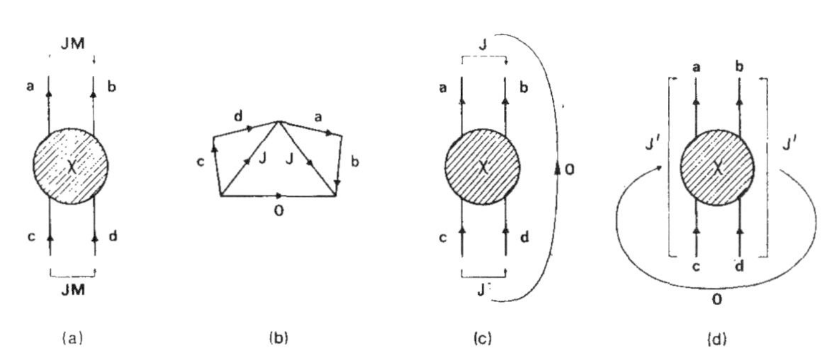

FIG. 6. Angular momentum couplings of a generalized vertex function.

On the right-hand side of Eq. (31) we have written a Wigner 6-j symbol. These expressions are useful in actual calculations. The angular momentum structure is most clearly displayed, however, by using the X-coefficient, so in most cases we shall not simplify the expressions we obtain.

Although we have only discussed the case of a single V interaction above, the same discussion can be applied to a product $\chi = VVV \cdots$, such as a $\hat{Q}$-box [12] composed of linked diagrams, since this also transforms as an angular momentum scalar. Thus for a given diagram, such as that shown in Fig. 6a, we first couple to zero resultant angular momentum as indicated in Figs. 6b and c, obtaining

$$\langle abJM \mid \chi \mid cdJM \rangle = \hat{J}^{-1} \langle \overset{J}{\overset{\urcorner\downarrow}{ab}} \mid \chi \mid \overset{J}{\overset{\urcorner\downarrow}{cd}} \rangle^{0}. \tag{33}$$

Hence, we need to evaluate the scalar coupled diagram of Fig. 6c and multiply by $\hat{J}^{-1}$. Having obtained the scalar coupled diagram, the cross coupling is easily performed. An example is shown in Fig. 6d corresponding to Eq. (28).

To summarize our developments thus far, we have associated factors

$$\langle \overset{JM}{\overset{\urcorner\downarrow}{ab}} \mid V \mid \overset{JM}{\overset{\urcorner\downarrow}{cd}} \rangle, \qquad \underset{\underset{JM}{\llcorner\!\uparrow\qquad\lrcorner}}{\overset{\overset{JM}{\downarrow\!\ulcorner\qquad\urcorner}}{\langle ab \mid V \mid cd \rangle}} \qquad \text{and} \qquad \underset{\underset{JM}{\llcorner\!\uparrow\qquad\lrcorner}}{\overset{\overset{JM}{\downarrow\!\ulcorner\qquad\urcorner}}{\langle ab \mid V \mid cd \rangle}}$$

with the diagrams in Figs. 5a, b and c, respectively. Further, as we have just seen, V may here be replaced by more complicated vertices χ containing any number of V

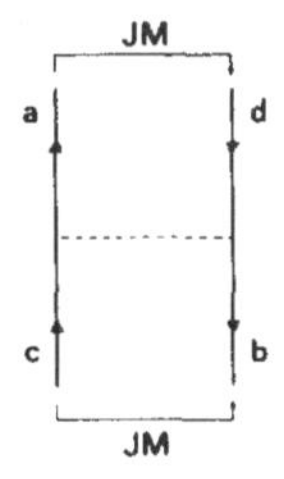

FIG. 7. Particle–hole diagram.

interactions. However, we have only considered particle–particle diagrams thus far. Then, turning to particle–hole diagrams, we may ask whether similar rules hold. Consider, for example, the particle–hole diagram shown in Fig. 7 corresponding to the matrix element $\langle ad^{-1}JM \mid V \mid cb^{-1}JM\rangle$. The angular momentum coupling is similar to that of the particle–particle diagram shown in Fig. 5c (except for the direction of the coupling between a and d). However, the detailed tensorial character of the coupling between a and d is different in the two cases, such that

$$\langle ad^{-1}JM \mid V \mid cb^{-1}JM\rangle = (-1)^{2d} \overbrace{\underbrace{\langle ab \mid V \mid cd\rangle}_{JM}}^{JM}. \tag{34}$$

Here we are only considering the angular momentum structure of the respective diagrams. The phase factors due to contractions of fermion operators have already been taken care of in Section 2.

The origin of the phase in Eq. (34) is seen by coupling the particle and hole spherical tensors T^a and $T^{d\dagger}$ according to

$$\overbrace{T^a T^{d\dagger}}^{JM} = \sum_{\alpha\delta} C^{a\ \ d\ \ J}_{\alpha\,-\delta\,M}(-1)^{d-\delta}\, T^a_\alpha T^{d\dagger}_\delta . \tag{35}$$

Then, taking the Hermitian conjugate of both sides, we obtain

$$(\overbrace{T^a T^{d\dagger}}^{JM})^\dagger = (-1)^{2d+J-M}\, \overbrace{T^d T^{a\dagger}}^{J-M}. \tag{36}$$

Here, the left-hand side describes the angular momentum properties of the bra vector $\langle ad^{-1}JM \mid$ appearing in the particle–hole matrix element on the left-hand side of Eq. (34). Further, except for the phase $(-1)^{2d}$, the right-hand side of Eq. (36) describes the angular momentum coupling between a and d in the cross-coupled matrix element on the right-hand side of Eq. (34). Hence, relation (34) results. It is instructive to contrast Eq. (36) with the relation

$$(\overbrace{T^a T^{d}}^{JM})^\dagger = (-1)^{J-M}\, \overbrace{T^{d\dagger} T^{a\dagger}}^{J-M}, \tag{37}$$

which describes the tensorial character of a particle–particle bra vector and is thus equivalent to Eq. (16), above.

The difference between the coupling schemes employed on the left- and right-hand sides of Eq. (34) may be further illustrated by considering the explicit expressions for the coupled matrix elements in terms of the uncoupled ones, namely,

$$\langle ad^{-1}JM \mid V \mid cb^{-1}JM\rangle = \sum_{\alpha\beta\gamma\delta} C^{a\ \ d\ \ J}_{\alpha\,-\delta\,M} C^{c\ \ b\ \ J}_{\gamma\,-\beta\,M}(-1)^{d-\delta}(-1)^{b-\beta}\langle(a\alpha)(b\beta) \mid V \mid (c\gamma)(d\delta)\rangle, \tag{38}$$

$$\overbrace{\underbrace{\langle ab \mid V \mid cd\rangle}_{JM}}^{JM} = (-1)^{J-M} \sum_{\alpha\beta\gamma\delta} C^{a\ \ d\ \ \ J}_{-\alpha\ \delta\ -M} C^{c\ \ \ b\ \ J}_{\gamma\ -\beta\ M} (-1)^{a-\alpha}(-1)^{b-\beta} \langle (a\alpha)(b\beta) \mid V \mid (c\gamma)(d\delta)\rangle. \tag{39}$$

Again, in writing Eq. (38) we have ignored the phase factors due to contractions of fermion operators, since we are only interested here in the angular momentum structure of the diagrams. We now observe that in Eq. (38) the phase factors $(-1)^{j-m}$ are always associated with *hole* lines, whereas in Eq. (39) they are always associated with *outgoing* lines. Thus, an outgoing hole line in the particle–hole coupling scheme (38) is treated similarly to an outgoing particle line in the particle–particle coupling scheme (39). On the other hand, an incoming hole line in the particle–hole scheme is treated differently from an incoming particle line in the particle–particle scheme. Now, our diagram rules will always be expressed in terms of angular momentum couplings among *particle* lines only, as was done, for example, in Eq. (39). Thus, in such a scheme we have to associate a phase factor $(-1)^{2h}$ with a particle–hole pair ph^{-1} on the top (bra) side of the diagram.

This can easily be generalized, so that for each complete diagram, which is expressed in terms of particle–particle coupled interaction vertices (matrix elements), we obtain a phase factor

$$(-1)^{n_{\text{tph}}}, \tag{40}$$

where n_{tph} is the number of particle-hole pairs on the top (bra) side of the diagram. We emphasize that this phase is due to the particular coupling scheme of Eq. (35) chosen for external particle and hole lines, namely,

$$A_i^\dagger = \sum_{m_p m_h} (-1)^{h-m_h} C^{p\ \ \ \ h\ \ \ \ J}_{m_p\ -m_h\ M} a^\dagger_{pm_p} a_{hm_h}$$

and similarly for $A_f^\dagger$. Then, A_f is defined according to Eq. (36) as $A_f = (A_f^\dagger)^\dagger$.[4] Further, we note that the phase arises from angular momentum considerations. Thus, the phase arises twice if we couple both angular momentum and isospin, i.e.,

$$(-1)^{2h} \to (-1)^{2(j_h+1/2)} = +1, \tag{41}$$

and in this case the phases in Eqs. (34) and (40) can simply be ignored.

Finally, let us point out that in actual calculations the coupled matrix elements needed are in fact of the following three basic types:

$$\langle \overset{JT}{\overline{ab}} \mid V \mid \overset{JT}{\overline{cd}}\rangle^{00}, \qquad \overbrace{\underbrace{\langle ab \mid V \mid cd\rangle}_{JT}}^{JT}{}^{00} \qquad \text{and} \qquad \overbrace{\underbrace{\langle ab \mid V \mid cd\rangle}_{JT}}^{JT}{}^{00}, \tag{42}$$

[4] Note that Eq. (36) implies that scalar coupled matrix elements be constructed according to our standard rule [see comments following Eqs. (24) and (25)], by coupling the external state with lines entering vertices to the external state with lines leaving vertices. If both particles and holes are present, particle–hole pairs may be ignored in determining the coupling order (this implies that if initial and final states involve only particle–hole pairs, the order of coupling is arbitrary).

where both ordinary angular momenta and isospins are coupled. The superscript 00 implies the coupling of both angular momenta and isospins to a total of zero. The transformation relations among the matrix elements (42) involve straightforward generalizations of Eqs. (28) and (29). For example, corresponding to Eq. (28) we have

$$\overset{JT}{\underset{JT}{\langle ab \mid V \mid cd\rangle^{00}}} = \sum_{J'T'} X\begin{pmatrix} c & a & J \\ d & b & J \\ J' & J' & 0 \end{pmatrix} X\begin{pmatrix} \frac{1}{2} & \frac{1}{2} & T \\ \frac{1}{2} & \frac{1}{2} & T \\ T' & T' & 0 \end{pmatrix} \langle \overset{J'T'}{ab} \mid V \mid \overset{J'T'}{cd}\rangle^{00}. \tag{43}$$

3.2. *Ladder Diagrams*

Our strategy is now to decompose a given diagram into products of ladders which have very simple angular momentum structure. An example of a particle–particle ladder is given in Fig. 8a. Figures 8b and c give alternative forms for a particle–hole ladder; the diagrams are actually equivalent, as can be seen by contracting the dashed line to a dot. Backward going vertices can also be included so that Fig. 8d is also a ladder. Hole–hole ladders can be treated similarly, although we do not give an explicit example.

Let us study the angular momentum structure of ladders in some detail. First, consider the simple particle–particle ladder shown in Fig. 9a. We note that angular momentum couplings are imposed on the external lines, in accordance with Eq. (17). For the internal lines we sum freely over all m-states. Then, keeping the p_1 and p_2 of the intermediate states fixed and disregarding the energy denominator (9) and overall factor (13) associated with the corresponding uncoupled diagram, we obtain

$$\begin{aligned} \text{Diagram 9a} &= \sum_{\alpha\beta\gamma\delta} (-1)^{a-\alpha}(-1)^{b-\beta}(-1)^{J-M} C^{\,a\ \ b\ \ J}_{-\alpha-\beta-M} C^{\,c\ d\ J}_{\gamma\ \delta\ M} \\ &\quad \times \sum_{m_1 m_2} \langle (a\alpha)(b\beta) \mid V \mid (p_1 m_1)(p_2 m_2)\rangle\langle (p_1 m_1)(p_2 m_2) \mid V \mid (c\gamma)(d\delta)\rangle \\ &= \hat{p}_1 \hat{p}_2 \langle \overset{JM}{ab} \mid V \mid \overset{00}{\underset{00}{p_1 p_2\rangle\langle p_1 p_2}} \mid V \mid \overset{JM}{cd}\rangle. \end{aligned} \tag{44}$$

In obtaining the coupling of the internal lines on the right-hand side of Eq. (44), we used the relation

$$\sum_m \mid jm\rangle\langle jm \mid = \hat{j} \sum_m (-1)^{j-m} C^{\,j\ \ j\ \ 0}_{m\,-m\,0} \mid jm\rangle\langle jm \mid = \hat{j} \mid j\rangle\langle j \mid^{0}_{0} = \hat{j}\, \overset{00}{\mid j\rangle\langle j \mid}. \tag{45}$$

Thus, each internal line may be regarded as being cut into two pieces which are coupled to zero resultant angular momentum, as shown in Fig. 9b. Further, the order of the coupling is such that the incoming piece of the internal line (i.e., the piece entering a vertex) is coupled to the outgoing piece (i.e., the piece leaving a vertex). Now, such a cross-channel coupling of the internal lines has the disadvantage

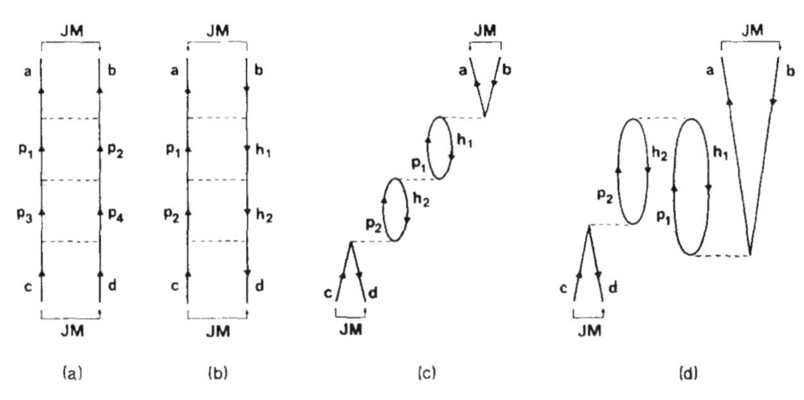

FIG. 8. Ladder diagrams.

that is connects *different* matrix elements, which is clearly inconvenient from a computational point of view. However, by a simple recoupling transformation we can change the scheme of coupling such that internal lines belonging to the same matrix element are coupled together. Using the transformation

$$\hat{r}\hat{s}\,\underset{00}{\overset{00}{|\,rs\rangle\langle rs\,|}} = \hat{r}\hat{s}\sum_{J'} X\begin{pmatrix} r & r & 0 \\ s & s & 0 \\ J' & J' & 0 \end{pmatrix} \{|\,\overset{J'}{rs}\rangle\langle \overset{J'}{rs}\,|\}^0_0 = \sum_{J'} \hat{J}'\{|\,\overset{J'}{rs}\rangle\langle \overset{J'}{rs}\,|\}^0_0 = \sum_{J'M'} |\,\overset{J'M'}{rs}\rangle\langle \overset{J'M'}{rs}\,| \quad (46)$$

in Eq. (44), we obtain the obvious and well-known result

$$\text{Diagram 9a} = \langle \overset{JM}{ab} \,|\, V \,|\, \overset{JM}{p_1 p_2}\rangle\langle \overset{JM}{p_1 p_2} \,|\, V \,|\, \overset{JM}{cd}\rangle. \quad (47)$$

This expression clearly shows the simplicity of the angular momentum structure of

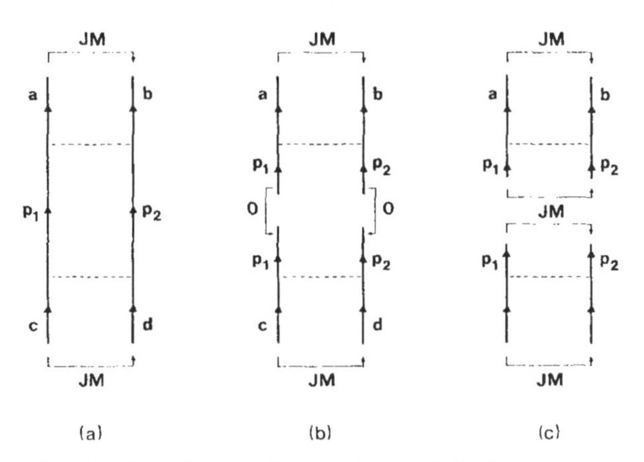

FIG. 9. Factorization of a particle–particle ladder diagram.

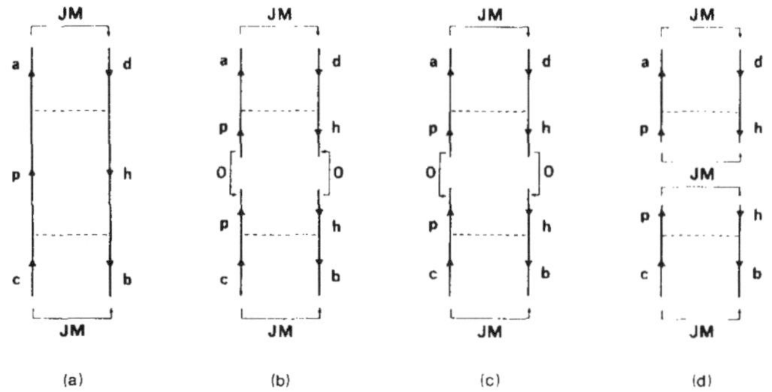

FIG. 10. Factorization of a particle–hole ladder diagram.

ladder diagrams. This simplicity lies in the fact that the J and M of the initial coupling imposed on the external lines is conserved throughout the sequence of interactions. Thus, a ladder diagram is expressed as a simple product of matrix elements of the interaction vertices, all with the same J and M, as illustrated in Fig. 9c.

In passing we note that the nuclear G-matrix, which is a basic concept in nuclear structure calculations with realistic forces, is simply given by the sum of all particle–particle ladders. Thus, the angular momentum structure of the G-matrix is uniquely determined by the total J and T of the external lines, which are conserved throughout the sequence of V interactions. We can express our diagrams in terms of G vertices by replacing V by G everywhere and, to avoid double counting, leaving out diagrams with particle–particle intermediate states between successive G vertices. However, it may be argued that *low-energy* intermediate states should be excluded from G and treated explicitly as particle–particle ladders in G (see, e.g., Ref. [4]).

It may seem that we have worked unnecessarily hard for the above trivial results. Our efforts are not wasted, however, as they have prepared us for treating the more subtle case of particle–hole ladders. Consider the simple second-order particle–hole ladder shown in Fig. 10a. The external lines are coupled to given J and M according to definition (38), while for the internal lines we sum freely over all the m-states. Thus, in a fashion similar to that used for the particle–particle ladder in Eq. (44), we obtain for the particle–hole ladder of Fig. 10a, keeping the p and h of the intermediate states fixed and disregarding the energy denominator (9) and the overall factor (13) of the corresponding uncoupled diagram

$$\text{Diagram 10a} = \sum_{\alpha\beta\gamma\delta} (-1)^{d-\delta}(-1)^{b-\beta}\, C^{a\ \ d\ \ J}_{\alpha\,-\delta\ M}\, C^{c\ \ b\ \ J}_{\gamma\,-\beta\ M}$$

$$\times \sum_{m_p m_h} \langle (a\alpha)(hm_h) \mid V \mid (pm_p)(d\delta)\rangle\langle (pm_p)(b\beta) \mid V \mid (c\gamma)(hm_h)\rangle$$

$$= (-1)^{2d}\, \hat{p}\hat{h}\, \underbrace{\langle \overbrace{ah \mid V \mid \underbrace{pd\rangle\langle p}_{00}}^{JM} \overbrace{b \mid V \mid ch}^{JM}\rangle}_{00}. \tag{48}$$

Here, we have used Eq. (34) for the coupling of the external lines; the phase $(-1)^{2d}$ is thus associated with the top particle–hole pair, as discussed in the previous subsection. The coupling scheme of the internal lines is illustrated in Fig. 10b. We note that here we have followed the convention for the order of coupling defined in Eq. (45), namely that the incoming part of an internal line is coupled to the outgoing part. This has the consequence that in Fig. 10b the upper part of the diagram is coupled to the lower part via the particle line, while the lower part is coupled to the upper part via the hole line. This "hybrid" coupling scheme is clearly not so convenient for making the recoupling transformation in the intermediate states necessary for factorization on the right-hand side of Eq. (48). In order to perform this transformation, which is similar to that of Eq. (46) employed for the particle–particle ladder, the direction of coupling must be the same for both internal lines. To achieve this, we choose to reverse the direction of coupling for the internal *hole* line. This reversal of the coupling, which is illustrated in Fig. 10c, clearly gives a phase factor $(-1)^{2h}$. Then, we can apply the transformation (46) to rewrite Eq. (48) as

$$\text{Diagram 10a} = (-1)^{2d}(-1)^{2h}\langle \overbrace{\underbrace{ah \mid V \mid pd}_{JM}}^{JM}\rangle\langle \overbrace{\underbrace{pb \mid V \mid ch}_{JM}}^{JM}\rangle. \tag{49}$$

This factorization, which is illustrated in Fig. 10d, is similar to that obtained for the particle–particle ladder. There are, however, phase differences. Firstly, there is a phase factor $(-1)^{2d}$ associated with the top particle–hole pair, as explained in the previous subsection. We emphasize that this phase only arises if the particle–hole ladder is a complete diagram by itself. As we shall eventually see, ladders may also be considered as internal building blocks of more complex diagrams. In such cases the external legs of the particle–hole ladder will be internal lines of the complete diagram and will furthermore be coupled according to Eq. (39), which is the appropriate coupling scheme for our diagram rules, and thus the phase $(-1)^{2d}$ will not arise. Further discussion of this point is given in Subsection 3.4. Secondly, there is a phase factor $(-1)^{2h}$ associated with the internal particle–hole pair cut open. Both these phase factors arise from angular momentum considerations. In addition there are phase and overall factors not shown here coming from the contractions of fermion operators in the corresponding uncoupled diagram, as discussed in Section 2.

To summarize, we have just shown the need for a phase factor

$$(-1)^{n_{ph}}, \tag{50}$$

where n_{ph} is the number of internal particle–hole pairs cut open for which the particle and hole lines are coupled in the *same* direction across the cut, e.g., $n_{ph} = 1$ in Fig. 10c. The above phase rule will be further discussed in Subsection 3.4. As was the case for the phase factor (40), the phase factor (50) is associated with the coupling of ordinary angular momenta only. If the isospins are coupled as well, this phase arises twice as shown in Eq. (41), and can thus simply be ignored.

We are now able to evaluate more complicated ladder diagrams merely by inspection. Consider for example the third-order ladders shown in Fig. 8. We shall not consider isospin and only be concerned with the angular momentum structure of the diagrams. Thus, as before we shall keep the j-values of the intermediate states fixed and ignore the energy denominators and the overall factors associated with the corresponding uncoupled diagrams. Then, applying the rules of this and the previous subsection we immediately have

$$\text{Diagram 8a} = \langle \overset{JM}{\overline{ab}} \mid V \mid \overset{JM}{\overline{p_1 p_2}} \rangle \langle \overset{JM}{\overline{p_1 p_2}} \mid V \mid \overset{JM}{\overline{p_3 p_4}} \rangle \langle \overset{JM}{\overline{p_3 p_4}} \mid V \mid \overset{JM}{\overline{cd}} \rangle, \tag{51}$$

$$\text{Diagram 8c} = (-1)^3 \underbrace{\overbrace{\langle h_1 a \mid V \mid p_1 b \rangle}^{JM}}_{JM} \underbrace{\overbrace{\langle h_2 p_1 \mid V \mid p_2 h_1 \rangle}^{JM}}_{JM} \underbrace{\overbrace{\langle d p_2 \mid V \mid c h_2 \rangle}^{JM}}_{JM}, \tag{52}$$

$$\text{Diagram 8d} = (-1)^3 \underbrace{\overbrace{\langle h_2 h_1 \mid V \mid p_2 p_1 \rangle}^{JM}}_{JM} \underbrace{\overbrace{\langle d p_2 \mid V \mid c h_2 \rangle}^{JM}}_{JM} \underbrace{\overbrace{\langle p_1 a \mid V \mid h_1 b \rangle}^{JM}}_{JM}. \tag{53}$$

The phases in Eqs. (52) and (53) arise from the fact that the corresponding diagrams each have one particle–hole pair on the top and two internal particle–hole pairs. We note that the order of coupling of the intermediate angular momenta must be consistent, implying for example that p_1 must couple to h_1 in both matrix elements. Further, the external legs are coupled to angular momentum J according to the chosen convention. For example, in diagram 8c the direction of the coupling for c and d is from c to d, while for a and b it is from b to a.

Note that this procedure can be applied to generalized ladder diagrams for which the rungs of the ladder may be of more complex structure than a single V interaction. We may use the notation χ of Fig. 6 and Eq. (33) for such complex rungs. The analysis

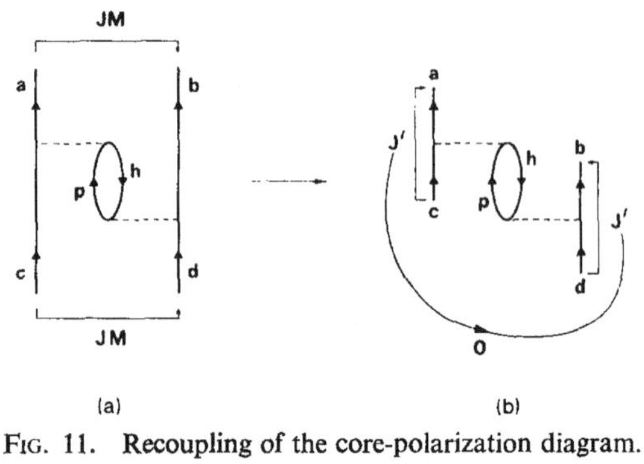

FIG. 11. Recoupling of the core-polarization diagram.

of this subsection then serves to break the diagram into a product of coupled matrix elements of χ, which must then be evaluated by the techniques described in the following subsections.

3.3. *Recognizing Ladders*

Sometimes in the original coupling scheme the diagram (or part of a diagram) is not in an obvious ladder form, but can be put into the desired ladder form by angular momentum recoupling. A simple example is the core-polarization diagram of Fig. 11a. The diagram is first coupled to a scalar and then transformed to the cross-coupled scheme indicated in Fig. 11b. This is evidently of ladder form so that the method of Subsection 3.2 can be used. Thus, we obtain, ignoring as before the phases and factors of the corresponding uncoupled diagram discussed in Section 2:

$$\text{Diagram 11a} = \hat{J}^{-1} \sum_{J'} X \begin{pmatrix} c & d & J \\ a & b & J \\ J' & J' & 0 \end{pmatrix} \langle ab \mid \chi \mid cd \rangle^{0}$$

$$= (-1)^{1} \hat{J}^{-1} \sum_{J'} X \begin{pmatrix} c & d & J \\ a & b & J \\ J' & J' & 0 \end{pmatrix} \hat{J}' \langle ah \mid V \mid cp \rangle \langle pb \mid V \mid hd \rangle. \quad (54)$$

In the second step of Eq. (54) we used the phase rule (50). This expression is identical to that derived by Osnes *et al.* [14]. Bertsch [19] has also obtained an expression of this form using essentially the same method that we present here. (He only discusses the evaluation of this particular diagram, however.) Further, it is easily shown by applying Eqs. (22), (28) and (31) that this expression is equivalent to those obtained by Barrett and Kirson [9] and by Ellis and Osnes [4]. The present expression, however, has the advantage that, once the cross-coupled matrix elements are calculated, a single summation over J' is required, whereas the other forms are written as a summation over three J's. Thus, our expression is not only easy to derive, but is immediately in a transparent form which is suitable for efficient computation.

3.4. *Decomposition of Diagrams into Ladders by Cutting Internal Lines*

There are diagrams (or parts of a diagram) which can be put into the form of ladders by judiciously cutting open one or more internal lines. This technique was in fact already used in Subsection 3.2 to factorize ladder diagrams. According to Eq. (45) an internal line consists of two pieces (an incoming one and an outgoing one) which are coupled together to a scalar in the cross channel. This coupling can then be broken by a suitable recoupling transformation. As an example, consider the diagram shown in Fig. 12a, which is apparently not of ladder form. First, we couple the external legs to a scalar. Then, if we apply Eq. (45) to the internal line p_1 (corresponding to cutting this line), we obtain the diagram shown in Fig. 12b. Here, both angular momentum couplings run from an incoming line to an outgoing line, in accordance with our

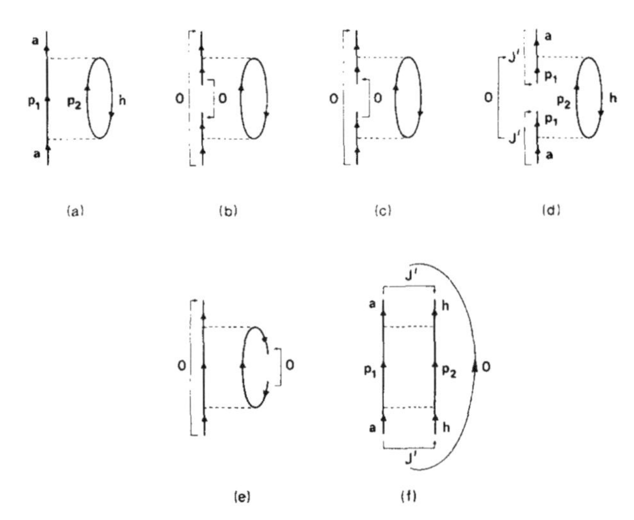

FIG. 12. Examples of cutting internal lines.

convention. Thus, the arrows of the two couplings point in opposite directions. This situation is similar to that encountered in Eq. (48) for the particle–hole ladder and is not convenient for performing the desired recoupling transformation leading to the diagram of Fig. 12d. To prepare for this recoupling, we reverse the direction of coupling for the internal line, as shown in Fig. 12c. This clearly gives rise to a phase $(-1)^{2p_1}$. Then, it is straightforward to make the recoupling transformation to diagram (d), which is of the desired ladder form. Following these steps, it is trivial to write down the expression for the diagram of Fig. 12a. We obtain, ignoring as before all the factors associated with the uncoupled diagram and keeping the j-values of the internal lines fixed:

$$\text{Diagram 12a} = \hat{a}^{-1}\hat{p}_1(-1)^{2p_1}\sum_{J'} X\begin{pmatrix} a & a & 0 \\ p_1 & p_1 & 0 \\ J' & J' & 0 \end{pmatrix} \overbrace{\langle ap_1 \mid \chi \mid p_1 a\rangle^{0}}^{J'}_{J'} \tag{55a}$$

$$= \hat{a}^{-1}\hat{p}_1(-1)^{2p_1}\sum_{J'} X\begin{pmatrix} a & a & 0 \\ p_1 & p_1 & 0 \\ J' & J' & 0 \end{pmatrix} \hat{J}'(-1)^{2h}\overbrace{\langle ah \mid V \mid p_1 p_2\rangle}^{J'M'}_{J'M'}\overbrace{\langle p_1 p_2 \mid V \mid ah\rangle}^{J'M'}_{J'M'} \tag{55b}$$

$$= \hat{a}^{-2}(-1)^{2p_1}(-1)^{2h}\sum_{J'} \hat{J}'^{2}\overbrace{\langle ah \mid V \mid p_1 p_2\rangle}^{J'M'}_{J'M'}\overbrace{\langle p_1 p_2 \mid V \mid ah\rangle}^{J'M'}_{J'M'}. \tag{55c}$$

In Eq. (55a) the factors $\hat{a}^{-1}$ and $\hat{p}_1$ are associated with the step from diagram (a) to diagram (b). The further step to diagram (c) gives the phase $(-1)^{2p_1}$, while the X-coefficient takes care of the transformation from diagram (c) to diagram (d). The matrix element of χ is simply short-hand notation for diagram (d) and is written out in detail in Eq. (55b). Here, the factor $\hat{J}$ follows from uncoupling the scalar coupled matrix element of χ, while the phase $(-1)^{2h}$ follows by the rule (50) from cutting the intermediate particle–hole pair of the ladder. Finally, by substituting expression (32) for the X-coefficient, we obtain Eq. (55c).

Now, the present technique is not unique as a given diagram can often be put into different ladder forms by cutting different internal lines. The final result must of course be the same in all cases, and thus one may choose the procedure leading to the simplest calculation. For example, the diagram of Fig. 12a can also be put into ladder form by cutting the internal hole line h, as shown in Fig. 12e. This diagram can then be transformed to the particle–particle ladder of Fig. 12f by straightforward angular momentum recoupling. Note that in Fig. 12f we have, for clarity, opened out the h lines; internal lines may be twisted at will when evaluating the angular momentum structure of a diagram, provided that the direction of the arrow (toward or away from the vertex) is maintained. By this procedure we readily obtain for the diagram of Fig. 12a

$$\text{Diagram 12a} = \hat{a}^{-1}\hat{h}\sum_{J'} X\begin{pmatrix} a & a & 0 \\ h & h & 0 \\ J' & J' & 0 \end{pmatrix} \langle \overset{J'}{ah} \mid \chi \mid \overset{J'}{ah} \rangle^{0}$$

$$= \hat{a}^{-2}\sum_{J'} \hat{J}'^{2} \langle \overset{J'M'}{ah} \mid V \mid \overset{J'M'}{p_1 p_2} \rangle \langle \overset{J'M'}{p_1 p_2} \mid V \mid \overset{J'M'}{ah} \rangle. \tag{56}$$

This expression must of course be equivalent to that of Eq. (55). This is in fact easily shown by applying Eqs. (23), (26), (28), (24) and the orthogonality relation for X-coefficients to Eq. (55). Comparing the two procedures, it is apparent that the procedure leading to Eq. (56) is slightly simpler and more convenient for computation than that leading to Eq. (55).

The present example serves to illustrate the meaning of the phase rule (50). In fact, the phase factor $(-1)^{n_{ph}}$ is associated with cut particle–hole pairs only. Thus, it is not unique for a given diagram, but depends on how it is evaluated. In Eq. (55) the original diagram was expressed in terms of a particle–hole ladder, while in Eq. (56) it was expressed in terms of a particle–particle ladder. Thus, the phase rule $(-1)^{n_{ph}}$ only applies to the evaluation of the particle–hole ladder in Eq. (55).

As another example, consider the simple third-order ring diagram shown in Fig. 13a. We cut the lines labelled p_1 and h_1, as shown in Fig. 13b. Since the coupling across p_1 is in the opposite direction to the coupling across h_1, we reverse the latter as shown in Fig. 13c, obtaining a phase $(-1)^{2h_1}$. Then we perform a straightforward recoupling transformation to obtain the diagram of Fig. 13d, which is in the form of a scalar coupled particle–hole ladder. Writing down these steps in algebraic form, we readily obtain

$$\text{Diagram 13a} = \hat{p}_1\hat{h}_1(-1)^{2h_1}\sum_{J'} X\begin{pmatrix} p_1 & p_1 & 0 \\ h_1 & h_1 & 0 \\ J' & J' & 0 \end{pmatrix} \langle h_1 p_1 \mid \chi \mid p_1 h_1 \rangle^{0}, \tag{57}$$

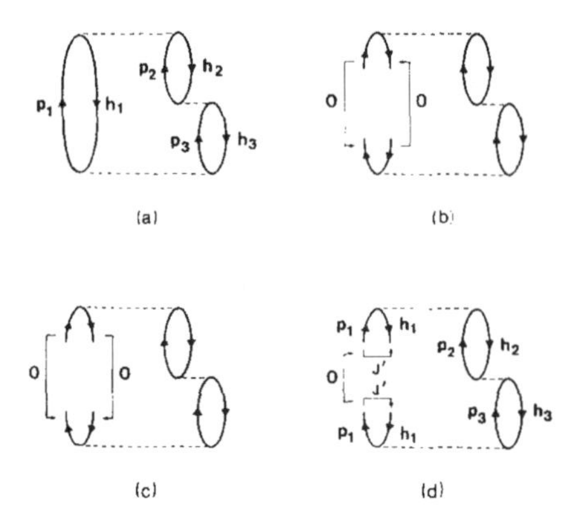

FIG. 13. Example of cutting internal lines.

where the scalar coupled particle–hole ladder is easily evaluated by the rules of Subsection 3.2:

$$\langle h_1 p_1 \mid \chi \mid p_1 h_1 \rangle^0 = (-1)^{2h_2+2h_3}\, \hat{J}' \langle h_1 h_2 \mid V \mid p_1 p_2 \rangle \langle p_2 h_3 \mid V \mid h_2 p_3 \rangle \langle p_1 p_3 \mid V \mid h_1 h_3 \rangle. \tag{58}$$

Note that h_1 is an internal hole line, so that the phase rule (40) does not apply to the

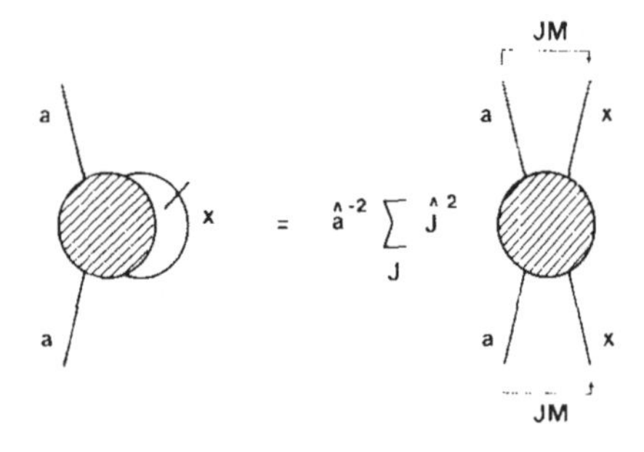

FIG. 14. Some useful formulae. The slashes indicate where we have cut the lines.

evaluation of the particle–hole ladder in Eq. (58). This is because the coupling between p_1 and h_1 is not externally imposed, but results from suitable recoupling of the internal lines. This coupling is automatically expressed in the appropriate scheme, and the phase rule (40) is not of relevance here. The phases on the right-hand sides of Eqs. (57) and (58) follow from the rule (50) since we cut in all three particle–hole pairs.

The present technique is schematically illustrated in Fig. 14. Note here that the lines may be either particle or hole lines. To obtain correct phases the lines on the right-hand side of the figure should be regarded as external lines. The rules of Fig. 14 are also obeyed if the single lines labelled a, x and y each actually represent several lines coupled to total angular momenta of a, x and y, respectively.

In the examples considered above, the lines were cut horizontally. However, we often wish to cut lines vertically, as indicated in Fig. 15. Note that the number of lines in the upper part of the cut must be the same as the number of lines in the lower part. A vertical cut can obviously be viewed as two horizontal cuts, and recoupling transformations can then be used to write the diagram in the desired form. As an example, consider the diagram shown in Fig. 16a. We first couple the diagram to a scalar and recouple it to cross-coupled form (Fig. 16b). Then, we cut the lines h_1 and h_2, as indicated in Fig. 16c. A final recoupling transformation serves to factorize the original diagram into the diagrams shown in Figs. 16d and e. Picking up the phases and factors from these operations we readily obtain

$$\text{Diagram 16a} = \hat{J}^{-1} \sum_{J'} X \begin{pmatrix} c & d & J \\ a & b & J \\ J' & J' & 0 \end{pmatrix} (-1)^{2h_1} \hat{h}_1 \hat{h}_2 X \begin{pmatrix} h_2 & h_2 & 0 \\ h_1 & h_1 & 0 \\ J' & J' & 0 \end{pmatrix} X \begin{pmatrix} J' & J' & 0 \\ J' & J' & 0 \\ 0 & 0 & 0 \end{pmatrix}$$

$$\times \langle \overbrace{\underbrace{a h_1 | \chi | c h_2}_{J'}}^{J'} \rangle^0 \langle \overbrace{\underbrace{h_2 b | V | h_1 d}_{J'}}^{J'} \rangle^0$$

$$= (-1)^{2h_1} \hat{J}^{-1} \sum_{J'} X \begin{pmatrix} c & d & J \\ a & b & J \\ J' & J' & 0 \end{pmatrix} \hat{J}'^{-1} \langle \overbrace{\underbrace{a h_1 | \chi | c h_2}_{J'}}^{J'} \rangle^0 \langle \overbrace{\underbrace{h_2 b | V | h_1 d}_{J'}}^{J'} \rangle^0 \qquad (59a)$$

$$= (-1)^{2h_1} \hat{J}^{-1} \sum_{J'J''} X \begin{pmatrix} c & d & J \\ a & b & J \\ J' & J' & 0 \end{pmatrix} X \begin{pmatrix} c & a & J' \\ h_2 & h_1 & J' \\ J'' & J'' & 0 \end{pmatrix} \hat{J}''$$

$$\times \langle \overset{J''M''}{a h_1} | V | \overset{J''M''}{p_1 p_2} \rangle \langle \overset{J''M''}{p_1 p_2} | V | \overset{J''M''}{c h_2} \rangle \langle \overbrace{\underbrace{h_2 b | V | h_1 d}_{J'M'}}^{J'M'} \rangle. \qquad (59b)$$

This agrees with the expression given by Barrett and Kirson [9] (diagram 5 of their Appendix) and is probably in a form which is a little more efficient to compute.

Although the vertical cut through the lines h_1 and h_2 in Fig. 16c could easily be

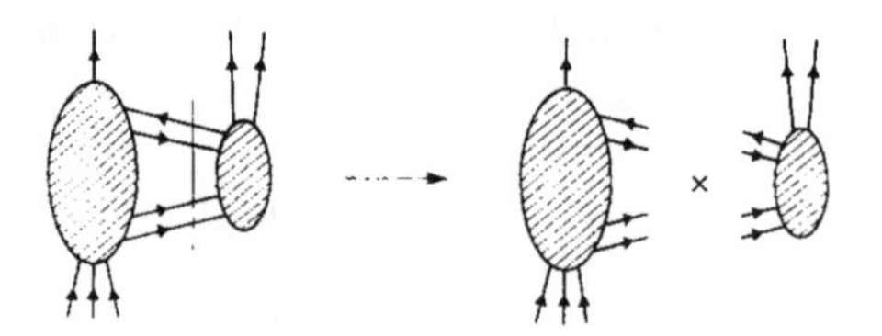

FIG. 15. Factorization by a vertical cut.

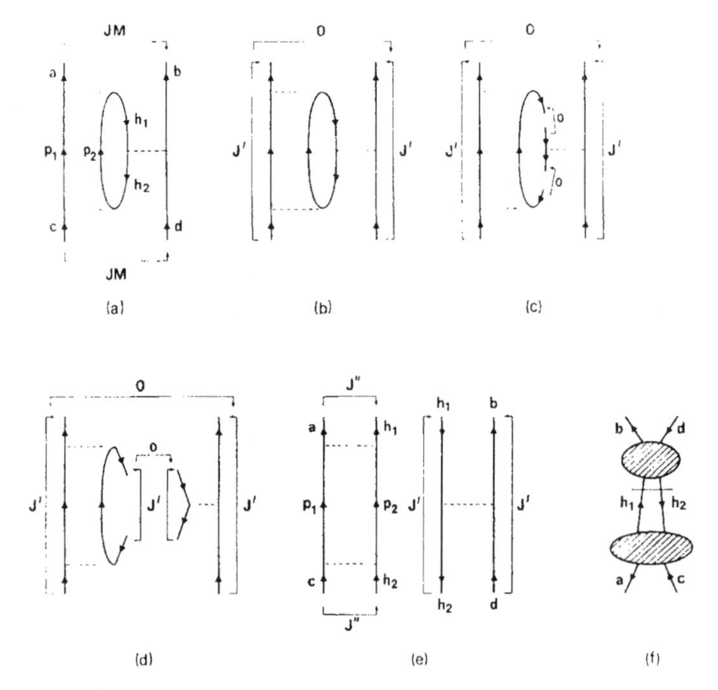

FIG. 16. Decomposition of a complicated diagram into a product of ladders.

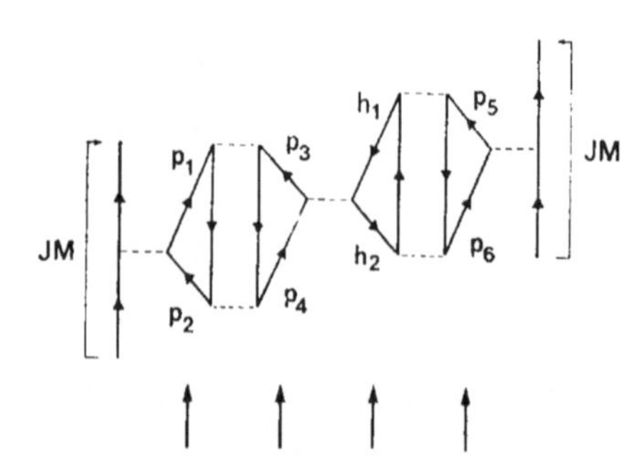

FIG. 17. Generalized particle–hole ladder in the cross channel.

treated as two horizontal cuts, the resulting calculation was somewhat lengthy. This tedious procedure can be circumvented, however, by observing that the diagram is of generalized ladder form when viewed sideways (i.e., in the cross channel). Thus, treating diagram (b) as a generalized particle–hole ladder, as shown in Fig. 16f, we immediately obtain Eq. (59a). In this case the phase $(-1)^{2h_1}$ is due to the phase rule (50), whereas in the former treatment it followed from cutting h_1 and reversing the coupling across the cut. The two approaches are of course equivalent, as the phase rule (50) was originally derived by cutting the internal particle–hole pair in a particle–hole ladder and reversing the coupling across the cut hole line.

To further illustrate the notion of a generalized ladder, we give another example, shown in Fig. 17. Here, the four vertical cuts, indicated by arrows, obviously give a phase factor

$$(-1)^{2p_2+2p_3+2h_1+2p_5}.$$

Note that the cut lines can be either holes or particles.

3.5. *One-Body Insertions*

In the previous subsections we showed that ladder diagrams are particularly simple to evaluate as they factorize into coupled matrix elements of the individual vertices. Also, we showed that more complicated diagrams may be decomposed into products of ladder diagrams by cutting internal lines. Another class of diagrams to which similar simplifications can be applied consists of diagrams with so-called one-body insertions.

A diagram with a one-body insertion has the general structure shown in Fig. 18a. The one-body insertion is denoted by A and may contain any number of V interactions. Since V is a scalar, the line entering and the line leaving A must carry the same j and m

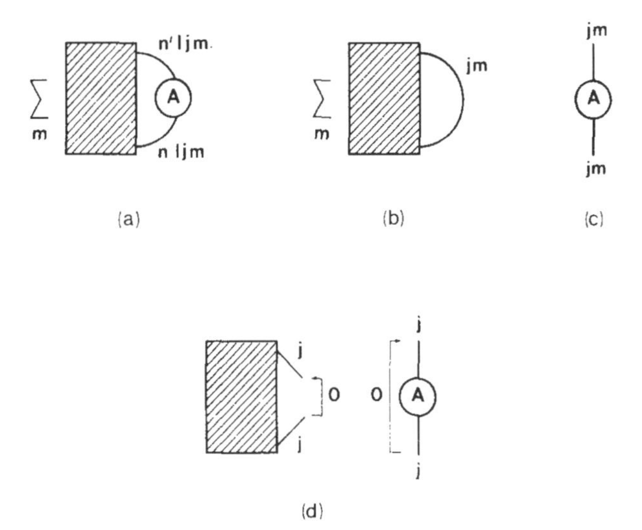

FIG. 18. Factorization of one-body insertions.

quantum numbers. In addition the orbital angular momentum quantum number l must be conserved since V conserves parity. However, the principal quantum numbers n and n' may differ. This should be borne in mind when writing down the complete expression for the diagram, but here we are only interested in the angular momentum properties. In this regard we can write

$$\text{Diagram 18a} = \text{Diagram 18b} \times \text{Diagram 18c}. \tag{60}$$

As indicated in Fig. 18, the z-component m is summed over in diagram 18a, and this summation is carried out in diagram 18b. We emphasize that there is no summation over m in diagram 18c which, of course, is independent of m.

An equivalent result is obtained by cutting off the one-body insertion A by means of a vertical cut, as described in the previous subsection. Since diagram 18a is a scalar coupled particle–hole ladder in the cross channel, it can be written as the product of the two diagrams shown in Fig. 18d, multiplied by the phase $(-1)^{2j}$ deriving from the phase rule (50). Then, by using Eq. (45), the product of Fig. 18d is easily seen to be equivalent to Eq. (60).

Thus we have shown that in determining the angular momentum factors of a diagram, the one-body insertions can be cut off and evaluated separately. Some specific examples will be presented in Section 4 (cases a, b, d and e).

3.6. *Effective Operators*

We now discuss the evaluation of diagrams which yield the effective matrix element of an operator T_μ^λ which transforms like a spherical tensor of rank λ, component μ. The operator is defined so that

$$(T_\mu^\lambda)^\dagger = (-1)^{\lambda-\mu} T_{-\mu}^\lambda, \tag{61}$$

consistent with our previous phase convention defined in Subsection 3.1. In a general diagram such as Fig. 19a, the operator T_μ^λ occurs *once* as indicated by the cross. (Note however that the perturbing potential may contain a one-body term U with

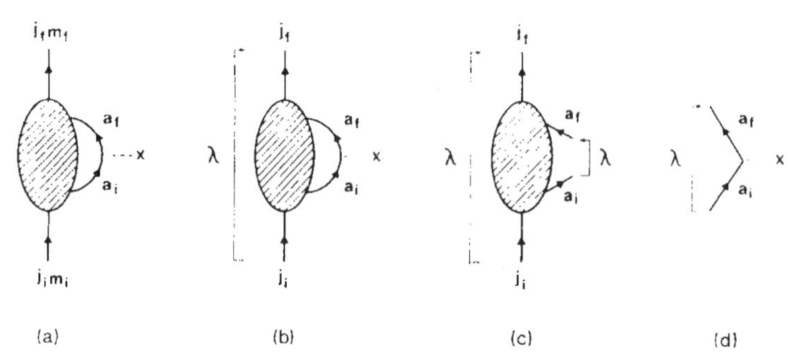

FIG. 19. Decomposition of effective operator diagrams. The operator vertex is denoted by ----×.

$\lambda = 0$, and this can occur *more* than once. This case can also be treated by the methods discussed here.) The blob in Fig. 19a indicates a sequence of V interactions and, of course, the diagram is linked. In Fig. 19a and most of our examples, the valence state consists of a single particle. Our method is general, however, and can be applied to valence states involving arbitrary numbers of particle and hole lines.

By using the Wigner–Eckart theorem we can express the transition matrix elements in terms of *reduced* transition matrix elements which are independent of the magnetic quantum numbers. We shall use the Wigner–Eckart theorem in the following form[5]

$$\langle j_f m_f | (V)^m T^\lambda_\mu (V)^n | j_i m_i \rangle = (-1)^{\lambda-\mu} (-1)^{j_f - m_f} C^{j_i \; j_f \; \lambda}_{m_i \, -m_f \, -\mu} \langle j_f \| (V)^m \mathbf{T}^\lambda (V)^n \| j_i \rangle, \quad (62)$$

implying that

$$\langle j_f \| \mathbf{T}^\lambda \| j_i \rangle = (-1)^{\lambda-\mu} \overbrace{\langle j_f | T^\lambda_\mu | j_i \rangle}^{\lambda-\mu} = \hat{\lambda}^{-1} \overbrace{\langle j_f | \mathbf{T}^\lambda | j_i \rangle}^{\lambda}{}^0, \quad (63)$$

where the last matrix element is explicitly coupled to a resultant angular momentum of zero, as indicated by the superscript. Note also that by using Eq. (61) we obtain

$$\langle j_2 \| \mathbf{T}^\lambda \| j_1 \rangle = (-1)^{j_2+\lambda-j_1} \langle j_1 \| \mathbf{T}^\lambda \| j_2 \rangle, \quad (64)$$

where we have taken the reduced matrix element to be real, as will always be the case here.

It is the reduced matrix element which we shall define to be the value of a given diagram and which we shall evaluate. To be more explicit, we define the value of the diagram in Fig. 19a to be that of Fig. 19b where the external lines i and f are coupled. Clearly, the diagram in Fig. 19b is just the reduced matrix element defined in Eq. (63). On the other hand, if we wish to compute the m-scheme matrix element corresponding to Fig. 19a, we can simply use Eq. (62). Note that if we are dealing with hole rather than particle states, we should use Eq. (63) with the labels i and f interchanged.

Now, to calculate the reduced matrix element of Fig. 19b, we notice that it has the form of a particle–hole ladder in the cross channel. Thus, we can use the techniques of the previous subsections to express it as the product of the two diagrams shown in Figs. 19c and d, multiplied by a phase $(-1)^{2a_f}$ coming from the phase rule (50). Firstly, by definition diagram (d) gives a contribution

$$\text{Diagram 19d} = \langle a_f \| \mathbf{T}^\lambda \| a_i \rangle. \quad (65)$$

Secondly, in diagram (c) we have coupled $j_i(a_i)$ and $j_f(a_f)$ to a resultant angular momentum λ, so that the diagram is in the cross-coupled form already discussed for the effective interaction. It is independent of μ, the z-component of λ, so that, if desired, we can further couple the two angular momenta λ to a resultant angular momentum of zero, provided that a factor $\hat{\lambda}^{-1}$ is introduced. This is sometimes convenient, as we shall see.

[5] The reduced matrix element defined here is $(-1)^{j_i+\lambda-j_f} \hat{\lambda}^{-1}$ times that of Edmonds [18].

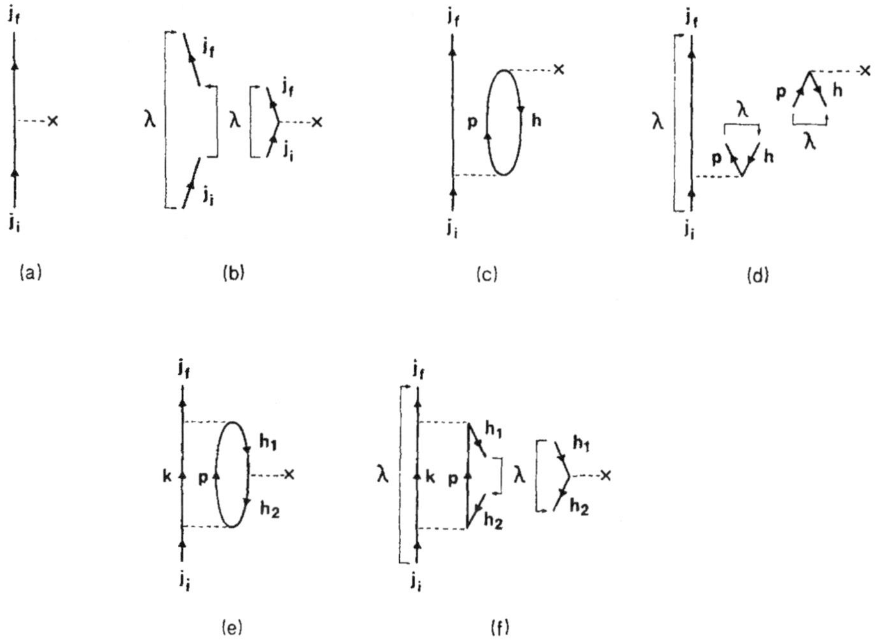

FIG. 20. Some low-order effective operator diagrams.

Our rule for obtaining the value of an effective operator diagram like Fig. 19b (i.e., the reduced matrix element) is therefore to write it as a product

$$\text{Diagram 19b} = (-1)^{2a_f} \times \text{Diagram 19c} \times \text{Diagram 19d}, \tag{66}$$

where diagram 19c can be evaluated by the techniques discussed previously and diagram 19d is given by Eq. (65).

Let us give a few examples. The most simple diagram is given in Fig. 20a, and this is pulled apart in Fig. 20b. Since we have two Clebsch–Gordan coefficients coupling j_i and j_f to λ, the sum over z-components gives unity. In addition there is a phase $(-1)^{2j_f}$ which cancels the phase $(-1)^{2a_f}$ in Eq. (66), and so we obtain the obvious result

$$\text{Diagram 20a} = \langle j_f \| \mathbf{T}^\lambda \| j_i \rangle. \tag{67}$$

The diagram of Fig. 20c is evaluated in the form indicated in Fig. 20d, giving

$$\text{Diagram 20c} = (-1)^{2h} \underset{\lambda\mu}{\overset{\lambda\mu}{\langle j_f p \mid V \mid j_i h \rangle \langle h \| \mathbf{T}^\lambda \| p \rangle}}, \tag{68}$$

where as usual the phases and factors associated with the corresponding uncoupled diagram have been left out. As a final example consider Fig. 20e, which is evaluated in the form of Fig. 20f:

$$\text{Diagram 20e} = (-1)^{2h_2}\langle j_f h_1 \mid \chi \mid j_i h_2\rangle\langle h_2 \| \mathbf{T}^\lambda \| h_1\rangle$$

$$= (-1)^{2h_2}(-1)^{h_1+h_2-\lambda}\hat{\lambda}^{-1}\sum_J X\begin{pmatrix} j_i & j_f & \lambda \\ h_2 & h_1 & \lambda \\ J & J & 0 \end{pmatrix}\hat{J}$$

$$\times \langle j_f h_1 \mid V \mid kp\rangle\langle kp \mid V \mid j_i h_2\rangle\langle h_2 \| \mathbf{T}^\lambda \| h_1\rangle. \tag{69}$$

Note the similarity here to the evaluation of Fig. 16a in Subsection 3.4. The above expressions agree with those given by Siegel and Zamick [15].

The present technique may be generalized to treat more complicated cases, such as the two-body effective operator diagrams. Consider as an example the diagram of Fig. 21a. We first cut off the operator vertex, and write the diagram as the product of Figs. 21b and c, multiplied by a phase deriving from the phase rule (50). The techniques described in Subsections 3.1 to 3.5 can be readily used to evaluate Fig. 21b.

3.7. *Further Remarks*

Our basic approach has been to break a given diagram into a product of ladder diagrams which are easily dealt with. Up to now, we have not considered cases with three or more valence lines. We shall now discuss the angular momentum factors associated with such diagrams. An example of an effective interaction diagram with three valence lines is shown in Fig. 22a. Such a diagram contributes to the effective interaction for nuclei with three nucleons outside closed shells, for instance ^{19}O. The diagram of Fig. 22a does not appear to have an obvious ladder form. It is composed of five ladders, but each has only one rung, i.e., one interaction vertex. The diagram of Fig. 22b has in fact the same basic structure, although it only has two valence lines. This is seen by cutting and pulling open the hole line. This type of diagram is usually called a three-body cluster.

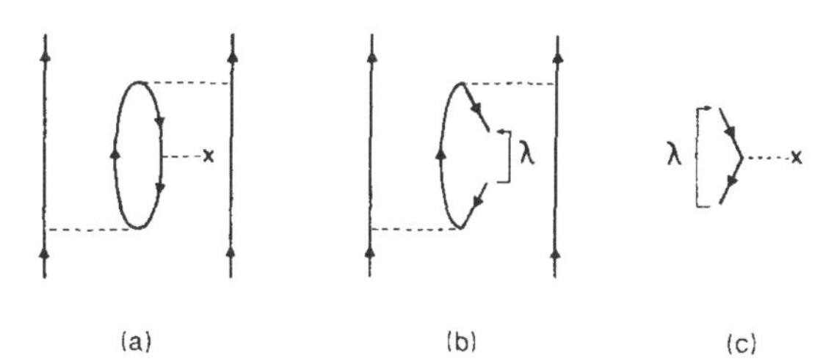

FIG. 21. Treatment of a two-body effective operator diagram.

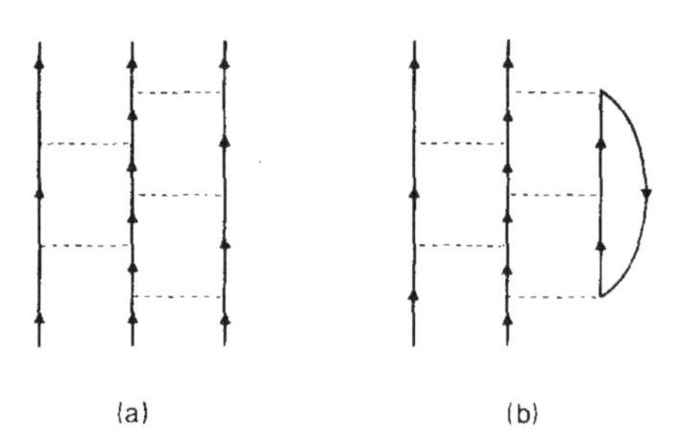

FIG. 22. Three-body cluster diagram.

The basic principle for determining the angular momentum structure of such diagrams remains the same, namely we break the diagrams into products of ladders (including those with one rung). By systematically recoupling and successively breaking off ladders, the procedure is indeed straightforward. This is clearly illustrated by the example shown in Fig. 23. It is readily seen that the lower two-body ladder can be broken off yielding

$$\text{Diagram 23a} = \sum_{J_{23}} X \begin{pmatrix} j_1 & j_2 & J_{12} \\ 0 & j_3 & j_3 \\ j_1 & J_{23} & J \end{pmatrix} \times \text{Diagram 23b} \times \text{Diagram 23c}, \qquad (70)$$

with the obvious restriction $J_{23} = J_{45}$. Clearly, the basic building blocks, such as diagrams (b) and (c) in this example, may appear in several diagrams, but they only need to be evaluated once. This procedure may be generalized in a straightforward manner to treat diagrams with more than three valence lines, such as the four-body clusters.

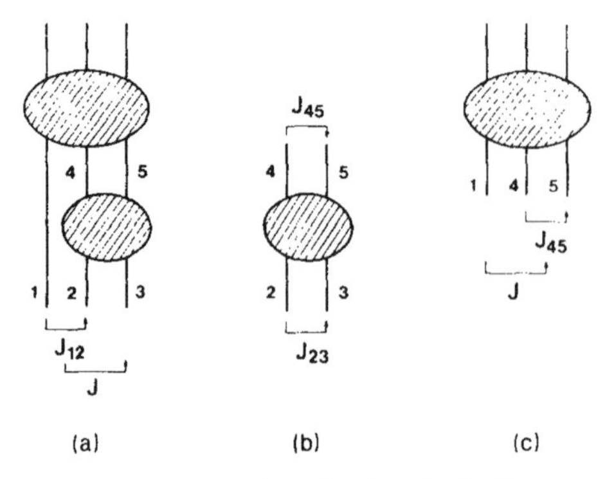

FIG. 23. Factorization of a three-body diagram.

4. Summary and Illustrative Examples

In this section we work out some examples which illustrate the power of the present technique. It is useful first to summarize the basic steps of our procedure:

(i) In the first step we use the diagram rules for the uncoupled representation, as discussed in Section 2, to obtain all factors aside from angular momentum coupling.

(ii) Then, the angular momentum factors are obtained by the following procedure:

(a) First, it is often convenient to couple the diagram explicitly to a scalar.

(b) Then, the diagram is decomposed into products of ladder diagrams, insofar as is possible, by angular momentum recoupling, by cutting internal lines, etc., as explained in Section 3. Since in general there are many different ways of cutting a diagram, we have not given explicit rules for the angular momentum factors and phases associated with a given cut, except for the simple cases shown in Fig. 14 and the phase rule (50) which applies to cuts of particle–hole pairs where the lines are coupled in the same order on each side of the cut. However, for all other cuts the angular momentum factors and phases can easily be worked out using standard angular momentum algebra.

(c) Each ladder is readily expressed in terms of standard coupled and cross-coupled matrix elements, keeping in mind rule (e), below.

(d) One-body insertions can be cut off and treated separately.

(e) There are special phase factors associated with particle–hole pairs, namely,

$$(-1)^{n_{tph}+n_{ph}}, \tag{71}$$

where n_{tph} is the number of particle–hole pairs on the top (bra) side of the diagram and n_{ph} is the number of internal particle–hole pairs cut open for which the particle and hole lines are coupled in the same direction across the cut. Recall that the second part of this phase rule $[(-1^{n_{ph}}]$ is operational in the sense that it may give different phases for different decompositions of a given diagram. Any such phase difference will of course be compensated by other phase differences in the components of the diagram. The phase (71) arises from angular momentum considerations alone. Thus, if isospin couplings are considered as well, this phase arises twice and can therefore be ignored.

(f) Finally, we sum over all intermediate-state quantum numbers.

We now use this technique to calculate the diagrams of Fig. 3; the slashes on these diagrams are to indicate where we have cut the lines. For each case, we briefly outline the steps that are needed to arrive at a given formula. We give the results in terms of the X-coefficients since this is most transparent, however the expressions can be simplified by using Eqs. (30)–(32). For simplicity we do not explicitly indicate the summation over intermediate particle and hole states. Further, only j out of the set of labels nlj

needed to specify a single-particle state is indicated in the matrix elements. (Note that if j_1 is required to be equal to j_2, this does not imply $n_1 = n_2$.)

We shall give both the angular momentum and the isospin factors; the latter are obtained from the former by replacing the single-particle angular momenta j by $\frac{1}{2}$ and substituting T labels for J labels. Thus the standard JT coupled matrix elements which we shall use are obtained by making these substitutions in Eq. (17), while the cross-coupled matrix elements are defined by Eq. (43). We also show explicitly the phase factor (71) even though it is the same for angular momentum and isospin and therefore contributes $+1$. Further, the normalization factors for external lines will be included. They are needed when the external lines are coupled to total angular momentum J and isospin T. For example, for the diagram of Fig. 3f we need an extra factor of $[(1 + \delta_{j_a j_b})(1 + \delta_{j_c j_d})]^{-1/2}$. We should point out that our standard coupled matrix elements of Eq. (17) do *not* include these factors. Thus, they are related to the usual two-particle matrix elements, say those of Kuo and Brown [16], by

$$\langle \overset{JT}{ab} \mid V \mid \overset{JT}{cd} \rangle = \sqrt{(1 + \delta_{ab})(1 + \delta_{cd})}\, \langle abJT \mid V_{\mathbf{KB}} \mid cdJT \rangle. \tag{72}$$

This applies to all of the examples given in this section. In Eq. (72) and throughout this section we have replaced $JMTM_T$ by the less cumbersome JT despite the fact that none of the matrix elements are scalar coupled.

Expressions for several of the diagrams considered here can be found in the works of Kassis [17], Barrett and Kirson [9] and Siegel and Zamick [15]. In many cases, our expressions appear to be somewhat more transparent in structure and efficient for computation.

EXAMPLE a. Cutting Fig. 3a as indicated, we couple the h_4 lines to a scalar and similarly for the p_2 lines, obtaining a factor $\hat{h}_4\hat{\tfrac{1}{2}}\hat{p}_2\hat{\tfrac{1}{2}}$. Coupling also p_1 and h_3 (necessarily to a scalar) we can recouple to obtain

$$\begin{aligned}
\text{Diagram 3a} &= -\tfrac{1}{2}[(\epsilon_{h_3} - \epsilon_{p_1})(\epsilon_{h_1} + \epsilon_{h_2} - \epsilon_{p_1} - \epsilon_{p_2})]^{-1} \sum_{\substack{JT\\J'T'}} (-1)^{2(h_3+1/2)}(-1)^{2(h_3+1/2)} \\
&\quad \times \hat{h}_4\hat{\tfrac{1}{2}}\hat{p}_2\hat{\tfrac{1}{2}}\hat{J}\hat{T}\hat{J}'\hat{T}' X\begin{pmatrix} p_1 & h_3 & 0 \\ h_4 & h_4 & 0 \\ J & J & 0 \end{pmatrix} X\begin{pmatrix} \frac{1}{2} & \frac{1}{2} & 0 \\ \frac{1}{2} & \frac{1}{2} & 0 \\ T & T & 0 \end{pmatrix} X\begin{pmatrix} h_3 & p_1 & 0 \\ p_2 & p_2 & 0 \\ J' & J' & 0 \end{pmatrix} X\begin{pmatrix} \frac{1}{2} & \frac{1}{2} & 0 \\ \frac{1}{2} & \frac{1}{2} & 0 \\ T' & T' & 0 \end{pmatrix} \\
&\quad \times \langle \overset{JT}{h_3 h_4} \mid V \mid \overset{JT}{p_1 h_4} \rangle \langle \overset{J'T'}{h_1 h_2} \mid V \mid \overset{J'T'}{h_3 p_2} \rangle \langle \overset{J'T'}{p_1 p_2} \mid V \mid \overset{J'T'}{h_1 h_2} \rangle \\
&= -\tfrac{1}{2}[(\epsilon_{h_3} - \epsilon_{p_1})(\epsilon_{h_1} + \epsilon_{h_2} - \epsilon_{p_1} - \epsilon_{p_2})]^{-1} \sum_{\substack{JT\\J'T'}} (-1)^{2(h_3+1/2)}(-1)^{2(h_3+1/2)} \\
&\quad \times [\hat{p}_1\hat{\tfrac{1}{2}}]^{-2}[\hat{J}\hat{T}\hat{J}'\hat{T}']^2 \langle \overset{JT}{h_3 h_4} \mid V \mid \overset{JT}{p_1 h_4} \rangle \langle \overset{J'T'}{h_1 h_2} \mid V \mid \overset{J'T'}{h_3 p_2} \rangle \langle \overset{J'T'}{p_1 p_2} \mid V \mid \overset{J'T'}{h_1 h_2} \rangle.
\end{aligned} \tag{73}$$

Here and in the following we have collected the phases and factors associated with the diagram in uncoupled form in front of the summation sign. Apart from the energy denominators these can be inferred from Table I. Recall also that the entire expression has to be summed over the intermediate particle and hole states. In Eq. (73) one of the phase factors $(-1)^{2(h_3+1/2)}$ comes from cutting the particle–hole pair p_1h_3, while the other comes from inverting the scalar coupling between p_1 and h_3 in the lower part of the diagram.

EXAMPLE b. We first cut off the one-body insertion of Fig. 3b using the technique of Fig. 18d. The remaining diagram is in the form of a ladder if we cut open the h_1 line. Noting that $h_2 = h_3$, we have

$$\begin{aligned}\text{Diagram 3b} = &-\frac{1}{2^2}[(\epsilon_{h_1}+\epsilon_{h_2}-\epsilon_{p_3}-\epsilon_{p_4})(\epsilon_{h_1}+\epsilon_{h_2}-\epsilon_{p_1}-\epsilon_{p_2})\\ &\times(\epsilon_{h_1}+\epsilon_{h_3}-\epsilon_{p_1}-\epsilon_{p_2})]^{-1}\sum_{\substack{JT\\J'T'}}(-1)^{2(h_2+1/2)}(-1)^{2(h_2+1/2)}[\hat{h}_2\hat{\tfrac{1}{2}}]^{-2}\\ &\times[\hat{J}\hat{T}\hat{J}'\hat{T}']^2\overset{JT}{\langle h_1h_2\mid V\mid p_3p_4\rangle}\overset{JT}{\langle p_3p_4\mid V\mid p_1p_2\rangle}\overset{JT}{\langle p_1p_2\mid V\mid h_1h_3\rangle}\\ &\times\overset{J'T'}{\langle h_3h_4\mid V\mid h_2h_4\rangle}.\end{aligned} \tag{74}$$

Here, a phase $(-1)^{2(h_2+1/2)}$ is obtained by cutting off the one-body insertion, since this amounts to cutting a particle–hole ladder in the cross channel. Further, another phase $(-1)^{2(h_2+1/2)}$ is obtained by inverting the scalar coupling between h_2 and h_3 in the one-body insertion so that the proper "in" coupled to "out" ordering is obtained.

EXAMPLE c. We cut open the h_1 line of Fig. 3c and couple the pieces to a scalar. Since we must have $a = c$, we also couple the complete diagram to a scalar. This allows us to recouple the diagram to ladder form, obtaining

$$\begin{aligned}\text{Diagram 3c} = &+\frac{1}{2^2}[(\epsilon_c+\epsilon_{h_1}-\epsilon_{p_1}-\epsilon_{p_2})(\epsilon_{h_3}+\epsilon_{h_2}-\epsilon_{p_1}-\epsilon_{p_2})]^{-1}\\ &\times\sum_{JT}[\hat{c}\hat{\tfrac{1}{2}}]^{-2}[\hat{J}\hat{T}]^2\overset{JT}{\langle ah_1\mid V\mid p_1p_2\rangle}\overset{JT}{\langle h_3h_2\mid V\mid ch_1\rangle}\overset{JT}{\langle p_1p_2\mid V\mid h_3h_2\rangle}.\end{aligned} \tag{75}$$

EXAMPLE d. We cut off the one-body insertion. The remaining diagram is already in ladder form and noting that $a = c$ and $p_1 = p_2$, we have

$$
\begin{aligned}
\text{Diagram 3d} &= -\tfrac{1}{2}[(\epsilon_{h_1}+\epsilon_{h_2}-\epsilon_a-\epsilon_{p_1})(\epsilon_{h_1}+\epsilon_{h_2}-\epsilon_a-\epsilon_{p_2})]^{-1} \\
&\quad\times \sum_{\substack{JT\\J'T'}} (-1)^{2(p_1+1/2)}(-1)^{2(p_1+1/2)}[\hat{c}\hat{\tfrac{1}{2}}\,\hat{p}_1\hat{\tfrac{1}{2}}]^{-2}[\hat{J}\hat{T}\hat{J}'\hat{T}']^2 \\
&\quad\times \langle \overset{JT}{h_1h_2} \mid V \mid \overset{JT}{cp_1}\rangle\langle \overset{JT}{ap_2} \mid V \mid \overset{JT}{h_1h_2}\rangle\langle \overset{J'T'}{p_1h_3} \mid V \mid \overset{J'T'}{p_2h_3}\rangle. \qquad (76)
\end{aligned}
$$

The two phases inside the summation have similar origin to example (b).

EXAMPLE e. We cut the line labelled p_2 in Fig. 3e and note that $p_2 = d$. The one-body part can then be treated as in Fig. 12 so we find

$$
\begin{aligned}
\text{Diagram 3e} &= -\tfrac{1}{2}[(\epsilon_c+\epsilon_b+\epsilon_{h_1}+\epsilon_{h_2}-\epsilon_d-\epsilon_a-\epsilon_{p_2}-\epsilon_{p_1}) \\
&\quad\times (\epsilon_{h_1}+\epsilon_{h_2}-\epsilon_d-\epsilon_{p_1})]^{-1}\sum_{J'T'}(-1)^{2(b+1/2)}(-1)^{2(p_2+1/2)}(-1)^{2(p_2+1/2)} \\
&\quad\times [\hat{d}\hat{\tfrac{1}{2}}]^{-2}[\hat{J}'\hat{T}']^2\langle \overset{JT}{\underset{JT}{ap_2 \mid V \mid bc}}\rangle\langle \overset{J'T'}{h_1h_2} \mid V \mid \overset{J'T'}{p_2p_1}\rangle\langle \overset{J'T'}{dp_1} \mid V \mid \overset{J'T'}{h_1h_2}\rangle. \qquad (77)
\end{aligned}
$$

Here, the phase $(-1)^{2(b+1/2)}$ comes from the phase rule (71) [or (40)] as there is one particle–hole pair on the top side of the diagram. Then, we obtain a phase $(-1)^{2(p_2+1/2)}$ by inverting the scalar coupling of the two pieces of the p_2 line at the cut and another phase $(-1)^{2(p_2+1/2)}$ by inverting the scalar coupling of p_2 and d obtained by angular momentum recoupling.

EXAMPLE f. We cross-couple the external legs of Fig. 3f and cut the p_1h_1 pair to obtain the product of two ladder diagrams. Care should be taken to keep track of the phases arising from changes in the coupling order. After performing the appropriate recouplings we obtain

$$
\begin{aligned}
\text{Diagram 3f} &= +[(1+\delta_{ab})(1+\delta_{cd})]^{-1/2}[(\epsilon_c+\epsilon_{h_1}-\epsilon_{p_1}-\epsilon_b) \\
&\quad\times (\epsilon_c+\epsilon_{h_2}-\epsilon_{p_1}-\epsilon_{p_2})]^{-1}\sum_{\substack{J'T'\\J''T''}}(-1)^{2(h_1+1/2)}(-1)^{2(h_2+1/2)}(-1)^{c+d-J+1-T} \\
&\quad\times [\hat{J}\hat{T}]^{-1}\,\hat{J}''\hat{T}''X\begin{pmatrix} d & c & J \\ a & b & J \\ J' & J' & 0\end{pmatrix}X\begin{pmatrix} \frac{1}{2} & \frac{1}{2} & T \\ \frac{1}{2} & \frac{1}{2} & T \\ T' & T' & 0\end{pmatrix}\langle \overset{J'T'}{\underset{J'T'}{ah_1 \mid V \mid dp_1}}\rangle \\
&\quad\times X\begin{pmatrix} c & b & J' \\ p_1 & h_1 & J' \\ J'' & J'' & 0\end{pmatrix}X\begin{pmatrix} \frac{1}{2} & \frac{1}{2} & T' \\ \frac{1}{2} & \frac{1}{2} & T' \\ T'' & T'' & 0\end{pmatrix}\langle \overset{J''T''}{\underset{J''T''}{h_2b \mid V \mid h_1p_2}}\rangle\langle \overset{J''T''}{\underset{J''T''}{p_1p_2 \mid V \mid h_2c}}\rangle.
\end{aligned}
$$

(78)

Here, the phase $(-1)^{c+d-J+1-T}$ obviously comes from inverting the coupling between the external lines c and d. Further, the phase $(-1)^{2(h_1+1/2)}$ is obtained by cutting the particle–hole pair p_1h_1, while the phase $(-1)^{2(h_2+1/2)}$ is due to the factorization of the remaining particle–hole ladder with intermediate state p_2h_2.

EXAMPLE g. We cross-couple the external legs of Fig. 3g and cut the p_1h_1 pair. The right-hand part of the diagram is tackled by cutting the p_2p_4 pair so as to give a one-rung ladder and a core-polarization diagram whose evaluation was discussed in Subsection 3.3. Explicitly we find

$$\begin{aligned}
\text{Diagram 3g} = {} & +[(1+\delta_{ab})(1+\delta_{cd})]^{-1/2}[(\epsilon_c+\epsilon_d+\epsilon_{h_1}-\epsilon_a-\epsilon_{p_2}-\epsilon_{p_4}) \\
& \times(\epsilon_c+\epsilon_{h_1}+\epsilon_{h_2}-\epsilon_a-\epsilon_{p_2}-\epsilon_{p_3})(\epsilon_c+\epsilon_{h_1}-\epsilon_a-\epsilon_{p_1})]^{-1} \\
& \times\sum_{\substack{J_1T_1\\ J_2T_2\\ J_3T_3}}(-1)^{2(h_1+1/2)}(-1)^{2(h_2+1/2)}[\hat{J}\hat{T}]^{-1}\,\hat{J}_3\hat{T}_3X\begin{pmatrix} c & d & J\\ a & b & J\\ J_1 & J_1 & 0\end{pmatrix}X\begin{pmatrix}\frac{1}{2} & \frac{1}{2} & T\\ \frac{1}{2} & \frac{1}{2} & T\\ T_1 & T_1 & 0\end{pmatrix} \\
& \times X\begin{pmatrix} p_1 & h_1 & J_1\\ d & b & J_1\\ J_2 & J_2 & 0\end{pmatrix}X\begin{pmatrix}\frac{1}{2} & \frac{1}{2} & T_1\\ \frac{1}{2} & \frac{1}{2} & T_1\\ T_2 & T_2 & 0\end{pmatrix}X\begin{pmatrix} p_1 & d & J_2\\ p_2 & p_4 & J_2\\ J_3 & J_3 & 0\end{pmatrix}X\begin{pmatrix}\frac{1}{2} & \frac{1}{2} & T_2\\ \frac{1}{2} & \frac{1}{2} & T_2\\ T_3 & T_3 & 0\end{pmatrix} \\
& \times\langle \overset{J_2T_2}{h_1b} \mid V\mid \overset{J_2T_2}{p_2p_4}\rangle\langle \overset{J_3T_3}{\underset{J_3T_3}{h_2p_4\mid V\mid p_3d}}\rangle\langle \overset{J_3T_3}{\underset{J_3T_3}{p_2p_3\mid V\mid p_1h_2}}\rangle\langle \overset{J_1T_1}{\underset{J_1T_1}{ap_1\mid V\mid ch_1}}\rangle.
\end{aligned} \tag{79}$$

Here, the phase $(-1)^{2(h_1+1/2)}$ comes from cutting the particle–hole pair p_1h_1, whereas the phase $(-1)^{2(h_2+1/2)}$ arises from factorization of the core-polarization "ladder" with intermediate state p_3h_2. The derivation of the above expression would probably become much more difficult without the present technique.

EXAMPLE h. The diagram of Fig. 3h is in ladder form already, provided that we cross-couple the external legs. We find

$$\begin{aligned}
\text{Diagram 3h} = {} & +\tfrac{1}{2}(1+\delta_{ab})^{-1/2}(\epsilon_c+\epsilon_d-\epsilon_{p_1}-\epsilon_{p_2})^{-1} \\
& \times(-1)^{2(d+1/2)}[\hat{c}\hat{\tfrac{1}{2}}]^{-1}\,\hat{J}\hat{T}X\begin{pmatrix} 0 & c & c\\ J & d & c\\ J & J & 0\end{pmatrix}X\begin{pmatrix} 0 & \frac{1}{2} & \frac{1}{2}\\ T & \frac{1}{2} & \frac{1}{2}\\ T & T & 0\end{pmatrix} \\
& \times\langle \overset{JT}{ab}\mid V\mid \overset{JT}{p_1p_2}\rangle\langle \overset{JT}{p_1p_2}\mid V\mid \overset{JT}{cd}\rangle.
\end{aligned} \tag{80}$$

Here, the phase $(-1)^{2(d+1/2)}$ arises from the top particle–hole pair.

EXAMPLE i. We take our tensor operator to be the product of an operator $T_\mu{}^\lambda$ acting in ordinary space and an operator $T_\nu{}^\tau$ acting in isospin space. Cutting off the operator in Fig. 3i as shown in Fig. 19 and recoupling to obtain a ladder, we find for the reduced matrix element (in both ordinary and isospin space)

$$\begin{aligned}\text{Diagram 3i} &= +[(\epsilon_a + \epsilon_{h_1} - \epsilon_{p_2} - \epsilon_{p_3})(\epsilon_c + \epsilon_{h_1} - \epsilon_{p_1} - \epsilon_{p_3})]^{-1} \\ &\times \sum_{JT} (-1)^{2(p_2+1/2)}(-1)^{2(h_1+1/2)} \langle p_2 \| \mathbf{T}^\lambda \| p_1 \rangle \langle \tfrac{1}{2} \| \mathbf{T}^\tau \| \tfrac{1}{2} \rangle [\hat{\lambda}\hat{\tau}]^{-1} \hat{J}\hat{T} \\ &\times X \begin{pmatrix} c & a & \lambda \\ p_1 & p_2 & \lambda \\ J & J & 0 \end{pmatrix} X \begin{pmatrix} \frac{1}{2} & \frac{1}{2} & \tau \\ \frac{1}{2} & \frac{1}{2} & \tau \\ T & T & 0 \end{pmatrix} \langle \overbrace{a\,\underbrace{h_1 \mid V \mid p_3}_{JT}\,p_2}^{JT} \rangle \langle \overbrace{p_3\,\underbrace{p_1 \mid V \mid c}_{JT}\,h_1}^{JT} \rangle .\end{aligned} \tag{81}$$

Here, the phase $(-1)^{2(p_2+1/2)}$ is obtained by cutting off the one-body operator, since this amounts to cutting a particle–hole pair in the cross channel. Further, there is a phase $(-1)^{2(h_1+1/2)}$ coming from the factorization of the remaining ladder diagram with the particle–hole intermediate state p_3h_1.

EXAMPLE j. The diagram of Fig. 3j is an example of a two-body effective operator diagram. By cutting off the operator and recoupling twice, we can express the diagram as the product of a two-body ladder diagram and a simple one-body operator diagram, obtaining

$$\begin{aligned}\text{Diagram 3j} &= +\tfrac{1}{2}[(\epsilon_c + \epsilon_d - \epsilon_{p_2} - \epsilon_{p_3})(\epsilon_d - \epsilon_{p_1})]^{-1} \langle p_1 \| \mathbf{T}^\lambda \| d \rangle \langle \tfrac{1}{2} \| \mathbf{T}^\tau \| \tfrac{1}{2} \rangle \\ &\times \hat{p}_1 \hat{\tfrac{1}{2}} \hat{J}_f \hat{T}_f X \begin{pmatrix} c & d & J_i \\ p_1 & p_1 & 0 \\ J_f & \lambda & J_i \end{pmatrix} X \begin{pmatrix} \frac{1}{2} & \frac{1}{2} & T_i \\ \frac{1}{2} & \frac{1}{2} & 0 \\ T_f & \tau & T_i \end{pmatrix} X \begin{pmatrix} J_f & \lambda & J_i \\ J_f & 0 & J_f \\ 0 & \lambda & \lambda \end{pmatrix} X \begin{pmatrix} T_f & \tau & T_i \\ T_f & 0 & T_f \\ 0 & \tau & \tau \end{pmatrix} \\ &\times \langle \overbrace{ab}^{J_fT_f} \mid V \mid \overbrace{p_2 p_3}^{J_fT_f} \rangle \langle \overbrace{p_2 p_3}^{J_fT_f} \mid V \mid \overbrace{cp_1}^{J_fT_f} \rangle .\end{aligned} \tag{82}$$

ACKNOWLEDGMENT

One of the authors (T.T.S.K.) would like to thank Professor A. Faessler for many helpful discussions and his hospitality at Jülich.

REFERENCES

1. T. T. S. KUO, S. Y. LEE, AND K. F. RATCLIFF, *Nucl. Phys. A* **176** (1971), 65.
2. T. T. S. KUO, *Ann. Rev. Nucl. Sci.* **24** (1974), 101.
3. B. H. BRANDOW, *Rev. Mod. Phys.* **39** (1967), 771.

4. P. J. ELLIS AND E. OSNES, *Rev. Mod. Phys.* **49** (1977), 777.
5. J. SHURPIN, D. STROTTMAN, T. T. S. KUO, M. CONZE, AND P. MANAKOS, *Phys. Lett. B* **69** (1977), 395.
6. J. SHURPIN, H. MÜTHER, T. T. S. KUO, AND A. FAESSLER, *Nucl. Phys. A* **293** (1977), 61.
7. E. M. KRENCIGLOWA AND T. T. S. KUO, *Nucl. Phys. A* **240** (1975), 195.
8. P. J. ELLIS, *in* "Proceedings of the International Topical Conference on Effective Interactions and Operators in Nuclei," Tuscon, 1975 (B. R. Barrett, Ed.), Lecture Notes in Physics No. 40, p. 296, Springer-Verlag, Berlin, 1975.
9. B. R. BARRETT AND M. W. KIRSON, *Nucl. Phys. A* **148** (1970), 145.
10. P. GOODE AND D. S. KOLTUN, *Nucl. Phys. A* **243** (1975), 44.
11. N. M. HUGENHOLTZ, *Physica* **23** (1957), 481.
12. T. T. S. KUO AND E. OSNES, "Many-Body Theory for Nuclear Structure," monograph in preparation.
13. M. BARANGER, *Phys. Rev.* **120** (1960), 957.
14. E. OSNES, T. T. S. KUO, AND C. S. WARKE, *Nucl. Phys. A* **168** (1971), 190.
15. S. SIEGEL AND L. ZAMICK, *Nucl. Phys. A* **145** (1970), 89.
16. T. T. S. KUO AND G. E. BROWN, *Nucl. Phys.* **85** (1966), 40.
17. N. I. KASSIS, *Nucl. Phys. A* **194** (1972), 205.
18. A. R. EDMONDS, "Angular Momentum in Quantum Mechanics," p. 75, Princeton Univ. Press, Princeton, N.J., 1957.
19. G. F. BERTSCH, "The Practitioner's Shell Model," p. 147, American Elsevier, New York, 1972.

PHYSICAL REVIEW C VOLUME 23, NUMBER 4 APRIL 1981

Contributions from high-momentum intermediate states to effective nucleon-nucleon interactions

H. M. Sommermann,* H. Müther, K. C. Tam,† T. T. S. Kuo,† and Amand Faessler
Institut für Theoretische Physik, Universität Tübingen, D-7400 Tübingen, Germany
(Received 22 September 1980)

We reexamine the intermediate state convergence problem for the core polarization diagram (G_{3p1h}) in the expansion of an effective shell-model interaction using a modern meson exchange Bonn-Jülich potential in comparison to the phenomenological Reid soft-core interaction. The Bonn-Jülich N-N potential contains a much weaker tensor force as compared with the Reid soft-core interaction. As a consequence its repulsive contributions to G_{3p1h} from high momentum intermediate particle states are much reduced and G_{3p1h} can be calculated in a good approximation with $2\hbar\omega$ particle-hole excitations alone. This opens up the possibility of calculating higher order processes in a reliable and expedient way. We have further determined the spectra of ^{18}O and ^{18}F using the folded diagram perturbation theory to determine the effective interaction. The energy levels are best reproduced by the Bonn-Jülich N-N potential.

[NUCLEAR STRUCTURE ^{18}O, ^{18}F; dependence of effective interaction on N-N tensor force. Convergence of intermediate state summation in core polarization.]

I. INTRODUCTION

Core polarization processes have played a very important role in the development of microscopic effective interactions. The significance of the second-order core polarization diagram (G_{3p1h}) was pointed out in the early works of Bertsch[1] and Kuo and Brown.[2] Its primary function is to provide the necessary long-range quadrupole-quadrupole component (P_2 force) in the empirical effective interaction between valence nucleons.[3] Conventionally, this core polarization diagram was calculated with a restricted sum over the intermediate particle-hole harmonic oscillator states: $\epsilon_p - \epsilon_h = 2\hbar\omega$. It was argued that the primary mechanism for core rearrangement should be the excitation of low-energy phonons of the core by the valence nucleons. Vary, Sauer, and Wong[4] (VSW) questioned the validity of this approximation. They included particle-hole excitations up to $22\hbar\omega$ for $A=18$ nuclei and found a rather slow rate of convergence, usually referred to as the VSW effect. Similar conclusions have also been reached by Kung, Kuo, and Ratcliff[5] (KKR). Both calculations employed the Reid soft-core (RSC) nucleon-nucleon potential.[6] Using this interaction, the contributions to G_{3p1h} from p-h excitations with energy greater than $2\hbar\omega$ were found to be predominantly repulsive, and thereby they almost counterbalanced the attractive contribution form G_{3p1h} with $2\hbar\omega$. In particular, the diagonal $J=0$, $T=1$ sd-shell matrix elements acquired, on the average, 500 keV repulsion from excitations with $\epsilon_p - \epsilon_h > 2\hbar\omega$. As a result, the ground state of ^{18}O now received an attractive shift of only about 350 keV over the bare interactions results. $2\hbar\omega$ excitations in the core polarization diagram calculated from RSC are therefore inadequate. The previous good agreement with experiment[2] could not be reproduced by calculations which took into account the excitations to high lying states. This convergence problem poses a major difficulty in the computation of the effective interaction from realistic nucleon-nucleon potentials. First of all, it is numerically quite difficult to calculate core polarization processes with intermediate particle-hole excitations up to $22\hbar\omega$. This puts severe limitations on the extent of such calculations. Moreover, even after having included all these higher excitations and having obtained converged values of G_{3p1h}, the results do not seem to yield adequate long-range multipole components, whose need has been established in the empirical effective interaction.

The slow rate of convergence in the intermediate-state summation can be traced to the tensor component of the Reid soft-core interaction. While central force components alone lead to rapid convergence, the strong tensor force requires one to carry the summation over intermediate p-h states up to very high excitation energies. The energy denominators provide little assistance for convergence.

In a meson exchange theory of the nucleon-nucleon interaction, the intermediate-range tensor force originates from the exchange of pseudoscalar pions and ρ vector-mesons. The ρ-exchange tensor interaction has the opposite sign as compared with the tensor part of the one-pion exchange potential. It is very important for a number of phenomena that the ρ-exchange tensor potential cuts off the one pion-exchange tensor potential at short distances. This is accomplished by choosing a relatively strong ρ-N coupling, which is in agreement with the analysis of Höhler and Pietarinen.[7] The resulting tensor interaction becomes much weaker than in earlier N-N potentials. Recent theoretical studies of the πNN vertex function[8] and new empirical evidence,[9] in fact, suggest a rather weak tensor force. The Reid soft-core interaction, however, contains a very strong tensor force component, which results in a D-state probability of $P_D \approx 6.4\%$ for the deuteron. At present a much smaller D-state probability, $P_D \approx 4\%$, seems favorable, although usual potential models with such a low D-state probability tend to overbind nuclear matter significantly. This discrepancy can be resolved by considering explicit isobar degrees of freedom in the two-nucleon force. This procedure yields modifications of the effective NN interaction in a nuclear medium due to the presence of other nucleons. The medium range attraction of the NN interaction originating from terms with intermediate isobars is reduced by Pauli and dispersion effects. Many-body corrections of this type play a large role, and even in light nuclei such as ^{16}O these effects are not negligible.[10] The Bonn-Jülich (MDFPΔ2) potential[11] is based on this meson theoretical framework and we have used it to investigate the so-called Vary-Sauer-Wong effect. Our main objective is to determine the influence of the tensor force, which is derived from a modern meson exchange interaction, on the convergence properties of the core polarization diagram. Based on our present knowledge of nucleon-nucleon potentials, are we allowed, in a good approximation, to consider only $2\hbar\omega$ intermediate excitations in the calculation of the core polarization diagram G_{3p1h}? Can we show that there is no Vary-Sauer-Wong effect? Is it possible to restore the much needed P_2 force which was thought to originate from G_{3p1h}? The answer to these questions will be crucial for further nuclear structure calculations.

In Sec. II we will formulate the problem of a converged calculation of the core polarization diagram and discuss our treatment of the intermediate state summation. We will briefly review the construction of the microscopic effective interaction and the evaluation of the folded diagram series, from which we obtain the spectra of ^{18}O and ^{18}F. Some interesting technical details are given in Sec. III, including a brief description of rather new momentum-space techniques. Finally, in Sec. IV, we give the results and discussion of our calculation. We will show in this paper the delicate dependence of the above slow convergence behavior on the N-N potential used in the calculation.

II. FORMULATION

In this section we outline our treatment of the core polarization process in a microscopic calculation of the effective interaction between valence nucleons.

In many body perturbation theory we start from a nuclear Hamiltonian H which is given as the sum of a single particle Hamiltonian H_0 plus the residual interaction H_1.

$$H = H_0 + H_1 , \quad (1a)$$

$$H_0 = T + U , \quad (1b)$$

$$H_1 = V - U . \quad (1c)$$

T and V denote, respectively, the kinetic and potential energy of the many body system. The auxiliary one-body potential U is chosen such as to "minimize" the remaining interaction, which consequently is treated by perturbation theory.

In the treatment of low-lying nuclear states the success of the nuclear shell model indicates that the choice of U as a spherical harmonic oscillator potential is appropriate. With this choice the core polarization diagram [Fig. 1(a) diagram (iv)] may be written[5] as

$$\langle abJT | G_{3p1h} | cdJT \rangle = (1+\delta_{ab})^{-1/2}(1+\delta_{cd})^{-1/2} \sum_{p\text{-}h} \sum_{J_1 J_2 T_1 T_2} (2J_1+1)(2J_2+1)(2T_1+1)(2T_2+1)$$

$$\times \begin{Bmatrix} J_2 & j_p & j_a \\ j_h & J_1 & j_b \\ j_c & j_d & J \end{Bmatrix} \begin{Bmatrix} T_2 & \frac{1}{2} & \frac{1}{2} \\ \frac{1}{2} & T_1 & \frac{1}{2} \\ \frac{1}{2} & \frac{1}{2} & T \end{Bmatrix} \langle hbJ_1T_1 | G(\omega_f) | pdJ_1T_1 \rangle$$

$$\times \left\langle p \left| \frac{1}{\omega_0 - H_0} \right| p \right\rangle \langle apJ_2T_2 | G(\omega_i) | chJ_2T_2 \rangle \quad (2a)$$

with

$$\omega_f = \epsilon_c + \epsilon_d + \epsilon_h - \epsilon_a , \quad (2b)$$

$$\omega_i = \epsilon_c + \epsilon_h , \quad (2c)$$

$$\omega_0 = \epsilon_c + \epsilon_d . \quad (2d)$$

For the example of ^{18}O or ^{18}F the model space is composed of single particle orbits from the *sd* shell for the two valence nucleons. The summation for the particle states p ranges over all harmonic oscillator states above the $0s$ and $0p$ shells.

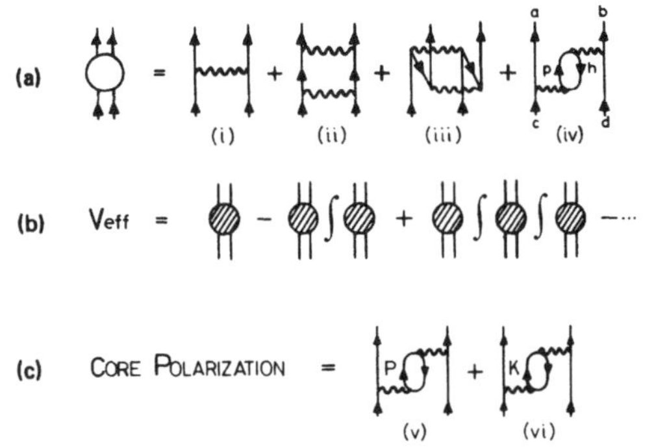

FIG. 1. (a) The particle-particle interaction. The valence particle interaction for $A = 18$ nuclei is given up to second order in terms of the Brueckner reaction matrix (denoted by wavy lines). (b) The folded diagram expansion for the effective interaction. The model space effective interaction series is indicated and terms with up to two folds are shown. The cross-hatched box consists of the one-particle Q boxes for each valence nucleon plus the particle-particle interaction of (a). (c) The core polarization process G_{3p1h}. The core polarization diagram has been calculated as the sum of diagrams (v) and (vi). The former contains low-lying intermediate particle states in harmonic oscillator representation while the latter includes the orthogonalized high-lying plane wave particle states.

While low-lying single particle states can be well approximated by harmonic oscillator functions, this clearly does not hold true for highly excited nucleons, and it may be more appropriate to use the choice $U = 0$ for these states. The question concerning the choice of the single particle spectrum in the calculation of G_{3p1h} is similar to that in the calculation of the bare Brueckner reaction matrix G,

$$G(\omega) = V + V\frac{Q}{\omega - H_0}G(\omega)\,. \tag{3}$$

Q represents the Pauli exclusion operator. Several authors[12,13] have emphasized the importance of a consistent treatment which leads to the inclusion of $-U$ insertions to all orders in the reaction matrix (remember that $H_1 = V - U$). In this section we follow closely the development given by Kung, Kuo, and Ratcliff.[5] The reaction matrix is then defined by

$$G^T(\omega) = V + VQ\frac{1}{\omega - QTQ}QG^T(\omega)\,. \tag{4}$$

The same arguments may be applied to the calculation of the core polarization diagram G_{3p1h}. Adding $-U$ insertions to all orders to the particle line p of G_{3p1h} we are led to a geometric series of the form

$$-\frac{\hat{q}}{\omega_0 - H_0}U\frac{\hat{q}}{\omega_0 - H_0} + \frac{\hat{q}}{\omega_0 - H_0}U\frac{\hat{q}}{\omega_0 - H_0}U\frac{\hat{q}}{\omega_0 - H_0} - \cdots\,. \tag{5}$$

Together with G_{3p1h} this series can be readily summed up and we obtain

$$\langle ab|G^T_{3p1h}|cd\rangle = \int dk \int dk' \sum_h \langle hb|G|kd\rangle \left\langle k\left|\hat{q}\frac{1}{\omega_0' - \hat{q}t\hat{q}}\hat{q}\right|k'\right\rangle \langle ak'|G|ch\rangle\,. \tag{6}$$

$\hat{q}$ projects onto one-particle oscillator states in the sd shell and above. $\omega_0' = (\epsilon_c + \epsilon_d) - (\epsilon_a + \epsilon_d - \epsilon_h)$. $-U$ insertions to the hole lines in G_{3p1h} will be canceled, essentially, by the corresponding Brueckner-Hartree-Fock (BHF) self-energy insertions, as the BHF self-consistent single particle wave functions for the hole states ($0s_{1/2}$, $0p_{3/2}$, and $0p_{1/2}$) are well represented by harmonic oscillator wave functions. The wave functions of particle states at higher excitation energies should not be expected to resemble harmonic oscillator functions. In fact, for very high excitation energy they should be well represented by free particle states. Thus there exist two different schemes for the evaluation of the core polarization process, denoted by G_{3p1h} and G^T_{3p1h}, respectively. In the first scheme, which corresponds to Eq. (2), no $-U$ insertions are considered. This approach was taken by Vary, Sauer, and Wong. In the second scheme $-U$ insertions are included for the high-lying single particle states. This method leads to Eq. (6) and has been used by Kung, Kuo, and Ratcliff. It may be pointed out that both results are quite similar.[5]

In the present work we will adopt a double partition approach which we regard as most appropriate. We have calculated G^T_{3p1h} for three compound spectra which consist of oscillator states up to $0p$, $1s0d$, and $1p0f$ shells, respectively, and orthogonalized plane waves above these states. This compound spectrum approach corresponds to the exclusion of $-U$ insertions from low-lying states where Brueckner-Hartree-Fock considerations suggest cancellation of such insertions with self-energy bubbles, but inclusion of $-U$ insertions to all orders at energies above which BHF consideration should no longer be applicable.

G^T_{3p1h} represents one term in the expansion of the

effective interaction. After we have studied the convergence problems in G^T_{3p1h}, which arise from the tensor components in the N-N interaction, we would like to evaluate the complete model space effective interaction. We can then determine the theoretical spectra of ^{18}O and ^{18}F and make a comparison with the experimental data. For the calculation of the microscopic effective interaction we employ the folded diagram perturbation theory of Kuo, Lee, and Ratcliff[14] and of Krenciglowa and Kuo.[15] The calculation is carried out in two steps.

We first calculate the Q box as a sum of some irreducible diagrams. Our model space for $A=18$ nuclei consists of two-particle states which are composed of the valence nucleons in the *sd* shell. The Q box is calculated in low-order perturbation theory up to second order in the reaction matrix [see Fig. 1(a)]. Diagram (i) represents the Brueckner reaction matrix. Diagram (ii) is included to treat the scattering into low-lying two-particle states separate from the excitation of high-lying intermediate two-particle states which are considered in solving the Bethe-Goldstone equation. For that purpose the Bethe-Goldstone equation has been solved using a projection operator P defined in the nomenclature of Ref. 16 by $P(3,10,21)$. Intermediate states in which both particles occupy a *sd*- or *pf*-shell orbit, for example, are not considered in the bare reaction matrix (i). Scattering to these states, which are best represented by oscillator wave functions, is treated by calculating diagram (ii). In addition to the core polarization diagram (iv) we also include the process (iii), which we call G_{4p2h}. Please note that there will be no convergence problem for the intermediate state summation except in G_{3p1h}. All other diagrams contain only finite sums over intermediate particles and holes. Diagrams containing bubble insertions have been eliminated since the BHF conditions are reasonably satisfied when using appropriate harmonic oscillator wave functions.

In the second step we evaluate the folded diagram series. Using the derivative method of Kuo, Lee, and Ratcliff[14] we include all terms involving up to four folds. The folded diagram series up to two folds is shown in Fig. 1(b). The Q box in this case is the sum of one-body and two-body irreducible diagrams. After folding, all one-body terms are removed. The complete Hamiltonian is obtained as the sum of a diagonal single particle Hamiltonian containing the empirical single particle energies plus the theoretical model space effective interaction.

III. METHOD OF CALCULATION

A special feature of the present calculation is the evaluation of the reaction matrix elements needed for G^T_{3p1h} in a "mixed representation." They are of the form $\langle n_1 n_2 | G | k_3 n_4 \rangle$, where n_1, n_2, and n_4 are oscillator states and k_3 is a plane-wave state with δ-function normalization. To calculate the G matrix in this mixed representation both Moshinsky brackets and the less known vector brackets are needed. The vector brackets $\langle klKL\lambda | k_3 l_3 k_4 l_4 \lambda \rangle$ facilitate the transformation of two nucleon laboratory frame plane-wave states into relative and center-of-mass coordinates.

$$|k_3 l_3 j_3 k_4 l_4 j_4; JT\rangle = \sum_{\lambda} \sum_{lLS\mathcal{J}} \int dK \int dk \frac{1-(-)^{l+S+T}}{\sqrt{2}} [(2j_3+1)(2j_4+1)(2\lambda+1)(2S+1)]^{1/2}$$
$$\times \begin{Bmatrix} l_3 & l_4 & \lambda \\ \frac{1}{2} & \frac{1}{2} & S \\ j_3 & j_4 & J \end{Bmatrix} \langle klKL\lambda | k_3 l_3 k_4 l_4 \lambda \rangle [(2\mathcal{J}+1)(2\lambda+1)]^{1/2} W(\mathcal{J}LS\lambda; Jl) | klS(\mathcal{J})KL; JT\rangle . \quad (7)$$

The vector bracket transformation has been extensively discussed by Wong and Clement.[17] The formulation by Balian and Brezin[18] is more convenient for the purpose of numerical calculations. A detailed description of this method in relation to the present problem can be found in Ref. 5.

The reaction matrix is obtained as the sum of the free two-nucleon G matrix G_F minus a Pauli correction term:

$$G(\omega) = V + VQ \frac{1}{\omega - QTQ} QG(\omega) = G_F - \Delta G . \quad (8a)$$

G is calculated according to the method of Krenciglowa, Kung, Kuo, and Osnes[16] using plane wave intermediate states. The Pauli exclusion operator Q is handled with the help of the Tsai-Kuo operator identity[19] which leads to the equations

$$G_F(\omega)=V+V\frac{1}{e}G_F(\omega)\,, \tag{8b}$$

$$\Delta G=G_F\frac{1}{e}P\left[1\Big/P\left(\frac{1}{e}+\frac{1}{e}G_F\frac{1}{e}\right)P\right]P\frac{1}{e}G_F\,. \tag{8c}$$

Similarly, the core polarization diagram G^T_{3p1h} can be expressed as the sum of two terms

$$\langle ab\,|\,G^T_{3p1h}\,|\,cd\rangle=(\text{free term})-(\text{Pauli term})\,, \tag{9a}$$

where

$$(\text{free term})=\sum_h\int dk\langle hb\,|\,G\,|\,kd\rangle\left\langle k\left|\frac{1}{\omega_0'-t}\right|k\right\rangle\langle ak\,|\,G\,|\,ch\rangle\,, \tag{9b}$$

$$(\text{Pauli term})=\sum_h\int dk\int dk'\langle hb\,|\,G\,|\,kd\rangle\left\langle k\left|\frac{1}{\omega_0'-t}\hat{p}\left[1\Big/\hat{p}\left(\frac{1}{\omega_0'-t}\right)\hat{p}\right]\hat{p}\,\frac{1}{\omega_0'-t}\right|k'\right\rangle\langle ak'\,|\,G\,|\,ch\rangle\,. \tag{9c}$$

Here $\hat{p}$ is a one-body operator $\hat{p}=1-\hat{q}$ projecting onto the harmonic oscillator single-particle states in the ^{16}O core. ω_0' has been defined in connection with Eq. (6). The core-polarization diagrams, as well as all other graphs, were derived with the help of diagrammatic rules developed by Kuo *et al.*[20]

After we have obtained the Q box as described above, the folded diagram series is obtained using the derivative method of Ref. 14 as previously mentioned. This approach is more suitable in the case of a two valence nucleon system than the iteration scheme of Krenciglowa.[15] We have used a harmonic oscillator shell spacing of $\hbar\omega=14$ MeV. To test the convergence properties of the folded diagram series, several *sd*-shell single particle energies were employed. This rather interesting point is discussed in detail in Sec. IV.

The calculations are performed using two different *NN* potentials V, the Reid soft core[6] potential and the potential MDFPΔ2 of the Bonn-Jülich group.[11] Both potentials were adjusted to describe the two nucleon data. While, however, the Reid potential to be used in a nuclear many-body calculation is identical to the one which is used in the description of the two nucleon data, this is not the case for the Bonn-Jülich potential. The Bonn-Jülich potential is derived in the framework of meson exchange theory using noncovariant time-dependent perturbation theory. It consists of the terms

$$V=V_{\text{OBE}}(\omega)+U_{N\Delta}(\omega)\frac{\tilde{Q}}{\omega-\tilde{H}_0}U_{N\Delta}(\omega)+V_{\Delta\Delta}(\omega)\frac{1}{\omega-\tilde{H}_0}V_{\Delta\Delta}(\omega)\,. \tag{10}$$

The first term of Eq. (10), V_{OBE}, is a one boson exchange part of the potential. Due to the treatment within noncovariant perturbation theory, it depends on the starting energy ω. The second and third terms of Eq. (10) describe contributions to the effective *NN* potential where one or both of the interacting nucleons are intermediately excited to the $\Delta(3,3)$ resonance. These terms are graphically represented in Fig. 2. They contribute significantly to the medium range attractive parts of the *NN* potential. While, however, for the two-nucleon system the operator $\tilde{Q}$ in Eq. (10) is set equal to one, for the interaction of two nucleons in a nucleus, the Pauli operator $\tilde{Q}$ prevents intermediate $N\Delta$ states with nucleons in occupied states. This Pauli quenching reduces the attractive second term of Eq. (10) and therefore yields less attractive matrix elements for *NN* states with isospin $T=1$. Also, the modification of the propagators of Eq. (10), which in a nuclear medium are defined in terms of self-consistent single particle rather than kinetic energies, yields a reduction of the terms with intermediate isobars. This effect is usually referred to as dispersion quenching. A more detailed description of the technical treatment of this effective consideration of mesonic and isobar degrees of freedom of the *NN* interaction is given in Ref. 11.

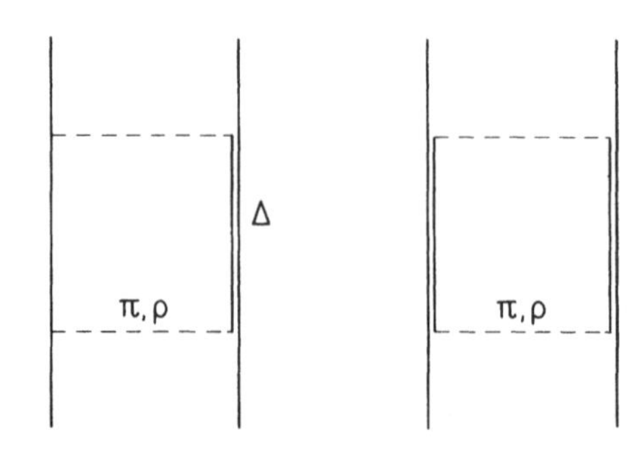

FIG. 2. Graphical representation of contributions to the Bonn-Jülich effective potential due to excitation of intermediate $\Delta(3,3)$ states. The diagrams correspond to the second and third terms of Eq. (10).

IV. RESULTS AND DISCUSSION

We begin the presentation of our results with a discussion of the momentum dependence (k_3) of the reaction matrix $\langle n_1 n_2 | G | k_3 n_4 \rangle$ and its influence on the convergence behavior of G^T_{3p1h}. We then give some numerical results for G^T_{3p1h} calculated from the Bonn-Jülich and Reid soft-core potentials and make a statement about their respective Vary-Sauer-Wong effects. Finally, the complete effecttive interaction of Figs. 1(a) and 1(b) is calculated and the resulting spectra of ^{18}O and ^{18}F are studied in detail. In this context we also investigate the convergence behavior of the effective interaction expansion.

In Fig. 3 we show two representative G-matrix elements plotted as a function of the intermediate particle momentum k. The matrix element in the lower part contains primarily contributions from the central components of the N-N potential. The Bonn-Jülich (MDFPΔ2) and RSC potentials show a very similar behavior in this case. The reaction matrix element which is displayed in the upper part of Fig. 3 is dominated by the tensor interaction in the N-N interaction. Note the large peak at high particle momenta in the case of the Reid potential. In a harmonic oscillator representation its contributions to G_{3p1h} are only included when summing over particle-hole excitations with energies much larger than $2\hbar\omega$. The same matrix element, calculated from the Bonn-Jülich potential, peaks at the same particle momentum. Its magnitude, however, is drastically reduced when compared with the corresponding RSC matrix element.

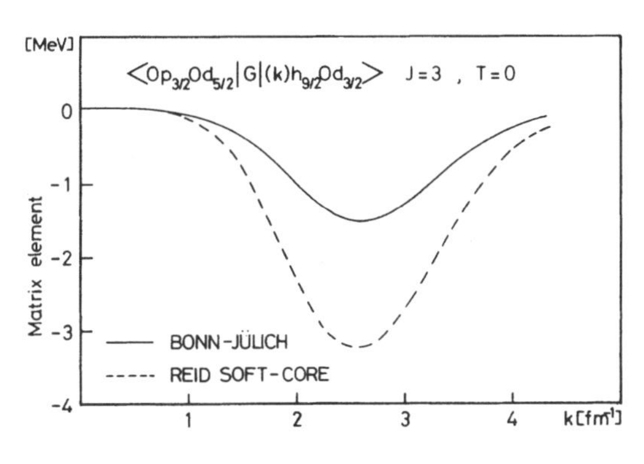

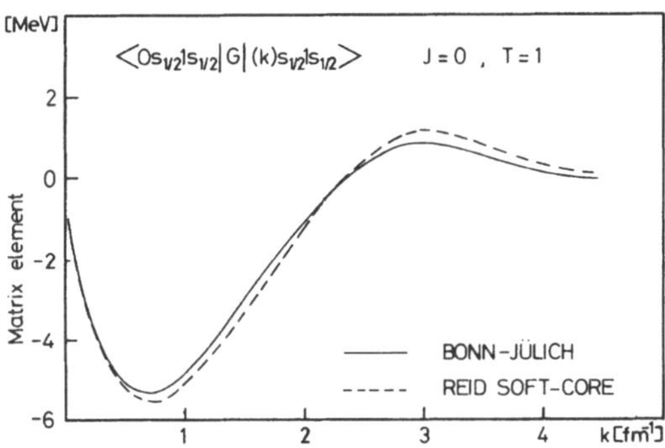

FIG. 3. Reaction matrix elements $\langle n_1 n_2 | G | k_3 n_4 \rangle$ in mixed representation. Two reaction matrix elements in the laboratory frame are plotted as a function of the particle momentum k_3. The matrix element in the lower figure is dominated by the central component of the N-N interaction, while the upper one contains mostly tensor contributions.

These results are, of course, an expression of the fact that the MDFPΔ2 potential contains a much weaker tensor force and correspondingly larger central components than the RSC interaction. Its consequences are of crucial importance in the calculation of G^T_{3p1h}. While there is no intermediate state convergence problem in G^T_{3p1h} from the central-component G matrices which tend to zero already for relatively small values of k, the contribution to G^T_{3p1h} from the tensor components depends sensitively on the magnitude of $G(k)$ for large momenta k. The core polarization processes with high momentum intermediate particles are repulsive and in the case of the Reid potential, have led to converged G^T_{3p1h} results which are markedly different from the conventional $2\hbar\omega$ calculations.

We have calculated the core polarization diagram G^T_{3p1h} for several compound spectra as indicated in Fig. 1(c). Let us concentrate here on the comparison of diagrams containing $2\hbar\omega$ excitations alone with those calculated from additional plane wave intermediate particle states above the pf shell. The latter choice is regarded to be physically most appropriate.

The results for the diagonal G^T_{3p1h} matrix elements with its valence nucleons coupled to $J=0$, $T=1$ are shown in Table I. The Bonn-Jülich po-

TABLE I. Diagonal matrix elements of G_{3p1h}. The following notation for the harmonic oscillator orbitals is used: $4=0d_{5/2}$, $5=0d_{3/2}$, $6=1s_{1/2}$. Further details are given in Sec. IV.

$J=0$, $T=1$	$\langle ab \mid G_{3p1h} \mid cd\rangle$		$2\hbar\omega$	$k>10$
	44	44	-1.039 [a]	0.476
			-1.575 [b]	0.107
	55	55	-0.304 [a]	0.477
			-0.770 [b]	0.183
	66	66	$+0.135$ [a]	0.515
			-0.464 [b]	0.171

[a] Reid soft core.
[b] Bonn-Jülich potential.

tential leads to much more attraction for the $2\hbar\omega$ contributions to G^T_{3p1h} as compared with RSC. In addition, a most important result emerges from the contributions of the orthogonalized plane wave particle states beyond the *pf* shell ($k > 10$). While they provide a substantial repulsion in the case of the Reid potential, this same repulsion is reduced by a factor of 4 when using the Bonn-Jülich potential. We therefore can make the important statement that the Vary-Sauer-Wong effect is of negligible importance if we work with a modern meson exchange potential having a weaker tensor force. Core polarization processes can then be calculated in an excellent approximation using $2\hbar\omega$ particle-hole excitations alone. The core polarization diagram seems to be again able to provide the quadrupole-quadrupole component, the P_2 force, of the empirical effective interaction, as had been suggested long ago by Kuo and Brown.[3]

A comprehensive presentation of our results for the Vary-Sauer-Wong effect is given in Figs. 4 and 5. In Fig. 4 we compare the ^{18}O spectra which have been calculated by VSW for the Reid soft-core interaction with the spectra that we have obtained from the Bonn-Jülich potential. In Fig. 5 we present the respective results for ^{18}F. The spectra denoted by I and I′ are calculated with only $2\hbar\omega$ particle-hole excitations. II and II′ are evaluated with converged values for G_{3p1h}; i.e., p-h excitations up to $22\hbar\omega$ are included in II, while we have summed over all plane wave intermediate particle states in II′. Notice the large shifts in the calculation of Vary, Sauer, and Wong. In our calculation the discrepancy between I′ and II′ becomes much smaller. The spectra calculated from core polarization diagrams with and without high momentum intermediate particles are almost identical when using the Bonn-Jülich potential. This is another way of stating that for modern N-N interactions the so-called Vary-Sauer-Wong effect is negligible.

We now turn to the comparison of our spectra for $A = 18$ nuclei with experiment. For that purpose we calculate the effective interaction V_{eff} by evaluating the folded diagram series as des-

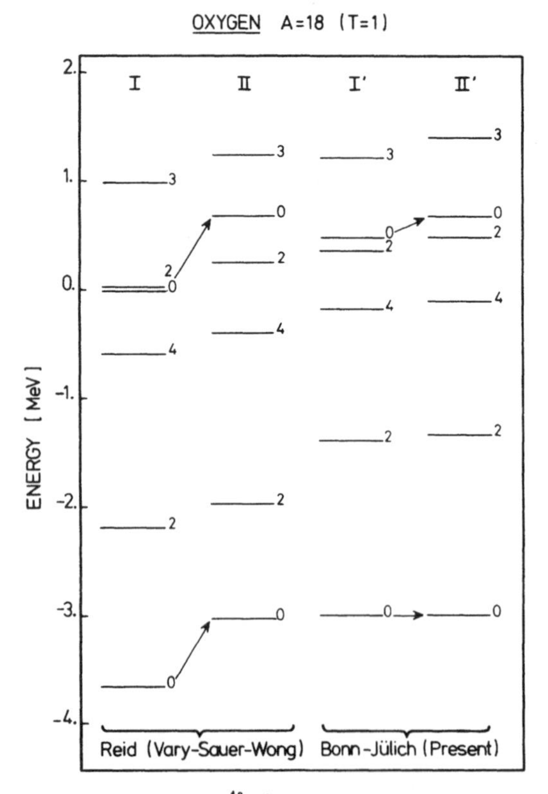

FIG. 4. Spectrum of ^{18}O ($T = 1$). No folded diagrams have been included in the calculation of the effective interaction.

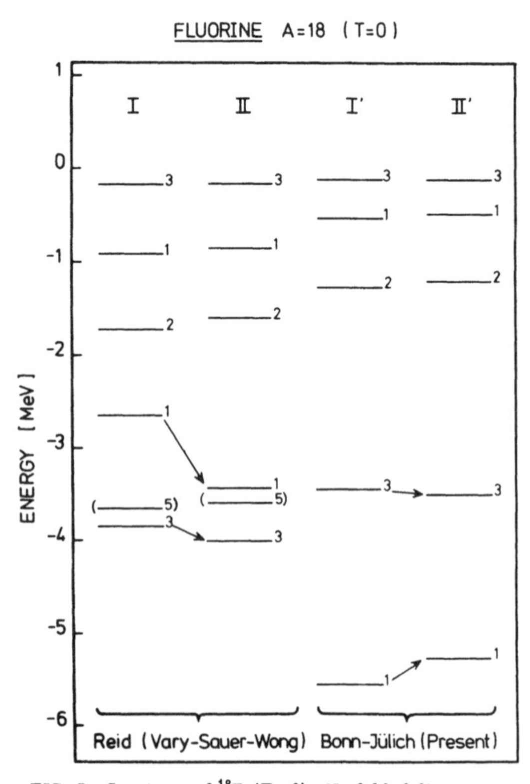

FIG. 5. Spectrum of ^{18}F ($T = 0$). No folded diagrams have been included, just as in Fig. 4.

cribed in Secs. II and III. Using the derivative method we follow closely the procedure of Shurpin *et al.*[22,23] It is then of interest to study the convergence behavior of this series as a function of the number of folds [see Fig. 1(b)]. An important feature which we have learned from our calculation is the rather sensitive dependence of the rate of convergence on the choice of the single particle spectrum. Strictly speaking, V_{eff} is independent of the single particle Hamiltonian H_0 as long as the folded diagram expansion converges. But, since we evaluate only terms with up to four folds, the result may, in fact, depend on our choice of H_0. In order to determine this dependence we have employed harmonic oscillator spectra ($\hbar\omega = 14$ MeV) with various constant shifts. According to the different shifts, we have changed the single particle energies in the *sd* shell between −5 and −13 MeV. In Table II we show the results for two matrix elements of the effective interaction, calculated from the Bonn-Jülich potential. Terms with certain numbers of folds indeed vary significantly with changes in the single particle spectrum. For example, the once-folded term for the $J=0$, $T=1$ case changes from 1.80 to 1.35 MeV when the *sd*-shell energy is shifted from −6 to −8 MeV. The corresponding change in the total V_{eff}, however, is only 0.08 MeV. We find optimal convergence behavior for *sd*-shell energies around −8 MeV. It is extremely nice to see that convergence of the folded diagram series can be obtained and that, moreover, the resulting effective interaction is rather insensitive to the choice of H_0. Tam *et al.*[21] have shown that the folded diagram series calculated with Brueckner-Hartree-Fock self-consistent spectra exhibits the same good convergence and leads to almost identical results. The folded diagrams introduce a remarkable self-correcting behavior and stability of V_{eff}. The above observations are equally true in calculations with the Reid soft-core interaction.

The $A=18$ spectra, including the contributions from high momentum particle states in G^T_{3p1h} as well as the folded diagrams, are displayed in Figs. 6 and 7 for ^{18}O and ^{18}F, respectively.

We notice the similarity of the spectra of ^{18}O derived from the meson exchange MDFPΔ2 potential and the phenomenological Reid soft-core interactions. Nevertheless, even though both theoretical ^{18}O spectra exhibit surprisingly good relative consistency, their respective 0^+ ground states are still underbound by about 1 MeV. We have seen in the case of the Bonn-Jülich potential that little repulsion comes from G^T_{3p1h} terms with particle-hole excitation energies larger than $2\hbar\omega$, resulting in a net attractive shift of roughly 800 keV with respect to the corresponding Reid G^T_{3p1h} contributions. But at the same time, the $T=1$ bare reaction matrix elements of the Bonn-Jülich potential are reduced by the quenching of the isobar terms in the potential as discussed in Sec. III. The overall situation, therefore, has not changed much and both calculations show the need of a better many-body treatment in order to explain the experimental levels.

In the case of ^{18}F the Reid soft-core potential again underbinds the low-lying levels. Historically, large deformed admixtures to the 1^+ ground

TABLE II. The convergence behavior of matrix elements of V_{eff} with respect to the choice of the single particle spectrum. ϵ_{sd} denotes the s-d shell single particle energy of our harmonic oscillator spectrum. See Sec. IV for further explanation.

	$\langle 0d_{5/2}0d_{5/2} \| V_{eff} \| 0d_{3/2}0d_{3/2}\rangle$; $J=0$, $T=1$ for the Bonn-Jülich potential					
ϵ_{sd}	Q	$-Q\int Q$	$Q\int Q\int Q$	$-Q\int Q\int Q\int Q$	$Q\int Q\int Q\int Q\int Q$	V_{eff}(sum)
−5	−4.48	2.04	−0.85	0.48	−0.46	−3.26
−6	−4.43	1.80	−0.44	0.11	−0.13	−3.09
−7	−4.39	1.57	−0.09	−0.10	−0.02	−3.03
−8	−4.35	1.35	0.20	−0.18	−0.04	−3.01
−10	−4.26	0.94	0.62	−0.04	−0.16	−2.90
−13	−4.13	0.39	0.93	0.59	0.07	−2.17
	$\langle 0d_{5/2}0d_{3/2} \| V_{eff} \| 0d_{5/2}0d_{3/2}\rangle$; $J=1$, $T=0$ for the Bonn-Jülich potential					
ϵ_{sd}	Q	$-Q\int Q$	$Q\int Q\int Q$	$-Q\int Q\int Q\int Q$	$Q\int Q\int Q\int Q\int Q$	V_{eff}(sum)
−5	−8.68	5.33	−3.96	3.84	−4.58	−8.06
−6	−8.56	4.75	−2.75	2.09	−2.15	−6.63
−7	−8.46	4.20	−1.70	0.86	−0.78	−5.87
−8	−8.35	3.67	−0.80	0.06	−0.13	−5.54
−10	−8.14	2.70	0.60	−0.52	−0.06	−5.42
−13	−7.84	1.40	1.86	0.26	−0.54	−4.86

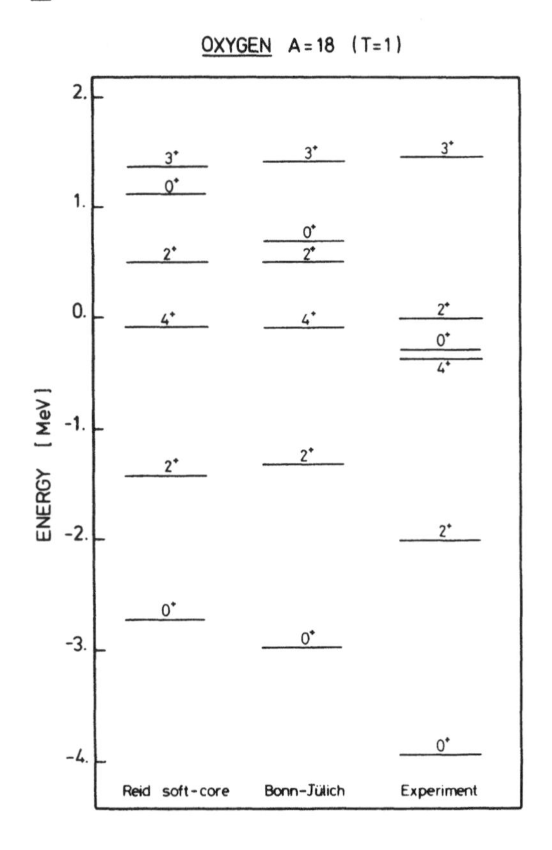

FIG. 6. Spectrum of ^{18}O ($T=1$). This spectrum was calculated with the inclusion of the folded diagrams as well as the contribution from G^{T}_{3p1h} with high momentum intermediate particle states.

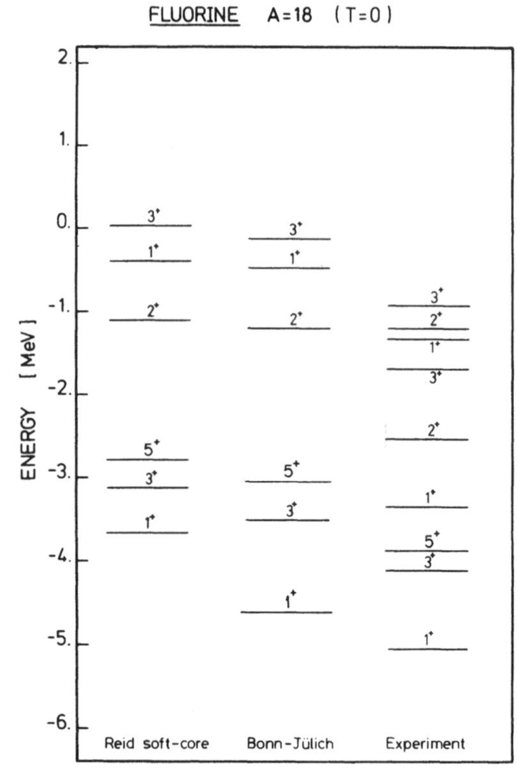

FLUORINE A=18 (T=0)

ENERGY [MeV]

Reid soft-core Bonn-Jülich Experiment

FIG. 7. Spectrum of ^{18}F ($T=0$). See explanation of Fig. 6.

state had to be invoked to explain the discrepancy between experiment and perturbation theory. As we can see in Fig. 7, however, this lowest 1^+ level received, compared with RSC, a large downward shift of about 1.0 MeV when we employed the MDFPΔ2 potential. This surprisingly larger attractiveness of the Bonn-Jülich potential in the $T=0$ channel, as compared with the Reid soft-core interaction, can be explained as follows. Both N-N interactions exhibit practically the same behavior in free space. When acting in a nuclear medium their tensor components like $\langle {}^3S_1|V|{}^3D_1\rangle (Q/e) \langle {}^3D_1|V|{}^3S_1\rangle$, which are dominant for $T=0$, are reduced due to the presence of the Pauli exclusion operator. The Bonn-Jülich potential, however, contains a much smaller tensor force to begin with and consequently loses less of its strength. This results in a more attractive nuclear reaction matrix. For both ^{18}O and ^{18}F the meson exchange Bonn-Jülich potential leads to theoretical spectra which are in better agreement with experiment than the phenomenological Reid soft-core interaction.

In summary, we have shown that for a modern meson exchange interaction of the Bonn-Jülich group (MDFPΔ2) the Vary-Sauer-Wong effect is of negligible importance. This has the important consequence that core polarization processes can be calculated in a good approximation by using $2\hbar\omega$ particle-hole excitations alone. This opens up the possibility of calculating in a reliable and expedient way higher order processes. Employing the folded diagram theory convergence for the effective interaction expansion in terms of the order of folding can be obtained if an appropriate single particle spectrum is used. While the Bonn-Jülich potential provides almost sufficient attraction for

TABLE III. Matrix elements of the effective interaction as derived from the Reid soft-core potential. Detailed explanations are given in the Appendix.

TJ	$abcd$	$\langle abJT \vert V_{\text{eff}} \vert cdJT \rangle$	TJ	$abcd$	$\langle abJT \vert V_{\text{eff}} \vert cdJT \rangle$
01	4444	−0.9428	03	5555	−1.8413
01	4445	2.5485	04	4545	−3.1864
01	4455	0.8749	05	4444	−2.7687
01	4456	0.2012	10	4444	−1.6489
01	4466	−0.4938	10	4455	−3.1541
01	4544	2.4597	10	4466	−0.9163
01	4545	−4.7293	10	5544	−2.8665
01	4555	0.3870	10	5555	−0.3763
01	4556	2.0179	10	5566	−0.6681
01	4566	1.4940	10	6644	−0.8513
01	5544	0.9889	10	6655	−0.7112
01	5545	0.3343	10	6666	−0.9664
01	5555	−0.2448	11	4545	−0.0601
01	5556	−0.7448	11	4556	0.1539
01	5566	−0.2184	11	5645	0.1457
01	5644	0.1354	11	5656	0.4900
01	5645	1.9809	12	4444	−0.6207
01	5655	−0.7390	12	4445	−0.2464
01	5656	−2.9651	12	4446	−0.7456
01	5666	−0.8258	12	4455	−0.9417
01	6644	−0.4642	12	4456	0.8159
01	6645	1.4457	12	4544	−0.2706
01	6655	−0.1157	12	4545	0.0283
01	6656	−0.8859	12	4546	−0.0627
01	6666	−2.2023	12	4555	−0.7727
02	4545	−3.6993	12	4556	0.6784
02	4546	−1.1745	12	4644	−0.7326
02	4556	−1.1111	12	4645	−0.0739
02	4645	−1.1837	12	4646	−0.8839
02	4646	−0.4453	12	4655	−0.4293
02	4656	−1.9956	12	4656	1.1647
02	5645	−1.0988	12	5544	−0.7773
02	5646	−1.9456	12	5545	−0.7119
02	5656	−1.1847	12	5546	−0.4540
03	4444	−0.6847	12	5555	0.3551
03	4445	1.4474	12	5556	0.2479
03	4446	−1.3565	12	5644	0.7219
03	4455	0.4658	12	5645	0.6456
03	4544	1.3526	12	5646	1.1474
03	4545	−0.8020	12	5655	0.2700
03	4546	0.8486	12	5656	0.0875
03	4555	1.7530	13	4545	0.2022
03	4644	−1.3123	13	4546	−0.0806
03	4645	0.8469	13	4645	−0.0817
03	4646	−2.9346	13	4646	0.4429
03	4655	0.2644	14	4444	0.1491
03	5544	0.3700	14	4445	−1.1536
03	5545	1.6544	14	4544	−1.1014
03	5546	0.2619	14	4545	−0.8725

the ground state of ^{18}F, still more attraction is needed in the case of ^{18}O. It may be necessary to include core deformed states and higher order processes in order to obtain the experimental spectra. However, overall the Bonn-Jülich potential is more successful than the Reid soft-core interaction in reproducing the experimental $A = 18$ levels.

TABLE IV. Matrix elements of the effective interaction as derived from the meson exchange Bonn-Jülich potential. Detailed explanations are given in the Appendix.

TJ	$abcd$	$\langle abJT \mid V_{eff} \mid cdJT \rangle$	TJ	$abcd$	$\langle abJT \mid V_{eff} \mid cdJT \rangle$
01	4444	−1.3513	03	5555	−2.1679
01	4445	2.8204	04	4545	−3.4190
01	4455	1.3951	05	4444	−3.0342
01	4456	0.4236	10	4444	−1.8819
01	4466	−0.8164	10	4455	−3.1784
01	4544	2.7246	10	4466	−0.8879
01	4545	−5.2713	10	5544	−2.8712
01	4555	0.0923	10	5555	−0.6995
01	4556	1.6269	10	5566	−0.6240
01	4566	1.6559	10	6644	−0.8207
01	5544	1.4246	10	6655	−0.6507
01	5545	0.0486	10	6666	−1.3200
01	5555	−0.6844	11	4545	0.2345
01	5556	−0.8270	11	4556	−0.0491
01	5566	0.0852	11	5645	−0.0516
01	5644	0.3537	11	5656	0.7137
01	5645	1.5991	12	4444	−0.4825
01	5655	−0.8024	12	4445	−0.2810
01	5656	−3.0076	12	4446	−0.7123
01	5666	−0.5222	12	4455	−0.9925
01	6644	−0.8005	12	4456	0.7916
01	6645	1.6139	12	4544	−0.3024
01	6655	0.1508	12	4545	0.0994
01	6656	−0.5755	12	4546	−0.1829
01	6666	−2.5338	12	4555	−0.8360
02	4545	−3.4732	12	4556	0.6153
02	4546	−1.0983	12	4644	−0.6921
02	4556	−1.1548	12	4645	−0.1942
02	4645	−1.0973	12	4646	−0.7608
02	4646	−0.4492	12	4655	−0.5025
02	4656	−2.0468	12	4656	1.1472
02	5645	−1.1375	12	5544	−0.8476
02	5646	−1.9904	12	5545	−0.7771
02	5656	−1.1167	12	5546	−0.5076
03	4444	−0.8179	12	5555	0.4384
03	4445	1.5905	12	5556	0.2280
03	4446	−1.4043	12	5644	0.7080
03	4455	0.7839	12	5645	0.5856
03	4544	1.4978	12	5646	1.1229
03	4545	−0.8359	12	5655	0.2457
03	4546	0.8738	12	5656	0.0533
03	4555	1.7446	13	4545	0.4295
03	4644	−1.3516	13	4546	−0.0981
03	4645	0.8569	13	4645	−0.1003
03	4646	−3.0314	13	4646	0.5888
03	4655	0.2780	14	4444	0.3159
03	5544	0.6847	14	4445	−1.2618
03	5545	1.6727	14	4544	−1.1983
03	5546	0.2752	14	4545	−0.9951

ACKNOWLEDGMENTS

The authors would like to thank Dr. J. Shurpin for making available part of the computer codes and Prof. K. F. Ratcliff for many stimulating discussions. This work has been supported by the Deutsche Forschungs Gemeinschaft (DFG), through the U. S. DOE under Contract No. EY-76-S-02-3001, and by an award to one of the authors (T.T.S.K.) from the Alexander von Humboldt Foundation.

APPENDIX

The matrix elements of the effective interactions for $A=18$ nuclei are given as derived from the

Reid soft-core potential (Table III) and the meson exchange Bonn-Jülich potential (Table IV). In this calculation we have considered the processes of Fig. 1(a) with the inclusion of high momentum intermediate states plus the effects of the folded diagrams. The matrix elements in the particle-particle coupling scheme are characterized by the total isospin T, total angular momentum J, and the single particle orbits a, b, c, and d. Our convention for the harmonic oscillator orbitals is $4 = 0d_{5/2}, 5 = 0d_{3/2}, 6 = 1s_{1/2}$.

---

*On leave from the State University of New York at Albany.

†On leave from the State University of New York at Stony Brook.

[1]G. F. Bertsch, Nucl. Phys. 74, 234 (1965).

[2]T. T. S. Kuo and G. E. Brown, Nucl. Phys. 85, 40 (1966).

[3]G. E. Brown and T. T. S. Kuo, Nucl. Phys. A92, 481 (1967).

[4]J. P. Vary, P. D. Sauer, and C. W. Wong, Phys. Rev. C 7, 1776 (1973).

[5]C. L. Kung, T. T. S. Kuo, and K. F. Ratcliff, Phys. Rev. C 19, 1063 (1979).

[6]R. V. Reid, Ann. Phys. (N.Y.) 50, 411 (1968).

[7]G. Höhler and E. Pietarinen, Nucl. Phys. B95, 210 (1975).

[8]J. W. Durso, A. D. Jackson, and B. J. Verwest, Nucl. Phys. A252, 404 (1977).

[9]H. Arenhövel and W. Fabian, Nucl. Phys. A282, 397 (1977); E. L. Lomon, Phys. Lett. 68B, 419 (1977).

[10]K. Holinde, R. Machleidt, A. Faessler, and H. Müther, Phys. Rev. C 15, 1432 (1977).

[11]K. Holinde, R. Machleidt, M. R. Anastasio, A. Faessler, and H. Müther, Phys. Rev. C 18, 870 (1978); M. R. Anastasio, A. Faessler, H. Müther, K. Holinde, and R. Machleidt, *ibid.* 18, 2416 (1978).

[12]C. W. Wong, Nucl. Phys. A91, 399 (1967).

[13]M. Baranger, in *Nuclear Structure and Nuclear Reactions*, Proceedings of the International School of Physics "Enrico Fermi," Course XI, edited by M. Jean and R. A. Ricci (Academic, New York, 1969), p. 511.

[14]T. T. S. Kuo, S. Y. Lee, and K. F. Ratcliff, Nucl. Phys. A176, 65 (1971); T. T. S. Kuo, Annu. Rev. Nucl. Sci. 24, 101 (1974).

[15]E. M. Krenciglowa and T. T. S. Kuo, Nucl. Phys. A235, 171 (1974).

[16]E. M. Krenciglowa, C. L. Kung, T. T. S. Kuo, and E. Osnes, Ann. Phys. (N.Y.) 101, No. 1, 154 (1976).

[17]C. W. Wong and D. M. Clement, Nucl. Phys. A183, 210 (1972).

[18]R. Balian and E. Brezin, Nuovo Cimento 61B, 403 (1969).

[19]S. F. Tsai and T. T. S. Kuo, Phys. Lett. 39B, 427 (1972).

[20]T. T. S. Kuo, J. Shurpin, K. C. Tam, E. Osnes, and P. J. Ellis, State University of New York at Stony Brook report (unpublished).

[21]K. C. Tam, H. Müther, M. Sommermann, T. T. S. Kuo, and A. Faessler, University of Tübingen report (unpublished).

[22]J. Shurpin, D. Strottmann, T. T. S. Kuo, M. Conze, and P. Manakos, Phys. Lett. B69, 395 (1977).

[23]J. Shurpin, H. Müther, T. T. S. Kuo, and A. Faessler, Nucl. Phys. A293, 61 (1977).

Nuclear Physics **A408** (1983) 310–358

FOLDED DIAGRAMS AND 1s-0d EFFECTIVE INTERACTIONS DERIVED FROM REID AND PARIS NUCLEON-NUCLEON POTENTIALS

J. SHURPIN[†] and T. T. S. KUO

Physics Dept., State University of New York at Stony Brook, Stony Brook, NY 11794, USA

and

D. STROTTMAN

Theoretical Division, Los Alamos National Laboratory, Los Alamos, NM 87544, USA

Received 18 January 1983
(Revised 20 April 1983)

Abstract: The sd-shell effective-interaction matrix elements are derived from the Paris and Reid potentials using a microscopic folded-diagram effective-interaction theory. A comparison of these matrix elements is carried out by calculating spectra and energy centroids for nuclei of mass 18 to 24. The folded diagrams were included by both solving for the energy-dependent effective interaction self-consistently and by including the folded diagrams explicitly. In the latter case the folded diagrams were grouped either according to the number of folds or as prescribed by the Lee and Suzuki iteration technique; the Lee-Suzuki method was found to converge better and yield the more reliable results. Special attention was given to the proper treatment of one-body connected diagrams in the calculation of the two-body effective interaction.

We first calculate the (energy-dependent) G-matrix appropriate for the sd-shell for both potentials using a momentum-space matrix-inversion method which treats the Pauli exclusion operator essentially exactly. This G-matrix interaction is then used to calculate the irreducible and non-folded diagrams contained in the $\hat{Q}$-box. The effective-interaction matrix elements are obtained by evaluating a $\hat{Q}$-box folded diagram series. We considered four approximations for the basic $\hat{Q}$-box. These were (C1) the inclusion of diagrams up to 2nd order in G, (C2) 2nd order plus hole-hole phonons, (C3) 2nd order plus (bare TDA) particle-hole phonons, and (C4) 2nd order plus both hole-hole and particle-hole phonons.

The contribution of the folded diagrams was found to be quite large, typically about 30%, and to weaken the interaction. Also, due to the greater energy dependence of higher-order diagrams, the effect of folded diagrams was much greater in higher orders. That is, the contribution from higher-order diagrams for most cases was greatly reduced by the folded diagrams. The convergence of the folded-diagram series deteriorates with the inclusion of higher-order $\hat{Q}$-box processes in the method which groups diagrams by the number of folds, but remains excellent in the Lee-Suzuki method.

Whereas the inclusion of the particle-hole phonon was essential to obtain agreement with experiment in earlier work, when the folded diagrams are included the effect of the particle-hole phonon is to reduce the amount of binding. All four approximations to both potentials produce interactions which badly underbind nuclei. The excitation spectra given by these interactions are, however, all rather similar to each other. The Paris interaction produces more binding than does the Reid, but differences between results obtained with the two interactions were often less than

[†] Present address: Telas Computer Systems, 71 West 35th St., NY, NY 10001.

differences obtained in the four approximations. Essentially no difference was found between the effective non-central interactions from the Reid and Paris potentials after including the folded diagrams, although these two potentials themselves are quite different, especially in the strength of the tensor force.

Comparisons between calculated spectra and experiment were done for ^{18}O, ^{18}F, ^{19}F, ^{20}O, ^{20}Ne, ^{22}Ne, ^{22}Na and ^{24}Mg.

1. Introduction

An outstanding problem in nuclear physics is to describe the properties of finite nuclei in terms of the interactions between their constituent nucleons. To describe low-energy phenomena, usually a non-relativistic approximation of the nucleon-nucleon (NN) interaction is carried out to construct a non-relativistic NN potential, V_{NN}. One must then solve the non-relativistic many-body Schrödinger equation. This problem has been the subject of numerous articles and several reviews [1,2]. As is well known, one cannot proceed directly with this ambitious prescription because of several major difficulties.

To begin with, the NN potential is itself not completely understood. Until recently, to circumvent this problem one constructed phenomenological potentials which were required to reproduce as much of the experimental data, such as low-energy phase shifts and deuteron properties, as possible. Examples of potentials of this type are the Yale potential [3]), the Hamada-Johnston potential [4]), and the Reid potential [5]), the last being by far the most widely used in recent years. This procedure, however, is not quite satisfactory because several different potentials can (and indeed do) yield the same phase shifts. Also, phenomenological potentials do not necessarily give a good theoretical basis for the interaction. A potential which is largely free of this latter shortcoming is the recently proposed Paris potential [6]) which is largely theoretically derived. The long- and medium-range parts ($r \lesssim 0.8$ fm) reflect π, 2π,and $3\pi(\omega)$ exchange. Only the short-range part, which should be important for high-energy phenomena, is constructed so as to fit the phase shifts. So far, the Paris potential has been tested mainly in nuclear matter where it gives reasonable results [7]). A major aim of the present work is to study the results the Paris potential gives for finite nuclei.

To solve the nuclear many-body problem, perturbation techniques adapted from quantum field theory and statistical mechanics were developed. A difficulty encountered immediately was that the matrix elements of V_{NN} are either infinite (hard-core potentials) or at least very large (soft-core potentials) when the relative wave functions are non-vanishing near the origin (e.g. in a harmonic oscillator basis). Brueckner and his collaborators [8]) summed the interactions between pairs of nucleons to all orders to obtain the reaction matrix, or G-matrix interaction, whose matrix elements are finite. The perturbation theories were then expressed in terms of G rather than V_{NN}. The earliest studies were carried out in infinite nuclear matter by Brueckner, Bethe and Goldstone [8,9]. The Goldstone linked-cluster non-degenerate perturbation expansion [10] can also be used for closed-shell nuclei such as ^{16}O and leads to the so-called Brueckner-Hartree-Fock (BHF) theory [11].

For open-shell nuclei, i.e. nuclei with valence nucleons such as ^{18}O, several approaches have been developed. There are the energy-dependent theories of Feshbach [12] and of Bloch and Horowitz [13]. In these theories, the effecttive interaction secular equation must be solved at the self-consistent energies. The energy-independent formulation, which leads to folded diagrams, has been developed via time-dependent perturbation theory by Morita [14], Oberlechner *et al.* [15], Johnson and Baranger [16], and by Kuo, Lee and Ratcliff [17], while Brandow [18] used time-independent perturbation methods. The various energy-independent effective interactions are consistent with each other [1]. In addition, their form is similar to that of the empirical effective interaction, which is also energy independent. In this work, we employ effective interactions of the type obtained by Kuo, Lee and Ratcliff. In this formulation the non-folded diagrams are first grouped together to form the $\hat{Q}$-box interaction, and then the folded diagrams are included in a very systematic way. Recent developments in this area have been made by Kuo and Krenciglowa [19]. These will be examined and applied here. In addition, the energy-dependent effective interaction has been shown to be equivalent to the folded-diagram effective interaction [20]. We will look at this point in actual calculations.

Another difficulty which had plagued calculations in finite nuclei is the presence of the Pauli exclusion operator in the equation for the G-matrix. This has been largely overcome in the method of Tsai and Kuo [21], which is used in conjunction with momentum-space matrix inversion (to find G_F, the G-matrix in free space, with no Pauli operator) as described by Krenciglowa, Kung, Kuo and Osnes [22]. This method is particularly suitable for the Paris potential (which is momentum dependent) and is the one we have used to construct our G-matrix.

Thus the general procedure is as follows: We cannot solve the nuclear many-body problem directly. Instead, we must solye a model-space problem of finite (low) dimension whose solutions are in principle, the same as the low-lying states of the complete (infinite dimensional) problem. For example, in our prototype nucleus ^{18}O we consider the shell-model problem of two sd-shell nucleons outside and ^{16}O closed core. The model space in which we seek to solve this problem consists of all states with two sd-shell particles. Clearly, there are many other possible configurations for ^{18}O, such as $2+n$ sd-shell (or higher-shell) particles and n holes in the ^{16}O core. What we must do is to find within the model space an *effective* interaction between the two particles which takes into account those configurations we do not include explicitly. This is done by the $\hat{Q}$-box formulation [17] of the folded-diagram theory.

There are still various remaining problems and open questions in the procedure referred to above, and we will address some of them here. The $\hat{Q}$-box contains diagrams to all orders in the interaction (V_{NN}, or equivalently G). Using existing methods it is quite difficult to evaluate even the third-order diagrams. This difficulty is twofold. First, the formulae for the diagrams require tedious and lengthy derivations. Second, the computation time required to numerically evaluate the diagrams using the formulae so obtained makes the calculations practically impossible. A new method was recently proposed by Kuo, Shurpin, Tam, Osnes and Ellis [23]) which gives a simple and systematic way of writing down the diagram formulae almost by inspection. The formulae so obtained are often easier to interpret physically than previous ones. The method has the added advantage that it yields diagram formulae which are usually much more efficient for numerical computation. This method will be used in the present work.

Perhaps the most important remaining question regarding the effective interaction is whether, and under what conditions, it converges. This question has received a great deal of attention in recent years [1,2]). In the early calculations of Kuo and Brown [24]) only two terms in the series were retained, the bare G-matrix and the 3p1h (second-order) core-polarization diagrams. Yet their results agreed fairly well with experiment, and subsequent, far more elaborate calculations, yielded no dramatic improvement. This seemed to suggest that the effective-interaction series converged very rapidly, or at least that the higher-order terms cancelled among themselves. Not long afterward, the validity of this was questioned by the work of Barrett and Kirson [25]). They showed that many of the third-order diagrams were of comparable size with the second-order core-polarization diagram. Moreover, the third order as a whole was generally of opposite sign to the second order. Thus truncating the series after the second order seemed not to be justified. In the folded-diagram theory the question of convergence can be divided into two parts; first, whether the $\hat{Q}$-box converges and, second, whether the folded-diagram series converges. In view of the Kuo and Brown results, which included no folded diagrams, there is also the question of why, or indeed whether, we need to include the folded diagrams at all. This work will attempt to address these and related matters. We now briefly outline the structure and content of the sections which follow.

We begin in sect. 2 with a brief review of the basic formalism of microscopic theories and model-space problems leading to the diagramatic expansion for the effective interaction. The notions of active and passive lines and linked and irreducible diagrams lead to a definition of the $\hat{Q}$-box, the basic entity in our expression for the effective interaction. The grouping of terms in the folded-diagram series is then examined, especially as it relates to one-body and two-body connected graphs. We briefly review also two techniques to obtain effective interactions which are also expressed in terms of the $\hat{Q}$-box. The first is the one derived by Kuo and Krenciglowa [19]) starting from the equation of motion. We show how this method allows us to include the "hole-hole phonon" which considers the interaction between two hole lines to all orders. The second is the iteration formula for the effective interaction of Lee and Suzuki [26]).

Sect. 3 contains a brief discussion of the NN interaction. Special attention is given to its momentum-dependent form. We carry out various comparisons between the Paris and Reid potentials. We also do a series of phase shift calculations using the momentum-space matrix-inversion method [27]. This is a necessary test of our computer code for our calculation of the Paris potential G-matrix which we will use in all subsequent nuclear structure calculations.

Sects. 4 and 5 describe an extensive series of calculations for sd-shell nuclei, based upon the ideas developed in the previous sections. Most of these are done in parallel for the Paris and Reid potentials. The hope is that we can perhaps trace the nuclear structure results back to the features of the free NN interaction, especially for the Paris potential. We study the effects of various refinements of the $\hat{Q}$-box, such the inclusion of hole-hole and TDA particle-hole phonons. Our calculation is done with an energy-dependent G-matrix and this enables us to include the off-energy-shell effects in the calculation of effective-interaction diagrams. Other components of the calculation such as our choice for the single-particle energies are also investigated. Our main effort is directed towards understanding the structure and function of the folded-diagram series and its convergence properties. The numerical results for the sd-shell matrix elements are tabulated. The spectra of several sd-shell nuclei (^{18}O, ^{18}F, ^{19}F, ^{22}Na, ^{22}Ne) are presented and discussed. We will compare the various spectra both among themselves and with experiment.

In sect. 6 we discuss the conclusions to be drawn from our various results and suggest several possibilities which may improve our results.

2. Folded-diagram effective interactions

2.1. INTRODUCTION

As outlined in sect. 1, our aim is to perform a microscopic nuclear-structure calculation based on a realistic NN potential $V = V_{NN}$. That is, for a nucleus with A nucleons we try to solve the non-relativistic Schrödinger equation

$$H\psi(1, 2, \ldots A) = E\psi(1, 2, \ldots A), \tag{2.1}$$

where

$$H = T + V, \tag{2.2}$$

T being the kinetic energy operator. It is clear that even for light nuclei we are unable to solve eq. (2.1), the A-body problem, exactly. The way to proceed is to project the wave function ψ of the complete Hilbert space onto a model space of low dimension in which we are able to solve the problem in a more physical way.

We first rewrite the hamiltonian as

$$H = H_0 + H_1, \tag{2.3a}$$

$$H_0 = T + U, \tag{2.3b}$$

$$H_1 = V - U, \tag{2.3c}$$

where V is the NN interaction and U is some single-particle potential. U is chosen so as to define a convenient set of unperturbed basis functions ϕ_n given by

$$H_0 \phi_n = \varepsilon_n \phi_n. \tag{2.4}$$

Frequent choices for U are a harmonic oscillator potential, which we will use in the present calculation, or a Hartree-Fock single-particle potential. We then divide the complete Hilbert space into two regions P (the model space) and Q (its complement) such that

$$P + Q = 1, \tag{2.5a}$$

$$P = \sum_i |i\rangle\langle i|, \tag{2.5b}$$

$$Q = \sum_\alpha |\alpha\rangle\langle\alpha|, \tag{2.5c}$$

$$\langle i|\alpha\rangle = 0, \tag{2.5d}$$

where both $|i\rangle$ and $|\alpha\rangle$ are eigenfunctions (i.e. ϕ_i and ϕ_α) of the unperturbed hamiltonian H_0. The division of the Hilbert space (i.e. what to choose as the model space) depends on the particular problem, and one is guided by physical intuition. For ^{18}O a reasonable choice would be a closed ^{16}O core with 2 particles in the sd shell. Thus in eqs. (2.5). $|i\rangle$ is any state with a closed ^{16}O core and 2 particles in the sd shell while $|\alpha\rangle$ represents the respective complement (in which all states have holes in the ^{16}O core and/or particles in shells above the sd shell).

We seek to reduce eq. (2.1) to a more tractable problem:

$$PH_{\text{eff}}P\psi = EP\psi, \tag{2.6a}$$

where

$$H_{\text{eff}} = H_0 + V_{\text{eff}}. \tag{2.6b}$$

Note that in eq. (2.6) E and ψ are the same as in eq. (2.1). For a model space (P-

space) of dimension d, H_{eff} will reproduce d of the eigenstates of H. One must obtain the form of V_{eff} which must account for all states belonging to Q that are now being left out of the problem.

There are several forms for V_{eff} to be found in the liteature. There are energy-dependent forms, $V_{eff}(E)$, which have been obtained by Feshbach [12]) and by Bloch and Horowitz [13]). In this work we will use the folded-diagram theory of V_{eff} in the form derived by Kuo, Lee and Ratcliff [17]) (henceforth referred to as KLR). This form of V_{eff} is independent of E. This is a desirable feature, since the empirical matrix elements (which are obtained by constructing a V_{eff} which can best reproduce the experimentally observed properties of various sd-shell nuclei) are also independent of E.

2.2. FOLDED-DIAGRAM EXPANSION OF V_{eff}

The folded-diagram expansion [17]) for H_1^{eff} (or V_{eff}) is

$$V_{eff} = \hat{Q} - \hat{Q}' \int \hat{Q} + \hat{Q}' \int \hat{Q} \int \hat{Q} - \hat{Q}' \int \hat{Q} \int \hat{Q} \int \hat{Q} + \ldots, \tag{2.8}$$

where $\hat{Q}$ is the $\hat{Q}$-box and $\int$ the generalized folding operation. The first $\hat{Q}$ in each term which is folded upon is written with a prime ($\hat{Q}'$) to indicate that the lowest-order diagram it contains is at least second order in the interaction. The $\hat{Q}$-box is composed of irreducible linked diagrams [1, 17]).

It can be shown [1]) that V_{eff} of eq. (2.7) can be written in terms of derivatives of the $\hat{Q}$-box (the so-called derivative formulation) as

$$V_{eff} = F_0 + F_1 + F_2 + F_3 + \ldots, \tag{2.8}$$

where F_n is the term with n foldings. As will be discussed later, a main part of our calculations will be carried out using this F_n expansion for V_{eff}.

The KLR-model-space eigenvalue problem, after cancellation of the unlinked diagrams, is

$$\sum_j \langle i|(H_0 + V_{eff})|j\rangle\langle j|\psi_n\rangle = (E_n - E_g^c)\langle i|\psi_n\rangle, \tag{2.9}$$

where E_g^c is the true ^{16}O ground-state energy and $\langle i|\psi\rangle$ is the projection of the true wave function ψ onto the model space. Note that the eigenvalue here is $(E_n - E_g^c)$ the energy relative to the ^{16}O ground state. Observe that V_{eff} contains 1-body diagrams such as diagram (b) in fig. 1. The sum of the 1-body contributions is actually what determines the ^{17}O energies. It is customary in eq. (2.9) to replace H_0 by H_0^{exp}, i.e. by the experimental sd-shell single-particle levels (which are taken as the low-lying ^{17}O states). As this accounts for the 1-body diagrams, we must not include them a second time in V_{eff}. That is, V_{eff} should contain only the 2-body connected diagrams. For the first term, F_0, this is easily done. We simply do not include the 1-body diagrams in the $\hat{Q}$-box. For the subsequent terms, this cannot be done in this way. Consider fig. 2, where in each case the dashed line represents a $\hat{Q}$-box. Diagrams (a) and (c) belong to F_0 and diagrams (b) and (d) to F_2. A "blob" attached to a line represents any sequence of interactions attached to that line. Thus (c) and (d) should be excluded from V_{eff}, since their contribution is already accounted for by H_0^{exp}. Diagram (b), however, should still be in V_{eff}, since it is 2-body connected. Indeed, its contribution can be quite large. It is only diagram (d) which we must remove. To this end, the $\hat{S}$-box is introduced [28]). The $\hat{S}$-box is defined as the 1-body part of the $\hat{Q}$-box (i.e. the set of all linked and irreducible diagrams in which all the interactions are connected to one line and the other line is merely a "spectator"). What must be removed from V_{eff} are all the terms in which the $\hat{S}$-box folds upon itself only. Thus we have the following:

$$F_0(\hat{Q}) = \hat{Q},$$

$$F_1(\hat{Q}) = \hat{Q}_1\hat{Q},$$

$$F_2(\hat{Q}) = \hat{Q}_2\hat{Q}\hat{Q}+\hat{Q}_1\hat{Q}_1\hat{Q},$$

$$F_3(\hat{Q}) = \hat{Q}_3\hat{Q}\hat{Q}\hat{Q}+\hat{Q}_2\hat{Q}_1\hat{Q}\hat{Q}+\hat{Q}_2\hat{Q}\hat{Q}_1\hat{Q}+\hat{Q}_1\hat{Q}_2\hat{Q}\hat{Q}+\hat{Q}_1\hat{Q}_1\hat{Q}_1\hat{Q}, \tag{2.10}$$

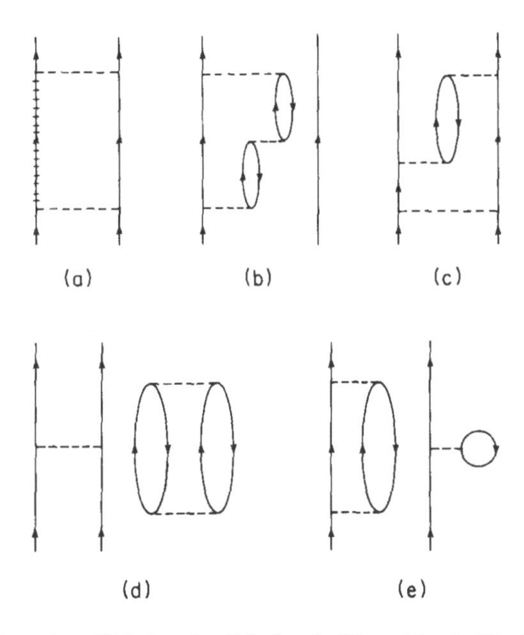

Fig. 1. Examples of linked and unlinked, reducible and irreducible diagrams.

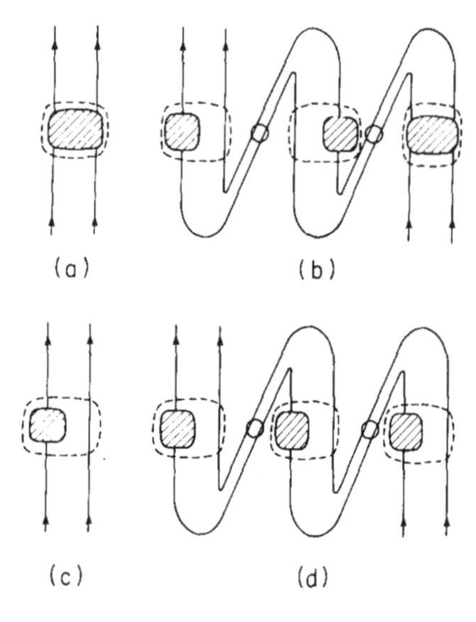

Fig. 2. One-body and two-body connected diagrams in the folded-diagram series for V_{eff}. The dashed line represents the Q-box interaction between the two lines. A "blob" represents a series of V (or U) interactions attached to the line.

and so forth, where

$$\hat{Q}_n \equiv \frac{1}{n!} \left. \frac{\mathrm{d}^n \hat{Q}}{\mathrm{d}\omega^n} \right|_{\omega = \omega_0}, \tag{2.11}$$

with ω_0 being the degenerate model-space energy ($PH_0 = P\omega_0$). We now define

$$\tilde{F}_n \equiv F_n(\hat{Q}) - F_n(\hat{S}), \tag{2.12}$$

where $\hat{S}$ is the 1-body box which we just introduced. Clearly the difference $\tilde{F}_n$ contains only the proper 2-body connected diagrams of V_{eff}. This gives us the equation we will actually use in our calculations,

$$V_{\text{eff}} = \tilde{F}_0 + \tilde{F}_1 + \tilde{F}_2 + \tilde{F}_3 + \tilde{F}_4 \ldots \tag{2.13}$$

We include terms up to $\tilde{F}_4$ only. The number of terms in F_n grows very rapidly with n, as can be seen from eq. (2.10). In sect. 4 we will explore the convergence properties of the folded-diagram series for V_{eff} when we group the terms according to the number of folds as in eq. (2.13).

2.3. THE EQUATION-OF-MOTION METHOD FOR V_{eff}

Recently, Kuo and Krenciglowa [19]) have proposed a somewhat different and more general method to obtain a folded-diagram expansion for the effective interaction. This method enables us to sum up a larger class of diagrams contained in V_{eff} as compared with the KLR method outlined in the previous section. The starting point of the KLR method is a model space whose states consist of two (sd-shell) particles outside the *unperturbed* ^{16}O core, $[a_i^+ a_j^+]|c\rangle$. Thus the eigenfunctions we obtain in solving eq. (2.9) are the projections of the true wave functions onto this "uncorrelated" model space. At the heart of the newly proposed method is to use as a starting point $|\psi_0^c\rangle$, the true (correlated) ground state of ^{16}O in place of $|c\rangle$. The model-space basis states are thus of the form $[a_i^+ a_j^+]|\psi_0^c\rangle$. An advantage of this is that when we obtain a secular equation of the form eq. (2.9) its eigenfunctions will be

$$\langle \psi_0^c | [a_i a_j] | \psi^{A+2} \rangle,$$

which can be related to experimentally observable spectroscopic factors. This cannot be done if we use $|c\rangle$.

We do not give any details of the method here, except to point out some of its specific implications for our calculations described in sect. 4. In this method we write V_{eff} as

$$V_{eff} = \bar{Q} - \bar{Q} \int \hat{Q} + \bar{Q} \int \hat{Q} \int \hat{Q} - \bar{Q} \int \hat{Q} \int \hat{Q} \int \hat{Q} + \ldots, \tag{2.14}$$

where the bar over the first $\hat{Q}$-box in each term indicates the last interaction (attached to the outgoing valence lines) must take place at $t = 0$. All the other interactions are now constrained by $\infty > t > -\infty$. As in eq. (2.9), the first $\bar{Q}$ in each folded term is at least second order in the interaction.

The value of any diagram (i.e. the diagram rules) will in general be different in the new and old schemes. We do not pursue this point except to evaluate in detail one diagram which will be important in our calculation. Consider the diagram in fig. 3. We suppress factors (such as V matrix elements) which do not depend upon which method we use and concentrate on the time integrations only. In the old scheme the time constraints are

$$t_0 = 0, \qquad 0 > t_1 > -\infty, \qquad 0 > t_2 > -\infty, \qquad t_2 > t_3 > -\infty,$$

whereas in the new scheme they are

$$t_2 = 0, \qquad 0 > t_3 > -\infty, \qquad \infty > t_0 > t_3, \qquad t_0 > t_1 > -\infty.$$

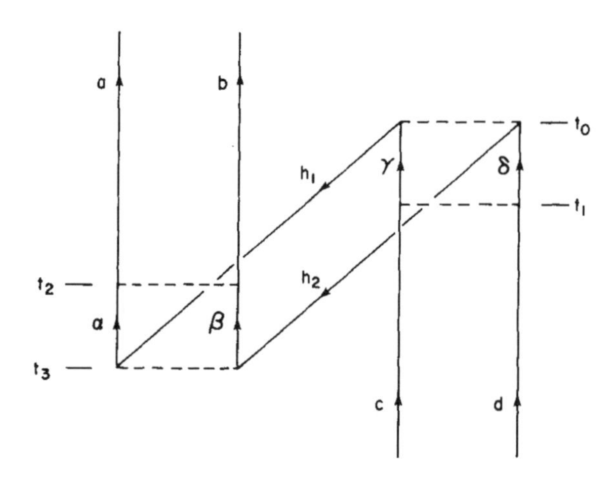

Fig. 3. The 4p2h diagram which is evaluated differently in the old and new schemes.

Thus in the old scheme we have for this diagram the energy factor

$$\frac{1}{\omega_0-(\varepsilon_\gamma+\varepsilon_\delta)}\,\frac{1}{\omega_0-(\varepsilon_a+\varepsilon_b+\varepsilon_c+\varepsilon_d-\varepsilon_{h_1}-\varepsilon_{h_2})}\,\frac{1}{(\varepsilon_{h_1}+\varepsilon_{h_2})-(\varepsilon_\alpha+\varepsilon_\beta)}, \tag{2.15}$$

while in the new scheme we have

$$\frac{1}{\omega_0-(\varepsilon_\gamma+\varepsilon_\delta)}\,\frac{1}{-\omega_0-(-\varepsilon_{h_1}-\varepsilon_{h_2})}\,\frac{1}{(\varepsilon_{h_1}+\varepsilon_{h_2})-(\varepsilon_\alpha+\varepsilon_\beta)}, \tag{2.16}$$

where we have defined the starting energy ω_0 as

$$\omega_0 \equiv \varepsilon_c+\varepsilon_d.$$

Note the on-energy shell expression for the middle energy denominator obtained in the new scheme. This allows us to sum up the 4p2h diagrams to infinite order as we will do in sect. 4. In the old scheme we could not do this. Note also that in fig. 3 the vertex a time t_2 is the "last" interaction attached to the (outgoing) valence line.

2.4. THE LEE-SUZUKI ITERATION METHOD FOR $V_{\rm eff}$

In our calculation of $V_{\rm eff}$, the first step is to calculate the $\hat{Q}$-box. Then we calculate the folded-diagram series of eq. (2.7) or (2.14) using either the F_n method of eq. (2.13) or the Lee-Suzuki iteration method [26]) which is here outlined.

The Lee-Suzuki method calculates the effective interaction $V_{\rm eff}$ in an iterative scheme. Denoting the nth iteration for the effective interaction as R_n, we have R_n

related to R_k $(k < n)$ by

$$R_n = (1-\hat{Q}_1 - \sum_{m=2}^{n-1} \hat{Q}_m \prod_{k=n-m+1}^{n-1} R_k)^{-1}\hat{Q}, \tag{2.17}$$

where $\hat{Q}$ is the $\hat{Q}$-box interaction defined in subsect. 2.2 and $\hat{Q}_n$ represents the energy derivative of the $\hat{Q}$-box as defined in eq. (2.14). An advantage of this method is that we only need to know $\hat{Q}$ and $\hat{Q}_n$ at a given starting energy ω_0. As we will discuss later in sect. 4, the calculated V_{eff} is essentially independent of ω_0. (Theoretically, V_{eff} is independent of ω_0.) The $\hat{Q}$-box here is the same as in the KLR theory or the method described in the preceeding section, and it enters in a perhaps more natural and convenient way. It contains all the linked and irreducible diagrams but no folded diagrams. The folded diagrams enter by way of eq. (2.17), which provides a way of regrouping the folded-diagram series.

We will examine how this grouping affects the convergence of V_{eff} in sect. 4, where V_{eff} will be calculated by first evaluating the Q-box and its derivatives and then calculating R_n according to eq. (2.17).

3. Method of calculation

In the last section, we obtained an expansion for the effective interaction in terms of matrix elements of the NN interaction V. It has been common practice to use for V phenomenological NN potentials such as the Reid potential [5]. In the present work we will use both this potential and the Paris potential [7] as V. We will first compare some essential features of these two potentials, and then discuss some details of the momentum-space matrix-inversion method on which our calculation will be based.

3.1. COMPARISON BETWEEN THE PARIS AND REID NN POTENTIAL

A main purpose of the present work is to compare the spectra of several sd-shell nuclei calculated from the Paris NN potential [7] with those from the Reid NN potential. It is important to note that these two potentials are in fact quite different, although they give quite similar NN phase shifts. The Reid NN potential is energy independent, and for each partial wave channel is given as a sum of a few Yukawa-type terms, i.e. $\sum_i c_i e^{-\mu_i r}/\mu_i r$, where c_i and μ_i are parameters and r is the internucleon distance. On the other hand, the Paris potential has a linear energy dependence which is, for convenience, transformed into a p^2 dependence. Then, the Paris potential, in its parametrized form [7], is given in r-space for each isospin T as

$$\begin{aligned} V(r,p^2) = {} & V_s(r,p^2)\Omega_s + V_t(r,p^2)\Omega_T \\ & + V_{LS}(r)\Omega_{LS} + V_T(r)\Omega_T + V_{SO_2}(r)\Omega_{SO_2}, \end{aligned} \tag{3.1}$$

where

$$V(r,p^2) = V^a(r)+\frac{p^2}{m}V^b(r)+V^b(r)\frac{p^2}{m}, \tag{3.2}$$

$$p^2 = -\hbar^2\left(\frac{1}{r}\frac{d^2}{dr^2}r-\frac{\hat{L}^2}{r^2}\right), \tag{3.3}$$

all in appropriate units. The labels s, t, LS, T, and SO_2 refer to the standard singlet, triplet, spin-orbit, tensor, and quadratic spin-orbit non-relativistic invariants, respectively. The $V(r)$'s are expressed as a sum of Yukawa-type terms. For example, for the central singlet component of the potential one has

$$\Omega_s = \tfrac{1}{4}(1-\boldsymbol{\sigma}_1\cdot\boldsymbol{\sigma}_2), \tag{3.4}$$

$$V_s(r) = \sum_{j=1}^{n} g_j\frac{e^{-m_jr}}{m_jr} \tag{3.5}$$

in eqs. (3.1) and (3.2), respectively. The potential is regularized at the origin ($r = 0$) by requiring

$$g_n = -m_n\sum_{j=1}^{n-1}\frac{g_j}{m_j}. \tag{3.6}$$

The form of the other components as well as the numerical values of the parameters g_j and m_j are found in ref. [7]). Although it is possible to work with this potential directly in configuration space, it is somewhat inconvenient to do so. Fortunately, the situation is much simpler if a plane-wave basis is used, as in the present work. Then the plane-wave matrix elements of the Paris potential are given simply as

$$\langle k|V|k'\rangle = \langle k|V^a|k'\rangle + \left(\frac{\hbar^2k^2}{m}+\frac{\hbar^2k'^2}{m}\right)\langle k|V^b|k'\rangle, \tag{3.7}$$

for any partial wave $^{2S+1}L_J$. Here k and k' are the magnitudes of the relative momenta of the outgoing and incoming nucleon pairs respectively. For convenience, numerical integration was used. This was checked against the analytic expression for the $^1S_0$ partial wave and found to be quite accurate. In fig. 4 we plot $V(k,k)$ for the Paris potential † (solid line) and for the Reid potential (dashed line), for partial waves $^1S_0$ to $^3D_1$. One is immediately struck by the fact that, although both potentials were made to fit the same scattering data, the two potentials are distinctly different. Similar differences have also been found for other partial waves.

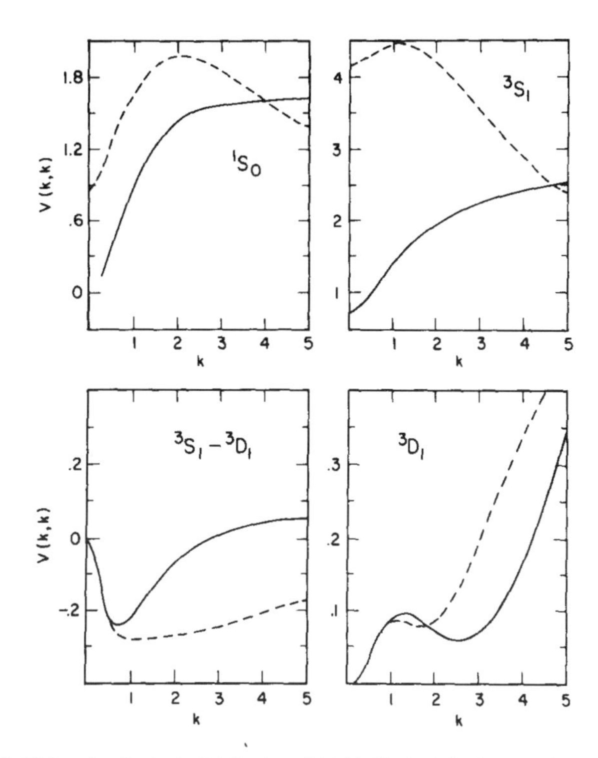

Fig. 4. Plot of $V(k, k)$ for the Paris (solid line) and Reid (dashed line) potentials for selected partial waves. The k are in fm^{-1} and V is in units of 41.47 MeV · fm^3.

In the Paris potential the components other than the central ones are not energy dependent and are functions of r alone. Thus they can be compared with their counterparts in the Reid potential directly in configuration space. We do this in fig. 5 for the important $T = 0$ $S = 1$ tensor force[†]. As shown, the tensor forces for these two potentials are quite different. It is of interest to see how this difference manifests itself in nuclear-structure calculations.

Before proceeding to the actual nuclear structure calculation we study the Paris potential NN phase shifts and compare with those of the Reid potential. A more compelling reason for looking at the phase shifts is that the method used to calculate them (momentum-space matrix inversion) is also used in the calculation of the G-matrix and thus provides a check of our computer programs.

[†] In the early stages of this work, the Paris potential parameters were taken from a preprint. In the published version [7]) the $T = 0$ parameters are somewhat different. Thus, in fig. 4 the preprint parameters were used, as they were in the phase-shift calculations of the following section. Fig. 5, however, used the published parameters. (The preprint had an even weaker tensor force.) Of course, for the G-matrix elements the published parameters were used.

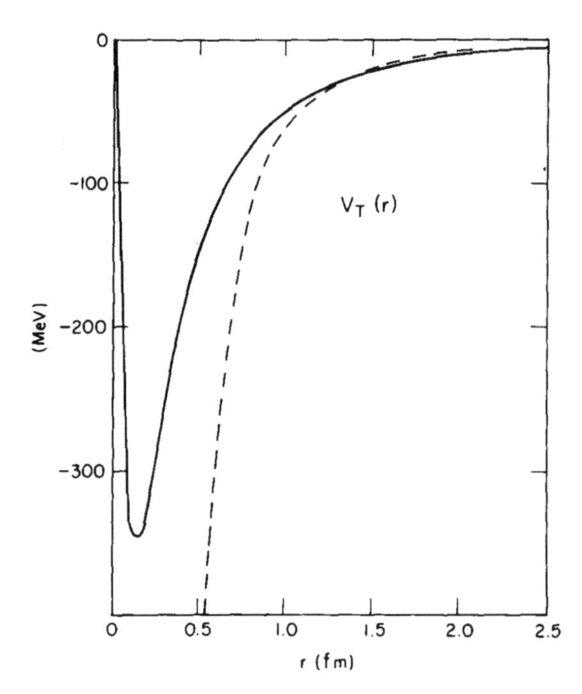

Fig. 5. Comparison of the $T = 0$, $S = 1$ tensor component between the Paris (solid line) and Reid (dashed line) potential.

3.2. PHASE-SHIFT CALCULATIONS

To calculate NN scattering phase shifts, we use the momentum-space matrix-inversion method as outlined in ref. [27]). Briefly, for an uncoupled partial wave, the phase shift is given by

$$\delta_l(k) = -\tan^{-1}[kR_l(k,k)], \tag{3.12}$$

where $R_l(k,k')$ is given by the principal value integral equation

$$R_l(k,k') = V_l(k,k') - \frac{2}{\pi}\mathrm{P}\int_0^\infty \frac{V_l(k,q)R_l(q,k')}{q^2-k'^2}q^2\mathrm{d}q. \tag{3.13}$$

This integral can be rewritten to avoid the principal-value singularity [27]). A similar expression exists for the coupled-channels case. To solve eq. (3.13) we use gausian integration. To determine the validity of this procedure, we chose several quite different sets of Gauss points and looked at the resulting phase shifts to see if and how they were affected.

The integral (3.13) from 0 to ∞ is broken up into several intervals. The intervals are specified by S_n, the start of the nth interval and by N_n, the number of Gauss points in the interval between S_{n-1} and S_n. For finite intervals, the Gauss points q_i (and weights) are mapped from the region $-1 \leqq x_i \leqq +1$ onto the interval. If the interval extends from S_n to infinity, we use a tangent mapping to obtain the Gauss points

$$q_i = S_n + C \tan \tfrac{1}{4}\pi(1+x_i). \tag{3.14}$$

N_t is the number of points for which the tangent mapping is used. The constant C is for adjusting the distribution of the tangent Gauss points. The total number of points is N_{tot}. The procedure is essentially the same as that of ref. [22]). The choices of Gauss points for several calculations are specified in table 1, all with $C = 1$.

Tables 2a and 2b show the phase shifts for the Paris potential calculated with Gauss point sets 1 and 2 of table 1. Note that they agree remarkably well with each other despite the large difference between the Gauss points used. We have found that the distribution of Gauss points is far more important than the number of points used. In table 2c we quote the phase shifts obtained in ref. [7]). Note that the phase shifts we calculate are the eigenphases whereas the ones from ref. [7]) are the nuclear bar phase shifts. The two can be related by equating the expressions for the S-matrix obtained from the two methods. [See e.g. ref. [29]).] We have done this and found the agreement with the published values to be very good for table 2b. We found it to be important to map the Gauss points onto several intervals rather than one interval covering the entire range from 0 to ∞. The phase shifts (and hence the R-matrix) were found to be quite stable with respect to reasonable variations of the Gauss points. For the G-matrix calculation we have used the Gauss points set (G) of table 1.

For comparison, the Reid-potential phase shifts are given in table 2d. [These are the ones we calculated using Gauss points set 3 of table 1, i.e. eigenphases, but when converted into the nuclear bar phases they agree closely with those given by Reid in his paper [5]).] The significant difference between the two potentials is in the

TABLE 1

Specification of the Gauss points

Set	S_0	S_1	S_2	S_3	N_1	N_2	N_3	N_t	N_{tot}
1	0	3.0	10.0	25.0	6	5	4	0	15
2	0	2.75	8.0	18.0	11	7	4	3	25
3	0	2.6	7.0	15.0	11	7	4	3	25
G	0	2.5	7.0	15.0	15	10	6	4	35

The boundaries of the integration intervals are the S_n (in fm^{-1}) and N_n and N_t are the number of points in the interval. The last row (G) refers to the Gauss points used in the G-matrix calculation of subsect. 3.4.

TABLE 2a

Results of phase-shift calculations (in degrees) for the Paris potential calculated with the momentum-space matrix inversion method using Gauss points set (1) of table 1

E_{lab} (MeV) =	25	50	95	142	210	330
$^1S_0$	48.68	38.42	25.25	14.92	3.18	−12.46
$^1D_2$	0.75	1.84	3.79	5.65	7.76	9.47
$^3P_0$	9.20	11.93	10.26	6.12	−0.45	−11.01
$^3P_1$	−5.24	−8.60	−12.91	−16.55	−21.21	−28.50
$^3P_2$	2.91	6.53	11.63	14.97	17.21	16.85
ε_2	−17.59	−16.53	−14.11	−12.02	−9.74	−6.94
$^3F_2$	−0.19	−0.18	0.10	0.44	0.62	−0.12
$^1P_1$	−6.99	−10.66	−14.75	−17.65	−20.81	−25.00
$^3S_1$	80.51	62.31	43.43	30.16	15.97	−1.46
ε_1	1.84	2.20	2.79	3.80	6.06	13.86
$^3D_1$	−2.94	−6.79	−12.22	−16.44	−20.81	−25.49
$^3D_2$	3.86	9.59	18.10	23.84	28.00	28.75

$^1P_1$ partial wave. Otherwise, the phase shifts are very similar. Of course it must be remembered that both potentials were constructed to reproduce the phase shifts. Still, the potentials themselves are markedly different and one cannot predict in advance their results in a nuclear structure calculation.

3.3. THE G-MATRIX

In sect. 2 the effective interaction is given as a series of diagrams, each containing matrix elements of the NN interaction V. In general, matrix elements of

TABLE 2b

Same as table 2a, but using Gauss points set 2 of table 1

E_{lab} (MeV) =	25	50	95	142	210	330
$^1S_0$	48.58	38.31	25.18	14.90	3.21	−12.48
$^1D_2$	0.75	1.84	3.79	5.64	7.75	9.44
$^3P_0$	9.19	11.93	10.25	6.10	−0.45	−11.04
$^3P_1$	−5.25	−8.60	−12.94	−16.61	−21.32	−28.75
$^3P_2$	2.91	6.51	11.59	14.90	17.11	16.74
ε_2	−17.60	−16.53	−14.12	−12.03	−9.75	−6.95
$^3F_2$	−0.19	−0.18	0.10	0.44	0.62	−0.12
$^1P_1$	−7.01	−10.67	−14.79	−17.74	−20.93	−25.21
$^3S_1$	80.25	62.04	43.16	29.95	15.91	−1.44
ε_2	1.80	2.18	2.78	3.81	6.09	13.83
$^3D_1$	−2.90	−6.77	−12.30	−16.54	−20.87	−25.70
$^3D_2$	3.87	9.59	18.03	23.73	27.85	28.48

TABLE 2c

Same as table 2a, but for the Paris potential quoted from ref. [7]); see text for explanation

E_{lab} (MeV) =	25	50	95	142	210	330
$^1S_0$	48.51	38.74	25.85	15.68	4.03	−11.80
$^1D_2$	0.75	1.81	3.72	5.56	7.66	9.37
$^3P_0$	8.87	11.82	10.35	6.31	−0.16	−10.79
$^3P_1$	−5.04	−8.41	−12.76	−16.47	−21.27	−28.90
$^3P_2$	2.44	5.73	10.66	14.06	16.49	16.44
ε_2	−0.85	−1.78	−2.67	−2.91	−2.71	−2.00
$^3F_2$	0.11	0.35	0.77	1.04	1.08	0.14
$^1P_1$	−7.03	−10.75	−14.87	−17.87	−21.13	−25.52
$^3S_1$	79.87	61.56	42.72	29.61	15.57	−2.61
ε_2	1.72	1.96	2.24	2.69	3.59	5.54
$^3D_1$	−2.95	−6.76	−12.29	−16.52	−20.73	−29.75
$^3D_2$	3.97	9.62	17.96	23.58	27.57	28.13

V are infinite or very large even for soft-core potentials. This is especially true in an oscillator basis with non-vanishing relative wave function at the origin. To overcome this, the reaction matrix, or G-matrix was introduced [8]). In operator language we have in our case [21])

$$G(\omega_0) = V + VQ\frac{1}{\omega_0 - QTQ}QG(\omega_0), \tag{3.15}$$

where Q is the Pauli exclusion operator and ω_0 is the starting energy. Diagrammatically, eq. (3.15) is expanded in fig. 6. Note that intermediate lines

TABLE 2d

Same as table 2a, but for the Reid potential using Gauss points 3 of table 1

E_{lab} (MeV) =	25	50	95	142	210	330
$^1S_0$	49.09	38.79	25.64	15.52	4.32	−10.14
$^1D_2$	0.68	1.66	3.39	5.11	7.18	9.31
$^3P_0$	8.60	11.55	10.22	5.98	−1.18	−13.29
$^3P_1$	−4.60	−8.09	−13.16	−17.39	−22.23	−28.46
$^3P_2$	3.02	7.10	13.10	16.97	19.51	19.78
ε_2	−15.86	−14.40	−12.09	−10.54	−9.11	−7.26
$^3F_2$	−0.15	−0.12	0.18	0.55	0.90	0.79
$^1P_1$	−1.94	−4.26	−10.67	−17.47	−26.07	−37.94
$^3S_1$	81.17	62.80	43.85	30.90	17.80	2.51
ε_1	1.90	2.58	3.97	5.98	9.72	18.05
$^3D_1$	−2.97	−6.91	−12.28	−16.15	−20.18	−26.04
$^3D_2$	4.13	10.10	17.80	22.13	24.98	26.08

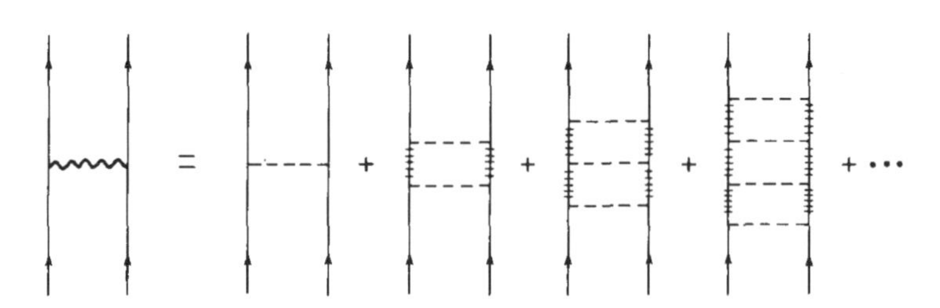

Fig. 6. Definition of the G-matrix between two particle lines.

must be passive lines – the railed lines which represent the single-particle states higher than the pf shell of fig. 7. Otherwise the diagram is not allowed in the $\hat{Q}$-box (recall subsect. 2.2). Thus all the diagrams in the $\hat{Q}$-box and hence all the diagrams contributing to V_{eff} have their V-vertices "replaced" by G-vertices in analogy with fig. 6. Since the matrix elements of G are finite we can proceed to calculate the matrix elements of V_{eff}.

We do not go into the details of the G-matrix calculation, as we closely follow the procedure of ref. [22]). We only make several notes in passing. As pointed out, eq. (3.15) is similar in structure to eq. (3.13) and we use the same momentum space method to solve it. We use 35 k-space Gauss points distributed as indicated by the last line of table 1. This choice should be adequate, judging by the phase-shift calculations described in the previous section. Also, in eq. (3.15) we use a two-particle Pauli operator Q specified by $(n_{\text{I}}, n_{\text{II}}, n_{\text{III}}) = (3, 10, 21)$. That is, the intermediate states included in G must lie outside the shaded region in fig. 7. The choice of Q is very important in determining which G-matrix diagrams are allowed in the $\hat{Q}$-box as well as in determining the allowed orbits for intermediate states of

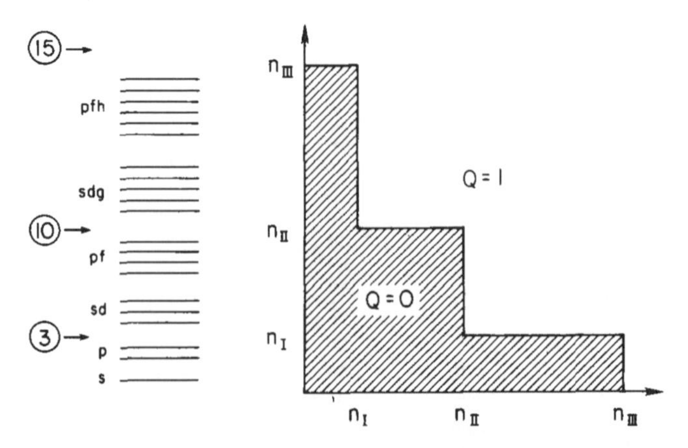

Fig. 7. Specification of the Pauli operator Q used in the calculation of the G-matrix. It is defined by $(n_{\text{I}}, n_{\text{II}}, n_{\text{III}}) = (3, 10, 21)$ with the orbit numbering as explained by the circled numbers.

$\hat{Q}$-box diagrams. [See e.g. ref. [28].] Our procedure was as follows. We started from the Reid or Paris Potential, obtained $V(k, k')$ using gaussian integration, and then $G_F(k, k')$, the "free" reaction matrix by gaussian integration and matrix inversion. Gaussian integration is again used to obtain $\langle NLnl|G_F(\omega)|N'L'n'l'\rangle$ in a rel-CM oscillator basis. The Pauli operator was treated by the Tsai-Kuo method [21], which expresses eq. (3.15) in terms of P, the complement of Q, and G_F. Finally, everything is gathered together and transformed to the laboratory frame oscillator basis by means of a Moshinsky transformation. We obtain

$$G(ab, cd; JT) \equiv \langle \overset{JT}{\overbrace{ab}} | G | \overset{JT}{\overbrace{cd}} \rangle, \tag{3.16}$$

where a, b, c, d refer to the oscillator orbits specified by (n_a, l_a, j_a), etc. J and T are total angular momentum and isospin, respectively, and the coupling is as indicated by the arrows in eq. (3.16). The matrix elements (3.16) are precisely what we need for nuclear-structure calculations. Extensive computer programs have been constructed for carrying out the above G-matrix calculation. Tables of G-matrix elements will be given later in sect. 4.

3.4. CALCULATION OF $\hat{Q}$-BOX DIAGRAMS

In the previous sections we have sketched the development of the effective interaction V_{eff} in terms of the $\hat{Q}$-box formulation of the folded-diagram theory. The $\hat{Q}$-box is the sum of all the linked and irreducible diagrams. Thus we need a set of rules to enable us to evaluate the diagrams in a simple systematic way. The diagram rules in the uncoupled representation or m-scheme are well known [2,30].

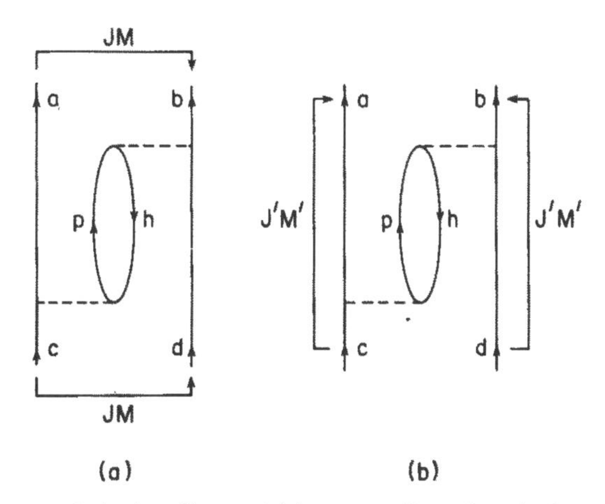

Fig. 8. The 3p1h core-polarization diagram (a) is more easily evaluated when recoupled as in (b).

For actual nuclear-structure calculations, we will need to evaluate diagrams in the angular-momentum-coupled representation, such as the core-polarization diagram of fig. 8. Recently, a set of simple diagram rules has been developed [23]) to evaluate these angular-momentum-coupled diagrams, and we will employ it to calculate all the $\hat{Q}$-box diagrams in the present work.

To illustrate the above method, we give in the following the formula for calculating diagram (a) of fig. 8. This diagram is first recoupled to become diagram (b), using the usual X or 9-j transformation coefficients. Then diagram (a) is readily given [23]) as

$$(\mathrm{a}) = \frac{1}{\sqrt{(2J+1)(2T+1)}} \sum_{J'T'} \times \begin{Bmatrix} c & d & j \\ a & b & j \\ J' & J' & 0 \end{Bmatrix} \times \begin{Bmatrix} \frac{1}{2} & \frac{1}{2} & T \\ \frac{1}{2} & \frac{1}{2} & T \\ T' & T' & 0 \end{Bmatrix} \sqrt{(2J'+1)(2T'+1)}$$

$$\times \underset{J'T'}{\overset{J'T'}{\langle ap|G(\omega_1)|ch\rangle}} \underset{J'T'}{\overset{J'T'}{\langle hb|G(\omega_2)|pd\rangle}}, \tag{3.17}$$

where the last matrix elements are of the cross-coupled type [23]) and the energy variables are $\omega_1 = \varepsilon_c + \varepsilon_h$ and $\omega_2 = \varepsilon_c + \varepsilon_d - \varepsilon_a + \varepsilon_h$, the ε's being the single-particle energies.

4. Results for two-body effective interactions

4.1. THE $\hat{Q}$-BOX DIAGRAMS

The next step in calculating an effective interaction is to evaluate the $\hat{Q}$-box (and its derivatives). The $\hat{Q}$-box is composed of all the linked and irreducible (G-matrix) diagrams. We take as a first approximation to the $\hat{Q}$-box all the diagrams up to second order in the G-matrix. The diagrams we include are D1 through D7 and U (the one-body potential) as shown in fig. 9.

In diagram D5 the intermediate particle states are restricted to the pf-shell because the G-matrix was calculated with a Puli operator defined by $(n_{\mathrm{I}}, n_{\mathrm{II}}, n_{\mathrm{III}}) = (3, 10, 21)$ which takes into account passive particle states of orbit 11 and above only. Orbits 7–10 (the pf-shell) must be included explicitly. For the same reason, particles in the pf-shell are also included in the intermediate states of diagram D2.

Our $\hat{Q}$-box, expressed in terms of the G-matrix, takes into account the particle-particle interaction (outside the model space) to all orders. If we use the equation of motion method of Kuo and Krenciglowa (subsect. 2.3) we can extend our $\hat{Q}$-box

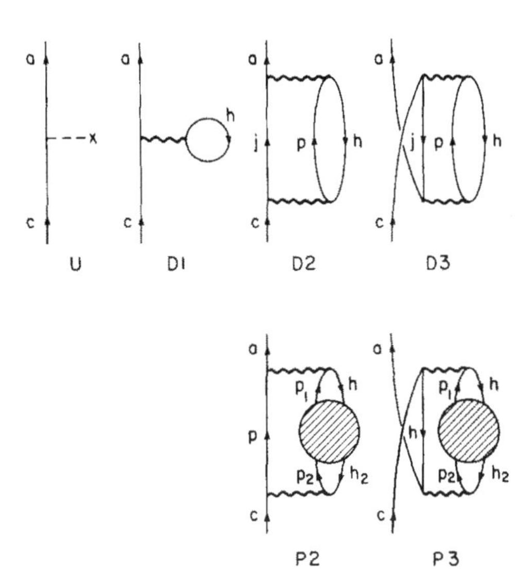

Fig. 9a. The one-body diagrams included in various approximations to the Q-box. The shaded circle represents a (bare) TDA particle-hole phonon in diagrams P2 and P3.

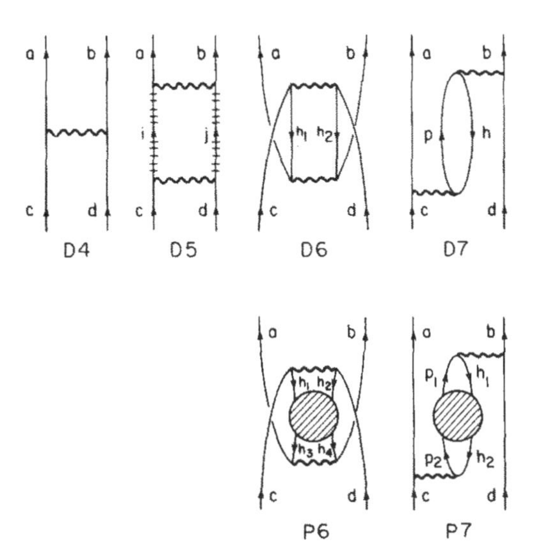

Fig. 9b. The two-body diagrams included in various approximations to the Q-box. The shaded circle represents a (bare) TDA particle-hole phonon in diagram P7 and a hole-hole phonon in P6.

to include the hole-hole interaction to all orders as well. That is, the two hole lines of diagram D6 are allowed to interact any number of times so that D6 is replaced by diagram P6. (See fig. 9.) In P6, the blob represents a hole-hole phonon. This refinement of the $\hat{Q}$-box approximation can have a significant effect on the matrix elements of V_{eff} and the resulting spectra. Another modification is in the treatment of the core-polarization processes. In place of the particle-hole pair in D2, D3 and D7 we can include a TDA particle-hole phonon. That is, D2, D3 and D7 are replaced in the $\hat{Q}$-box by P2, P3 and P7, respectively. In obtaining the TDA eigenvalues and eigenvectors, we include only the bare (first order) G-matrix particle-hole interaction vertex as shown by diagram (a) of fig. 10. Because we include the TDA phonons only as a "correction" to the particle-hole intermediate states, inclusion of only this bare vertex should be sufficient. Also, it has been demonstrated that the inclusion of all the second order particle-hole interaction vertices, such as diagram (b) of fig. 10, in RPA-phonon calculations gives results very similar to the ones obtained in bare TDA only [2,32]).

One result of including the TDA states rather than simple particle-hole states is that we can now isolate the spurious 1^- state in diagrams P2 and P3. This cannot be done if we include only the particle-hole intermediate state by second-order perturbation theory because then all the 1^- states (there are 7 of them) are mixed. The TDA phonon intermediate state allows us to isolate the spurious 1^- state and remove it in lowest order. This is done by explicitly removing the spurious 1^- contributions in the calculation of diagrams P2 and P3. The removal of the spurious state changes P2 and P3 by about 0.6 MeV. The results we will quote here are from calculations in which the spurious state is removed. However, to gauge the effect of including TDA phonons in place of simple particle-hole states, we really should leave the spurious state in for the comparison to be valid, since we do not remove it in the simple particle-hole case.

The calculation was done in a harmonic oscillator basis with $\hbar\Omega = 14$ MeV. The unperturbed starting energy ω_0, the point at which we calculated the effective interaction, was taken to be -10 MeV (for two sd-shell particles). The different

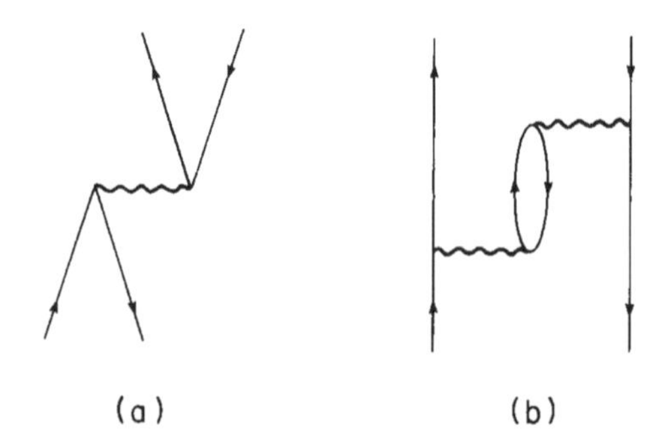

Fig. 10. The bare TDA vertex (a), and one of the second-order TDA vertex (b).

approximations to the $\hat{Q}$-box are referred to as C1, C2, C3 and C4 for the $\hat{Q}$-box with 2nd order diagrams only, 2nd order plus hole-hole phonon, and 2nd order plus particle-hole phonon, and 2nd order plus both hole-hole and particle-hole phonons, respectively. Thus, the C1 interaction includes diagrams D4, D5, D6 and D7 of fig. 9b; for the C2 interaction, D6 is replaced by P6, etc. Effective interactions obtained with these four methods will be compared in sect. 5.

4.2. INCLUSION OF THE FOLDED DIAGRAMS

In the early microscopic calculations of this type [2, 24]) the folded diagrams were generally left out and the effective interaction was taken to be the $\hat{Q}$-box alone (in some low order approximation such as the bare G plus 3p1h core-polarization diagrams). Barrett and Kirson [25]) have investigated several once-folded diagrams and found their effects important. In the present work, we shall investigate the effect of folded diagrams in a more extensive and systematic way. When more processes are included in the $\hat{Q}$-box, it will become apparent that the $\hat{Q}$-box alone gives substantial overbinding as well as poor level spacing. (See e.g. fig. 11 for $n = 1$, which corresponds to $V_{\mathrm{eff}} = \hat{Q}$.) Thus we must include the folded diagrams to renormalize the effective interaction. This we can do in one of the three ways described below.

It has been shown [20]) that the $\hat{Q}$-box evaluated at a self-consistent starting energy is equivalent to the effect of V_{eff} with all the folded diagrams included. That is,

$$[H_0 + V_{\mathrm{eff}}]|\Psi_n\rangle = [H_0 + \hat{Q}(E_n)]|\Psi_n\rangle = E_n|\Psi_n\rangle, \tag{4.1}$$

where E_n is the total energy of the nuclear system relative to the exact ground-state energy of the core system. For example, E_n is $E_n(^{19}\mathrm{O}) - E_0(^{16}\mathrm{O})$ when the above equation is used to calculate the energy spectrum of $^{19}\mathrm{O}$. As discussed earlier, V_{eff} is related to the $\hat{Q}$-boxes by, for example, eqs. (2.13)–(2.16) and in calculating V_{eff} we need to evaluate the folded diagrams involving the $\hat{Q}$-boxes. Eq. (4.1) is in some aspects more convenient for calculation because we just evaluate the $\hat{Q}$-box at the self-consistent energy without the need for calculating any $\hat{Q}$-box folded diagrams. For a given $\hat{Q}$-box, this self-consistent energy can usually be obtained by a graphical method. There is, however, a rather important disadvantage associated with the above self-consistent equation. Namely, $\hat{Q}(E_n)$ is dependent on the total energy of the nuclear system and for different nuclei and different nuclear states the $\hat{Q}$-boxes will need to be recalculated. In other words, this formulation does not lead to a set of matrix elements of V_{eff} which can be used to treat the low-lying states of *all* the sd-shell nuclei, unlike the case of the empirical shell model. For example, we would like to have a set of two-body matrix elements of V_{eff} which can be used to calculate both the ground state of $^{18}\mathrm{O}$ and the excited states of

^{20}Ne. The matrix elements of $\hat{Q}(E_n)$ of eq. (4.1) do not meet this requirement. Another inconvenience of $\hat{Q}(E_n)$ is that it contains unlinked valence diagrams, whereas V_{eff} contains only linked valence diagrams. Thus in the present work we will not use the self-consistent formulation of eq. (4.1) for major calculations except for checking the convergence of the folded-diagram series for simple cases ^{18}O and ^{18}F.

A second method is to sum the folded-diagram series explicitly as described in sect. 2. Using the method of grouping terms according to the number of folds we can proceed in one of two ways. We can use eqs. (2.12) and (2.13) to obtain what are usually called the sd-shell effective-interaction matrix elements. In this case, the energy levels obtained are relative to ^{17}O since the one-body terms are removed from V_{eff} and are replaced by the experimental ^{17}O low-lying states. Alternatively, we can retain the one-body diagrams in V_{eff} (i.e. use $F_n = F_n(\hat{Q})$) in which case the energy levels obtained are relative to the ^{16}O ground state. In either case, all the $\hat{Q}_n$ (see eqs. (2.10)–(2.13)) are evaluated at the unperturbed starting energy ω_0, at which point all the matrix multiplication is carried out.

A third way to include the folded diagrams is to group them as prescribed by the Lee-Suzuki iteration method. Here one also uses the $\hat{Q}_n$ evaluated at ω_0 but the terms are grouped differently. One solves for the eigenvalues of the H_0+R_n, where R_n is defined by eq. (2.17), increasing n until the eigenvalues converge. At that point one has essentially included all of the folded diagrams and

$$V_{\text{eff}} = R_\infty \tag{4.2}$$

is obtained. As written, eq. (2.17) contains the one-body graphs in the $\hat{Q}$-boxes so that the eigenvalues obtained are the energy levels relative to the ^{16}O ground state. However, one may obtain the usual sd-shell matrix elements (i.e. two-body-connected only) by removing the one-body graphs from V_{eff}. This is done by doing two parallel calculations

$$R_n^Q = (1-\hat{Q}_1 - \sum_{m=2}^{n-1} Q_m \prod_{k=n-m+1}^{n-1} R_k^Q)^{-1}\hat{Q}, \tag{4.3a}$$

$$R_n^S = (1-\hat{S}_1 - \sum_{m=2}^{n-1} \hat{S}_m \prod_{k=n-m+1}^{n-1} R_n^S)^{-1}\hat{S}, \tag{4.3b}$$

and then, after convergence is reached, taking the eigenvalues of $(H_0+R_n^Q-R_n^S)$. In this case, of course, the one-body part of V_{eff} is taken from experiment.

We thus have three ways to take into account the folded diagrams: (i) the self-consistent graphical method (SC), with an energy-dependent effective interaction, and (ii) the method of grouping according to the number of folds (F_n), and (iii) the Lee-Suzuki iteration method (LS), where V_{eff} from either method (ii) or (iii) is energy-independent. Before going on to compare the results of the three methods in the following sections, we make several observations. For the energy-dependent effective interaction (SC method) we need to know (only) the $\hat{Q}$-box, but at all energies. For the energy-independent interaction (F_n or LS method) we need to know the $\hat{Q}$-box and all its derivatives, but at a single energy only. As mentioned earlier, the F_n and LS method directly lead to a set of energy-independent matrix elements of V_{eff} which can be used for any sd-shell nuclei, but not for the SC method.

4.3. CONVERGENCE

Before stating the results of the calculations we have outlined above, we first investigate the reliability of the methods used to obtain them. The most important questions at this point are whether the diagramatic series for V_{eff} converges and what are some of the factors which influence convergence. In the previous section, we described three ways in which we can take into account the contribution from the folded diagrams. In the SC method, there is no problem of convergence (unless perhaps at points where $\hat{Q}(\omega)$ is singluar. [See ref. [1]).]

We now examine the structure of the folded-diagram series in the F_n method as it is embodied in eqs. (2.10) through (2.13). The energy (ω) dependence of the $\hat{Q}$-box is of central importance. The $\hat{Q}$-box is defined as a sum of diagrams, so that its energy dependence is the sum of the energy dependences of the individual diagrams. The energy dependence of the first-order diagrams (D1 and D4) is practically linear. The second-order diagrams have a more pronounced energy dependence. When phonons are included, the energy dependence is even stronger because most of the energy dependence comes from the energy denominators associated with low-lying intermediate states whereas the G-matrix itself has a relatively weak energy dependence. Thus, the higher order the process, the stronger the energy dependence.

The result of this is that as one goes from C1 to C4 (i.e. includes higher-order processes as discussed in subsect. 4.1) the size of $\hat{Q}^{(n)}$, the nth derivative of $\hat{Q}$, increases. The higher derivatives show an especially large increase. The effect of this on the folded-diagram series terms F_n is easily predicted by eqs. (2.10)–(2.13). As seen in table 3, as one goes from C1 to C4 the size of $\tilde{F}_n$ increases. Also, the relative increase is larger for higher n.

The above discussion of how well V_{eff} converges when the terms are grouped according to the number of foldings is clearly reflected in the resulting eigenvalues (spectra). In fig. 11 we show the spectra (relative to ^{17}O) of V_{eff} where

$$V_{eff} = \sum_{n=0}^{N} \tilde{F}_n, \tag{4.4}$$

TABLE 3

Representative V_{eff} matrix elements in the F_n method using the Paris potential with starting energy $\omega_0 = -10$ MeV

T	J	a	b	c	d	F_0	F_1	F_2	F_3	F_4	V_{eff}
1	0	4	4	4	4	−2.9467	0.9698	−0.3173	0.0611	−0.0060	−2.2391
1	0	4	4	5	5	−4.2159	0.8920	0.0841	0.0032	−0.0362	−3.2728
1	0	4	4	6	6	−1.1982	0.3679	−0.0913	0.0109	0.0009	−0.9098
1	0	5	5	4	4	−4.2159	1.3098	0.0232	−0.0408	−0.0490	−2.9726
1	0	5	5	5	5	−0.9605	−0.0552	0.0206	0.0535	0.0124	−0.9293
1	0	5	5	6	6	−0.9140	0.2817	−0.0054	−0.0084	−0.0077	−0.6539
1	0	6	6	4	4	−1.1982	0.4570	−0.1436	0.0285	−0.0031	−0.8594
1	0	6	6	5	5	−0.9140	0.2052	0.0055	0.0050	−0.0098	−0.7080
1	0	6	6	6	6	−2.3319	0.8047	−0.0935	−0.0436	0.0184	−1.6485
1	0	4	4	4	4	−3.0753	1.1547	−0.3888	−0.0648	0.0287	−2.3455
1	0	4	4	5	5	−4.3978	0.8670	0.4197	0.3979	−0.2491	−2.9623
1	0	4	4	6	6	−1.3937	0.4764	0.1102	−0.0219	0.0170	−1.0324
1	0	5	5	4	4	−4.3978	1.4623	0.4520	0.3467	0.1094	−2.0275
1	0	5	5	5	5	−1.0489	−0.1873	−0.1352	−0.0785	−0.0324	−1.4823
1	0	5	5	6	6	−0.9222	0.3616	0.0153	0.0222	0.0174	−0.5057
1	0	6	6	4	4	−1.3937	0.8599	−0.491	0.0921	0.0125	−0.9533
1	0	6	6	5	5	−0.9222	0.5364	0.1605	0.1410	−0.0518	−0.1361
1	0	6	6	6	6	−2.9024	1.3358	−0.0541	−0.2569	0.0378	−1.8399

The upper half is for calculation C1 and the lower half for C4. The orbits (4, 5, 6) represent respectively ($0d_{5/2}$, $0d_{3/2}$, $1s_{1/2}$). All entries are in MeV.

for $N = 0$ to 4 using the Paris interaction. For the C1 (i.e. second-order diagrams only) calculation the convergence is quite satisfactory and is not shown. For the C4 calculation, however, convergence has markedly deteriorated.

To study the convergence of V_{eff} when the folded diagrams are grouped according to the LS scheme we look at fig. 12 where we give the spectra of $V_{eff} = R_n$ (actually, of $R_n^Q - R_n^S$) for $n = 1$ to 6. Fig. 12 shows the "worst case" calculation C4 for $T = 1$ and $T = 0$. It is clear from the figure that even for this case the LS series converges very rapidly for all the states.

It appears, on the basis of its relatively rapid and smooth convergence, that the LS method is preferred over the F_n method. The argument in favor of LS over F_n is strengthened by the following. Consider table 4 in which we compare the eigenvalues of all three methods, SC, LS, and F_n for all four $\hat{Q}$-box approximations. (All are for the Paris potential.) Since we include the SC method in the comparison, the eigenvalues in table 4 are relative to ^{16}O. The agreement between the SC and LS eigenvalues is striking. (Note that our numbers are really only valid in the interval −20 to −6 MeV because we have calculated the $\hat{Q}$-box only in this energy interval.) Actually, the F_n numbers are also in good agreement generally, except for the $TJ = 01$ ground state for which the F_n method gives a rather poor result. Thus we can say that the LS iteration formula reaches the "exact result" (i.e.

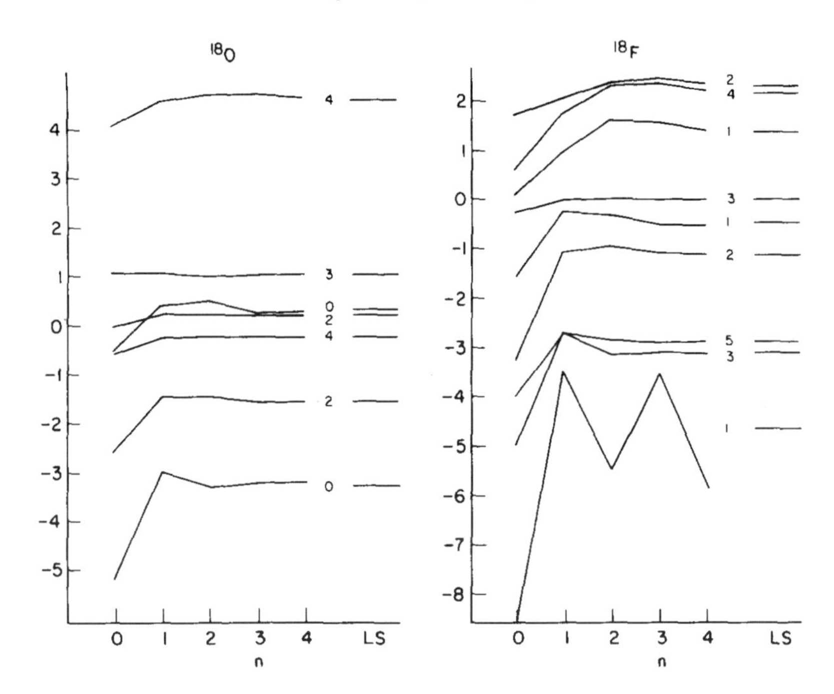

Fig. 11. The $A = 18$ spectra as a function of the number of terms n in the F_n method. The LS results (after six iterations) are included for comparison. These results are for the Paris potential in approximation C4 which converged the most poorly of those considered.

the self-consistent solution) after only 6 iterations. [In the SC calculation, (i) self-consistency was considered to have been reached with agreement to within 0.00005 MeV, and (ii) we know the $\hat{Q}$-box at 8 points only so that the higher derivatives are less reliable. In view of these two limitations, the agreement attained is indeed remarkable.] We therefore use the LS spectra as the standard against which we measure the convergence of the F_n series in fig. 11. If we know the $\hat{Q}$-box and its derivatives, we can find V_{eff} with only a small number of iterations when using the LS method for calculating the folded diagrams. The other half of the problem, what to include in the basic $\hat{Q}$-box, still remains.

4.4. THE EFFECTIVE INTERACTION

We now present results of the various calculations we have been describing.

We first examine the importance of the spurious particle-hole $TJ^{\pi} = 01^{-}$ state which enters into diagrams D2(P2) and D3(P3). As pointed out earlier we can isolate its effect when we include diagrams P2 and P3 in place of D2 and D3. The spurious state's contribution to the diagrams is very large, often more than 10%. The reason, of course, is that because it has a particle-hole eigenvalue ω_{ph} close to

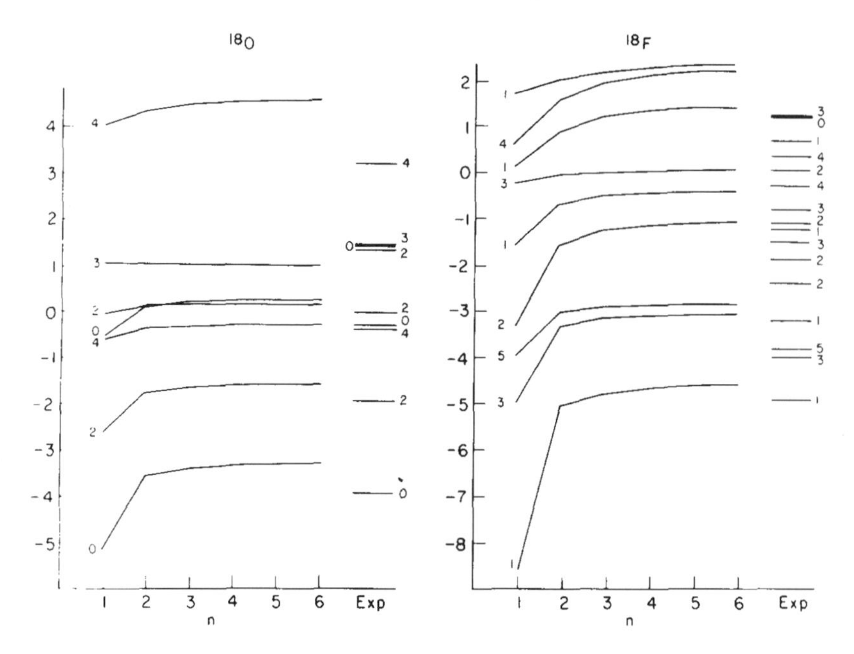

Fig. 12. The $A = 18$ spectra after n iterations in the LS method for calculation C4 using the Paris potential. The experimental spectra are included for comparison. The ground-state energies, relative to ^{17}O, are from ref. [33]) and the excitation energies are from ref. [34]).

zero the energy denominator $(\omega_0 - \omega_{ph})$ is relatively small. However, P2 and P3 enter with opposite sign so that the spurious state contribution to the effective interaction tends to cancel out. When the one-body diagrams are included (i.e. if we do not use the experimental ^{17}O levels so that the spectra are relative to ^{16}O) the effect can be as large as 0.1 MeV or more. When the one-body part of V_{eff} is taken from experiment (i.e. diagrams P2 and P3 contribute only in the folded terms which are two-body connected) the effect is quite small, 20 keV or less. Thus we can proceed to make valid comparisons between the different $\hat{Q}$-box approximations (calculations C1, C2, C3 and C4) even though the spurious state is not treated consistently (C1 and C2 include it while C3 and C4 remove it in lowest order).

Unperturbed single-particle energies ε were used in both the hole-hole and particle-hole phonon calculations. Our results are not to be compared with some of those to be found in the literature [see e.g. ref. [31])] which use the experimental single-particle energies for the ε. The Paris and Reid results are very similar with no surprises. The important thing is that both the hole-hole and particle-hole energies are significantly renormalized from their unperturbed values when we allow the two hole lines or particle and hole lines to interact. The resulting smaller denominators (i) make the $\hat{Q}$-box "larger" (i.e. $|P2| > |D2|$ and $|P3| > |D3|$), and (ii)

TABLE 4

Low-lying $A = 18$ levels

TJ	C1	C2	C3	C4
10	−10.59920	−10.71672	−10.35508	−10.46468
	−10.59920	−10.71670	−10.35502	−10.46462
	−10.59920	−10.727	−10.371	−10.490
2	−8.88340	−8.89500	−8.75340	−8.76300
	−8.88341	−8.89499	−8.75343	−8.76306
	−8.882	−8.894	−8.756	−8.766
4	−7.44860	−7.44860	−7.08640	−7.08640
	−7.44858	−7.44858	−7.08653	−7.08653
	−7.450	−7.450	−7.083	−7.083
01	−12.10408	−12.73150	−11.86850	−12.46800
	−12.10405	−12.73148	−11.86847	−12.46801
	−12.197	−13.223	−12.287	−13.786
3	−10.49140	−10.52120	−10.45800	−10.48412
	−10.49140	−10.52123	−10.45797	−10.48408
	−10.496	−10.526	−10.478	−10.506
5	−10.11280	−10.11280	−9.63784	−9.63784
	−10.11280	−10.11280	−9.63786	−9.63786
	−10.113	−10.113	−9.637	−9.637
2	−9.16022	−9.25420	−9.03048	−9.10274
	−9.16024	−9.25423	−9.03052	−9.10277
	−9.158	−9.252	−9.029	−9.101

All entries are in MeV and are relative to the ^{16}O ground state. In each case the three rows correspond to the SC, LS and F_m, respectively. The Paris potential were used with a starting energy of −10 MeV.

make the $\hat{Q}$-box more energy dependent, so that the folded diagrams become much more important. As we shall see, these two effects tend to cancel each other.

The sd-shell matrix elements of V_{eff} for $I = 1$, $J = 0$ are given in table 5 for calculations C1 through C4 for both the Paris and Reid potentials. As more higher-order processes are introduced into the $\hat{Q}$-box, V_{eff} becomes more non-hermitian. This is entirely due to the contribution of the folded terms.

Although the effective interaction obtained by this manner is not hermitian, the eigenvalues are all real. The non-hermitian matrix V_{eff} may be transformed to an hermitian form by use of a similarity transformation, although the columns of the transformation matrix will not be mutually orthogonal. Alternatively, the two-body interaction may be forced to be hermitian by the simple expedient of using for the matrix element V_{ik} the average of V_{ik} and V_{ki} [ref. [18])]. This method was used for the calculations described in the next section. The error introduced by this

TABLE 5

The sd-shell effective-interaction matrix elements for $T = 1$, $J = 0$

TJ	$abcd$	C1	C2	C3	C4
10	4444	−2.2360	−2.2925	−2.1822	−2.2464
		−2.0899	−2.1472	−1.9961	−2.0583
10	4455	−3.2670	−3.3960	−3.1239	−3.2296
		−3.1363	−3.2610	−3.2353	−3.3460
10	4466	−0.9096	−0.9304	−1.0031	−1.0227
		−0.8976	−0.9193	−0.9940	−1.0144
10	5544	−2.9536	−3.0050	−2.5074	−2.5609
		−2.8348	−2.8866	−2.5420	−2.5960
10	5555	−0.9409	−1.0253	−1.1018	−1.1576
		−0.8450	−0.9273	−0.9555	−1.0171
10	5566	−0.6505	−0.6669	−0.5517	−0.5663
		−0.6428	−0.6598	−0.5591	−0.5746
10	6644	−0.8596	−0.8801	−0.8594	−0.8809
		−0.8327	−0.8533	−0.8523	−0.8733
10	6655	−0.7021	−0.7357	−0.3381	−0.3584
		−0.6893	−0.7216	−0.4075	−0.4305
10	6666	−1.6478	−1.6575	−1.7779	−1.7863
		−1.4433	−1.4532	−1.5039	−1.5126

In each case the upper row is for the Paris potential and the lower for the Reid potential. The full non-hermitian V_{eff} is given. The orbits (4, 5, 6) represent respectively ($0d_{5/2}$, $0d_{3/2}$, $1s_{1/2}$). All entries are in MeV. In all cases, the Lee-Suzuki (LS) iteration method was used.

approximation is easily checked for the $A = 18$ case. The largest array occurs for $JT = 10$; in this instance the difference between the lowest eigenvalue of the hermitian and non-hermitian array is 0.4 keV when using the Paris potential. The greatest error for this JT is only 2.4 keV.

5. Results for several-particle systems

In this section we shall discuss results obtained using the newly derived matrix elements as shown in table 5 for several-nucleon systems. We feel this to be important for several reasons. First of all the matrix elements were derived to predict nuclear spectra, and their success or failure to do so can only be found by actual comparison with experiment. In cases where agreement is poor (and there shall be many such cases) one may perhaps obtain clues to where improvements may be made. A comparison between results obtained using the different approximations will be possible. Another point is that the matrix elements of table 5 are only the two-body parts of the model-space effective interaction. Their results for the spectra of several-nucleon systems may serve to indicate the potential importance of the many-body effective interactions. Finally, from a knowledge of the nature of the nuclear wave functions, one may establish the relative importance of the differing spin-orbit and tensor components of the two interactions.

Before discussing the results obtained with the Paris and Reid interactions, we briefly recall the four approximations used in obtaining the effective interaction:

C1: $\hat{Q}$-box with 2nd order diagrams;

C2: 2nd order plus the hole-hole phonon;

C3: 2nd order plus the particle-hole phonon;

C4: 2nd order plus hole-hole and particle-hole phonon.

For convenience we shall refer to the interaction obtained using the Paris interaction with approximation C1 as P1, etc.

5.1. CENTROIDS

A global assessment of an interaction may be obtained by calulating energy centroids of an interaction defined in jj-coupling by

$$\varepsilon_{ab} = \frac{\sum_{J,T} [J][T] \langle j_a j_b JT | V | j_a j_b JT \rangle}{\sum_{J,T} [J][T]}, \tag{5.1}$$

and in LS-coupling by

$$\varepsilon(\lambda\mu) = \frac{\sum [J][T] \langle (\lambda\mu) LSTJ | V | (\lambda\mu) LSTJ \rangle}{g(\lambda\mu)(2S+1)(2T+1)}. \tag{5.2}$$

In eq. (5.2) $(\lambda\mu)$ labels the possible SU(3) representations and $g(\lambda\mu)$ is the dimension of $(\lambda\mu)$. In the sd-shell the three possible SU(3) representations for two particles are (40) and (02) which are spatially symmetric and (21) which is spatially antisymmetric. The quantity $[k] \equiv 2k+1$.

If the two-body interaction V were written as

$$V = \sum_{k=0}^{2} (R^{(k)} \cdot S^{(k)}(\sigma_1 \sigma_2)), \tag{5.3}$$

then a tensor decomposition of two-body matrix elements is obtained from

$$\langle L\alpha || R^{(k)} || L'\alpha' \rangle \langle ST\beta || S^{(k)} || S'T'\beta' \rangle$$
$$= (-)^{L+S} \sum_J (-)^J \begin{Bmatrix} L & S & J \\ S' & L' & k \end{Bmatrix} \langle LST\alpha\beta J | V | L'S'T\alpha'\beta' J \rangle. \quad (5.4)$$

In eq. (5.4) α and β are any labels necessary to completely specify the matrix element; we shall take them to be irreducible representations of SU(3) and SU(4), respectively. We shall refer to the $k = 0$, 1 and 2 interactions as central, spin-orbit and tensor interactions, although their actual origin may be from other components.

By applying eq. (5.4) and then transforming the two-body matrix elements to *jj*-coupling, one may determine the amount of effective central, spin-orbit and tensor interactions contributing to each two-body matrix element. It is trivial to prove that non-central components of the interaction will not contribute to an SU(3) centroid; hence, an SU(3) energy centroid is a measure of the effective central interaction. An analogous statement is *not* true for a *jj*-coupling centroid.

In fig. 13 are shown contributions to the energy centroids $\varepsilon_1 = \varepsilon_{\frac{1}{2}\frac{1}{2}}$ and $\varepsilon_5 \equiv \varepsilon_{\frac{5}{2}\frac{5}{2}}$ ($\frac{5}{2}$ for $d_{\frac{5}{2}}$ and $\frac{1}{2}$ for $s_{\frac{1}{2}}$) for several interactions. Only a central or rank-zero interaction can contribute to ε_1.

The twelve interactions shown in fig. 13 include the eight interactions obtained in this work and four interactions which have been widely used in shell-model calculations. The interaction "KB" is the Kuo-Brown interaction [24]) and "K" is the Kuo interaction [35]), both derived from the Hamada-Johnston potential [4]). The interactions "K12" and "R" are interactions obtained in ref. [36]), namely "Kuo plus 12 free parameters" or K12fp and "RIP". The former interaction is essentially the Kuo interaction, but with those diagonal matrix elements involving only the $d_{\frac{5}{2}}$ or $s_{\frac{1}{2}}$ particles being varied to produce a better fit to spectra. The RIP interaction is a purely central interaction obtained by varying the values of radial integrals.

From fig. 13 one finds the spin-orbit (ε^{V}) and tensor (ε^{T}) contributions to ε_5 are small and of opposite sign with the obvious exception of RIP. Further, the magnitude is essentially the same for the eight interactions obtained in the present work. The contribution to the ε_5 of the central (ε^{C}) effective interaction does vary slightly from interaction to interaction and is appreciably different for the eight new interactions from the Kuo and K12fp interactions. Since K12fp was fit to *inter alia* binding energies, this suggests the new interactions may underbind nuclei.

The $s_{\frac{1}{2}}$ centroid, ε_1, does depend upon the original two-body interaction with the Paris potential producing more attraction as expected. However, again the three Kuo interactions are more attractive.

It is of interest to compare these results with those obtained for the $\hat{Q}$-box or unfolded interaction. The $s_{\frac{1}{2}}$ centroid is -3.71 MeV and -3.16 MeV for the Paris

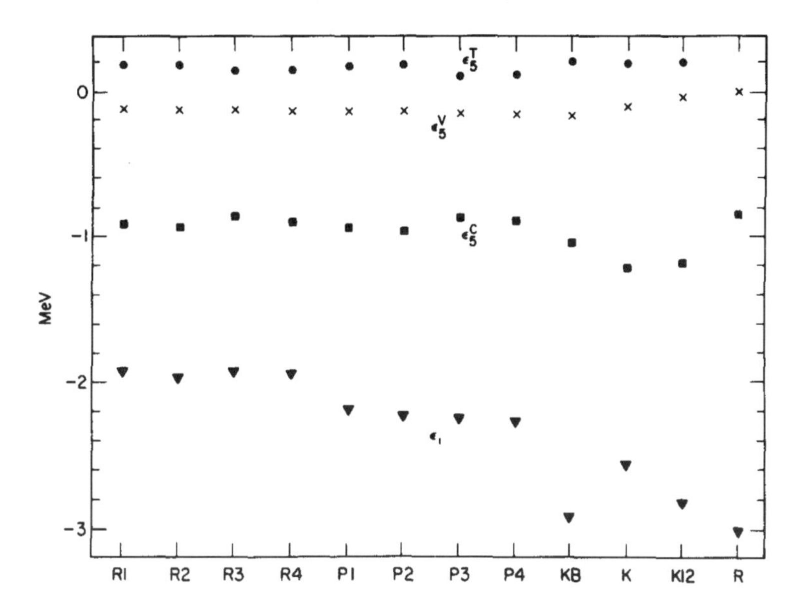

Fig. 13. The centroids of the $0d_{5/2}$ matrix elements, ε_5, and $1s_{1/2}$ matrix elements, ε_1 for 12 interactions. The quantities ε_5^C, ε_5^V and ε_5^T are centroids of the rank-0, rank-1 and rank-2 parts of the interaction, respectively. The four interactions R1, R2, R3 and R4 are derived from the Reid potential with the four approximations C1, C2, C3 and C4, respectively. The interactions derived from the Paris interaction are similarly labelled, P1, P2, P3 and P4. The remaining interactions are KB: Kuo-Brown [24]), K: Kuo [35]), K12: "Kuo+12 free parameters" and R: RIP [36]).

and Reid interaction, respectively. (These numbers are for the $\hat{Q}$-box with phonons, i.e. diagrams D4, D5, P6 and P7 and assuming a starting energy of $\omega_0 = -10$ MeV.) Hence, the inclusion of folded diagrams results in less attraction of 1.5 MeV! The contribution to ε_5 of the central, spin-orbit and tensor components of the Paris interaction is -1.41, -0.12, and 0.25 MeV and for the Reid interaction -1.42, -0.17 and 0.26 MeV. The difference between the two interactions is the effective two-body spin-orbit interaction. The main effect of including folded diagrams is on the central components: there is less attraction of approximately 0.5 MeV. The non-central contributions are almost unaffected.

In the beginning of the sd-shell, the SU(3) model provides a much more viable coupling scheme than does *jj*-coupling as well as providing some physical insight into the nuclear structure calculations. In fig. 14 are plotted energy centroids of the three SU(3) representations of two sd particles. As remarked above, the (40) and (02) are spatially symmetric, or belong to the SU(6) representation [2], and (21) belongs to [11]. The (21) centroid is positive (repulsive) as expected from phenomenological potentials.

Although no great differences were discernable in *jj*-coupling amongst the four methods of calculating an interaction, substantial differences do appear in SU(3).

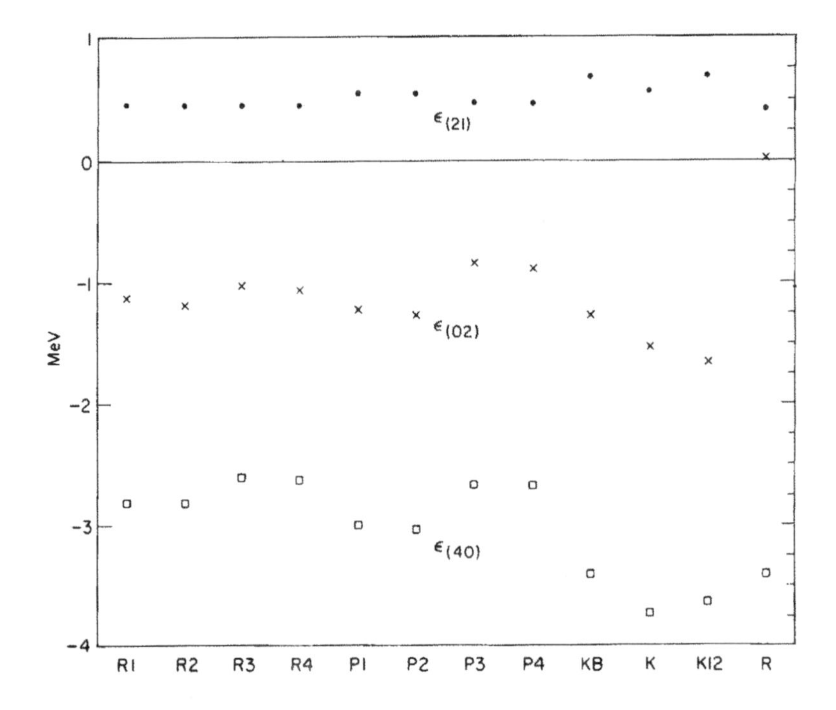

Fig. 14. Energy centroids of the three SU(3) representations occuring in the two-particle system in the sd shell. The labelling of the 12 interactions is explained in the caption of fig. 13.

Calculations C3 and C4 are for both two-body potentials less attractive than for C1 and C2. From the centroids one may establish that for the Paris interaction the inclusion of the hole-hole phonon produces an added attraction of 0.038, 0.049 and 0.042 MeV for (40), (02) and [2], respectively, and the particle-hole phonon produces less attraction of 0.355, 0.148 and 0.296 MeV. The hole-hole phonon does not affect [11] and the particle-hole phonon moves the [11] centroid up by 0.053 MeV. Similar, albeit smaller, numbers result from the Reid potential.

The effect of the 3p1h diagram in the original work of Kuo and Brown [24]) was to lower the 0^+ state and raise the 4^+ state of ^{18}O, thereby producing better agreement between theory and experiment. An examination of fig. 14 as well as the calculated spectra of ^{18}O in fig. 16 gives rise to an apparent paradox in that by adding the particle-hole phonon to the Q-box produces less attraction for both the spatially symmetric centroids and the ground state of ^{18}O. The behavior of the (40) energy centroid may be easily explained since the $J = 4$ matrix element enters with a weight of nine greater than for $J = 0$ and the $J = 4$ state is raised by the particle-hole renormalization.

However, the puzzle of less binding for the 0^+ of ^{18}O remains. In table 6 are listed diagonal matrix elements of the (40) and (02) representations; these two

TABLE 6

$(\lambda\mu)$	L	D7	P7	P1-P3
(40)	0	−0.983	−1.335	0.058
	2	−0.210	−0.267	0.420
	4	0.273	0.339	0.313
(02)	0	−0.432	−0.525	−0.079
	2	0.168	0.155	0.075
$C(40)$		0.038	−0.017	0.319
$C(02)$		0.031	0.028	0.099

The contributions to the diagonal two-body $T = 1$ matrix elements and SU(3) centroids arising from diagrams D7 and P7 of fig. 9b using the Paris potential and $\omega_0 = -10$ MeV (columns headed by D7 and P7) and the difference of the matrix elements obtained from interaction P1 and P3 (last column).

representations account for approximation 90% of the wave function of the low-lying states. The results are for the Paris potential. As in the earlier work the contributions from the simple 3p1h graphs are to lower the 0^+ by almost one MeV and to raise the 4^+ by 0.3 MeV. The results for the particle-hole phonon are even larger. The net effect on the centroids are virtually nil.

The diagonal (40) $L = 0$ matrix elements for the full $\hat{Q}$-box interaction (D4+D5+D6+D7) is −5.67 MeV and −6.02 MeV when D7 is replaced by P7. Renormalizing the interaction by including the folded diagrams produces an appreciably less attractive interaction: the matrix element becomes −4.14 and −4.08 MeV for P1 and P4, respectively. Further, the energy separation between the $L = 0$ and $L = 4$ states of the (40) representation is 2.36 MeV (P1) and 2.61 MeV (P4) rather than 3.20 MeV or 3.61 MeV for the $\hat{Q}$-box interaction containing D7 or P7, respectively. Thus, an effect of including folded diagrams is to obtain less binding energy and a more compact spectrum; much of the effect of the phonon on the $\hat{Q}$-box interaction disappears. A similar, though less dramatic, effect occurs for hole-hole phonon.

The above discussion may be summarized as follows. As one goes from C1 to C4 the strength of the $\hat{Q}$-box increases. When one also includes the folded terms in V_{eff} it is found that the matrix elements are typically weaker for C3 than C1, for example. For C4 the spectra obtained in lowest order (i.e. from $V_{\mathrm{eff}} = \hat{Q}$) are noticeably lower than the C1 spectra but when the folded terms are included in V_{eff} the C4 spectra are quite similar to those of C1, while the C3 spectra actually are less bound than the C1. Thus, when we include higher-order processes in the $\hat{Q}$-box the folded diagrams become more important.

The core renormalization due to 4p2h states was characterized in ref. [38]) as similar to a pairing interaction. The hole-hole phonon also exhibits some characteristics of pairing interaction in that [11] states are essentially unaffected

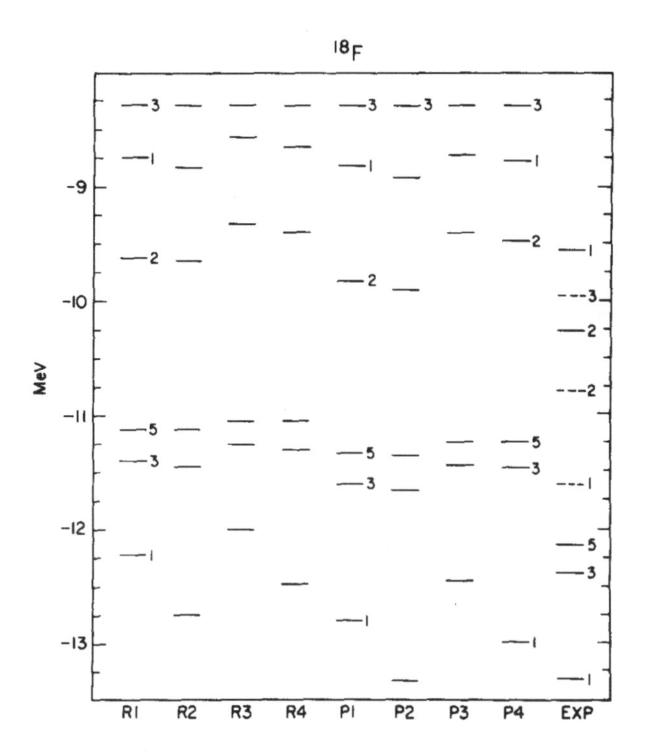

Fig. 15. Energy spectra of ^{18}F as calculated with the eight interactions derived in this work. The labelling of the eight interactions is explained in the caption of fig. 13.

and low angular momentum states of the two particle system are affected the most (see figs. 15 and 16). However, a pairing interaction would affect (40) more than (02) in the ratio of 5:1, unlike the hole-hole phonon.

The inclusion of the particle-hole phonon makes the relative separation of (40) and (02) centroids smaller – 0.207 MeV less for the Paris potential – which has unfortunate consequences in several-nucleon nuclei. Bands built from SU(3) representations other than the ground state or leading SU(3) representation will now lie at a lower excitation energy. The energy of an SU(3) centroid in a n-particle system is [39])

$$
\begin{aligned}
E(n(\lambda\mu)) = {} & (2n-n^2)E(20) \\
& +\tfrac{1}{36}(-15n+7n^2-3\langle C^{(4)}\rangle+2\langle C^{(3)}\rangle)E(40) \\
& +\tfrac{1}{18}(21n+n^2-3\langle C^{(4)}\rangle-\langle C^{(3)}\rangle)E(02) \\
& +\tfrac{1}{4}(n^2-5n+\langle C^{(4)}\rangle)E(21),
\end{aligned}
\tag{5.5}
$$

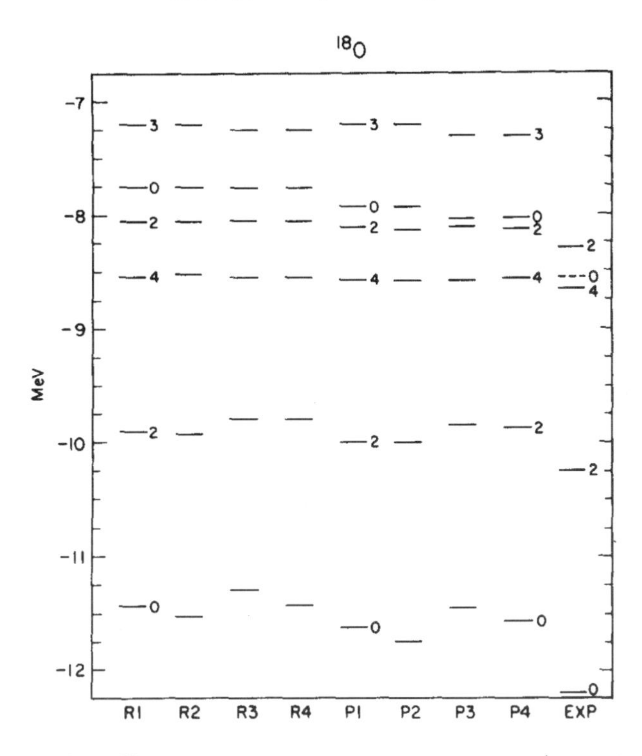

Fig. 16. Energy spectra of ^{18}O as calculated with the eight interactions derived in this work. The labelling of the eight interactions is explained in the caption of fig. 13.

with the expectation values of the Casimir operators $C^{(k)}$ of SU(k) given by [40])

$$\langle C^{(3)} \rangle = \lambda^2 + \mu^2 + \lambda\mu + 3(\lambda + \mu),$$

$$\langle C^{(4)} \rangle = \sum f_i^2 + 3(f_1 - f_4) + f_2 - f_3,$$

for representations $(\lambda\mu)$ and $[f]$, respectively. The SU(4) representation $[f]$ must be used in evaluating eq. (5.5) which is congruent to the SU(6) representation $[\tilde{f}]$.

From eq. (5.5) the energy difference between (80), the ground-state band of ^{20}Ne, and (42), the first excited band, is

$$E(80) - E(42) = \tfrac{7}{3}[E(40) - E(02)].$$

Hence, including the particle-hole phonon moves the two centroids 0.48 MeV closer together for the Paris potential. Though this may seem small, the two centroids are already too near each other as compared to experiment. Further, the changes will be larger in nuclei with more particles such as ^{28}Si.

A comparison with the centroids of the K12fp interaction is useful to predict possible consequences in heavier nuclei and, since the twelve parameters were varied to produce a better fit with experiment, will indicate some problems which will arise. The K12fp interaction has a separation of the (40) and (02) energy centroids 0.22 MeV larger than the C2 approximation to the Paris interaction (P2). There is also a larger separation between [2] and [11]. Hence, the interactions obtained in this work will have too many states of low energy and both SU(3) and SU(4) will be more badly violated than with K12fp. Further, the energy centroid of K12fp for [2] is lower representing more binding. Hence, we may anticipate the current interactions will badly underbind nuclei throughout the sd-shell.

The SU(3) centroids obtained for the $\hat{Q}$-box interaction suggests that results obtained using the $\hat{Q}$-box or unfolded interaction would be very similar to that obtained using the earlier Kuo or Kuo-Brown interactions. The centroids for the (40), (02) and (21) representations are -4.03, -1.71 and 0.46 MeV for the Reid interaction and -4.17, -1.73 and 0.54 MeV for the Paris interaction. (These results are for a $\hat{Q}$-box which includes phonons, i.e. diagrams D4, D5, P6 and P7 with $\omega_0 = -10$ MeV.)

These centroids are not appreciably different from those of the Kuo interaction. The inclusion of the folded diagrams results in less attraction for (40) of 1.35 MeV and 0.55 MeV for (02) whereas the (21) centroid is essentially unaffected. One may tentatively conclude that the folded diagrams do not affect the spatially antisymmetric states and the more correlated a state, the greater the effect of folded diagrams.

5.2. SPECTRA

We now turn to a consideration of the predictions of the matrix elements in nuclei. Although all nuclei with A equal to 18 to 22 and ^{24}Mg were calculated, we shall for brevity present results only for the more illuminating cases. In all cases the single-particle energies were taken from experiment. The calculations were performed using a modified version of the Glasgow shell-model code[41]).

In fig. 15 are shown results for ^{18}F and experiment. Although many of the features anticipated from an examination of the centroids appear, the spectra are obviously more detailed. The inclusion of the phonons affect the low-spin states (1^+ and 2^+) more than the 3^+ and 5^+; in particular, the second 3^+ level is essentially unaffected. Although the P2 interaction reproduces the ground-state binding energy, the other interactions do rather worse. All interactions fail to produce the proper energy for 3_1^+ and 5_1^+, although P3 reproduces the excitation spectrum reasonably well. Since these energies enter the evaluation of the centroid – and position of levels in heavier nuclei – with more weight than the 1^+, the expectations based on the centroids appear valid.

In fig. 16 are shown results for ^{18}O. Again, the limits provided by the centroids are validated by the detailed calculation of structure. The Paris and Reid interactions do not provide sufficient binding although the Paris interaction is somewhat stronger than the Reid. Again, the higher-spin states appear to be relatively unaffected by the inclusion of phonons. Although the 4_1^+ level is correctly predicted, the 0_1^+ and 2_1^+ are too high, resulting in a compressed spectrum. This will also occur in heavier nuclei.

In fig. 17 are results for ^{19}F. The wave functions of the ground-state band are dominated by the SU(3) representation (60) and have S equal to one-half. Hence, as in ^{18}O, the energies depend primarily on the central interactions with small contributions from the spin-orbit interaction.

The error in the predicted binding energy ranges from approximately 1 MeV for the P2 interaction to 2.7 MeV for the R3 interaction. Inclusion of the hole-hole phonon not only shifts the centroids down, but also expands the spectrum; the position of the $\frac{13}{2}^+$ is essentially unaffected by inclusion of phonons. However, the particle-hole phonon tends to contract the spectrum. The final result with both hole-hole and particle-hole phonons included produces a spectrum which has the correct level ordering but with insufficient binding.

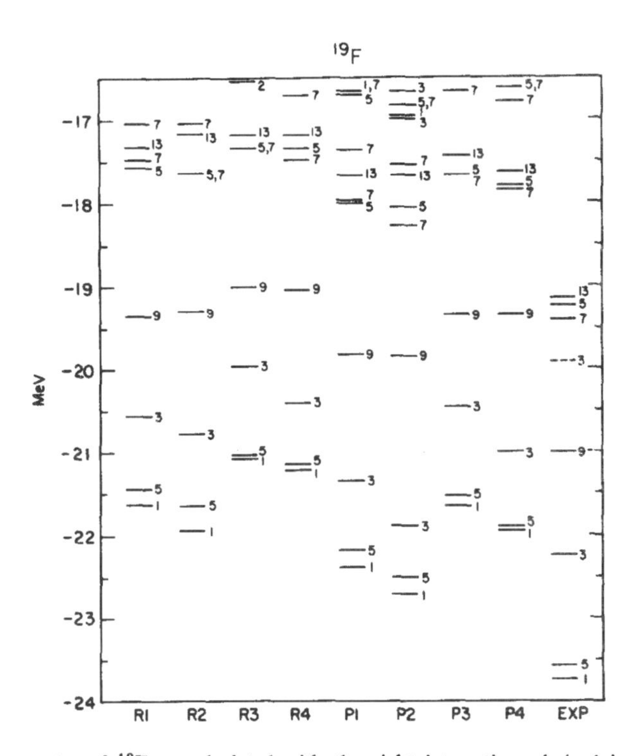

Fig. 17. Energy spectra of ^{19}F as calculated with the eight interactions derived in this work. The labelling of the eight interactions is explained in the caption of fig. 13. The dashed line represents a $\frac{3}{2}^+$ state dominantly 7p4h.

The calculation of levels in ^{20}O is of interest because only the $T = 1$ interaction may contribute. The results are not shown. The second 0^+ and 2^+ states both lie about one MeV lower than the experimental position. Again the spectrum is too compressed and the amount of binding is about one MeV too little.

The calculated spectra of ^{20}Ne may be found elsewhere[42]). Briefly, deficiencies found in other nuclei also appear in ^{20}Ne: there is too little binding energy, 3.2 MeV for P1 and 4.8 MeV for R1; the spectrum is too compressed with the 8^+ appearing 0.5 MeV for P1 and 1.4 MeV for the R1 interaction too low relative to the ground state; the (42) excited band lies at too low an energy relative to the ground state.

The tensor interaction of the Paris potential is appreciably weaker than that of the Reid interaction, but the contribution from the tensor components to the $\hat{Q}$-box two-body matrix elements are only 2% larger for the Reid potential than for the Paris potential. However, the process of including the folded diagrams may alter this situation: the many angular momentum recouplings and matrix products may cause the effective non-central components to change or conceivably even disappear. Since it is in general difficult to choose an appropriate weighting of the many two-body matrix elements (recall the non-central components do not contribute to SU(3) energy centroids), the expectation value was evaluated between several wave functions generated for mass 20 nuclei using the C1 interactions. The non-central components were obtained using eq. (5.4). The results are given in table 7.

The lowest $T = 0$ wave functions are dominated by $S = 0$ components for which the non-central forces cannot contribute. However, the lowest $T = 1$ wave functions have $S = 1$ and both the two-body spin-orbit and tensor force may contribute. An examination of table 7 demonstrates several points. First, the effective tensor – and also spin-orbit – force of the interaction obtained from the Reid potential is not very different from that of the Paris potential. This is somewhat surprising. Second, the contribution from the two-body effective spin-orbit interaction is appreciably larger than that from the effective tensor interaction. Third, in most instances the effective tensor interaction is repulsive. Fourth, essentially all the differences between results obtained using the two interactions arise from the effective central interaction. Finally, the contribution from the non-central components to the energy of the $T = 0$, $S = 0$ states is approximately the same as that to the $T = 1$ states.

As remarked in subsect. 5.1 the inclusion of the particle-hole phonon and then folding results in less binding, even for the low-spin states, contrary to expectation based on previous interactions which included only the $\hat{Q}$-box with no folded diagrams. In table 7 are also listed the contributions to the energy of mass-20 levels of three diagrams D4, P6 and P7. The quantities listed are the contribution to the interaction from only the $\hat{Q}$-box and does not include any effects such diagrams have on the interaction after including folded diagrams.

TABLE 7

Contributions to the energy of $T = 0$ and $T = 1$ states in mass 20 from specified parts of the interaction

J	C	V	T	D4	P6	P7
$T = 0$						
0_1	−24.68	−0.24	−0.07	−26.76	−2.71	−1.37
	−23.01	−0.19	−0.05	−25.55	−2.88	−2.29
0_2	−18.21	−0.16	0.15	−20.18	−2.01	−2.13
	−16.82	−0.12	0.20	−19.00	−1.91	−1.41
2	−22.97	−0.33	0.05	−25.45	−1.90	−0.80
	−21.52	−0.27	0.07	−24.39	−2.23	−2.03
4	−20.60	−0.41	−0.03	−23.53	−1.58	−0.39
	−19.27	−0.37	−0.02	−22.59	−1.94	−1.27
6	−15.91	−0.65	0.16	−19.76	−0.13	0.59
	−14.99	−0.61	0.18	−18.99	−0.62	−0.48
$T = 1$						
1	−13.52	−0.88	0.06	−16.10	−1.36	0.23
	−12.60	−0.86	0.07	−15.48	0.40	−0.69
2	−13.97	−0.89	0.09	−16.64	−1.07	−0.07
	−13.04	−0.82	0.11	−15.93	0.08	−0.91
3	−13.33	−0.83	0.26	−16.10	−0.91	0.21
	−12.39	−0.80	0.29	−15.34	0.45	−0.80
4	−12.84	−0.77	0.36	−15.31	−1.20	−0.02
	−11.97	−0.72	0.38	−14.60	0.06	−0.55

The columns labelled "C", "V" and "T" contain the expectation values of the effective centroid, spin-orbit and tensor interactions; the columns headed "D4", "P6" and "P7" contain the expectation values of diagram D4, P6 and P4 of fig. 9. The wave functions used were generated using the C1 interaction with ^{17}O single-particle energies. For each J the first line contains results for the Paris interactions and the second for Reid interactions.

From the table it is seen the contribution from the particle-hole phonon diagram (P7) is much larger for the Reid interaction than for the Paris interaction. This should be related to the relative strength of the tensor force of these two interactions. For both interactions the effect is to lower the energy of the low-spin states and increase the energy of the high-spin state. This reinforces our earlier conclusion that the apparent repulsion resulting from the inclusion of the particle-hole phonon was a manifestation of the folded diagrams.

One may also investigate the extent the hole-hole phonon contribution simulates a pairing interaction. One expects the pairing interaction to be attractive and, in ^{20}Ne it can contribute to states of $J \leqq 4$. From table 7 one observes the contribution from P6 obeys these expectations.

Finally, we note that if one were to take as the interaction just the contributions from D4 and P7 using a starting energy of $\omega_0 = -10$ MeV [similar to the original work of Kuo and Brown [24])], agreement with experiment would be quite good. Using the P1 interaction, the calculated binding energy would be in error by only

70 keV and the excitation energy of the 8^+ is only 0.5 MeV too small. The R1 interaction underbinds ^{20}Ne by 170 keV but produces an 8^+ within 100 keV of experiment.

In fig. 18 are results for ^{22}Na. Since ^{22}Na is odd-odd, the low-lying states have $S = 1$ and all parts of the interaction may contribute. The level structure of ^{22}Na has routinely defied successful calculation with both phenomenological interactions [43]) and realistic interactions [36]). The first 1^+ (and sometimes also the second 1^+) states tend to lie lower than the experimental 3^+ ground state. The present calculations have the same failings. Interactions which are fit to the structure of mass-22 nuclei can produce the correct ordering by either using a repulsive triplet-even tensor interaction [44]) or by varying a number of two-body matrix elements [37]). Because of the number of matrix elements in the latter approach, it is difficult to establish the physical origin of the discrepancy.

From fig. 18 it is apparent the spectra is too compressed. Experimentally there are seven levels below 3 MeV excitation; the R1 interaction produces 12 levels and P4 produces 11 levels below this energy. The 2^+ level appears 1.5 to 2 MeV too low.

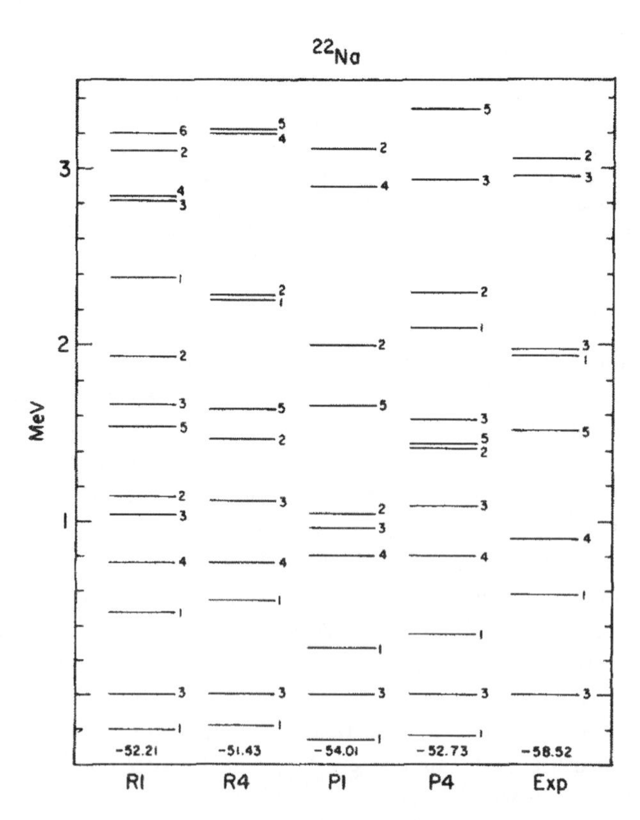

Fig. 18. Energy spectra of ^{22}Na calculated with four of the interactions derived in this work.

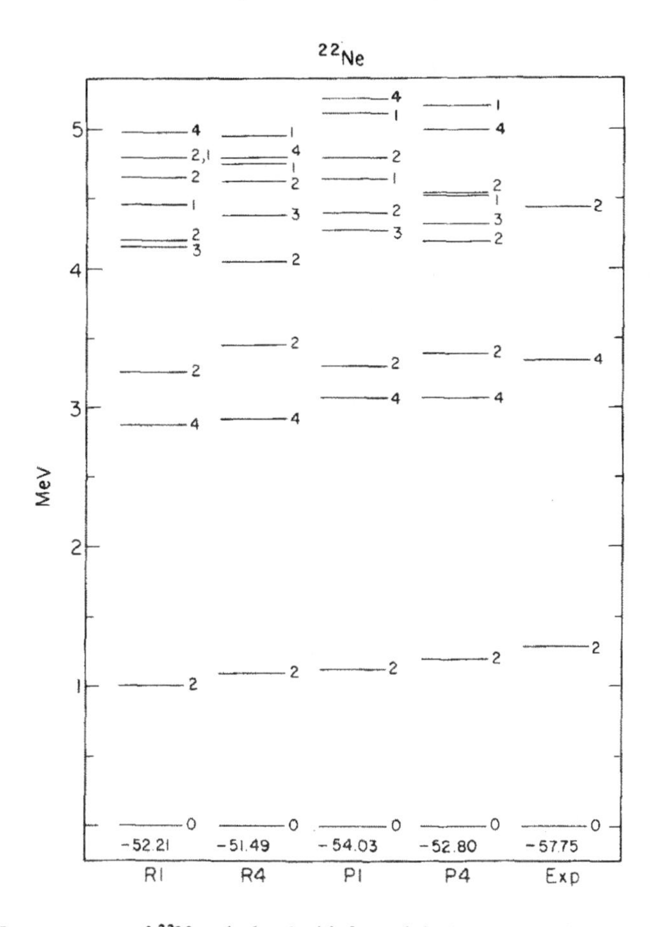

Fig. 19. Energy spectra of ^{22}Ne calculated with four of the interactions derived in this paper.

In fig. 19 are shown results for ^{22}Ne. The low-lying states are dominated by (82) $S = 0$. Again the spectra is too compact with the 2_1^+ and 4_1^+ having too small an excitation energy and the 6_1^+ which is not shown, even worse, although the excitation spectrum for the 2_1^+ and 4_1^+ states is acceptable. The most obvious failure is the position of the 2_2^+ which is a member of the (82) $K = 2$ band. Its incorrect position is now almost traditional [36,43]. Although its energy may be corrected by explicit inclusion of pf excitations [45], interactions may be found which do predict its energy correctly [37,42] although the physical origin of the modifications is not known. Modifying the matrix elements which violate S – which are substantial in the present interactions – do not substantially modify the results [46].

Finally, we present results for ^{24}Mg. The amount of underbinding is substantial, ranging from 5.5 MeV for the P1 interaction to over 10 MeV for R4. The $K = 2$

band again appears much too low. As predicted by the centroids, the excitation energy of the 0_2^+ is too low when the particle-hole phonon is included.

6. Discussion and conclusion

The aim of this paper was to perform within the framework of the shell model a microscopic nuclear-structure calculation, the starting point of which was the free NN interaction. The first step in this procedure, the calculation of the G-matrix, is probably quite sound. The methods used, momentum space matrix inversion to obtain G_F and the Tsai-Kuo method to treat that part of G which depends on the Pauli operator Q, are very accurate. The second step is the calculation of the $\hat{Q}$-box, and it is in this step that we have to make some approximations. Four approximation schemes, denoted as C1, C2, C3 and C4 have been used. We will return to discuss them later. In the final step, we obtain V_{eff} by "folding" the $\hat{Q}$-boxes together, i.e. evaluating the folded diagram $\hat{Q}$-box series of V_{eff}. For a given $\hat{Q}$-box, this step can be quite accurately carried out by using the Lee-Suzuki iteration method. This method was shown to yield essentially exact results after only a small number of iterations. We have found that the effect of higher-order $\hat{Q}$-box diagrams is suppressed significantly by the inclusion of the corresponding folded diagrams. Generally speaking, the effect of the folded diagrams is quite significant in the calculation of V_{eff}.

Another aim of these calculations was to see how the Paris potential would fare in finite nuclei. To our knowledge, this is the first microscopic calculation of finite open-shell nuclei for which the Paris potential was used. Our results indicate that the Paris potential is as good as, if not better than, the Reid potential. Indeed, in all nuclei we considered, the Paris potential seemed to give better, albeit inadequate, ground-state binding energy. However, in most cases differences in spectra produced with or without the inclusion of phonons were as great as those between the Paris and Reid potential.

We demonstrated that in mass-20 nuclei the contributions to the energy from the effective non-central components of the Reid and Paris interaction were essentially the same. Differences in spectra and binding energies arose mainly from the effective central interaction, although there were appreciable differences in the contribution to the energies from diagrams P6 and P7. However, the Reid interaction tended to produce wave functions with a slightly larger occupation of the $d_{\frac{3}{2}}$ than did the Paris interaction.

The third aim of these calculations was to do a systematic study of the contribution of the folded diagrams to V_{eff}. It is quite clear that the inclusion of the folded diagrams have a much greater effect on the $A = 18$–24 spectra than do various infinite-order partial summations (hole-hole and particle-hole phonons) that this calculation and others include. Until now, it has been practically taken for granted that the inclusion of TDA (or RPA, for that matter) phonons would give increased binding. Our calculations have shown that when the folded diagrams are included, this may not be the case. The C3 spectra (for which the TDA particle-hole phonons were included) show appreciably less binding than do the C1 spectra (for which we included no phonons). By explicitly calculating the expectation value of the particle-hole diagram for mass-20 states, we demonstrated this apparent anomaly was a result of the folded diagrams.

Why did the early calculations, starting with those of Kuo and Brown, reproduce the experimental spectra as well as they did, even though all diagrams above second order were ignored? The answer suggested by the current work seems to be that although the $\hat{Q}$-box they used contain only some low-order diagrams, the effect of the higher-order $\hat{Q}$-box diagrams would have been cancelled in large part if the folded diagrams were included in V_{eff} as well. This is because the higher-order diagrams have a relatively greater energy dependence so that they significantly renormalize the $\hat{Q}$-box interaction. Thus if higher-order diagrams are included (in the basic $\hat{Q}$-box), the folded diagrams must be included as well (in the expansion for V_{eff}). Indeed, using an effective interaction defined by the sum of contributions from diagram D4, the bare G-matrix, and P7, the particle-hole phonon, of fig. 9, with no folded diagrams included, much better agreement with experiment was obtained.

When compared with experiments the excitation spectra of several nuclei are reproduced reasonably well but our calculated binding energies are generally too small by several MeV, especially for nuclei in the middle of the sd-shell. It is not clear how to remedy this, and will be an interesting problem for further study. Several directions may be pursued. In the present work we have included only the two-body components of V_{eff}. V_{eff} clearly has many-body components. For example, ^{19}O will have 3-body and ^{22}Ne should have 6-body effective interactions. Calculation of such many-body components is rather complicated.

In fig. 20 we give some representative diagrams which contribute to 3-body effective interactions. Diagram (i) belongs to the non-folded term $\hat{Q}$ of eq. (2.9); it originates from our model-space restriction that nucleons are confined to certain low-lying shell-model states. (Note that a railed line represents a particle outside the model space.) Diagram (ii) is a 3-body effective interaction diagram due to non-nucleonic degrees of freedom. Here nucleons 1 and 2 interact by exchanging a pion, exciting nucleon 2 into an isobar. Then it is deexcited back to a nucleon. Three-body forces can also arise from folding two 2-body irreducible boxes together, as shown by diagram (iii). Polls *et al.*[47]) have recently investigated the contribution of these types of diagrams to the three-body effective interaction for the sd-shell nuclei using the Reid [5]) and Jülich-Bonn [48]) NN potentials. They found that the explicit inclusion of Δ degrees of freedom (diagram (ii) of fig. 20) has rather a negligible contribution to the three-body effective interaction. But the three-body effective interaction due to the truncation of the shell-model space

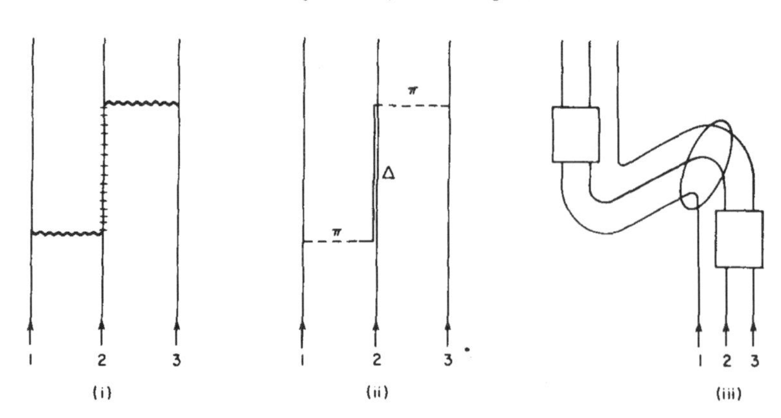

Fig. 20. Some of the diagrams which contribute to the three-body effective interaction.

(diagrams (i) and (iii) is significant, although not overwhelmingly large. They give reduced binding energy of approximately 0.5 MeV to the $A = 19$ nuclei. [A similar result has been obtained by Huang *et al.* [49]) who performed a folded-diagram calculation for the low-lying states of ^{19}O using mainly the M3Y [50]) potential as the underlying two-body effective interaction without considering the effect due to the Δ-degrees of freedom.]

Another possibility to improve the present calculation is to include still more irreducible diagrams in the $\hat{Q}$-box than were included in the present work. In fact, one would like at this point to have the exact $\hat{Q}$-box. One possible step toward this goal would be the inclusion of the 3rd order diagrams in the $\hat{Q}$-box. Such a calculation is facilitated by the new techniques [23]) developed to evaluate angular-momentum-coupled diagrams. However, the net result of a systematic 3rd order calculation is not clear. Without the folded diagrams the spectra would presumably show less binding than for 2nd order. [See e.g. ref. [25]).] However, the derivative (slope) of the 3rd order contribution would also be positive, and moreover, large compared to the 2nd order contribution so that the net result upon inclusion of the folded diagrams is not at all clear.

By solving a set of self-consistent integral equations, Chakravarty *et al.* [51]) have calculated several large classes of $\hat{Q}$-box diagrams to all orders. It is interesting that their results of the $\hat{Q}$-box interaction for the 1s0d shell are generally not very different from ours with the C4 approximation. They have not calculated the folded $\hat{Q}$-box series for V_{eff}. It would be of interest to do so to compare the V_{eff} so obtained with ours.

Contributions from highly excited states to the $\hat{Q}$-box due to the tensor interaction, the so-called Vary-Sauer-Wong effect [51]), were not evaluated in the present work. However, recent work [52]) using the Bonn potential [48]) showed that two-body potentials with a weak tensor interaction give rise to a negligible Vary-Sauer-Wong effect. The Paris potential has a weak tensor component.

We should like to thank J. M. Richard for providing the routines used to evaluate the Paris potential.

This work was supported in part by the DOE under contract DE-AC02-76ER13001.

Note added in proof: The coupled-cluster method of Kümmel and his collaborators has been recently extended and applied to open-shell nuclei [54-46]. It will be of interest to compare their method and results with those of the present work. We are very grateful to Prof. Kümmel for discussions in this regard.

References

1) T. T. S. Kuo, Ann. Rev. Nucl. Sci. **24** (1974) 101, and references therein
2) P. J. Ellis and E. Osnes, Rev. Mod. Phys. **49** (1977) 777, and references therein
3) G. Breit, Rev. Mod. Phys. **34** (1962) 776;
K. E. Lassila *et al.*, Phys. Rev. **126** (1962) 881
4) T. Hamada and I. D. Johnston, Nucl. Phys. **34** (1962) 382
5) R. V. Reid, Ann. of Phys. **50** (1968) 411
6) R. Vinh Mau, in Mesons in nuclei, ed. M. Rho and D. Wilkinson (North-Holland, Amsterdam, 1979) p. 151
7) M. Lacombe *et al.*, Phys. Rev. **C21** (1980) 861
8) K. A. Brueckner, C. A. Levinson and H. M. Mahmoud, Phys. Rev. **95** (1954) 217
9) H. A. Bethe, Phys. Rev. **103** (1956) 1353; Ann. Rev. Nucl. Sci. **21** (1971) 93;
H. A. Bethe and J. Goldstone, Proc. Roy. Soc. **A238** (1957) 551
10) J. Goldstone, Proc. Roy. Soc. **A239** (1957) 267
11) K. T. R. Davies, M. Baranger, R. M. Tarbutton and T. T. S. Kuo, Phys. Rev. **177** (1969) 1519
12) H. Feshbach, Ann. of Phys. **19** (1962) 287
13) C. Bloch and J. Horowitz, Nucl. Phys. **8** (1958) 51
14) T. Morita, Prog. Theor. Phys. **29** (1963) 351
15) G. Oberlechner *et al.*, Nuovo Cim. **B68** (1970) 23
16) M. B. Johnson and M. Baranger, Ann. of Phys. **62** (1971) 172
17) T. T. S. Kuo, S. Y. Lee and K. F. Ratcliff, Nucl. Phys. **A176** (1971) 65
18) B. H. Brandow, Rev. Mod. Phys. **39** (1967) 771; Int. J. Quantum Chem. **15** (1979) 207
19) T. T. S. Kuo and E. M. Krenciglowa, Nucl. Phys. **A342** (1980) 454
20) E. M. Krenciglowa and T. T. S. Kuo, Nucl. Phys. **A235** (1974) 171
21) S. F. Tsai and T. T. S. Kuo, Phys. Lett. **39B** (1972) 427
22) E. M. Krenciglowa, C. L. Kung, T. T. S. Kuo and E. Osnes, Ann. of Phys. **101** (1976) 154
23) T. T. S. Kuo, J. Shurpin, K. C. Tam, E. Osnes and P. J. Ellis, Ann. of Phys. **132** (1981) 237
24) T. T. S. Kuo and G. E. Brown, Nucl. Phys. **A85** (1966) 40
25) B. R. Barrett and M. W. Kirson, Nucl. Phys. **A148** (1970) 145
26) S. Y. Lee and K. Suzuki, Phys. Lett. **91B** (1980) 173; Prog. Theor. Phys. **64** (1980) 2091
27) G. E. Brown, A. D. Jackson and T. T. S. Kuo, Nucl. Phys. **A133** (1969) 481
28) J. Shurpin, D. Strottman, T. T. S. Kuo, M. Conze and P. Manakos, Phys. Lett. **69B** (1977) 395;
J. Shurpin, H. Muther, T. T. S. Kuo and A. Faessler, Nucl. Phys. **A293** (1977) 61
29) H. P. Stapp, T. J. Ypsilantis and N. Metropolis, Phys. Rev. **105** (1957) 302
30) T. T. S. Kuo and E. Osnes, Many-body theory for nuclear structure, monograph in preparation
31) E. Osnes, T. T. S. Kuo and C. S. Warke, Nucl. Phys. **A168** (1971) 190
32) T. T. S. Kuo and E. Osnes, Nucl. Phys. **A226** (1974) 204
33) A. H. Wapstra and K. Bos, Atom. Nucl. Data Tables **19** (1977) 175
34) F. Ajzenberg-Selove, Nucl. Phys. **A300** (1978) 1
35) T. T. S. Kuo, Nucl. Phys. **A103** (1967) 71

36) E. C. Halbert, J. B. McGrory, B. H. Wildenthal and S. P. Pandya in Advances in nuclear physics, vol. 4, ed. M. Baranger and E. Vogt (Plenum, New York, 1970)
37) B. M. Preedom and B. H. Wildenthal, Phys. Rev. **C6** (1972) 1633
38) T. T. S. Kuo and G. E. Brown, Nucl. Phys. **A92** (1967) 481
39) J. C. Parikh, Phys. Lett. **41B** (1972) 468
40) S. I. So and D. Strottman, J. Math. Phys. **20** (1970) 153
41) R. R. Whitehead, A. Watt, D. J. Cole and I. Morrison, in Advances in nuclear physics, vol. 10, ed. J. Negele and E. Voyt (Plenum, New York, 1938)
42) J. Shurpin, Ph. D. thesis, State University of New York at Stony Brook (1980), unpublished
43) Y. Akiyama, A. Arima and T. Sebe, Nucl. Phys. **59** (1964) 1; **85** (1966) 184
44) N. Anyas-Weiss and D. Strottman, Nucl. Phys. **A306** (1978) 201
45) A. Arima and D. Strottman, Nucl. Phys. **A162** (1971) 605
46) M. Conze, H. Feldmeier and P. Manakos, Phys. Lett. **43B** (1973) 101; private communication
47) A. Polls, H. Müther, A. Faessler, T. T. S. Kuo and E. Osnes, preprint (University of Tubingen, 1982)
48) K. Holinde, R. Machleidt, M. R. Anastasio, A. Faessler and H. Müther, Phys. Rev. **C18** (1978) 870
49) W. Z. Huang, H. Q. Song, Z. X. Wang and T. T. S. Kuo, Chinese J. Nucl. Phys., in press (1982)
50) G. Bertsch, J. Borysowich, H. McManus and W. G. Love, Nucl. Phys. **A284** (1977) 399
51) S. Chakravarty, P. J. Ellis, T. T. S. Kuo and E. Osnes, Phys. Lett. **109B** (1982) 141
52) J. P. Vary, P. U. Sauer and C. W. Wong, Phys. Rev. **C7** (1973) 1776
53) H. M. Sommermann, H. Müther, K. C. Tam, T. T. S. Kuo and A. Faessler, Phys. Rev. **C23** (1981) 1765
54) R. Offerman, W. Ey and H. Kümmel, Nucl. Phys. **A273** (1976) 349
55) W. Ey, Nucl. Phys. **A296** (1978) 189
56) J. G. Zabolitzky and W. Ey, Nucl. Phys. **A328** (1979) 507

Nuclear Physics **A401** (1983) 124–142

THREE-BODY FORCES IN sd-SHELL NUCLEI

A. POLLS[†], H. MÜTHER and A. FAESSLER
Institut für Theoretische Physik, Universität Tübingen, D-7400 Tübingen, W. Germany

T. T. S. KUO
Physics Department, State University of New York at Stony Brook, Stony Brook, NY, 11794, USA

and

E. OSNES
Institute of Physics, University of Oslo, Norway

Received 17 November 1982

Abstract: The influence of three-body forces on the excitation spectra of nuclei with 3 valence nucleons in the sd-shell is investigated. Three-body forces are considered, which arise from an intermediate excitation of the interacting nucleons to the $\Delta(3, 3)$ resonance. Besides these real three-nucleon forces, effective three-body interactions are taken into account which are due to the restriction of the nuclear structure calculation to sd-shell configurations. Significant cancellations are observed between the different contributions to the effective three-nucleon force. The resulting three-body matrix elements yield only a small influence on the spectrum of the $A = 19$ systems. The typical size of the matrix elements, however, is large enough to expect a serious influence on the results of shell-model calculations with more than three valence nucleons.

1. Introduction

During the last few years it has been increasingly realized that non-nucleonic degrees of freedom can be important in microscopic nuclear structure calculations. This means that for certain investigations on the structure of nuclei the nucleons cannot be considered as elementary particles, but effects, which are due to possible excitations of the nucleons for example to the $\Delta(3, 3)$ resonance have to be taken into account. As an example of the importance of such isobar degrees of freedom one may think of the attempts to calculate the binding energy of nuclear systems microscopically. Here one realized that the attractive components of the nucleon-nucleon (NN) interaction at medium range are to a considerable extent due to intermediate excitation of the interacting nucleons to the $\Delta(3, 3)$ resonance. A corresponding diagram is displayed in fig. 1a. If now such terms are treated

[†] Supported by the DFG.

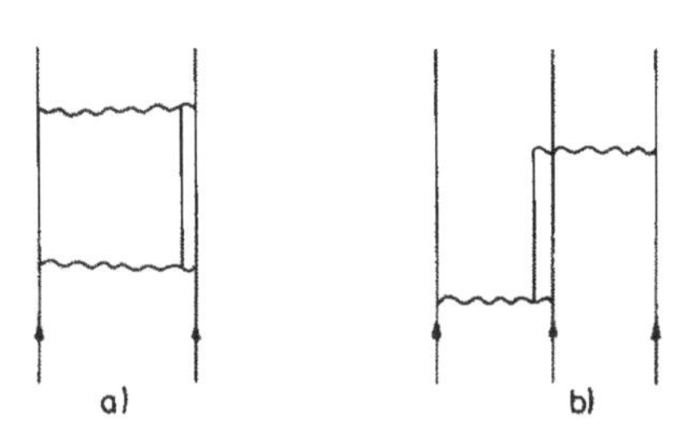

Fig. 1. Isobar contributions to the effective 2N (a) and 3N interaction (b).

explicitly in a nuclear structure calculation, one finds that the attractive components of these isobar terms are reduced in a nuclear medium since the intermediate nucleon of fig. 1a can only be scattered into a state above the Fermi surface and also because the energy denominators in calculating these contributions are typically larger in a nuclear medium than in the vacuum. This quenching of the effective NN interaction has been investigated by several groups in calculating binding energies of nuclear matter [1-4]) and finite nuclei [4-6]).

If, however, isobar degrees of freedom are taken into account in a nuclear structure calculation, one should not only consider their effect on the NN interaction but also realize, that there exist terms with intermediate isobar excitations involving three nucleons (see fig. 1b). From the point of view of classical nuclear physics, which does not allow for excitations of the nucleons, such contributions can be incorporated by assuming an empirical three-body force. The contribution of such three-body forces to the binding energy of nuclei has been studied [6-9]) and it was found that they yield attraction, which is comparable in size with the contributions from three-nucleon correlations [10,11]). This shows that three-body forces of this kind can have a significant influence on nuclear structure calculations. The evaluation of binding energies alone, however, gives only very global information on the structure of the three-body force to be studied. Therefore it is the aim of this work to investigate the influence of three-body forces, which are determined microscopically, on the results of shell-model calculations.

The effective hamiltonian to be used in such shell-model calculations not only contains a "real" three-body force as discussed above. The restriction of the nuclear structure calculation to the shell-model configurations of some valence nucleons outside an inert core makes it necessary to renormalize the hamiltonian for the valence nucleons. With such a renormalization one tries to incorporate in a perturbative way the effects of the configurations which are not treated in the shell-model calculation explicitly. This procedure leads to an effective two-body interaction between the valence nucleons which has widely been discussed in the literature [see e.g. refs. [12-14])]. The renormalization of the hamiltonian, however, also yields effective three-body forces for nuclei with three or more valence nucleons. The diagrams for the lowest order contributions are displayed in fig. 2.

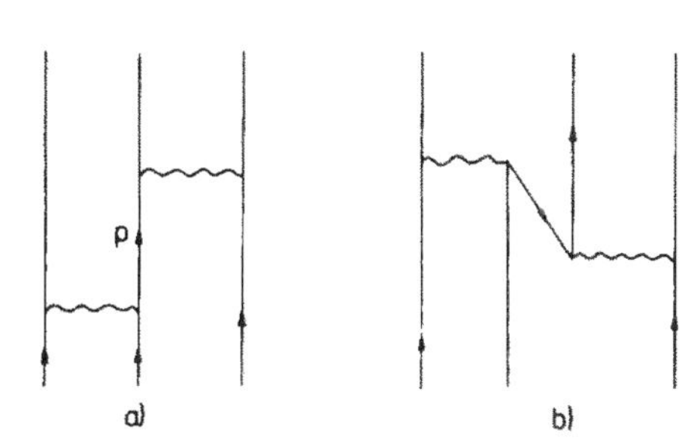

Fig. 2. Lowest order contribution to an effective three-body force between valence nucleons in a shell-model calculation. Since the intermediate-particle line of fig. 2a stands for a state above the valence shell, the diagram of fig. 2a accounts for excitations of one of the valence nucleons to higher shells, whereas the diagram of fig. 2b represents an intermediate core excitation.

Such effective three-body forces have been calculated already for the p-shell [15,16]), the $d_{\frac{5}{2}}s_{\frac{1}{2}}$ shell [31]) and the $f_{\frac{7}{2}}$ shell [17,24,32,33]), assuming oscillator states for intermediate particle spectrum and summing all contributions up to 6 $\hbar\omega$ [ref. [16])] or even up to 0 $\hbar\omega$ [refs. [17,31])] only. From investigations on the convergence of the renormalized effective two-body interaction one knows that such a restriction to low-lying particle states can yield misleading results [18,19]). Therefore, in this work a technique is used which sums over all particle states in a plane-wave basis orthogonalized to the hole and valence states [19,20]). As a final remark in this introduction we could like to discuss the question of the experimental evidence for three-body forces in a shell-model analysis. One may argue, that it is possible to adjust an effective hamiltonian, which contains one- and two-body terms only, in such a way, that a diagonalisation of this hamiltonian in a shell-model basis yields excellent agreement with experimental data [21]). This may be interpreted as an indication that there is no need for three-body forces. If one uses, however, the Preedom-Wildenthal interaction, fitted to $A = 17$–29 nuclei, to estimate the binding energy of ^{40}Ca, an overbinding of about 30 MeV is obtained [22]). This could be corrected by a repulsive three-body force [23]) with an average matrix element of 15 keV. Similar estimates were made also for the p-shell [16]) and pf-shell [17,24,30,32-35]). Such estimates only yield a lower bound for the size of the three-body matrix elements. Microscopically derived three-body forces may be larger and their effects could be hidden in the empirically adjusted two-body part of the effective hamiltonian. Nevertheless, these estimates indicate, that the matrix elements of a three-body force are probably small and a large influence on the calculated spectra should only be expected for systems with a large (> 3) number of valence nucleons.

In sect. 2 we describe the technique to calculate the sd-shell matrix elements of the "real" and "effective" three-body force. Results for the matrix elements and their influence on the calculated spectra of $A = 19$ nuclei are presented in sect. 3. The main conclusions of this work are summarized in sect. 4.

2. Calculation of three-body matrix elements

An effective hamiltonian for a shell-model calculation may be written in the form

$$H_{\text{eff}} = \sum_i \varepsilon_i a_i^+ a_i + \tfrac{1}{4} \sum_{ijlm} \langle ij|V_2|lm\rangle a_i^+ a_j^+ a_m a_l$$

$$+ \tfrac{1}{36} \sum_{ijklmn} \langle ijk|V_3|lmn\rangle a_i^+ a_j^+ a_k^+ a_n a_m a_l, \tag{1}$$

assuming that effective n-body interactions V_n for $n > 3$ can be neglected. In this equation the $a_i^+(a_i)$ stand for creation (annihilation) operators for nucleons in the states $i, j, \ldots$ of the valence shell, which corresponds to the sd-shell in the present work. The single-particle energies ε_i are normally extracted from the energies of the $A = 17$ nuclei assuming that the lowest states with spin $\frac{5}{2}$, $\frac{1}{2}$ and $\frac{3}{2}$ can be considered as states with one valence nucleon outside an ^{16}O core. This is justified in a folded-diagram expansion of H_{eff} [ref. [13]]. Also here we use these experimental single-particle energies which are $\varepsilon_i = -4.15$, -3.28 and 0.93 for the $d_{\frac{5}{2}}$, $1s_{\frac{1}{2}}$ and $d_{\frac{3}{2}}$ states, respectively. The matrix elements of the two-body part V_2 of the hamiltonian of eq. (1) are taken from ref. [19]. In this paper two effective shell-model interactions are derived from two realistic NN interactions, the Reid soft-core potential [25] and the potential MDFPΔ2 of the Bonn-Jülich collaboration [4]. While the Reid potential is a purely phenomenological potential, the MDFPΔ2 is based on the one-boson-exchange model, with the extension that intermediate excitations of the interacting nucleons to the $\Delta(3, 3)$ resonance (see discussion in the introduction) are treated explicitly. For both potentials an effective shell-model interaction has been evaluated in ref. [19] using the folded diagram expansion of Kuo, Lee and Ratcliff [26].

Matrix elements for the three-body term in eq. (1) are evaluated between antisymmetrized and normalized three-particle states

$$|lmn\rangle = |(lm)J_{lm}nJ\rangle. \tag{2}$$

This notation means, that the angular momenta of states l and m (j_l and j_m) are coupled to $J_{l,m}$ leading together with the angular momentum j_n to a total angular momentum J of the three-particle state. Note, that for the sake of abbreviation all isospin quantum numbers are omitted. If now one restricts the calculation to three-body terms with only two two-body interaction lines (see figs. 1b and 2) the contributions of the nine topologically distinct Hugenholtz diagrams of fig. 3 have to be considered. In the schematic diagrams of fig. 3, the dots stand for antisymmetrized two-body matrix elements and the intermediate crossed lines represent isobar excitation, particle-states above the valence shell or a hole line depending on whether we are evaluating terms of figs. 1b, 2a or 2b, respectively. To

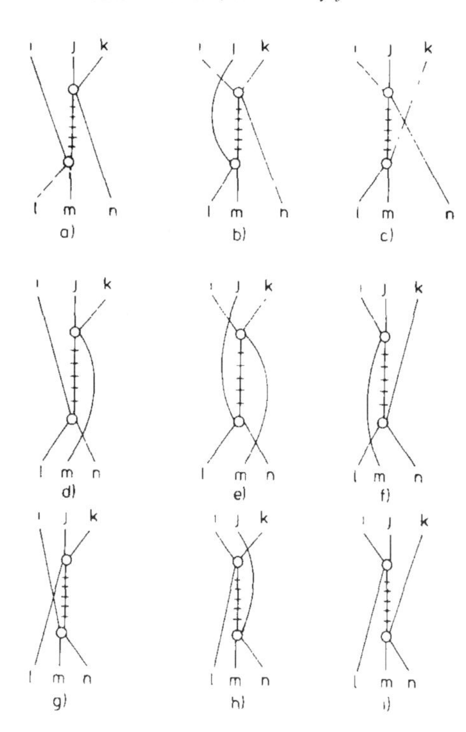

Fig. 3. Topologically distinct Hugenholtz diagrams for second-order three-body terms. In these diagrams the dots stand for antisymmetrized two-body matrix elements and the crossed line for an intermediate state. The specific form of the two-body matrix elements and the intermediate state depends on the contribution which is actually considered.

calculate the different kinds of diagrams of fig. 3 in terms of angular momentum coupled two-body matrix elements one in general has to recouple the angular momenta of the single-particle states. As an example we show this recoupling procedure for the diagram of fig. 3a:

$$\langle ijk|V_3(\text{fig. 3a})|lmn\rangle = \sum_{J_{jk}}\sum_{p}(-)^{j_i+j_k+j_p+j_n}\hat{J}_{ij}\hat{J}_{lm}\hat{J}^2_{jk}$$

$$\times\begin{Bmatrix} j_i & j_j & J_{ij} \\ j_k & J & J_{jk}\end{Bmatrix}\begin{Bmatrix} j_i & j_p & J_{lm} \\ j_n & J & J_{jk}\end{Bmatrix}\langle jk|\tilde{V}|pn\rangle_{J_{jk}}\frac{1}{e}\langle ip|\tilde{V}|lm\rangle_{J_{lm}}. \qquad (3)$$

In this equation $\hat{J}$ stands for $\sqrt{2J+1}$ and the curly brackets denote 6-j symbols. The summation over intermediate states p, the energy denominator e and the two-body matrix elements will be discussed below. The three-body matrix element is written in the shorthand notation of eq. (2) without indicating the intermediate and total angular momenta. The corresponding equation for the other diagrams of fig. 3 are given in the appendix.

We first concentrate on the "real" three-body force graphically displayed in fig. 1b. For this case the intermediate states p denote $\Delta(3, 3)$ excitations and therefore the two-body matrix elements in eq. (3) and in the appendix are transition matrix elements for processes NN → NΔ. These matrix elements are derived in the OBE model considering π- and ρ-exchange. Other mesons do not contribute since they cannot yield a change of the spin and isospin from the nucleon value $\frac{1}{2}$ to the Δ-isobar value $\frac{3}{2}$. Effects of NN correlations are taken into account in the very same way as in the Brueckner G-matrix for the NN interaction. This means that the transition matrix elements are defined by a "transition G-matrix" $G_{\Delta\mathrm{N}}$ using e.g.

$$\begin{aligned}\langle jk|\tilde{V}|pn\rangle &= \langle jk|G_{\Delta\mathrm{N}}|\Delta n\rangle \\ &= \langle jk|V_{\Delta\mathrm{N}}|\Delta n\rangle + \sum_{N_1N_2>V} \frac{\langle jk|G|N_1N_2\rangle\langle N_1N_2|V_{\Delta\mathrm{N}}|\Delta n\rangle}{\varepsilon_j+\varepsilon_k-(\varepsilon_{N_1}+\varepsilon_{N_2})},\end{aligned}$$

which is graphically displayed in fig. 4. The bare transition potential $V_{\Delta\mathrm{N}}$ of eq. (4) is the same transition potential which is also contained in MDFPΔ2 [see ref. [4])] and also the NN → NN G-matrix has been calculated for this potential and therefore G contains also intermediate NΔ and ΔΔ states itself. The sum over intermediate particle states N_1N_2 above the valence shell $N_1N_2 > V$ has been performed in an oscillator basis for the relative motion including states up to $2n+1 = 24$.

As a complete set of single-particle wave functions for the intermediate isobar excitation one could take the set of harmonic oscillator waves using the same oscillator constant as for the states of the bound nucleons. Since in a numerical calculation one only can treat a few low-lying oscillator states, such a representation of the isobar wave function seems not to be very appropriate. Also one knows, that two-body potentials and the transition potential of eq. (4), containing strong tensor components, tend to excite states of relatively high momentum [18,19]). Therefore a representation of the isobar wave function in terms of oscillator waves would probably converge only very slowly, especially since the energy denominator contained e.g. in eq. (3) hardly supports such a convergence. Therefore we decided to represent the intermediate isobar states in terms of plane

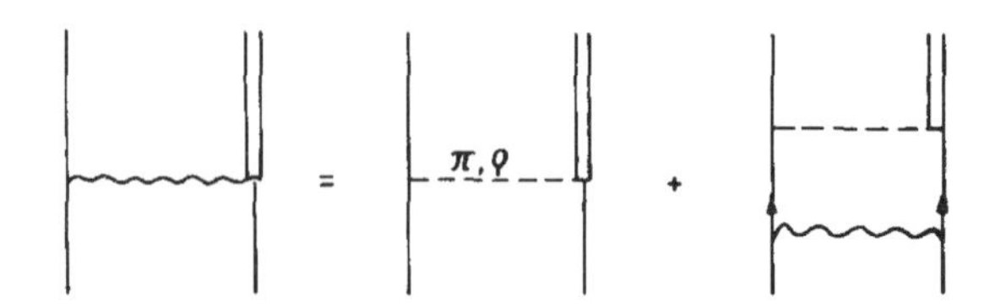

Fig. 4. Graphical representation of eq. (4). The wiggly lines denote G-matrix elements, while the dashed line stand for a bare π- and ρ-exchange.

waves. In this case a numerical integration over the single-particle momentum of the isobar state can be performed easily, to include a complete set of intermediate states.

The price one has to pay is that e.g. the ket of the transition potential of eq. (4) contains one plane wave state with momentum p for the isobar Δ and one oscillator sate for the nucleon N:

$$|\Delta \mathrm{N}\rangle = |pl_\Delta j_\Delta, nl_\mathrm{N} j_\mathrm{N}, J_{\Delta\mathrm{N}}\rangle. \tag{5}$$

Such a two-body wave function in a mixed representation can be calculated from a state of two plane waves by normal numerical integration with the oscillator function for the nucleon:

$$|pl_\Delta j_\Delta, nl_\mathrm{N} j_\mathrm{N}, J_{\Delta\mathrm{N}}\rangle = \int \mathrm{d}q R_{nl_\mathrm{N}}(q) |pl_\Delta j_\Delta, ql_\mathrm{N} j_\mathrm{N}, J_{\Delta\mathrm{N}}\rangle. \tag{6}$$

Finally, the vector brackets [27]) $\langle klKL|pl_\Delta ql_\mathrm{N}\lambda\rangle$ facilitate the transformation from the laboratory plane-wave states into relative (kl) and c.m. (KL) free waves.

$$\begin{aligned} |pl_\Delta j_\Delta, ql_\mathrm{N} j_\mathrm{N}, J_{\Delta\mathrm{N}}\rangle = \sum_{\lambda lLSJ} \int \mathrm{d}K \int \mathrm{d}k (-)^{J+L+S+\lambda} \\ \times \hat{j}_\Delta \hat{j}_\mathrm{N} \hat{\lambda} \hat{S} \begin{Bmatrix} l_\Delta & l_\mathrm{N} & \lambda \\ S_\Delta & \frac{1}{2} & S \\ j_\Delta & j_\mathrm{N} & J_{\Delta\mathrm{N}} \end{Bmatrix} \langle klKL|pl_\Delta ql_\mathrm{N}\lambda\rangle \hat{J}\hat{\lambda} \\ \times \begin{Bmatrix} J & L & J_{\Delta\mathrm{N}} \\ \lambda & S & l \end{Bmatrix} |k(lS)J, KL, J_{\Delta\mathrm{N}}\rangle. \end{aligned} \tag{7}$$

The matrix elements of the transition potential $V_{\Delta\mathrm{N}}$ are conveniently calculated in a basis of relative and c.m. plane-wave states. The NN state of the matrix element can then be transformed into an oscillator representation for relative and c.m. motion, followed by a normal Moshinsky transformation into two-nucleon laboratory oscillator states. The ΔN state of the transition matrix element is obtained applying the transformations of eq. (5) to (7).

The energy denominator e of eq. (3) and of the corresponding equations in the appendix is obtained as the sum of the single-particle energies of the initial state $(\varepsilon_l + \varepsilon_m + \varepsilon_n)$ minus the sum of the single-particle energies for the intermediate state. For eq. (3), i.e. the diagram of fig. 3a, specifying the intermediate state as an isobar excitation, this energy denominator can therefore be written

$$e = e_\Delta = \varepsilon_l + \varepsilon_m - \varepsilon_i - \varepsilon_\Delta, \tag{8}$$

with

$$\varepsilon_\Delta = \frac{p^2}{2m_\Delta} + (m_\Delta - m_N), \tag{9}$$

where m_Δ and m_N stand for the mass of the Δ-resonance and the nucleon, respectively.

Next we consider the contribution from intermediate particle states to the effective three-body force as displayed in fig. 2a. Also in this case one can expect bad convergence properties, if a set of oscillator states is used for the intermediate-particle states [18]. Therefore we also use here a plane-wave representation for the particle states using the same technique as in ref. [19] for the evaluation of the core-polarisation diagram G_{3p1h} in the effective two-body interaction. The method is in straight analogy to the one described above for the isobar excitation. For the two-body interaction $\tilde{V}$ in eq. (3) the Brueckner G-matrix elements calculated with the Reid potential or the MDFPΔ2 of the Bonn-Jülich collaboration [4]) are used. The plane-wave states of the intermediate particle states have to be orthogonal to the states of the occupied and valence shells. This can be achieved by using for the propagator in eq. (3)

$$\frac{1}{e} = \frac{1}{e_N} = \hat{Q}\frac{1}{\varepsilon_1 + \varepsilon_m - \varepsilon_i - \hat{Q}p^2/2m_N\hat{Q}}\hat{Q}, \tag{10}$$

with a projection operator $\hat{Q}$ which acting on the intermediate state p projects on single-particle states above the valence shell. With the help of an operator identity [28]) this propagator can be rewritten as

$$\frac{1}{e_N} = \frac{1}{e_{\text{free}}} + \frac{1}{e_{\text{free}}}\hat{P}\left[\hat{P}\frac{1}{e_{\text{free}}}\hat{P}\right]^{-1}\hat{P}\frac{1}{e_{\text{free}}}, \tag{11}$$

where $\hat{P}$ is the projection operator on states of the core and valence shells and

$$e_{\text{free}} = \varepsilon_l + \varepsilon_m - \varepsilon_i - p^2/2m_N. \tag{12}$$

For the contributions listed in the appendix only the single-particle energies have to be modified.

Finally, evaluating the contributions of the kind displayed in fig. 2b involves a summation over intermediate hole states h only. This can conveniently be done in an oscillator basis. In this case the energy denominator consists of single-particle energies of hole and valence states. It is given as the sum of the single-particle energies of the initial state plus the energy of the hole state minus the sum of all intermediate particle states. For the contribution of diagram a of fig. 3 (see also eq. (3)) this yields

$$e_H = \varepsilon_m + \varepsilon_h - \varepsilon_k - \varepsilon_j, \tag{13}$$

and similar expressions are obtained for the other contributions. In the actual calculation single-particle energies are used which are degenerate in the major shells. For the valence shell the average of the experimental single-particle energies was used ($\varepsilon_{val} = -2.31$) and the hole state energies were considered to be below this value by $\hbar\omega$ and $2\hbar\omega$ for the p-shell and s-shell, respectively, assuming $\hbar\omega = 14$ MeV.

3. Results and discussions

To evaluate the "real" three-body force originating from intermediate isobar excitations, one needs the G-matrix for the transition potential $G_{\Delta N}$ (see eq. (4)). Typical matrix elements of $G_{\Delta N}$ are presented in figs. 5 and 6. While the upper half of these figures shows matrix elements using the mixed representation of the ΔN state (see eq. (5)) as a function of the momentum k of the isobar state, the lower half displays results if also the isobar state is represented by oscillator states using the same oscillator

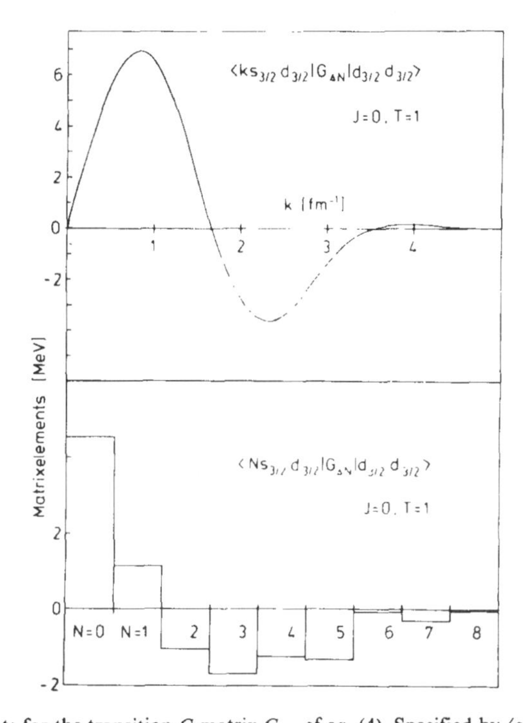

Fig. 5. Matrix elements for the transition G-matrix $G_{N\Delta}$ of eq. (4). Specified by $\langle s_{3/2}(\Delta)d_{3/2}|G_{N\Delta}|d_{3/2}d_{3/2}$ $J = 0, T = 1\rangle$. The upper half displays results for this matrix element using plane-wave representation for the isobar state as a function of the momentum k. In the lower half of this figure the corresponding matrix elements are given using oscillator representation also for the isobar for different radial quantum numbers N. The oscillator parameter is $b = 1.72$ fm.

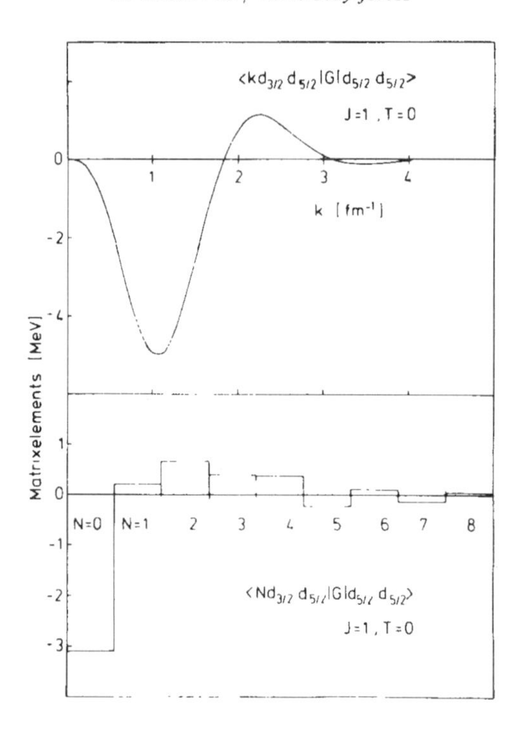

Fig. 6. Matrix elements for the transition matrix $G_{N\Delta}$ $\langle i_{9/2}(\Delta)d_{5/2}|G_{\Delta N}|d_{5/2}d_{5/2}$ $J = 2$, $T = 1\rangle$. For further explanation see caption to fig. 5.

length as for the nucleons. From these figures one can see, that using the mixed representation isobar states should be taken into account up to a momentum of 4 to 5 fm^{-1}. For higher momenta the matrix elements of the transition potential $G_{\Delta N}$ are so small that these states can safely be ignored. Fig. 5 gives results for a matrix element with an isobar in an $s_{\frac{1}{2}}$ state. Using the oscillator representation also for the isobar state, one finds that the largest matrix element is obtained for the radial quantum number $N = 0$. (Counting the nodes not including the ones at zero and infinity.) This corresponds to an excitation of a valence nucleon to an isobar state with orbital wave function the same as for the $0s_{\frac{1}{2}}$ nucleon state in the ^{16}O core. But also the matrix elements for an isobar state with larger radial oscillator quantum number are non-negligible. In this case one should take into account excitations up to $N = 5$ which corresponds to 10 $\hbar\omega$ additional excitation energy relative to the $(N, l) = (0, 0)$ state. In the example of fig. 6 the isobar is in an $i_{\frac{9}{2}}$ state and the size of the matrix elements shows that also isobar states with a high orbital angular momentum $(l = 6)$ have to be considered. Also here oscillator excitations up to at least 12 $\hbar\omega$ should be considered which shows again the efficiency of the plane-wave representation for intermediate particle states.

Using these $G_{\Delta N}$ matrix elements with a plane-wave representation for the isobar, the "real" three-body force of fig. 1b has been calculated as described in sect. 2. The result for the average of all diagonal matrix elements for antisymmetrized three-particle states in the sd-shell,

$$\bar{V}_3 = \sum_{\substack{\alpha\beta\gamma \\ JT}} (2J+1)(2T+1)\langle\alpha\beta\gamma|V_3|\alpha\beta\gamma\rangle_{JT} \Big/ \left(\sum_{\alpha\beta\gamma JT} \langle\alpha\beta\gamma|\alpha\beta\gamma\rangle_{JT}(2J+1)(2T+1) \right), \tag{14}$$

is given in table 1. This average value is repulsive and very small ($\bar{V}_3 = 0.014$ MeV). Therefore its sign and size is an agreement with the empirical estimate [22,23]), which is discussed at the end of sect. 1 of this paper. It is worth mentioning, that in the calculation of binding energies of closed-shell nuclei the three-body terms with intermediate isobars yield a relatively large attractive contribution [8]), while the corresponding three-body force between valence nucleons tends to give a repulsive contribution. This difference may be due to the larger average distance of the interacting valence nucleons as compared to the interaction distances in the core.

Besides the average value of the three-body matrix elements $\bar{V}_3$, we also give the root mean square deviation $\Delta\bar{V}_3$ in the averaging procedure of eq. (14). The value of $\Delta\bar{V}_3$, which is larger than the average itself, clearly indicates that individual matrix elements can be very different from the average value. This means that the inclusion of the three-body force in a shell-model calculation should not only lead to a global shift of all energies for a given nucleus, but that quite different shifts for the individual states in the spectrum could be obtained.

To consider the "effective" three-body force displayed in fig. 2 we first discuss

TABLE 1

Average 3-body matrix elements $\bar{V}_3$ (see eq. (14)) and the root mean square deviation $\Delta\bar{V}_3$ in the averaging procedure are given for the "real" three-body force (see fig. 1b) which is due to intermediate isobar excitation and for the "effective" three-body forces of figs. 2a and 2b

		$\bar{V}_3$ (MeV)	$\Delta\bar{V}_3$ (MeV)
	fig. 1b	0.014	0.037
MDFPΔ2:	fig. 2a	0.043	0.081
	fig. 2b	−0.006	0.057
	total	0.051	0.098
Reid:	fig. 2a	0.034	0.057
	fig. 2b	−0.005	0.043
	total	0.043	0.078

The results for the "effective" 3N forces and for the total sum (including the term of fig. 1b) have been evaluated starting from the NN potential MDFPΔ2 of ref. [4]) and the Reid soft-core potential.

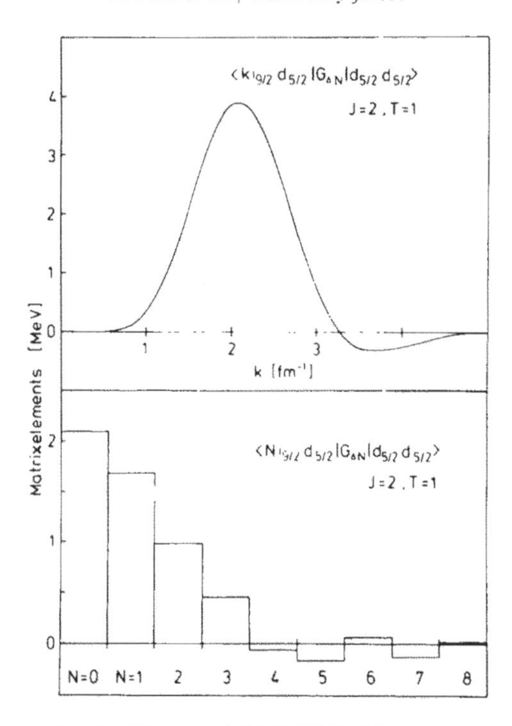

Fig. 7. G-matrix elements for the NN potential MDFPΔ2. The $d_{3/2}$ state of the matrix element $\langle d_{3/2}d_{5/2}|G|d_{5/2}d_{5/2}\, J = 1, T = 0\rangle$ is represented by a plane wave and oscillator state for the upper and lower part of the figure respectively. For further explanation see caption to fig. 5.

some relevant matrix elements of the Brueckner G-matrix which plays here the same role as $G_{\Delta N}$ did for the 3N term with an intermediate isobar. Therefore, in analogy to figs. 5 and 6, fig. 7 displays typical results for a G-matrix describing the excitation of a valence nucleon to an intermediate particle state. Representing this particle state in a plane-wave representation, without requesting orthogonality to valence and hole states (see eq. (11)), one sees that states up to a momentum k of 5 fm^{-1} have to be considered. (In the specific example of fig. 7 it would probably be sufficient to take into account states up to 3 fm^{-1}). In the lower half of fig. 7, the corresponding oscillator matrix elements are given. One sees that by far the largest matrix element is obtained for the $d_{\frac{3}{2}}$ state with $N = 0$. For a shell-model calculation in the complete sd-shell, however, the $0d_{\frac{3}{2}}$ state is a valence state and therefore it should not be contained in the summation over intermediate particle states. The calculation of effective 3N forces presented in ref. [29]), was restricted to excited particle states in an oscillator basis with $N = 1$. So this typical example gives a hint about two important points. First of all it shows, that if a plane-wave representation of the particle states is used, a careful orthogonalisation to the hole and valence states is required to avoid over-counting. In the present calculation this is done as explained in sect. 2 of this paper. Secondly, since the $0d_{\frac{3}{2}}$ state would not belong to the valence states for a shell-model calculation restricted to the $0d_{\frac{3}{2}}$ subshell, this example also illustrates that effective three-body forces can be much stronger, if the shell-model space is restricted to one subshell only.

Results for the average 3N matrix elements $\bar{V}_3$ of the "effective" three-body forces of figs. 2a, b are also listed in table 1 together with the root mean square deviations. Also for the "effective" three-body force similar conclusions can be drawn as for the "real" 3N force discussed above. The matrix elements are small and comparable in size to those from the isobar terms. The average value is repulsive, but the deviation of individual matrix elements from this average is relatively large so that for certain states also some attraction could be obtained from 3N forces.

Results of shell-model calculations for $A = 19$, $T = \frac{1}{2}$ states with and without inclusion of the total 3N force, which is the sum of the "real" and the "effective" 3N force, are displayed in fig. 8. One can hardly see any difference between the

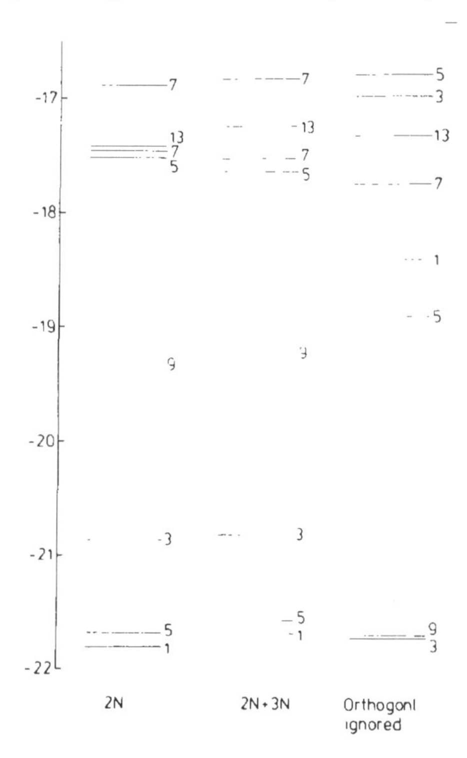

Fig. 8. Results of shell-model calculations for $A = 19$, $T = \frac{1}{2}$ states without (1st column) and with (2nd column) inclusion of 3N forces. The effective 2N and 3N interactions are derived from the NN potential MDFPΔ2 of the Bonn-Jülich collaboration. The last column is obtained if in the calculation of the "effective" 3N forces the orthogonalisation of the intermediate particle states to the core and valence states is ignored. For this third spectrum 2 states at lower energies ($J = \frac{5}{2}$ at -24.785 MeV and $J = \frac{1}{2}$ at -25.48 MeV) are not displayed. The numbers at the energy levels denote $2J$ of that state.

TABLE 2

Results of shell-model calculations for $A = 19$ states with isospin $T = \frac{1}{2}$

J	MDFPΔ2		Reid	
	2N	2N+3N	2N	2N+3N
$\frac{1}{2}$	−21.801	−21.676	−21.192	−21.037
$\frac{3}{2}$	−20.863	−20.812	−20.117	−20.085
$\frac{5}{2}$	−21.681	−21.558	−20.920	−20.790
	−17.516	−17.629	−17.264	−17.318
$\frac{7}{2}$	−17.459	−17.516	−17.092	−17.157
	−16.887	−16.819	−16.864	−16.764
$\frac{9}{2}$	−19.328	−19.217	−19.146	−19.058
$\frac{13}{2}$	−17.420	−17.231	−17.086	−16.944

For the columns labeled 2N only an effective two-body interaction has been considered while the columns labeled 2N+3N contain results of calculations for which also the total 3N contributions, discussed in this paper, are taken into account. Results are displayed starting from NN potentials MDFPΔ2 and Reid soft core. Details of the derivation of the 2N part of the hamiltonian can be found in ref. [19]).

spectra of the first and second column of this figure. Only if one makes a mistake and ignores the orthogonalisation of the intermediate particle states of fig. 2a to the core and valence states, a sizable effect is obtained. This demonstrates again that one has to be very careful treating this overcounting problem.

To make the effects of the 3N force visible, the results of fig. 8 are given numerically in table 2. From these values one can see that the addition of the 3N interaction yields small attractive or repulsive contributions, depending on the state considered. The largest energy shift for these low-lying states is about 190 keV. The results only depend slightly on the NN interaction used. Similar results were also obtained for the $T = \frac{3}{2}$ states of 3 valence nucleons. (See fig. 9 and table 3.) In this case the energy differences due to the 3N terms are slightly smaller (only up to 60 keV).

At first sight these results seem to indicate that the effects of 3N forces can safely be ignored in shell-model calculations. This is of course true for systems with only 3 valence nucleons. Following the analysis of Quesne [30]), however, one estimates the average energy shift of states with n valence particles due to a 3N force as

$$\Delta E_3(n) = \binom{n}{3} \bar{V}_3. \tag{15}$$

In this equation $\bar{V}_3$ stands for the average matrix element of eq. (14) and $\binom{n}{3}$ is the binomial coefficient. This means that already for $n = 4$ (e.g. ^{20}Ne) the effects could be about a factor 4 larger, in the middle of the shell (e.g. ^{28}Si) the binomial factor is already 220. This shows that the 3N terms discussed in this work might cause drastic changes in shell-model investigations for nuclei with more than 3 valence nucleons.

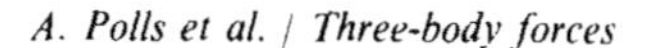

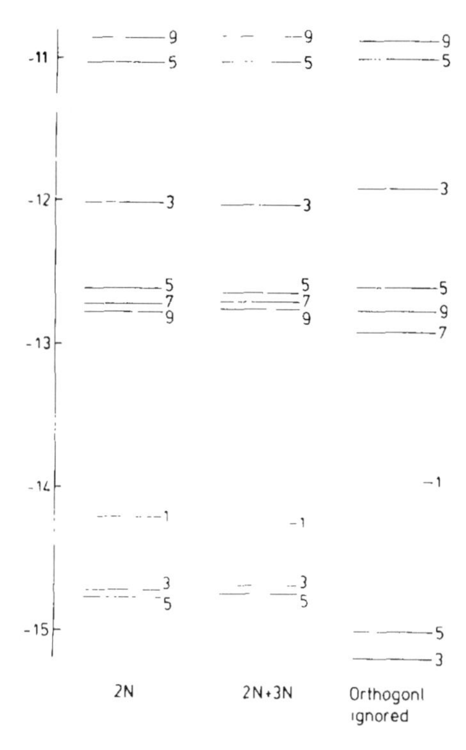

Fig. 9. Results of shell-model calculations for ^{19}O ($T = \frac{3}{2}$). For further explanation see caption to fig. 8.

TABLE 3

Results of shell-model calculations for ^{19}O ($T = \frac{3}{2}$)

J	MDFPΔ2		Reid	
	2N	2N+3N	2N	2N+3N
$\frac{1}{2}$	−14.214	−14.255	−14.239	−14.268
$\frac{3}{2}$	−14.730	−14.692	−15.189	−15.153
	−12.017	−12.028	−12.314	−12.314
$\frac{5}{2}$	−14.779	−14.752	−14.955	−14.921
	−12.615	−12.641	−12.807	−12.810
	−11.032	−11.025	−11.114	−11.124
$\frac{7}{2}$	−12.721	−12.703	−13.103	−13.081
$\frac{9}{2}$	−12.778	−12.752	−13.152	−13.122
	−10.857	−10.843	−11.206	−11.182

For further explanation see caption to table 2.

4. Conclusions

The aim of this contribution is an investigation of the effect of 3N forces on shell-model calculations for finite nuclei choosing as an example for numerical studies the sd-shell. Besides the effects from "real" 3N forces, which are due to intermediate excitation of isobar ($\Delta(3,3)$) configurations (see fig. 1b) we also take into account the effects of "effective" 3N interactions which originate from the elimination of nuclear configurations (see fig. 2). It turns out that the effects from the "effective" 3N force are of the same size as those of the "real" three-body interaction. The values of the matrix elements are relatively small and of the order of only 50 keV to 100 keV. The average over all matrix elements is repulsive. It is interesting to note, that this repulsive tendency and the order of magnitude of the matrix elements is in agreement with the empirical estimates [14,22,23]) for the average size of a 3N force in the sd-shell (see also discussion in the introduction). Furthermore we would like to recall that the contribution of three-body terms with intermediate $\Delta(3,3)$ excitations to the binding energy of closed-shell nuclei or nuclear matter is attractive [8]), whereas the corresponding 3N interaction between valence nucleons tends to give repulsion. Individual matrix elements of the 3N interaction, however, can be very different from the repulsive average. This leads to shifts in the calculated energies for the low lying states of nuclei with 3 valence nucleons in the sd-shell ranging from -0.11 to $+0.19$ MeV. Typically the effects are slightly larger for $T=\frac{1}{2}$ than for $T=\frac{3}{2}$ states. Of course this effect of the 3N force is relatively small. Since however, the importance of the 3N terms increases with the number of triplets which can be formed by the valence nucleons, we expect much larger effects in the middle or at the end of the sd-shell. Such investigations are in preparation.

This work has been supported by the Deutsche Forschungsgemeinschaft. This support is gratefully acknowledged.

Appendix

In this appendix we show the angular momentum recoupling which is needed to calculate the contributions of the different diagrams of fig. 3 in terms of angular momentum coupled two-body matrix elements. The corresponding recoupling for the isospin is not shown explicitly. The result for fig. 3a is already given in the text (eq. (3)). There we also explain the nomenclature. Note that the explicit minus signs in the expressions for figs. 3c, 3e and 3g are due to the interchange of two fermions.

The remaining phase factors are due to angular momentum recoupling and can readily be generalized (together with the other factors) to include isospin recoupling:

$$\langle ijk|V_3(\text{fig. 3b})|lmn\rangle = \sum_{J_{ik}p} (-)^{j_k-j_n+J_{ij}-J_{lm}} \hat{J}_{ij}\hat{J}_{lm}\hat{J}^2_{ik}$$

$$\times \begin{Bmatrix} j_i & j_j & J_{ij} \\ J & j_k & J_{ik} \end{Bmatrix} \begin{Bmatrix} j_p & j_j & J_{lm} \\ J & j_n & J_{ik} \end{Bmatrix} \langle ik|\tilde{V}|pn\rangle_{J_{ik}} \frac{1}{e} \langle pj|\tilde{V}|lm\rangle_{J_{lm}};$$

$$\langle ijk|V_3(\text{fig. 3c})|lmn\rangle = -\sum_{p} (-)^{j_k+j_n+J_{ij}+J_{lm}} \hat{J}_{ij}\hat{J}_{lm}$$

$$\times \begin{Bmatrix} j_p & j_n & J_{ij} \\ J & j_k & J_{lm} \end{Bmatrix} \langle ij|\tilde{V}|pn\rangle_{J_{ij}} \frac{1}{e} \langle pk|\tilde{V}|lm\rangle_{J_{lm}};$$

$$\langle ijk|V_3(\text{fig. 3d})|lmn\rangle = \sum_{pJ_{jk}J_{ln}} (-)^{j_j+j_k+j_m+j_n+J_{lm}+J_{jk}+J_{ln}}$$

$$\times \hat{J}_{ij}\hat{J}_{lm}\hat{J}^2_{jk}\hat{J}^2_{ln} \begin{Bmatrix} j_i & j_j & J_{ij} \\ j_k & J & J_{jk} \end{Bmatrix} \begin{Bmatrix} j_l & j_m & J_{lm} \\ J & j_n & J_{ln} \end{Bmatrix} \begin{Bmatrix} j_i & j_p & J_{ln} \\ j_m & J & J_{jk} \end{Bmatrix}$$

$$\times \langle jk|\tilde{V}|mp\rangle_{J_{jk}} \frac{1}{e} \langle ip|\tilde{V}|ln\rangle_{J_{ln}};$$

$$\langle ijk|V_3(\text{fig. 3e})|lmn\rangle = -\sum_{pJ_{ik}J_{ln}} (-)^{j_k+j_n+J_{ij}+J_{lm}}$$

$$\times \hat{J}_{ij}\hat{J}_{lm}\hat{J}^2_{ik}\hat{J}^2_{ln} \begin{Bmatrix} j_i & j_j & J_{ij} \\ J & j_k & J_{ik} \end{Bmatrix} \begin{Bmatrix} j_l & j_m & J_{lm} \\ J & j_n & J_{ln} \end{Bmatrix} \begin{Bmatrix} j_p & j_m & J_{ik} \\ J & j_j & J_{ln} \end{Bmatrix}$$

$$\times \langle ik|\tilde{V}|pm\rangle_{J_{ik}} \frac{1}{e} \langle pj|\tilde{V}|ln\rangle_{J_{ln}};$$

$$\langle ijk|V_3(\text{fig. 3f})|lmn\rangle = \sum_{pJ_{ln}} (-)^{j_k-j_n+J_{ij}-J_{lm}} \hat{J}_{ij}\hat{J}_{lm}\hat{J}^2_{ln}$$

$$\times \begin{Bmatrix} j_p & j_m & J_{ij} \\ J & j_k & J_{ln} \end{Bmatrix} \begin{Bmatrix} j_l & j_m & J_{lm} \\ J & j_n & J_{ln} \end{Bmatrix} \langle ij|\tilde{V}|pm\rangle_{J_{ij}} \frac{1}{e} \langle pk|\tilde{V}|ln\rangle_{J_{ln}};$$

$$\langle ijk|V_3(\text{fig. 3g})|lmn\rangle = -\sum_{pJ_{jk}J_{mn}} (-)^{j_j+j_k+j_m+j_n+J_{jk}+J_{mn}+2J}$$

$$\times \hat{J}_{ij}\hat{J}_{lm}\hat{J}^2_{jk}\hat{J}^2_{mn} \begin{Bmatrix} j_i & j_j & J_{ij} \\ j_k & J & J_{jk} \end{Bmatrix} \begin{Bmatrix} j_l & j_m & J_{lm} \\ j_n & J & J_{mn} \end{Bmatrix} \begin{Bmatrix} j_i & j_p & J_{mn} \\ j_l & J & J_{jk} \end{Bmatrix}$$

$$\times \langle jk|\tilde{V}|lp\rangle_{J_{jk}} \frac{1}{e} \langle ip|\tilde{V}|mn\rangle_{J_{mn}};$$

$$\langle ijk|V_3(\text{fig. 3h})|lmn\rangle = \sum_{pJ_{ik}J_{mn}} (-)^{j_j+j_k+j_m+j_n+J_{ij}+J_{ik}+J_{mn}}$$

$$\times \hat{J}_{ij}\hat{J}_{lm}\hat{J}^2_{ik}\hat{J}^2_{mn} \begin{Bmatrix} j_i & j_j & J_{ij} \\ J & J_k & J_{jk} \end{Bmatrix} \begin{Bmatrix} j_l & j_m & J_{lm} \\ j_n & J & J_{mn} \end{Bmatrix} \begin{Bmatrix} j_l & j_p & J_{ik} \\ j_j & J & J_{mn} \end{Bmatrix}$$

$$\times \langle ik|\tilde{V}|lp\rangle_{J_{ik}} \frac{1}{e} \langle jp|\tilde{V}|mn\rangle_{J_{mn}};$$

$$\langle ijk|V_3(\text{fig. 3i})|lmn\rangle = \sum_{p,J_{mn}} (-)^{j_p+j_k+j_m+j_n} \hat{J}_{ij}\hat{J}_{lm}\hat{J}^2_{mn}$$

$$\times \begin{Bmatrix} j_l & j_p & J_{ij} \\ j_k & J & J_{mn} \end{Bmatrix} \begin{Bmatrix} j_l & j_m & J_{lm} \\ j_n & J & J_{mn} \end{Bmatrix} \langle ij|\tilde{V}|lp\rangle_{J_{ij}} \frac{1}{e} \langle pk|\tilde{V}|mn\rangle_{J_{mn}}.$$

References

1) A. M. Green, Rep. Prog. Phys. **39** (1976) 1109
2) A. M. Green and J. A. Niskanen, Nucl. Phys. **A249** (1975) 493
3) B. D. Day and F. Coester, Phys. Rev. **C13** (1976) 1720
4) M. R. Anastasio, A. Faessler, H. Müther, K. Holinde and R. Machleidt, Phys. Rev. **C18** (1978) 2416
5) K. Holinde, R. Machleidt, A. Faessler and H. Müther Phys. Rev. **C15** (1977) 1932
6) H. Müther, Nucl. Phys. **A328** (1979) 429
7) C. Hajduk and P. U. Sauer, Nucl. Phys. **A322** (1974) 324
8) A. Faessler, H. Müther, K. Shimizu and W. Wadia, Nucl. Phys. **A333** (1980) 428
9) K. Shimizu, A. Polls, H. Müther and A. Faessler, Nucl. Phys. **A364** (1981) 461
10) B. D. Day, Phys. Rev. **C24** (1981) 1203
11) H. Kümmel, K. H. Lührmann and J. G. Zabolitzky, Phys. Reports **36C** (1978) 1
12) B. R. Barret, ed., Effective interactions and operators in nuclei (Springer, Berlin, 1975)
13) T. T. S. Kuo, Ann. Rev. Nucl. Sci. **24** (1974) 101
14) A. Arima, Nucl. Phys. **A354** (1981) 19c
15) B. Singh, Nucl. Phys. **A219** (1974) 621
16) H. Dirim, J. P. Elliott and J. A. Evans, Nucl. Phys. **A244** (1975) 301
17) A. Poves, in The many-body problem, Jastrow correlations versus Brueckner theory, ed. R. Guardiola and J. Ros (Springer, Berlin, 1981) p. 282
18) J. P. Vary, P. U. Sauer and C. W. Wong, Phys. Rev. **C7** (1973) 1776
19) H. M. Sommermann, H. Müther, K. C. Tam, T. T. S. Kuo and A. Faessler, Phys. Rev. **C23** (1981) 1765
20) C. L. Kung, T. T. S. Kuo and K. F. Ratcliff, Phys. Rev. **C19** (1979) 1063
21) B. M. Preedom and B. H. Wildenthal, Phys. Rev. **C6** (1972) 1633
22) M. Nomura, Phys. Lett. **85B** (1979) 187
23) B. J. Cole, A. Watt and R. R. Whitehead, Phys. Lett. **57B** (1975) 24
24) E. Osnes, Phys. Lett. **26B** (1968) 274)
25) R. V. Reid, Ann. of Phys. **50** (1968) 411
26) T. T. S. Kuo, S. Y. Lee and K. F. Ratcliff, Nucl. Phys. **A176** (1971) 65
27) R. Balian and E. Brezin, Nuovo Cim. **61B** (1969) 403
28) S. F. Tsai and T. T. S. Kuo, Phys. Lett. **39B** (1972) 427
29) W. Huang, H. Song, Z. Wang and T. T. S. Kuo, Chinese Nuclear Physics, to be published

30) C. Quesne, Phys. Lett. **31B** (1970) 7
31) B. R. Barrett, E. C. Halbert and J. B. McGrory, Ann. of Phys. **90** (1975) 321
32) Y. Yarid, Nucl. Phys. **A225** (1974) 382
33) F. Andreozzi and G. Sartoris, Nucl. Phys. **A270** (1976) 388
34) G. F. Bertsch, Phys. Rev. Lett. **21** (1968) 1694
35) I. Eisenstein and M. W. Kirson, Phys. Lett. **B47** (1973) 315

PHYSICAL REVIEW C VOLUME 33, NUMBER 2 FEBRUARY 1986

Model-space nuclear matter calculations with the Paris nucleon-nucleon potential

T. T. S. Kuo, Z. Y. Ma,* and R. Vinh Mau†
Department of Physics, State University of New York at Stony Brook, Stony Brook, New York 11794
(Received 2 August 1985)

Using a model-space Brueckner-Hartree-Fock approach, we have carried out nuclear matter calculations using the Paris nucleon-nucleon potential. The self-consistent single particle spectrum from this approach is continuous for momentum up to k_M, where $k_M \approx 2k_F$ is the momentum space boundary of our chosen model space. The nuclear matter average binding energy and saturation Fermi momentum given by our calculations are ~ 15.6 MeV and ~ 1.56 fm^{-1}, respectively. When using the conventional Brueckner-Hartree-Fock approach with a spectrum which has a gap at k_F, the corresponding results are ~ 11.5 MeV and ~ 1.50 fm^{-1}. The gain of approximately 4 MeV in binding energy between the two calculations comes mainly from the $^3S_1$ and $^1S_0$ partial wave channels. We have investigated the effect of adding an empirical density dependent central potential to the Paris potential. It is found that the addition of such a potential whose strength is $\sim 10\%$ of the central component of the Paris potential is adequate in making the nuclear matter binding energy and saturation density in simultaneous agreement with the empirical values.

I. INTRODUCTION

In recent years nucleon-nucleon potentials derived from meson and isobar degrees of freedom have been able to describe rather satisfactorily experimental nucleon-nucleon phase shifts and deuteron properties. The next question to be answered is whether such potentials are able to predict nuclear many-body properties. As is well known, the simplest many-body system as far as theoretical calculations are concerned is the infinitely large and homogeneous nuclear matter. Many methods have been proposed for carrying out nuclear matter calculations, such as the Brueckner-Hartree-Fock (BHF),[1] the Fermi hypernetted chain,[2] the e^S (Ref. 3), and the model-space BHF (Refs. 4 and 5) methods. From the Weiszäcker empirical nuclear mass formula, the binding energy per nucleon (BE/A) in nuclear matter is deduced to be ~ 16 MeV, and from electron scattering experiments of nuclei, the nuclear matter saturation density (ρ_0) is deduced to be ~ 0.17 nucleon per fm^3. Theoretical derivations using the methods mentioned above have, however, never been able to reproduce the values of the binding energy per nucleon (BE/A) and ρ_0 which are in simultaneous agreement with the corresponding empirical values; the calculated values of BE/A and ρ_0 using various nucleon-nucleon potentials generally lie on a band—the Coester band[6]—which deviates significantly from the corresponding empirical values. This deviation can be attributed to different sources. It could be due to the inadequacy of the many-body techniques used so far in the calculations. It could also be that the assumption that two-body forces are dominant in nuclear matter is not accurate enough and that three or more body forces are not negligible. This point was, in fact, suggested by several recent studies,[8,9] although, even in the case of three-body forces, the derivation of these forces is still very ambiguous.

In this paper, we report our results of several model-space BHF (MBHF) calculations of nuclear matter, using the Paris nucleon-nucleon (NN) potential.[7] This method was introduced by Ma and Kuo.[4] They have applied it using the Reid NN potential. They have also carried out some preliminary MBHF nuclear matter calculations using the Paris NN potential, including only partial waves with $l \leq 5$. Results of these calculations have been very briefly reported.[5] The present paper carries out more extensive MBHF calculations using the Paris NN potential and will report the results in more detail. In addition, we will study the effect of adding an empirical density dependent two-body effective interaction to the Paris potential. This effective density dependent interaction is assumed to represent all effects due to the modification of the two-body forces by the medium, three-body forces, etc. We will show indeed that we only need to add a fairly small density dependent component to the Paris potential so as to make the calculated nuclear matter saturation density in good agreement with the experimental one.

We will first, in Sec. II, briefly describe the MBHF method for nuclear matter calculations. Mahaux and his collaborators[10] have pointed out that the discontinuous single particle spectrum used in conventional BHF calculations is unsatisfactory on several fundamental grounds, mainly because this spectrum has an artificial energy gap of ~ 60 MeV at the Fermi surface k_F. These authors therefore proposed a continuous single particle spectrum based on a Green's function method. The MBHF method is derived from a model-space approach, which leads to a single particle spectrum which is continuous within the chosen model space. Hence if one chooses a model space which extends beyond k_F, one will obtain a single particle spectrum which is continuous is k_F. As discussed later, one may choose to treat the hole-line spectrum slightly different when carrying out nuclear matter calculations within the model space. Then the resulting single particle spectrum will have a small gap at k_F. An important feature of our spectrum is that its potential energy is generally attractive in the momentum region k_F to $\sim 2k_F$. In

Sec. III we will report on our results of nuclear matter calculations by using the Paris nucleon-nucleon potential as well as describing some details of our computational methods, such as the Born approximation for calculating the potential energy contribution from partial waves with $l \geq 5$. In Sec. IV, we will describe and discuss several MBHF nuclear matter calculations using the Paris nucleon-nucleon potential with its central component modified by an empirical density dependent factor. A discussion and a conclusion are presented in Sec. V.

II. THE MBHF METHOD FOR NUCLEAR MATTER

In this section we briefly describe the MBHF method for nuclear matter.[4,5] In treating nuclear many-body problems, one usually introduces a one-body auxiliary potential U to the nuclear Hamiltonian $H=T+V$, where V is a chosen nucleon-nucleon (NN) potential, and rewrite it as

$$H=(T+U)+(V-U)\equiv H_0+H_1 \ . \quad (1)$$

The exact solutions of the Schrödinger equation

$$H\psi_n=E_n\psi_n \quad (2)$$

are, of course, independent of the choice of U. But when solving Eq. (2), using some approximation methods as invariably done in practice, the choice of U can play a very important role. The MBHF method is basically a method for the choice of U. For nuclear matter calculations, we begin with choosing a model space P defined by

$$P\equiv\{k\leq k_M\} \ , \quad (3)$$

where all nucleons are restricted to have momentum k less than k_M, the momentum space boundary of P. Typical values for k_M are $\sim 3 \text{ fm}^{-1}$, as will be discussed later. Using effective interaction theories,[11,12] we can transform Eq. (2) into a model space equation

$$H_{\text{eff}}P\psi_m=E_mP\psi_m \ , \quad (4)$$

with

$$H_{\text{eff}}=P(H_0+V_{\text{eff}})P \ , \quad (4a)$$

where H_{eff} and V_{eff} are, respectively, the model-space effective Hamiltonian and interaction. Clearly V_{eff} itself is dependent on U.

The effective interaction V_{eff} generally contains many-body components, i.e.,

$$V_{\text{eff}}=\overline{V}^{(0)}+\overline{V}^{(1)}+\overline{V}^{(2)}+\cdots \ , \quad (5)$$

where $\overline{V}^{(n)}$ denotes the n-body components of V_{eff}. This is so even when the original interaction V is taken merely as a two-body interaction, such as the Paris potential.[17] A basic step of the MBHF method is to choose U such that

$$P\overline{V}^{(1)}P=0 \ . \quad (6)$$

This is in fact a model-space Hartree-Fock condition, as it is equivalent to requiring $\langle 1\text{p}1\text{h} \,|\, PH_{\text{eff}}P \,|\, 0\text{p}0\text{h}\rangle=0$ where 1p1h and 0p0h are the familiar one-particle—one-hole and zero-particle—zero-hole eigenstates of H_0, respectively. From Eq. (6) we can only determine PUP, and we have chosen $PUQ=QUP=QUQ=0$.[4,5] Let us mention that the unknown to be solved for from Eq. (6) is PUP, and in doing so a self-consistent procedure must be employed. This is because, briefly speaking, we need to know H_0 and P in order to calculate the matrix elements of U from Eq. (6), while H_0 and P themselves are also dependent on U. For nuclear matter, this self-consistent procedure is simplified because our P, given in terms of plane-wave single particle states, is independent of U. As in Ref. 4, we use a one-G-matrix approximation in solving Eq. (6), and this leads to the following self-consistent equations for PUP:

$$\langle k \,|\, U^{\text{MHF}} \,|\, k\rangle=\sum_{h\leq k_F}\langle kh \,|\, \overline{G}(\omega) \,|\, kh\rangle \ , \quad (7)$$

$$\langle kh \,|\, \overline{G}(\omega) \,|\, kh\rangle = \langle kh \,|\, V \,|\, kh\rangle + \sum_{mn}\frac{\langle kh \,|\, V \,|\, mn\rangle\overline{Q}(mn)\langle mn \,|\, \overline{G}(\omega) \,|\, kh\rangle}{\omega-\epsilon_n^M-\epsilon_m^M} \ , \quad (7a)$$

$$\omega=\epsilon_k^M+\epsilon_h^M \ , \quad (7b)$$

$$\epsilon_l^M=t_l+\langle k_l \,|\, u^{\text{MHF}} \,|\, k_l\rangle, \ k_l\leq k_M \ ,$$
$$=t_l, \ k_l>k_M \ , \quad (7c)$$

where t_l is the kinetic energy $\hbar^2k_l^2/2m$. U and U^{MHF} are related by $U=\sum_{i=1}^{A}u^{\text{MHF}}(i)$. Note that the intermediate states of the $\overline{G}$ matrix must belong to the Q space, and this is ensured by using in Eq. (7a)

$$\overline{Q}(mn)=1, \text{ if } \max(k_m,k_n)>k_M \text{ and } \min(k_m,k_n)>k_F \ ,$$
$$=0, \text{ otherwise} \ . \quad (7d)$$

An angle-average approximation for $\overline{Q}$ (Ref. 4) has been used in our calculations. In this way, $\overline{Q}$ is dependent only on the magnitudes of the center of mass and relative momenta **K** and **k**. Then, as shown in Fig. 1, the values of $\overline{Q}$ are calculated depending on which regions the magnitudes of K and k belong to. These regions (a to f) are di-

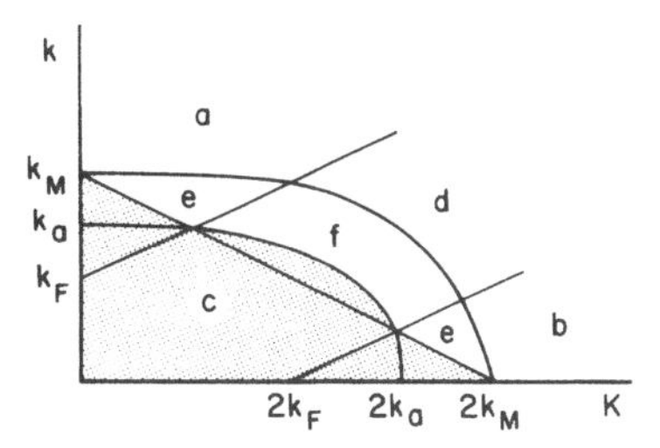

FIG. 1. Angle averaged projection operator $\overline{Q}(m,n)$ of Eq. (7). k_a^2 is $(k_F^2+k_M^2)/2$.

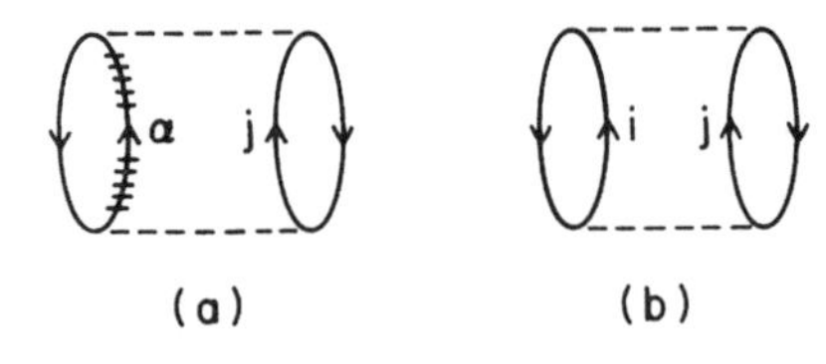

FIG. 2. Structure of $\overline{G}$ and G_M matrices. Particles with momentum $>k_M$ are denoted by railed lines. Bare particle lines are those with momentum $\leq k_M$.

vided by three lines, $k-K/2=\pm k_F$ and $k+K/2=k_M$, and two ellipses $k^2+K^2/4=k_M^2$ and $(k_F^2+k_M^2)/2$. In regions a and b we have $\overline{Q}=1$, and in the shaded region c we have $\overline{Q}=0$. The angle-averaged approximations are used only for regions d, e, and f, and the values of $\overline{Q}$ in these regions are, respectively, $(k^2-k_F^2-K^2/4)/kK$, $[(k+K/2)^2-k_M^2]/kK$, and $(2k^2-k_F^2-k_M^2+K^2/2)/kK$.

The $\overline{G}$ interaction defined above contains only those diagrams whose intermediate states have at least one particle with momentum $>k_M$. In nuclear matter calculations, particle-hole excitations with particle momentum $<k_M$ should also be included, and this can be done by including processes corresponding to repeated $\overline{G}$ interactions within the model space P. Let us give some examples. As shown in Fig. 2, diagram (a) is contained in $\overline{G}$ but not (b). This is because the bare particle lines i and j both have momentum $<k_M$, while the railed line a is a line with momentum $>k_M$. To include diagram (b) in the nuclear matter calculation, we need to calculate diagrams second order in $\overline{G}$. To include this type of diagram more completely, we adopt the MBHF method.[4] Briefly speaking, we first calculate the model-space two-body effective interaction by including $\overline{G}$ and all the two-body folded diagrams generated by $\overline{G}$, i.e.,

$$V_{\text{eff}}^{(2)} \approx \overline{G} - \overline{G} \int \overline{G} + \overline{G} \int \overline{G} \int \overline{G} - \cdots \equiv \overline{G}_F . \tag{8}$$

We then carry out BHF nuclear matter calculations within the model space, using the $\overline{G}_F$ as the effective nucleon-nucleon interaction within the model space. This leads to the following self-consistent equations:

$$\langle h_1h_2 | G_M(\omega) | h_1h_2 \rangle = \langle h_1h_2 | \overline{G}_F | h_1h_2 \rangle + \sum_{mn} \frac{\langle h_1h_2 | \overline{G}_F | mn \rangle Q(mn) \langle mn | G_M(\omega) | h_1h_2 \rangle}{\omega - \epsilon_m - \epsilon_n} , \tag{9}$$

where $\omega=\epsilon_{h_1}+\epsilon_{h_2}$ and

$$Q(m,n)=1, \text{ if } k_F<(k_m,k_n)\leq k_M ,$$
$$=0, \text{ otherwise} , \tag{9a}$$

$$\epsilon_m = t_m + \sum_{h<k_F} \langle mh | G_M(\epsilon_m+\epsilon_h) | mh \rangle \text{ if } k_m \leq k_F ,$$
$$=\epsilon_m^M \text{ if } k_m > k_F , \tag{9b}$$

where ϵ_m^M was given in Eq. (7b). The potential energy (PE) per particle in nuclear matter is given in terms of G_M by

$$\langle \text{PE} \rangle = \frac{1}{A} \sum_{h_1,h_2\leq k_F} \langle h_1h_2 | G_M(\epsilon_{h_1}+\epsilon_{h_2}) | h_1h_2 \rangle \tag{9c}$$

with the single particle energies ϵ given by Eq. (9b).

We can simplify the above calculations. By substituting $\overline{G}_F$ of Eq. (8) into Eq. (9) and making use of Eqs. (7a)–(7d), we can rewrite Eq. (9) as

$$\langle h_1h_2 | G_M(\omega) | h_1h_2 \rangle = \langle h_1h_2 | V | h_1h_2 \rangle + \sum_{mn} \frac{\langle h_1h_2 | V | mn \rangle [Q(mn)+\overline{Q}(mn)] \langle mn | G_M(\omega) | h_1h_2 \rangle}{\omega - \epsilon_m - \epsilon_n} \tag{10}$$

with $\omega=\epsilon_{h_1}+\epsilon_{h_2}$ and Q and $\overline{Q}$ given, respectively, by Eqs. (7d) and (9a). Equation (10) is more convenient for calculation than Eq. (9), because we now calculate G_M directly from V whereas for Eq. (9) we need to first calculate $\overline{G}_F$ and then calculate G_M from $\overline{G}_F$. From Eq. (10) we can see rather clearly the connection between the MBHF method and the conventional BHF method. The essential difference is in the treatment of the single particle spectrum for $k>k_F$. In the BHF method, free-particle spectrum is used for particles with $k>k_F$. In the present method, free particle spectrum is used only for $k>k_M$, whereas the model-space HF spectrum of Eq. (7c) is used for $k_F<k<k_M$. It is easily seen that the above MBHF method reduces to the conventional BHF method if we choose $k_M=k_F$. How should we choose k_M? If the calculations are carried out exactly, the results should be independent of k_M. In practice, we must make some approximations, and therefore the choice of k_M will affect our results. We have found that for $k_M \sim 2k_F$ or ~ 3 fm^{-1} the results of our nuclear matter binding energy calculations are quite stable with respect to small variations of k_M.[5] We have therefore chosen $k_M=2k_F$ in our calculations.

III. NUCLEAR MATTER CALCULATIONS USING PARIS POTENTIAL

The Paris potential reproduces the low-energy ($E \lesssim 330$ MeV) two-nucleon scattering data and deuteron properties very well, and its long- and medium-range parts are field theoretically derived including components from one-, two-, and three-pion exchanges. The short-range ($r<0.8$ fm) part of this potential is determined phenomenologically. In our nuclear matter calculations, we have used the parametrized form of this potential as the NN potential V of Eq. (1). This form of the Paris NN potential has a significant momentum dependent component which has been shown[13] to have important effects on the nuclear matter single-particle spectrum, effective mass, and binding energy. It should be pointed out that in numerical calculations special care must be given to the treatment of this momentum dependent component, as was found in coordinate space phase-shift and nuclear matter calculations.[13] We calculate the nuclear matter G matrix using momentum space integral equation methods, and have found that it is very important to treat the momentum space mesh points at high momentum (~ 30 fm^{-1}) with great care. A fairly high concentration of momentum space Gaussian points must be placed in this region in order to obtain numerical stability.

A first step in our calculation is the evaluation of the partial wave matrix element of the form

$$\langle kl \mid V \mid k'l' \rangle = \int_0^\infty r^2 dr\, j_l(kr) V(r) j_{l'}(k'r) \,, \tag{11}$$

where $V(r)$ is the Paris potential. This matrix element may be evaluated using numerical integration, but in this way high accuracy is difficult to obtain; this is because of the strong short-range components contained in $V(r)$ and the rapid oscillations of the Bessel functions when k and/or k' become large. The parametrized form of $V(r)$ is particularly convenient because it is composed of a sum of Yukawa terms of the form e^{-mr}/r and their derivatives. Then the matrix elements of Eq. (11) can be analytically evaluated by way of the integration formula

$$\int_0^\infty r^2 dr\, j_l(kr) \frac{e^{-mr}}{r} j_l(k'r) dr' = \frac{1}{2kk'} Q_l(z) \,, \tag{12}$$

where $z=(k^2+k'^2+m^2)/2kk'$ and $Q_l(z)$ is the Legendre function of the second kind. Thus the matrix elements $\langle kl \mid V \mid k'l' \rangle$ can be calculated either numerically or analytically. We have used both methods, and obtained satisfactory agreement between their results; this serves to check our computer programs. Results reported in this work were all carried out using analytically calculated $\langle kl \mid V \mid k'l' \rangle$.

The G_M matrix of Eq. (10) is then calculated in a partial wave basis, using angle average approximations for Q and $\bar{Q}$, namely

$$\langle k\alpha \mid G_M(\omega) \mid k'\alpha' \rangle = \langle k\alpha \mid V \mid k'\alpha' \rangle + \frac{2}{\pi} \int_0^\infty p^2 dp \sum_\beta \frac{\langle k\alpha \mid V \mid p\beta \rangle (Q+\bar{Q}) \langle p\beta \mid G_M(\omega) \mid k'\alpha' \rangle}{\omega - \epsilon_{k_1'} - \epsilon_{k_2'}} \,, \tag{13}$$

where α and β denote the two-nucleon partial wave quantum numbers ($lSTj$), k and k' the relative momenta, and $\epsilon_{k_1'}$ and $\epsilon_{k_2'}$ are the single particle energies given by Eq. (9b). The average potential energy per nucleon as given by Eq. (9c) is also calculated in terms of the partial wave G_M matrices, namely

$$\langle \mathrm{PE} \rangle = \sum_{k_1,k_2 \le k_F} \frac{4\pi}{A\Omega} \sum_\alpha (2T+1)(2j+1) \times \langle k\alpha \mid G_M(K^2) \mid k\alpha \rangle \,, \tag{14}$$

where A and Ω are, respectively, the mass number and volume of nuclear matter. The relative and center-of-mass momenta k and K are integrated over under the constraint that k_1 and k_2 are both less than k_F. ($\mathbf{k}_1=\mathbf{k}+\mathbf{K}/2$, $\mathbf{k}_2=\mathbf{k}-\mathbf{K}/2$.) Furthermore, an angle average approximation[1] for K^2 has been used in our calculation. For high partial waves ($l \ge 5$), short-range correlations between nucleons in nuclear matter are not important. Hence for these partial waves we have replaced G_M by V in Eq. (14). As shown in Table I, this replacement is judged to be a very accurate approximation for evaluating the nuclear matter potential energy for partial waves with $l \ge 4$. Here we see that the short-range correlations are important only for $l \le 2$ partial waves. For example, the $l=0$ contributions to U from V are generally repulsive. The main effect of including the short-range correlations is the conversion of V into G_M, and we see that the contributions from G_M are mostly attractive.

TABLE I. Average nuclear matter potential energies, in MeV, for various l values. The entries headed by G_M are calculated according to Eq. (14), while those under V are calculated in the same way except that G_M is replaced by V.

k_F (fm^{-1})	1.2		1.4		1.6	
l	V	G_M	V	G_M	V	G_M
0	29.97	−29.44	50.16	−37.66	77.99	−44.91
1	5.19	2.26	9.27	4.04	15.49	7.00
2	−2.91	−3.24	−5.56	−6.16	−9.54	−10.57
3	0.87	0.86	1.64	1.61	2.71	2.64
4	−0.48	−0.48	−1.09	−1.08	−2.12	−2.10

We have found that the high ($l > 5$) partial waves have the effect of reducing the saturation density of nuclear matter. Their contribution to U at saturation density is found to be ~ 0.6 MeV per nucleon. In a previous calculation,[5] the $l > 5$ partial waves were not included and the resulting saturation density was slightly larger than the one found in this work.

In Fig. 3, we show our results for the nuclear matter saturation curves, using the MBHF method as described above [mainly Eqs. (10), (13), and (14)] and elsewhere.[4,5] The average binding energy and saturation density are found to be 15.6 MeV and $k_F = 1.56$ fm^{-1}, respectively. (The calculations were performed using $k_M = 2k_F$.) We have also performed the usual BHF calculations, which correspond to the MBHF calculations with the special choice of $k_M = k_F$. As shown, saturation k_F and average binding energy for BHF are 1.5 fm^{-1} and 11.5 MeV, respectively. When compared with BHF, our MBHF calculations give an additional binding energy of ~ 4 MeV per nucleon. The nuclear matter incompressibility coefficient

$$\kappa = 9\rho^2 \frac{d^2}{d\rho^2} \left[\frac{E}{A} \right]_{k_F} = k^2 \frac{d^2}{dk^2} \left[\frac{E}{A} \right]_{k_F} \tag{15}$$

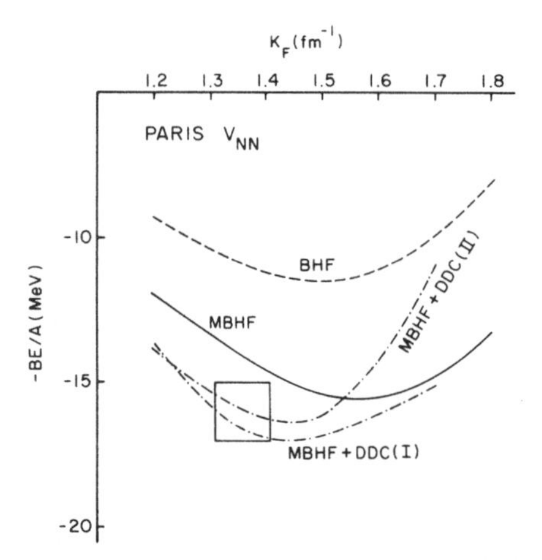

FIG. 3. Nuclear matter saturation curves. Curves BHF and MBHF are calculated using the Paris NN potential. The other two curves are both MBHF calculations using the Paris potential supplemented by density dependent central components. The binding energy per nucleon BE/A, saturation k_F and incompressibility coefficient κ for these curves are

Curve	BE/A	k_F	κ
BHF	11.5	1.50	153
MBHF	15.5	1.56	188
DDC(I)	17.1	1.45	231
DDC(II)	16.4	1.45	336

The empirical values for BE/A and saturation density are indicated by the box.

can be deduced from the saturation curves of Fig. 3. It is deduced to be approximately 150 MeV for BHF and 190 MeV for MBHF. Both are somewhat smaller than the empirical value of $\kappa = 220 \pm 20$ MeV.

In Table II, contributions to average potential energies from individual partial waves are tabulated, for two k_F values for both BHF and MBHF. As shown, the increase in binding energy from BHF to MBHF mainly comes from the $^3S_1$-$^3D_1$ and $^1S_0$ channels. (We have checked our computer programs by comparing our BHF results with those of Day.[14] For example, our BHF binding energy per nucleon at $k_F = 1.4$ fm^{-1} is 11.19 MeV while his value is 11.15 MeV. In general, satisfactory agreement between our results and those of Day has been obtained.) As discussed in Sec. II and elsewhere,[4,5] an essential difference between BHF and MBHF is the choice or determination of the single particle spectrum ϵ_k used in nuclear matter calculations. Let us divide ϵ_k into three regions and express it as

$$\begin{aligned} \epsilon_k &= \frac{\hbar^2 k^2}{2m_2^*} - \Delta_2 \quad \text{for } k < k_F , \\ &= \frac{\hbar^2 k^2}{2m_1^*} - \Delta_1 \quad \text{for } k_F < k < k_M , \\ &= \frac{\hbar^2 k^2}{2m} \quad \text{for } k > k_M . \end{aligned} \tag{16}$$

TABLE II. Decomposition of average nuclear matter potential energies (PE), in MeV, calculated from the Paris NN potential. See the text for other explanations.

k_F	1.4 (fm^{-1})		1.5 (fm^{-1})	
Channel	BHF	MBHF	BHF	MBHF
$^1S_0$	−16.97	−17.26	−19.16	−19.57
$^3S_1$-$^3S_1$	−17.99	−20.40	−19.29	−21.96
$^3D_1$-$^3D_1$	1.69	1.68	2.22	2.21
$^1P_1$	4.87	4.81	6.14	6.05
$^3P_0$	−3.84	−3.86	−4.52	−4.57
$^3P_1$	11.96	11.46	15.03	14.72
$^3P_2$-$^3P_2$	−8.14	−8.35	−10.65	−10.90
$^3F_2$-$^3F_2$	−0.69	−0.69	−0.95	−0.95
$^1D_2$	−3.21	−3.22	−4.30	−4.31
$^3D_2$	−4.65	−4.70	−6.13	−6.18
$^3D_3$-$^3D_3$	0.13	0.07	0.18	0.11
$^3G_3$-$^3G_3$	0.25	0.25	0.36	0.36
$^1F_3$	0.97	0.97	1.31	1.30
$^3F_3$	1.84	1.84	2.50	2.50
$^3F_4$-$^3F_4$	−0.51	−0.51	−0.76	−0.76
$^3H_4$-$^3H_4$	−0.11	−0.11	−0.17	−0.17
$^1G_4$	−0.56	−0.56	−0.79	−0.79
$^3G_4$	−0.87	−0.86	−1.24	−1.24
$^3G_5$-$^3G_5$	0.09	0.08	0.13	0.13
$^3I_5$-$^3I_5$	0.04	0.04	0.07	0.07
$l > 4$	0.31	0.31	0.43	0.43
PE	−35.57	−38.94	−39.48	−43.41
KE	24.38	24.38	27.99	27.99
E/A	−11.19	−14.55	−11.49	−15.42

TABLE III. Single particle spectra derived from BHF and MBHF, using the Paris NN potential.

k_F	1.4		1.5	
	BHF	MBHF	BHF	MBHF
m/m_1^*	1	1.28	1	1.31
Δ_1	0	66.74	0	74.40
m/m_2^*	1.52	1.49	1.56	1.53
Δ_2	83.79	89.87	94.58	101.77

In MBHF, one first determines m_1^* and Δ_1 based on Eqs. (7). Then within the model space ($k<k_M$) we can further include the hole-line self-energy insertions. This leads to the (m_2^*,Δ_2) spectrum for $k<k_F$, as shown in Eq. (16). For $k>k_M$ we use the free particle spectrum. In Table III, we give some typical values for m_1^*, Δ_1, m_2^*, and Δ_2. In BHF, the self-consistent spectrum is used only for $k<k_F$, while the free particle is used for all other momenta. The values of m_2^* and Δ_2 for BHF are also shown in Table III. Clearly the MBHF and BHF spectrum are quite different for $k_F<k<k_M$. For $k<k_F$, the two are rather similar except that the MBHF spectrum is about 5 MeV lower than the BHF spectrum.

IV. DENSITY DEPENDENT CENTRAL POTENTIAL

We have seen that the binding energy per nucleon given by the MBHF nuclear matter calculations using the Paris NN potential is rather close to the empirical value of ~ 16 MeV/A. The saturation density given by such calculations is, however, larger than the empirical value corresponding to $k_F \sim 1.36$ fm^{-1}. Day[15] has pointed out that by using only a two-body NN interaction that fits the low energy scattering data and the deuteron properties, it is difficult to reproduce the empirical nuclear matter saturation properties. One can therefore conjecture that a density dependent component in the bare NN interaction for nucleons in nuclear matter may be needed. Based on our calculations, we would like this component to have a net repulsive effect in nuclear matter binding energy calculations for $k_F \geq 1.5$ fm^{-1}, while for $k_F \lesssim 1.5$ fm^{-1} it should have a net attractive effect. Recently there has been much discussion on the effect of the three-body NN interaction in nuclear matter calculations. Although the effect of three-body and higher-body NN interactions to nuclear matter binding energy calculations may be small as compared to the contribution from the two-body NN interaction, their effect on nuclear matter saturation properties may, however, not be negligible.

In this work, we would like to investigate the effect of an empirical density dependent force in nuclear matter saturation density calculations. Our purpose is mainly to estimate the general strength of such a density dependent force so that its addition to the Paris NN potential will shift the nuclear matter saturation density given by MBHF to $k_F \approx 1.36$ fm^{-1}. We regard this density dependent central (DDC) piece as an empirical device to account for effects other than the two-body forces. It is well-known that a three-body force in nuclear matter can be generally represented by a two-body density dependent force. Thus we introduce a parametrized density dependent two-body force

$$V_{\rm DDC}(k_F)=\alpha e^{-\beta(k_F-k_0)^2}V_C$$

where V_C is the isospin independent central component of the Paris NN potential. α, β, and k_0 are parameters. We have performed MBHF calculations using various values for α, β, and k_0. In Fig. 3, the curve labeled MBHF + DDC(I) is obtained using $\alpha=0.1$, $\beta=20$ fm^2, and $k_0=1.30$ fm^{-1}. The NN interaction used in this MBHF calculation is given by the sum of $V_{\rm Paris}$ and $V_{\rm DDC}$. As shown, the resulting saturation Fermi momentum and binding energy per nucleon are, respectively, ~ 1.45 fm^{-1} and ~ 17 MeV, in good agreement with the empirical values. The resulting incompressibility coefficient is ~ 230 MeV, which is also in good agreement with the empirical value of $\sim 200 \pm 20$ MeV. The above results indicate clearly that we only need a rather weak density-dependent two-body central interaction, whose strength is of the order of 10% of that of the Paris central potential, in order to bring the calculated nuclear matter saturation density and binding energy in simultaneous agreement with the empirical values. Based on a σ model, Jackson, Rho, and Krotscheck[9] have investigated the three-body forces for nucleons in nuclear matter. They suggested that the main effect of such forces may be represented by a two-body effective central interaction due to one σ-meson exchange with effective mass m'_σ, which is related to the bare mass m_σ by $m'_\sigma \approx m_\sigma(1-\alpha\rho+\beta\rho^{5/3})$, where ρ is the nuclear matter density. The constants α and β were given as ~ 0.5 fm^3 and ~ 1.2 fm^5, respectively. The medium range attraction of the NN potential comes mainly from the σ exchange. Thus, the renormalized m'_σ makes this part of the NN interaction density dependent. It is difficult to rigorously incorporate this effect into the Paris NN potential, because its medium range attractive part is due to the $\pi\pi$ S wave interaction rather than a "σ meson." As a preliminary investigation, we have simply modified the central part of the Paris potential V_C by a similar density dependent factor, converting it into $V_C/(1-\alpha'\rho+\beta'\rho^{5/3})$. When using this modified potential with $\alpha'=1.7$ fm^3 and $\beta'=4.3$ fm^5 in our MBHF calculation of nuclear matter, the resulting binding energy and saturation density are both in reasonably good agreement with the empirical values, as shown by the curve MBHF + DDC(II) of Fig. 3. The incompressibility coefficient obtained from this curve is ~ 336 MeV, which is somewhat too large as compared with the empirical value.

V. DISCUSSION AND CONCLUSION

We have carried out MBHF nuclear matter calculations using the Paris NN potential. Comparing with the corresponding BHF results, MBHF gives an additional binding energy of about 4 MeV per nucleon while slightly increasing the saturation density. The trend of these results is approximately the same as that observed in a MBHF calculation of nuclear matter using the Reid NN potential.[4,5]

The main difference between MBHF and BHF is the use of the single particle potential. Using a model-space HF approach, the self-consistent single particle spectrum given by MBHF is a continuous one for momentum $0<k<k_M$ where k_M is the chosen momentum-space model-space boundary. If one chooses $k_M>k_F$, then one has a continuous single particle spectrum extended beyond k_F. If one chooses $k_M=k_F$, then MBHF reduces to BHF whose single particle spectrum has a huge discontinuity at k_F; this is rather unphysical, as has been pointed by Mahaux and his collaborators[10,13] some time ago. Nuclear matter calculations are numerically rather complicated, and it will be very helpful to have checks with independent calculations. The present calculation is rather similar to a recent nuclear matter calculation using the Paris NN potential carried out by Lejeune, Martzolff, and Grange.[13] They also used a continuous single particle spectrum, but theirs is derived from a Green's function approach while ours is from a model space HF approach. Their single particle potential is generally complex, while ours is real. Nevertheless, the real parts of their single particle spectrum and our spectrum are numerically very close to each other for momentum $\lesssim 3$ fm^{-1}, with difference ≈ 10 MeV or less. (They used the real part of their single particle potential in their nuclear matter calculation.) Note that their spectrum is continuous for all momenta while ours has a small gap at k_M. It is rather satisfactory to note that the resulting BE/A and saturation k_F for their and our calculations are, respectively, (16.0±2 MeV, 1.62 fm^{-1}) and (15.5 MeV and 1.56 fm^{-1}). They are in remarkably good agreement. (Note that in addition to the above difference in the single particle spectrum, these two calculations also differ in methods of calculations. Their reaction matrix was obtained by solving differential equations in the coordinate space, while we have calculated our reaction matrix using the momentum space matrix inversion method.) The above confirms the general trend that nuclear matter calculations using a continuous single particle similar to the one derived in this work or that of Ref. 13 can increase the average nuclear matter binding energy by about 4 MeV, as compared with the conventional BHF results.

As discussed elsewhere,[5] the gain in BE/A from BHF to MBHF nuclear matter calculations is primarily due to the difference in the single particle spectrum used in these two calculations. The particle-hole gap in the single particle spectrum of MBHF is considerably smaller than that of BHF. Consequently, the energy denominators for low energy particle-hole excitations in nuclear matter are significantly reduced. This increases the contribution from these excitations to the nuclear matter binding energy, particularly for the $^3S_1$-$^3D_1$ channel (see Table II) where the NN tensor interaction is important.

The above gain in BE/A can also be explained from a different viewpoint. In BHF, the potential energy for a nucleon in the momentum region k_F to k_M is zero, while in MBHF it is generally attractive and has an average value of about -40 MeV. The probability of having a nucleon excited from $<k_F$ to the above momentum region is approximately given by the familiar wound integral whose value is found to be ~ 0.1. The potential energy for a nucleon below k_F in BHF and MBHF are approximately equal to each other. Hence the gain in BE/A from BHF to MBHF should be approximately $0.1\times(-40)=-4$ MeV. This estimate agrees well with the result shown in Fig. 3.

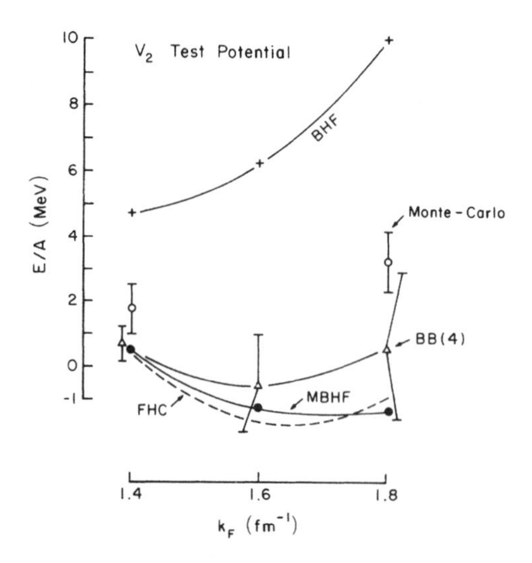

FIG. 4. Nuclear matter calculations using the V_2 test potential. Results of our calculations are denoted by BHF and MBHF. Those from the Green's function Monte-Carlo calculations (Ref. 17), the Fermi hypernetted chain calculations (Ref. 18), and Day's four-hole-line calculations (Ref. 19) are denoted by Monte-Carlo, FHC, and BB(4), respectively. Results for Monte Carlo and BB(4) are given with error estimates.

We have found that the contribution from the high-order partial waves ($l\gtrsim 4$) to the nuclear matter binding energy to be generally not important as indicated by Table II. This is consistent with the results of Sprung *et al.*[17] who estimated the contributions from high-order partial waves to nuclear matter binding energy directly from phase shifts, and the results of Grangé *et al.*[18] who investigated these contributions for nuclear matter calculations using continuous single particle spectra.

Although the calculated BE/A is in fairly good agreement with the empirical value, the calculated saturation density, however, is too high. We have investigated the effect of adding a weak density dependent central potential to the Paris NN potential. Although our investigation in this regard is rather preliminary, its results do indicate that we only need a rather *weak* (about 10% of the Paris central potential) additional density dependent central force to make the calculated BE/A and saturation k_F in simultaneous agreement with the empirical values.

To further present calculations, it seems to be of the highest priority to calculate some higher-order diagrams within the framework of MBHF. In MBHF, one essentially includes only the two-hole-line diagrams. Hence it is basically the same as BHF except for the use of the

MBHF continuous single particle spectrum as mentioned above. An important question to be answered is the following: Now that the MBHF BE/A is already in fairly good agreement with the empirical value, will this good agreement remain when one further includes some higher-order diagrams such as the ring diagrams within the model space? If it is so, then the net effect of all the higher-order diagrams must be small. It will be of much interest to find out if this turns out to be true. Ring diagram nuclear matter calculations using the Paris potential and the MBHF approach are now being carried out.[16]

To test the accuracy of the MBHF approach, we have carried out model nuclear matter calculations using the V_2 test potential. This potential is just the central part of the $^3S_1$ Reid NN potential, and for this potential highly accurate and elaborate nuclear matter calculations are available, namely the Monte-Carlo calculation,[19] the Fermi hypernetted chain calculation,[20] and Day's four-hole-line BHF calculation.[21] As shown in Fig. 4, our BHF results largely deviate from the results of these calculations. But our MBHF results agree with the latter two remarkably well. This is certainly an encouraging agreement.

Note added in proof. M. A. Matin and M. Dey [Phys. Rev. C **27**, 2356 (1983); **29**, 344 (1984)] have performed similar nuclear matter calculations and obtained BE/$A \approx 21$ MeV and a saturation point at $k_F \approx 1.6$ fm^{-1}.

ACKNOWLEDGMENTS

The authors are very grateful to Dr. B. D. Day for many stimulating discussions and advice, and for providing them with his unpublished results. Without them, this work would have not been possible. The authors are also very grateful to Prof. G. E. Brown for encouragement and many enlightening discussions, and to Dr. C. Hajduk for kindly providing them with his elaborate computer programs for computing the momentum-space matrix elements of the Paris potential. Finally Z.Y.M. and R.V.M. wish to thank Prof. G. E. Brown for his warm hospitality during their stay at Stony Brook. This work was supported in part by the U.S. Department of Energy Contract No. DE-AC02-76ER13001.

*Permanent address: Institute of Atomic Energy, P.O. Box 275(41), Beijing, The People's Republic of China.

†Permanent address: Division de Physique Theorique, Institut de Physique Nucleaire, 91406 Orsay, France.

[1] H. A. Bethe, Annu. Rev. Nucl. Sci. **21**, 93 (1971); D. W. L. Sprung, Adv. Nucl. Phys. **5**, 225 (1972); B. D. Day, Rev. Mod. Phys. **50**, 495 (1978).

[2] R. Jastrow, Phys. Rev. **98**, 1479 (1955); S. Fantoni and S. Rosati, Nuovo Cimento **A20**, 179 (1974).

[3] H. Kümmel, K. H. Lührmann, and J. G. Zabolitzky, Phys. Rep. **36C**, 1 (1978).

[4] Z. Y. Ma and T. T. S. Kuo, Phys. Lett. **B127**, 137 (1983).

[5] T. T. S. Kuo and Z. Y. Ma, in *Nucleon-Nucleon Interaction and Nuclear Many Body Problems,* edited by S. S. Wu and T. T. S. Kuo (World-Scientific, Singapore, 1984), p. 178.

[6] F. Coester, S. Cohen, B. Day, and C. M. Vincent, Phys. Rev. C **1**, 769 (1970).

[7] M. Lacombe, B. Loiseau, J. M. Richard, R. Vinh Mau, J. Coté, P. Pires, and R. de Tourreil, Phys. Rev. C **21**, 861 (1980).

[8] S. Barshay and G. E. Brown, Phys. Rev. Lett. **34**, 1106 (1975).

[9] A. D. Jackson, M. Rho, and E. Krotscheck, Nucl. Phys. **A407**, 495 (1983).

[10] J. P. Jeukenne, A. Lejeune, and C. Mahaux, Phys. Rep. **25**, 83 (1975).

[11] T. T. S. Kuo, S. Y. Lee, and K. F. Ratcliff, Nucl. Phys. **A176**, 65 (1971); T. T. S. Kuo, Annu. Rev. Nucl. Sci. **24**, 101 (1974); T. T. S. Kuo and E. M. Krenciglowa, Nucl. Phys. **A342**, 454 (1980).

[12] P. J. Ellis and E. Osnes, Rev. Mod. Phys. **49**, 777 (1977).

[13] A. Lejeune, M. Martzolff, and P. Grangé, *Lecture Notes in Physics* (Springer, Berlin, 1984), Vol. 198, p. 36.

[14] B. D. Day, private communication.

[15] B. D. Day, Phys. Rev. Lett. **47**, 226 (1981).

[16] H. Q. Song, S. D. Yang, and T. T. S. Kuo, private communication.

[17] D. W. L. Sprung, P. K. Banerjee, A. M. Jopko, and M. K. Scrivastava, Nucl. Phys. **A144**, 245 (1970).

[18] P. Grangé, A. Lejeune, and C. Mahaux, Nucl. Phys. **A319**, 50 (1979).

[19] D. Ceperley, G. V. Chester, and M. H. Kalos, Phys. Rev. B **16**, 3081 (1977).

[20] K. E. Schmidt and V. R. Pandharipande, Nucl. Phys. **A328**, 240 (1979).

[21] B. D. Day, Nucl. Phys. **A328**, 1 (1979).

Nuclear Physics **A462** (1987) 491-526
North-Holland, Amsterdam

INFINITE ORDER SUMMATION OF PARTICLE-PARTICLE RING DIAGRAMS IN A MODEL-SPACE APPROACH FOR NUCLEAR MATTER*

H.Q. SONG[1], S.D.YANG[2] and T.T.S. KUO

Department of Physics, The State University of New York at Stony Brook, Stony Brook, NY 11794, USA

Received 28 February 1986
(Revised 25 July 1986)

Abstract: A ring diagram model-space nuclear matter theory is formulated and applied to the calculation of the binding energy per nucleon (BE/A), saturation Fermi momentum (k_F) and incompressibility coefficient (K) of symmetric nuclear matter, using the Paris and Reid nucleon-nucleon potentials. A model space is introduced where all nucleons are restricted to have momentum $k \leq k_M$, typical values of k_M being $\sim 3.2\ \mathrm{fm}^{-1}$. Using a model-space Hartree-Fock approach, self-consistent single particle spectra are derived for holes ($k \leq k_F$) and particles with momentum $k_F < k \leq k_M$. For particles with $k > k_M$, we use a free particle spectrum. Within the model space we sum up the particle-particle ring diagrams (both forward- and backward-going) to all orders. A rather simple expression for the energy shift ΔE_0^{pp} is obtained, namely ΔE_0^{pp} is expressed as integrals involving the trace of $Y(\lambda)Y^{+}(\lambda)G^{M}$ where G^{M} is the model-space reaction matrix, the Y's are transition amplitudes obtained from solving RPA-type secular equations and λ is a strength parameter to be integrated from 0 to 1. We have used angle-average approximations in our calculations, and in this way different partial wave channels are decoupled. For the $^3S_1$-$^3D_1$ channel, the effect of the ring diagrams is found to be particularly important. The inclusion of the ring diagrams has largely increased the role of the tensor force in determining the nuclear matter saturation properties, and consequently we obtain saturation densities which are significantly lower than those given by most other calculations. For the Paris potential, our results for BE/A, k_F and K are respectively 17.38 MeV, 1.42 fm^{-1} and 96.3 MeV. For the Reid potential, the corresponding results are 15.15 MeV, 1.30 fm^{-1} and 110.7 MeV. Our calculated values for the binding energy per nucleon and saturation density are *both* in rather satisfactory agreement with the corresponding empirical values.

1. Introduction

A primary aim of nuclear matter theory is to provide a microscopic derivation of the empirical nuclear matter properties such as the binding energy per nucleon BE/A, saturation density ρ_0 (or saturation Fermi momentum k_F) and the incompressibility coefficient K. In fact there have been a very large amount of works [1-6]) in

* Supported in part by the United States Department of Energy under contract number DE-AC02-76ER13001.

[1] Permanent address: Institute of Nuclear Research, Academia Sinica, P.O. Box 8204, Shanghai, the People's Republic of China.

[2] Permanent address: Dept. of Physics, Jilin University, Changchun, the People's Republic of China.

0375-9474/87/$03.50 © Elsevier Science Publishers B.V.
(North-Holland Physics Publishing Division)

this area, and certainly nuclear matter theory has played a very important role in nuclear physics. A long standing difficulty in nuclear matter theory has been the inability of almost all theoretical calculations in obtaining BE/A and ρ_0 which are in simultaneous agreement with the corresponding empirical values. [7]) (The accepted empirical values for BE/A, ρ_0 and K are respectively ~16 MeV, ~0.17 fm^{-1} and ~200 MeV. Recently Brown and Osnes [8]) have suggested that K should be considerably smaller, being ~110 MeV.) Frequently, if the calculated BE/A is satisfactory, then the calculated ρ_0 is too high. And if the calculated ρ_0 is more or less correct, then the calculated BE/A is usually far from being adequate.

There have been a number of developments aimed at resolving the above discrepancy. Many believe that a major rethinking of the two-body nucleon-nucleon(NN) potential is in order [9]). The concern here is primarily that the usual practice of employing only a two-body NN potential which fits the NN scattering data in nuclear matter calculations is indequate in reproducing the empirical properties of nuclear matter. In fact, a recent nuclear matter calculation by Kuo, Ma and Vinh Mau [10]) indicated that a small three-body interaction, cast in the form of a density dependent two-body interaction, was quite helpful in bringing the calculated values of both the BE/A and ρ_0 to reasonable agreement with the empirical results. Another line of developments is in the direction of refining the many-body methods used in nuclear matter calculations. Many elaboorate many-body methods have been advocated, such as the exp S approach [11]), the Green function approach [3]), the hyper-netted chain approach [12]), and the model-space approach [13]). All these approaches are formulated to give a more accurate nuclear matter theory than the prototypical lowest-order BHF theory with a discontinuous single-particle spectrum. The above theories are all non-relativistic. It should be noted that Shakin and his collaborators [14]) have recently proposed a highly promising relativistic nuclear matter theory, suggesting that relativistic effects may be very important in reproducing the correct nuclear matter saturation density.

The main purpose of the present study is a rather confined one. Recently, Yang, Heyer and Kuo [YHK] [15]) have proposed a convenient and rigorous method for summing up certain classes of ring diagrams to all orders for the calculation of ground state energies of general many-body systems. It seems to be of much interest to investigate the application of this method to nuclear matter calculations. In particular, the long range correlations arising from ring diagrams correlations in nuclear matter seem not to have been studied in a general way in existing nuclear matter calculations, and we feel that they may play an important role in determining the nuclear matter bulk properties BE/A, ρ_0 and K and the inter-relation among them. The combination of the YHK ring diagram method and the model-space nuclear matter approach [13]) may provide a new and suitable formulation for studying such long range correlations in nuclear matter. In this paper, we formulate and carry out model-space nuclear matter calculations where within a chosen model space the particle-particle ring diagrams of nuclear matter are summed to all orders. The effect of these ring diagrams on BE/A, ρ_0 and K are, in particular, examined.

Our calculation is also motivated by the following physical intuition. In treating many-body problems, it is definitely of great importance to choose a "good" single particle (s.p.) potential, in the sense that this potential should be a good representation of the actual average nuclear field felt by a nucleon in the nuclear medium under consideration. From this view point, the conventional BHF s.p. spectrum is certainly unsatisfactory because it has a large artificial discontinuity at the Fermi surface k_F [3]. We will adopt the model-space approach of Ma and Kuo [13] for determining the s.p. potential. This approach advocates the following partition concept. A model space P is defined as a configuration space where all nucleons are restricted to have momentum $k \leq k_M$, k_M being the momentum space boundary of P. As is discussed later, a typical value for k_M is $\sim 3\ \text{fm}^{-1}$. It is then conjectured that nuclear correlations with their intermediate states outside P can be adequately treated with the usual G-matrix approach. In fact, these correlations are primarily due to the short-range repulsion of the NN potential. Nuclear correlations whose intermediate states are inside P should be treated more thoroughly, such as including the ring diagrams to all orders. Based on the above concept our nuclear matter calculations are carried out in the following two-step approach. First, we choose a model space P and calculate self-consistently the model-space reaction matrix and s.p. spectrum. Then, within P, we sum up the particle-particle ring diagrams to all orders using the YHK method. Calculations are carried out using both the Paris [16] and the Reid [17] NN potentials.

The present work is a continuation of the model-space BHF (MBHF) nuclear matter calculations of Ma and Kuo [13], and Kuo, Ma and Vinh Mau [10]*. In their calculations, a space P was first chosen and within P a continuous s.p. spectrum was determined self-consistently. Then the MBHF calculations were carried out by summing up only the upward-going G-matrix ladder diagrams within P to all orders. The binding energies given by these MBHF calculations are already fairly satisfactory. Hence there is the concern that nuclear matter may become overly bound when we include ring diagrams to all orders in our calculations. This concern in fact appears to be a very cogent one because in our model space approach the s.p. spectrum is continuous. Thus there is either no gap between particle and hole states, or a relatively small gap at k_F if one chooses to treat the hole s.p. spectrum slightly differently from the particle one [13]. Consequently it is relatively easy to have particle-hole excitations, and the effect of the ring diagrams for such model-space calculations may be particularly large or even overly so. Needless to say, it is this concern which has provided us with additional motivation for carrying out the present nuclear matter calculations.

* Several calculational details of the MBHF method for nuclear matter calculations were not given in refs. [10,13]. We will describe them in this paper.

The YHK ring diagram method, as outlined in sect. 2, can be used to sum up both the particle-particle and the particle-hole ring diagrams. In the present paper, we limit ourselves to the study of the particle-particle ring diagrams only*. To our knowledge, the effect of this type of ring diagrams has not been studied in nuclear matter calculations before. Our formulation for calculating such ring diagrams is described in sect. 3, where we show that the YHK ring diagram method can be rigorously reformulated in terms of the model-space reaction matrix and in the binding energy calculation the trace is taken entirely within the model space. A desirable feature of our formulation is that the final calculational procedure is indeed relatively simple. Basically, we just solve an RPA type secular equation and calculate the transition amplitude Y's. Then the nuclear matter binding energy calculation is reduced to simple integrals involving the trace of $YY^{+}G^{M}$ where G^{M} is the model space reaction matrix. The normalization of the Y-amplitudes is determined self-consistently, and its role in determining the nuclear matter saturation density is discussed.

Results of our ring diagram calculations and some calculational details, such as those for the angle-average approximation for the RPA equation, are presented in sect. 4. Although we do not calculate the particle-hole ring diagrams in the present work, we give arguments that, within the present formulation, the effect of the nuclear matter particle-hole ring diagrams may be considerably less important than that of the corresponding particle-particle ring diagrams. This is discussed in sect. 5 where we present our discussions and conclusions of the present calculation.

2. Summation of ring diagrams

Since the basic formulation for the summation of ring diagrams has been given in detail elsehwere [15]), here we will just describe some of its essential features which are necessary in modifying it so as to treat the G-matrix ring diagrams of nuclear matter. We start from a hamiltonian $H = T + V$. Introducing a s.p. potential U, we rewrite it as $H = H_0 + H_1 \equiv (T+U)+(V-U)$. Here V is the NN potential, such as the Paris potential [16]). The s.p. wave functions and energies defined by H_0 are denoted by ϕ_i and ε_i, respectively. As described later, we have used a model-space approach [13]) for determining U.

Let E_0 and W_0 denote, respectively, the true and unperturbed ground state energy of nuclear matter. Then the binding energy per nucleon BE/A for nuclear matter is just $-E_0/A$ in the limit of $A\to\infty$, A being the number of nucleons. Thus a basic quantity to be calculated in nuclear matter theory is $-E_0/A$. As is well known, the energy shift $\Delta E_0 = E_0 - W_0$ can be calculated using standard linked diagram expansions [18]). Some low order diagrams of ΔE_0 are given in fig. 1.

* We plan to investigate the particle-hole ring diagrams in a future publication. As we will discuss later, there are indications that the effect of the particle-hole ring diagrams may be smaller than that of the particle-particle ring diagrams.

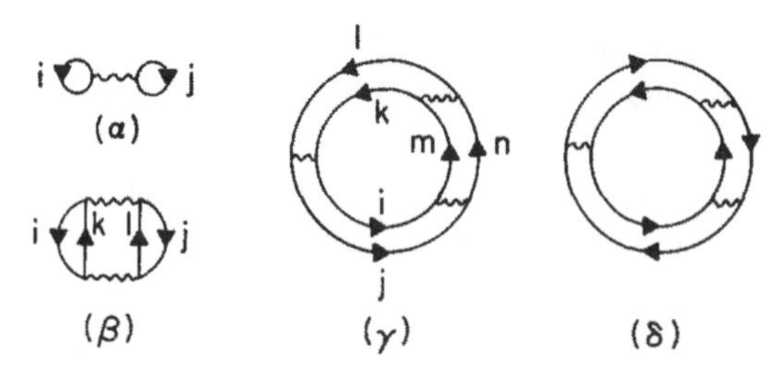

Fig. 1. Low-order linked diagrams of the energy shift $\Delta E_0 = E_0 - W_0$. Each wave line vertex denotes an anti-symmetrized vertex of V.

The contributions to ΔE_0 from diagrams (α), (β) and (γ) of fig. 1 are [15]) respectively

$$\Delta E_0^{\mathrm{pp}}(1) = \frac{-1}{2\pi i}\int_{-\infty}^{\infty} \mathrm{d}\omega\, \mathrm{e}^{i\omega 0^+} F_{ij}(\omega)\bar{V}_{ijij}, \tag{2.1}$$

$$\Delta E_0^{\mathrm{pp}}(2) = \frac{-1}{2\pi i}\int_{-\infty}^{\infty} \mathrm{d}\omega\, \mathrm{e}^{i\omega 0^+} \tfrac{1}{2} F_{ij}(\omega)\bar{V}_{ijkl}F_{kl}(\omega)\bar{V}_{klij}, \tag{2.2}$$

$$\Delta E_0^{\mathrm{pp}}(3) = \frac{-1}{2\pi i}\int_{-\infty}^{\infty} \mathrm{d}\omega\, \mathrm{e}^{i\omega 0^+} \tfrac{1}{3} F_{ij}(\omega)\bar{V}_{ijkl}F_{kl}(\omega)\bar{V}_{klmn}F_{mn}(\omega)\bar{V}_{mnij}, \tag{2.3}$$

where repeated indices are summed over all the s.p. states without restriction (e.g., we sum over both $i>j$ and $i<j$), and F_{ij} is the unperturbed pair propagator given by

$$F_{ij}(\omega) = \frac{\bar{n}_i\bar{n}_j}{\omega-(\varepsilon_i+\varepsilon_j)+i0} - \frac{n_i n_j}{\omega-(\varepsilon_i+\varepsilon_j)-i0}. \tag{2.4}$$

The matrix elements of $\bar{V}$ are defined as

$$\bar{V}_{ijkl} = \tfrac{1}{2}(V_{ijkl} - V_{ijlk}). \tag{2.5}$$

Note that V_{ijkl} is the simple product matrix element $\int \mathrm{d}^3r_1\, \mathrm{d}^3r_2\, \phi_i^*(\boldsymbol{r}_1)\phi_j^*(\boldsymbol{r}_2)\cdot V(r_{12})\phi_k(\boldsymbol{r}_1)\phi_l(\boldsymbol{r}_2)$. In eq. (2.4), $n_i=1$ if $k_i<k_F$ and $=0$ if $k_i>k_F$, and $\bar{n}_i \equiv 1-n_i$. In addition, the factors $\pm i0$ in the denominators are abbreviations for $\pm i\eta$ in the limit of $\eta\to 0^+$. (0^+ here and in eqs. (2.1) to (2.3) denotes an infinitesimally small positive quantity.) This notation will be used from now on.

In this way, the particle-particle (pp) ring diagrams such as those shown by diagrams (α), (β) and (γ) of fig. 1 can be summed up to all orders, leading to

$$\Delta E_0^{\mathrm{pp}} = \frac{-1}{2\pi i}\int_{-\infty}^{\infty} \mathrm{d}\omega\, \mathrm{e}^{i\omega 0^+}\, \mathrm{tr}\,(F\bar{V} + \tfrac{1}{2}(F\bar{V})^2 + \tfrac{1}{3}(F\bar{V})^3 + \cdots\}, \tag{2.6}$$

where the series inside the curly brackets form a logarithmic function. However, this is not so convenient for calculation. Let us introduce a strength parameter λ, and convert eq. (2.6) into

$$\Delta E_0^{\mathrm{pp}} = \frac{-1}{2\pi i} \int_0^1 \frac{\mathrm{d}\lambda}{\lambda} \int_{-\infty}^{\infty} \mathrm{d}\omega \; \mathrm{e}^{i\omega 0^+} \operatorname{tr}\{\lambda F\bar{V} + (\lambda F\bar{V})^2 + (\lambda F\bar{V})^3 + \cdots\}. \tag{2.7}$$

It is of interest to note that the introduction of the strength parameter λ makes the quantity inside the curly brackets a simple geometric series. This is not only computationally convenient but also physically desirable - it tells us that the ground state energy shift ΔE_0^{pp} is dependent on how the many-body system evolves from a non-interacting system ($\lambda = 0$) to a fully intereacting system ($\lambda = 1$).

We now introduce a λ-dependent particle-particle Green function $G^{\mathrm{pp}}(\omega, \lambda)$ defined by

$$G_{ijkl}^{\mathrm{pp}}(\omega, \lambda) = F_{ij}(\omega)\delta_{ij,kl} + F_{ij}(\omega)\lambda \bar{V}_{ijmn}(G_{mnkl}^{\mathrm{pp}}(\omega, \lambda). \tag{2.8}$$

Then ΔE_0^{pp} of eq. (2.7) can be cast into the compact form

$$\Delta E_0^{\mathrm{pp}} = \frac{-1}{2\pi i} \int_0^1 \frac{\mathrm{d}\lambda}{2\pi i} \int_{-\infty}^{\infty} \mathrm{d}\omega \; \mathrm{e}^{i\omega 0^+} \operatorname{tr}\{G^{\mathrm{pp}}(\omega, \lambda)\bar{V}\lambda\}. \tag{2.9}$$

An alternative approach is via the λ-dependent generalized reaction matrix $K^{\mathrm{pp}}(\omega, \lambda)$ defined by

$$K_{ijkl}^{\mathrm{pp}}(\omega, \lambda) = \lambda \bar{V}_{ijkl} + \lambda \bar{V}_{ijmn} F_{mn}(\omega) K_{mnkl}^{\mathrm{pp}}(\omega, \lambda). \tag{2.10}$$

Then ΔE_0^{pp} is given by

$$\Delta E_0^{\mathrm{pp}} = \frac{-1}{2\pi i} \int_0^1 \frac{\mathrm{d}\lambda}{\lambda} \int_{-\infty}^{\infty} \mathrm{d}\omega \; \mathrm{e}^{i\omega 0^+} \operatorname{tr}\{F(\omega) K^{\mathrm{pp}}(\omega, \lambda)\}. \tag{2.11}$$

The task now confronting us is how to actually calculate ΔE_0^{pp} for nuclear matter. The expressions (2.9) and (2.11) are both compact and appealing but it is more important to investigate whether they provide a convenient and practical scheme for actual calculations. Before investigating this question for nuclear matter, let us first point out that our ΔE_0^{pp} sums up the particle-particle ring diagrams to all orders, including diagrams (α), (β) and (γ) of fig. 1, and the general particle-particle ring diagram (i) of fig. 2. Thus we draw the general ring diagram in the circular form shown in fig. 2 to emphasize the fact that the two lines in each pair propagator can be either both particles ($> k_{\mathrm{F}}$) or both holes ($< k_{\mathrm{F}}$). Hence our present formulation includes contributions to ΔE_0 from repeated interactions between two hole lines (i.e. hole-hole correlations) to all orders and also ground-state correlation interactions connecting two particle to two hole lines to all orders. In contrast, only the repeated interactions between a pair of particle lines are included in the usual BHF type nuclear matter calculations, as illustrated by diagram (ii) of fig. 2.

We have found that it is more convenient in several aspects to use eq. (2.9) than eq. (2.11) for calculating ΔE_0^{pp}. This is because the frequency integration (i.e. $\int \mathrm{d}\omega$)

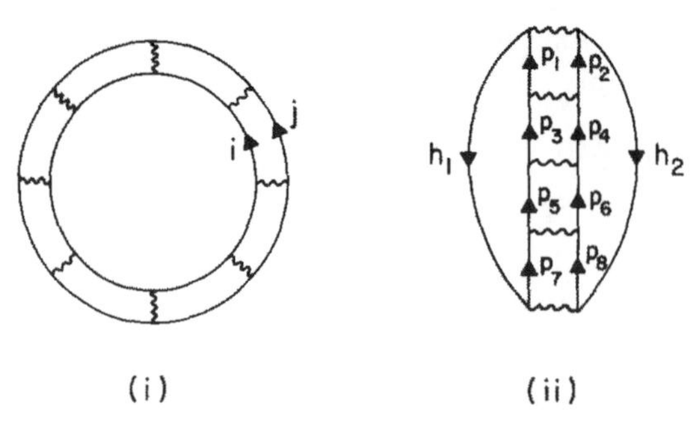

Fig. 2. (i) A general ring diagram included in the present work. Note that i and j can be either both particles ($>k_F$) or both holes ($<k_F$). (ii) A general diagram included in the usual BHF type nuclear matter calculations. Note that all p's are restricted to be $>k_F$, and h's$<k_F$.

in the former is considerably easier to carry out than in the latter. [15]) Hence, in the present work, we use the former scheme. To carry on the calculation, we first need to calculate the Green function G^{pp} and also find a way to perform the ω integration. A convenient way to proceed is to write G^{pp} in its Lehmann's representation, namely

$$G^{pp}_{ijkl}(\omega,\lambda)=\sum_n \frac{X_n(ij,\lambda)X_n^*(kl,\lambda)}{\omega-\omega_n^+(\lambda)+i0}-\sum_m \frac{Y_m(ij,\lambda)Y_m^*(kl,\lambda)}{\omega-\omega_m^-(\lambda)-i0}, \tag{2.12}$$

where

$$\omega_n^+(\lambda)=E_n^{A+2}(\lambda)-E_0^A(\lambda), \tag{2.12a}$$

$$\omega_m^-(\lambda)=E_0^A-E_m^{A-2}(\lambda), \tag{2.12b}$$

$$X_n(ij,\lambda)=\langle\Psi_0^A(\lambda)|a_ja_i|\Psi_n^{A+2}(\lambda)\rangle, \tag{2.12c}$$

$$Y_m(ij,\lambda)=\langle\Psi_m^{A-2}(\lambda)|a_ja_i|\Psi_0^A(\lambda)\rangle. \tag{2.12d}$$

Here the eigenfunctions and eigenvalues are defined by $(H_0+\lambda H_1)\Psi_0^A(\lambda)=E_0^A(\lambda)\Psi_0^A(\lambda)$, $(H_0+\lambda H_1)\Psi_n^{A+2}(\lambda)=E_n^{A+2}(\lambda)\Psi_n^{A+2}(\lambda)$, and similarly for $\Psi_m^{A-2}(\lambda)$ and $E_m^{A-2}(\lambda)$. Substituting the above into eq. (2.9) leads to the result [15])

$$\Delta E_0^{pp}=\int_0^1\frac{d\lambda}{\lambda}\sum_{\substack{m\\(A-2)}}\sum_{ijkl}Y_m(ij,\lambda)Y_m^*(kl,\lambda)\bar{V}_{klij}\lambda, \tag{2.13}$$

where the amplitudes Y_m are calculated from the RPA-type equation

$$\sum_{ef}\{(\varepsilon_i+\varepsilon_j)\delta_{ij,ef}+(\bar{n}_i\bar{n}_j-n_in_j)\lambda\bar{V}_{ijef}\}Y_m(ef,\lambda)=\omega_m^-(\lambda)Y_m(ij,\lambda). \tag{2.14}$$

Note that the factor $(\bar{n}_i\bar{n}_j-n_in_j)$ can also be written as $(1-n_i-n_j)$. It should be emphasized that in eqs. (2.13) and (2.14) the summations are unrestricted, namely the indices for the summations are in fact $i>j$ and $i<j$, and similarly for k, l, e and f. Some simplification may be effected by making use of the symmetry properties of $\bar{V}$ and Y_m as depicted by eqs. (2.5) and (2.12d). The convergence factor $e^{i\omega 0^+}$ in

eq. (2.9) has played an important role in obtaining eq. (2.13) from eqs. (2.9) and (2.12). This factor enables us to convert the ω integral in eq. (2.9) into a contour integral closed in the upper ω plane. This point will be discussed in more detail when we introduce the model-space G-matrix. The normalization of the Y_m vectors of eqs. (2.12d) and (2.14) will play an important role in our calculation and we will discuss this matter in sects. 3 and 4.

For nuclear matter calculations, $\bar{V}$ in eqs. (2.13) and (2.14) represents the matrix elements of a NN potential such as the Paris or Reid potential. Both have very strong short range repulsions and hence their $\bar{V}$ matrix elements are generally positive and very large. ($\bar{V}$ becomes infinite, if the NN potential employed has a hard-core short range repulsion.) Thus eqs. (2.13) and (2.14) are clearly not suitable for applying to many-body problems with interactions having strong short range repulsions. (In fact these equations are undefined in the case of hard-core NN potentials.) Some type of G-matrix partial summations are needed to take care of the strong short-range NN correlations. There is another difficulty. As it stands, eq. (2.14) is a matrix equation of infinite dimension. This is because its indices i, j, e and f run over all the s.p. states, of which there are infinitely many. (For nuclear matter, its s.p. states are continuum plane-wave states with momentum from 0 to ∞.) Thus eq. (2.14), as it stands, is not convenient for actual calculations. It would be very helpful if some form of truncation of it can be effected. Hence to carry out nuclear matter calculations based on the formulation outlined by eqs. (2.13) and (2.14), the two above mentioned difficulties must first be overcome.

3. Model space approach

We introduce here a model-space approach in order to reformulate the ring diagram method described in sect. 2 so that it can be expressed in terms of the model-space reaction matrix G^{M}. In this way, as will be seen shortly, we can overcome the two difficulties just mentioned. Let us divide the s.p. states into two groups: those with momentum $k > k_{\mathrm{M}}$ and those with $k \leqslant k_{\mathrm{M}}$. In drawing diagrams, the propagator corresponding to the former will be denoted by upward-going railed lines (see fig. 3). Our model space P is defined as a configuration space where all nucleons are restricted to have momentum $k \leqslant k_{\mathrm{M}}$, k_{M} being the momentum space

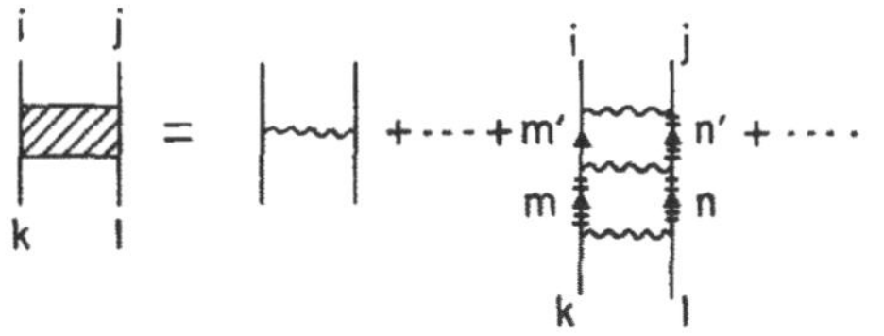

Fig. 3. Structure of the model space G^{M} matrix. Here the railed lines m, n and n' denote s.p. states with momentum $> k_{\mathrm{M}}$. m' denotes a s.p. state with momentum between k_{F} and k_{M}.

boundary of P. As mentioned earlier, typical values for k_M are $\sim 3\ \text{fm}^{-1}$. This point will be further discussed in sect. 4.

We define a model-space reaction matrix G^M by

$$G^{M}_{ijkl}(\omega) = V_{ijkl} + \sum_{mn} V_{ijmn} \frac{Q^{M}(mn)}{\omega - \varepsilon_m - \varepsilon_n + i0} G^{M}_{mnkl}(\omega)\,, \tag{3.1}$$

where the exclusion operator $Q^M(mn)$ is given by

$$Q^{M}(mn) = \begin{cases} 1, & \text{if } \max(k_m, k_n) > k_M \text{ and } \min(k_m, k_n) > k_F\,, \\ 0, & \text{otherwise}\,. \end{cases} \tag{3.2}$$

Hence the intermediate s.p. states m and n of G^M must both be particles ($>k_F$), and at least one of them must be outside the model space P. The diagramatic structure of G^M is shown in fig. 3 where we note that the intermediate states must have at least one railed line (momentum $> k_M$). It is important to note the $+i0$ factor in the propagator of G^M. Comparing with eq. (2.4), we see that only the retarded part of $F(\omega)$ (i.e. two lines propagating upward) can take part in the intermediate states of G^M. Another consequence of this $+i0$ factor is that the poles of $G^M(\omega)$ are all in the *lower* half of the complex ω-plane, as may be readily verified. This is an essential point in our method for summing up the $\bar{G}$-matrix ring diagrams, as we will soon discuss. Similar to eq. (2.5), we define the antisymmetrized reaction matrix as

$$\bar{G}^{M}_{ijkl}(\omega) = \tfrac{1}{2}\{G^{M}_{ijkl}(\omega) - G^{M}_{ijlk}(\omega)\}\,. \tag{3.3}$$

We now discuss how to regroup the particle-particle ring diagram with $\bar{V}$ vertices, as shown in fig. 1 (α) to (γ) and fig. 2(i), into ring diagrams with $\bar{G}^M$ vertices. Let us consider the example shown in fig. 4. Here diagram (i) is a particle-particle ring diagram which is third order in $\bar{V}$. Diagram (ii) has the same (ij), (kl) and (mn) propagators which are all within the model space P. But the propagators (uv), (pq) and (rs) are only the retarded propagators. For example, the (uv) propagator is $(1-n_u)(1-n_v)/(\omega - \varepsilon_u - \varepsilon_v + i0)$. In contrast, the (ij), (kl) and (mn) propagators are composed of both the retarded and the advanced parts as indicated by eq. (2.4).

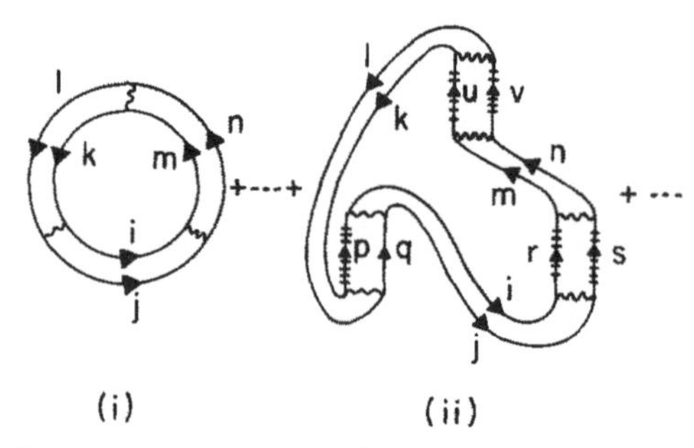

Fig. 4. Partial summation of ring diagrams with $\bar{V}$ vertices into ring diagrams with $\bar{G}^M$ vertices.

Furthermore, at least one line of each of the (u, v), (p, q) and (r, s) propagators must be outside the P space. In this way, we assure that there will be no double counting in combining particle-particle ring diagrams with $\bar{V}$ vertices into those with $\bar{G}^{\mathrm{M}}$ vertices. For example, any particle-particle ring diagram with three P-space propagators is included in the series shown in fig. 4. The sum of this entire series is a particle-particle ring diagram which is third order in $\bar{G}^{\mathrm{M}}$. Similarly all of the ring diagrams with four segments of P-space propagators are grouped together to form a corresponding ring diagram which is fourth order in $\bar{G}^{\mathrm{M}}$ vertex.

Thus we can now write ΔE_0^{pp} in two equivalent ways, either

$$\Delta E_0^{\mathrm{pp}} = \Delta E_0^{\mathrm{pp}}(1) + \Delta E_0^{\mathrm{pp}}(2) + \Delta E_0^{\mathrm{pp}}(3) + \cdots \tag{3.4}$$

or

$$\Delta E_0^{\mathrm{pp}} = \Delta E_0^{\mathrm{pp}}(1') + \Delta E_0^{\mathrm{pp}}(2') + \Delta E_0^{\mathrm{pp}}(3') + \cdots . \tag{3.5}$$

The various terms in eq. (3.4) have been given by eqs. (2.1) to (2.3). For example, $\Delta E_0^{\mathrm{pp}}(3)$ is third order in $\bar{V}$. In contrast, the various terms in eq. (3.5) are now expressed in terms of the $\bar{G}^{\mathrm{M}}$ interactions. For example, $\Delta E_0^{\mathrm{pp}}(3')$ is third order in $\bar{G}^{\mathrm{M}}$. $\Delta E_0^{\mathrm{pp}}(n')$ differs from $\Delta E_0^{\mathrm{pp}}(n)$ only in the replacement of the $\bar{V}$'s of the latter by the corresponding $\bar{G}^{\mathrm{M}}$'s and in the restriction on the summation of the indices. For example, $\Delta E_0^{\mathrm{pp}}(3')$ is given by

$$\begin{aligned}\Delta E_0^{\mathrm{pp}}(3') = \frac{-1}{2\pi i}\int_{-\infty}^{\infty} \mathrm{d}\omega\, \mathrm{e}^{i\omega 0^+}\tfrac{1}{3} \sum_{\substack{ijklmn \\ \epsilon P}} F_{ij}(\omega)\bar{G}^{\mathrm{M}}_{ijkl}(\omega) \\ \times F_{kl}(\omega)\bar{G}^{\mathrm{M}}_{klmn}(\omega)F_{mn}(\omega)\bar{G}^{\mathrm{M}}_{mnij}(\omega)\,,\end{aligned} \tag{3.6}$$

where we note that the indices $(i, j, \ldots n)$ are summed within the model space only.

Based on the above results, the steps leading to the eqs. (2.8) and (2.9) can be easily repeated using the $\bar{G}^{\mathrm{M}}$ vertices. Thus corresponding to eq. (2.9) we now have

$$\Delta E_0^{\mathrm{pp}} = \frac{-1}{2\pi i}\int_0^1 \frac{\mathrm{d}\lambda}{\lambda}\int_{-\infty}^{\infty} \mathrm{d}\omega\, \mathrm{e}^{i\omega 0^+}\, \mathrm{tr}_{\mathrm{P}}\{G^{\mathrm{pp}}(\omega,\lambda)\bar{G}^{\mathrm{M}}(\omega)\lambda\}\,, \tag{3.7}$$

where the subscript P denotes that the trace is to be taken only within the model space P, and the model space reaction matrix $\bar{G}^{\mathrm{M}}$ has been given by eqs. (3.3) and (3.1). The Green function G^{pp} is now given in terms of $\bar{G}^{\mathrm{M}}$, namely

$$G^{\mathrm{pp}}_{ijkl}(\omega,\lambda) = F_{ij}(\omega)\delta_{ij,kl} + F_{ij}(\omega)\lambda\bar{G}^{\mathrm{M}}_{ijmn}(\omega)G^{\mathrm{pp}}_{mnkl}(\omega,\lambda)\,, \tag{3.8}$$

where note that the indices $i, j, \ldots$ and n are all within the model space P, and repeated indices are summed over freely (i.e. we sum over $m \gtrless n$). Thus eq. (3.8) is a model-space effective Dyson's equation. General properties of this type of the model-space effective Dyson's equations for the Green functions have been studied by Wu and Kuo. [19])

So far the result, eq. (3.7), was obtained by way of a diagramatic analysis ΔE_0^{pp}. It can also be derived algebraically, which serves to double check our result. Let us do this. We start from the general expression (2.6) for ΔE_0^{pp}. Let us divide the two-particle subspace into two groups, denoted respectively by the projection operatos $\hat{P}$ and $\hat{Q}$. To be more explicit, we choose $\hat{P}$ to be $\hat{P}=\sum_{ab}|ab\rangle\langle ab|$ where the s.p. states a and b both have momentum $\leq k_{\text{M}}$, $\hat{Q}$ is the complement of $\hat{P}$, i.e. for the two-particle subspace we have $\hat{P}+\hat{Q}=1$. We introduce the notations tr_{P} and tr_{Q} to denote the trace for the $\hat{P}$- and $\hat{Q}$-space, respectively. Then the various terms in the integrand of eq. (2.6) can be decomposed as

$$\begin{aligned}
&\text{tr}\{F\bar{V}\}=\text{tr}_{\text{P}}\{F\bar{V}\}+\text{tr}_{\text{Q}}\{F\bar{V}\},\\
&\tfrac{1}{2}\text{tr}\{(F\bar{V})^2\}=\tfrac{1}{2}\text{tr}_{\text{P}}\{F\bar{V}\hat{P}F\bar{V}\}+\tfrac{1}{2}\text{tr}_{\text{Q}}\{F\bar{V}\hat{Q}F\bar{V}\}+\text{tr}_{\text{P}}\{F\bar{V}\hat{Q}F\bar{V}\},\\
&\tfrac{1}{3}\text{tr}\{(F\bar{V})^3\}=\tfrac{1}{3}\text{tr}_{\text{P}}\{F\bar{V}\hat{P}F\bar{V}\hat{P}F\bar{V}\}+\tfrac{1}{3}\text{tr}_{\text{Q}}\{F\bar{V}\hat{Q}F\bar{V}\hat{Q}F\bar{V}\}+\text{tr}_{\text{P}}\{F\bar{V}\hat{Q}F\bar{V}\hat{Q}F\bar{V}\}\\
&\qquad+\text{tr}_{\text{P}}\{F\bar{V}\hat{P}F\bar{V}\hat{Q}F\bar{V}\}
\end{aligned}\tag{3.9}$$

and similarly for the other terms. In deriving the above, we have used the cyclic invariance property of the trace. Note that we have split ΔE_0^{pp} into two general types of terms, one preceded by tr_{P} and the other preceded tr_{Q}.

It is important to note that all the tr_{Q} terms do *not* contribute to the integral $\int \text{d}\omega\, \text{e}^{i\omega 0^+}\dots$ of eq. (3.6). This is because of the convergence factor $\text{e}^{i\omega 0^+}$. As indicated in fig. 5, this factor enables us to convert the above integral into a contour integral closed in the upper ω-plane. Note that all the propagators of the tr_{Q} terms are of the form QF. From eq. (2.4), we know that QF has poles only in the lower ω-plane. Hence all these tr_{Q} terms do not contribute to ΔE_0^{pp}, giving us the rather interesting result that ΔE_0^{pp} can be calculated entirely from the tr_{P} terms.

A straightforward regrouping of the tr_{P} terms yields

$$\Delta E_0^{\text{pp}}=\frac{-1}{2\pi i}\int_{-\infty}^{\infty}\text{d}\omega\,\text{e}^{i\omega 0^+}\,\text{tr}_{\text{P}}\{PF\bar{K}+\tfrac{1}{2}(PF\bar{K})^2+\tfrac{1}{3}(PF\bar{K})^3+\cdots\},\tag{3.10}$$

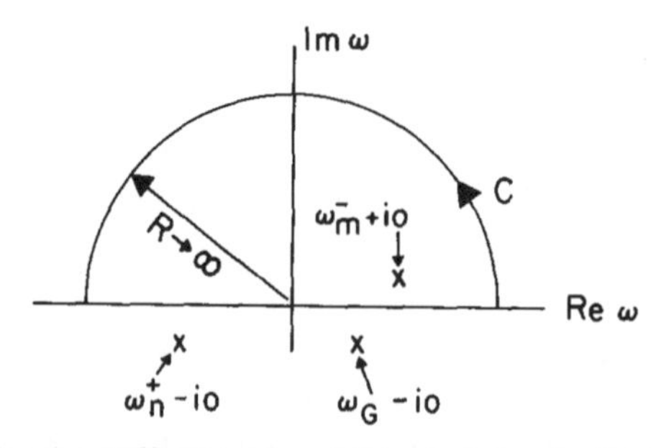

Fig. 5. Contour integration for ΔE_0^{pp}. The poles of the retarded and advanced Green's functions are indicated by (ω_0^+-i0) and (ω_m^-+i0), respectively. The poles of the model-space reaction matrix are indicated by $(\omega_{\text{G}}-i0)$.

where $\bar{K}$ is defined as

$$\bar{K} = \bar{V} + \bar{V}QFQ\bar{K}. \tag{3.11}$$

Comparing with eqs. (3.1) and (3.3), we see that the above $\bar{K}$ is just the $\bar{G}^{\mathrm{M}}$ matrix. Hence eq. (3.10) gives the same result as that given by eq. (3.5) and (3.6). This concludes our algebraic proof for eq. (3.7).

Now we discuss how to calculate ΔE_0^{pp} as given by eq. (3.7). By writing $G^{\mathrm{pp}}(\omega, \lambda)$ in its Lehmann representation, the steps leading to eq. (2.13) from eq. (2.12) to (2.12d) can now be repeated. The result is, however, interestingly different. Let us denote the ω integral of eq. (3.7) by I. Then from eq. (2.12) we have

$$I = \int_{-\infty}^{\infty} \mathrm{d}\omega\, \mathrm{e}^{i\omega 0^+} \left\{ \sum_n \frac{X_n(\lambda) X_n(\lambda)^+}{\omega - \omega_0^+ + i0} - \sum_m \frac{Y_m(\lambda) Y_m(\lambda)^+}{\omega - \omega_m^- - i0} \right\} \bar{G}^{\mathrm{M}}(\omega). \tag{3.12}$$

In fig. 5, we discuss the locations of the poles of the above integrand. For the retarded Green function, the poles $(\omega_n^+ - i0)$ are in the lower ω-plane. For the advanced Green function, the poles $(\omega_m^- + i0)$ are in the upper ω-plane. When we defined our model-space reaction matrix G^{M} by eq. (3.1), we have judiciously chosen its propagator to be $(\omega - \varepsilon_m - \varepsilon_n + i0)^{-1}$. Hence the poles of $\bar{G}^{\mathrm{M}}(\omega)$, denoted by $\omega_G - i0$, are in the lower ω-plane. The convergence factor $\mathrm{e}^{i\omega 0^+}$ in eq. (3.12) dictates that we can replace the integral $\int_{-\infty}^{\infty} \mathrm{d}\omega(\ldots)$ of eq. (3.9) by the contour integral $\oint_{\mathrm{C}} \mathrm{d}\omega(\ldots)$ where C is the upper-plane contour shown in fig. 5. Thus we have the rather nice result that the poles of $\bar{G}^{\mathrm{M}}(\omega)$ and of the retarded Green function both do not contribute to ΔE_0^{pp} of eq. (3.7). This leads to the result

$$\Delta E_0^{\mathrm{pp}} = \int_0^1 \frac{\mathrm{d}\lambda}{\lambda} \sum_{\substack{m \\ (A-2)}} \sum_{\substack{ijkl \\ \in \mathrm{P}}} Y_m(ij, \lambda) Y_m^*(kl, \lambda) \bar{G}_{klij}^{\mathrm{M}}(\omega_m^-(\lambda))\lambda. \tag{3.13}$$

The above is a main result of the present work, and it is the primary equation on which our ring-diagram nuclear matter calculations will be based. Let us now discuss some specific features of this equation. First let us compare eq. (3.13) with eq. (2.13). It is interesting to note that ΔE_0^{pp} given by these two equations are mathematically equivalent to each other, although in appearance they look rather different. Clearly eq. (3.13) is considerably more convenient for numerical calculations than eq. (2.13). As discussed earlier, the matrix elements $\bar{V}_{klij}$ of eq. (2.13) are not suitable for direct numerical calculations because they are usually the plane-wave matrix elements of a singular NN potential. Furthermore, the indices i, j, k and l in eq. (2.13) are summed over the entire s.p. space. In contrast, the $\bar{G}^{\mathrm{M}}$ reaction matrix in eq. (3.13) is a numerically well-behaved quantity and, in addition, the indices i, j, k and l are now summed within the model space P, namely their momentum values are restricted to be $\leqslant k_{\mathrm{M}}$, the momentum space boundary of P. Hence, eq. (3.13) is definitely more convenient for actual nuclear matter calculations. It should be pointed out that the Y-amplitudes of eq. (3.13) are no longer calculated from eq. (2.14). Instead, they are now calculated from a model space self-consistent equation of the RPA type namely

$$\sum_{ef}\{(\varepsilon_i+\varepsilon_j)\delta_{ij,ef}+(\bar{n}_i\bar{n}_j-n_in_j)\lambda L_{ijef}(\omega)\}Y_m(ef,\lambda)=\mu_m(\omega,\lambda)Y_m(ij,\lambda) \quad (3.14)$$

with the self-consistent condition for ω

$$\mu_m(\omega,\lambda)=\omega\equiv\omega_m^-(\lambda)\,, \quad (3.14a)$$

$$L_{ijef}(\omega)\equiv\bar{G}^{M}_{ijef}(\omega)\,. \quad (3.14b)$$

Since the derivation and general properties of the above type of self-consistent equations have been given in detail elsewhere [19,20]), we will not give the details of the derivation of the above equations except to mention that they are obtained from eqs. (3.8), (2.12) and (2.4). In eq. (3.14), L is in general an irreducible vertex function containing both two-body and one-body terms (self-energy insertions). Here, as indicated by eq. (3.14b), we include in L only the two-body reaction matrix $\bar{G}^M$. As we discuss later, it is important to also include one-body terms in L when s.p. self-energy insertions are taken into account.

It should be emphasized that the normalization condition for the Y-amplitudes is [20])

$$\sum_{\substack{p_1>p_2>k_F\\ \in P}}|Y_m(p_1p_2,\lambda)|^2-\sum_{\substack{k_F>h_1>h_2\\ \in P}}|Y_m(h_1h_2,\lambda)|^2=-Z_m^\lambda \quad (3.15)$$

with

$$Z_m^\lambda=\left\{1-\left.\frac{\partial\mu_m(\omega,\lambda)}{\partial\omega}\right|_{\omega=\mu_m(\omega,\lambda)}\right\}^{-1}. \quad (3.16)$$

In the above, p_1 and p_2 are obviously the particle s.p. states within the model space P and h_1 and h_2 are the hole s.p. states within P.

A brief summary of our calculational procedure is in order. The model space reaction matrix $\bar{G}^M$ is calculated from eqs. (3.1) to (3.3). Then we solve the self-consistent equations (3.14) and (3.14a). This may be done by a graphical method where we plot $\mu_m(\omega,\lambda)$ as a function of ω. The self-consistent solution is given by the point where $\omega=\mu_m(\omega,\lambda)$ is satisfied. The slope of the $\mu_m(\omega,\lambda)$ curve at this point is then used in eq. (3.16) to determine the normalization constant Z_m^λ. In this way, we can calculate Y_m and ω_m^-. Using these and $\bar{G}^M$, the energy shift is then calculated using eq. (3.13). Recall that Y_m is the $A\to(A-2)$ transition amplitude defined in eq. (2.12d). If these amplitudes can be determined experimentally, then we can use them directly in eq. (3.13). We don't know how realistic it is to do so, but this seems to be an interesting prospect for further study.

How does the present ring-diagram approach compare with the usual BHF [1,2]) and the model-space BHF (MBHF) approaches [10,13]) for nuclear matter? It may be most convenient to explain this by first examining their diagrammatic structures. As shown in fig. 6, diagram (α) is the typical two-hole-line diagram included in the usual BHF theory. (h_1 and h_2 are hole lines.) Here the box B represents a BHF G-matrix using discontinuous s.p. spectrum, namely a self-consistent spectrum for the hole lines but free-particle spectrum for the particle lines. Diagram (β) is a typical diagram included in the MBHF theory of nuclear matter [10,13]. Here each M-box denotes a model-space reaction matrix G^M defined in eq. (3.1) and p_1 to p_4 are all particle lines (with momentum between k_M and k_F), and h_1 and h_2 are hole lines. The main difference between BHF and MBHF is in the treatment of the s.p. spectrum between k_F and k_M (the momentum space boundary of the model space). In BHF, one uses a free particle spectrum in this region, while in MBHF one uses a self-consistent spectrum in this region as we will discuss later. The present calculation goes considerably further. Here we sum up the ring diagrams of the type shown by diagram (γ) of fig. 6 to all orders. Again, each M-box represents a G^M vertex whose intermediate states are all particle states outside the model space P. A specific feature is that the intermediate states $i, j, k, \ldots$ to s are all within P and each pair of them, such as the pair (rs), can be either both particles or both holes. Hence the present method sums up the particle–particle ring diagrams of all possible zig-zag shapes to all orders. Clearly the present method sums up a much larger class of diagrams than BHF and MBHF.

It is of interest to note the following feature of eq. (3.13). The expression ΔE_0^{pp} given by eq. (3.13) reduces to the corresponding expression for the BHF case if we set $k_M = k_F$ and, in addition, set the Y_m amplitudes equal to their unperturbed values (i.e. in eq. (2.12d) we replace the true wave functions Ψ_m^{A-2} and Ψ_0^A by the respective unperturbed wave functions Φ_m^{A-2} and Φ_0^A). In eq. (3.13), the Y_m amplitudes serve as a kind of distribution function. The effect of the particle-particle ring diagrams is exhibited by the deviation of these amplitudes from their respective unperturbed values. Clearly the magnitudes and distributions of the Y-amplitudes will play an instrumental role in our present calculation.

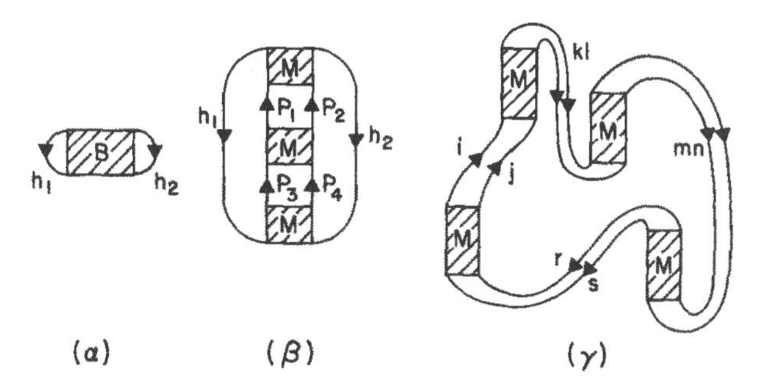

Fig. 6. Typical diagrams included in the usual BHF (α), the MBHF (β), and the present ring-diagram (γ) approaches for nuclear matter. Box B denotes a BHF reaction matrix, and box M denotes a model-space reaction matrix of eq. (3.1).

4. Calculation and results

4.1. SINGLE PARTICLE POTENTIAL AND SPECTRUM

An important step in our nuclear matter calculation is the determination of the model-space s.p. potential U. In our formulation described so far, we have used $H=(T+U)+(V-U)\equiv H_0+H_1$, and the s.p. wave functions ϕ_k and energies ε_k are defined by H_0. To carry on the calculation, we must know ϕ_k and ε_k, namely we must know U. For nuclear matter, we use plane waves for ϕ_k and, hence, the quantity to be determined is the s.p. spectrum ε_k.

We adopt a model-space Hartree-Fock method [13]) for determining ε_k. The principle of this method is briefly described below. Using effective interaction theories [21]), one can formally transform the nuclear hamiltonian H into a model-space (P) effective hamiltonian $PH_{\text{eff}}P$, where $H_{\text{eff}}=H_0+PV_{\text{eff}}P$. The effective interaction V_{eff} has in general many-body components such as the one-body part $V^{(1)}_{\text{eff}}$, two-body part $V^{(2)}_{\text{eff}}$, etc. The model-space Hartree-Fock method for determining U is to choose a U such that the condition $PV^{(1)}_{\text{eff}}P=0$ is satisfied. Using a reaction matrix approximation [13]), this condition leads to the following set of self-consistent equations for determining the s.p. spectrum ε_k:

$$\varepsilon_{k_1}=t_{k_1}+\Gamma_{k_1}(\varepsilon_{k_1})\,, \tag{4.1}$$

$$\Gamma_{k_1}(\omega)=2\sum_{h\leq k_F}\langle k_1 h|\bar{G}^{\text{M}}(\omega+\varepsilon_h)|k_1 h\rangle \qquad k_1\leq k_{\text{M}}\,, \tag{4.1a}$$

$$\Gamma_{k_1}(\omega)\equiv 0 \qquad k_1>k_{\text{M}}\,. \tag{4.1b}$$

where t_{k_1} is the s.p. kinetic energy $\hbar^2k_1^2/2m$, $\bar{G}^{\text{M}}$ is the reaction matrix defined in eqs. (3.3) amd (3.1), and k_{M} is $\sim 3\ \text{fm}^{-1}$ which is the momentum space boundary of the model space P. The s.p. potential is the one-body vertex function Γ evaluated at the self-consistent energy $\omega=\varepsilon_{k_1}$, namely

$$U(k_1)=\Gamma_{k_1}(\varepsilon_{k_1})\,. \tag{4.2}$$

A special feature of the above approach for the s.p. potential is that $U(k_1)$ is determined self-consistently for $k_1\leq k_{\text{M}}$, and for $k_1>k_{\text{M}}$ we set $U(k_1)=0$. We will use an effective mass description for the s.p. spectrum. Then the s.p. spectrum is given as

$$\varepsilon_{k_1}=\begin{cases}\dfrac{\hbar^2k_1^2}{2m^*}+\Delta & k_1\leq k_{\text{M}}\\[2ex] \dfrac{\hbar^2k_1^2}{2m} & k_1>k_{\text{M}}\,.\end{cases} \tag{4.3}$$

The effective mass m^* and zero-point energy Δ are determined self-consistently, as described below.

It is convenient to carry out the calculation of eq. (4.1a) in the relative and center-of-mass (RCM) frame. Similarly, we also need to calculate G^M of eq. (3.1) in terms of the RCM momentum variables. An essential step in doing so is to replace the projection operator Q^M of eq. (3.2) by its angle-average approximation [13] $\bar{Q}^M$ given by

$$\bar{Q}^M(k, K, k_F, k_m) = \begin{cases} 1 & \text{in regions a, b} \\ 0 & \text{c} \\ [k^2 - k_F^2 + \frac{1}{4}K^2]/kK & \text{d} \\ [(k+\frac{1}{2}K)^2 - k_M^2]/kK & \text{e} \\ [2k^2 - k_F^2 + \frac{1}{2}K^2 - k_M^2]/kK & \text{f}, \end{cases} \tag{4.4}$$

where the values of k and K for the various regions are defined in fig. 7. $\boldsymbol{k}$ and $\boldsymbol{K}$ are related to $\boldsymbol{k}_m$ and $\boldsymbol{k}_n$ by $\boldsymbol{k} = \frac{1}{2}(\boldsymbol{k}_m - \boldsymbol{k}_n)$ and $\boldsymbol{K} = \boldsymbol{k}_m + \boldsymbol{k}_n$. The above result is obtained from eq. (3.2) by assuming that all relative directions between $\boldsymbol{k}$ and $\boldsymbol{K}$ are equally likely and thereby we can average over these directions.

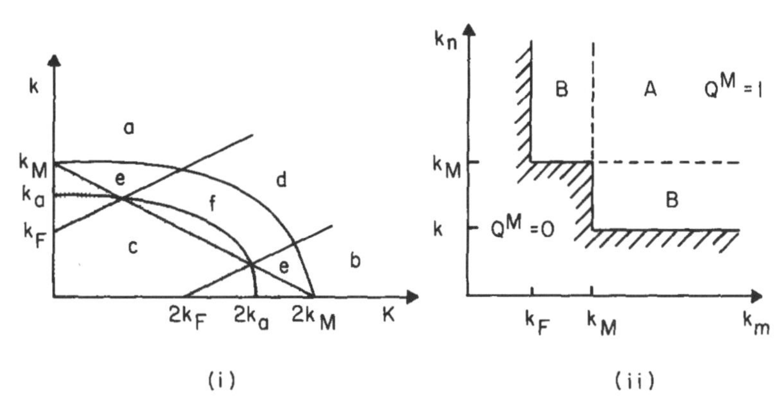

Fig. 7. (i) Momentum regions for the angle-averaged projection operator $\bar{Q}^M$ of eq. (4.4). Note that $k_a^2 = \frac{1}{2}(k_F^2 + k_M^2)$ and the two curves are ellipses. (ii) The projection operator Q^M of eq. (3.2).

Using the above averaged $\bar{Q}^M$, eq. (3.1) for the model space reaction matrix can be decomposed into separate partial wave channels. Namely for each partial wave channel α such as the ${}^3S_1$-${}^3D_1$ channel, we have

$$\langle kl|G^M(\omega, K\alpha)|k'l'\rangle = \langle kl|V|k'l'\rangle + \frac{2}{\pi}\int_0^\infty k''^2\,dk'' \sum_{l''} \frac{\langle kl|V|k''l''\rangle \bar{Q}^M(k'', K, k_F, k_M)\langle k''l''|G^M(\omega, K\alpha)|k'l'\rangle}{\omega - H_0(k''K)}, \tag{4.5}$$

where α stands for the partial wave quantum numbers $(ll'SJT)$, and K the center-of-mass momentum. For simplicity, the K and (SJT) quantum numbers associated with the bra and ket vectors have been suppressed. For example, $\langle kl|$ should in fact be $\langle klSJT, K|$. $H_0(k'', K)$ is the unperturbed energy of the intermediate state (k'', K);

and its determination is discussed later. We note that our convention for plane waves is

$$\langle \boldsymbol{r}|klSJ\rangle = j_l(kr)\mathscr{Y}_{lSJ}(\hat{\boldsymbol{r}}) \tag{4.6}$$

where $j_l(kr)$ is the spherical Bessel function, and $\mathscr{Y}$ is the vector spherical harmonics corresponding to $\boldsymbol{l}+\boldsymbol{S}=\boldsymbol{J}$. We also note that G^{M} is diagonal in K and α, as indicated by eq. (4.5). This is a consequence of using the angle averaged projection operator $\bar{Q}^{\mathrm{M}}$ of eq. (4.4).

Angle-average approximations [22]) are also used in eqs. (4.2) and (4.1a). This leads to the familiar equation for the s.p. potential; e.g. we have for $k_1 < k_{\mathrm{F}}$

$$U(k_1) = \sum_{\alpha}(2T+1)(2J+1)\left\{\frac{8}{\pi}\int_0^{k_-} k^2\,\mathrm{d}k\, G^{\mathrm{M}}_{lSJT}(k,\bar{K}_1)\right.$$
$$\left. + \frac{1}{\pi k_1}\int_{k_-}^{k_+} k\,\mathrm{d}k[k_{\mathrm{F}}^2 - k_1^2 + 4k(k_1-k)]G^{\mathrm{M}}_{lSJT}(k,\bar{K}_2)\right\}, \tag{4.7}$$

where

$$k_- = \tfrac{1}{2}(k_{\mathrm{F}} - k_1)$$
$$k_+ = \tfrac{1}{2}(k_{\mathrm{F}} + k_1)$$
$$\bar{K}_1^2 = 4(k_1^2 + k^2)$$
$$\bar{K}_2^2 = 4(k_1^2+k^2) - (2k+k_1-k_{\mathrm{F}})(2k+k_1+k_{\mathrm{F}}) \tag{4.7a}$$

and the partial wave G^{M} matrix elements are given by

$$G^{\mathrm{M}}_{lSJT}(k,K) \equiv \langle kl|G^{\mathrm{M}}(\omega, K\alpha)|kl\rangle \tag{4.7b}$$

with

$$\omega = \frac{\hbar^2 k^2}{m^*} + \frac{\hbar^2 K^2}{4m^*} + 2\Delta\,. \tag{4.7c}$$

In the above, we have used the effective mass description for the s.p. spectrum as given by eq. (4.3). Now we return to the intermediate-state spectrum $H_0(k'', K)$ of eq. (4.5). Referring to fig. 7(ii), the momentum variables k_m and k_n corresponding to k'' and K may be in either region A or B. In region A, both particles have momentum $> k_{\mathrm{M}}$. Then according to eq. (4.3) we have for this region

$$H_0(k'', K) = \frac{\hbar^2 k''^2}{m} + \frac{\hbar^2 K^2}{4m}\,. \tag{4.8}$$

But in regions B, we have one nucleon with momentum $> k_{\mathrm{M}}$ while the other with momentum between k_F and k_M, and the situation is complicated because now the two nucleons have different m^* and Δ. We have used for this case an angle average approximation for the nucleon in region B, denoting its momentum by k_m, that k_m^2 is replaced by its angle average $\langle k_m^2 \rangle$ which is equal to $\frac{1}{2}(k_F^2+k_M^2)$., Then for regions B we have

$$H_0(k'', K)=\frac{\hbar^2 k''^2}{m}+\frac{\hbar^2 K^2}{4m}+\frac{\hbar^2}{4m}\left(\frac{m}{m^*}-1\right)(k_F^2+k_M^2)+\Delta\,. \tag{4.8a}$$

The s.p. potential $U(k_1)$ and spectrum ε_{k_1} are then calculated self-consistently: We first assume some initial values $m^*(i)$ and $\Delta(i)$ for the effective mass and zero-point energy. Then $U(k_1)$ is calculated for various k_1 values $\leq k_M$ using eqs. (4.4) to (4.8a). This leads to a new s.p. spectrum $\varepsilon_{k_1}(i+1)$ from which the new values for $m^*(i+1)$ and $\Delta(i+1)$ are determined by means of a least $-\chi^2$ fit. Iteration calculations are continued until $m^*(n+1)=m^*(n)$ and $\Delta(n+1)=\Delta(n)$ within a small percentage deviation. (In our calculations, the deviation for Δ is about 0.1% and for m^* about 0.5%.)

In fig. 8, we show a typical s.p. spectrum ε calculated using the above model-space approach and using the Paris NN interaction [16]). As shown, a characteristic feature of our spectrum is that it is continuous from momentum 0 to k_M, unlike the usual BHF s.p. spectrum which has a large discontinuity at k_F. Beyond k_M, we use the free s.p. spectrum because our method is designed to determine only the s.p. potential within the model space. [13]) Thus our spectrum has a small discontinuity of ~10 MeV at k_M. We have chosen $k_M=3.2\,\text{fm}^{-1}$. [Ma and Kuo [13]) have found that at $k_M \approx 2k_F$ their MBHF calculations are rather stable with respect to small variations of k_M.

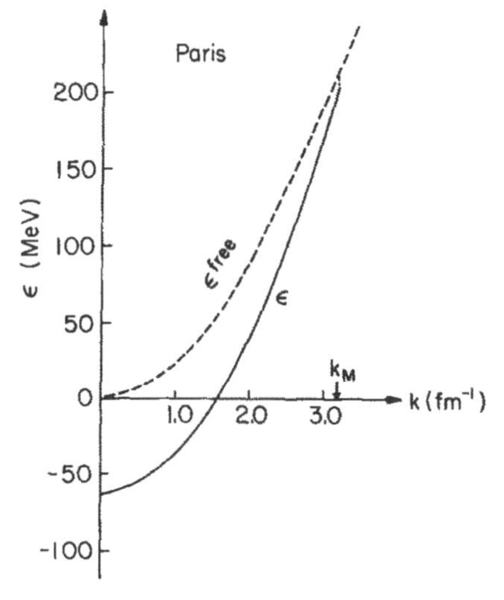

Fig. 8. Single particle spectra ε and ε_H calculated from our model space approach and the Paris NN interaction. It is for $k_F=1.4\,\text{fm}^{-1}$ and $k_M=3.2\,\text{fm}^{-1}$. The free particle spectrum is denoted by $\varepsilon^{\text{free}}$.

Based on this, we have made our choice of k_M as stated above. Note that our choice of K_M is different from theirs: we use a fixed k_M irrespective of the k_F values while their k_M varies with k_F. It is difficult to assess precisely which of these two schemes is better. For $k_F \approx 1.5\ \mathrm{fm}^{-1}$, those two choices are clearly very similar to each other. In the regions of very large and very small k_F, we feel that our choice is more reasonable. For example, if $k_F = 0.5\ \mathrm{fm}^{-1}$ then $k_M = 2k_F = 1.0\ \mathrm{fm}^{-1}$ would certainly seem to be too small.] In fig. 9 we show a s.p. spectrum calculated from the Reid NN interaction [17]). Clearly it is rather similar to the spectrum calculated using the Paris NN interaction; the Paris spectrum being a few MeV lower. We note that the numerical values of our model-space s.p. spectrum are in fact quite close to those of the continuous s.p. spectrum of Mahaux and his collaborators [3]), although these two spectra employ different methods of derivation.

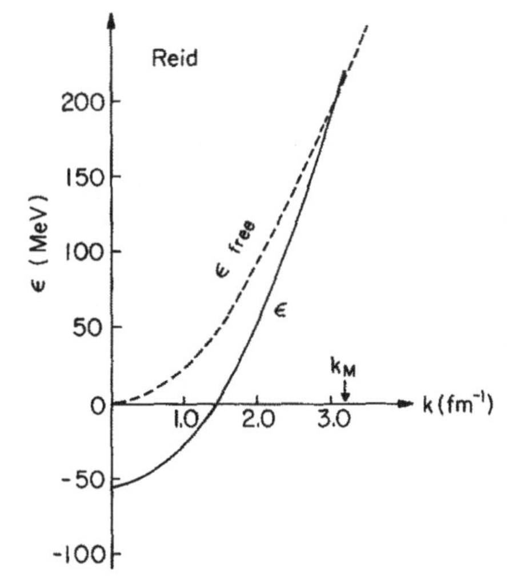

Fig. 9. Same as fig. 8 except for the Reid NN interaction.

Remember that we have used an effective mass description for the s.p. spectrum as shown in eq. (4.3). In table 1, we give our results for the effective mass m^* and the zero-point energy Δ for various k_F values and for both the Paris and Reid NN interactions. When plotted as a function of k_F, they form rather smooth curves. It should be pointed out that it is often not very accurate to fit our calculated spectrum by a parabola with two parameters m^* and Δ. (We have used a six point fit for the momentum range 0 to k_M, three points below k_F and three above. Typical differences between calculated and fitted spectra are ~1-2 MeV. In order to carry out calculations in the RCM coordinates, it is however necessary to use the m^* and Δ description for the s.p. spectrum.) As shown in the table, values of m^* calculated from the two NN potentials are quite similar to each other, but there are considerable differences between the two sets of Δ values. To see the trend of the dependence of m^* and Δ

TABLE 1

Self-consistent effective mass m^* and zero-point energy Δ for various k_F values, calculated with $k_M = 3.2\ \text{fm}^{-1}$ and the Paris and Reid NN interactions

	$k_F(\text{fm}^{-1})$	m/m^*	Δ (MeV)
Paris	1.20	1.190	−44.55
	1.30	1.234	−53.77
	1.36	1.262	−59.66
	1.40	1.275	−63.50
	1.50	1.327	−73.71
	1.60	1.375	−84.23
	1.70	1.435	−95.22
	1.80	1.504	−106.28
Reid	1.00	1.120	−24.51
	1.10	1.155	−31.09
	1.20	1.196	−38.42
	1.30	1.244	−46.22
	1.36	1.275	−51.06
	1.40	1.289	−54.20
	1.50	1.347	−62.45
	1.60	1.402	−70.43

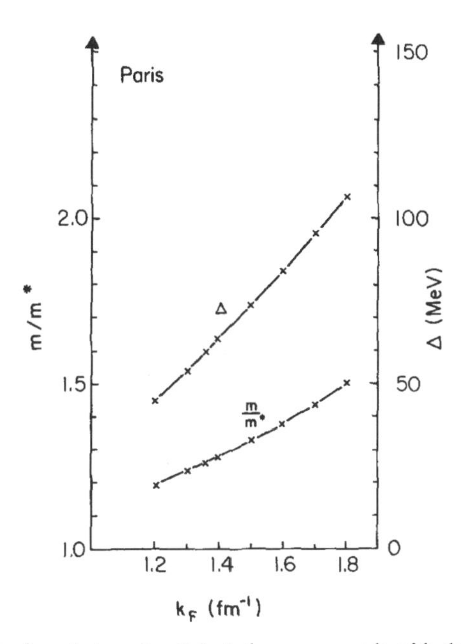

Fig. 10. Dependence of m^*, and Δ on k_F. Calculations were made with the Paris NN potential and $k_M = 3.2\ \text{fm}^{-1}$.

on the values of k_F, we plot our results for them in fig. 10 for the Paris potential. The corresponding curves for the Reid potential are rather similar.

4.2. ANGLE-AVERAGE APPROXIMATION FOR RPA EQUATION

We now turn to the calculation of the energy shift ΔE_0^{pp} of eq. (3.13). To do so, we need to know the transition amplitudes Y and the model-space reaction matrix $\bar{G}^M$. We have just described how to calculate $\bar{G}^M$ in the previous sub-section. Now we discuss how to calculate the transition amplitudes Y from the RPA-type secular equation (3.14).

As it stands, eq. (3.14) is not convenient for computation. It is expressed in terms of the laboratory momentum variables; its indices $(i, j. e, f)$ are in fact the laboratory s.p. momenta $(\boldsymbol{k}_i, \boldsymbol{k}_j, \boldsymbol{k}_e, \boldsymbol{k}_f)$. Our plan is to use a momentum-space discretization method [22-24]) to convert eq. (3.14) into a finite dimensional matrix equation. To do this in terms of these laboratory momentum variables will lead to matrix equations of impractically large dimensions. It may become greatly simplified if we can transform this equation to its RCM representation. For the calculation of $\bar{G}^M$, we have already used the angle-average approximation as discussed in sect. 4.1. Hence $\bar{G}^M$ is diagonal in its center-of-mass momentum variable. This simplifies eq. (3.14) slightly but is still not adequate. It seems to be indispensable that we also have to make an angle-average approximation for the occupation factor $(\bar{n}_i\bar{n}_j - n_i n_j) = 1 - (n_i + n_j)$ of eq. (3.14).

We define a function $Q_R(\boldsymbol{k}_i, \boldsymbol{k}_j) = 1-(n_i+n_j) = 1$ or -1 depending on the values of k_i and k_j. This is seen clearly in fig. 11(i) where Q_R is equal to 1 in regions A and B (i and j are both particles) and -1 in region C (i and j are both holes). Q_R is equal to zero for all other regions. The values of $Q_R(\boldsymbol{k}, \boldsymbol{K})$ clearly depends on

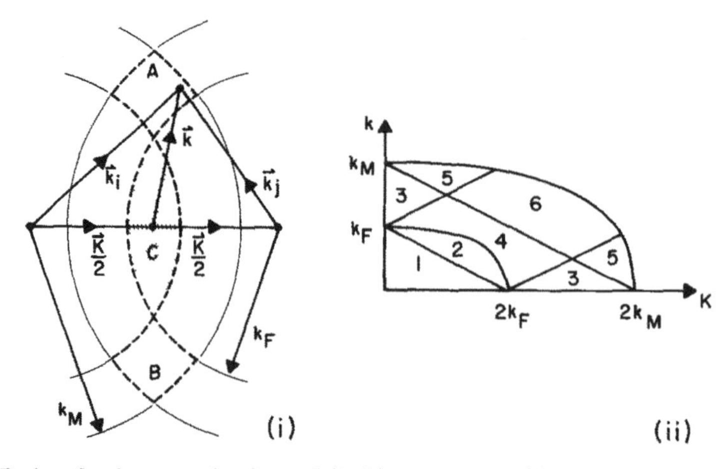

Fig. 11. Regions for the occupation factor $Q(\boldsymbol{k}_i, \boldsymbol{k}_j) = 1 - n_i - n_j$ and its angle-average approximation $\bar{Q}_R(k, K, k_F, k_M)$. In (ii) the lines are $(k - \frac{1}{2}K) = \pm k_F$ and $k + \frac{1}{2}K = k_F$ and k_M. The curves are ellipses.

the angle between $\boldsymbol{k}$ and $\boldsymbol{K}$. Assuming that all values for this angle are equally likely, we can replace $Q_R(\boldsymbol{k}, \boldsymbol{K})$ by its angle-average approximation $\bar{Q}_R(k, K, k_F, k_M)$. With the help of fig. 11(i), the values of $\bar{Q}_R$ are obtained as

$$\bar{Q}_R(k, K, k_F, k_M) = \begin{cases} -1 & \text{region } 1 \\ -|x_1| & 2 \\ 1 & 3 \\ |x_1| & 4 \\ |x_2| & 5 \\ \min(|x_1|, |x_2|) & 6 \end{cases} \tag{4.9}$$

where

$$\begin{aligned} x_1 &\equiv (k_F^2 - k^2 - \tfrac{1}{4}K^2)/kK, \\ x_2 &\equiv (k_M^2 - k^2 - \tfrac{1}{4}K^2)/kK. \end{aligned} \tag{4.9a}$$

The regions 1 to 6 of eq. (4.9) refer to the regions in the (k, K) plane shown in fig. 11(ii).

The replacement of Q_R by its angle-average approximation $\bar{Q}_R$ greatly simplifies eq. (3.14); it can now be decomposed into separate partial-wave equations. Namely, it becomes for partial wave channel α

$$\sum_{l'} \int dk' \left\{ \varepsilon_{kK} \delta(k-k')\delta_{ll'} + \lambda \frac{2k'^2}{\pi} \bar{Q}_R(k, k)\langle kl|L(\omega, K)|k'l'\rangle \right\} Y_m(k'l'k, \lambda)$$
$$= \mu_m(\omega, \lambda) Y_m(klK, \lambda), \tag{4.10}$$

where $\bar{Q}_R(k, K)$ is an abbreviation for $\bar{Q}_R(k, K, k_F, k_M)$ of eq. (4.9). ε_{kK} is the unperturbed energy $\hbar^2k^2/m^* + \hbar^2K^2/4m^* + 2\Delta$. The wave function (kl) stands for the RCM partial wave function $(klSJT, K)$ and similarly for (k', l'), as explained in sect. 4.1. Clearly, eq. (4.10) is to be solved together with the self-consistent condition (3.14a), giving the self-consistent solution $\omega_m^-(\lambda)$. We emphasize that eq. (4.10) is much simpler for calculation than eq. (3.14), and this is made possible by the introduction of the angle-average approximation for $1-(n_i+n_j)$. This enables us to solve eq. (3.14) separately for each partial wave channels, such as the $^3S_1$-$^3D_1$ channel. It should be pointed out that this approximation is quire similar to the angle-average approximations for the Pauli exclusion operators and potential energy calculations which have been used in almost all existing nuclear matter calculations. [1-4,22,25]

We now turn to the vertex function $L(\omega, K)$ of eq. (4.10), which is a standard equation for determining the poles and residues of the particle-particle Green functions [19,20]. L is the irreducible vertex function which has, in general, both two-body and one-body terms. For simplicity, we included in eq. (3.14b) only a two-body term, $\bar{G}^M$, in L. The following consideration is important. As shown by eqs. (3.15) and (3.16), the normalization constant Z_m for the Y_m amplitudes depend on the energy derivatives of the self-consistent energy μ_m. Hence, Z_m in fact depends on the energy derivative $dL/d\omega$. As described in sect. 4.1, we have introduced a s.p. potential U defined by the self-energy vertex function Γ (see eqs. (4.1) to (4.2)). Thus the one-body vertex functions have already entered our calculations. Therefore our L is in fact

$$\langle ij|L(\omega)|ef\rangle = \bar{G}^{M}_{ijef} + \delta_{ie}\delta_{jf}\{S_i(\omega_1)+S_j(\omega_2)\} \tag{4.11}$$

with the one-body vertex function given as

$$S_i(\omega_1) \equiv \Gamma_i(\omega_1) - U(i) \tag{4.11a}$$

and similarly for $S_j(\omega_2)$. It is readily seen that $\omega_1 = \omega - \varepsilon_j$ and $\omega_2 = \omega - \varepsilon_i$. For clarity, the above equations are expressed in the laboratory frame where the indices $i, j, \ldots$ stand for the laboratory momenta $k_i, k_j, \ldots$. Note that our choice of U, as discussed in sect. 4.1, is to make $S_i(\omega = \varepsilon_i) = 0$ (and $S_j(\omega = \varepsilon_j) = 0$). This certainly does *not* imply at all that $S_i(\omega_1) = S_j(\omega_2) = 0$ and $dS_i(\omega_1)/d\omega = dS_j(\omega_2)/d\omega = 0$. In fact these energy derivatives will be rather large and will play an important role in determining Z_m. With $L(\omega)$ given by eq. (4.11), the ring diagrams included in our nuclear matter calculations are in fact of the general structure shown in fig. 12.

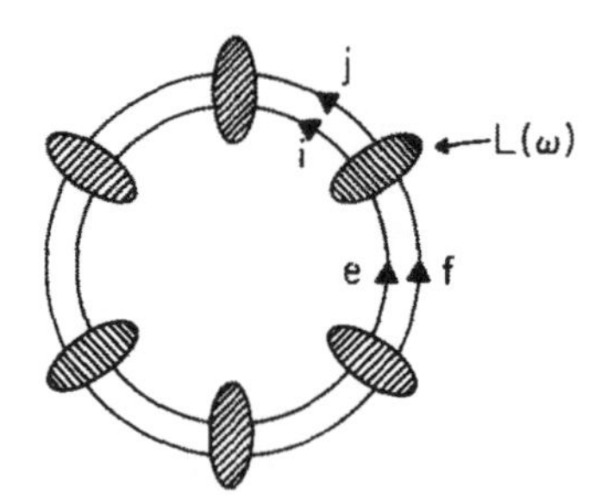

Fig. 12. A general sign diagram included in our nuclear matter calculations. The vertex function $L(\omega)$ is given by eq. (4.11).

As discussed earlier, in our calculations we need the RCM matrix elements $\langle kl|L(\omega, K)|k'l'\rangle$. For given RCM momenta k and K, the angle-averaged values for k_i^2 and k_j^2 being both particles (regions A and B of fig. 11(i)) or both holes (region C of fig. 11(i)) can be readily found. Namely both are given by $(k^2+\frac{1}{4}K^2)$. It has been necessary at several stages of our calculations to use the angle-average approximations. The same is also true for treating S_i and S_j of eq. (4.11). Thus the matrix element of $L(\omega, K)$ of eq. (4.10) is obtained as

$$\langle kl|L(\omega, K)|k'l'\rangle = \langle kl|\bar{G}^{M}(\omega, K\alpha)|k'l'\rangle + 2\delta_{kk'}\delta_{ll'}\{\Gamma_{\bar{k}_1}(\omega - \varepsilon_{\bar{k}_1}) - U(\bar{k}_1)\}, \tag{4.12}$$

where $\bar{G}^{M}$ has been given by eq. (3.3), and Γ and U have been given by eqs. (4.1) to (4.2). Recall that $\bar{k}_1^2 = k^2 + \frac{1}{4}K^2$.

We now proceed to solve eqs. (4.10) and (4.12). First, we recall that they are to be solved together with the self-consistent condition (3.14a), i.e. $\mu_m(\omega, \lambda) = \omega \equiv \omega_m^-(\lambda)$. It is certainly numerically possible to carry out such self-consistent solutions by way of graphical methods. In so doing, one needs a rather large amount of computer time. It is more convenient, as well as physically more meaningful, to carry out the following perturbative solution. We have found that the dependence of $\bar{G}^M$ on ω is quite weak for ω in the vicinity of ω_m^-. (Note that only ω_m^- and the corresponding Y_m amplitudes enter in our calculation of ΔE_0^{pp}, as shown by eq. (3.13).) Thus the dependence of $L(\omega, K)$ on ω comes almost entirely from that of $\Gamma_{\bar{k}_1}(\omega)$. As is well known, $-d\Gamma/d\omega$ is closely related to the two-body wound integral $\kappa^{(2)}$ whose values are ~ 0.10.

Let us first calculate $d\Gamma/d\omega$. From eqs. (4.1a) and (3.1) we have

$$\frac{d}{d\omega}\Gamma_{k_1}(\omega) = -\frac{2}{A}\sum_{k_h \leq k_F}\langle k_1 k_h|\bar{G}^M\left(\frac{Q^M}{e}\right)^2\bar{G}^M|k_1 k_h\rangle\,, \tag{4.13}$$

where $\bar{G}^M$ and e are both dependent on ω, i.e. $\bar{G}^M = \bar{G}^M(\omega + \varepsilon_h)$ and $e = (\omega + \varepsilon_h) - H_0$ (see eqs. (4.8) and (4.8a) for H_0). At the self-consistent energy $\omega = \varepsilon_{k_1}$ we have $k^{(2)}(k_1) = -(d/d\omega)\Gamma_{k_1}(\omega)$ which is the familiar two-body wound integral. Using angle average approximations quite similar to those used for the derivation of $U(k_1)$ of eq. (4.7), we obtain from eq. (4.13)

$$\begin{aligned}\kappa^{(2)}(k_1, \omega) &\equiv -\frac{d}{d\omega}\Gamma_{k_1}(\omega)\\ &= \frac{8}{\pi}\sum_{\alpha}(2T+1)(2J+1)\int_0^{k_-} k^2\, dk\langle k\alpha|G\left(\frac{Q}{e}\right)^2 G(\bar{K}_1)|k\alpha\rangle\\ &\quad + \frac{1}{2\pi k_1}\sum_{\alpha}(2T+1)(2J+1)\\ &\quad \times\int_{k_-}^{k_+} dk[k_F^2 - k_1^2 + 4k(k_1 - k)]k\langle k\alpha|G\left(\frac{Q}{e}\right)^2 G(\bar{K}_2)|k\alpha\rangle\,,\end{aligned} \tag{4.14}$$

where $k_\pm$, $\bar{K}_1$ and $\bar{K}_2$ have been given in eq. (4.7a). The matrix elements of $G(Q/e)^2 G(K)$ are abbreviations for

$$\langle k\alpha|G\left(\frac{Q}{e}\right)^2 G(k)|k\alpha\rangle \equiv \sum_{l'}\frac{2}{\pi}\int_0^\infty k'^2\, dk'\left\{\frac{\bar{Q}^M(k', K)\langle k'l'|G^M(\omega, k\alpha)|kl\rangle}{\omega - H_0(k', K)}\right\}^2\,, \tag{4.15}$$

where $\bar{Q}^{M}$ is the $\bar{Q}^{M}(k, K, k_F, k_M)$ of eq. (4.4) and H_0 has been given by eqs. (4.8) and (4.8a). Recall that α stands for the quantum numbers $(ll'SJT)$ and (kl) represents $(klSJT, K)$.

In fig. 13, we present some representative values of $\kappa^{(2)}(k_1, \omega)$. As shown, they are generally rather small. (This is mainly because we have used a model space with $k_M = 3.2\ \text{fm}^{-1}$.) We have found that $\kappa^{(2)}$ generally varies rather slowly with ω and k_1 in the energy and momentum regions important for our present calculation. Consequently, $\mathrm{d}\Gamma/\mathrm{d}\omega$ has the same behavior and it should be a good approximation to solve eqs. (4.10), (4.12) and (3.14) with their ω-dependent terms treated by first order perturbation theory*, as is done in the present calculation.

4.3. NUCLEAR MATTER BINDING ENERGIES

Having calculated the reaction matrix $\bar{G}^{M}$ and the transition amplitudes Y in the previous two subsections, we are now ready to calculate the energy shift ΔE_0^{pp} of eq. (3.13). Using angle-average approximations similar to those used in the derivation of eqs. (4.4) and (4.9), we can express ΔE_0^{pp} as

$$\frac{\Delta E_0^{pp}}{A} = \frac{3}{\pi^2 k_F^3} \sum_{\alpha} (2J+1)(2T+1) \int_0^1 \mathrm{d}\lambda \int_0^{2k_M} K^2\,\mathrm{d}K \sum_m \sum_{ll'}$$
$$\times \int_0^{k_M} k^2\,\mathrm{d}k \int_0^{k_M} k'^2\,\mathrm{d}k'$$
$$\times\, Y_m(klK, \lambda)^* \langle kl| G^M(\omega_m^-(\lambda), K\alpha)|k'l'\rangle Y_m(k'l'K, \lambda)\,, \qquad (4.16)$$

which is the potential energy per nucleon (PE/A) of nuclear matter. Let us briefly

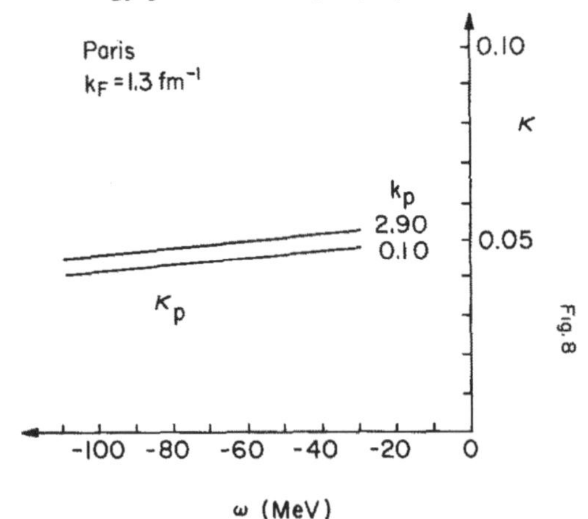

Fig. 13. Variation of $\kappa^{(2)}(k_1, \omega)$ with k_1 and ω, calculated with $k_F = 1.3\ \text{fm}^{-1}$ and the Paris NN interaction. $k_M = 3.2\ \text{fm}^{-1}$ is used.

* Briefly speaking, we consider $[H_0 + A(\omega)]\psi(\omega) = E(\omega)\psi(\omega)$ with $\omega = E(\omega)$. H_0 is ω-independent. We first solve $[H_0 + A(\omega_0)]\phi_0 = E_0\phi_0$, and then treat $(\omega - \omega_0)\,\mathrm{d}A/\mathrm{d}\omega$ by first order perturbation theory (for ω not very far away from ω_0.) This gives, for example, $E(\omega) \approx E_0 + (\omega - \omega_0)\langle\tilde{\phi}_0|\,\mathrm{d}A/\mathrm{d}\omega|\phi_0\rangle$.

explain the above. α denotes the partial wave quantum numbers ($ll'SJT$). For each α, λ and K we solve the secular equation (4.10) to obtain $\omega_m^-(\lambda)$ and $Y_m(\ldots,\lambda)$. Recall that, as shown by eq. (3.13), ΔE_0^{pp} was originally expressed in terms of the laboratory frame momentum variables $(\boldsymbol{k}_i, \boldsymbol{k}_j)$ and $(\boldsymbol{k}_k, \boldsymbol{k}_l)$. We have now transformed it to the RCM frame with momentum variables k, k' and K. This is made possible by replacing the occupation factor $(\bar{n}_i\bar{n}_j - n_i n_j)$ by its angle-average approximation $\bar{Q}_R$ of eq. (4.9) in the derivation of the RPA type seculation equation (4.10). Thus the amplitudes Y's are already angle averaged, and therefore in eq. (4.16) the upper integration limits for k and K are respectively k_M and $2k_M$. We note that the Y's are normalized to Z_m^λ according to eq. (3.15) which is expressed in terms of the laboratory momentum variables. In actual calculations this equation is transformed into the RCM frame also with the approximation that the occupation factor $(\bar{n}_i\bar{n}_j - n_i n_j)$ is replaced by its angle average $\bar{Q}_R$. Clearly, the binding energy per nucleon (BE/A) is given by $-\mathrm{BE}/A = 3\hbar^2k_F^2/10m + \Delta E_0^{pp}/A$.

The integrations in eq. (4.16) are carried out numerically using gaussian mesh points and weights. We have found that satisfactory accuracy is obtained when using approximately (3, 3, 30) mesh points for the (λ, K, k) integrations, respectively. It is interesting to note that when setting $\lambda = 1$ (i.e. $\int_0^1 d\lambda$ is removed), $k_M = k_F$, and $L = 0$ in eq. (4.10), eq. (4.16) becomes the usual BHF formula for calculating PE/A. We have used this limit for performing BHF nuclear matter calculations and obtained results in very good agreement with results of other BHF calculations*. This serves as a check of our computer programs. Note that when L is set to 0 in eq. (4.10), the amplitudes Y_m reduce to their respective unperturbed values (i.e. the true wave functions Ψ_m^{A-2} and Ψ_0^A in eq. (2.12d) are replaced by the respective unperturbed wave functions Φ_m^{A-2} and Φ_0^A).

In figs. 14 and 15 we show the results of our nuclear matter BE/A calculations as a function of k_F (or density $\rho = 2k_F^3/3\pi^2$) for both the Paris and the Reid NN interactions. The curves labelled RING are our ring diagram calculations outlined by eq. (4.16). The results from BHF calculations, labelled BHF, are also given for comparison. Let us discuss fig. 14 first. Clearly, the RING calculations give an additional binding energy of about 4 MeV per nucleon, as compared with the BHF result. This gain in binding energy is perhaps attributable, to a large extent, to the use of a "continuous" s.p. spectrum which has an attractive potential energy in the momentum region k_F to k_M. Ma *et al.* [10,13]) have carried out model-space BHF (MBHF) nuclear matter calculations which are essentially BHF calculations using a "continuous" s.p. spectrum for momentum $k \leq k_M$. They have found that MBHF nuclear matter calculations using the Paris NN interaction gives an additional BE/A of about 4 MeV compared with the corresponding BHF calculations. Lejeune, Mattzolff and Grange [27]) have performed BHF nuclear matter calculations using the Paris NN interactions and a continuous s.p. spectrum. Their results and those of Ma *et al.* are in fact in very good agreement with each other; their calculations

* We have compared our BHF results with those of refs. [10,13]) using the Paris and Reid potentials, of ref. [22]) using the Reid potential, and of Day and Wiringa's calculation [26]) using the Paris potential.

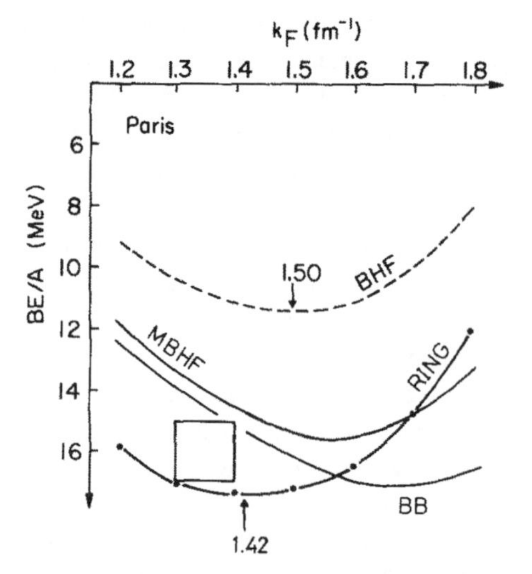

Fig. 14. Results of ring diagram nuclear matter calculations (RING) obtained with eq. (4.16), $k_M = 3.2\,\text{fm}^{-1}$, and the Paris NN interaction. Results for the BHF calculations are also shown (BHF). The arrows indicate the saturation Fermi momenta. The box indicates the empirical nuclear matter properties. The curves labelled BB and MBHF are respectively Day and Wiringa's [26]) and Kuo, Ma and Vinh Mau's [10]) results. See text for further explanation.

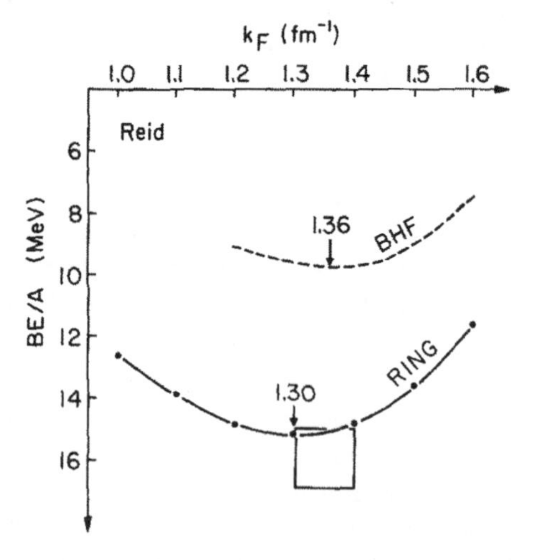

Fig. 15. Same as fig. 14 except for the Reid NN interaction.

using a continuous s.p. spectrum give an additional BE/A of about 4 MeV compared with those using a conventional discontinuous s.p. spectrum. Similar calculations have also been carried out by Dey and Matin [28]). They have, however, found that the gain in BE/A due to the use of a continuous s.p. spectrum is about 7 MeV. All these calculations indicate that the use of "continuous" s.p. spectrum of the form shown in fig. 8 is very likely to give an extra BE/A of several MeV compared with the case of using a conventional discontinuous BHF s.p. spectrum.

Day and Wiringa [26]) have performed both variational and Brueckner-Bethe calculations of nuclear matter using the Paris potential. The latter included two-, three- and four-hole line contributions. They have found that the results given by these two approaches generally agree with each other rather well. In fig. 14, we include their Brueckner-Bethe result (labelled BB) for comparison. As shown, their binding energy agrees with our RING result rather well, but their saturation density is significantly higher. Kuo, Ma and Vinh Mau [10]) have performed model-space Brueckner Hartree-Fock calculations of nuclear matter using the Paris potential. Their results (labelled MBHF) are also included in fig. 14 for comparison.

The most interesting feature of fig. 14 is therefore not that our RING calculations give satisfactory BE/A. It is, instead, that they give rather satisfactory saturation density. Most likely, this is due to the inclusion of the particle-particle ring diagrams to all orders in our calculations. Let us trace our calculations. The potential energy per nucleon of our calculations is evaluated according to eq. (4.16). Let us decompose $\Delta E_0^{\mathrm{pp}}/A$ of this equation into its λ-integrands for each partial waves:

$$\frac{\Delta E_0^{\mathrm{pp}}}{A}=\sum_\alpha \Delta E_\alpha=\sum_\alpha \int_0^1 \mathrm{d}\lambda\, I_\alpha(\lambda)\,. \tag{4.17}$$

We show in table 2 a set of typical values for $I_\alpha(\lambda)$ and ΔE_α. For most partial wave channels, $I_\alpha(\lambda)$ does not vary much from $\lambda=0$ to $\lambda=1$. But for the $^3S_1$-$^3D_1$ channel, we notice that there is a large change of $I_\alpha(\lambda)$; it changes from -12.7 to -29.9 when λ changes from ~ 0.11 to ~ 0.89. It is easily seen that $\lambda=0$ corresponds to the case in which only terms of first order in $\bar{G}^{\mathrm{M}}$ are included in ΔE_0^{pp} of eq. (4.16). Hence we have the rather interesting and important observation that it is the $^3S_1$-$^3D_1$ channel where the effect of ring diagrams is most pronounced. Since it is this channel where the NN tensor interaction is most important, it may be inferred from our calculations that the importance of the ring diagram correlations is closely related to the NN tensor interaction.

Fig. 14 seems to indicate that the effect of the ring diagrams is rather sensitive to the nuclear matter density. Let us use the BHF curve as a reference. With respect to it, we see that the ring diagrams give a considerably larger gain in BE/A for low density than for high density. Thus the inclusion of the ring diagrams shifts the saturation to a lower value ($k_F \cong 1.4\ \mathrm{fm}^{-1}$ as shown by the figure). Let us investigate this point more closely.

In fig. 16 we show the contributions to PE/A from the $^3S_1$-$^3D_1$ partial wave channels for various k_F values. For the Paris NN interaction, the ring diagram calculations are represented by curves 1 and the BHF calculation by curves 2. For the Reid NN interaction, the former are represented by curves 3 and the latter by

TABLE 2

A decomposition of $\Delta E^{pp}/A$ of eq. (4.16)

α	0.1127	0.5000	0.8873	$\int_0^1 d\lambda \ldots$
$^1S_0$	−16.762	−16.974	−17.212	−16.981
$^3S_1$-$^3D_1$	−12.719	−20.708	−29.867	−21.033
$^1P_1$	4.922	4.543	4.229	4.562
$^3P_0$	−3.767	−3.749	−3.726	−3.748
$^3P_1$	11.972	10.726	9.755	10.802
$^3P_2$-$^3F_2$	−8.527	−8.980	−9.491	−8.996
$^1D_2$	−3.187	−3.133	−3.085	−3.135
$^3D_2$	−4.607	−4602	−4.632	−4.612
$^3D_3$-$^3G_3$	0.483	0.287	0.102	0.290
$^1F_3$	0.980	0.949	0.920	0.950
$^3F_3$	1.857	1.801	1.748	1.802
$^3F_4$-$^3H_4$	−0.625	−0.622	−0.618	−0.622
$^1G_4$	−0.567	−0.552	−0.538	−0.553
$^3G_4$	−0.878	−0.858	−0.839	−0.858
$^3G_5$-$^3I_5$	0.143	0.134	0.125	0.134
$L>4$				0.309
PE/A				−41.688
KE/A				24.384
BE/A				−17.304

curves 4. Consider the $^3S_1$-$^3D_1$ channel first, and we note that in this channel the NN tensor interaction plays an important role. It is well known that the intermediate states induced by the NN tensor interaction are mainly of moderately high excitation energy (~300 MeV) [ref. [29])] At high densities, these intermediate states are to a large extent blocked (because k_F is large). At lower densities, these states are relatively empty. Thus if the ring diagrams are induced mainly by the NN tensor interaction, then we expect their effect to be more pronounced at low densities than at high densities. This conjecture is clearly supported by our results, as shown by the $^3S_1$-$^3D_1$ curves of fig. 16. Here curves 1 and 2 are rather close to each other at $k_F \sim 1.6\,\text{fm}^{-1}$, but at $k_F \sim 1.2\,\text{fm}^{-1}$ curve 1 becomes considerably lower. Similar behavior is also observed for curves 3 and 4.

Turning to the $^1S_0$ curves of fig. 16, we see that curves (1, 3) are respectively higher than curves (2, 4) in the region of high densities. This suggests that the effect of the short-range repulsion of the NN interaction is also enhanced by the ring diagrams. (This enhancement is also evident for the $^3S_1$-$^3D_1$ curves of fig. 16.) That the difference between curves 3 and 4 is larger than the difference between 1 and 2 is perhaps a reflection that the Reid NN interaction has a stronger short-range repulsion than the Paris NN interaction.

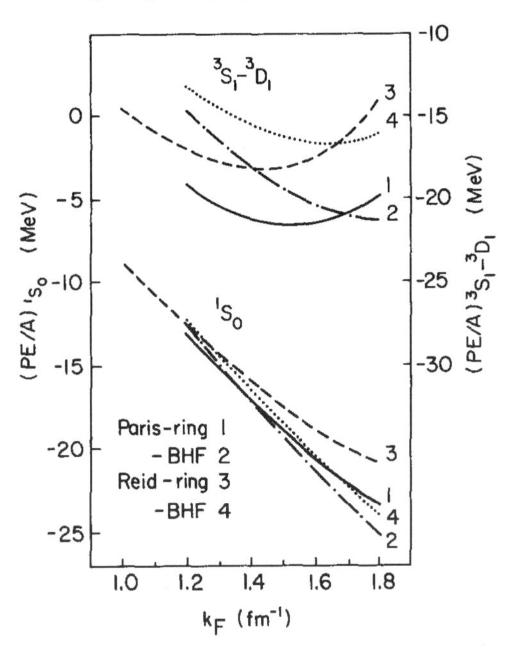

Fig. 16. Contributions to the potential energy per nucleon (PE/A) from the $^1S_0$ and $^3S_1$-$^3D_1$ channels, for the Paris and Reid NN interactions. Calculations including ring diagrams are denoted by "RING", and the BHF calculations are denoted by "BHF".

Let us now go back to fig. 15 where nuclear matter saturation curves calculated with the Reid NN interaction are shown. The BHF calculation gives BE/$A \sim 9.5$ MeV with a saturation Fermi momentum $k_F \sim 1.36\ \mathrm{fm}^{-1}$. Just like the case with the Paris NN interaction of fig. 14, the ring diagrams shift the saturation density to a much lower value ($k_F \sim 1.30\ \mathrm{fm}^{-1}$). It is interesting to note that this k_F is in fact lower than the empirical value of $k_F \sim 1.35\ \mathrm{fm}^{-1}$.

In figs. 17a and b, we give the major partial wave contributions to PE/A of our ring diagram nuclear matter calculations using the Paris NN interaction. In figs. 17c and d the corresponding contributions are given for the Reid NN interaction. It is interesting to note that only the $^3S_1$-$^3D_1$ curves depict a saturation behavior. The overall nuclear matter saturation is governed by a delicate balance among these PE/A contributions and the kinetic energies at various k_F values. We note that the PE/A curves for the $^1S_0$, $^3S_1$-$^3D_1$, $^1P_1$ and $^3P_1$ of the Reid NN interaction are all significantly different from those of the Paris NN interaction.

The normalization condition of the Y-amplitudes, as given by eqs. (3.15) and (3.16), has played an important role in our nuclear matter calculations. As shown by eq. (4.16), $\Delta E_0^{pp}/A$ depends directly upon the Y-amplitudes, and the overall magnitudes of Y depend on the normalization constant Z_m of eqs. (3.15) and (3.16). From our discussion in sect. 4.2, it is easy to see that Z_m^λ of eq. (3.16) is approximately

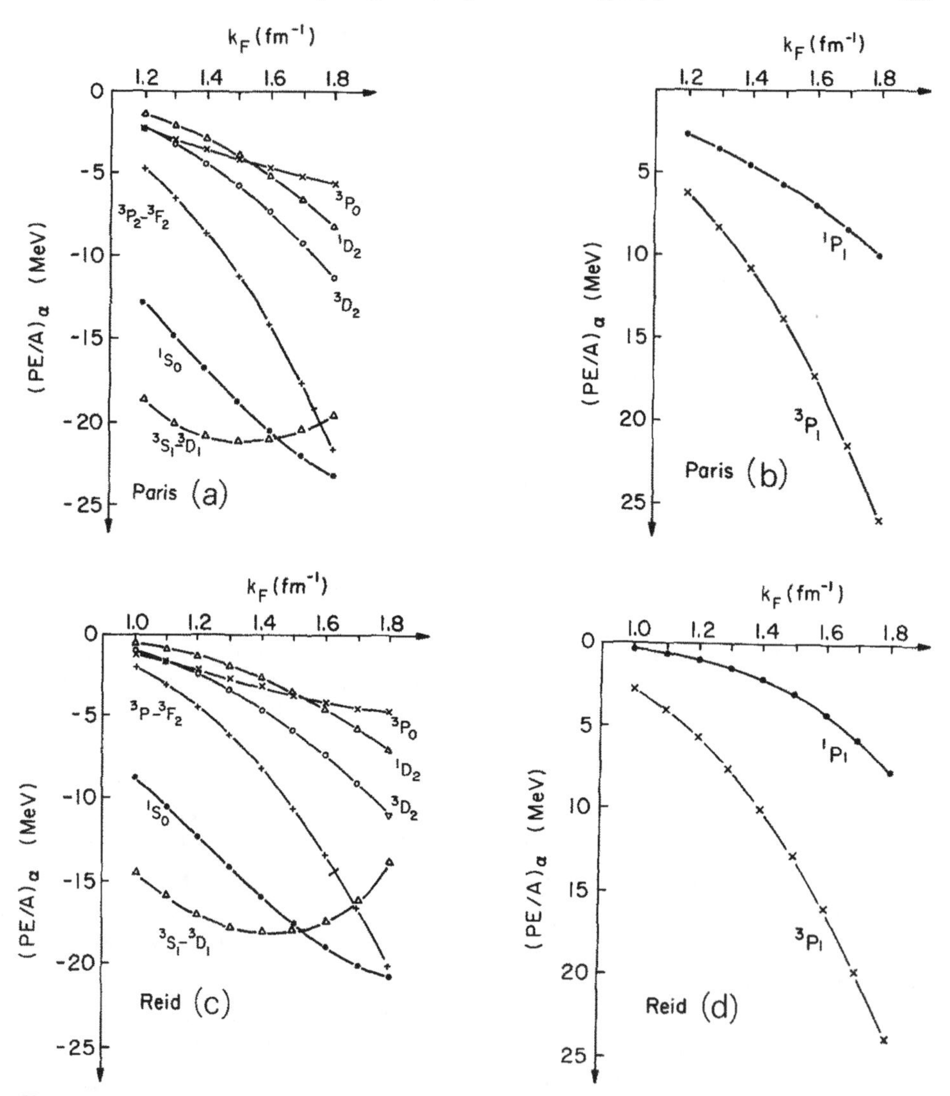

Fig. 17. Partial wave contributions to the potential energy per nucleon (PE/A) in our ring diagram nuclear matter calculations for the Paris (a, b) and Reid (c, d) NN interactions. $k_M = 3.2\ \text{fm}^{-1}$ is used.

equal to $(1+2\kappa^{(2)})^{-1}$ for $\lambda=0$ and 1 for $\lambda=0$, where typical values of $\kappa^{(2)}$ have been given in fig. 13. Note that λ is integrated from 0 to 1 as indicated by eq. (4.16). As expected, we have found that $\kappa^{(2)}$ generally increases with k_F. Namely, at higher k_F (i.e. higher density) $\kappa^{(2)}$ becomes larger and, consequently, Z_m^λ becomes smaller in general. Hence in eq. (4.16) the magnitudes of the Y-amplitudes become generally smaller, by a rather significant factor, as k_F increases. This has been an important factor in determining the saturation density of our nuclear matter calculations.

From the saturation curves of figs. 14 and 15, we have deduced the incompressibility coefficients

$$K = k_F^2 \frac{d^2}{dk_F^2}\left(-\frac{BE}{A}\right). \tag{4.18}$$

They are 96.3 and 110.7 MeV for our ring diagram calculations using, respectively, the Paris and Reid NN potentials. Again, both the ring diagrams and the tensor force have played important roles in determining their values. We note that our value for the Reid case is in good agreement with the value recently suggested by Brown and Osnes [8]).

Finally, let us discuss an essential feature of our present nuclear matter theory. Our calculation is based on eq. (4.16), and to carry out its calculation we must first solve the RPA equation (4.10). Clearly this equation involves a non-Hermitian matrix and consequently its energy solutions may become complex. When this takes place, ΔE_0^{pp} of eq. (4.16) becomes generally complex and consequently nuclear matter becomes unstable. We have indeed found such complex solutions at $k_F = 1.2\,\mathrm{fm}^{-1}$ for the Paris potential case (fig. 14) and at $k_F = 1.0\,\mathrm{fm}^{-1}$ for the Reid potential case (fig. 15). (No complex solutions are found for all the other points of the RING curves of figs. 14 and 15.) Hence our calculation predicts nuclear matter instability at the above two situations. In the present work we have simply discarded the complex solutions of eq. (4.10), and the $k_F = 1.0$ point of fig. 15 are obtained in this way. Nuclear matter instability is an important problem. [5]) It appears that our ring-diagram nuclear matter theory may provide a useful tool for studying this problem. We plan to carry out a more thorough investigation of the instability situation associated with complex solutions of eq. (4.10) in a future publication.

5. Discussion and conclusion

We have developed a new method for summing up the particle-particle ring diagrams of nuclear matter to all orders. This type of ring diagrams has not been investigated before. A desirable feature of our method is its relative simplicity for making calculations. One first chooses a model space, and calculates the model space reaction matrix G^M and s.p. spectrum. Using these, one solves a RPA-type secular equation to obtain the Y transition amplitudes. Then the potential energy of nuclear matter is given by simple integrals of the product YY^+G^M.

Using the Paris and Reid NN potentials, we have applied the above method to nuclear matter calculations. The effects of the particle-particle ring diagrams are found to be rather important. Compared to the results of conventional BHF nuclear matter calculations, our calculations with the inclusion of the ring diagrams give a larger binding energy per nucleon as well as a *smaller* saturation density. It should be emphasized that we use a model-space s.p. spectrum which has self-energy insertions to particle lines (with momentum $k_F < k < k_M$). This enhances the effect of our ring diagrams as compared with the case where ring diagrams are calculated using a conventional discontinuous BHF s.p. spectrum.

Our calculations indicate that the ring diagrams are particularly important for the treatment of the NN tensor force. With these diagrams included, the NN tensor force has the clear tendency to make nuclear matter saturate at lower densities (as compared with nuclear matter calculations without the inclusion of the ring diagrams). This is a rather interesting result; it enables our nuclear matter result to move laterally away from the well-known Coester band [8]). In fact our calculated binding energy per nucleon and saturation density of nuclear matter, for both the Paris and Reid NN potentials, are both in rather satisfactory agreement with the corresponding empirical values as shown in figs. 14 and 15.

Our calculations can still be improved upon in several areas. Consequently, we do not want to emphasize the above good agreement between our results and the empirical nuclear matter properties. Instead, we wish to emphasize the trend provided by the inclusion of our ring diagrams in lowering the nuclear matter saturation densities. Shakin and his collaborators [14]) have pointed out that relativistic corrections are very important in lowering the nuclear matter saturation densities. Our calculations do not include relativistic corrections, yet our results have shown that the inclusion of our ring diagrams seems to have a rather strong effect in lowering the nuclear matter saturation density. This raises the intriguing possibility that the calculated nuclear matter saturation density may become too low if ring diagrams and relativistic corrections are both included. Further study in this direction is certainly needed. In fact, we are planning to carry out a ring diagram nuclear matter calculation with the inclusion of relativistic effects. Several authors [5,9,10]) have suggested the need of three-body effective interactions in lowering the calculated nuclear matter saturation density. Our calculations indicate that this need may be reduced by the inclusion of the ring diagrams.

We now discuss several aspects where the present calculation can be improved upon and where further investigations are needed. First, an important step of our calculation is the choice of k_M, the momentum space boundary of our model space. We have used $k_M = 3.2\ \mathrm{fm}^{-1}$. Our results clearly depend on the choice of k_M. Ma and Kuo [13]) have investigated the dependence of their MBHF calculations on the choice of k_M, and found that the dependence of their results on k_M was rather smooth. (They found a local minimum for BE/A at $k_M \approx 2k_F$.) Their calculations did not include the ring diagrams. Using their results as a guide line, we have used $k_M = 3.2\ \mathrm{fm}^{-1}$ in our present calculation. Limitations in computer time has prevented us from carrying out the present calculation using a series of different k_M values. This should be done in the future. In short, the choice of k_M is an uncertainty in the present calculation and should be further investigated.

Next, we turn to the angle-average approximations. As is well known, these approximations are standard (and indispensible) in treating the Pauli exclusion operators in the usual BHF nuclear matter calculations and are generally considered to be fairly accurate [1,2]). We have now generalized them to the treatment of the particle-particle RPA secular equations. This is done primarily because by doing so different partial waves are decoupled. Thus the calculation is greatly simplified. Although the approximations used in the RPA equations are very similar to those used in treating the Pauli operators in BHF calculations and we may expect them to have similar accuracies, it should still be very desirable to actually check the accuracy of the former. This may be done for restricted cases by solving the RPA equations directly in the laboratory frame by way of a vector-bracket transformation method [30]), and comparing the results so obtained with those obtained with the angle-average approximations.

Finally, we come to the particle-hole ring diagrams. As mentioned earlier, the YHK ring diagram method is applicable to the summation of the particle-hole ring diagrams. But their numerical calculation is actually more complicated than the particle-particle ring diagrams. This is because to calculate the particle-hole matrix element we need to make a transformation between the particle-hole channels and the particle-particle channels. Thus the various partial wave channels of the particle-particle interactions are generally coupled. The representation advocated by Dickoff, Faessler and Müther [31]) and by Müther [32]) appears to be a convenient way for studying the particle-hole ring diagrams. In fact they have calculated the particle-hole ring diagrams in nuclear matter using the Reid NN potential. There are, however, differences between their calculations and what we would like to calculate within the present framework.

As indicated in fig. 1, diagram (δ) is a third order particle-hole ring diagram. Diagram (β) may be treated as either a second-order particle-particle or as a second-order particle-hole ring diagram. We have used the former scheme and included diagram (β) in ΔE_0^{pp}. Thus, as discussed in ref. [15]), the lowest order particle-hole ring diagram in ΔE_0^{ph} - the potential energy contribution from the particle-hole ring diagrams - is *third* order in G^M. In contrast, the particle-hole ring diagrams of refs. [31,32]) and of Day's calculation [33]) begin with diagrams second order in the reaction matrix. There are indications that the particle-hole ring series converge rather rapidly and may not be very important for nuclear matter binding energy calculations [33,31,32]). Hence we expect that our ΔE_0^{ph} would be considerably less important than our ΔE_0^{pp}. This behavior has in fact been observed in some Lipkin model calculations [15]) and preliminary G-matrix calculations of the binding energy of ^{16}O [ref. [34])]. Whether the above expectation will be realized or not for nuclear matter remains to be investigated. Hence it will be worthwhile and interesting to actually calculate ΔE_0^{ph} for nuclear matter based on the framework of the present work and ref. [15]). Two further differences between ΔE_0^{ph} so calculated and earlier particle-hole ring diagram calculations [31-33]) are worth noting. First we employ a model-space approach which employs a self-consistent particle ($k_F < k < k_M$) spectrum whereas the earlier calculations employed a conventional BHF s.p. spectra. Second, our method for summing up the particle-hole ring diagrams will be similar to that given by eq. (3.13), and it will be considerably different from the methods used in earlier calculations. It will be of interest to investigate the effect of these differences on the role the particle-hole ring diagrams play in nuclear matter calculations.

The authors wish to thank Prof. G.E. Brown for many helpful discussions and encouragment throughout the course of this work. They are very grateful to Z.Y. Ma for several helpful correspondences and for making her MBHF computer programs available to them. They also wish to thank Prof. S.S. Wu for a number of helpful discussions and correspondences, to J. Heyer for a careful reading of the manuscript, and to Diane Siegel for her excellent and pleasant help in preparing the manuscript. Last but not least, H.Q.S. and S.D.Y. wish to thank Profs. G.E. Brown and A.D. Jackson for hospitality and support during their visit at Stony Brook.

References

1) H.A. Bethe, Ann. Rev. Nucl. Sci. **21** (1971) 93
2) D.W.L. Sprung, Adv. Nucl. Phys. **5** (1972) 225
3) J.P. Jeukenne, A. Lejeune and C. Mahaux, Phys. Reports **25C** (1976) 83
4) B.D. Day, Rev. Mod. Phys. **50** (1978) 495
5) A.D. Jackson, Ann. Rev. Nucl. Part. Sci. **33** (1983) 105
6) S.-O. Bäckman, G.E. Brown and J.A. Niskanen, Phys. Reports **124** (1985) 1
7) F. Coester, S. Cohen, B. Day and C.M. Vincent, Phys. Rev. **C1** (1970) 769
8) G.E. Brown and E. Osnes, Phys. Lett. **159B** (1985) 223
9) B.D. Day, Phys. Rev. Lett. **47** (1981) 226
10) T.T.S. Kuo, Z.Y. Ma and R. Vinh Mau, Phys. Rev. C **33** (1986) 717
11) H. Kümmel, K.H. Lührmann and J.G. Zabolitsky, Phys. Reports **36C** (1978) 1
12) R. Jastrow, Phys. Rev. **98** (1955) 1479; S. Fantoni and S. Rosati, Nuovo Cim. **A20** (1974) 179
13) Z.Y. Ma and T.T.S. Kuo, Phys. Lett. **B127** (1983) 137; T.T.S. Kuo and Z.Y.Ma, in Nucleon-nucleon interaction and nuclear many body problems, ed. S.S. Wu and T.T.S. Kuo (World Scientific, Singapore 1984), p. 178
14) M.R. Anastasio, L.S. Celenza, W.S. Pong and C.M. Shakin, Phys. Reports **100** (1983) 327
15) S.D. Yang, J. Heyer and T.T.S. Kuo, Nucl. Phys. **A448** (1986) 420
16) M. Lacombe, B. Loiseau, J.M. Richard, R. Vinh Mau, J. Coté, P. Pires and R. de Tourreil, Phys. Rev. **C21** (1980) 861
17) R.V. Reid, Ann. of Phys. **50** (1968) 411
18) See, for example, A.L. Fetter and J.D. Walecka, Quantum theory of many-particle systems (McGraw-Hill, 1971)
19) S.S. Wu and T.T.S. Kuo, Nucl. Phys. **A430** (1984) 110
20) S.D. Yang and T.T.S. Kuo, Nucl. Phys. **A456** (1986) 413
21) T.T.S. Kuo, Ann. Rev. Nucl. Sci. **24** (1974) 101; P.J. Ellis and E. Osnes, Rev. Mod. Phys. **49** (1977) 777
22) M. Haftel and F. Tabakin, Nucl. Phys. **A158** (1970) 1
23) G.E. Brown, A.D. Jackson and T.T.S. Kuo, Nucl. Phys. **A133** (1969) 481
24) E.M. Krenciglowa, C.L. Kung, T.T.S. Kuo and E. Osnes, Ann. of Phys. **101** (1976) 154
25) H.A. Bethe, B.H. Brandow and A.G. Petschek, Phys. Rev. **129** (1963) 225
26) B.D. Day and R.B. Wiringa, Phys. Rev. **C32** (1985) 1057
27) A. Lejeune, M. Martzolff and P. Grange, Lecture Notes in Physics (Springer) vol. 198 (1984) 36
28) M.A. Matin and M. Dey, Phys. Rev. **C27** (1983) 2356: **C29** (1984) 344(E)
29) T.T.S. Kuo and G.E. Brown, Phys. Lett. **18** (1965) 54

30) See, for example, C.L. Kung, T.T.S. Kuo and K.F. Ratcliff, Phys. Rev. **C19** (1979) 1063
31) W.H. Dickoff, A. Faessler and H. Müther, Nucl. Phys. **A389** (1982) 492
32) H. Müther, Nucleon-nucleon interaction and nuclear many body problem ed. S.S. Wu and T.T.S. Kuo (World Scientific, 1984) p. 490.
33) B.D. Day, Phys. Rev. **C24** (1981) 1203
34) H. Müther and T.T.S. Kuo, private communication

CHAPTER III

Low-Momentum NN Interactions in a Renormalization Group Approach

Inspired by ideas from effective field theory and the renormalization group, we proposed a method for integrating out the strongly repulsive and experimentally unconstrained short-distance components of the free-space nucleon–nucleon potential. The resulting low-momentum interaction, V_{low-k}, was found to be nearly unique with respect to the underlying interaction, thereby yielding for the first time a model-independent picture of the nuclear force at low-momenta. In this chapter we have collected the foundational papers and subsequent applications on this subject.

[S. Bogner, T.T.S. Kuo and L. Coraggio (2001); S. Bogner, T.T.S. Kuo, L. Coraggio, A. Covello and N. Itaco (2002)]: These two papers reported the T-matrix equivalence method for deriving the low-momentum V_{low-k} interaction from an underlying realistic NN potential V_{NN}. V_{low-k} preserves the deuteron properties given by V_{NN} as well as its phase shifts with $E_{\rm lab} \leq \Lambda^2$, Λ being the decimation scale. It may be mentioned, however, that for a long time our computer codes were not cooperative; the phase shifts of V_{low-k} given by the codes did not match those of V_{NN}. Finally in the summer of 1998, Harry Lee of Argonne found the defect in our codes; it was our big joy. (At that time Harry Lee and TTSK were both visiting the Institute of Physics in Taipei, Taiwan.) Let us thank him again. These papers also pointed out that results of shell model calculations using V_{low-k} were in good agreement with experiments.

[A. Schwenk, G.E. Brown and B. Friman (2002)]: In this paper V_{low-k} was used within the framework of Landau Fermi liquid theory to study the interaction of quasiparticles near the Fermi surface of symmetric nuclear matter. The renormalization of the effective particle-hole interaction through polarization of the surrounding medium is included through the induced interaction formalism developed many years previously by S.V. Babu and G.E. Brown in their study of liquid ^{3}He. Starting with a novel set of sum rules for the Fermi liquid parameters characterizing the quasiparticle interaction and taking the lowest-order spin-independent ones from experiment, a set of spin-dependent parameters was obtained. Furthermore, the renormalization group equations in the particle-hole channel were derived, and it was found that the sum rules relating Fermi liquid parameters are renormalization group invariant.

[S.K. Bogner, T.T.S. Kuo and A. Schwenk (2003); S.K. Bogner, T.T.S. Kuo, A. Schwenk, D.R. Entem and R. Machleidt (2003)]: The main result of these papers was that for a decimation scale $\Lambda \leq 2.1$ fm^{-1}, the low-momentum interactions derived from various NN potentials (CD-Bonn, Argonne, Nijmegen, Idaho, ...) were practically all identical to each other. Thus for momenta below this scale, the V_{low-k} interaction was essentially unique. In retrospect, this is a reasonable result. The various NN potential models are all contrained by the same low-energy scattering phase shifts up to $E_{\rm lab} \simeq 350$ MeV, which corresponds to the above value of Λ. There are no experimental constraints beyond this momentum scale. It is then not surprising that the V_{low-k} parts of these interactions are nearly the same. When a larger Λ was employed, these papers showed that the various potentials were significantly different from each nother. It is ironic that beyond the above scale Λ, the NN interaction is by and large 'unknown'; one can only employ certain models in this region.

[J.D. Holt, T.T.S. Kuo and G.E. Brown (2004)]: The V_{low-k} interactions derived from the folded-diagram or Lee-Suzuki methods are by construction non-Hermitian. This paper showed that the above non-Hermitian V_{low-k} interactions could be transformed into Hermitian ones by way of a family of similarity transformations. The well-known Okubo Hermitian effective interaction corresponds to a special choice for the similarity transformation.

[J.D. Holt, T.T.S. Kuo, G.E. Brown and S.K. Bogner (2004)]: This paper found that the difference $(V_{NN} - V_{low-k})$ can be highly accurately represented by a momentum expansion of the form $(C_0 + C_2 q^2 + C_4 q^4)$, with C_0 being the dominant term and C_4 negligibly small. This indicates that V_{NN} and V_{low-k} differ from each other mainly in their short-range parts, the latter corresponding to the former with its short-range parts largely integrated out. (Note that C_0 is a delta function in r-space.)

[J.D. Holt, J.W. Holt, T.T.S. Kuo, G.E. Brown and S.K. Bogner (2005)]: This paper reported an all-order calculation for the core polarization effect using the V_{low-k} interaction. By solving the coupled self-consistent Kirson-Babu-Brown equations for the induced interactions, several classes of core polarization planar diagrams were summed to all orders. The use of the energy-independent V_{low-k} interaction played an important role in making the solution of these coupled equations possible. Our main result was that the all-order core polarization results were remarkably close to those given by the 2nd-order core polarization diagram alone, providing a belated support to the early Kuo–Brown interactions where the core polarization effect was evaluated using 2nd-order perturbation theory.

[B.-J. Schaefer, M. Wagner, J. Wambach, T.T.S. Kuo and G.E. Brown (2006)]: Low-momentum hyperon-nucleon (YN) potentials were derived from several Nijmegen YN potentials. The YN low-momentum interactions derived from them, however, exhibit significant differences among them, lacking the 'uniqueness' feature of the NN low-momentum interactions. This is largely because the YN scattering data are much fewer and also more difficult to determine (as compared to the NN scattering data). More and better experimental data on YN scatterings are needed.

[J.W. Holt and G.E. Brown (2006)]: This paper reviewed the Bethe-Brueckner nuclear matter theory, and in particular investigated an interesting connection between the well-known Moszkowski-Scott (MS) separation method and the low-momentum V_{low-k} interaction. The MS effective interaction is $V_{\rm long}(r)$ which is equal to V_{NN} for $r > d$ and equal to 0 otherwise, d being the separation distance. This paper found that $V_{\rm long}$ and V_{low-k} were nearly equivalent. In terms of the RG language, $V_{\rm long}$ is an r-space renormalized interaction with the hard-core (small-r) part of V_{NN} integrated out, while V_{low-k} is its k-space counterpart with the high-momentum part of V_{NN} similarly integrated out.

[L. Coraggio, A. Covello, A. Gargano, N. Itaco and T.T.S. Kuo (2009)]: A comprehensive and detailed review on shell model nuclear structure calculations based on the renormalized V_{low-k} interactions was presented. The shell-model calculations in the region around ^{132}Sn were remarkably successful and interesting. One of the authors (TTSK) has visited Naples many times to work together with the coauthors, and is very grateful to them, particularly Aldo, for their unsurpassed hospitality over the past 15 years. This has been a most pleasant collaboration.

ELSEVIER

Nuclear Physics A684 (2001) 432c–436c

www.elsevier.nl/locate/npe

Low momentum nucleon-nucleon potentials with half-on-shell T-matrix equivalence

Scott Bogner [a] T. T. S. Kuo[a] L. Coraggio [b] *

[a] Department of Physics and Astronomy, SUNY, Stony Brook, New York 11794, USA

[b] Dipartimento di Scienze Fisiche and Istituto Nazionale di Fisica Nucleare, Università di Napoli Federico II, I-80126 Napoli, Italy

We study a method by which realistic nucleon-nucleon potentials V_{NN} can be reduced, in a physically equivalent way, to an effective low-momentum potential V^{low-k} confined within a cut-off momentum k_{cut}. Our effective potential is obtained using the folded-diagram method of Kuo, Lee and Ratcliff, and it is shown to preserve the half-on-shell T-matrix. Both the Andreozzi-Lee-Suzuki and the Andreozzi-Krenciglowa-Kuo iteration methods have been employed in carrying out the reduction. Calculations have been performed for the Bonn-A and Paris NN potentials, using various choices for k_{cut} such as $2\ \mathrm{fm}^{-1}$. The deuteron binding energy, low-energy NN phase shifts, and the low-momentum half-on-shell T-matrix given by V_{NN} are all accurately reproduced by V^{low-k}. Possible applications of V^{low-k} directly to nuclear matter and nuclear structure calculations are discussed.

1. Introduction

Recently there has been much interest in studying nuclear physics problems, particularly the two-nucleon problem, using effective field theory (EFT) [1,2]. The basic idea of the EFT approach is to shrink the full-space theory to a small-space one which contains only the low-momentum modes. This is accomplished by integrating out the high momentum modes, thus generating effective couplings which implicitly contain the effects of the high-momentum modes. In this way, one can derive a low-momentum nucleon-nucleon (NN) potential, which is specifically designed for low-energy nuclear physics. The low-momentum NN potentials so constructed have indeed been quite successful in describing the two-nucleon system at low energy . [2]

In the present work, we would like to derive also a low-momentum NN potential, although along a different direction. There are a number of realistic nucleon-nucleon potentials V_{NN}, such as the Bonn [3] and Paris [4] potentials, which all describe the observed deuteron and NN scattering data very well. They have both low- and high-momentum components. In fact because of the strong short range repulsion contained in them, their momentum space matrix elements $V(k,k')$ are still significant at large momentum. We would like to reduce such realistic potentials, in a physically equivalent way, to certain

*Talk presented by TTSK at FB16 International Conference, Taipei, Taiwan March 6-10, 2000

0375-9474/01/$ – see front matter
PII S0375-9474(01)00449-3

effective low-momentum NN potentials, V^{low-k}, which have only slow momentum components, below a chosen cut-off momentum k_{cut}.

2. Transformation method

The Schroedinger equation for the full-space two-nucleon problem is written as

$$H \mid \Psi\rangle = E \mid \Psi\rangle;\; H = H_0 + V_{NN}, \tag{1}$$

where H_0 is the kinetic-energy operator for the two-nucleon system. A model space P is defined as a momentum subspace with $k \leq k_{cut}$, k being the two-nucleon relative momentum. We want to transform the above equation to a model-space one

$$PH_{eff}P \mid \Psi\rangle = EP \mid \Psi\rangle;\; H_{eff} = H_0 + V^{low-k}, \tag{2}$$

where the low-momentum effective interaction is denoted as V^{low-k}. As far as the low-energy physics is concerned, we would like to have H_{eff} to be physically equivalent to H. Specifically, this means the requirement that the deuteron binding energy, low-energy NN phase shifts and the low-momentum half-on-shell T-matrix of H are all reproduced by H_{eff}. The full-space half-on-shell T-matrix for V_{NN} is defined as

$$\langle p' \mid T(\omega) \mid p\rangle = \langle p' \mid V_{NN} \mid p\rangle + \int_0^{\infty} k^2 dk \langle p' \mid V_{NN} \mid k\rangle \frac{1}{\omega - H_0(k)} \langle k \mid T(\omega) \mid p\rangle;\; \omega = \varepsilon_p, \tag{3}$$

where ε_p is the unperturbed energy for state $\mid p\rangle$. The corresponding model-space T-matrix given by V_{low-k} is

$$\langle p' \mid T_{eff}(\omega) \mid p\rangle = \langle p' \mid V^{low-k} \mid p\rangle + \int_0^{k_{cut}} k^2 dk \langle p' \mid V^{low-k} \mid k\rangle \frac{1}{\omega - H_0(k)} \langle k \mid T_{eff}(\omega) \mid p\rangle;\; \omega = \varepsilon_p. \tag{4}$$

Note for T_{eff} the intermediate states are integrated up to k_{cut}. The boundary conditions associated with the free Green's function are not written out, for simplicity. For p and p' both belonging to P (i.e. both $\leq k_{cut}$), we require

$$\langle p' \mid T(\omega = \varepsilon_p) \mid p\rangle = \langle p' \mid T_{eff}(\omega = \varepsilon_p) \mid p\rangle. \tag{5}$$

Base on the Kuo-Lee-Ratcliff (KLR) folded-diagram method [5,6], Bogner, Kuo and Coraggio [7] have recently shown that the above requirements can be satisfied when the low-momentum effective interaction is given by the folded-diagram series

$$V^{low-k} = \hat{Q} - \hat{Q}' \int \hat{Q} + \hat{Q}' \int \hat{Q} \int \hat{Q} - \hat{Q}' \int \hat{Q} \int \hat{Q} \int \hat{Q} + \ \dots\ , \tag{6}$$

where $\hat{Q}$, often referred to as the $\hat{Q}$-box, is an irreducible vertex function in the sense that its intermediate states must be outside the model space P. The integral sign appearing above represents a generalized folding operation [5,6]. $\hat{Q}'$ is obtained from $\hat{Q}$ by removing terms of first order in the interaction V_{NN}.

Let us outline their proof. A general term of the above T-matrix can be written as $\langle p' \mid (V + V\frac{1}{e(p)}V + V\frac{1}{e(p)}V\frac{1}{e(p)}V + \cdots) \mid p\rangle$ where $e(p) \equiv (\varepsilon_p - H_o)$. Note that the intermediate states cover the entire space (1=P+Q where P denotes the model space and Q its complement). Writing it out in terms of P and Q, a typical term of T is $V\frac{Q}{e}V\frac{Q}{e}V\frac{P}{e}V\frac{Q}{e}V\frac{P}{e}V$. Note it has three segments partitioned by two $\frac{P}{e}$ propagators. Let us define a $\hat{Q}$-box as $\hat{Q} = V + V\frac{Q}{e}V + V\frac{Q}{e}V\frac{Q}{e}V + \cdots$, where all intermediate states belong to Q. One readily sees that the previous term is just a part of the three-$\hat{Q}$-box term, and in general we have $T = \hat{Q} + \hat{Q}\frac{P}{e}\hat{Q} + \hat{Q}\frac{P}{e}\hat{Q}\frac{P}{e}\hat{Q} + \cdots$. By performing a folded-diagram factorization of each term with more than one $\hat{Q}$-box, one can rewrite the T-matrix as $T = V^{low-k} + V^{low-k}\frac{P}{e}V^{low-k} + V^{low-k}\frac{P}{e}V^{low-k}\frac{P}{e}V^{low-k} + \cdots$. The above result then follows.

3. Results and discussion

The above V^{low-k} may be calulated using iteration methods. We have done so using both the Andreozzi-Lee-Suzuki (ALS) and Andreozzi-Krenciglowa-Kuo (AKK) iteration methods[8], for both the Bonn-A and Paris potentials. We note that V^{low-k} is energy independent, and it contains less information than the full-space V_{NN}. The ALS method converges to the lowest (in energy) d states of H, d being the dimension of the model space. (Here we have discretized the momentum space, writing H as a finite matrix.) In contrast, the AKK method converges to the d states of H with maximum P-space overlaps. We have found that the V^{low-k} given by both methods are very close to each other, an indication that the intruder-state problem [9] does not seem to be present in our present calculation. The deuteron binding energy given by V_{NN} is very accurately reproduced by V^{low-k}, for a wide range of k_{cut}. In Fig.1, we compare the phase shifts given by V_{NN} and those by V^{low-k}. They agree quite well. Empirical phase shifts are determined up to $E_{lab} \approx 300$MeV, and they are given by the fully-on-shell T-matrix. Hence we need to use $k_{cut} \sim 2\ fm^{-1}$, if we want V^{low-k} to reproduce the phase shifts up to this energy. In Fig. 2, we compare the half-on-shell T-matrices (calculated with the principal-value boundary condition) given by V_{NN} and V^{low-k}, they also agree quite well. Note that plotted are the (k',k) matrix elements with $k^2 = E_{lab}M/2\hbar^2$, M being the nucleon mass. There are a number of similarities between the model-space reduction method used here and the renormaliztion group method employed in effective field theory, and it would be useful to elucidate the connection between them. Since our method exactly preserves the half-on-shell T-matrix, it may provide a convenient way to study the flow equation which describes the change of the low-momentum effective interaction with respect to the momentum cutoff.

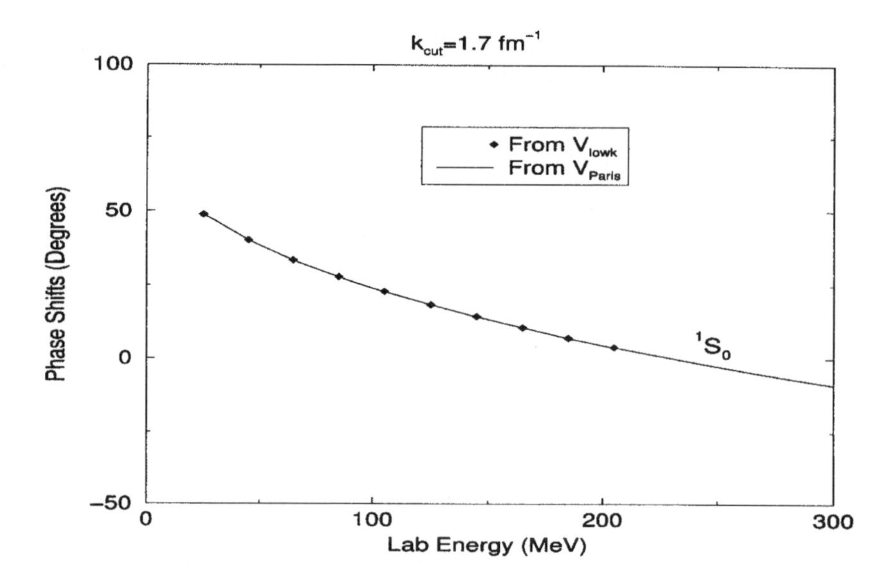

Figure 1. Comparison of phase shifts from V_{NN} and V^{low-k}.

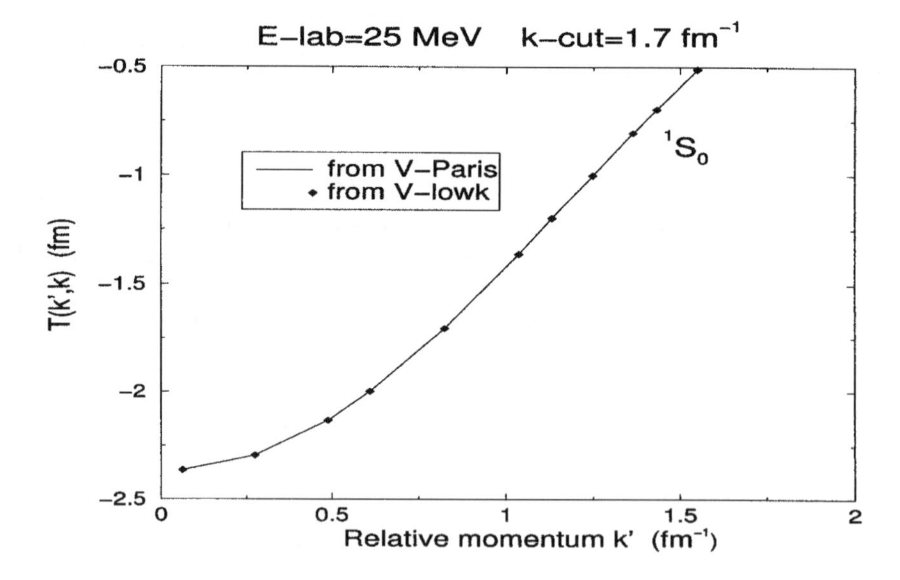

Figure 2. Comparison of half-on-shell T-matrix from V_{NN} and V^{low-k}.

To summarize, we have reduced realistic NN potentials V (Bonn-A and Paris) to corresponding effective low-momentum potentials V^{low-k} for a model space of $k \leq k_{cut}$. The deuteron binding energy, low-energy phase shifts and the low-momentum half-on-shell T-matrix given by V_{NN} are all reproduced by V^{low-k}. Because of the strong short-range repulsion contained in V_{NN}, it is well known that we can not use it directly in shell-model calulations of nuclei and/or in Hartree-Fock calculations of nuclear matter; we need first to convert V_{NN} into a G-matrix, to take care of the short-range correlations. The G-matrix so obtained is energy dependent. We have found that our V^{low-k} is a generally smooth potential (without strong short range repulsion), and it is energy independent. It may be suitable to use V^{low-k} directly in the above calculations, without the need of first calculating the usual Brueckner G-matrix. This would be of interest and desirable. We have done some shell model calculations [7] in this direction, and obtained rather encouraging results. Nuclear matter calculations using V^{low-k} are in progress.

Acknowledgement: We thank Prof. G.E. Brown for many stimulating discussions. This work was supported in part by the U.S. DOE Grant No. DE-FG02-88ER40388, and the Italian Ministero dell'Università e della Ricerca Scientifica e Tecnologica (MURST). TTSK is particularly grateful to T.S.H. Lee for several helpful discussions.

REFERENCES

1. D. B. Kaplan, "Effective Field Theories", Lectures given at 7th Summer School in Nuclear Physics Symmetries, Seattle, Washington, 1995 (e-print archive nucl-th/9506035).
2. E. Epelbaum, W. Glöckle, and Ulf-G. Meissner, Nucl. Phys. **A671** 295 (2000).
3. R. Machleidt, Adv. Nucl. Phys. **19**, 189 (1989).
4. M. Lacombe et al., Phys. Rev. **C21**, 861 (1980).
5. T.T.S. Kuo, S.Y. Lee and K.F. Ratcliff, Nucl. Phys. **A176** 65 (1971).
6. T.T.S. Kuo and E. Osnes, Springer Lecture Notes of Physics, **Vol. 364**, p.1 (1990).
7. S. Bogner, T.T.S. Kuo and L. Coraggio, preprint (LANL, nucl-th/9912056, December 1999).
8. F. Andreozzi, Phys. Rev. **C54**, 684 (1996).
9. K. Suzuki et al., Phys. Lett. **B308**, 1 (1993).

PHYSICAL REVIEW C, VOLUME 65, 051301(R)

Low momentum nucleon-nucleon potential and shell model effective interactions

Scott Bogner,[1] T. T. S. Kuo,[1] L. Coraggio,[2] A. Covello,[2] and N. Itaco[2]
[1]*Department of Physics, SUNY, Stony Brook, New York 11794*
[2]*Dipartimento di Scienze Fisiche, Università di Napoli Federico II and Istituto Nazionale di Fisica Nucleare, I-80126 Napoli, Italy*

(Received 11 June 2001; revised manuscript received 19 February 2002; published 16 April 2002)

A low momentum nucleon-nucleon (*NN*) potential $V_{low\text{-}k}$ is derived from meson exchange potentials by integrating out the model dependent high momentum modes of V_{NN}. The smooth and approximately unique $V_{low\text{-}k}$ is used as input for shell model calculations instead of the usual Brueckner *G* matrix. Such an approach eliminates the nuclear mass dependence of the input interaction one finds in the *G* matrix approach, allowing the same input interaction to be used in different nuclear regions. Shell model calculations of ^{18}O, ^{134}Te, and ^{135}I using the *same* input $V_{low\text{-}k}$ have been performed. For cut-off momentum Λ in the vicinity of 2 fm^{-1}, our calculated low-lying spectra for these nuclei are in good agreement with experiments, and are weakly dependent on Λ.

DOI: 10.1103/PhysRevC.65.051301 PACS number(s): 21.60.Cs, 21.30.Fe, 27.20.+n, 27.60.+j

A fundamental problem in nuclear physics has been the determination of the effective nucleon-nucleon (*NN*) interaction used in the nuclear shell model, which has been successful in describing a variety of nuclear properties. There have been a number of successful approaches [1–4] for this determination, ranging from empirical fits of experimental data, to deriving it microscopically from the bare *NN* potential. Despite impressive quantitative successes, the traditional microscopic approach suffers the fate of being "model dependent" owing to the fact that there is no unique V_{NN} to start from. Moreover, as the Brueckner *G* matrix has traditionally been the starting point, one obtains different input interactions for nuclei in different mass regions as a result of the Pauli blocking operator.

In this work, we propose a different approach to shell model effective interactions that is motivated by the recent applications of effective field theory (EFT) and the renormalization group (RG) to low energy nuclear systems [5–8]. Our aim is to remove some of the model dependence that arises at short distances in the various V_{NN} models, and also to eliminate the mass dependence one finds in the *G* matrix approach, thus allowing the same interaction to be used in different nuclear regions such as those for ^{18}O and ^{134}Te. A central theme of the RG-EFT approach is that physics in the infrared region is insensitive to the details of the short distance dynamics. One can therefore have infinite theories that differ substantially at small distances, but still give the same low energy physics if they possess the same symmetries and the "correct" long-wavelength structure [5,8]. The fact that the various meson models for V_{NN} share the same one pion tail, but differ significantly in how they treat the shorter distance pieces illustrates this explicitly as they give the same phase shifts and deuteron binding energy. In RG language, the short distance pieces of V_{NN} are like irrelevant operators since their detailed form cannot be resolved from low energy data.

Motivated by these observations, we would like to derive a low-momentum *NN* potential $V_{low\text{-}k}$ by integrating out the high momentum components of different models of V_{NN} in the sense of the RG [5,8], and investigate its suitability of being used directly as a model independent effective interaction for shell model calculations. We shall use in the present work the CD-Bonn *NN* potential [9] for V_{NN}. In the following, we shall first describe our method for carrying out the high-momentum integration. Shell model calculations for ^{18}O, ^{134}Te, and ^{135}I using $V_{low\text{-}k}$ will then be performed. Our results will be discussed, especially about their dependence on the cut-off momentum Λ.

The first step in our approach is to integrate out the model dependent high momentum components of V_{NN}. In accordance with the general definition of a renormalization group transformation, the decimation must be such that low energy observables calculated in the full theory are exactly preserved by the effective theory. We turn to the model space methods of nuclear structure theory for guidance, as there has been much work in recent years discussing their similarity to the Wilson RG approach [4,10,11]. While the technical details differ, both approaches attempt to thin out, or limit the degrees of freedom one must explicitly consider to describe the physics in some low energy regime. Once the relevant low energy modes are identified, all remaining modes or states are "integrated" out. Their effects are then implicitly buried inside the effective interaction in a manner that leaves the low energy observables invariant. One successful model-space reduction method is the Kuo-Lee-Ratcliff (KLR) folded diagram theory [12,13]. For the nucleon-nucleon problem in vacuum, the RG approach simply means that the low momentum *T* matrix and the deuteron binding energy calculated from V_{NN} must be reproduced by $V_{low\text{-}k}$, but with all loop integrals cut off at some Λ. Therefore, we start from the half-on-shell *T* matrix

$$T(k',k,k^2)=V_{NN}(k',k)+\int_0^\infty q^2dq\,V_{NN}(k',q)\frac{1}{k^2-q^2+i0^+}T(q,k,k^2). \tag{1}$$

We then define an effective low-momentum *T* matrix by

$$T_{low\text{-}k}(p',p,p^2)=V_{low\text{-}k}(p',p)+\int_0^{\Lambda}q^2dqV_{low\text{-}k}(p',q)\times\frac{1}{p^2-q^2+i0^+}T_{low\text{-}k}(q,p,p^2), \qquad (2)$$

where Λ denotes a momentum space cut-off (such as $\Lambda=2\ \mathrm{fm}^{-1}$) and $(p',p)\leqslant\Lambda$. We require the above T matrices satisfying the condition

$$T(p',p,p^2)=T_{low\text{-}k}(p',p,p^2);\ (p',p)\leqslant\Lambda. \qquad (3)$$

The above equations define the effective low momentum interaction $V_{low\text{-}k}$. In the following, let us show that the above equations are satisfied by the solution

$$V_{low\text{-}k}=\hat{Q}-\hat{Q}'\int\hat{Q}+\hat{Q}'\int\hat{Q}\int\hat{Q}-\hat{Q}'\int\hat{Q}\int\hat{Q}\int\hat{Q}+\ldots, \qquad (4)$$

which is just the KLR folded-diagram effective interaction [12,13]. A preliminary account of this result has been reported as a work in progress at a recent conference [14].

In time dependent formulation, the T matrix of Eq. (1) can be written as $\langle k'|VU(0,-\infty)|k\rangle$, U being the time evolution operator. In this way we can readily perform a diagrammatic analysis of the T matrix. A general term of it may be written as $\langle k'|(V+V1/e(k)V+V1/e(k)V1/e(k)V+\ldots)|k\rangle$ where $e(k)\equiv(k^2-H_0)$, H_0 being the unperrurbed Hamiltonian. Note that the intermediate states (represented by 1 in the numerator) cover the entire space, and $1=P+Q$ where P denotes the model space (momentum $\leqslant\Lambda$) and Q its complement. Expanding it out in terms of P and Q, a typical term of T is of the form $VQ/eVQ/eVP/eVQ/eVP/eV$. Let us define a $\hat{Q}$ box as $\hat{Q}=V+VQ/eV+VQ/eVQ/eV+\ldots$, where all intermediate states belong to Q. One readily sees that the T matrix can be regrouped as a $\hat{Q}$ box series, namely $\langle p'|T|p\rangle=\langle p'|[\hat{Q}+\hat{Q}P/e\hat{Q}+\hat{Q}P/e\hat{Q}P/e\hat{Q}+\ldots]|p\rangle$. Note that all the $\hat{Q}$ boxes have the same energy variable, namely p^2.

This regrouping is depicted in Fig. 1, where each $\hat{Q}$ box is denoted by a circle and the solid line represents the propagator P/e. The diagrams A, B, and C are, respectively, the one-and two- and three-$\hat{Q}$-box terms of T, and clearly $T=\mathrm{A}+\mathrm{B}+\mathrm{C}+\ldots$. Note the dashed vertical line is not a propagator; it is just a "ghost" line to indicate the external indices. We now perform a folded-diagram factorization for the T matrix, following closely the KLR folded-diagram method [12,13]. Diagram B of Fig. 1 is factorized into the product of two parts (see B1) where the time integrations of the two parts are independent from each other, each integrating from $-\infty$ to 0. In this way we have introduced a time-incorrect contribution which must be corrected. In other words B is not equal to B1, rather it is equal to B1 plus the folded-diagram correction B2. Note that the integral sign represents a generalized folding [12,13].

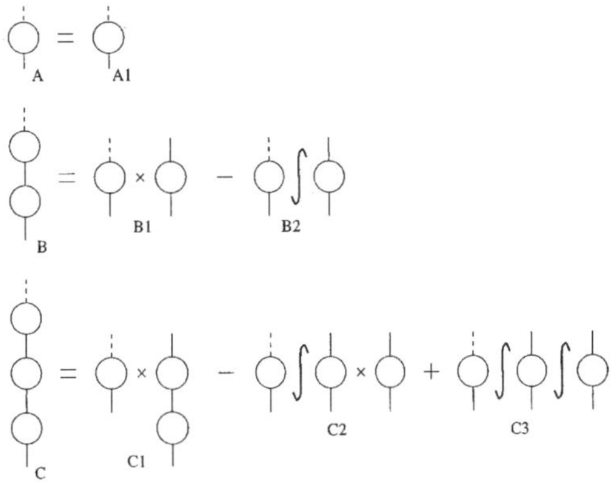

FIG. 1. Folded-diagram factorization of the half-on-shell T matrix.

Similarly we factorize the three-$\hat{Q}$-box term C as shown in the third line of Fig. 1. Higher-order $\hat{Q}$-box terms are also factorized following the same folded-diagram procedure. Let us now collect terms in the figure in a "slanted" way. The sum of terms A1, B2, C3,... is just the low-momentum effective interaction of Eq. (4). (Note that the leading $\hat{Q}$ box of any folded term must be at least second order in V_{NN}, and hence it is denoted as $\hat{Q}'$ box which equals to $\hat{Q}$ box with terms first-order in V_{NN} subtracted.) The sum B1, C2, D3,... is $V_{low\text{-}k}P/e\hat{Q}$. Similarly the sum $\mathrm{C1}+\mathrm{D2}+\mathrm{E3}+\ldots$ is just $V_{low\text{-}k}P/e\hat{Q}P/e\hat{Q}$. (Note diagrams D1, D2,..., E1, E2,... are not shown in the figure.) Continuing this way, it is easy to see that Eqs. (1)–(3) are satisfied by the low momentum effective interaction of Eq. (4).

The effective interaction of Eq. (4) can be calculated using interaction methods. A number of such iteration methods have been developed; the Krenciglowa-Kuo [15] and the Lee-Suzuki iteration methods [16] are two examples. These methods were formulated primarily for the case of degenerate PH_0P, H_0 being the unperturbed Hamiltonian. For our present two-nucleon problem, PH_0P is obviously nondegenerate. Nondegenerate iteration methods [17] are more complicated. However, a recent iteration method developed by Andreozzi [18] is particularly efficient for the nondegenerate case. This method shall be referred to as the Andreozzi-Lee-Suzuki (ALS) iteration method, and has been employed in the present work.

We have carried out numerical checks to ensure that certain low-energy physics of V_{NN} are indeed preserved by $V_{low\text{-}k}$. We first check the deuteron binding energy BE_d. We have calculated BE_d using $V_{low\text{-}k}$ for many values of Λ, and for all cases the BE_d given by $V_{low\text{-}k}$ agrees very accurately (to four places after the decimal) with that given by V_{NN}. (Note that when Λ approaches ∞ $V_{low\text{-}k}$ is the same as V_{NN}.) In Fig. 2, we present some $^1S_0$ and $^3P_0$ phase shifts calculated from the CD-Bonn V_{NN} (dotted line) and the $V_{low\text{-}k}$

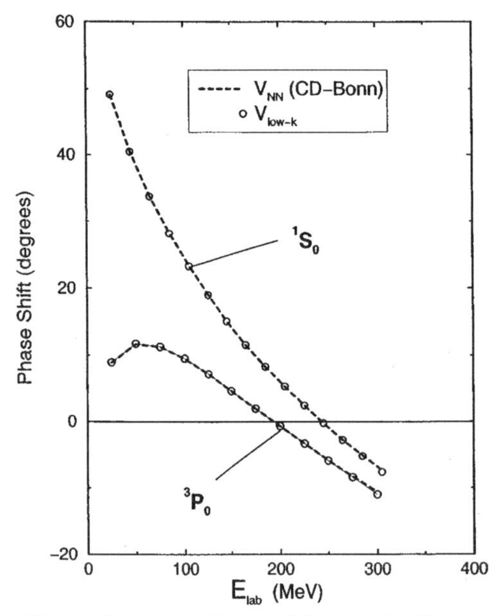

FIG. 2. Comparison of phase shifts given by $V_{low\text{-}k}$ and V_{NN}.

(circles) derived from it, using a momentum cut-off $\Lambda = 2.0\ \mathrm{fm}^{-1}$. As seen, the phase shifts from the former are well reproduced by the latter. We have also checked the half-on-shell T matrix given by V_{NN} and by $V_{low\text{-}k}$, and found very good agreement between them [14]. The above agreements are expected, as we have shown that the T-matrix equivalence of Eq. (3) holds for any Λ. In short, our numerical checks have reaffirmed that the deuteron binding energy, low energy phase shifts and low momentum half-on-shell T matrix of V_{NN} are all preserved by $V_{low\text{-}k}$. As far as those physical quantities are concerned, $V_{low\text{-}k}$ and V_{NN} are equivalent.

Having proven the "physical equivalence" of $V_{low\text{-}k}$ and V_{NN} in the sense of the RG, we turn now to microscopic shell model calculations in which we use $V_{low\text{-}k}$ as the input interaction. A folded-diagram formulation [13,2,3] is employed. An important feature here is that this formalism allows us to calculate the energy differences of neighboring many-body systems. For example, we can calculate the energy difference of ^{18}O and the ground state energy of ^{16}O, starting from the experimental ^{17}O single particle energies (s.p.e.) and a shell model effective interaction V_{eff} derived microscopically from an underlying NN potential. This folded diagram method has been rather successfully applied to many nuclei using G-matrix interactions [2,3]. There the basic inputs to the calculation are the matrix elements $\langle n_1 n_2 | G | n_3 n_4 \rangle$ where G is the Brueckner G matrix and the n's are harmonic oscillator wave functions.

Since $V_{low\text{-}k}$ is already a smooth potential, it is no longer necessary to first calculate the G matrix. Thus in our present work, the starting basic input are just the matrix

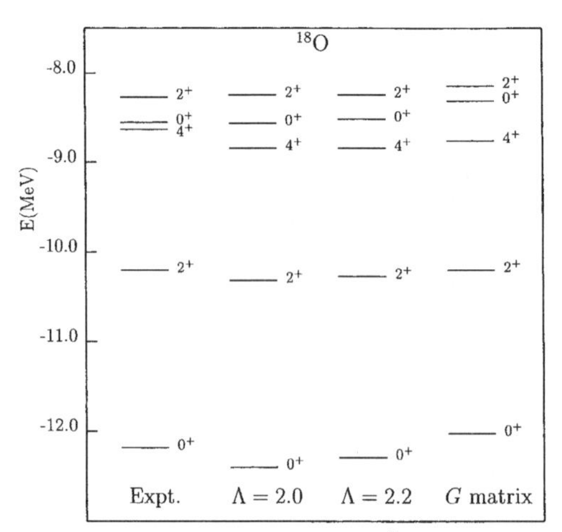

FIG. 3. Low-lying states of ^{18}O.

elements $\langle n_1 n_2 | V_{low\text{-}k} | n_3 n_4 \rangle$, and thereafter our calculation procedures are exactly the same as described in Refs. [2], [3]. A model space with two valence neutrons in the $(0d_{5/2}, 0d_{3/2}, 1s_{1/2})$ shell is used for ^{18}O, and one with two and three valence protons in the $(0g_{7/2}, 1d_{5/2}, 1d_{3/2}, 2s_{1/2}, 0h_{11/2})$ shell for ^{134}Te and ^{135}I, respectively. As is customary, we use s.p.e. extracted from the experimental spectra of the corresponding single-particle valence nuclei, ^{17}O and ^{133}Sb [19]. For the absolute scaling of the sets of s.p.e., the mass-excess values for ^{17}O and ^{133}Sb have been taken from Refs. [20, 21]. For ^{134}Te and ^{135}I, we assume that the contribution of the Coulomb interaction between valence protons is equal to the matrix element of the Coulomb force between the states $(g_{7/2})^2_{J^\pi=0^+}$.

As shown in Fig. 3, our calculated low-lying J^π states of ^{18}O agree highly satisfactorily with experiments [19]. In the same figure, results of the corresponding G-matrix calculations are also shown; the $V_{low\text{-}k}$ results are just as good or slightly better. It may be mentioned that our $V_{low\text{-}k}$ is slightly non-Hermitian. A Hermitian $V_{low\text{-}k}$ can be obtained using the Okubo transformation [7]. We have constructed such a Hermitian $V_{low\text{-}k}$ using the Suzuki-Okamoto method [22]. We have found that the shell model energy levels given by the two $V_{low\text{-}k}$'s are very similar, probably because our $V_{low\text{-}k}$ is only slightly non-Hermitian. In a concurrent paper [11] we have found that $V_{low\text{-}k}$ is almost independent of the underlying V_{NN} for the values of Λ considered here. Therefore, although the CD-Bonn potential [9] is used in our present calculations, we stress that similar results will be obtained if we calculate $V_{low\text{-}k}$ from other models such as the Paris or Argonne V-18 potentials.

It may be mentioned that the G matrix is energy dependent and Pauli blocking dependent, while $V_{low\text{-}k}$ is not. This is a desirable feature, indicating that $V_{low\text{-}k}$ may be suitable also for other nuclear regions. To study this point, we have used the same $V_{low\text{-}k}$ in a shell model calculation of ^{134}Te as mentioned earlier. It is encouraging that our results for ^{134}Te also agree well with experiments

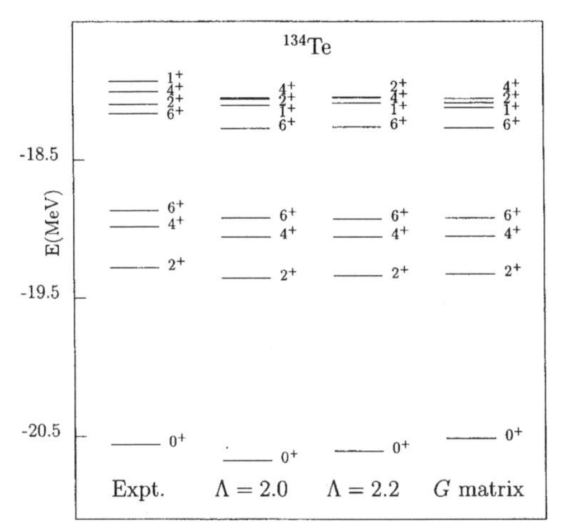

FIG. 4. Low-lying states of ^{134}Te.

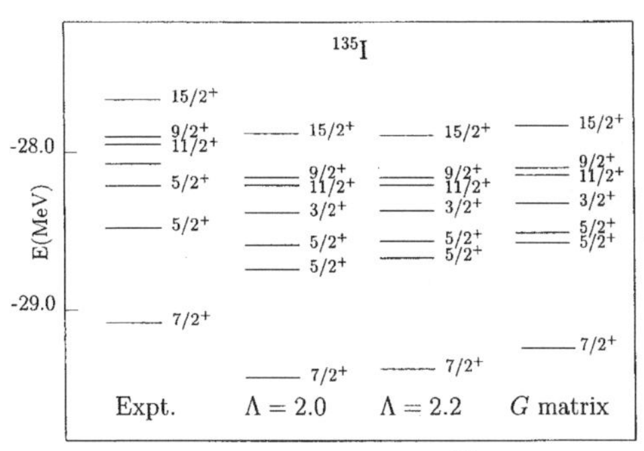

FIG. 5. Low-lying states of ^{135}I.

[19] as shown in Fig. 4. Again the $V_{low\text{-}k}$ results are just as good or slightly better than the G-matrix results. We emphasize that we have used *the same* $V_{low\text{-}k}$ interaction in both ^{18}O and ^{134}Te calculations, and it appears to work equally well for both nuclei. This is in marked contrast to the traditional approach in which one has to use different G matrices for different mass regions, as the associated Pauli blocking operators are different. This is an appealing result, as it suggests the possibility for a common shell-model interaction that attenuates much of the dependence on the V_{NN} model and is suitable for a wide range of nuclei.

It is of interest to investigate if the same $V_{low\text{-}k}$ is suitable for nuclei with more than two valence nucleons. It is primarily for this purpose we have carried out the shell model calculation of ^{135}I mentioned earlier, using the same $V_{low\text{-}k}$. Our results are shown in Fig. 5. It is gratifying that the calculated excitation spectra are in very good agreement with experiments [19]. We note that the valence interaction energy for the three valence nucleons given by our calculation is slightly overbound, by about 0.3 MeV. This may be an indication of the need of a weak three-body force for this nucleus, which has three valence nucleons. (Our $V_{low\text{-}k}$ is a two-body interaction.) We plan to study this topic in a future work.

An important issue is what value one should use for Λ. Guided by general EFT arguments, the minimum value for Λ must be large enough so that $V_{low\text{-}k}$ explicitly contains the necessary degrees of freedom for the physical system. For example, 2π exchanges are important for low energy nuclear physics, and to adequately include the corresponding degree of freedom we need to have $\Lambda_{\min}$ larger than $\sim m_{2\pi}$, i.e., $\sim 1.4\ \text{fm}^{-1}$. In fact we have found that $V_{low\text{-}k}$ varies strongly with Λ when it is smaller than that value. A general signal for $\Lambda_{\min}$ is when the calculated physical quantities first become insensitive to Λ [5]. Conversely, we want Λ to be smaller than the short distance scale $\Lambda_{\max}$ at which the model dependence of the different V_{NN} starts to creep in [5]. Systems in which these two constraints are consistent with each other (i.e., $\Lambda_{\min} < \Lambda_{\max}$) are amenable to EFT-RG inspired effective theories, as they possess a clear separation of scales between the relevant long wavelength modes and the model dependent short distance structure. We have found [11] that $\Lambda_{\max}$ should not be much greater than 2.0–2.5 fm^{-1} as this is the scale after which $V_{low\text{-}k}$ first becomes strongly dependent on the particular V_{NN} used. There is another consideration: Most *NN* potentials are constructed to fit empirical phase shifts up to $E_{lab} \approx 350$ MeV [9]. Since $E_{lab} \lesssim 2\hbar^2\Lambda^2/M$, M being the nucleon mass, and one should require $V_{low\text{-}k}$ to reproduce the same empirical phase shifts, a choice of Λ in the vicinity of 2 fm^{-1} would seem to be appropriate.

Guided by the above considerations, we have used in our calculations two values for the momentum cut-off, namely $\Lambda = 2.0$ and 2.2 fm^{-1} as shown in Figs. 3–5. It is satisfying to see that the results are rather insensitive to the choice of Λ, in harmony with the EFT philosophy mentioned earlier. Perhaps more importantly, both are in satisfactory agreement with experiments.

In summary, we have investigated a RG-EFT inspired approach to shell model calculations that is a "first step" towards a model independent calculation that uses one common interaction over a wide range of nuclei. Using the KLR folded diagram approach in conjunction with the ALS iteration method, we have performed a RG decimation where the model dependent pieces of V_{NN} models are integrated out to obtain a nearly unique low momentum potential $V_{low\text{-}k}$. This $V_{low\text{-}k}$ preserves the deuteron pole as well as the low energy phase shifts and half-on-shell T matrix. We have used $V_{low\text{-}k}$, which is a smooth potential, directly in shell model calculations of ^{18}O, ^{134}Te, and ^{135}I without first calculating the G matrix. The results are all in satisfactory agreement with experiment, and they are insensitive to Λ in the neighborhood of $\Lambda \approx 2$ fm^{-1}. We do feel that $V_{low\text{-}k}$ may become a promising and reliable effective interaction for shell model calculations of few valence nucleons, over a wide range of nuclear regions.

We thank Professor G. E. Brown and A. Schwenk for many discussions. This work was supported in part by the U.S. DOE Grant No. DE-FG02-88ER40388, and the Italian Ministero dell'Università e della Ricerca Scientifica e Tecnologica (MURST).

[1] P. J. Ellis and E. Osnes, Rev. Mod. Phys. **49**, 777 (1977).
[2] A. Covello *et al.*, Prog. Part. Nucl. Phys. **38**, 165 (1997).
[3] M. F. Jiang, R. Machleidt, D. B. Stout, and T. T. S. Kuo, Phys. Rev. C **46**, 910 (1992).
[4] W. Haxton and C. L. Song, Phys. Rev. Lett. **84**, 5484 (2000).
[5] P. Lepage, nucl-th/9706029.
[6] D. B. Kaplan, M. J. Savage, and M. B. Wise, Phys. Lett. B **424**, 390 (1998).
[7] E. Epelbaum, W. Glöckle, and Ulf-G. Meissner, Nucl. Phys. **A637**, 107 (1998).
[8] *Nuclear Physics with Effective Field Theory II*, edited by P. Bedaque *et al.* (World Scientific, Singapore, 1999).
[9] R. Machleidt, Phys. Rev. C **63**, 024001 (2001).
[10] S. K. Bogner and T. T. S. Kuo, Phys. Lett. B **500**, 279 (2001).
[11] S. K. Bogner, T. T. S. Kuo, A. Schwenk, D. R. Entem, and R. Machleidt, nucl-th/0108041.
[12] T. T. S. Kuo, S. Y. Lee, and K. F. Ratcliff, Nucl. Phys. **A176**, 65 (1971).
[13] T. T. S. Kuo and E. Osnes, Springer Lecture Notes of Physics, 1990, Vol. 364, p. 1.
[14] S. Bogner, T. T. S. Kuo, and L. Coraggio, Nucl. Phys. **A684**, 432c (2001).
[15] E. M. Krenciglowa and T. T. S. Kuo, Nucl. Phys. **A235**, 171 (1974).
[16] K. Suzuki and S. Y. Lee, Prog. Theor. Phys. **64**, 2091 (1980).
[17] T. T. S. Kuo *et al.*, Nucl. Phys. **A582**, 205 (1995), and references quoted therein.
[18] F. Andreozzi, Phys. Rev. C **54**, 684 (1996).
[19] Data extracted using the NNDC On-Line Data Service from the ENSDF database, files revised as of December 5, 2001, M. R. Bhat, in *Evaluated Nuclear Structure Data File (ENSDF), Nuclear Data for Science and Technology*, edited by S. M. Quaim (Springer-Verlag, Berlin, 1992), p. 817.
[20] G. Audi and A. H. Wapstra, Nucl. Phys. **A565**, 1 (1993).
[21] B. Fogelberg, K. A. Mezilev, H. Mach, V. I. Isakov, and J. Slivova, Phys. Rev. Lett. **82**, 1823 (1999).
[22] T. T. S. Kuo, P. J. Ellis, Jifa Hao, Zibang Li, K. Suzuki, R. Okamoto, and H. Kumagai, Nucl. Phys. **A560**, 621 (1993).

ELSEVIER

Nuclear Physics A 703 (2002) 745–769

www.elsevier.com/locate/npe

Low-momentum nucleon–nucleon interaction and Fermi liquid theory

Achim Schwenk [a,*], Gerald E. Brown [a], Bengt Friman [b]

[a] *Department of Physics and Astronomy, State University of New York, Stony Brook, NY 11794-3800, USA*
[b] *Gesellschaft fur Schwerionenforschung, Planckstr. 1, 64291 Darmstadt, Germany*

Received 4 October 2001; accepted 17 December 2001

We dedicate this paper to the memory of Sven-Olof Bäckman.

Abstract

We use the induced interaction of Babu and Brown to derive two novel relations between the quasiparticle interaction in nuclear matter and the unique low-momentum nucleon–nucleon interaction $V_{\text{low }k}$ in vacuum. These relations provide two independent constraints on the Fermi liquid parameters of nuclear matter. We derive the full renormalization group equations in the particle–hole channels from the induced interaction. The new constraints, together with the Pauli principle sum rules, define four combinations of Fermi liquid parameters that are invariant under the renormalization group flow. Using empirical values for the spin-independent Fermi liquid parameters, we are able to compute the major spin-dependent ones by imposing the new constraints and the Pauli principle sum rules. The effects of tensor forces are discussed. © 2002 Elsevier Science B.V. All rights reserved.

PACS: 21.65.+f; 71.10.Ay; 21.30.Fe; 11.10.Hi

Keywords: Nuclear matter; Fermi liquid theory; Nucleon–nucleon interactions; Effective interactions; Renormalization group

1. Introduction

This work was motivated by the results of Bogner, Kuo and Corragio [1], who have constructed a low-momentum nucleon–nucleon potential $V_{\text{low }k}$ using folded-diagram

* Corresponding author.
E-mail addresses: aschwenk@nuclear.physics.sunysb.edu (A. Schwenk), popenoe@nuclear.physics.sunysb.edu (G.E. Brown), b.friman@gsi.de (B. Friman).

PII: S0375-9474(01)01673-6

techniques. The starting point of their procedure is a realistic nucleon–nucleon interaction, which is reduced to a low-momentum potential by integrating out relative momenta higher than a cutoff Λ, in the sense of the renormalization group (RG) [2]. The hard momenta larger than Λ renormalize $V_{\text{low }k}$, such that the low-momentum half-on-shell T matrix and bound-state properties of the underlying theory remain unchanged. Consequently, the physics at relative momenta smaller than Λ is preserved.

Bogner et al. find that various, very different bare interactions, such as the Paris, Bonn, and Argonne potential and a chiral effective field theory model, flow to the same $V_{\text{low }k}$ for $\Lambda \lesssim 2\ \text{fm}^{-1}$ [2]. All the nuclear force models are constructed to fit the experimentally available nucleon–nucleon phase shifts up to momenta $k \sim 2\ \text{fm}^{-1}$. However, they differ substantially in their treatment of the short-range parts of the interaction, since these effects cannot be pinned down uniquely by the scattering data. Therefore the work of Bogner et al. demonstrates that one can isolate the physics of the nucleons at low momenta from the effects probed by high momenta and in this way obtain a unique low-momentum nucleon–nucleon potential $V_{\text{low }k}$. When one compares the low momentum part of the bare potentials with $V_{\text{low }k}$, one observes that for reasonable values of the cutoff the main effect of the RG decimation to a unique $V_{\text{low }k}$ is a constant shift in momentum space corresponding to a delta function in coordinate space.[1] This is in keeping with the ideas of effective field theory, where one projects $V_{\text{low }k}$ on one- and two-pion exchange terms plus contact terms, the latter resulting from the exchange of the heavy mesons. The nonpionic contact term contributions flow to "fixed point" values for $\Lambda \lesssim 2\ \text{fm}^{-1}$. Therefore, the most important feature of the unique low-momentum interaction is its value at zero initial and final relative momenta $V_{\text{low }k}(0,0)$, since it directly incorporates the largest effect of the RG decimation—the removal of the model-dependent short-range core by a smeared delta function.

Moreover, it is seen [2] that in the $^1S_0$ channel, $V_{\text{low }k}(0,0)$ is almost independent of the cutoff Λ for $1 \lesssim \Lambda \lesssim 3\ \text{fm}^{-1}$, while in the $^3S_1$ channel only a weak linear dependence on Λ remains in the same range of momenta. For Λ in this momentum range, the contribution of the short-range repulsion, which is peaked around approximately $4\ \text{fm}^{-1}$ [3], is already integrated out, while the common one-pion exchange long-range tail remains basically unchanged until $\Lambda \sim m_\pi$. The residual dependence on the cutoff in the $^3S_1$ channel is due to higher-order tensor contributions, which are peaked at an intermediate-momentum transfer of approximately $2\ \text{fm}^{-1}$ [4]. The weak cutoff dependence of $V_{\text{low }k}$ around $\Lambda \sim 2\ \text{fm}^{-1}$ is characteristic of effective field theories, where the dependence on the cutoff is expected to be weak, provided the relevant degrees of freedom—here nucleons and pions—are kept explicitly. The separation of scales implied by the exchanged meson masses is complicated, however, by the higher-order tensor interactions.

Diagrammatically $V_{\text{low }k}$ sums all ladders with bare potential vertices and intermediate momenta greater than the cutoff. Subsequently, the energy dependence of the ladder sum is removed in order to obtain an energy-independent $V_{\text{low }k}$. This is achieved by means of folding, which can be regarded as averaging over the energy-dependent effective interaction weighted by the low-momentum components of the low-energy scattering

[1] Due to the cutoff employed, the constant shift within the model space corresponds to a smeared delta function.

states. Therefore, it is intuitive to use $V_{\mathrm{low}\,k}$ for $\Lambda = k_{\mathrm{F}}$ as the Brueckner G matrix. This identification is approximative, since self-energy insertions and the dependence on the center-of-mass momentum are ignored in $V_{\mathrm{low}\,k}$. However, it has been argued that the self-energy insertions, which must be evaluated off-shell, are small [4]. Hence, we expect that $V_{\mathrm{low}\,k}$ reproduces the G matrix reasonably well. Furthermore, Bogner et al. [5] argue that $V_{\mathrm{low}\,k}$ may be used directly as a shell-model effective interaction instead of the Brueckner G matrix, since $V_{\mathrm{low}\,k}$ includes the effects of the repulsive core and is generally smooth. They find very good agreement for the low-lying states of core nuclei with two valence nucleons such as ^{18}O and ^{134}Te.

Our second motivation is the work of Birse et al. on the Wilsonian renormalization group treatment of two-body scattering [6,7], where the existence of a unique low-momentum potential is addressed. By demanding that the physical T matrix be independent of the cutoff,[2] they obtain a RG flow equation for the effective potential. After rescaling all dimensionful quantities with the cutoff, they find a trivial fixed point corresponding to zero scattering length and a nontrivial one corresponding to an infinite scattering length. The expansion around the nontrivial fixed point yields the effective range expansion. This demonstrates that the s-wave nucleon–nucleon potential, where the scattering length is large, must lie in the vicinity of the nontrivial fixed point. It would be of interest to clarify the role of this fixed-point structure in the RG flow to $V_{\mathrm{low}\,k}$ and, in particular, whether this can be used to understand why a *unique* potential is obtained already for $\Lambda \lesssim 2\ \mathrm{fm}^{-1}$.

In normal Fermi systems, the low-momentum quasiparticle interaction, which is characterized by the Fermi liquid parameters, is determined by a RG fixed point. In this paper we derive a relation between the Fermi liquid parameters of nuclear matter and the s-wave low-momentum nucleon–nucleon interaction $V_{\mathrm{low}\,k}(0,0)$ at $\Lambda = k_{\mathrm{F}}$. This relation connects the fixed point of the quasiparticle interaction to $V_{\mathrm{low}\,k}$ in the region where it depends only weakly on the cutoff. The existence of such a relation is supported by the success of the model space calculations of Bogner et al. [5], where $V_{\mathrm{low}\,k}$ is used as the shell-model effective interaction. These calculations are in spirit very similar to Fermi liquid theory. In both cases one uses an empirical single-particle spectrum and the energy is measured with respect to a filled Fermi sea. In the case of ^{18}O, the zero of the energy corresponds to the ground state of ^{16}O.

We start by giving a brief introduction of Landau's theory of normal Fermi liquids. We then review the induced interaction introduced by Babu and Brown [9], which will be used to derive the two new constraints. We give a diagrammatically motivated heuristic derivation of the induced interaction, which demonstrates that the induced interaction generates the complete particle–hole parquet for the scattering amplitude, i.e., all fermionic planar diagrams except for particle–particle loops. The latter should be included in the driving term. We then derive two new constraints that relate the Fermi liquid parameters to the low-momentum nucleon–nucleon interaction $V_{\mathrm{low}\,k}$, by solving the integral equation for the scattering amplitude and the induced interaction in a particular limit simultaneously.

[2] Their analysis was carried out for the reaction matrix, but a similar analysis holds for the T matrix. In [8] the RG equation for $V_{\mathrm{low}\,k}$ is derived from the Lippmann–Schwinger equation for the half-on-shell T matrix and it is shown that the same flow equation can be equivalently obtained from the Kuo–Lee–Ratcliff folded diagram series and the Lee–Suzuki similarity transformation.

Making contact with the RG approach to Fermi liquid theory, we derive the coupled RG equations for the particle–hole channels from the induced interaction. Within the particle–hole parquet, the particular combinations of Fermi liquid parameters that appears in these constraints, as well as the Pauli principle sum rules, are invariant under the (in medium) RG flow towards the Fermi surface. Using empirical values for the spin-independent Fermi liquid parameters, we are able to compute the major spin-dependent parameters by imposing the new constraints and the Pauli principle sum rules. Finally, we include tensor interactions in the constraints and demonstrate the necessity of a self-consistent treatment within the induced interaction.

2. Fermi liquid theory

Fermi liquid theory was invented by Landau [10] to describe strongly interacting normal Fermi systems at low temperatures. Landau introduced the quasiparticle concept to describe the elementary excitations of the interacting system. For low excitation energies, the corresponding quasiparticles are long lived and in a sense weakly interacting. One can think of the ground state of the system as a filled Fermi sea of quasiparticles, while quasiparticles above and quasiholes below the Fermi surface correspond to low-lying excited states. The quasiparticles can be thought of as free particles dressed by the interactions with the many-body medium.

When quasiparticles or quasiholes are added to the interacting ground state, the energy of the system is changed by

$$\delta E = \sum_{p\sigma} \epsilon_{p}^{(0)} \delta n_{p\sigma} + \frac{1}{2V} \sum_{p\sigma, p'\sigma'} f_{\sigma,\sigma'}(\boldsymbol{p}, \boldsymbol{p}') \delta n_{p\sigma} \delta n_{p'\sigma'} + \mathcal{O}(\delta n^3), \tag{1}$$

where V is the volume of the system, $\delta n_{p\sigma}$ the change in the quasiparticle occupation number and $\epsilon_{p}^{(0)} - \mu = v_{\mathrm{F}}(p - k_{\mathrm{F}})$ the quasiparticle energy expanded around the Fermi surface. The Fermi momentum is denoted by k_{F}, the Fermi velocity by v_{F} and the chemical potential by μ. The quasiparticle lifetime in normal Fermi systems at zero temperature, is very large close to the Fermi surface ($\tau \sim (p - k_{\mathrm{F}})^{-2}$). Consequently, the quasiparticle concept is useful for describing long wavelength excitations, where the corresponding quasiparticles are restricted to momenta $|\boldsymbol{p}| \approx k_{\mathrm{F}}$. When studying such excitations, one can set $|\boldsymbol{p}| = |\boldsymbol{p}'| = k_{\mathrm{F}}$ in the effective interaction $f_{\sigma,\sigma'}(\boldsymbol{p}, \boldsymbol{p}')$. In a rotationally invariant system, the only remaining spatial variable of f is then the angle θ between $\boldsymbol{p}$ and $\boldsymbol{p}'$. The dependence of $f_{\sigma,\sigma'}(\boldsymbol{p}, \boldsymbol{p}')$ on this angle reflects the nonlocality of the quasiparticle interaction.

It follows from Eq. (1) that the effective interaction is obtained from the energy by varying twice with respect to the quasiparticle occupation number. An illustrative example is the Hartree–Fock approximation, where one finds that the Landau f function is simply given by the direct and exchange terms of the bare interaction:

$$\begin{aligned} \frac{\delta E^{\mathrm{HF}}}{\delta n_{p\sigma} \delta n_{p'\sigma'}} &= \frac{f_{\sigma,\sigma'}^{\mathrm{HF}}(\boldsymbol{p}, \boldsymbol{p}')}{V} \\ &= \langle \boldsymbol{p}\sigma, \boldsymbol{p}'\sigma' | V | \boldsymbol{p}\sigma, \boldsymbol{p}'\sigma' \rangle - \langle \boldsymbol{p}\sigma, \boldsymbol{p}'\sigma' | V | \boldsymbol{p}'\sigma', \boldsymbol{p}\sigma \rangle. \end{aligned} \tag{2}$$

As in effective field theories, the functional form of the spin- and isospin-dependence of the Landau function is determined by the symmetries of the system only—in the case of symmetric nuclear matter these are invariance under spin and isospin rotations.[3] The dependence of f on the angle θ is expanded in Legendre polynomials:

$$\begin{aligned} f(\theta) &= \frac{1}{N(0)}\mathcal{F}(\theta) \\ &= \frac{1}{N(0)}\sum_l \left(F_l + F_l'\boldsymbol{\tau}\cdot\boldsymbol{\tau}' + G_l\,\boldsymbol{\sigma}\cdot\boldsymbol{\sigma}' + G_l'\,\boldsymbol{\tau}\cdot\boldsymbol{\tau}'\boldsymbol{\sigma}\cdot\boldsymbol{\sigma}'\right)P_l(\cos\theta) \\ &\quad + \frac{1}{N(0)}\frac{(\boldsymbol{p}-\boldsymbol{p}')^2}{k_{\mathrm{F}}^2}S_{12}(\boldsymbol{p}-\boldsymbol{p}')\sum_l \left(H_l + H_l'\boldsymbol{\tau}\cdot\boldsymbol{\tau}'\right)P_l(\cos\theta) \\ &\quad + \mathcal{O}\left(A^{-1/3}\right). \end{aligned} \tag{3}$$

Here $\boldsymbol{\sigma}$ and $\boldsymbol{\tau}$ are spin and isospin operators respectively, $S_{12}(\boldsymbol{k}) = 3\boldsymbol{\sigma}\cdot\hat{\boldsymbol{k}}\,\boldsymbol{\sigma}'\cdot\hat{\boldsymbol{k}} - \boldsymbol{\sigma}\cdot\boldsymbol{\sigma}'$ is the tensor operator and we have pulled out a factor $N(0) = 2m^\star k_{\mathrm{F}}/\pi^2$, the density of states at the Fermi surface, in order to make the Fermi liquid parameters F_l, F_l', G_l, G_l', H_l and H_l' dimensionless. The effective mass of the quasiparticles is defined as $m^\star = k_{\mathrm{F}}/v_{\mathrm{F}}$. We will discuss tensor interactions in Section 6, but in order to simplify the discussion, we suppress them in the derivation of the constraints. It is straightforward to generalize the derivation and include them. Finally, since we consider infinite nuclear matter, the spin–orbit interaction can be neglected.

As in effective field theories, the Fermi liquid parameters are determined by comparison with experiments. For nuclear matter we have the following relations for the incompressibility, the effective mass and the symmetry energy [10,11]:

$$K = \frac{3\hbar^2 k_{\mathrm{F}}^2}{m^\star}(1+F_0), \qquad \frac{m^\star}{m} = 1 + F_1/3, \quad \text{and} \quad E_{\mathrm{sym}} = \frac{\hbar^2 k_{\mathrm{F}}^2}{6m^\star}\left(1+F_0'\right). \tag{4}$$

In order to establish the connection between the quasiparticle interaction and the quasiparticle scattering amplitude, we consider the leading particle–hole reducible contributions to the full vertex function. We denote the bare particle–hole vertex by $B(p, p'; q)$, where the momenta p, p' etc. and q are 4-momenta, $p = (\varepsilon, \boldsymbol{p})$ and $q = (\omega, \boldsymbol{q})$:

$p' + \frac{q}{2}$ $\quad$ $p' - \frac{q}{2}$

$$B(p, p'; q) \quad = \quad [\text{vertex diagram}] \quad . \tag{5}$$

$p + \frac{q}{2}$ $\quad$ $p - \frac{q}{2}$

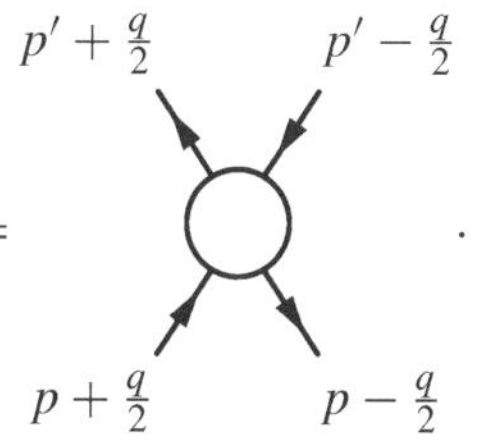

There are two possible ways to join two particle–hole vertices with a particle–hole loop:[4]

[3] When tensor forces are considered the quasiparticle interaction is not invariant under rotations in spin space, but under combined spin and spatial rotations.

[4] Note that in Eq. (7) we have used the antisymmetry of the bare vertex: $B(1+2, 3+4; 1-2) = -B(1+4, 3+2, 1-4)$.

$$-\mathrm{i}\int \frac{\mathrm{d}^4p''}{(2\pi)^4} B(p, p''; q) G\left(p'' + \frac{q}{2}\right) G\left(p'' - \frac{q}{2}\right) B(p'', p'; q)$$

$p' + \frac{q}{2}$ $p' - \frac{q}{2}$

$$= \quad p'' + \tfrac{q}{2} \qquad p'' - \tfrac{q}{2}, \tag{6}$$

$p + \frac{q}{2}$ $p - \frac{q}{2}$

$$-\mathrm{i}\int \frac{\mathrm{d}^4p''}{(2\pi)^4} B\left(\frac{p + p' + q}{2}, p''; p - p'\right) G\left(p'' + \frac{p - p'}{2}\right) G\left(p'' - \frac{p - p'}{2}\right)$$

$p' + \frac{q}{2}$ $p'' - \frac{p-p'}{2}$ $p' - \frac{q}{2}$

$$\times B\left(p'', \frac{p + p' - q}{2}; p - p'\right) = \quad . \tag{7}$$

$p'' + \frac{p-p'}{2}$

$p + \frac{q}{2}$ $p - \frac{q}{2}$

In the recent literature, the first channel, Eq. (6), is referred to as the zero-sound channel (ZS), while the second one is called ZS$'$. The ZS$'$ diagram, Eq. (7), is the exchange diagram to the ZS graph. Landau wrote down a Bethe–Salpeter equation, which sums the particle–hole ladders in the ZS channel. This equation relates the full particle–hole vertex $\Gamma(p, p'; q)$ to the ZS particle–hole irreducible one $\widetilde{\Gamma}(p, p'; q)$:

$p' + \frac{q}{2}$ $p' - \frac{q}{2}$

$p' + \frac{q}{2}$ $p' - \frac{q}{2}$ $p' + \frac{q}{2}$ $p' - \frac{q}{2}$

$$\Gamma \quad = \quad \widetilde{\Gamma} \quad + \quad p'' + \tfrac{q}{2} \;\; \Gamma,\ \widetilde{\Gamma} \;\; p'' - \tfrac{q}{2} \quad . \tag{8}$$

$p + \frac{q}{2}$ $p - \frac{q}{2}$ $p + \frac{q}{2}$ $p - \frac{q}{2}$

$p + \frac{q}{2}$ $p - \frac{q}{2}$

The Bethe–Salpeter equation in the ZS channel reads:

$$\Gamma(p, p'; q) = \widetilde{\Gamma}(p, p'; q) - \mathrm{i}\int \frac{\mathrm{d}^4p''}{(2\pi)^4} \widetilde{\Gamma}(p, p''; q) G\left(p'' + \frac{q}{2}\right) G\left(p'' - \frac{q}{2}\right) \times \Gamma(p'', p'; q). \tag{9}$$

As argued above, we set p and p' on the Fermi surface and let $q \to 0$. In finite nuclei, typical momentum transfers $|\boldsymbol{q}|$ are of the order of the inverse size of the nucleus. Therefore, on physical grounds, $|\boldsymbol{q}| \sim 1/R \sim A^{-1/3}$ vanishes in nuclear matter [11]. Landau noticed that the product of propagators $G(p'' + q/2)G(p'' - q/2)$ is singular in the limit $|\boldsymbol{q}| \to 0$ and $\omega \to 0$ (see, e.g., [12]) and therefore $\widetilde{\Gamma}$ is by construction finite as $q \to 0$. The singularity is due to the quasiparticle poles in the propagators:

$$
\begin{aligned}
&G\left(p'' + \frac{q}{2}\right)G\left(p'' - \frac{q}{2}\right) \\
&\quad = \frac{z}{\varepsilon'' + \omega/2 - v_{\mathrm{F}}(|\boldsymbol{p}'' + \boldsymbol{q}/2| - k_{\mathrm{F}}) + \mathrm{i}\delta_{p''+\frac{q}{2}}} \\
&\quad\quad \times \frac{z}{\varepsilon'' - \omega/2 - v_{\mathrm{F}}(|\boldsymbol{p}'' - \boldsymbol{q}/2| - k_{\mathrm{F}}) + \mathrm{i}\delta_{p''-\frac{q}{2}}} + \text{multi-pair background} \\
&\quad = \frac{2\pi \mathrm{i} z^2}{v_{\mathrm{F}}} \frac{v_{\mathrm{F}} \hat{\boldsymbol{p}}'' \cdot \boldsymbol{q}}{\omega - v_{\mathrm{F}} \hat{\boldsymbol{p}}'' \cdot \boldsymbol{q}} \delta(\varepsilon'')\delta(|\boldsymbol{p}''| - k_{\mathrm{F}}) + \text{nonsingular } \phi(p''),
\end{aligned}
\tag{10}
$$

where the quasiparticle energy is measured relative to the Fermi energy μ. We note that the singular part, which is due to the quasiparticle piece of the Green functions, vanishes in the limit $|\boldsymbol{q}| \to 0$ and $\omega \to 0$ with $|\boldsymbol{q}|/\omega \to 0$. Therefore, one can eliminate all quasiparticle–quasihole reducible contributions in a given channel by taking this limit.

The singularity of the ZS particle–hole propagator is reflected in the dependence of the coefficient of the delta functions in Eq. (10) on the order of the limits $|\boldsymbol{q}| \to 0$ and $\omega \to 0$. The $|\boldsymbol{q}|$ and ω limits of the particle–hole vertex are defined as

$$
\Gamma^{\omega}(p, p') = \lim_{\omega \to 0} \left(\Gamma(p, p'; q)|_{|\boldsymbol{q}|=0}\right) \quad \text{and} \tag{11}
$$

$$
\Gamma^{q}(p, p') = \lim_{|\boldsymbol{q}| \to 0} \left(\Gamma(p, p'; q)|_{\omega=0}\right). \tag{12}
$$

In the ω limit the singular part in Eq. (10) vanishes. Thus, from Eq. (8) it follows that Γ^{ω} itself is obtained by solving a Bethe–Salpeter equation, which sums the ZS particle–hole ladders with the nonsingular part ϕ only. Consequently, Γ^{ω} is *quasiparticle–quasihole* irreducible in the ZS channel.

With this at hand, one can eliminate $\widetilde{\Gamma}$ and the nonsingular ϕ to obtain the following quasiparticle–quasihole analogue of Eq. (8) for Γ, at $T = 0$ [12]:

$$
\begin{aligned}
&\Gamma_{\sigma\cdot\sigma', \tau\cdot\tau'}(p, p'; q) \\
&\quad = \Gamma^{\omega}_{\sigma\cdot\sigma', \tau\cdot\tau'}(p, p') \\
&\quad\quad + N(0)\, z^2 \frac{1}{4} \operatorname{Tr}_{\sigma''\tau''} \int \frac{\mathrm{d}\Omega_{p''}}{4\pi} \Gamma^{\omega}_{\sigma\cdot\sigma'', \tau\cdot\tau''}(p, p'') \frac{v_{\mathrm{F}} \hat{\boldsymbol{p}}'' \cdot \boldsymbol{q}}{\omega - v_{\mathrm{F}} \hat{\boldsymbol{p}}'' \cdot \boldsymbol{q}} \\
&\quad\quad \times \Gamma_{\sigma''\cdot\sigma', \tau''\cdot\tau'}(p'', p'; q).
\end{aligned}
\tag{13}
$$

Diagrammatically this equation corresponds to

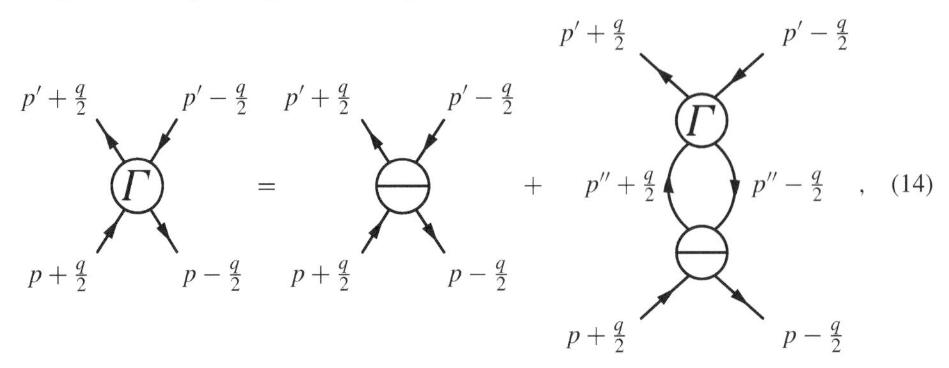

where Γ^{ω} is denoted by a blob with a line across. The line is drawn perpendicular to the channel, in which Γ^{ω} is quasiparticle–quasihole irreducible.[5]

The $|\boldsymbol{q}|$ limit Γ^{q} corresponds to the full particle–hole vertex for $q=0$, i.e., scattering of quasiparticles strictly on the Fermi surface with vanishing momentum transfer $|\boldsymbol{q}|\to 0$. Thus, Eq. (13) can be used to relate the two limits:

$$\Gamma^{q}_{\sigma\cdot\sigma',\tau\cdot\tau'}=\Gamma^{\omega}_{\sigma\cdot\sigma',\tau\cdot\tau'}-N(0)\,z^{2}\frac{1}{4}\int\frac{\mathrm{d}\Omega_{p''}}{4\pi}\,\mathrm{Tr}_{\sigma''\tau''}\,\Gamma^{\omega}_{\sigma\cdot\sigma'',\tau\cdot\tau''}\Gamma^{q}_{\sigma''\cdot\sigma',\tau''\cdot\tau'}\,. \tag{15}$$

The quasiparticle–quasihole irreducible vertex can be identified with the quasiparticle interaction introduced above [12], $N(0)\,z^{2}\Gamma^{\omega}(p,p')=\mathcal{F}(\theta)$, while $N(0)\,z^{2}\Gamma^{q}(p,p')=\mathcal{A}(\theta)$ is the quasiparticle forward scattering amplitude. The factor z is the spectral strength at the quasiparticle pole. By inserting this into Eq. (15) and expanding the angular dependence of $\mathcal{F}(\theta)$ and $\mathcal{A}(\theta)$ on Legendre polynomials, we arrive at a set of algebraic equations for the scattering amplitude with the solution:

$$\begin{aligned}\mathcal{A}(\theta)&=\sum_{l}\biggl(\frac{F_{l}}{1+F_{l}/(2l+1)}+\frac{F'_{l}}{1+F'_{l}/(2l+1)}\boldsymbol{\tau}\cdot\boldsymbol{\tau}'\\&\quad+\frac{G_{l}}{1+G_{l}/(2l+1)}\boldsymbol{\sigma}\cdot\boldsymbol{\sigma}'+\frac{G'_{l}}{1+G'_{l}/(2l+1)}\boldsymbol{\tau}\cdot\boldsymbol{\tau}'\boldsymbol{\sigma}\cdot\boldsymbol{\sigma}'\biggr)P_{l}(\cos\theta).\end{aligned} \tag{16}$$

The antisymmetry of the quasiparticle scattering amplitude implies two Pauli principle sum rules [10,13] for the Fermi liquid parameters, corresponding to scattering at vanishing relative momentum in singlet-odd and triplet-odd states:

$$\begin{aligned}&\sum_{l}\biggl(\frac{F_{l}}{1+F_{l}/(2l+1)}+\frac{F'_{l}}{1+F'_{l}/(2l+1)}\\&\quad+\frac{G_{l}}{1+G_{l}/(2l+1)}+\frac{G'_{l}}{1+G'_{l}/(2l+1)}\biggr)=0,\end{aligned} \tag{17}$$

[5] We use the notation that the particle–hole propagators in diagrams with the crossed blob correspond to the singular quasiparticle–quasihole part only.

$$\sum_l \left(\frac{F_l}{1 + F_l/(2l+1)} - 3\frac{F'_l}{1 + F'_l/(2l+1)} \right.$$
$$\left. - 3\frac{G_l}{1 + G_l/(2l+1)} + 9\frac{G'_l}{1 + G'_l/(2l+1)} \right) = 0. \tag{18}$$

It is important to note that the quasiparticle interaction is strictly speaking defined only in the Landau limit $q = 0$. This is reflected in the one-pion exchange (OPE) contribution (direct and exchange) to Γ^{ω}, where the direct tensor interaction, which is proportional to $\boldsymbol{q}^2$, vanishes in the Landau limit. For later use we give the one-pion exchange contribution to Γ^{ω}:

$$\Gamma^{\mathrm{OPE}}_{\boldsymbol{\sigma}\cdot\boldsymbol{\sigma}',\boldsymbol{\tau}\cdot\boldsymbol{\tau}'}(p, p'; q)$$
$$= -\frac{f^2}{3m_\pi^2}\boldsymbol{\tau}\cdot\boldsymbol{\tau}'\left\{\boldsymbol{q}^2\frac{S_{12}(\boldsymbol{q})}{\boldsymbol{q}^2 + m_\pi^2} - \frac{m_\pi^2\boldsymbol{\sigma}\cdot\boldsymbol{\sigma}'}{\boldsymbol{q}^2 + m_\pi^2}\right\}$$
$$+ \frac{f^2}{3m_\pi^2}\frac{3 - \boldsymbol{\tau}\cdot\boldsymbol{\tau}'}{2}\left\{(\boldsymbol{p} - \boldsymbol{p}')^2\frac{S_{12}(\boldsymbol{p} - \boldsymbol{p}')}{(\boldsymbol{p} - \boldsymbol{p}')^2 + m_\pi^2} - \frac{1}{2}\frac{m_\pi^2(3 - \boldsymbol{\sigma}\cdot\boldsymbol{\sigma}')}{(\boldsymbol{p} - \boldsymbol{p}')^2 + m_\pi^2}\right\}. \tag{19}$$

3. The induced interaction

The quasiparticle scattering amplitude includes particle–hole diagrams in the ZS channel to all orders. Therefore, if one were to use a finite set of diagrams for the quasiparticle–quasihole irreducible vertex Γ^{ω}, e.g., the Hartree–Fock approximation, Eq. (2), then the corresponding quasiparticle scattering amplitude, obtained by solving Eq. (13), would not obey the Pauli principle. This is because the particle–hole diagrams in the ZS$'$ channel, which, as discussed above are the exchange diagrams to those in the ZS channel, are not iterated. Thus, in order to obey the Pauli principle, it is necessary to iterate the ZS$'$ channel to all orders as well. This is done by the induced interaction, which was invented by Babu and Brown [9] and applied to nuclear matter by Sjöberg [14]. Here we give a diagrammatically motivated heuristic derivation of the induced interaction.

The integral equation for Γ^{ω} must be constructed in such a way that it generates all possible ZS and ZS$'$ joined diagrams for the quasiparticle scattering amplitude. To third order, these are given in Fig. 1.

Using an antisymmetric, particle–hole irreducible vertex function in Fig. 1 guarantees the antisymmetry of the quasiparticle scattering amplitude. We have marked the propagators $G(p'' + q/2)G(p'' - q/2)$, that are generated by solving the Bethe–Salpeter equation, (8), with thick lines. The diagrams with only thin lines are contained in $\widetilde{\Gamma}$. All of these can be constructed from a ZS$'$ ladder sum, where the vertices are lower order diagrams of $\widetilde{\Gamma}$ rotated by 90 degrees. To second order $\widetilde{\Gamma}$ consists of the one ZS$'$ bubble only, diagram (c), to third-order $\widetilde{\Gamma}$ also includes the two ZS$'$ bubble string, diagram (i), and the diagrams (d) and (e). The latter are constructed by taking a second (lower) order diagram, the one ZS$'$ bubble, diagram (c), rotating it by 90 degrees, and then inserting it as left or right vertex into the one ZS$'$ bubble. Thus, the integral equation for $\widetilde{\Gamma}$, for a system with spin only, reads:

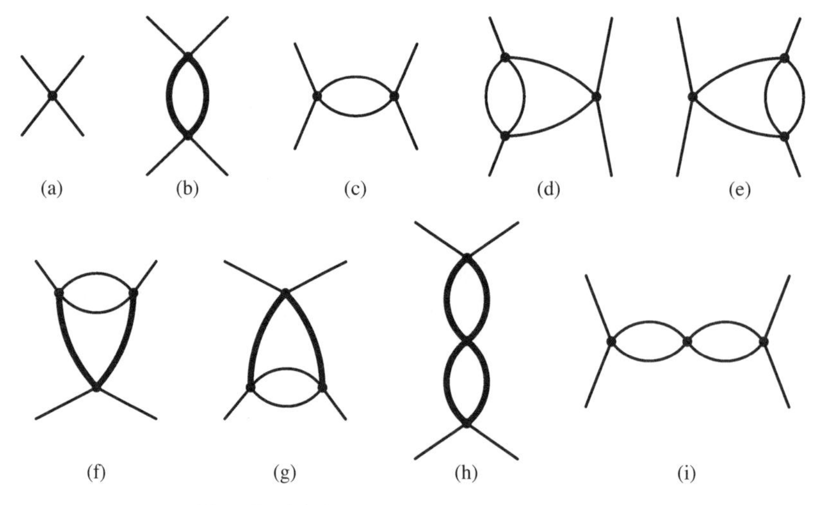

Fig. 1. Particle–hole parquet diagrams to third order.

$$
\begin{aligned}
&\widetilde{\Gamma}_{\sigma\cdot\sigma'}(p,p';q) \\
&\quad = I_{\sigma\cdot\sigma'}(p,p';q) - \frac{1}{2}(1+\boldsymbol{\sigma}\cdot\boldsymbol{\sigma}') \\
&\quad \times \left\{ \frac{1}{2}\mathrm{Tr}_{\sigma''} \int \frac{-\mathrm{i}\,\mathrm{d}^4 p''}{(2\pi)^4} \widetilde{\Gamma}_{\sigma\cdot\sigma''}\left(\frac{p+p'+q}{2}, p''; p-p'\right) G\left(p'' + \frac{p-p'}{2}\right) \right. \\
&\quad \times G\left(p'' - \frac{p-p'}{2}\right) \widetilde{\Gamma}_{\sigma''\cdot\sigma'}\left(p'', \frac{p+p'-q}{2}; p-p'\right) \\
&\quad \left. + \widetilde{\Gamma} G^2 \widetilde{\Gamma} G^2 \widetilde{\Gamma} + \cdots \right\},
\end{aligned}
\tag{20}
$$

where we have denoted the antisymmetric, ZS and ZS′ particle–hole irreducible vertex with I. The spin operator $\boldsymbol{\sigma}$ in the brackets of Eq. (20) is contracted with the spinors of the left particle–hole pair with momenta $p+q/2$ and $p'+q/2$, whereas $\boldsymbol{\sigma}$ in the left-hand side and in I is contracted with the bottom particle–hole pair spinors with momenta $p \pm q/2$. The recoupling between the two particle–hole channels is accounted for by including the spin exchange operator $P_\sigma = 1/2(1+\boldsymbol{\sigma}\cdot\boldsymbol{\sigma}')$. Since $\widetilde{\Gamma}$ is finite, we can take the limit $q \to 0$ in Eq. (20) and obtain for $\boldsymbol{p} \approx \boldsymbol{p}'$ [6]:

$$
\begin{aligned}
\widetilde{\Gamma}_{\sigma\cdot\sigma'}(p,p') &= I_{\sigma\cdot\sigma'}(p,p') - \frac{1}{2}(1+\boldsymbol{\sigma}\cdot\boldsymbol{\sigma}') \\
&\quad \times \left\{ \frac{1}{2}\mathrm{Tr}_{\sigma''}\left(N(0)\, z^2 \int \frac{\mathrm{d}\Omega_{\boldsymbol{p}''}}{4\pi} \widetilde{\Gamma}_{\sigma\cdot\sigma''}\left(\frac{p+p'}{2}, p''; p-p'\right) \right.\right.
\end{aligned}
$$

[6] To guarantee continuity in the forward scattering amplitude the limit $q \to 0$ has to be performed before taking $\boldsymbol{p} \to \boldsymbol{p}'$ [15].

$$
\begin{aligned}
&\times \frac{v_{\mathrm{F}} \hat{\boldsymbol{p}''} \cdot (\boldsymbol{p} - \boldsymbol{p}')}{\varepsilon - \varepsilon' - v_{\mathrm{F}} \hat{\boldsymbol{p}''} \cdot (\boldsymbol{p} - \boldsymbol{p}')} \widetilde{\Gamma}_{\boldsymbol{\sigma}'' \cdot \boldsymbol{\sigma}'} \left(\frac{p+p'}{2}; p - p' \right) \\
&+ \int \frac{-\mathrm{i}\, \mathrm{d}^4 p''}{(2\pi)^4} \widetilde{\Gamma}_{\boldsymbol{\sigma} \cdot \boldsymbol{\sigma}''} \left(\frac{p+p'}{2}, p''; p - p' \right) \phi(p'') \\
&\times \widetilde{\Gamma}_{\boldsymbol{\sigma}'' \cdot \boldsymbol{\sigma}'} \left(p'', \frac{p+p'}{2}; p - p' \right) \Bigg) \\
&+ \widetilde{\Gamma} \big((\mathcal{G}\mathcal{G})_{\mathrm{ZS}'} + \phi \big) \widetilde{\Gamma} \big((\mathcal{G}\mathcal{G})_{\mathrm{ZS}'} + \phi \big) \widetilde{\Gamma} + \cdots \Bigg\},
\end{aligned}
\tag{21}
$$

where $(\mathcal{G}\mathcal{G})_{\mathrm{ZS}'}$ denotes the quasiparticle–quasihole part of the propagators in the ZS′ channel. To both sides of Eq. (21) we add the series $\widetilde{\Gamma}\phi\widetilde{\Gamma} + \widetilde{\Gamma}\phi\widetilde{\Gamma}\phi\widetilde{\Gamma} + \cdots$ and obtain

$$
\begin{aligned}
\Gamma^{\omega} = {} & I + \widetilde{\Gamma}\phi\widetilde{\Gamma} + \widetilde{\Gamma}\phi\widetilde{\Gamma}\phi\widetilde{\Gamma} + \cdots - \frac{1}{2}(1 + \boldsymbol{\sigma} \cdot \boldsymbol{\sigma}') \big\{ \widetilde{\Gamma} \big((\mathcal{G}\mathcal{G})_{\mathrm{ZS}'} + \phi \big) \widetilde{\Gamma} \\
& + \widetilde{\Gamma} \big((\mathcal{G}\mathcal{G})_{\mathrm{ZS}'} + \phi \big) \widetilde{\Gamma} \big((\mathcal{G}\mathcal{G})_{\mathrm{ZS}'} + \phi \big) \widetilde{\Gamma} + \cdots \big\}.
\end{aligned}
\tag{22}
$$

By regrouping the terms we find

$$
\begin{aligned}
\Gamma^{\omega} = {} & I + \left(1 - \frac{1}{2}(1 + \boldsymbol{\sigma} \cdot \boldsymbol{\sigma}') \right) \widetilde{\Gamma}\phi \frac{1}{1 - \widetilde{\Gamma}\phi} \widetilde{\Gamma} \\
& - \frac{1}{2}(1 + \boldsymbol{\sigma} \cdot \boldsymbol{\sigma}') \Gamma^{\omega} (\mathcal{G}\mathcal{G})_{\mathrm{ZS}'} \frac{1}{1 - \Gamma^{\omega} (\mathcal{G}\mathcal{G})_{\mathrm{ZS}'}} \Gamma^{\omega}.
\end{aligned}
\tag{23}
$$

The first term $I_{qp} = I + \big(1 - \frac{1}{2}(1 + \boldsymbol{\sigma} \cdot \boldsymbol{\sigma}')\big) \widetilde{\Gamma}\phi (1 - \widetilde{\Gamma}\phi)^{-1} \widetilde{\Gamma}$ is *quasiparticle–quasihole* irreducible both in the ZS and ZS′ channel. Due to the identity $P_{\sigma}(1 - P_{\sigma}) = -(1 - P_{\sigma})$ and the antisymmetry of I, I_{qp} is also antisymmetric. For $\boldsymbol{p} = \boldsymbol{p}'$ the nonsingular parts of the ZS and the ZS′ graphs differ only in the spin dependence. This is reflected in the factor $(1 - P_{\sigma})$, which vanishes for a Fermi liquid of say spin up species only.

Eq. (23) is an integral equation for Γ^{ω}, which diagrammatically is of the form:

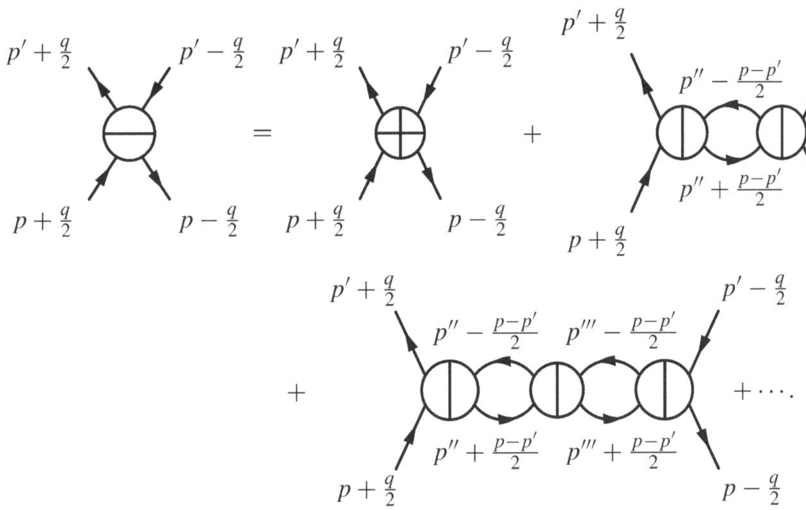

(24)

The diagram with the crossed lines denotes the driving term I_{qp}, which consists of all quasiparticle–quasihole irreducible diagrams (in both the ZS and ZS′ channels). The series of all ZS′ bubble diagrams corresponding to the remaining terms in Eq. (23) is called the induced interaction. It may be regarded as the linear response of the system to the presence of the quasiparticle. Due to the exchange of the external lines, it is explicit that all diagrams in the induced interaction are irreducible in the ZS channel. The limit $q \to 0$ is to be taken only after the iteration of the induced interaction. In order to illustrate this, consider the one-pion exchange vertex function, Eq. (19), as driving term. The momentum transfers in $\Gamma^{\mathrm{OPE}}(p, p'; q)$ are $\boldsymbol{q}$ and $\boldsymbol{p} - \boldsymbol{p}'$. However, due to the exchange character of the induced interaction, the corresponding momentum transfers in the vertices of the one ZS′ bubble are $\boldsymbol{p} - \boldsymbol{p}'$ and $\frac{1}{2}(\boldsymbol{p} + \boldsymbol{p}' + \boldsymbol{q}) - \boldsymbol{p}''$. Although terms proportional to $\boldsymbol{q}^2$ in the driving term vanish in the Landau limit, they appear in the induced interaction. Thus, in general the induced interaction requires input beyond Fermi liquid theory, since the Landau parameters are defined only in the $|\boldsymbol{q}| \to 0$ limit. In the limit $\boldsymbol{p} = \boldsymbol{p}'$, the induced interaction expressed solely in terms of Landau parameters is exact. Nevertheless, applications to nuclear matter [14,16,17], neutron matter [18,19] and liquid ^{3}He [20–22] have shown that the induced interaction is a very powerful approximation even for nonvanishing angles θ, i.e., $\boldsymbol{p} \neq \boldsymbol{p}'$.

The one ZS′ bubble contribution to the induced interactions is given by

$$\begin{aligned}\Gamma^{\mathrm{ind}(2)}_{\sigma\cdot\sigma',\tau\cdot\tau'}(p, p') &= -\frac{1}{4}(1 + \boldsymbol{\sigma}\cdot\boldsymbol{\sigma}')(1 + \boldsymbol{\tau}\cdot\boldsymbol{\tau}')N(0)z^2 \\ &\quad \times \frac{1}{4}\mathrm{Tr}_{\sigma''\tau''}\int \frac{\mathrm{d}\Omega_{p''}}{4\pi}\Gamma^{\omega}_{\sigma\cdot\sigma'',\tau\cdot\tau''}\left(\frac{p+p'}{2}, p''; p - p'\right) \\ &\quad \times \frac{v_{\mathrm{F}}\hat{\boldsymbol{p}}''\cdot(\boldsymbol{p} - \boldsymbol{p}')}{\varepsilon - \varepsilon' - v_{\mathrm{F}}\hat{\boldsymbol{p}}''\cdot(\boldsymbol{p} - \boldsymbol{p}')}\Gamma^{\omega}_{\sigma''\cdot\sigma',\tau''\cdot\tau'}\left(p'', \frac{p+p'}{2}; p - p'\right).\end{aligned} \tag{25}$$

In the extrapolation away from $\boldsymbol{p} = \boldsymbol{p}'$, the initial and final momenta are treated symmetrically. This yields the correct result, e.g., for a current–current coupling. Using the bare direct and exchange interaction as driving term, i.e.,

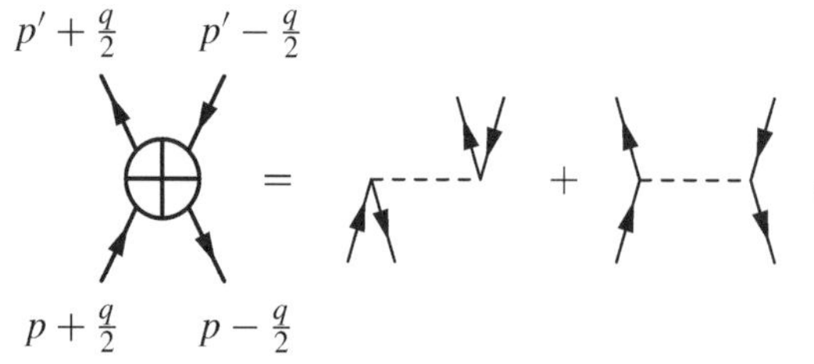

$p' + \frac{q}{2}$ $\quad p' - \frac{q}{2}$

[diagram] = [diagram] + [diagram] , (26)

$p + \frac{q}{2}$ $\quad p - \frac{q}{2}$

one finds that the lowest order contributions to Eq. (25) correspond to the following diagrams:

$\Gamma^{\mathrm{ind}(2)} =$ [diagram] + [diagram] + [diagram] + [diagram] . (27)

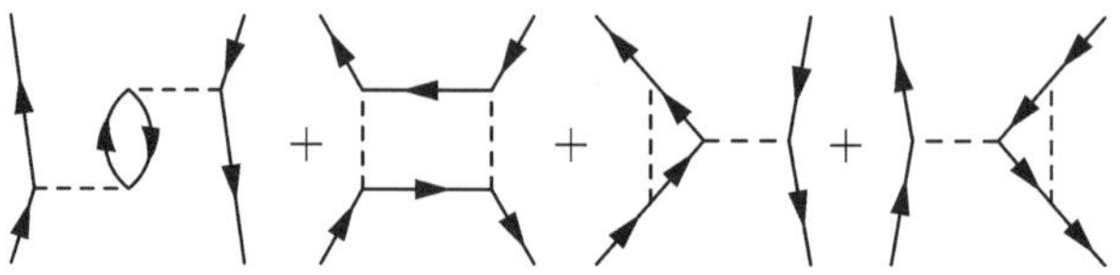

We expand the angular dependence of the quasiparticle interaction Γ^ω on Legendre polynomials, $\Gamma^\omega = \sum_l \Gamma_l^\omega P_l(\cos\theta)$. After inserting this in Eq. (25), we find

$$
\begin{aligned}
&\Gamma^{\mathrm{ind}(2)}_{\sigma\cdot\sigma',\tau\cdot\tau'}(p,p') \\
&\quad = -\frac{1}{4}(1+\boldsymbol{\sigma}\cdot\boldsymbol{\sigma}')(1+\boldsymbol{\tau}\cdot\boldsymbol{\tau}')N(0)\,z^2 \\
&\quad \times \frac{1}{4}\,\mathrm{Tr}_{\sigma''\tau''}\sum_{l,l'}\Gamma^\omega_{\sigma\cdot\sigma'',\tau\cdot\tau'',l}\,\Gamma^\omega_{\sigma''\cdot\sigma',\tau''\cdot\tau',l'} \\
&\quad \times \int \frac{\mathrm{d}\Omega_{p''}}{4\pi} P_l\left(\widehat{\frac{\boldsymbol{p}+\boldsymbol{p}'}{2}}\cdot\hat{\boldsymbol{p}}''\right) P_{l'}\left(\hat{\boldsymbol{p}}''\cdot\widehat{\frac{\boldsymbol{p}+\boldsymbol{p}'}{2}}\right)\frac{v_{\mathrm{F}}\hat{\boldsymbol{p}}''\cdot(\boldsymbol{p}-\boldsymbol{p}')}{\varepsilon-\varepsilon'-v_{\mathrm{F}}\hat{\boldsymbol{p}}''\cdot(\boldsymbol{p}-\boldsymbol{p}')}.
\end{aligned}
\tag{28}
$$

In order to cover all possible combinations of $\boldsymbol{p}$ and $\boldsymbol{p}'$, the induced interaction is needed for momentum transfer $\boldsymbol{q}' = \boldsymbol{p} - \boldsymbol{p}'$ up to $2k_{\mathrm{F}}$. This is done by extrapolating the quasiparticle–quasihole propagator in Eq. (10) to large q using the particle–hole propagator of a free Fermi gas with an effective mass $m^\star$. Furthermore, the external quasiparticles are assumed to be on the Fermi surface, so that $\varepsilon = \varepsilon' = 0$. For $l, l' = 0, 1$, the resulting integrals in Eq. (28) are given in [14]. We introduce the notation $\mathcal{F}_{\mathrm{ind}} = N(0)\,z^2\Gamma_{\mathrm{ind}}$ and decompose the induced interaction into its scalar, spin, isospin and spin–isospin components:

$$
\mathcal{F}_{\mathrm{ind}} = F_{\mathrm{ind}} + F'_{\mathrm{ind}}\boldsymbol{\tau}\cdot\boldsymbol{\tau}' + G_{\mathrm{ind}}\boldsymbol{\sigma}\cdot\boldsymbol{\sigma}' + G'_{\mathrm{ind}}\boldsymbol{\tau}\cdot\boldsymbol{\tau}'\boldsymbol{\sigma}\cdot\boldsymbol{\sigma}'. \tag{29}
$$

The resulting expression for the scalar induced interaction, F_{ind}, including $l = 0, 1$ terms, is [14,16]

$$
\begin{aligned}
4F_{\mathrm{ind}} = {} & 1\cdot\left(\frac{F_0^2\alpha_0(q'/k_{\mathrm{F}})}{1+F_0\alpha_0(q'/k_{\mathrm{F}})} + \left(1-\frac{q'^2}{4k_{\mathrm{F}}^2}\right)\frac{F_1^2\alpha_1(q'/k_{\mathrm{F}})}{1+F_1\alpha_1(q'/k_{\mathrm{F}})}\right) \\
& + 3\cdot\left(\frac{F_0'^2\alpha_0(q'/k_{\mathrm{F}})}{1+F_0'\alpha_0(q'/k_{\mathrm{F}})} + \left(1-\frac{q'^2}{4k_{\mathrm{F}}^2}\right)\frac{F_1'^2\alpha_1(q'/k_{\mathrm{F}})}{1+F_1'\alpha_1(q'/k_{\mathrm{F}})}\right) \\
& + 3\cdot\left(\frac{G_0^2\alpha_0(q'/k_{\mathrm{F}})}{1+G_0\alpha_0(q'/k_{\mathrm{F}})} + \left(1-\frac{q'^2}{4k_{\mathrm{F}}^2}\right)\frac{G_1^2\alpha_1(q'/k_{\mathrm{F}})}{1+G_1\alpha_1(q'/k_{\mathrm{F}})}\right) \\
& + 9\cdot\left(\frac{G_0'^2\alpha_0(q'/k_{\mathrm{F}})}{1+G_0'\alpha_0(q'/k_{\mathrm{F}})} + \left(1-\frac{q'^2}{4k_{\mathrm{F}}^2}\right)\frac{G_1'^2\alpha_1(q'/k_{\mathrm{F}})}{1+G_1'\alpha_1(q'/k_{\mathrm{F}})}\right),
\end{aligned}
\tag{30}
$$

where $q' = |\boldsymbol{q}'|$, $\alpha_0(x)$ and $\alpha_1(x)$ are the Lindhard (or density–density) and current–current correlation functions, respectively. The factor $(1 - q'^2/4k_{\mathrm{F}}^2)$ guarantees that the current response vanishes for back-to-back scattering.

$$
\alpha_0(x) = \frac{1}{2} + \frac{1}{2}\left(\frac{x}{4} - \frac{1}{x}\right)\ln\frac{1-x/2}{1+x/2}, \tag{31}
$$

$$
\alpha_1(x) = \frac{1}{2}\left[\frac{3}{8} - \frac{1}{2x^2} + \left(\frac{1}{2x^3} + \frac{1}{4x} - \frac{3x}{32}\right)\ln\frac{1+x/2}{1-x/2}\right]. \tag{32}
$$

Table 1
Spin–isospin recoupling coefficients

	F	F'	G	G'
F_{ind}	1	3	3	9
F'_{ind}	1	−1	3	−3
G_{ind}	1	3	−1	−3
G'_{ind}	1	−1	−1	1

For the spin, isospin and spin–isospin induced parts, the coefficients in (30) have to be changed according to Table 1. These coefficients follow from the recoupling of spin and isospin between the two particle–hole channels.

By construction, the induced interaction with the bare direct and exchange interaction as driving term generates the complete particle–hole parquet for the scattering amplitude. The particle–hole parquet are all planar fermionic diagrams except those that are joined by the particle–particle (BCS) channel. This corresponds to the solution to the fermionic parquet equations of Lande and Smith ignoring the coupling to the s channel [23]. The s channel diagrams are particle–hole irreducible in both the ZS and ZS$'$ channels. Hence, they should be included in the driving term. Traditionally the driving term has been computed within Brueckner theory by varying the energy twice with respect to the occupation number and removing all contributions that are included in the induced interaction [14,24]. If two hole contributions, which are expected to be small, are neglected, one can express the driving term as the direct and exchange Brueckner G matrix multiplied by the renormalization factor z^2. The factor z^2 accounts for some of the higher-order completely particle–hole irreducible diagrams. Diagrams involving, e.g., particle–particle ladders with a screened interaction are neglected. In this work we identify the G matrix in the driving term with the low-momentum nucleon–nucleon interaction $V_{\text{low }k}$ for $\Lambda = k_{\text{F}}$ [2] and consequently employ $z^2 V_{\text{low }k}$ as the driving term. As discussed in the introduction, $V_{\text{low }k}$ is smooth and includes the effects of the short range repulsion.

4. Relation between the Fermi liquid parameters and the low-momentum nucleon–nucleon interaction

For $\boldsymbol{p} = \boldsymbol{p}'$ the induced interaction expressed in terms of Fermi liquid parameters is exact and one can derive general constraints for these parameters. In this limit the integral in Eq. (28) simplifies to $-\delta_{l,l'}/(2l+1)$ and all higher ZS$'$ bubble terms are easily summed. Thus, Eq. (24) can be written as follows:

$$F_s + F_a \boldsymbol{\sigma}\cdot\boldsymbol{\sigma}' = F_s^d + F_a^d \boldsymbol{\sigma}\cdot\boldsymbol{\sigma}' + \int \frac{d\Omega_{p''}}{4\pi} \left\{ F_s(p, p'') A_s(p'', p) \frac{1+\boldsymbol{\sigma}\cdot\boldsymbol{\sigma}'}{2} + F_a(p, p'') A_a(p'', p) \frac{3-\boldsymbol{\sigma}\cdot\boldsymbol{\sigma}'}{2} \right\}, \tag{33}$$

where we again consider a system with spin only. For $\boldsymbol{p} = \boldsymbol{p}'$ the series of ZS$'$ bubbles is equivalent to the series of ZS bubbles summed by the scattering amplitude up to a sign and the spin exchange operator for the exchange of the external lines in the induced interaction. The equation for the scattering amplitude reads

$$A_s + A_a \boldsymbol{\sigma}\cdot\boldsymbol{\sigma}' = F_s + F_a \boldsymbol{\sigma}\cdot\boldsymbol{\sigma}' - \int \frac{d\Omega_{p''}}{4\pi}\big\{F_s(p,p'')A_s(p'',p) + F_a(p,p'')A_a(p'',p)\boldsymbol{\sigma}\cdot\boldsymbol{\sigma}'\big\}. \quad (34)$$

We have introduced the notation

$$\mathcal{F} = F_s + F_a\boldsymbol{\sigma}\cdot\boldsymbol{\sigma}', \qquad \mathcal{A} = A_s + A_a\boldsymbol{\sigma}\cdot\boldsymbol{\sigma}', \qquad \mathcal{F}_{\text{driving}} = F_s^d + F_a^d\boldsymbol{\sigma}\cdot\boldsymbol{\sigma}'. \quad (35)$$

It is easy to solve the integral equations for the driving term. In the $S=1$ channel the sum and in the $S=0$ channel the difference of the two integral equations, Eqs. (33) and (34), leads to

$$\mathcal{F}_{\text{driving}}(S=1) = \mathcal{A} = 0, \quad (36)$$

$$\mathcal{F}_{\text{driving}}(S=0) = 2\mathcal{F} - \mathcal{A}, \quad (37)$$

where we have used the Pauli principle sum rule in the case $S=1$. The first case, Eq. (36), projects on odd partial waves, while the second case, Eq. (37), projects on even partial waves. In symmetric nuclear matter, there are two spin–isospin states corresponding to odd partial waves ($S=T=0$ and $S=T=1$) and two corresponding to even states ($S=0, T=1$ and $S=1, T=0$). We thus obtain two new constraints on the Fermi liquid parameters of nuclear matter:

$$\sum_l\Big\{2F_l - \frac{F_l}{1+F_l/(2l+1)} + 2F_l' - \frac{F_l'}{1+F_l'/(2l+1)} - 3\Big(2G_l - \frac{G_l}{1+G_l/(2l+1)}\Big) - 3\Big(2G_l' - \frac{G_l'}{1+G_l'/(2l+1)}\Big)\Big\} = \mathcal{F}_{\text{driving}}(S=0, T=1), \quad (38)$$

$$\sum_l\Big\{2F_l - \frac{F_l}{1+F_l/(2l+1)} - 3\Big(2F_l' - \frac{F_l'}{1+F_l'/(2l+1)}\Big) + 2G_l - \frac{G_l}{1+G_l/(2l+1)} - 3\Big(2G_l' - \frac{G_l'}{1+G_l'/(2l+1)}\Big)\Big\} = \mathcal{F}_{\text{driving}}(S=1, T=0). \quad (39)$$

These are general constraints, which however are useful only if the driving term is known. Such a constraint was first derived by Bedell and Ainsworth [21] for paramagnetic Fermi liquids, like liquid ^{3}He or ^{3}He–^{4}He mixtures, and employed to extract the effective scattering length. As reasoned above, we approximate the driving term with $z^2 V_{\text{low }k}$. We need the matrix elements of $V_{\text{low }k}$ in the basis of total spin S and total isospin T. By summing over M_S, we project onto the central components of the forward scattering amplitude [17]:

$$\frac{1}{V}\mathcal{F}_{\text{driving}}(S,T) = \frac{z^2}{2S+1}N(0)\sum_{M_S}\big(\langle \boldsymbol{p}\boldsymbol{p}' ST|V_{\text{low }k}|\boldsymbol{p}\boldsymbol{p}' ST\rangle - \text{exchange}\big). \quad (40)$$

Transforming to relative momentum $\boldsymbol{q}' = \boldsymbol{p} - \boldsymbol{p}'$ and coupling angular momentum and total spin leads to

$$\mathcal{F}_{\text{driving}}(S,T) = z^2 N(0) \frac{4\pi}{2S+1} \sum_{J,l} (2J+1)\big(1-(-1)^{l+S+T}\big) \times \left\langle k = \frac{q'}{2} lSJT \middle| V_{\text{low }k} \middle| k = \frac{q'}{2} lSJT \right\rangle. \tag{41}$$

At vanishing relative momentum there are only s-wave contributions to the driving term due to the rotational invariance. Since the driving term is antisymmetric, these contributions are in the $S=0$, $T=1$ and $S=1$, $T=0$ channels, consistent with the two Pauli principle sum rules. Thus, with the input $z^2 V_{\text{low }k}$ for the driving term, the two relations, Eqs. (38) and (39), constrain the dimensionless Fermi liquid parameters of nuclear matter in a nontrivial way to

$$\mathcal{F}_{\text{driving}}(S=0,T=1) = z^2 \frac{16 m_{\text{N}} k_{\text{F}} (1+F_1/3)}{\pi} V_{\text{low }k}\big(0,0;\Lambda = k_{\text{F}}, {}^1\text{S}_0\big), \tag{42}$$

$$\mathcal{F}_{\text{driving}}(S=1,T=0) = z^2 \frac{16 m_{\text{N}} k_{\text{F}} (1+F_1/3)}{\pi} V_{\text{low }k}\big(0,0;\Lambda = k_{\text{F}}, {}^3\text{S}_1\big). \tag{43}$$

The dimension of the potential is absorbed by the density of states. As explained in the introduction and in [2], $V_{\text{low }k}$ is obtained from a RG decimation of various nuclear force models. Bogner et al. [2] find that the $V_{\text{low }k}$ obtained from various bare potentials at $\Lambda = k_{\text{F}}$ are identical. Moreover, when one compares the low-momentum part of the bare interaction models with $V_{\text{low }k}$, one observes that the main effect of the renormalization is a constant shift in momentum space. This correspond to a smeared delta function in coordinate space and accounts for the removal of the model-dependent short-range core. Thus, the two constraints, which use as dynamical input $V_{\text{low }k}(0,0)$, connect the pivotal matrix element of the RG decimation to the unique set of Fermi liquid parameters of nuclear matter. As the Fermi liquid parameters are fixed points under the RG flow towards the Fermi surface, the constraints relate $V_{\text{low }k}$ to these fixed points.

5. Renormalization group with the induced interaction

In the microscopic derivation of Fermi liquid theory, one isolates the quasiparticle part of the full propagator from the pair background. We have shown that, for $\boldsymbol{p} \approx \boldsymbol{p}'$, this is rigorously possible also when both particle–hole channels are taken into account. This is necessary in order to preserve the Pauli principle and leads to the induced interaction. Having reduced the theory to interactions among quasiparticles, we now separate the soft modes of the quasiparticle–quasihole propagators from the hard ones. To this end we introduce a momentum cutoff at $k_{\text{F}} \pm \Lambda_{\text{F}}$.[7] In this way we arrive at a theory of quasiparticles interacting in a model space of slow modes exclusively. For a discussion of the RG approach to Fermi liquid theory see [25–27].

[7] We denote with Λ_{F} the cutoff in medium, which is not to be confused with the cutoff Λ for $V_{\text{low }k}$.

In a shorthand notation we write $(\mathcal{G}\mathcal{G})_{\rm ZS} = (\mathcal{G}\mathcal{G})^{\rm S}_{\rm ZS} + (\mathcal{G}\mathcal{G})^{\rm H}_{\rm ZS}$ for the ZS propagators (at finite q) and with analogous expressions for the ZS′ channel. The indices S and H denote integrations over the soft (inside the shell) and hard (outside) momenta, respectively. We define the vertices $\gamma^q(p, p'; q, \Lambda_{\rm F})$ and $\gamma^\omega(p, p'; q, \Lambda_{\rm F})$ by

$$\gamma^q_{\rm ZS}(\Lambda_{\rm F}) = \gamma^\omega_{\rm ZS}(\Lambda_{\rm F}) + \gamma^\omega_{\rm ZS}(\Lambda_{\rm F})(\mathcal{G}\mathcal{G})^{\rm H}_{\rm ZS}\gamma^q_{\rm ZS}(\Lambda_{\rm F}), \tag{44}$$

$$\begin{aligned}\gamma^\omega_{\rm ZS}(\Lambda_{\rm F}) &= I_{qp} - \big\{\gamma^\omega_{\rm ZS'}(\Lambda_{\rm F})(\mathcal{G}\mathcal{G})^{\rm H}_{\rm ZS'}\gamma^\omega_{\rm ZS'}(\Lambda_{\rm F}) \\ &\quad + \gamma^\omega_{\rm ZS'}(\Lambda_{\rm F})(\mathcal{G}\mathcal{G})^{\rm H}_{\rm ZS'}\gamma^\omega_{\rm ZS'}(\Lambda_{\rm F})(\mathcal{G}\mathcal{G})^{\rm H}_{\rm ZS'}\gamma^\omega_{\rm ZS'}(\Lambda_{\rm F}) + \cdots\big\} \qquad (45) \\ &= I_{qp} - \gamma^\omega_{\rm ZS'}(\Lambda_{\rm F})(\mathcal{G}\mathcal{G})^{\rm H}_{\rm ZS'}\gamma^q_{\rm ZS'}(\Lambda_{\rm F}), \qquad (46)\end{aligned}$$

where I_{qp} denotes the quasiparticle–quasihole irreducible driving term defined above. Furthermore, we introduce the shorthand notation $\gamma^q_{\rm ZS} = \gamma^q(p, p'; q, \Lambda_{\rm F})$ and the exchange thereof $\gamma^q_{\rm ZS'} = \gamma^q((p + p' + q)/2, (p + p' - q)/2; p - p', \Lambda_{\rm F})$, where for simplicity the spin- and isospin-dependence is suppressed. Analogous expressions hold for $\gamma^\omega_{\rm ZS}$ and $\gamma^\omega_{\rm ZS'}$. Due to phase-space restrictions, the running of γ^q and γ^ω at $T = 0$ starts at $\Lambda_{\rm F} = \max(|\boldsymbol{q}|/2, |\boldsymbol{p} - \boldsymbol{p}'|/2)$. With a weakly energy-dependent driving term I_{qp}, we can set $\omega = 0$ in the flow equations. For $\Lambda_{\rm F} = 0$, the quasiparticle scattering amplitude Γ^q and the quasiparticle interaction Γ^ω are obtained as the $|\boldsymbol{q}| \to 0$ limit of $\gamma^q_{\rm ZS}$ and $\gamma^\omega_{\rm ZS}$, respectively. On the other hand, for $\Lambda_{\rm F} \geqslant k_{\rm F}$, the particle–hole contributions vanish in the momentum range of interest $|\boldsymbol{q}|, |\boldsymbol{p} - \boldsymbol{p}'| \leqslant 2k_{\rm F}$, so that $\gamma^q_{\rm ZS}(k_{\rm F}) = \gamma^\omega_{\rm ZS}(k_{\rm F}) = I_{qp}$.

We differentiate Eqs. (44) and (45) with respect to $\Lambda_{\rm F}$ and require $\mathrm{d}I_{qp}/\mathrm{d}\Lambda_{\rm F} = 0$. This corresponds to ignoring the flow from the particle–particle (BCS) channel. The coupled RG equations then read:

$$\begin{aligned}\frac{\mathrm{d}\gamma^q_{\rm ZS}}{\mathrm{d}\Lambda_{\rm F}} &= \gamma^q_{\rm ZS}\frac{\mathrm{d}(\mathcal{G}\mathcal{G})^{\rm H}_{\rm ZS}}{\mathrm{d}\Lambda_{\rm F}}\gamma^q_{\rm ZS} + \frac{\mathrm{d}\gamma^\omega_{\rm ZS}}{\mathrm{d}\Lambda_{\rm F}} + \frac{\mathrm{d}\gamma^\omega_{\rm ZS}}{\mathrm{d}\Lambda_{\rm F}}(\mathcal{G}\mathcal{G})^{\rm H}_{\rm ZS}\gamma^q_{\rm ZS} + \gamma^q_{\rm ZS}(\mathcal{G}\mathcal{G})^{\rm H}_{\rm ZS}\frac{\mathrm{d}\gamma^\omega_{\rm ZS}}{\mathrm{d}\Lambda_{\rm F}} \\ &\quad + \gamma^q_{\rm ZS}(\mathcal{G}\mathcal{G})^{\rm H}_{\rm ZS}\frac{\mathrm{d}\gamma^\omega_{\rm ZS}}{\mathrm{d}\Lambda_{\rm F}}(\mathcal{G}\mathcal{G})^{\rm H}_{\rm ZS}\gamma^q_{\rm ZS}, \qquad (47)\end{aligned}$$

$$\begin{aligned}\frac{\mathrm{d}\gamma^\omega_{\rm ZS}}{\mathrm{d}\Lambda_{\rm F}} &= -\Bigg\{\frac{1}{1 - \gamma^\omega_{\rm ZS'}(\mathcal{G}\mathcal{G})^{\rm H}_{\rm ZS'}}\frac{\mathrm{d}\gamma^\omega_{\rm ZS'}}{\mathrm{d}\Lambda_{\rm F}} + \frac{1}{1 - \gamma^\omega_{\rm ZS'}(\mathcal{G}\mathcal{G})^{\rm H}_{\rm ZS'}}\Bigg(\gamma^\omega_{\rm ZS'}\frac{\mathrm{d}(\mathcal{G}\mathcal{G})^{\rm H}_{\rm ZS'}}{\mathrm{d}\Lambda_{\rm F}} \\ &\quad + \frac{\mathrm{d}\gamma^\omega_{\rm ZS'}}{\mathrm{d}\Lambda_{\rm F}}(\mathcal{G}\mathcal{G})^{\rm H}_{\rm ZS'}\Bigg)\frac{1}{1 - \gamma^\omega_{\rm ZS'}(\mathcal{G}\mathcal{G})^{\rm H}_{\rm ZS'}}\gamma^\omega_{\rm ZS'} - \frac{\mathrm{d}\gamma^\omega_{\rm ZS'}}{\mathrm{d}\Lambda_{\rm F}}\Bigg\}. \qquad (48)\end{aligned}$$

Using the notation

$$\begin{aligned}\delta_{\rm ZS}(\Lambda_{\rm F}) &= \frac{\mathrm{d}\gamma^\omega_{\rm ZS}}{\mathrm{d}\Lambda_{\rm F}}(\mathcal{G}\mathcal{G})^{\rm H}_{\rm ZS}\gamma^q_{\rm ZS} + \gamma^q_{\rm ZS}(\mathcal{G}\mathcal{G})^{\rm H}_{\rm ZS}\frac{\mathrm{d}\gamma^\omega_{\rm ZS}}{\mathrm{d}\Lambda_{\rm F}} \\ &\quad + \gamma^q_{\rm ZS}(\mathcal{G}\mathcal{G})^{\rm H}_{\rm ZS}\frac{\mathrm{d}\gamma^\omega_{\rm ZS}}{\mathrm{d}\Lambda_{\rm F}}(\mathcal{G}\mathcal{G})^{\rm H}_{\rm ZS}\gamma^q_{\rm ZS} \qquad (49)\end{aligned}$$

and the analogous expression for the ZS′ channel, we write the RG equations in the compact form:

$$\frac{d\gamma^q_{ZS}}{d\Lambda_F} = \gamma^q_{ZS}\frac{d(\mathcal{G}\mathcal{G})^H_{ZS}}{d\Lambda_F}\gamma^q_{ZS} + \frac{d\gamma^\omega_{ZS}}{d\Lambda_F} + \delta_{ZS}(\Lambda_F), \tag{50}$$

$$\frac{d\gamma^\omega_{ZS}}{d\Lambda_F} = -\left\{\gamma^q_{ZS'}\frac{d(\mathcal{G}\mathcal{G})^H_{ZS'}}{d\Lambda_F}\gamma^q_{ZS'} + \delta_{ZS'}(\Lambda_F)\right\}. \tag{51}$$

In the limit $\boldsymbol{p} = \boldsymbol{p}'$ we can replace $\gamma^q_{ZS'}$ in Eq. (51) by γ^q_{ZS} and obtain

$$\frac{d\gamma^\omega_{ZS}}{d\Lambda_F} = -P_\sigma\left\{\gamma^q_{ZS}\frac{d(\mathcal{G}\mathcal{G})^H_{ZS}}{d\Lambda_F}\gamma^q_{ZS} + \delta_{ZS}(\Lambda_F)\right\}, \tag{52}$$

where the spin structure in the exchange channel is accounted for by P_σ. This implies that for $\boldsymbol{p} = \boldsymbol{p}'$ we have $d\gamma^q_{ZS}|_{|\boldsymbol{q}|=0}/d\Lambda_F = 0$ in singlet-odd and triplet-odd states, while $d(2\gamma^\omega_{ZS} - \gamma^q_{ZS})|_{|\boldsymbol{q}|=0}/d\Lambda_F = 0$ in singlet-even and triplet-even states. Thus, the Pauli principle sum rules and the new constraints are invariant under the RG flow. The coupled RG equations, (47) and (48) are nonperturbative. To lowest order, where δ in Eqs. (50) and (51) is neglected, these agree with the perturbative one-loop RG equations of Dupuis [27].

6. Fermi liquid parameters and tensor interactions

The aim of this section is to study whether phenomenological values for the Fermi liquid parameters are consistent with the sum rules as well as the constraints. For this purpose we approximate the quasiparticle interaction by the $l = 0$ and $l = 1$ terms. As additional input we take the phenomenological values for the scalar and isospin Fermi liquid parameters. The central spin and spin–isospin Fermi liquid parameters are then obtained from the sum rules and the constraints. By taking linear combinations of the sum rules, (17) and (18), and the constraints, Eqs. (38) and (39), the equations for G_l and G'_l decouple:

$$\sum_l\left\{\frac{F_l}{1+F_l/(2l+1)} + 3\frac{G'_l}{1+G'_l/(2l+1)}\right\} = 0, \tag{53}$$

$$\begin{aligned}\sum_l\bigg\{2F_l - \frac{F_l}{1+F_l/(2l+1)} - 2\left(2F'_l - \frac{F'_l}{1+F'_l/(2l+1)}\right)\\ - 3\left(2G'_l - \frac{G'_l}{1+G'_l/(2l+1)}\right)\bigg\}\\ = z^2\frac{16m_N k_F(1+F_1/3)}{\pi}V_{\text{low }k}\left(0,0;\Lambda = k_F, \frac{{}^1S_0 + 3\cdot{}^3S_1}{4}\right),\end{aligned} \tag{54}$$

$$\sum_l\left\{\frac{F_l}{1+F_l/(2l+1)} + \frac{3}{2}\frac{F'_l}{1+F'_l/(2l+1)} + \frac{3}{2}\frac{G_l}{1+G_l/(2l+1)}\right\} = 0, \tag{55}$$

$$\begin{aligned}\sum_l\bigg\{2F'_l - \frac{F'_l}{1+F'_l/(2l+1)} - \left(2G_l - \frac{G_l}{1+G_l/(2l+1)}\right)\bigg\}\\ = z^2\frac{16m_N k_F(1+F_1/3)}{4\pi}V_{\text{low }k}\left(0,0;\Lambda = k_F, {}^1S_0 - {}^3S_1\right).\end{aligned} \tag{56}$$

We note that the relevant input for the spin–isospin Fermi liquid parameters G'_l, Eq. (54), is the spin averaged s-wave low-momentum potential, whereas the one for the spin Fermi liquid parameters G_l, Eq. (56), is the difference of the spin-singlet and spin-triplet s-wave low-momentum potentials. Since the $^3S_1$ channel is only slightly more attractive than the $^1S_0$ channel, the right-hand side of Eq. (56) is small. Consequently, this constraint is not very sensitive to the precise value of the renormalization factor z^2.

The quasiparticle strength z was recently computed in a self-consistent description of the nucleon spectral functions. Roth [28] finds $z = 0.76$ at the Fermi surface for $k_F = 1.35\ \mathrm{fm}^{-1}$. However, there is a systematic uncertainty on the value of the z factor, since the relevance of experimental constraints from (e, e′p) knockout reactions on the jump in the occupation number at the Fermi surface is questionable. Furnstahl and Hammer [29] have recently shown that within the rigorous effective field theory for the interacting dilute Fermi gas the occupation numbers are not observable. We use $z = 0.8$ and for the nucleon effective mass at the saturation point we use $m^\star/m = 0.72$ corresponding to

$$F_1 = -0.85. \tag{57}$$

The empirical value for the anomalous orbital gyromagnetic ratio provides a constraint on F'_1. For a proton in the Pb region [30], $\delta g_l = 0.23 \pm 0.03$. In Fermi liquid theory [11] $\delta g_l = (1/3)(F'_1 - F_1)/(1 + F_1/3)$, which for $F_1 = -0.85$ yields

$$F'_1 = 0.14. \tag{58}$$

The incompressibility of nuclear matter is experimentally best constrained by the isoscalar giant monopole resonance and by fitting binding energies and the diffuseness of the nuclear surface. Microscopic calculations by Blaizot et al. [31] and Youngblood et al. [32] and the Thomas–Fermi equation of state of Myers and Swiatecki [33] give an incompressibility of $K = 230 \pm 20$ MeV. The Thomas–Fermi equation of state gives a symmetry energy of $E_{\mathrm{sym}} = 32.7$ MeV. The empirical value of the symmetry energy is limited by various fits to nuclear masses, resulting in $E_{\mathrm{sym}} = 31 \pm 5$ MeV [34]. Thus, we find

$$F_0 = -0.27, \qquad F'_0 = 0.71. \tag{59}$$

In units where $m = 1$, the matrix elements of the low-momentum nucleon–nucleon interaction are given by [2]

$$V_{\mathrm{low}\,k}\big(0, 0; \Lambda = k_F, {}^1S_0\big) = -1.95\ \mathrm{fm}, \tag{60}$$

$$V_{\mathrm{low}\,k}\big(0, 0; \Lambda = k_F, {}^3S_1\big) = -2.51\ \mathrm{fm}. \tag{61}$$

These are identical for the Bonn-A, Paris, and Argonne-V18 potential as well as a chiral model.

Very similar results are obtained by Feldmeier et al. [35,36], who introduce a unitary correlation operator including central and tensor correlations. For both the Bonn-A and Argonne-V18 potentials, they find $V_{\mathrm{UCOM}}(0, 0; {}^1S_0) = -1.88$ fm and $V_{\mathrm{UCOM}}(0, 0; {}^3S_1) = -2.86$ fm. The value in the $^3S_1$ channel depends on the range of the tensor correlations, which for the value quoted here is chosen to reproduce the d-state admixture when the uncorrelated deuteron trial wave function contains only an s-wave component. The dependence on the range of the tensor correlation corresponds to the cutoff dependence of $V_{\mathrm{low}\,k}(0, 0; \Lambda, {}^3S_1)$ discussed in the introduction.

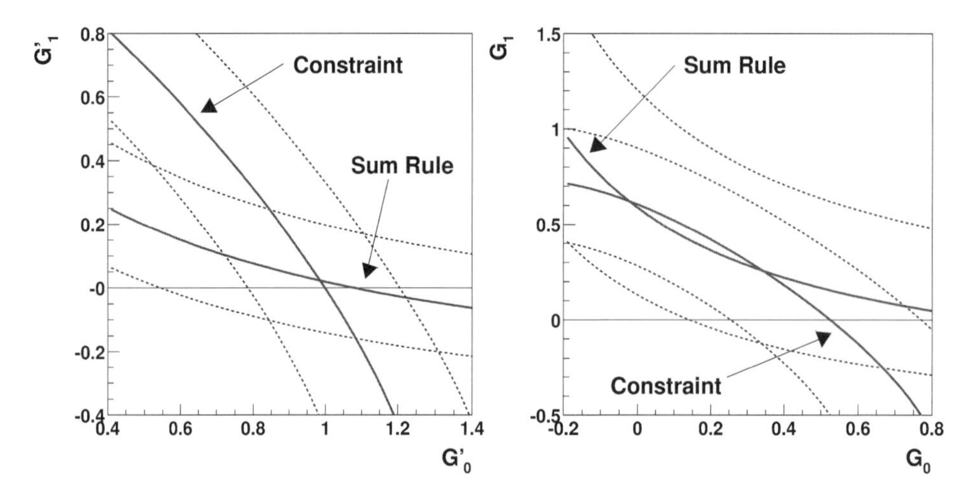

Fig. 2. The solution for the spin-dependent Fermi liquid parameters (solid lines) with error bands limited by the dashed lines. Here the effect of tensor parameters is neglected.

Epelbaoum et al. [37] constructed an effective potential from a s-wave Malfliet–Tjon type potential. The transformation method of Okubo used in their work is similar to the RG decimation employed for $V_{\text{low }k}$. They find $V_{\text{eff}}(0,0;{}^1\text{S}_0) = -1.94$ fm for a cutoff of $\Lambda = 300$ MeV. Their results are in a good agreement with $V_{\text{low }k}$.

In Fig. 2 we show the solution to the Eqs. (53)–(56) without tensor Fermi liquid parameters. In the error estimates we include the uncertainties in the input Fermi liquid parameters, the uncertainties due to the truncation of the Legendre series as well as the uncertainties in the driving term. The latter include only the estimated error of the renormalization factor z, since the effects of the neglected higher-order contributions are difficult to appraise. We thus find

$$G_0 = 0.15 \pm 0.3, \quad G_1 = 0.45 \pm 0.3, \tag{62}$$

$$G_0' = 1.0 \pm 0.2, \quad G_1' = 0 \pm 0.2. \tag{63}$$

The relative errors of G_0 and G_1 are large, because the corresponding bands are almost parallel. Nevertheless, this calculation demonstrates that $V_{\text{low }k}$ is a very promising starting point for calculations of Fermi liquid parameters.

The mean value of $G_0' = 1$ should be confronted with the experimental constraints imposed by the energy of the giant Gamow–Teller resonance. Since the Fermi liquid parameters embody the effective interaction in the nucleon subspace, the empirical value of g'_{NN}, obtained in a model that includes Δ-isobar degrees of freedom, must be corrected for the screening due to Δ–hole excitations. Including Δ–hole excitations to all orders, we find

$$G_0' = N(0)\frac{f_{\pi\text{NN}}^2}{m_\pi^2}\left\{ g'_{\text{NN}} - \frac{\dfrac{f_{\pi\text{N}\Delta}^2}{m_\pi^2} g'^2_{\text{N}\Delta}\dfrac{8}{9}\dfrac{\rho_0}{m_\Delta - m_\text{N}}}{1 + \dfrac{f_{\pi\text{N}\Delta}^2}{m_\pi^2} g'_{\Delta\Delta}\dfrac{8}{9}\dfrac{\rho_0}{m_\Delta - m_\text{N}}} \right\}, \tag{64}$$

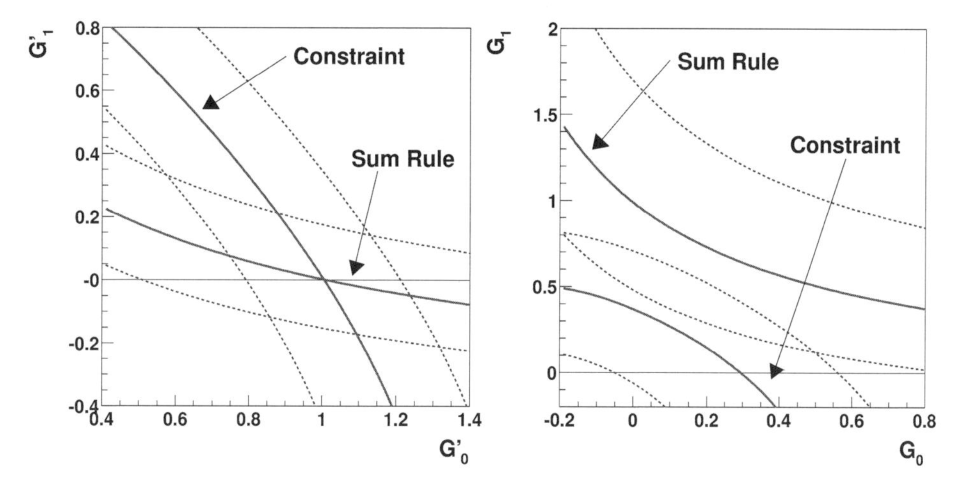

Fig. 3. The same as Fig. 2 but including tensor interactions.

where ρ_0 denotes the nuclear matter density. Furthermore, g'_{NN} is the short-range part of the spin–isospin-dependent effective nucleon–nucleon interaction in pionic units, while $g'_{\mathrm{N}\Delta}$ and $g'_{\Delta\Delta}$ are the corresponding $\mathrm{NN} \to \mathrm{N}\Delta$ and $\mathrm{N}\Delta \to \Delta\mathrm{N}$ interaction strengths. Kawahigashi et al. [38] find $g'_{\mathrm{NN}} = 0.6$, $g'_{\mathrm{N}\Delta} = 0.3$, while Körfgen et al. [39,40] obtain $g'_{\mathrm{N}\Delta} = 0.3$, $g'_{\Delta\Delta} = 0.3$. Using these values, we find $G'_0 = 1.0$ in good agreement with our result. The Δ–hole polarization reduces the value of G'_0 by about 10%.

The discussion presented above is easily generalized to include the effects of the tensor force. For $\boldsymbol{p} = \boldsymbol{p}'$ the tensor components of the quasiparticle interaction $\mathcal{F}$, Eq. (3), and the quasiparticle scattering amplitude $\mathcal{A}$ vanish and the tensor force enters only together with the spin-dependent parameters in the $\boldsymbol{\sigma} \cdot \boldsymbol{\sigma}'$ and $\boldsymbol{\tau} \cdot \boldsymbol{\tau}'\boldsymbol{\sigma} \cdot \boldsymbol{\sigma}'$ components of the scattering amplitude [13]. The coupling of spin and Landau l to good total angular momentum J was carried out by Bäckman et al. [17]. We use the tensor Fermi liquid parameters obtained from a G-matrix calculation using Reid's soft core potential for $l \leqslant 4$ and from the one pion exchange potential for higher l (see Table 2 of [17]). To account for the effects of tensor forces, we replace the spin-dependent parameters of the scattering amplitude $C_l = G_l/(1 + G_l/(2l + 1))$ and $C'_l = G'_l/(1 + G'_l/(2l + 1))$ in the constraints up to $l = 4$ with the corresponding expressions including tensor interactions [13]. Since the expansion of the tensor interaction in Landau l is poorly convergent, we include terms up to $l = 4$. We have checked that the contributions of higher l are negligible. We note that the tensor parameters of Ref. [17] are given for $z = 1$ and $m^\star/m = 1$. Consequently, these parameters should be reduced by the factor $z^2 m^\star/m$.

In Fig. 3 we show the solution to the Eqs. (53)–(56) including tensor Fermi liquid parameters. We have not included errors for the tensor parameters. Since the isospin tensor parameters H'_l are small (one-pion and one-rho exchange yields $H'_l = -H_l/3$ [16]), the solution for G'_0 and G'_1 in the spin–isospin sector is basically unaffected by the presence of tensor interactions. However, the solution for the spin Fermi liquid parameters G_0 and G_1 is strongly modified. In fact, the error bands overlap only in a small region, when we include tensor interactions. The reason is that in this sector the tensor parameters of Ref. [17] are quite large. We note that this may change, when the contribution of the induced interaction to the tensor parameters is included. In order to illustrate the possible effects of this type, we compute the leading contribution to the tensor Fermi liquid parameters from the one bubble polarization in the induced interaction using the one-pion exchange interaction. The lowest-order contribution to the tensor Fermi liquid parameters H_l from the one-pion exchange driving term, Eq. (19), is given by

$$H(\theta) = N(0) z^2 \frac{f^2}{3m_\pi^2} \frac{3}{2} \frac{k_F^2}{(\boldsymbol{p}-\boldsymbol{p}')^2 + m_\pi^2}. \tag{65}$$

The dominant tensor contribution from the one bubble term in the induced interaction is obtained by employing the direct tensor part π_T of the one-pion exchange potential as vertices in the induced interaction. This corresponds to the first and in part the third and the last diagrams of Eq. (27). More explicitly, we compute the diagrams

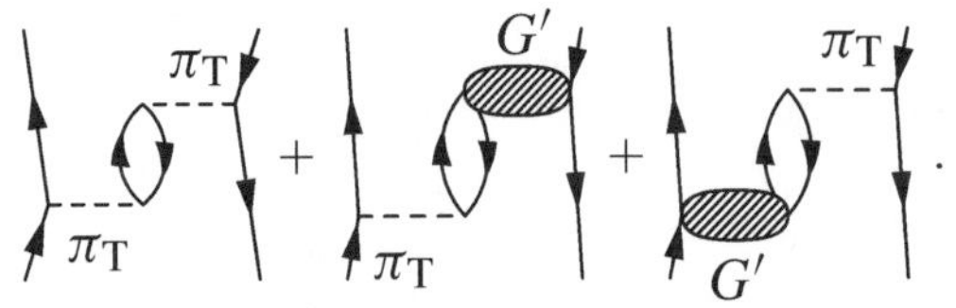

π_T π_T + G' π_T + π_T G' . (66)

For the long-range part of G' we include the momentum dependence by splitting the interaction into a one-pion exchange piece and a short-ranged piece:[8]

$$G' = N(0)\, z^2 \frac{f^2}{3m_\pi^2}\left(\frac{m_\pi^2}{\boldsymbol{q}^2 + m_\pi^2} + \Delta g'\right). \tag{67}$$

Using Eqs. (19) and (25) we then find

$$\begin{aligned}
&\Gamma^{\mathrm{ind(2)dir.OPE}}_{\boldsymbol{\sigma}\cdot\boldsymbol{\sigma}',\boldsymbol{\tau}\cdot\boldsymbol{\tau}'}(p,p') \\
&\quad = -\frac{1}{4}(1+\boldsymbol{\sigma}\cdot\boldsymbol{\sigma}')(3-\boldsymbol{\tau}\cdot\boldsymbol{\tau}')N(0)z^2\left(\frac{f^2}{3m_\pi^2}\right)^2 \\
&\quad\times \frac{1}{2}\mathrm{Tr}_{\sigma''}\int \frac{d\Omega_{p''}}{4\pi}(\boldsymbol{p}-\boldsymbol{p}')^2 \frac{S_{12''}(\boldsymbol{p}-\boldsymbol{p}')}{(\boldsymbol{p}-\boldsymbol{p}')^2+m_\pi^2}\left\{(\boldsymbol{p}-\boldsymbol{p}')^2\frac{S_{2''2}(\boldsymbol{p}-\boldsymbol{p}')}{(\boldsymbol{p}-\boldsymbol{p}')^2+m_\pi^2}\right. \\
&\quad\left. -2\frac{m_\pi^2\boldsymbol{\sigma}''\cdot\boldsymbol{\sigma}'}{(\boldsymbol{p}-\boldsymbol{p}')^2+m_\pi^2} - 6\Delta g'\right\}\frac{v_F\hat{\boldsymbol{p}}''\cdot(\boldsymbol{p}-\boldsymbol{p}')}{\varepsilon-\varepsilon'-v_F\hat{\boldsymbol{p}}''\cdot(\boldsymbol{p}-\boldsymbol{p}')}.
\end{aligned} \tag{68}$$

The integral over $\Omega_{p''}$ yields the Lindhard function $-\alpha_0(q'/k_F)$. By exploiting the following identities for the tensor operator,

$$\frac{1}{2}\mathrm{Tr}_{\sigma''}\, S_{12''}(\boldsymbol{p}-\boldsymbol{p}')S_{2''2}(\boldsymbol{p}-\boldsymbol{p}') = S_{12}(\boldsymbol{p}-\boldsymbol{p}') + 2\boldsymbol{\sigma}\cdot\boldsymbol{\sigma}', \tag{69}$$

$$\frac{1}{2}\mathrm{Tr}_{\sigma''}\, S_{12''}(\boldsymbol{p}-\boldsymbol{p}')\boldsymbol{\sigma}''\cdot\boldsymbol{\sigma}' = S_{12}(\boldsymbol{p}-\boldsymbol{p}'), \tag{70}$$

$$\frac{1}{2}(1+\boldsymbol{\sigma}\cdot\boldsymbol{\sigma}')S_{12}(\boldsymbol{p}-\boldsymbol{p}') = S_{12}(\boldsymbol{p}-\boldsymbol{p}'), \tag{71}$$

[8] This is justified since the q dependence of the exchanged heavy mesons is weak.

we finally arrive at the second order correction to the tensor Fermi liquid parameters

$$\Delta H(\theta) = H(\theta) N(0)\, z^2 \frac{f^2}{3m_\pi^2} \alpha_0(q'/k_F) \times \left\{ \frac{(\boldsymbol{p}-\boldsymbol{p}')^2}{(\boldsymbol{p}-\boldsymbol{p}')^2+m_\pi^2} - 2\frac{m_\pi^2}{(\boldsymbol{p}-\boldsymbol{p}')^2+m_\pi^2} - 6\Delta g' \right\}. \tag{72}$$

In order to reproduce the empirical value for $G_0' = 1$ with the direct one-pion exchange contribution plus $\Delta g'$, we need $\Delta g' = 0.5$.[9] The resulting corrections to the tensor Fermi liquid parameters $H_0 = 0.35$ and $H_1 = 0.43$ are $\Delta H_0 = -0.40$ and $\Delta H_1 = -0.69$. Thus, we find that the induced interaction tends to reduce the tensor Fermi liquid parameters, in agreement with the results of Dickhoff et al. [41]. The very large effects show that the tensor interactions must be treated self-consistently within the induced interaction. Finally, we note that about 60% of the left-hand side of Eq. (54) is due to the Landau parameter G_0'. Thus, there is a close connection between the spin-averaged s-wave low-momentum interaction $V_{\text{low }k}(0,0;\Lambda = k_F)$ and the local spin–isospin-dependent part of the quasiparticle interaction.

7. Summary and conclusions

In this paper we presented two new algebraic constraints that relate the Landau Fermi liquid parameters in nuclear matter to the driving term of the induced interaction. By identifying the driving term with the s-wave low-momentum nucleon–nucleon interaction $V_{\text{low }k}$ at $\Lambda = k_F$, including some straightforward in-medium effects, we obtained an intriguing relation between the effective interaction *in vacuum* and *in nuclear matter*.

The resulting constraints on the Fermi liquid parameters were used in conjunction with the Pauli principle sum rules to compute the major spin-dependent parameters, given the phenomenological values for the spin-independent parameters. We find good agreement with empirically determined parameters. The present calculation indicates that a good approximation to the driving term of the induced interaction can be obtained from $V_{\text{low }k}$ in a straightforward manner, by including minimal in-medium corrections, the wave-function renormalization factors and the nucleon effective mass in the density of states. A full calculation of the induced interaction, including a self-consistent treatment would be needed to firmly establish this identification. In such a calculation, the spin, isospin and velocity dependence of $V_{\text{low }k}$ would be reflected in the corresponding Fermi liquid parameters. A comparison with empirical parameters would then provide a test of, e.g., the velocity dependence of the low-momentum nucleon–nucleon interaction $V_{\text{low }k}$. In $V_{\text{low }k}$, the role of the (local) short-range repulsion of the bare interactions is taken over by a nonlocality, which interpolates between a weak repulsion at low energies and a stronger one at higher energies.

[9] Note that the value of g'_{NN} here is larger than in [38], because we use $z < 1$. The physics is determined by G_0', not by g'_{NN}.

The effects of tensor forces are also studied, using the tensor parameters obtained in a G-matrix calculation. We find a fairly large effect of the tensor force on the isoscalar spin-dependent parameters. However, as indicated by a simple estimate, this effect will probably be reduced when the tensor parameters are computed self-consistently by including the tensor force in the induced interaction.

Moreover, we derive the flow equations for the renormalization group decimation of the quasiparticle scattering amplitude and the quasiparticle interaction in the two particle–hole channels starting from the induced interaction. A solution of these equations would provide the scattering amplitude also for nonforward scattering, which is of high interest for the calculation of superfluid gaps and transport processes, e.g., in neutron star interiors. In condensed matter systems an ab initio RG analysis of this type [42] applied to the 2D Hubbard model has successfully established the existence of d-wave superconductivity. The RG equations for the quasiparticle scattering amplitude and the quasiparticle interaction we obtained from the induced interaction are nonperturbative. Existing RG studies in Fermi systems have been restricted to one-loop approximations.

Acknowledgement

We thank Scott Bogner and Tom Kuo for helpful discussions. A.S. acknowledges the kind hospitality of the Theory Group at GSI, where part of this work was carried out. The work of A.S. and G.E.B. was supported by the US-DOE grant No. DE-FG02-88ER40388.

References

[1] S.K. Bogner, T.T.S. Kuo, L. Coraggio, nucl-th/9912056.
[2] S.K. Bogner, T.T.S. Kuo, A. Schwenk, D.R. Entem, R. Machleidt, nucl-th/0108041.
[3] H.A. Bethe, Annu. Rev. Nucl. Sci. 21 (1971) 93.
[4] G.E. Brown, Unified Theory of Nuclear Models and Forces, third edition, Amsterdam, North-Holland, 1971.
[5] S.K. Bogner, T.T.S. Kuo, L. Coraggio, A. Covello, N. Itaco, nucl-th/0108040.
[6] M.C. Birse, J.A. McGovern, K.G. Richardson, Phys. Lett. B 464 (1999) 169.
[7] T. Barford, M.C. Birse, Talk presented at Mesons and Light Nuclei, Prague, 2001, nucl-th/0108024.
[8] S.K. Bogner, A. Schwenk, T.T.S. Kuo, G.E. Brown, nucl-th/0111042.
[9] S. Babu, G.E. Brown, Ann. Phys. 78 (1973) 1.
[10] L.D. Landau, Sov. Phys. JETP 3 (1957) 920;
L.D. Landau, Sov. Phys. JETP 5 (1957) 101;
L.D. Landau, Sov. Phys. JETP 8 (1959) 70.
[11] A.B. Migdal, Theory of Finite Fermi Systems and Applications to Atomic Nuclei, Interscience, London, 1967.
[12] A.A. Abrikosov, L.P. Gor'kov, I.E. Dzyaloshinski, Methods of Quantum Field Theory in Statistical Physics, Dover, New York, 1963.
[13] B.L. Friman, A.K. Dhar, Phys. Lett. B 85 (1979) 1.
[14] O. Sjöberg, Ann. Phys. 78 (1973) 39.
[15] N.D. Mermin, Phys. Rev. 159 (1967) 161.
[16] S.-O. Bäckman, G.E. Brown, J.A. Niskanen, Phys. Rep. 124 (1985) 1.
[17] S.-O. Bäckman, O. Sjöberg, A.D. Jackson, Nucl. Phys. A 321 (1979) 10.
[18] J. Wambach, T.L. Ainsworth, D. Pines, Nucl. Phys. A 555 (1993) 128.
[19] H.-J. Schulze, J. Cugnon, A. Lejeune, M. Baldo, U. Lombardo, Phys. Lett. B 375 (1996) 1.

[20] T.L. Ainsworth, K.S. Bedell, G.E. Brown, K.F. Quader, J. Low Temp. Phys. 50 (1983) 317.
[21] K.S. Bedell, T.L. Ainsworth, Phys. Lett. A 102 (1984) 49.
[22] T.L. Ainsworth, K.S. Bedell, Phys. Rev. B 35 (1987) 8425.
[23] A. Lande, R.A. Smith, Phys. Rev. A 45 (1992) 913.
[24] S.-O. Bäckman, Nucl. Phys. A 120 (1968) 593;
S.-O. Bäckman, Nucl. Phys. A 130 (1969) 427.
[25] R. Shankar, Rev. Mod. Phys. 66 (1994) 129.
[26] J. Polchinski, in: J. Harvey, J. Polchinski (Eds.), Proc. of the 1992 Theoretical Advanced Studies Institute in Elementary Particle Physics, World Scientific, Singapore, 1993.
[27] N. Dupuis, Eur. Phys. J. B 3 (1998) 315.
[28] E.P. Roth, Self-Consistent Green's Functions in Nuclear Matter, PhD thesis, Washington University, St. Louis, 2000.
[29] R.J. Furnstahl, H.-W. Hammer, nucl-th/0108069.
[30] R. Nolte, A. Baumann, K.W. Rose, M. Schumacher, Phys. Lett. B 173 (1986) 388.
[31] J.P. Blaizot, J.F. Berger, J. Dechargé, M. Girod, Nucl. Phys. A 591 (1995) 435.
[32] D.H. Youngblood, H.L. Clark, Y.-W. Lui, Phys. Rev. Lett. 82 (1999) 691.
[33] W.D. Myers, W.J. Swiatecki, Nucl. Phys. A 601 (1996) 141;
W.D. Myers, W.J. Swiatecki, Phys. Rev. C 57 (1998) 3020.
[34] P.E. Haustein, At. Data Nucl. Data Tables 39 (1988) 185.
[35] T. Neff, PhD thesis, Darmstadt University of Technology, 2001, in preparation.
[36] H. Feldmeier, T. Neff, R. Roth, J. Schnack, Nucl. Phys. A 632 (1998) 61.
[37] E. Epelbaoum, W. Glöckle, A. Krüger, Ulf-G. Meissner, Nucl. Phys. A 645 (1999) 413.
[38] K. Kawahigashi, K. Nishida, A. Itabashi, M. Ichimura, Phys. Rev. C 63 (2001) 044609.
[39] B. Körfgen, F. Osterfeld, T. Udagawa, Phys. Rev. C 50 (1994) 1637.
[40] B. Körfgen, P. Oltmanns, F. Osterfeld, T. Udagawa, Phys. Rev. C 55 (1997) 1819.
[41] W.H. Dickhoff, A. Faessler, H. Muether, S.-S. Wu, Nucl. Phys. A 405 (1983) 534.
[42] C.J. Halboth, W. Metzner, Phys. Rev. B 61 (2000) 7364;
C.J. Halboth, W. Metzner, Phys. Rev. Lett. 85 (2000) 5162.

Available online at www.sciencedirect.com

PHYSICS REPORTS

Physics Reports 386 (2003) 1–27

www.elsevier.com/locate/physrep

Model-independent low momentum nucleon interaction from phase shift equivalence

S.K. Bogner[a,b,*], T.T.S. Kuo[b], A. Schwenk[c,b]

[a] *Institute for Nuclear Theory, Box 351550, University of Washington, Seattle, WA 98195, USA*
[b] *Department of Physics and Astronomy, State University of New York, Stony Brook, NY 11794-3800, USA*
[c] *Department of Physics, The Ohio State University, Columbus, OH 43210, USA*

Accepted 10 July 2003
editor: G.E. Brown

Abstract

We present detailed results for the model-independent low momentum nucleon–nucleon interaction $V_{\mathrm{low}\,k}$. By introducing a cutoff in momentum space, we separate the Hilbert space into a low momentum and a high momentum part. The renormalization group is used to construct the effective interaction $V_{\mathrm{low}\,k}$ in the low momentum space, starting from various high precision potential models commonly used in nuclear many-body calculations. With a cutoff in the range of $\Lambda \sim 2.1\ \mathrm{fm}^{-1}$, the new potential $V_{\mathrm{low}\,k}$ is independent of the input model, and reproduces the experimental phase shift data for corresponding laboratory energies below $E_{\mathrm{lab}} \sim 350$ MeV, as well as the deuteron binding energy with similar accuracy as the realistic input potentials. The model independence of $V_{\mathrm{low}\,k}$ demonstrates that the physics of nucleons interacting at low momenta does not depend on details of the high momentum dynamics assumed in conventional potential models. $V_{\mathrm{low}\,k}$ does not have momentum components larger than the cutoff, and as a consequence is considerably softer than the high precision potentials. Therefore, when $V_{\mathrm{low}\,k}$ is used as microscopic input in the many-body problem, the high momentum effects in the particle–particle channel do not have to be addressed by performing a Brueckner ladder resummation or short-range correlation methods. By varying the cutoff, we study how the model independence of $V_{\mathrm{low}\,k}$ is reached in different partial waves. This provides numerical evidence for the separation of scales in the nuclear problem, and physical insight into the nature of the low momentum interaction.

PACS: 13.75.Cs; 21.30. – x; 11.10.Hi

Keywords: Nucleon–nucleon interactions; Effective interactions; Renormalization group

* Corresponding author. Institute for Nuclear Theory, Box 351550, University of Washington, Seattle, WA 98195, USA.
E-mail addresses: bogner@phys.washington.edu (S.K. Bogner), kuo@nuclear.physics.sunysb.edu (T.T.S. Kuo), aschwenk@mps.ohio-state.edu (A. Schwenk).

doi:10.1016/j.physrep.2003.07.001

Contents

1. Introduction

A major challenge of nuclear physics lies in the description of finite nuclei and nuclear matter on the basis of a microscopic theory. By microscopic theory, it is understood that the input nuclear force is, as best as possible, based on the available free-space data, e.g., the elastic nucleon–nucleon and nucleon–deuteron scattering phase shifts. The theoretical predictions are then obtained using a systematic many-body approach, which is reliable for strongly interacting systems.

At low energies and densities, one typically starts by treating protons and neutrons as nonrelativistic point-like fermions. However, as opposed to the Coulomb interaction in electronic systems, the nucleon–nucleon interaction remains unknown from the underlying theory of the strong interactions, quantum chromodynamics (QCD). Thus, phenomenological meson-exchange models of the two-nucleon force are commonly used as input in many-body calculations. Moreover, three- (and possibly four-) body forces are needed to obtain accurate results for few-body systems. Recent Green's function Monte Carlo calculations using a realistic two-body interaction, the Argonne v_{18} potential, have shown that three-body forces typically contribute a net attraction of $\approx 20\%$ to the low-lying states of light nuclei, $A \leqslant 10$ [1,2].

However, already at the lowest level of the two-body interaction, there are several quite different potential models commonly used. Therefore, it is a natural first step to ask whether it is possible to remove or to minimize this model dependence. In this work, we expand on previous results [3,4] and demonstrate that this is indeed possible, as long as one restricts the interaction to momenta which are constrained by the experimental scattering data, and explicitly includes the long-range part of the interaction.

There are number of high precision models of the nucleon–nucleon force $V_{\rm NN}$, which we will refer to as the "bare" interactions. All realistic potentials are given by the one-pion exchange (OPE) interaction at long distances, but vary substantially in their treatment of the intermediate-range attraction and the short-range repulsion. We briefly summarize the different frameworks used to generate the intermediate and short-range parts of the realistic potential models [5,6]. The Paris potential

[7] calculates the intermediate-range contributions to the nucleon–nucleon scattering amplitude from two-pion exchange using dispersion theory, and then assumes a local potential representation consisting of several static Yukawa functions. The ω-meson exchange at short distances is included as part of the three-pion exchange, and a repulsive core is introduced by sharply cutting off these parts at the internucleon distance $r \sim 0.8$ fm. At short distances, the interaction is simply given by a constant (but energy-dependent) soft core. The Bonn potentials [8] are based on multiple one-boson exchange interactions and the two-pion exchange potential calculated in perturbation theory. The two-pion exchange contribution is then approximated by an energy-independent σ-meson exchange term. Smooth form factors (with typical form factor masses $\approx$1–2 GeV) cut off the potential at short distances. The short-range repulsion originates from the ω exchange. The second generation CD-Bonn potential [9,10] refines the treatment of the one-boson exchange amplitudes to take into account the non-local structure arising from the covariant amplitudes. The Nijmegen potentials [11] consist of multiple one-boson exchange parts (with separate local and non-local versions), where the interaction parameters of the heavier mesons depend on the partial waves. At very short distances, the potentials are regularized by exponential form factors. Finally, the local Argonne potential [12] consists of the OPE interaction regularized at short distances, and a phenomenological parametrization at short and intermediate distances. The core is provided by Woods–Saxon functions that are effective at a distance $r \sim 0.5$ fm.

Over the past decade, there has also been a great effort in deriving low-energy nucleon–nucleon interactions in the framework of effective field theory (EFT) [13–19] (for a review see [20]). The EFT approach is based on local Lagrangian field theory with low-energy degrees of freedom, constrained by the symmetries of QCD. The Lagrangian contains the nucleon and pion fields and all possible interactions consistent with chiral symmetry. At low energies, the heavy mesons and nucleon resonances can be integrated out of the theory. Their effects are encoded in the renormalized pion exchange and scale-dependent coupling constants of model-independent contact interactions. By formulating a power counting scheme, the number of couplings in the effective Lagrangian can be truncated and fitted to a set of low-energy data. Once the low-energy couplings are determined, the power counting scheme enables the EFT to predict other processes with controlled error estimates.

In addition to the high precision phenomenological interactions, we study the Idaho potential [19], which is a chiral model inspired by EFT. The Idaho potential introduces some model dependence in contrast to the rigorous EFT interactions by, e.g., including several terms in the expansion which are subleading in the power counting with respect to omitted ones. This leads to a better description of the experimental phase shifts, comparable in accuracy to the realistic models. Since we will argue that our results are dependent on the accurate reproduction of the scattering phase shifts, the results for the Idaho EFT potential are presented here as well.

The couplings and parameters of the different potentials described above are fitted to the elastic nucleon–nucleon scattering phase shifts over laboratory energies $E_{\rm lab} \lesssim 350$ MeV and the low-energy deuteron properties. As a consequence, the bare interactions are constrained by the experimental data up to a corresponding relative momentum scale of[1]

$$\Lambda_{\rm data} \sim 2.1 \text{ fm}^{-1} . \qquad (1)$$

[1] Here, we use conventional scattering units where $c = \hbar = \hbar^2/m = 1$, i.e., the laboratory energy is related to the relative momentum through $E_{\rm lab} = 2k^2$.

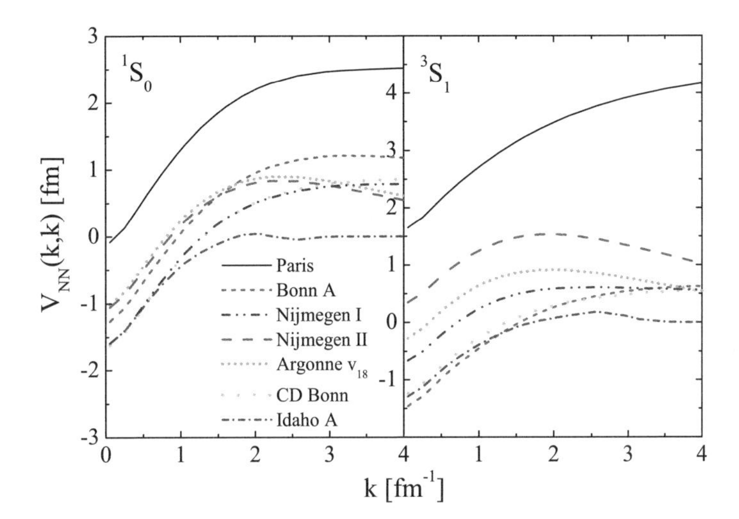

Fig. 1. The diagonal momentum-space matrix elements of the different high precision potentials V_{NN} versus relative momentum in the $^1S_0$ and $^3S_1$ partial wave.

This is manifest in Fig. 1, where we observe that various high precision potential models of V_{NN} have quite different momentum-space components, despite their common treatment of the OPE interaction and the reproduction of the same low-energy data. This indicates that the low-energy observables are not sensitive to the details or different assumptions of the short-distance physics. This insensitivity is a consequence of the separation of long- and short-distance scales in the nuclear force, and implies that the nucleon–nucleon interaction is amenable to an effective theory or renormalization group treatment.

1.1. The two-body problem in effective theories

In this work, we propose a renormalization group (RG) approach to the nucleon–nucleon interaction which removes the short-distance model dependencies of the high precision potentials, while preserving their high accuracy description of the nucleon–nucleon scattering data and deuteron properties. As our approach is a compromise between an EFT treatment and the high precision potentials, we first discuss in general terms how effective theory methods can be applied to the two-nucleon problem.

The fundamental principle underlying modern effective theory methods is simple: the low-energy physics is not sensitive to the details of the high-energy dynamics. In other words, one can construct arbitrarily many theories, which have the same long-wavelength structure and lead to identical low-energy observables, but differ at short distances and higher-energy scales. This ambiguity can be used constructively. Namely, it is possible to replace the detailed short-distance dynamics by simple effective interactions, which preserve the low-energy symmetries of the full theory.

For example, a forerunner to the modern EFT approach is the well-known effective range theory of Bethe and Longmire [21]. This allowed a quantitative description of low-energy nuclear phenomena without a detailed knowledge of the nuclear force. The low-energy scattering amplitude $T(k,k;k^2)$ is given by two phenomenological low-energy constants independent of an assumed nuclear force model [22]

$$\frac{1}{T(k,k;k^2)} = \frac{1}{a_s} - \frac{1}{2} r_0 k^2 + \mathcal{O}(k^4) , \tag{2}$$

where a_s denotes the scattering length and r_0 is the effective range parameter. The effective range theory encompasses the fundamental principle of modern effective theories, as any force models with different short-distance details can be tuned to give the same low-energy physics, which is captured in the effective range parameters.

More generally, this insensitivity can be exploited when the short-distance dynamics are either poorly understood, or are too complicated for calculation. Consider a quantum system described by the Hamiltonian

$$H = H_0 + V_L + V_H , \tag{3}$$

where V_L primarily couples to low-energy states, in our case the long-range part of the OPE interaction, and V_H includes the remaining complicated or unknown short-distance part of the interaction.

A first step in building an effective theory is to impose a cutoff Λ on the intermediate state energies and momenta. In the modern viewpoint, the cutoff takes on a physical meaning. It divides the low-energy states, which are essential to the low-energy physics, from the high-energy modes. In this way, one explicitly maintains only dynamics that are well understood. It is, however, not possible to simply neglect the remaining V_H, since its effects on the low-energy physics need to be included in the form of correction, or so-called counterterms. The correction terms are constrained by demanding that all low-energy spectra, low-energy amplitudes, etc., calculated using the bare Hamiltonian H are exactly reproduced using an effective Hamiltonian $H_{\text{low}\,k}$ in the truncated Hilbert space.

Schematically, in perturbation theory all amplitudes are of the form

$$\langle f|\mathscr{A}|i\rangle = \langle f|V_{\text{bare}}|i\rangle + \sum_{n=0}^{\infty} \frac{\langle f|V_{\text{bare}}|n\rangle\langle n|V_{\text{bare}}|i\rangle}{E_i - E_n} + \mathcal{O}(V_{\text{bare}}^3) , \tag{4}$$

where the intermediate state summations are over both low and high-energy states. We now define an effective low momentum potential $V_{\text{low}\,k}$ by demanding

$$\langle f|\mathscr{A}|i\rangle = \langle f|V_{\text{low}\,k}|i\rangle + \sum_{n=0}^{\Lambda} \frac{\langle f|V_{\text{low}\,k}|n\rangle\langle n|V_{\text{low}\,k}|i\rangle}{E_i - E_n} + \mathcal{O}(V_{\text{low}\,k}^3) , \tag{5}$$

where we introduce the correction terms through

$$V_{\text{low}\,k} = V_L + \delta V_{\text{ct}} . \tag{6}$$

The physical amplitudes cannot depend on the way in which we split the Hilbert space and must therefore be independent of the cutoff. Consequently, $V_{\text{low}\,k}$ changes or "runs" with the cutoff in order to cancel the cutoff dependence arising from the truncated intermediate state summations. This means that the effective theory must be RG invariant. Demanding RG invariance implies a RG equation for the low momentum interaction

$$\frac{d}{d\Lambda}\langle f|\mathscr{A}|i\rangle = 0 \;\Rightarrow\; \frac{d}{d\Lambda}V_{\mathrm{low}\,k} = \beta([V_{\mathrm{low}\,k}],\Lambda)\ , \tag{7}$$

where the β function of the problem depends on the cutoff-dependent low momentum potential, as well as explicit cutoff dependence from the regularization. Starting from a bare interaction defined in the full Hilbert space, i.e., with a large cutoff, the RG equation can be solved to obtain the physically equivalent effective potential. This process is referred to as integrating out, or decimating the high-energy degrees of freedom. It can be viewed as filtering out the details of the assumed high-energy dynamics, while implicitly incorporating their detail-independent effects on the low-energy physics.

Since the virtual high-energy states propagate over distances of the order $1/\Lambda$ or shorter, the correction terms are well approximated by contact interactions, regardless of the detailed form of V_{H}. The general form of δV_{ct} should therefore be given by a sum over all local contact operators consistent with the symmetries of the problem. In the specific case of the two-body system, we have schematically (for an excellent discussion of the details of the approach, see [23]):

$$\begin{aligned}\delta V_{\mathrm{ct}}(\mathbf{r}) &= C_0\,\frac{1}{\Lambda^2}\,\delta^3(\mathbf{r}) + C_2\,\frac{1}{\Lambda^4}\,\nabla^2\,\delta^3(\mathbf{r}) + C_2'\,\frac{1}{\Lambda^4}\,\nabla\cdot\delta^3(\mathbf{r})\,\nabla + \cdots \\ &\quad + C_{2n}\,\frac{1}{\Lambda^{2n+2}}\,\nabla^{2n}\,\delta^3(\mathbf{r}) + \ldots\end{aligned} \tag{8}$$

with dimensionless couplings C_{2n}. To a low-energy probe, the effective theory is indistinguishable from the underlying theory, provided the decimation has been performed exactly.

The power counting scheme of the EFT enables one to reliably truncate the correction terms, Eq. (8), and fit the couplings directly to a set of low-energy data, since a RG decimation of the underlying fundamental theory is at present not feasible. In our current approach, we will assume that the nucleon–nucleon interaction in the full Hilbert space is given by a high precision potential model, and then perform the RG decimation to low momenta exactly. In this way, the resulting low momentum potential contains all necessary counterterms to maintain exact RG invariance of the theory, and therefore reproduces the experimental phase shift data and the deuteron properties over the same kinematic range as the conventional models.

1.2. A schematic model

Before considering realistic nucleon–nucleon potential models, we apply the RG approach to a simplified schematic model that allows for a straightforward solution of the RG equation to construct $V_{\mathrm{low}\,k}$. In addition, the schematic model nicely illustrates the main point of our current approach. Independent of the schematic model constructed in the full Hilbert space, the RG decimation leads to a model-independent low momentum interaction $V_{\mathrm{low}\,k}$.

We consider a separable interaction, which we construct to approximately fit the experimental neutron–proton phase shift data in the $^1\mathrm{S}_0$ channel. For this purpose, we found the following separable

Table 1
The parameter sets of two separable models of type Eq. (10)

	I	II
n	1/2	1
α	1.8199	4.5354
g ($\mathrm{fm}^{(1-4n)/2}$)	3.1429	3.3542
m (fm^{-1})	1.0962	1.3892

interaction to be particularly useful [2]

$$V_{\mathrm{bare}}(k',k) = L(k')R(k) \tag{9}$$

$$= \frac{g}{(k'^2+m^2)^n}\left(-\frac{g}{(k^2+m^2)^n} + \frac{\alpha g(k^2+m^2)^n}{(k^2+\eta^2 m^2)^{2n}}\right) . \tag{10}$$

For the diagonal matrix elements, the analogy to the nucleon–nucleon interaction can be seen better

$$V_{\mathrm{bare}}(k,k) = -\frac{g^2}{(k^2+m^2)^{2n}} + \frac{\alpha g^2}{(k^2+\eta^2 m^2)^{2n}} . \tag{11}$$

The separable model consists of an attractive part and a short-range, $\eta > 1$, repulsive part. The strength of the repulsion, for given η, can be constructed to fit the $^1S_0$ phase shift data reasonably.

For separable potentials, the Lippmann–Schwinger equation can be resummed explicitly and one finds [22]

$$T_{\mathrm{bare}}(k',k;k^2) = \frac{L(k')R(k)}{1 + (2/\pi)\mathscr{P}\int_0^\infty [L(p)R(p)/(p^2-k^2)]p^2\,\mathrm{d}p} . \tag{12}$$

We choose a typical $\eta = 2$, which approximately reflects the ratio of the repulsive to attractive meson contributions $m_\omega/2m_\pi$. The remaining parameters of our model, g, m and α, are then fitted to reproduce the effective range expansion,

$$T_{\mathrm{bare}}(k,k;k^2) = \frac{1}{1/a_s - \frac{1}{2}r_0 k^2} \tag{13}$$

and the change in sign of the experimental phase shifts $T_{\mathrm{bare}}(\kappa,\kappa;\kappa^2) = 0$ at $\kappa = 1.79\ \mathrm{fm}^{-1}$ for a given power n. We constructed two exemplary separable models with the parameters given in Table 1. The separable models are fitted to a neutron–proton $^1S_0$ scattering length, $a_s = -23.73$ fm, and effective range $r_0 = 2.70$ fm.

In Fig. 2, we show the phase shifts calculated from the models of V_{bare}. We observe that the separable models I and II can achieve a reasonably realistic description of the nucleon–nucleon phase shifts for laboratory energies below $E_{\mathrm{lab}} \lesssim 350$ MeV, i.e., for relative momenta $k \lesssim 2.1\ \mathrm{fm}^{-1}$.

As discussed in the previous section, the low momentum amplitudes can be preserved with an effective interaction $V_{\mathrm{low}\,k}$ acting in the Hilbert space of low momentum modes, $k < \Lambda$, exclusively.

[2] For simplicity, we have chosen different functions for $L(k')$ and $R(k)$, in order to include a repulsive part in the potential without having to introduce several separable terms, which would complicate the discussion.

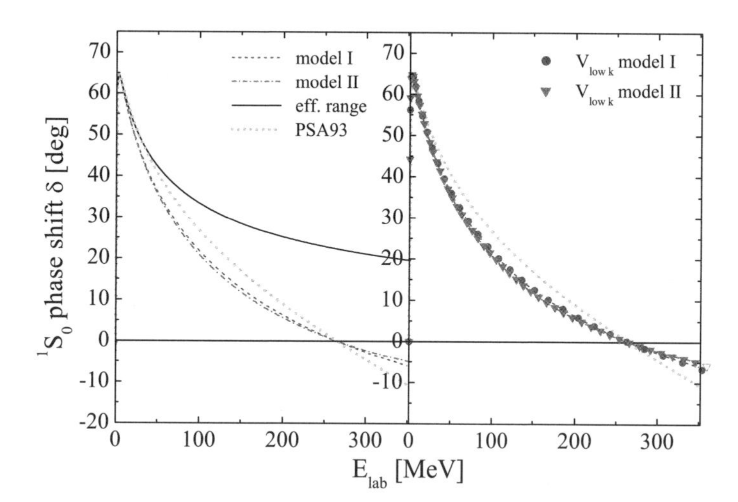

Fig. 2. The left figure shows the phase shifts calculated from the separable models I and II in comparison with the effective range expansion and the Nijmegen multi-energy phase shift analysis (PSA93) [24,25]. In the right figure, the phase shifts calculated from the corresponding low momentum interactions are given. The low momentum interactions are obtained by solving the RG equation, Eq. (14), to a scale $\Lambda = 2.1\ \mathrm{fm}^{-1}$.

The low momentum potential is renormalized by the high momentum modes according to an RG equation, of general form given by Eq. (7). In the current approach, we demand that the low momentum half-on-shell T matrix (and consequently the phase shifts) are RG invariant, see also [3,4]. The RG equation which follows from this requirement is derived in [26]. For separable interactions, the RG equation for the diagonal components of the low momentum interaction simplifies and is given in closed form by [3]

$$\frac{\mathrm{d}}{\mathrm{d}\Lambda} V_{\mathrm{low}\,k}(k,k;\Lambda) = \frac{2/\pi}{1-(k/\Lambda)^2} \frac{V_{\mathrm{low}\,k}(k,k;\Lambda)\, V_{\mathrm{low}\,k}(\Lambda,\Lambda;\Lambda)}{1 + \dfrac{2}{\pi}\mathscr{P}\displaystyle\int_0^{\Lambda} \frac{V_{\mathrm{low}\,k}(p,p;\Lambda)}{p^2-\Lambda^2}\, p^2\,\mathrm{d}p} . \tag{14}$$

The initial condition of the RG equation is given by the bare interaction,

$$V_{\mathrm{low}\,k}(k,k;\Lambda_0) = V_{\mathrm{bare}}(k,k) , \tag{15}$$

at an initial scale Λ_0 where the high momentum modes are negligible,

$$V_{\mathrm{bare}}(\Lambda_0,\Lambda_0) \approx 0 . \tag{16}$$

The RG equation is then used to evolve the bare potentials to low momenta, thus removing the model-dependent realization of the high momentum parts. We solve the separable RG equation to a cutoff $\Lambda = 2.1\ \mathrm{fm}^{-1}$, which corresponds to a laboratory energy scale up to which the realistic

[3] The RG equation is regularized by imposing a small gap between the low momentum Hilbert space, $k \leqslant \Lambda - \varepsilon$, and the cutoff Λ in the integration over intermediate momenta, $p_{\mathrm{loop}} \leqslant \Lambda$. In this way the T matrix can be calculated to chosen accuracy. In other words, the effect of this regularization on the scattering amplitude is $\mathcal{O}(\varepsilon)$.

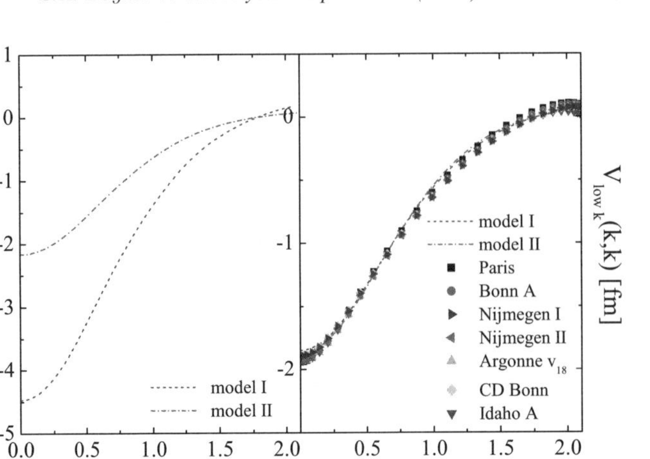

Fig. 3. The diagonal momentum-space matrix elements of V_{bare} and $V_{\text{low}\,k}$ derived from the separable models I and II. As a comparison, we show the results for the diagonal matrix elements of the $V_{\text{low}\,k}$ from Section 3 derived from the high precision nucleon–nucleon interactions.

nucleon–nucleon interactions are constrained by experiment, $E_{\text{lab}} \sim 350$ MeV. First, we compare the phase shifts calculated in the low momentum space

$$\tan\delta = -kT_{\text{low}\,k}(k,k;k^2)\ , \tag{17}$$

$$T_{\text{low}\,k}(k,k;k^2) = \frac{V_{\text{low}\,k}(k,k;\Lambda)}{1 + (2/\pi)\mathscr{P}\displaystyle\int_0^{\Lambda}[V_{\text{low}\,k}(p,p;\Lambda)/p^2 - k^2]\,p^2\,\mathrm{d}p} \tag{18}$$

and verify in Fig. 2 that the phase shifts are indeed reproduced by $V_{\text{low}\,k}$. In the RG approach, we have by construction

$$T_{\text{low}\,k}(k',k;k^2) = T_{\text{bare}}(k',k;k^2)\ . \tag{19}$$

In Fig. 3, we show the resulting diagonal matrix elements of $V_{\text{low}\,k}$ in comparison to the $V_{\text{low}\,k}$ derived from the high precision nucleon–nucleon interactions. Although the bare separable models I and II differ substantially, we find that the low momentum interactions in the truncated Hilbert space are identical. Furthermore, we observe that, to a surprisingly good accuracy, these simple models can in fact reproduce the diagonal matrix elements of the low momentum interaction obtained from the realistic potential models. In the RG equation for separable models, the matrix element corresponding to zero phase shift remains invariant,

$$V_{\text{low}\,k}(\kappa,\kappa;\Lambda) = V_{\text{bare}}(\kappa,\kappa) = T(\kappa,\kappa;\kappa^2) = 0\ . \tag{20}$$

Interestingly, we find that the diagonal matrix element of the realistic $V_{\text{low}\,k}$ also changes sign approximately at relative momentum κ.

1.3. Organization

This work is organized as follows. In the next section we discuss the calculational methods used for constructing the effective low momentum interaction $V_{\mathrm{low}\,k}$. In particular, we show how $V_{\mathrm{low}\,k}$ is related to the solution of the Bloch–Horowitz equation. Since the equivalence of the model space effective interaction methods and the RG approach is discussed in detail in [26], we will focus on the discussion of the effective interaction methods used in this work. In Section 3, we present the results for the low momentum interactions derived from the realistic potential models. We give results for all partial waves with $J \leqslant 4$, which provides ample evidence for the claimed model independence of $V_{\mathrm{low}\,k}$. By studying specific partial waves, we will show how the model independence is reached as the cutoff is varied, and also demonstrate its relation to the phase shift equivalence and the common long-range pion physics of the different potential models. Finally, we verify that the low momentum interaction indeed reproduces the scattering data below $E_{\mathrm{lab}} \lesssim 350$ MeV and the deuteron binding energy with similar accuracy as the high precision potentials. After summarizing our results, we elaborate on the advantages of using $V_{\mathrm{low}\,k}$ as the microscopic input interaction for nuclear many-body calculations.

2. Calculational methods

The schematic model provides a nice example illustrating the use of nonperturbative RG techniques to construct the low momentum effective theory. One could therefore apply the same direct method as for the schematic model. However, in the general non-separable case, the direct solution of the RG equation would be more involved. Therefore, we have adapted conventional effective interaction techniques, also referred to as model space methods, for the calculation of the low momentum interaction in the general case. These methods have been successfully used to derive nuclear shell model interactions in a truncated Hilbert space, see e.g., the review [27]. Both the RG and model space techniques are concerned with integrating out the high-energy states, such that the low-energy observables remain invariant in the simpler effective theory. In spite of their similarities, the formal equivalence of conventional effective interaction theory and the RG was only recently demonstrated for the two-body problem [26]. There, we have shown that the low momentum Hamiltonian obtained from the solution of the RG equation is equivalent to the effective theory derived using Bloch–Horowitz or Lee–Suzuki projection methods. Since our current work makes use of this equivalence for the calculation of $V_{\mathrm{low}\,k}$, we briefly discuss the relevant techniques below.

2.1. Integrating out the high momentum modes by similarity transformations

Consider a physical system, for which we are only interested in its low-energy properties. The first step in effective interaction theory is to define Feshbach projection operators onto the physically important low-energy model space, the so-called P-space, and the high-energy complement, the Q-space,

$$P = \sum_{i=1}^{d} |i\rangle\langle i| = \begin{pmatrix} 1 & 0 \\ 0 & 0 \end{pmatrix} \quad \text{and} \quad Q = \sum_{i=d+1}^{\infty} |i\rangle\langle i| = \begin{pmatrix} 0 & 0 \\ 0 & 1 \end{pmatrix} . \tag{21}$$

The projection operators satisfy $P+Q=1$, $PQ=QP=0$, $P^2=P$ and $Q^2=Q$, and we have written the projection operators in matrix form as well. In most applications, the eigenstates of the unperturbed Hamiltonian H_0 are taken as basis states $|i\rangle$.

The Schrödinger equation can then be written in block form as

$$\begin{pmatrix} PHP & PHQ \\ QHP & QHQ \end{pmatrix} \begin{pmatrix} P|\Psi\rangle \\ Q|\Psi\rangle \end{pmatrix} = E \begin{pmatrix} P|\Psi\rangle \\ Q|\Psi\rangle \end{pmatrix} . \tag{22}$$

By using the second block row of Eq. (22), the Q-space projection $Q|\Psi\rangle$ can be eliminated and one obtains the fundamental equations of Bloch–Horowitz effective interaction theory [28,29]

$$P\mathscr{H}^{\rm BH}_{{\rm low}\,k}(E)P|\Psi\rangle = EP|\Psi\rangle , \tag{23}$$

where the effective Hamiltonian $\mathscr{H}^{\rm BH}_{{\rm low}\,k}(E)$ is obtained by the solution of the Bloch–Horowitz equation

$$\mathscr{H}^{\rm BH}_{{\rm low}\,k}(E) = H + H\frac{1}{E - QHQ}H . \tag{24}$$

The Bloch–Horowitz equation generates an effective Hamiltonian, whose P-space projection $H^{\rm BH}_{{\rm low}\,k}(E)=P\mathscr{H}^{\rm BH}_{{\rm low}\,k}(E)P$ is operative only within the low-energy model space and, e.g., enables an exact diagonalization in shell model applications. The Q-space states have been decoupled from the problem in a way that preserves the low-energy spectrum exactly, i.e., the d lowest-lying eigenvalues. The eigenstates of $H^{\rm BH}_{{\rm low}\,k}(E)$ are simply given by the P-space projections of the exact eigenstates $P|\Psi_n\rangle$. The effective Hamiltonian $H^{\rm BH}_{{\rm low}\,k}(E)$ depends on the exact energy eigenvalue one is solving for, and thus necessitates a self-consistent treatment. Although the self-consistency problem can be solved nicely using the Lanczos algorithm [30], the use of energy-dependent two-body interactions in many-body calculations can pose computational and conceptual difficulties.

A refinement of the Bloch–Horowitz theory that avoids these difficulties is the Lee–Suzuki similarity transformation [31,32]. The Lee–Suzuki method constructs a similarity transformation that brings the Hamiltonian to the following block structure:

$$\Theta^{-1}H\Theta = \mathscr{H}^{\rm LS}_{{\rm low}\,k} = \begin{pmatrix} P\mathscr{H}P & P\mathscr{H}Q \\ 0 & Q\mathscr{H}Q \end{pmatrix} . \tag{25}$$

Since the determinant of a block-triangular matrix factorizes into the determinants of the independent P- and Q-space blocks, the low-energy spectrum can be calculated exactly from the smaller effective model space Hamiltonian $H^{\rm LS}_{{\rm low}\,k} = P\mathscr{H}^{\rm LS}_{{\rm low}\,k}P$. As in the Bloch–Horowitz approach, the eigenstates of $H^{\rm LS}_{{\rm low}\,k}$ are given by $P|\Psi_n\rangle$. Making an ansatz for the Lee–Suzuki similarity transformation, one parametrizes Θ as a non-orthogonal transformation in terms of the so-called wave operator ω

$$\Theta = 1+\omega = \begin{pmatrix} 1 & 0 \\ \omega & 1 \end{pmatrix} \quad \text{and} \quad \Theta^{-1} = 1-\omega = \begin{pmatrix} 1 & 0 \\ -\omega & 1 \end{pmatrix} , \tag{26}$$

where $\omega=Q\,\omega\,P$ connects the Q- and P-space. Inserting the ansatz into Eq. (25) leads to a non-linear constraint on ω, the decoupling equation,

$$Q\,\mathscr{H}^{\rm LS}_{{\rm low}\,k}P = 0 \;\Rightarrow\; QHP + QHQ\,\omega - \omega\,PHP - \omega\,PHQ\,\omega = 0 . \tag{27}$$

Once a solution to this non-linear operator equation is found, the energy-independent effective Hamiltonian is given by

$$H^{\mathrm{LS}}_{\mathrm{low}\,k} = PHP + PHQ\,\omega P = PHP + PVQ\,\omega P \;, \tag{28}$$

where we have assumed that P and Q commute with the unperturbed Hamiltonian.

The relation to the Bloch–Horowitz theory can be clarified by inserting the Lee–Suzuki solution, Eq. (28), into the decoupling equation, Eq. (27), and solving for the wave operator. The resulting formal expression for the wave operator allows one to express the energy-independent Lee–Suzuki $H^{\mathrm{LS}}_{\mathrm{low}\,k}$ as a summation of the lowest d, self-consistent Bloch–Horowitz solutions $H^{\mathrm{BH}}_{\mathrm{low}\,k}(E)$, where each term is weighted by the projection of the corresponding eigenstate of the effective theory $P\,|\Psi_n\rangle(\langle\widetilde{\Psi_n}|\,P)$ [4]

$$\begin{aligned} H^{\mathrm{LS}}_{\mathrm{low}\,k} &= \sum_{n=1}^{d} \left(PHP + PH\,\frac{1}{E_n - QHQ}\,HP \right) P\,|\Psi_n\rangle(\langle\widetilde{\Psi_n}|\,P) \\ &= \sum_{n=1}^{d} H^{\mathrm{BH}}_{\mathrm{low}\,k}(E_n)\,P\,|\Psi_n\rangle(\langle\widetilde{\Psi_n}|\,P) \;. \end{aligned} \tag{29}$$

In this work, we apply the Lee–Suzuki formalism to the free-space nucleon–nucleon problem to derive the energy-independent low momentum interaction

$$V_{\mathrm{low}\,k} = H^{\mathrm{LS}}_{\mathrm{low}\,k} - PH_0P = PVP + PVQ\,\omega P \;. \tag{30}$$

This approach has been proven to be equivalent to the solution of the RG equation that results from imposing a momentum-space cutoff and demanding RG invariant half-on-shell (HOS) T matrices [26]. For a given partial wave, the P and Q operators are defined in a continuous plane wave basis as

$$P = \frac{2}{\pi}\int_0^{\Lambda} p^2\,\mathrm{d}p\,|p\rangle\langle p| \quad \text{and} \quad Q = \frac{2}{\pi}\int_{\Lambda}^{\infty} q^2\,\mathrm{d}q\,|q\rangle\langle q| \;. \tag{31}$$

There are several methods to solve the non-linear decoupling equation for the wave operator. For the two-body problem, e.g., the exact solutions for the eigenstates can be used, and $V_{\mathrm{low}\,k}$ can be calculated directly using Eq. (29). For more complex applications, one solves the decoupling equation using iterative techniques. Since the iterative methods are numerically very robust, we have utilized the iterative algorithm of Andreozzi [33] to calculate $V_{\mathrm{low}\,k}$. The defining equations of this iteration scheme are given by

$$\mathscr{X}_0 = -(QHQ)^{-1}\,QHP \;, \tag{32}$$

$$\mathscr{X}_n = (\mathscr{Q}(\omega_{n-1}))^{-1}\,\mathscr{X}_{n-1}\,H^{\mathrm{LS}}_{\mathrm{low}\,k}(\omega_{n-1}) \quad \text{for } n \geqslant 1 \;, \tag{33}$$

where

$$\omega_n = \mathscr{X}_0 + \mathscr{X}_1 + \cdots + \mathscr{X}_n \;, \tag{34}$$

$$\mathscr{Q}(\omega_n) = QHQ - Q\,\omega_n\,PHQ \tag{35}$$

[4] The tilde denotes the bi-orthogonal complement, since projections of orthogonal $|\Psi_n\rangle$ are not necessarily orthogonal.

and

$$H^{\mathrm{LS}}_{\mathrm{low}\,k}(\omega_n) = PHP + PHQ\,\omega_n P \; . \tag{36}$$

The iteration converges when $H^{\mathrm{LS}}_{\mathrm{low}\,k}(\omega_n) \approx H^{\mathrm{LS}}_{\mathrm{low}\,k}(\omega_{n-1})$.

A convenient feature of the algorithm is that each iteration only requires one matrix inversion in the Q-space. The Q-space matrix inversion is manageable, typically requiring a set of ≈ 60 Gauss–Legendre mesh points extending to $\approx 25\ \mathrm{fm}^{-1}$ to include the high momentum behavior of the input interaction. Finally, we mention that variations of the algorithm exist in which all matrix inversions occur in P-space, which can offer significant computational advantages for large-scale problems [33].

2.2. A refinement: hermitian low momentum interactions

By construction, the Lee–Suzuki $H^{\mathrm{LS}}_{\mathrm{low}\,k}$ is non-hermitian because the eigenstates of the effective theory are projections of the full eigenstates onto the low momentum subspace. The non-hermiticity is a result of eliminating the energy-dependence of the Bloch–Horowitz theory by means of the non-orthogonal similarity transformation.[5] However, one can perform an additional similarity transformation on $H^{\mathrm{LS}}_{\mathrm{low}\,k}$ to obtain an energy-independent and hermitian Hamiltonian [34,35], which we denote by $\bar{H}^{\mathrm{LS}}_{\mathrm{low}\,k}$ and correspondingly, the hermitian low momentum interaction by $\bar{V}_{\mathrm{low}\,k}$. Both similarity transformations combine to give a unitary transformation, which is equivalent to the well-known Okubo transformation [36].

Here, we follow the method of Andreozzi [33] in constructing the second similarity transformation. Using the definition of the Lee–Suzuki transformation, Eq. (25), one finds

$$\Theta^\dagger \Theta\, \mathscr{H}^{\mathrm{LS}}_{\mathrm{low}\,k} = (\mathscr{H}^{\mathrm{LS}}_{\mathrm{low}\,k})^\dagger\, \Theta^\dagger \Theta \; , \tag{37}$$

where we have used $H = H^\dagger$. Projecting onto the low momentum space, the P-space block can be written as

$$(P + \omega^\dagger\, \omega)\, H^{\mathrm{LS}}_{\mathrm{low}\,k} = (H^{\mathrm{LS}}_{\mathrm{low}\,k})^\dagger\, (P + \omega^\dagger \omega) \; . \tag{38}$$

The operator $P + \omega^\dagger\, \omega$ is hermitian and positive definite. Therefore, it can be written by means of a Cholesky decomposition as a product of a lower-triangular matrix L and its adjoint $L^\dagger$. Writing

$$(P + \omega^\dagger\, \omega) = LL^\dagger \tag{39}$$

in Eq. (38) and multiplication by L^{-1} from the left and $(L^\dagger)^{-1}$ from the right, one finds

$$L^{-1}\, (\mathscr{H}^{\mathrm{LS}}_{\mathrm{low}\,k})^\dagger\, L = L^\dagger\, \mathscr{H}^{\mathrm{LS}}_{\mathrm{low}\,k}\, (L^\dagger)^{-1} = (L^{-1} (\mathscr{H}^{\mathrm{LS}}_{\mathrm{low}\ k})^\dagger L)^\dagger \; . \tag{40}$$

Therefore, we can write the hermitian $\bar{H}^{\mathrm{LS}}_{\mathrm{low}\,k}$ directly in terms of the Lee–Suzuki transformation and the Cholesky decomposition as

$$\bar{H}^{\mathrm{LS}}_{\mathrm{low}\,k} = L^\dagger\, H^{\mathrm{LS}}_{\mathrm{low}k}\, (L^\dagger)^{-1} \tag{41}$$

and the combined transformation $U = \Theta\, (L^\dagger)^{-1}$ is unitary.

---

[5] Although the Bloch–Horowitz Hamiltonian $H^{\mathrm{BH}}_{\mathrm{low}\,k}(E_n)$ is hermitian for a given energy E_n, one encounters the same non-orthogonality problem between eigenstates, since the Hamiltonian changes self-consistently depending on the energy E_n.

We mention that the differences between the non-hermitian $V_{\text{low}\,k}$ and the hermitian $\bar{V}_{\text{low}\,k}$ can be understood in the RG approach. The non-hermitian effective theory is obtained by introducing a cutoff in momentum space and imposing RG invariance for the HOS T matrix with an energy-independent low momentum potential. Since the initial conditions of the RG equation are given by the bare interaction, this implies that the effective theory preserves the low momentum components of the full wave functions, in addition to the phase shifts and the bound state poles. The choice of HOS T matrix invariance is motivated by the physical observation that the low momentum components of the wave functions mainly probe the long-range OPE part of the interaction. Conversely, the hermitian effective theory results from RG invariant on-shell T matrices. Thus, the resulting hermitian $\bar{V}_{\text{low}\,k}$ no longer preserves the low momentum components of the full wave functions, although the observable phase shifts and bound state poles are preserved. In practice, we will find that the numerical differences between the hermitian and non-hermitian interactions are quite small.

Finally, we note that the Lee–Suzuki similarity transformation method is an algebraic realization of the more general folded diagram method of Kuo, Lee and Ratcliff (see [27,37,38]). In this connection, one can recast the results of this section in a diagrammatic formalism, which can be useful in many-body systems. Moreover, the folded diagram formulation has been applied to the two-body problem to prove diagrammatically that the low momentum effective theory preserves the low-energy scattering observables of the input interaction [3].

3. Results

We now discuss the main results for the low momentum interaction $V_{\text{low}\,k}$, which we have obtained by integrating out the high momentum components of different realistic potential models. Most of the results shown here are for a momentum cutoff of $\Lambda = 2.1\ \text{fm}^{-1}$, which corresponds to a distance scale $1/\Lambda \sim 0.5$ fm. With cutoffs chosen in this range, the model-dependent parameterization of the repulsive core at $r \lesssim 0.5$ fm in the bare interactions is not resolved in the effective theory. Consequently, one is able to reproduce the deuteron pole and scattering observables below $E_{\text{lab}} =$ 350 MeV without introducing a specific short-distance model. Moreover, the RG invariance of the decimation guarantees that our conclusions are independent of the precise value of the cutoff.

Before we present our results in detail, we summarize our main findings and refer the reader to the key figures. The most important properties of $V_{\text{low}\,k}$ are:

1. $V_{\text{low}\,k}$ renormalizes to a nearly universal interaction as the cutoff is lowered to the scale of $\Lambda \sim 2.1\ \text{fm}^{-1}$. We will argue that the model independence of the diagonal matrix elements of $V_{\text{low}\,k}$, as shown in Figs. 4 and 5, is driven by the phase shift equivalence of the input models. The correlation between the phase shift equivalence and the collapse of the low momentum interactions is clearly demonstrated in Fig. 8. We also find practically identical results for the hermitian $\bar{V}_{\text{low}\,k}$, see Figs. 18 and 19.
2. $V_{\text{low}\,k}$ reproduces the phase shifts and the deuteron binding energy with similar accuracy as the high precision models. The phase shifts calculated from $V_{\text{low}\,k}$ are shown in Figs. 16 and 17.
3. In the effective theory one does not have to make model-dependent assumptions on the short-distance interaction. The RG evolution filters out the short-distance details of the input potentials, while preserving the model-independent effects of the high momentum modes on the low

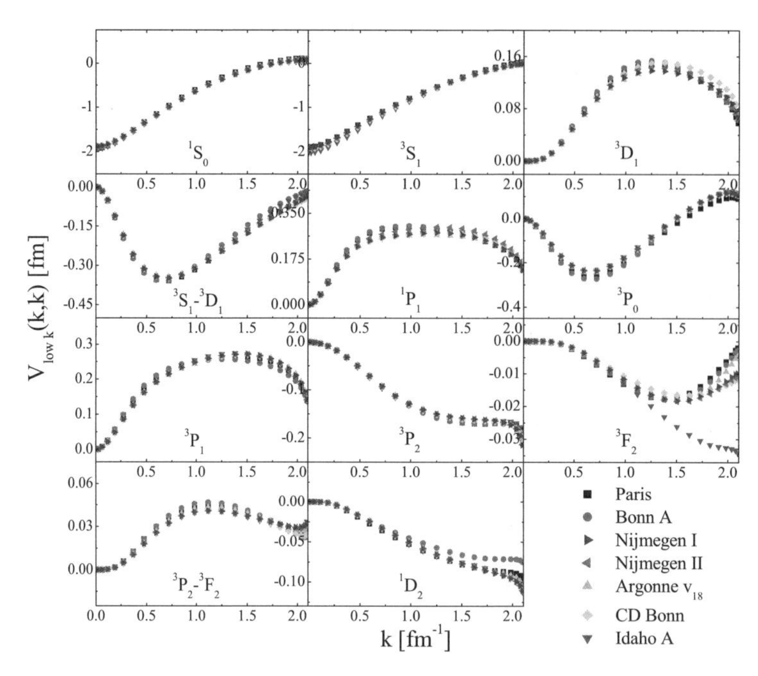

Fig. 4. Diagonal momentum-space matrix elements of the $V_{\mathrm{low}\,k}$ obtained from the different potential models for a cutoff $\Lambda = 2.1\ \mathrm{fm}^{-1}$. Results are shown for the partial waves $J \leqslant 4$.

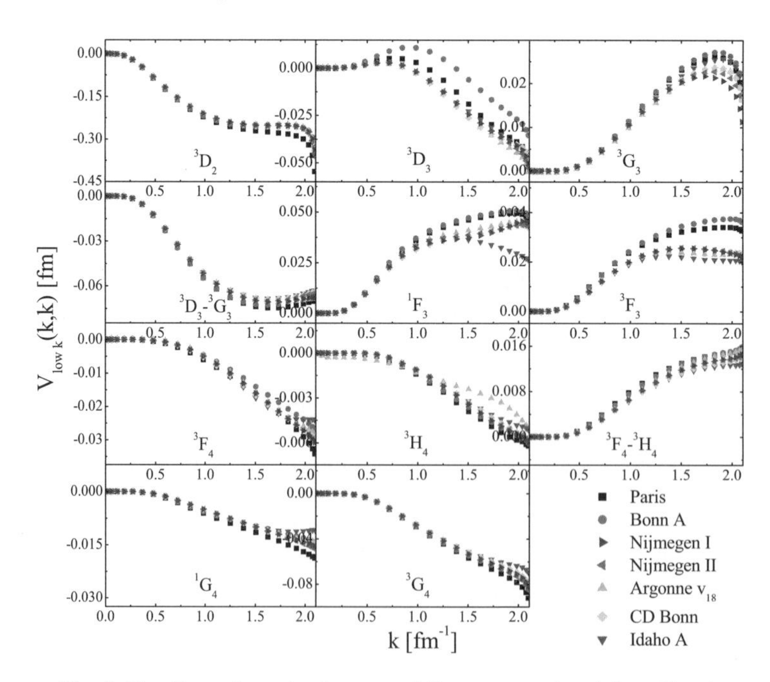

Fig. 5. The diagonal matrix elements of $V_{\mathrm{low}\,k}$ are continued from Fig. 4.

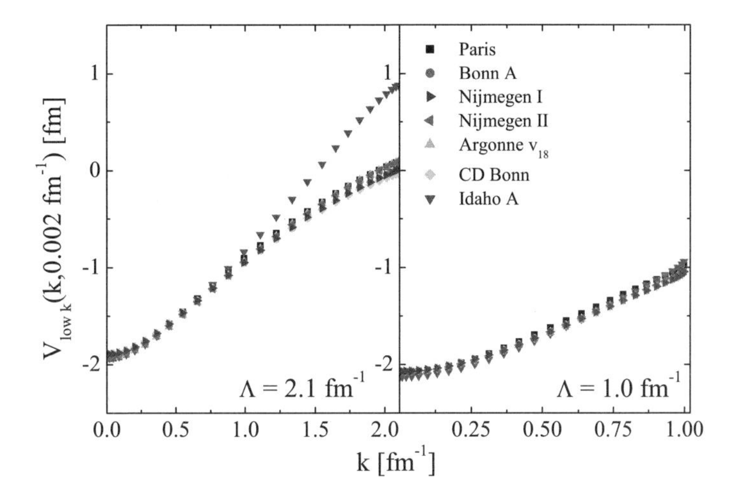

Fig. 6. The off-diagonal momentum space matrix elements of the $V_{\mathrm{low}\,k}$ obtained for a cutoff $\Lambda = 2.1\ \mathrm{fm}^{-1}$ (left) and $\Lambda = 1.0\ \mathrm{fm}^{-1}$ (right) in the $^1S_0$ partial wave.

momentum observables, and leaving the long-range part (OPE) of the interaction intact. This can be seen most clearly in Figs. 11–13.

We also point out that the numerical differences between $V_{\mathrm{low}\,k}$ and the hermitian $\bar{V}_{\mathrm{low}\,k}$ are quite small, see Fig. 15. For brevity, we mostly present results for $V_{\mathrm{low}\,k}$ in this work, since the $\bar{V}_{\mathrm{low}\,k}$ is easily derived from $V_{\mathrm{low}\,k}$ using the transformation described in Section 2.2.

3.1. Model independence of the low momentum interaction

For cutoffs around $\Lambda \sim 2.1\ \mathrm{fm}^{-1}$, the effective theory preserves the nucleon–nucleon phase shifts up to laboratory energies of $E_{\mathrm{lab}} \sim 350$ MeV. This corresponds to the scale of the scattering data that all the high precision potential models are fitted to. We have argued that, based on the separation of mass scales, the model dependence of the realistic interactions should be reduced as we integerate out the ambiguous high momentum modes. In Figs. 4 and 5, we present the results of the RG decimation. We observe that the diagonal matrix elements of the low momentum interactions derived from the different potential models are nearly identical, and thus insensitive to the input potential model for cutoffs $\Lambda \lesssim 2.1\ \mathrm{fm}^{-1}$. The largest relative differences between the $V_{\mathrm{low}\,k}$ are found in the $^3F_2$, $^3D_3$, $^1F_3$ and $^3F_3$ partial waves. These differences are correlated with the deviations in the phase shifts, i.e., the different accuracies, of the potential models. We will discuss this point in detail below.

We note that if we perform the additional similarity transformation to obtain the resulting hermitian $\bar{V}_{\mathrm{low}\,k}$, we find the same model independence and practically identical diagonal matrix elements, as shown in Figs. 18 and 19 in the appendix. Since the hermitian $\bar{V}_{\mathrm{low}\,k}$ is obtained by symmetrizing $V_{\mathrm{low}\,k}$, it is expected that the effects on the diagonal matrix elements are not significant.

One obtains further insight into the collapse of $V_{\mathrm{low}\,k}$ by studying the off-diagonal matrix elements. These are shown in the $^1S_0$ partial wave in Fig. 6. We find a similar nearly universal behavior for the off-diagonal matrix elements of $V_{\mathrm{low}\,k}$, although those calculated from the Idaho potential differ at $k' \approx 1.2\ \mathrm{fm}^{-1}$, which is approximately the mass scale given by the 2π exchange. The conventional

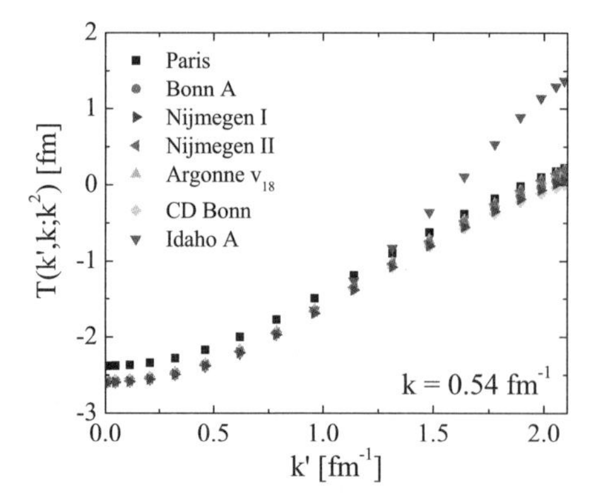

Fig. 7. Comparison of the HOS T matrices calculated from the different potential models in the $^1S_0$ partial wave.

models include 2π contributions in which non-nucleonic intermediate states are explicitly taken into account. In contrast, the Idaho potential starts from a local, effective Lagrangian, in which nucleons and pions are the only explicit degrees of freedom. For example, contributions from Δ intermediate states between two successive pion exchanges are contracted to a renormalized $\mathscr{L}^{\rm eff}_{\pi{\rm NN}}$ vertex. Moreover, the conventional potential models all contain a strong, short-range repulsion associated with the exchange of heavy mesons, mostly originating from the ω exchange. In the Idaho potential, the repulsive core is not explicitly present. However, the effects of the short-range repulsion on low-energy observables are contained in contact terms that are fitted to the low-energy data [19]. Therefore, in momentum space, the Idaho potential changes over to contact terms at momentum transfers above the range given by $2m_\pi$.

While the Idaho potential is phase shift equivalent to the conventional models, the HOS T matrices, i.e., the low momentum components of the wave functions, differ significantly from the conventional models at energies and momenta larger than the above scale of $2m_\pi$. In Fig. 7, we find that the conventional models give remarkably similar HOS T matrices over their range of phase shift equivalence. This approximate HOS T matrix equivalence can be understood by expanding the HOS T matrix around the fully on-shell part. The difference can be written as a wound integral that contains the off-shell behavior,

$$T(k',k;k^2) = T(k,k;k^2) + \frac{k^2 - k'^2}{k'} \int_0^\infty {\rm d}r \sin(k'r)(u_k(r) - v_k(r)) \,, \tag{42}$$

where $u_k(r) = r\,\Psi_k(r)$ denotes the exact and $v_k(r)$ is the asymptotic S-wave wave function

$$\lim_{r\to\infty} u_k(r) \to v_k(r) \sim \sin(kr + \delta) \,. \tag{43}$$

We can use Eq. (42) to express the difference of the HOS T matrices for two different, but phase shift equivalent bare interactions. This leads to

$$\Delta T(k',k;k^2) = \frac{k^2 - k'^2}{k'} \int_0^R {\rm d}r \sin(k'r)\Delta u_k(r) \,, \tag{44}$$

where R is the distance over which the two models differ substantially at short distances, which we can take for simplicity as the size of the repulsive core. Regardless of the details of the differences, at

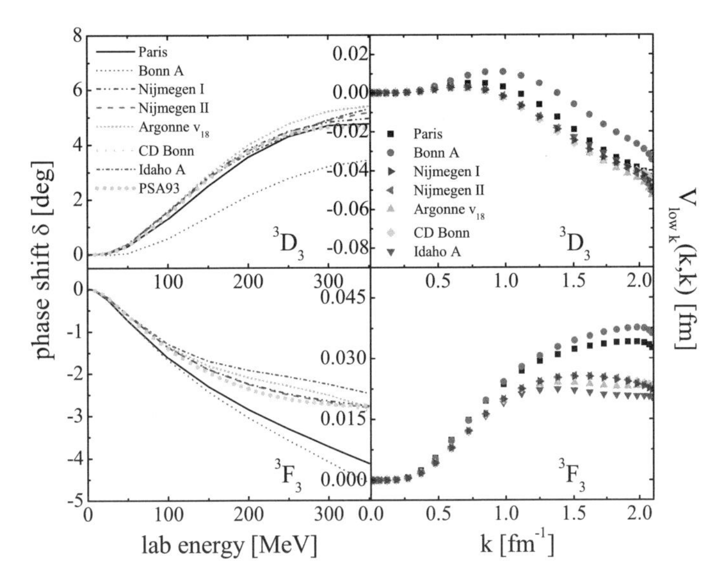

Fig. 8. Comparison of the phase shifts given by the V_{NN} models and the diagonal matrix elements of $V_{\mathrm{low}\,k}$ in the $^3D_3$ and $^3F_3$ partial waves. Results for $V_{\mathrm{low}\,k}$ are shown with a cutoff $\Lambda = 2.1\ \mathrm{fm}^{-1}$.

scattering energies well below the strength of the short-range repulsion, the differences between the wave functions in the core are suppressed by strong two-body correlations. Therefore, it is physically reasonable that the conventional realistic potentials give similar HOS T matrices at low energies.

The deviations in the off-diagonal matrix elements of $V_{\mathrm{low}\,k}$ derived from the Idaho potential are obviously removed, if we integrate out further below $2m_\pi$. As shown in Fig. 6, we find that the off-diagonal matrix elements in the $^1S_0$ partial wave collapse onto one curve as well for $\Lambda \lesssim 1.2\ \mathrm{fm}^{-1}$.

The preceding observations strongly suggest that the collapse of the diagonal matrix elements of $V_{\mathrm{low}\,k}$ is driven by the phase shift equivalence of the input models, while the collapse of the off-diagonal matrix elements is controlled by approximate HOS T matrix equivalence at low energy and momentum. The latter can in turn be mainly attributed to a common off-shell behavior in the OPE interaction, with some differences arising from the two-pion exchange and the crossover to contact interactions. The correlation between the diagonal matrix elements and the phase shifts is nicely shown in the $^3D_3$ and $^3F_3$ partial waves, where we observe the largest relative deviations from model independence. The left panels of Fig. 8 clearly show that the older generation Paris and Bonn potential models give significantly different phase shifts in these partial waves. This leads to the observed differences in the $V_{\mathrm{low}\,k}$ derived from these bare interactions, as can be seen in the right panels of Fig. 8. For comparison, we have also shown the results of the Nijmegen multi-energy phase shift analysis (PSA93) [24,25].

A further example of this correlation can be seen in Fig. 9, where the Idaho potential leads to considerably different phase shifts in the $^3F_2$ partial wave above laboratory energies $E_{\mathrm{lab}} \gtrsim 100$ MeV. This corresponds to a relative momentum of $k \gtrsim 1.0\ \mathrm{fm}^{-1}$. Accordingly in Fig. 10, we find that the diagonal matrix elements of $V_{\mathrm{low}\,k}$ derived from the Iadaho potential deviate from the others

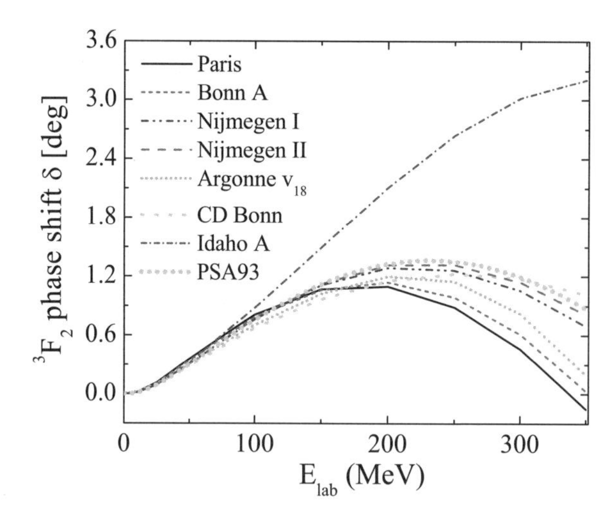

Fig. 9. Phase shifts in the $^3F_2$ partial wave calculated from different V_{NN} models.

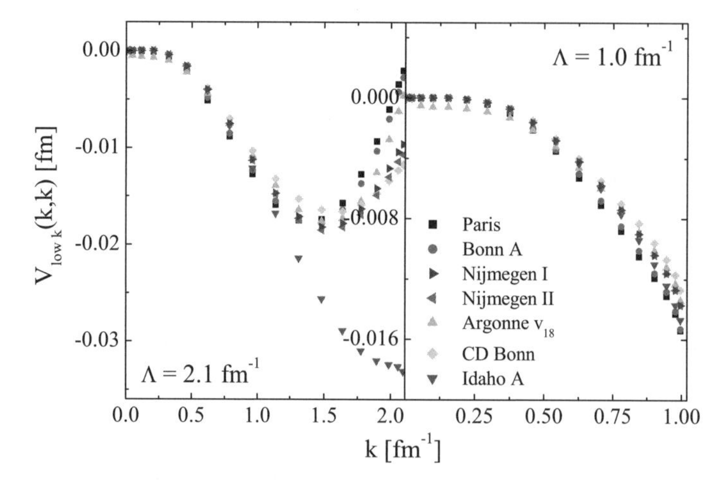

Fig. 10. The diagonal matrix elements of $V_{\mathrm{low}\,k}$ for $\Lambda = 2.1\ \mathrm{fm}^{-1}$ (left) and $\Lambda = 1.0\ \mathrm{fm}^{-1}$ (right) in the $^3F_2$ partial wave.

for momenta larger than this scale. As for the off-diagonal matrix elements in the $^1S_0$ partial wave, these differences vanish as the cutoff is lowered to $\Lambda \lesssim 1.0\ \mathrm{fm}^{-1}$.

In the N^{2}LO version of the Idaho potential used here, the contact interactions do not contribute to the $L \geqslant 3$ partial waves. Thus, in the $^3F_2$ partial wave the Idaho potential is completely given by pion exchange. Our results demonstrate that the effects of the short-distance physics cannot simply be ignored. Rather, their renormalization effects on the low momentum physics has to be taken into account. If we lower the cutoff to $\Lambda \lesssim 1.0\ \mathrm{fm}^{-1}$, the effects of the high momentum modes is sufficiently small, and one finds that indeed the physics is given adequately

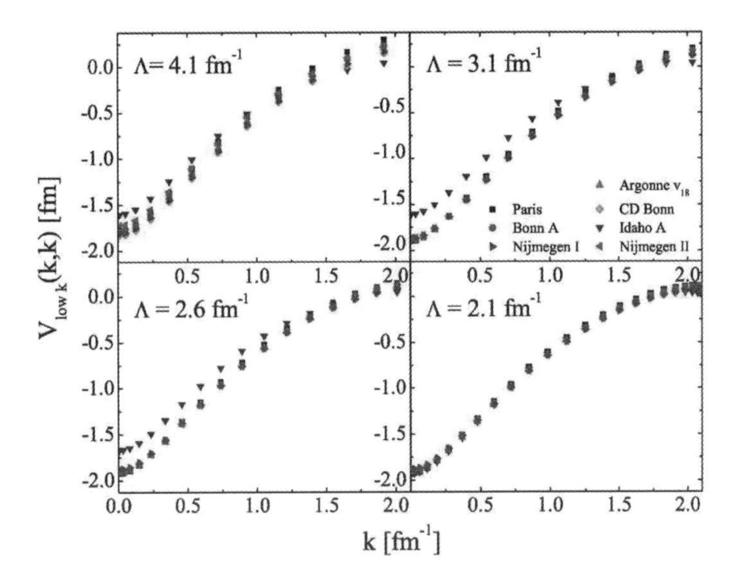

Fig. 11. The collapse of the diagonal momentum-space matrix elements of $V_{\mathrm{low}\,k}$ as the cutoff is lowered to $\Lambda = 2.1\ \mathrm{fm}^{-1}$ in the $^1S_0$ partial wave.

by pion exchange. This can be understood in the Born approximation, which is a reasonable approximation for the high partial waves, especially at low energies. In the Born approximation, the integral term of the Lippmann–Schwinger equation is neglected, and thus the low momentum interaction in the cutoff Hilbert space is given by simply cutting off the bare interaction.

The RG approach has the advantage that the nature of the low momentum interaction can be revealed by studying the change of $V_{\mathrm{low}\,k}$ with the cutoff scale. In this way, the RG evolution clearly provides evidence for the separation of scales in the two-nucleon problem, see also [3]. Furthermore, we will show that it also provides physical insight into the low momentum interaction.

In the evolution of the $^1S_0$ matrix elements shown in Fig. 11, we find that the model independence of $V_{\mathrm{low}\,k}$ is reached at somewhat larger values of the cutoff around $\Lambda \sim 3.5\ \mathrm{fm}^{-1}$. This scale lies below the ω and ρ meson masses, and consequently the details of the repulsive core are not resolved in the effective theory. Also note that for $\Lambda \gtrsim 3.0\ \mathrm{fm}^{-1}$, the Idaho potential is not renormalized, since it is an EFT potential without high momentum components by construction.

Conversely, we find in Fig. 12 that the collapse in the $^3S_1$ partial wave occurs at a comparatively lower value for the cutoff of $\Lambda \sim 2.1$–$2.6\ \mathrm{fm}^{-1}$. This arises from the fact that the tensor force is operative in the triplet S-wave. Since $V_{\mathrm{low}\,k}$ acquires a large second-order renormalization from the tensor force, one indeed expects that $V_{\mathrm{low}\,k}$ becomes model-independent at a lower cutoff in the $^3S_1$ partial wave.

The nature of the tensor interaction of $V_{\mathrm{low}\,k}$ can be studied by considering the couping between the $^3S_1$ and $^3D_1$ partial wave. In this channel, only the tensor force enters. The results for the RG evolution in the $^3S_1$–$^3D_1$ block are presented in Fig. 13. We observe that the realistic models are practically identical for momenta below $k \lesssim 0.7\ \mathrm{fm}^{-1}$, which corresponds to the pion mass. This clearly demonstrates that the tensor part of the input models differs significantly for momenta above

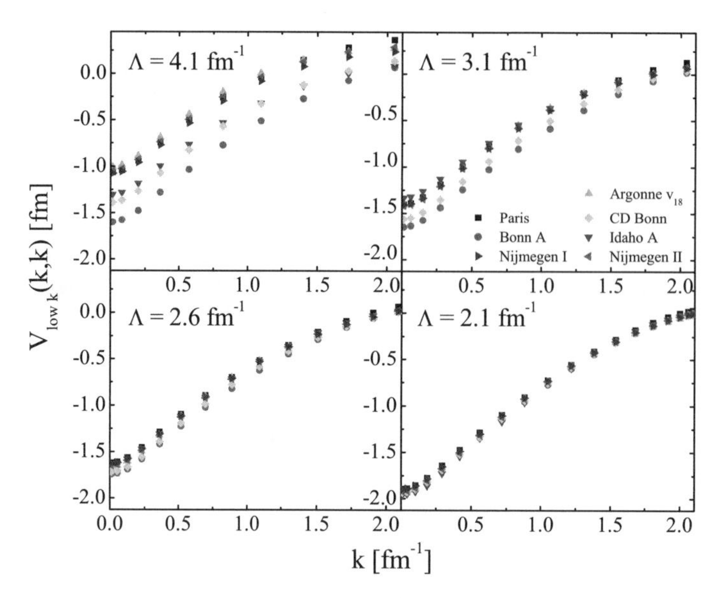

Fig. 12. The collapse of the diagonal momentum-space matrix elements of $V_{\mathrm{low}\,k}$ as the cutoff is lowered to $\Lambda = 2.1\ \mathrm{fm}^{-1}$ in the $^3S_1$ partial wave.

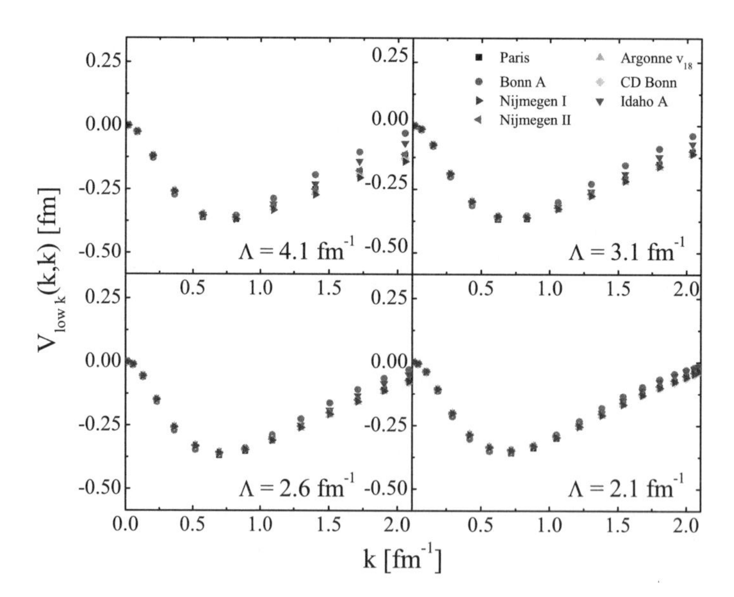

Fig. 13. The collapse of the diagonal momentum-space matrix elements of $V_{\mathrm{low}\,k}$ as the cutoff is lowered to $\Lambda = 2.1\ \mathrm{fm}^{-1}$ in the coupled $^3S_1$–$^3D_1$ block.

m_π. However, the effects of the different high momentum tensor components on the low momentum physics can be taken into account by employing a renormalized interaction, which in turn becomes model independent at a cutoff of $\Lambda \sim 2.1\ \mathrm{fm}^{-1}$.

Table 2
Comparison of the deuteron binding energy E_D for $V_{\mathrm{low}\,k}$ and the bare interactions. We note that all $V_{\mathrm{low}\,k}$ are derived with the same momentum mesh. For the values reported here, we have chosen not to optimize the mesh for each bare interaction separately

E_D (MeV)	Paris	Bonn A	Nijm. I	Nijm. II	Arg. v_{18}	CD Bonn	Idaho A
V_{NN}	−2.2218	−2.2242	−2.2246	−2.2242	−2.2247	−2.2238	−2.2242
Quoted	−2.2249	−2.22452	−2.224575	−2.224575	−2.224575	−2.224575	−2.2242
$V_{\mathrm{low}\,k}$	−2.2218	−2.2242	−2.2246	−2.2242	−2.2247	−2.2238	−2.2242

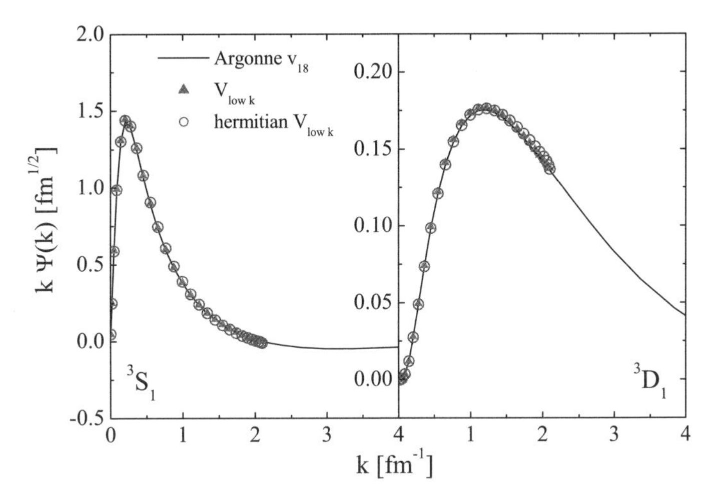

Fig. 14. Comparison of the momentum-space S- and D-state deuteron wave functions calculated from $V_{\mathrm{low}\,k}$, the hermitian $\bar{V}_{\mathrm{low}\,k}$ and the bare V_{NN}. Results are shown for a cutoff of $\Lambda = 2.1\ \mathrm{fm}^{-1}$ and the Argonne v_{18} potential.

The RG decimation to $V_{\mathrm{low}\,k}$ preserves the deuteron binding energies of the bare interactions. In Table 2, we give the binding energies obtained from the $V_{\mathrm{low}\,k}$, and verify that the deuteron poles are in fact reproduced in the effective theory. We note that we have not optimized the momentum mesh for each input V_{NN}, which would further improve the agreement with the quoted values. Referring to Fig. 14, the low momentum components of the bare deuteron S- and D-state wave functions are RG invariant as well. Moreover, although the hermitian $\bar{V}_{\mathrm{low}\,k}$ does not exactly preserve the low momentum projections of the wave functions, we find in Fig. 14 that the deuteron wave functions calculated from $V_{\mathrm{low}\,k}$ and the hermitian $\bar{V}_{\mathrm{low}\,k}$ are practically identical.

Finally, we note that the same general results hold for both $V_{\mathrm{low}\,k}$ and $\bar{V}_{\mathrm{low}\,k}$, and that the numerical differences are quite small. This is demonstrated as an example for the $^1S_0$ partial wave in Fig. 15. It can also be seen that the differences are as expected most noticeable for the far off-shell parts of the low momentum interaction.

3.2. Phase shift equivalence of the low momentum interaction

The model-independent low momentum interaction reproduces the experimental phase shift data with similar accuracy as the high precision nucleon–nucleon potentials. Therefore, $V_{\mathrm{low}\,k}$ fits the

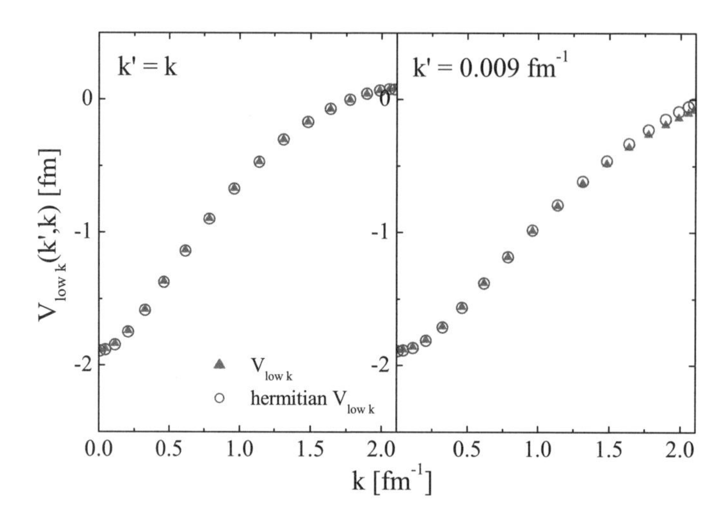

Fig. 15. Comparison of the diagonal (left) and off-diagonal (right) matrix elements of $V_{\mathrm{low}\,k}$ and the hermitian $\bar{V}_{\mathrm{low}\,k}$ in the $^1S_0$ partial wave. Results are shown for a cutoff of $\Lambda = 2.1$ fm^{-1} and the Argonne v_{18} potential.

measured phase shifts for laboratory energies below $E_{\mathrm{lab}} = 350$ MeV with a similar χ^2/datum $\approx$ 1–2, depending on which conventional potential model V_{NN} is used. Since the hermitian $\bar{V}_{\mathrm{low}\,k}$ preserves this content as well as the hermiticity of the Hamiltonian, we demonstrate the high accuracy of the hermitian $\bar{V}_{\mathrm{low}\,k}$ here. In Figs. 16 and 17, the neutron–proton phase shifts and the mixing parameters in coupled channels for all partial waves $J \leqslant 4$ are given. For clarity of the figures, we show these for the low momentum $\bar{V}_{\mathrm{low}\,k}$ derived from the CD Bonn potential only. Since both the hermitian and non-hermitian low momentum interactions preserve the phase shifts of the input models, results of the same accuracy are derived from the various other models. For comparison, the phase shifts and mixing parameters calculated from the bare CD Bonn potential and the results of the Nijmegen multi-energy phase shift analysis [24,25] are included in Figs. 16 and 17.

We conclude that the low-energy scattering data can be equally well reproduced by an effective low momentum interaction $V_{\mathrm{low}\,k}$ restricted to the Hilbert space of low momentum modes, $k < \Lambda_{\mathrm{data}}$. This is achieved with a model-independent interaction, without introducing further assumptions on the details of the short-distance dynamics.

4. Summary and advantages

We have shown that the RG decimation of different high precision nucleon–nucleon interactions to low momenta leads to a unique interaction, called $V_{\mathrm{low}\,k}$. This result is obtained by truncating the full Hilbert space to an effective space, which consists of momentum components up to the scale of the constraining low-energy data. The differences in the realistic interactions arise from the assumed high momentum dynamics. By restricting the interaction to the low momentum space, the details of the short-distance physics are not resolved, and the detail-independent effects of the high momentum modes on the low momentum observables are included in the renormalized effective interaction. The existence of a nearly universal $V_{\mathrm{low}\,k}$ demonstrates that it is possible to separate the physics of

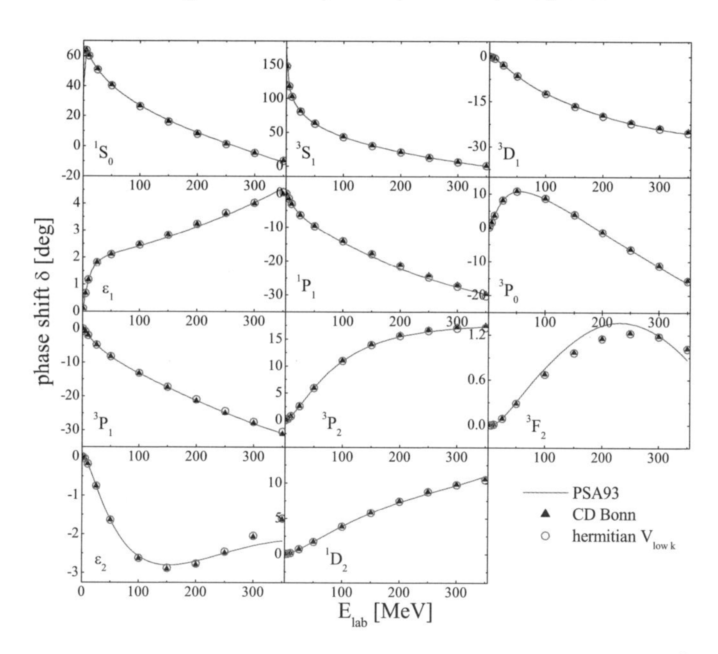

Fig. 16. The neutron–proton phase shifts and mixing parameters calculated from the hermitian $\bar{V}_{\mathrm{low}\,k}$ in the partial waves $J \leqslant 4$ are shown. The $\bar{V}_{\mathrm{low}\,k}$ is derived from the CD Bonn potential with a cutoff of $\Lambda = 2.1\ \mathrm{fm}^{-1}$. In comparison, the phase shifts and mixing parameters calculated from the bare CD Bonn potential as well as the results of the Nijmegen multi-energy phase shift analysis [24,25] are included.

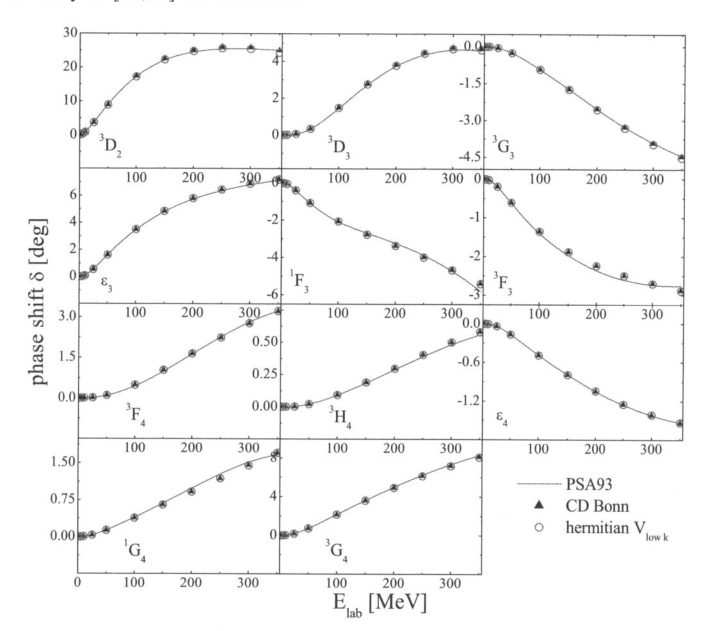

Fig. 17. The neutron–proton phase shifts and mixing parameters are continued from Fig. 16, for details we refer to the text of Fig. 16.

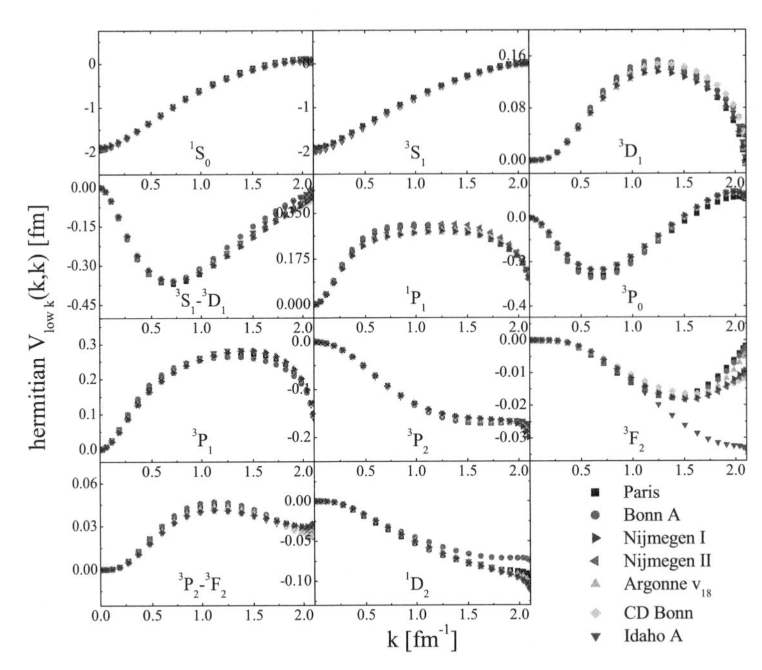

Fig. 18. Diagonal momentum-space matrix elements of the hermitian $\bar{V}_{\mathrm{low}\,k}$ obtained from the different potential models for a cutoff $\Lambda = 2.1\ \mathrm{fm}^{-1}$. Results are shown for the partial waves $J \leqslant 4$.

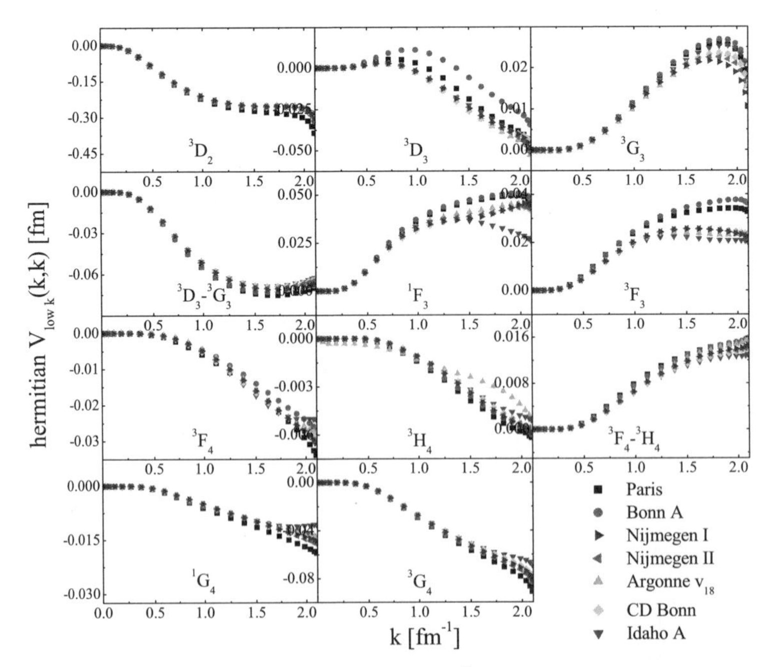

Fig. 19. The diagonal elements of the hermitian $\bar{V}_{\mathrm{low}\,k}$ are continued from Fig. 18.

low momentum nucleons from the ambiguous short-distance parametrizations used in the realistic potential models.

The RG invariance of the effective theory guarantees that the HOS T matrix and the deuteron properties of the full theory are exactly preserved in the model space. Consequently, $V_{\text{low}\,k}$ is phase shift equivalent to the high precision potential models and reproduces the experimental elastic scattering data with similar accuracy. A further refinement transforms $V_{\text{low}\,k}$ into a hermitian low momentum interaction $\bar{V}_{\text{low}\,k}$, which remains model independent and preserves the low-energy on-shell amplitudes of the bare interactions. In this manner, the resulting RG decimation incorporates the model-independent effects of the short-distance physics in the low momentum interaction, while the model-dependent effects are filtered out.

We believe that the low momentum interaction $V_{\text{low}\,k}$ satisfies the first of the two requirements of a microscopic many-body theory given in the Introduction. Namely, that the input interaction is, as best as possible, based on the available low-energy data. A necessary condition for the collapse of the low momentum interactions is the separation of long from short-distance scales in the nuclear problem. In addition, we have argued that the accuracy of the reproduction of the phase shifts drives the collapse of the on-shell components of the low momentum interactions. Furthermore, the evolution of $V_{\text{low}\,k}$ with the cutoff provides physical insight into the separation of scales in the nuclear problem, the effects of the short-range repulsion in the renormalization, and the tensor content of the low momentum interaction.

$V_{\text{low}\,k}$ sums high momentum ladders in free space. Therefore, the low momentum interaction is considerably softer than the bare interactions and does not have a repulsive core. As a consequence, when used as the microscopic input in the many-body problem, $V_{\text{low}\,k}$ does not lead to strong high momentum scattering effects in the particle–particle channel, which necessitates a Brueckner resummation or short-range correlation methods for the conventional models.

The low momentum interaction should be regarded as a new input potential for many-body calculations. However, we emphasize that in contrast to the Brueckner G matrix, $V_{\text{low}\,k}$ does not have high momentum components above the cutoff and therefore one does not have to be concerned about double counting the high momentum contributions. In EFT, this is understood easily, since the ultraviolet divergences can be renormalized in free space and the many-body dynamics does not lead to new ultraviolet divergences. More precisely, in any many-body diagram, the excitations to high momentum states that are absorbed in $V_{\text{low}\,k}$ vanish due to the momentum space cutoff by construction. Using $V_{\text{low}\,k}$ is simply analogous to using the relatively soft CD Bonn interaction, instead of the stronger Argonne v_{18} potential.

Since $V_{\text{low}\,k}$ accounts for a large part of the phase space in the many-body system, it will be interesting to explore whether calculations starting with $V_{\text{low}\,k}$ could be organized in a more perturbative fashion. Encouraging results for a perturbative expansion of the shell model effective Hamiltonian in few-body systems has been demonstrated for the deuteron in [39]. For larger systems, this idea more relies on a geometrical phase space argument.

Acknowledgements

We thank Gerry Brown, Bengt Friman and Dick Furnstahl for helpful discussions. This work was supported by the US-DOE Grant DE-FG02-88ER40388, the US-DOE Grant DE-FG03-00ER41132, the NSF under Grant No. PHY-0098645 and by an Ohio State University Postdoctoral Fellowship.

Appendix

Diagonal matrix elements of the hermitian $\bar{V}_{\text{low } k}$ (see Figs. 18 and 19).

References

[1] S.C. Pieper, V.R. Pandharipande, R.B. Wiringa, J. Carlson, Phys. Rev. C 64 (2001) 014001.
[2] S.C. Pieper, K. Varga, R.B. Wiringa, nucl-th/0206061.
[3] S.K. Bogner, T.T.S. Kuo, L. Coraggio, Nucl. Phys. A 684 (2001) 432c.
[4] S.K. Bogner, T.T.S. Kuo, A. Schwenk, D.R. Entem, R. Machleidt, nucl-th/0108041.
[5] R. Machleidt, I. Slaus, J. Phys. G 27 (2001) R69.
[6] R. Vinh Mau, An advanced course in modern nuclear physics, in: J.M. Arias, M. Lozano (Eds.), Lecture Notes in Physics, Vol. 581, Springer, Berlin, 2001, p. 1.
[7] M. Lacombe, B. Loiseau, J. M. Richard, R. Vinh Mau, J. Côté, P. Pirès, R. de Tourreil, Phys. Rev. C 21 (1980) 861.
[8] R. Machleidt, K. Holinde, C. Elster, Phys. Rep. 149 (1987) 1.
[9] R. Machleidt, F. Sammarruca, Y. Song, Phys. Rev. C 53 (1996) 1483.
[10] R. Machleidt, Phys. Rev. C 63 (2001) 024001.
[11] V.G.J. Stoks, R.A.M. Klomp, C.P.F. Terheggen, J.J. de Swart, Phys. Rev. C 49 (1994) 2950.
[12] R.B. Wiringa, V.G.J. Stoks, R. Schiavilla, Phys. Rev. C 51 (1995) 38.
[13] S. Weinberg, Phys. Lett. B 251 (1990) 288.
[14] S. Weinberg, Nucl. Phys. B 363 (1991) 3.
[15] C. Ordonez, L. Ray, U. van Kolck, Phys. Rev. Lett. 72 (1994) 1982.
[16] D.B. Kaplan, M.J. Savage, M.B. Wise, Nucl. Phys. B 534 (1998) 329.
[17] T.-S. Park, K. Kubodera, D.-P. Min, M. Rho, Phys. Rev. C 58 (1998) 637.
[18] E. Epelbaum, W. Glockle, U.-G. Meissner, Nucl. Phys. A 671 (2000) 295.
[19] D.R. Entem, R. Machleidt, Phys. Lett. B 524 (2001) 93.
[20] S.R. Beane, P.F. Bedaque, W.C. Haxton, D.R. Phillips, M.J. Savage, in: M. Shifman (Ed.), At the Frontier of Particle Physics, Vol. 1, World Scientific, Singapore, p. 133, nucl-th/0008064.
[21] H.A. Bethe, C. Longmire, Phys. Rev. 77 (1950) 647.
[22] G.E. Brown, A.D. Jackson, The Nucleon–Nucleon Interaction, North-Holland, Amsterdam, 1976.
[23] G.P. Lepage, "how to renormalize the Schrödinger equation", Lectures given at 9th Jorge Andre Swieca Summer School: Particles and Fields, Sao Paulo, Brazil, February, 1997, nucl-th/9706029.
[24] V.G.J. Stoks, R.A.M. Klomp, M.C.M. Rentmeester, J.J. de Swart, Phys. Rev. C 48 (1993) 792.
[25] NN-OnLine, http://nn-online.sci.kun.nl.
[26] S.K. Bogner, A. Schwenk, T.T.S. Kuo, G.E. Brown, nucl-th/0111042.
[27] M. Hjorth-Jensen, T.T.S. Kuo, E. Osnes, Phys. Rept. 261 (1995) 125.
[28] C. Bloch, Nucl. Phys. 6 (1958) 329.
[29] C. Bloch, J. Horowitz, Nucl. Phys. 8 (1958) 91.
[30] W.C. Haxton, C.-L. Song, Phys. Rev. Lett. 84 (2000) 5484.
[31] S.Y. Lee, K. Suzuki, Phys. Lett. B 91 (1980) 173.
[32] K. Suzuki, S.Y. Lee, Prog. Theor. Phys. 64 (1980) 2091.
[33] F. Andreozzi, Phys. Rev. C 54 (1996) 684.
[34] K. Suzuki, Prog. Theor. Phys. 68 (1982) 246.
[35] K. Suzuki, R. Okamoto, Prog. Theor. Phys. 70 (1983) 439.
[36] S. Okubo, Prog. Theor. Phys. 12 (1954) 603.
[37] T.T.S. Kuo, S.Y. Lee, K.F. Ratcliff, Nucl. Phys. A 176 (1971) 65.
[38] T.T.S. Kuo, E. Osnes, Folded-diagram theory of the effective interaction in nuclei, atoms and molecules, Lecture Notes in Physics, Vol. 364, Springer, Berlin, 1990, p. 1.
[39] W.C. Haxton, T. Luu, Phys. Rev. Lett. 89 (2002) 182503.

ELSEVIER

Available online at www.sciencedirect.com

PHYSICS LETTERS B

Physics Letters B 576 (2003) 265–272

www.elsevier.com/locate/physletb

Towards a model-independent low momentum nucleon–nucleon interaction

S.K. Bogner [a], T.T.S. Kuo [a], A. Schwenk [a], D.R. Entem [b], R. Machleidt [b]

[a] *Department of Physics and Astronomy, State University of New York, Stony Brook, NY 11794-3800, USA*
[b] *Department of Physics, University of Idaho, Moscow, ID 83844-0903, USA*

Received 1 May 2003; received in revised form 7 August 2003; accepted 4 October 2003

Editor: J.-P. Blaizot

Abstract

We provide evidence for a high precision model-independent low momentum nucleon–nucleon interaction. Performing a momentum-space renormalization group decimation, we find that the effective interactions constructed from various high precision nucleon–nucleon interaction models, such as the Paris, Bonn, Nijmegen, Argonne, CD Bonn and Idaho potentials, are identical. This model-independent low momentum interaction, called $V_{\mathrm{low\,k}}$, reproduces the same phase shifts and deuteron pole as the input potential models, without ambiguous assumptions on the high momentum components, which are not constrained by low energy data and lead to model-dependent results in many-body applications. $V_{\mathrm{low\,k}}$ is energy-independent and does not necessitate the calculation of the Brueckner G matrix.

PACS: 21.30.Cb; 21.60.-n; 21.30.Fe; 11.10.Hi

Keywords: Nucleon–nucleon interaction; Effective interactions; Renormalization group

1. Introduction

In low energy nuclear systems such as finite nuclei and nuclear matter, one can base complicated many-body calculations on a simple picture of point-like nucleons interacting by means of a two-body potential and three-body forces when needed. Unlike for electronic systems where the low energy Coulomb force is unambiguously determined from quantum electrodynamics, there is much ambiguity in nuclear physics owing to the non-perturbative nature of quantum chromodynamics (QCD) at low energy scales. Consequently, there are a number of high precision, phenomenological meson exchange models of the two-nucleon force V_{NN}, such as the Paris [1], Bonn [2], Nijmegen [3], Argonne [4] and CD Bonn [5,6] potentials, as well as model-independent but less accurate treatments based on chiral effective field theory (EFT) [7–10], for a review see [11]. We also study the Idaho potential [12], which is based on the EFT framework, but some model dependence is introduced in order to achieve similarly high precision as compared to the other interaction models.

E-mail addresses: bogner@phys.washington.edu (S.K. Bogner), kuo@nuclear.physics.sunysb.edu (T.T.S. Kuo), aschwenk@mps.ohio-state.edu (A. Schwenk).

0370-2693/$ – see front matter
doi:10.1016/j.physletb.2003.10.012

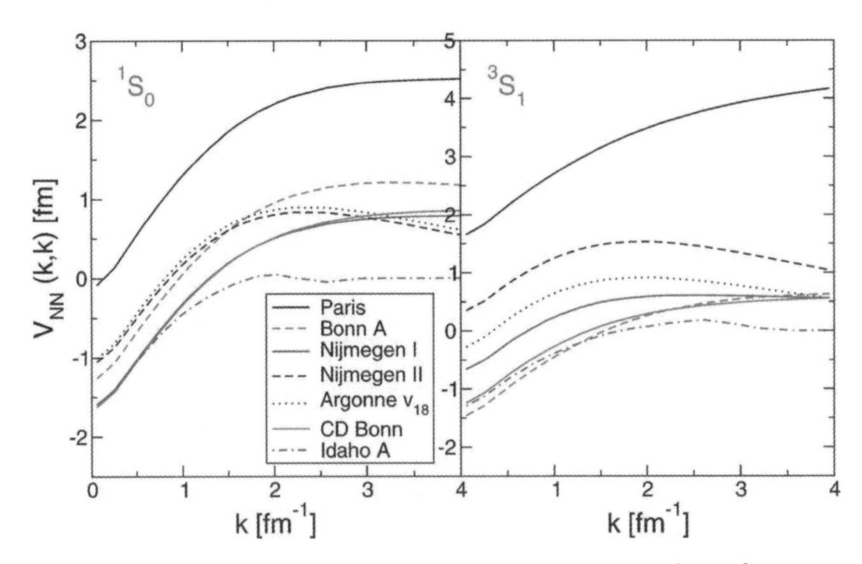

Fig. 1. Momentum-space matrix elements of V_{NN} for different bare potentials in the $^1S_0$ and $^3S_1$ channels.

These nuclear force models incorporate the same one-pion exchange interaction (OPE) at long distance, but differ in their treatment of the intermediate and short-range parts (e.g., parameterization of the repulsive core compared to ω exchange, different form factors, dispersive or field theoretical treatment of the 2π exchange). The fact that the different short distance constructions reproduce the same phase shifts and deuteron properties indicates that low energy observables are insensitive to the details of the short distance dynamics. The EFT approach exploits this insensitivity by explicitly keeping only the pion and nucleon degrees of freedom (in accordance with the spontaneous breaking of chiral symmetry in the QCD vacuum) and encoding the effects of the integrated heavy degrees of freedom in the form of couplings which multiply model-independent delta functions and their derivatives, see, e.g., [13]. The EFT thus provides a model-independent description of the two-nucleon system. However, the high precision, i.e., $\chi^2/\text{datum} \approx 1$, description of the nucleon–nucleon scattering data provided by conventional models is at present not achieved by the rigorous EFT potentials.

In Fig. 1, we observe that the realistic models of V_{NN} have quite different momentum-space matrix elements despite their common OPE parts and reproduction of the same low energy data. It demonstrates that the low energy phase shifts and deuteron properties cannot distinguish between the models used for the short distance parts. However, in many-body calculations the assumed short-distance structure of a particular V_{NN} enters by means of virtual nucleon states extending to high momentum, which will lead to model-dependent results, e.g., in Brueckner Hartree–Fock calculations of the binding energy of nuclear matter. It is clearly of great interest to remove such model dependence from microscopic nuclear many-body calculations.

Motivated by these observations, we perform a renormalization group (RG) decimation and integrate out the ambiguous high momentum components of the realistic interactions. Due to the separation of scales in the nuclear problem, it is reasonable to expect that the model dependence of the input potentials will be largely removed as the high momentum components are excluded from the Hilbert space. Starting from any of the V_{NN} in Fig. 1, we integrate out the momenta above a cutoff Λ to obtain an effective low momentum potential, called $V_{\text{low k}}$. The physical condition is that the effective theory reproduces the deuteron pole and the half-on-shell (HOS) T matrices of the input V_{NN} model (i.e., the observable scattering phase shifts and the low momentum components of the two-body wave functions, which probe the OPE part of the interaction), but with all loop momenta cutoff at Λ. We will find the striking result of Fig. 2 that the $V_{\text{low k}}$

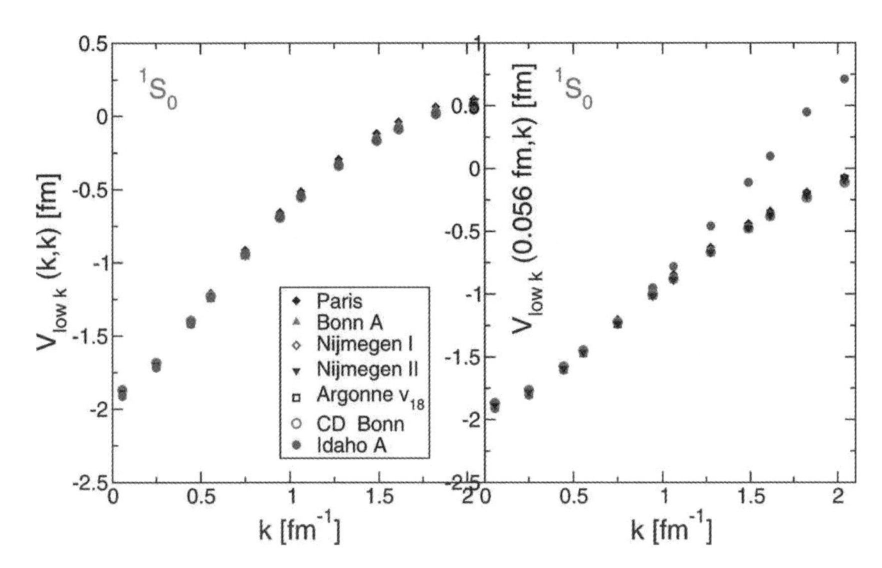

Fig. 2. Diagonal and off-diagonal momentum-space matrix elements of the $V_{\text{low k}}$ obtained from the different bare potentials in the $^1S_0$ channel for a cutoff $\Lambda = 2.1\ \text{fm}^{-1}$.

becomes independent of the particular input model for $\Lambda \lesssim 2.1\ \text{fm}^{-1}$, which corresponds to laboratory energies $E_{\text{lab}} \lesssim 350$ MeV. The latter is the energy scale over which the high precision potentials are constrained by experimental data.

Our strategy thus is a compromise between conventional models and EFT treatments. The resulting $V_{\text{low k}}$ reproduces the nucleon–nucleon scattering phase shifts with similar accuracy as the high precision potentials, but without making further assumptions on the detailed high momentum structure, which cannot be resolved by fitting to the low energy data only.

2. Renormalization group decimation

The first step in the RG decimation is to define an energy-dependent effective potential (called the $\hat{Q}$ box or the Bloch–Horowitz potential in effective interaction theory), which is irreducible with respect to cutting intermediate low momentum propagators. The $\hat{Q}$ box resums the effects of the high momentum modes,

$$\hat{Q}(k', k; \omega) = V_{\text{NN}}(k', k) + \frac{2}{\pi}\mathcal{P}\int_{\Lambda}^{\infty} \frac{V_{\text{NN}}(k', p)\hat{Q}(p, k; \omega)}{\omega - p^2} p^2\, dp, \tag{1}$$

where k', k, and p denote the relative momentum of the outgoing, incoming, and intermediate nucleons. The principal value HOS T matrix for a given partial wave is obtained by solving the Lippmann–Schwinger equation

$$T\big(k', k; k^2\big) = V_{\text{NN}}(k', k) + \frac{2}{\pi}\mathcal{P}\int_{0}^{\infty} \frac{V_{\text{NN}}(k', p)T(p, k; k^2)}{k^2 - p^2} p^2\, dp. \tag{2}$$

In terms of the effective cutoff-dependent $\hat{Q}$ potential, the scattering equation can be expressed as

$$T\big(k', k; k^2\big) = \hat{Q}\big(k', k; k^2\big) + \frac{2}{\pi}\mathcal{P}\int_{0}^{\Lambda} \frac{\hat{Q}(k', p; k^2)T(p, k; k^2)}{k^2 - p^2} p^2\, dp. \tag{3}$$

The low momentum effective theory defined by Eqs. (1) and (3) preserves the low energy scattering amplitudes and bound states independently of the chosen model space, i.e., the value of the cutoff Λ. However, the energy dependence of the effective $\hat{Q}$ potential is inconvenient for practical calculations. In order to eliminate the energy dependence, we introduce the so-called Kuo–Lee–Ratcliff folded diagrams, which provide a way of reorganizing the Lippmann–Schwinger equation, Eq. (3), such that the energy dependence of the $\hat{Q}(k', k; \omega)$ box is converted to a purely momentum dependent interaction, $V_{\mathrm{low\,k}}(k', k)$ [14,15]. The folded diagrams are correction terms one must add to Eq. (3), if one were to set all $\hat{Q}$ box energies right side on-shell. This explicitly leads to (for details see [15,16])

$$V_{\mathrm{low\,k}}(k', k) = \hat{Q}(k', k; k^2) + \frac{2}{\pi}\mathcal{P}\int_0^{\Lambda} p^2\,dp\,\hat{Q}(p, k; k^2) \times \frac{\hat{Q}(k', p; k^2) - \hat{Q}(k', p; p^2)}{k^2 - p^2} + \mathcal{O}(\hat{Q}^3). \quad (4)$$

The folded diagram resummation indicated in Eq. (4) can be carried out to all orders using the similarity transformation method of Lee and Suzuki [17,18]. By construction, the resulting energy-independent $V_{\mathrm{low\,k}}$ preserves the low momentum HOS T matrix of the input V_{NN} model [19],

$$T(k', k; k^2) = V_{\mathrm{low\,k}}(k', k) + \frac{2}{\pi}\mathcal{P}\int_0^{\Lambda} \frac{V_{\mathrm{low\,k}}(k', p)T(p, k; k^2)}{k^2 - p^2} p^2\,dp, \quad (5)$$

where all momenta are constrained to lie below the cutoff Λ. As our RG decimation preserves the half-on-shell T matrix, we have $dT(k', k; k^2)/d\Lambda = 0$, which implies a RG equation for $V_{\mathrm{low\,k}}$ [16]

$$\frac{d}{d\Lambda}V_{\mathrm{low\,k}}(k', k) = \frac{2}{\pi}\frac{V_{\mathrm{low\,k}}(k', \Lambda)T(\Lambda, k; \Lambda^2)}{1 - (k/\Lambda)^2}. \quad (6)$$

Similarly, a scaling equation is obtained for the $\hat{Q}$ box by integrating out an infinitesimal momentum shell, and one finds

$$\frac{d}{d\Lambda}\hat{Q}(k', k; p^2) = \frac{2}{\pi}\frac{\hat{Q}(k', \Lambda; p^2)\hat{Q}(\Lambda, k; p^2)}{1 - (p/\Lambda)^2}. \quad (7)$$

This equation was obtained previously by Birse et al. by requiring the invariance of the full-off-shell T matrix, $dT(k', k; p^2)/d\Lambda = 0$ [20].

The RG equation, Eq. (6), lies at the heart of the approach presented here: given a microscopic input model with a large cutoff, one can use the RG equation to evolve the bare interaction to a physically equivalent, but simpler effective theory valid for energies below the cutoff. The RG evolution separates the details of the assumed short distance dynamics, while incorporating their detail-independent effects on low energy phenomena through the running of the effective interaction [16,21].

The principal difference between the presented RG approach and the standard EFT one is that we do not expand the interaction in powers of local operators. This implies that, first, one does not make assumptions on locality, and that second, we do not truncate after a certain power included in the low momentum interaction. In the EFT approach, power-counting is used to calculate observables to a given order with controllable errors. In contrast, we start from a Hamiltonian in a large space, which reproduces the low energy observables with high accuracy and then decimate to a smaller, low momentum space so that the observables are reproduced. Since the exact RG equation is solved without truncation, the invariance of the T matrices is guaranteed.

Nevertheless, there are close similarities between the presented RG approach and the standard EFT one. For example, at cutoffs above the pion mass, $V_{\mathrm{low\,k}}$ keeps the pion exchange explicit, while the unresolved short distance physics could be encoded in a series of contact terms. Moreover, we find below that $V_{\mathrm{low\,k}}$ seems only to depend on the fact that all potential models have the same pion tail, and fit the same phase shifts. This is similar to EFT treatments, where the interaction is constrained by pion exchange, phase shifts and the choice of regulator.

3. Results

Referring to Fig. 2, we find the central result of this Letter. The diagonal matrix elements of $V_{\mathrm{low\,k}}$ obtained from the different V_{NN} of Fig. 1 collapse onto the same curve for $\Lambda \lesssim 2.1\ \mathrm{fm}^{-1}$. Similar results are found in all partial waves and will be reported elsewhere in detail [23]. We emphasize that $V_{\mathrm{low\,k}}$ reproduces the experimental phase shift data and the deuteron pole with similar accuracy as compared with

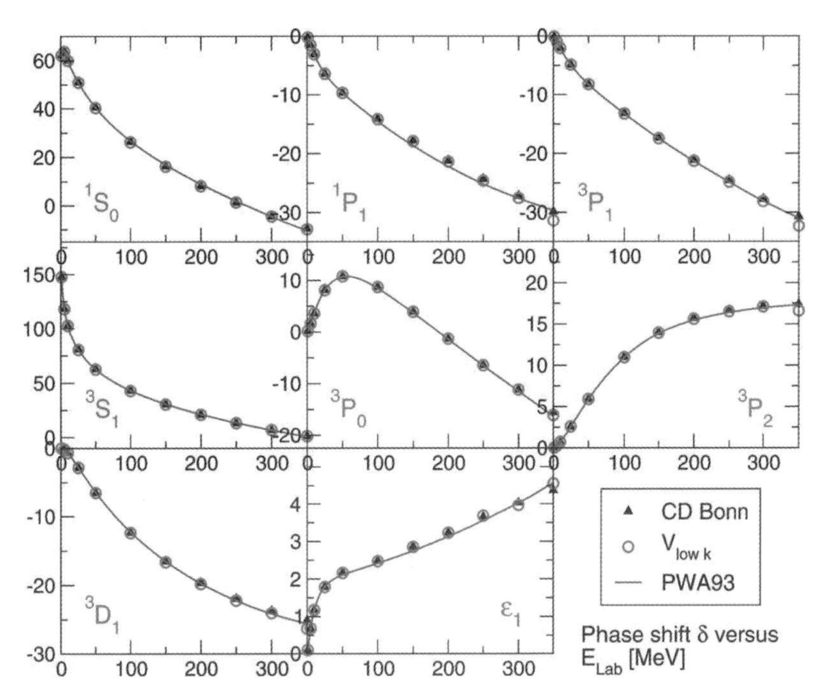

Fig. 3. S-wave (singlet and triplet with mixing parameter) and P-wave phase shifts of $V_{\mathrm{low\,k}}$ for a cutoff $\Lambda = 2.1\ \mathrm{fm}^{-1}$ compared to the input V_{NN} model used, here the CD Bonn potential. We also show the results of the multi-energy phase shift analysis (PWA93) of the Nijmegen group [22]. For $V_{\mathrm{low\,k}}$ the agreement of the calculated phase shifts with the PWA93 is determined by the quality of the fit for the bare interaction.

the high precision potential models, as we show in Fig. 3. The reproduction of the phase shifts with the renormalized $V_{\mathrm{low\,k}}$ does however not require ambiguous high momentum components, which are not constrained by the low energy scattering data. We note that the change in sign of the $^1\mathrm{S}_0$ and the $^3\mathrm{P}_0$ phase shifts indicates that the effects of the repulsive core are properly encoded in $V_{\mathrm{low\,k}}$.

Intriguingly, the $V_{\mathrm{low\,k}}$ mainly differs from the bare potential by a constant shift. This was previously observed by Epelbaum et al. for a toy two-Yukawa bare potential [24]. The constant shift in momentum space corresponds to a smeared delta function in coordinate space and accounts for the renormalization of the repulsive core from the bare interactions, see also [25].

In Fig. 2, we find a similar behaviour for the off-diagonal matrix elements, although the $V_{\mathrm{low\,k}}$ derived from the Idaho potential begins to differ from the others at approximately $2m_\pi = 1.4\ \mathrm{fm}^{-1}$. This discrepancy in the off-diagonal matrix elements arises from the fact that the Idaho potential used here treats the 2π exchange differently than the meson models do. We can integrate out further and lower the cutoff to $\Lambda \lesssim 1.4\ \mathrm{fm}^{-1}$, then the off-diagonal elements collapse as well.

These results can be understood from T matrix preservation. The HOS T matrix determines the phase shifts as well as the low momentum components of the low energy scattering and bound state wave functions. Using the spectral representation of the T matrix, $V_{\mathrm{low\,k}}$ can be expressed as

$$V_{\mathrm{low\,k}}(k',k) = T\big(k',k;k^2\big) + \frac{2}{\pi}\mathcal{P}\int_0^\Lambda T\big(k',p;p^2\big) \times \frac{1}{p^2-k^2}T\big(p,k;p^2\big). \tag{8}$$

The V_{NN} models give the same on-shell T matrices over the phase shift equivalent kinematic range of $k \lesssim 2.1\ \mathrm{fm}^{-1}$, but their off-shell behavior is a priori unconstrained. In practice, however, one observes that

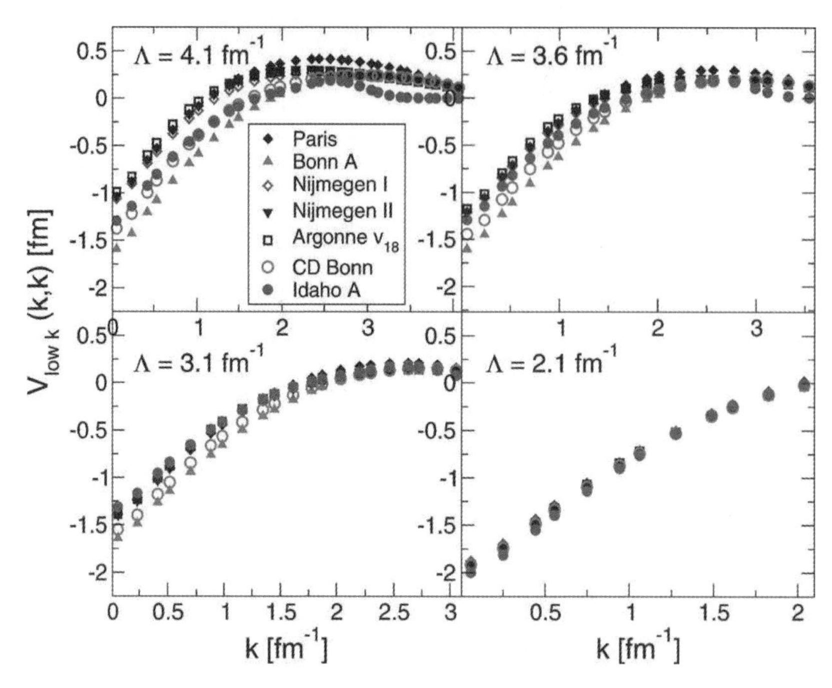

Fig. 4. Evolution of the diagonal momentum-space matrix elements for the $V_{\mathrm{low\,k}}$ derived from the different bare potentials from larger values of the cutoff to $\Lambda = 2.1\ \mathrm{fm}^{-1}$ in the $^3\mathrm{S}_1$ partial wave.

the realistic potential models result in similar HOS T matrices at low energies and momenta. This is understood from the fact that these models share the same long-range OPE interaction and differ most noticeably on short-distance scales set by the repulsive core, $r \sim 0.5$ fm and smaller. Consequently, one expects that the off-shell differences are suppressed at energies and momenta below a corresponding scale of $\Lambda \sim 1/r \sim 2.0\ \mathrm{fm}^{-1}$. It is clear from Eq. (8) that the approximate HOS T matrix equivalence is a sufficient condition for the $V_{\mathrm{low\,k}}$ to be independent of the various potential models. Moreover, the spectral representation clarifies that the off-diagonal matrix elements of $V_{\mathrm{low\,k}}$ are more sensitive to a particular off-shell behavior as observed in Fig. 2, and thus deviate at a lower cutoff than the diagonal matrix elements. The collapse of the diagonal momentum-space matrix elements at the scale set by the constraining scattering data, $\Lambda \approx 2.1\ \mathrm{fm}^{-1}$, is nicely illustrated in Fig. 4, which shows the RG evolution as the cutoff is successively lowered in the $^3\mathrm{S}_1$ partial wave.

Next, we analyze the scaling properties of $V_{\mathrm{low\,k}}$. For this purpose, we show matrix element $V_{\mathrm{low\,k}}(0,0)$ versus cutoff in Fig. 5 for the $^1\mathrm{S}_0$ and $^3\mathrm{S}_1$ partial waves. The main results are the following. T matrix preservation guarantees that $V_{\mathrm{low\,k}}(0,0)$ flows toward the scattering length as the cutoff is taken to zero, these are $a_{^1\mathrm{S}_0} = -23.73$ fm and $a_{^3\mathrm{S}_1} = 5.42$ fm and are well reproduced with $V_{\mathrm{low\,k}}$.

For cutoffs $\Lambda > m_\pi$, $V_{\mathrm{low\,k}}$ is nearly independent of the cutoff in the $^1\mathrm{S}_0$ channel and weakly linearly dependent in the $^3\mathrm{S}_1$ channel. According to EFT principles, the couplings are nearly independent of the cutoff as long as Λ is large enough to explicitly include the relevant degrees of freedom needed to describe the scale one is probing. Thus, the running of $V_{\mathrm{low\,k}}$ initiated at $\Lambda \sim m_\pi$ is a result of integrating out the pion. The rapid changes at very small Λ are the result of the large scattering length. To illustrate the effects of the large scattering length, we solve the RG equation for $V_{\mathrm{low\,k}}(0,0)$ for small cutoffs, with the $^1\mathrm{S}_0$ scattering length as the $\Lambda = 0$ boundary condition. One finds

$$V_{\mathrm{low\,k}}(0,0) \approx \frac{1}{1/a_{^1\mathrm{S}_0} - 2\Lambda/\pi}, \tag{9}$$

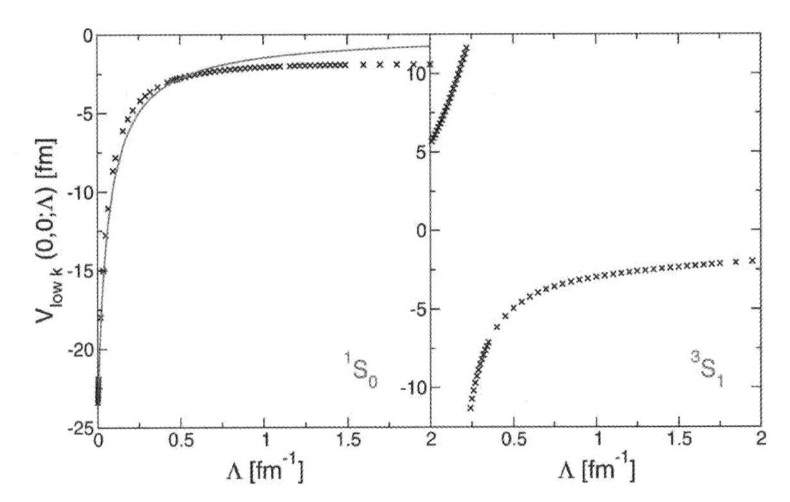

Fig. 5. RG flow of $V_{\mathrm{low\,k}}(0,0;\Lambda)$ versus cutoff Λ in the $^1\mathrm{S}_0$ and $^3\mathrm{S}_1$ partial waves. The solid line represents the solution of the RG equation for small Λ as discussed in the text.

which is shown as the solid curve in Fig. 5. Clearly, the agreement for small $\Lambda < m_\pi/2$ is convincing. Eq. (9) is identical to the Kaplan–Savage–Wise solution for the renormalization of the momentum-independent contact term in the pionless EFT, upon identifying Λ with the regulator mass μ used in dimensional regularization [8]. The weak dependence on the cutoff in the $^3\mathrm{S}_1$ partial wave results from the dominantly second order tensor contributions, which are peaked at relative momenta $k \approx 2\ \mathrm{fm}^{-1}$ [26].

Finally, we note that $V_{\mathrm{low\,k}}$ preserves all scattering and bound states with energy $|E| < \Lambda^2$. Therefore, the jump in the $^3\mathrm{S}_1$ channel is the bound state contribution to the T matrix and occurs at $\Lambda = k_{\mathrm{D}} = \sqrt{m E_{\mathrm{D}}/\hbar^2} = 0.23\ \mathrm{fm}^{-1}$. From effective range theory and the Low equation [27], we have for the bound state contribution and thus the discontinuity

$$\Delta V_{\mathrm{low\,k}}(0,0;k_{\mathrm{D}})_{^3\mathrm{S}_1} = \frac{4}{\pi\, k_{\mathrm{D}}(1 - k_{\mathrm{D}} r_0)} = 22.8\ \mathrm{fm}, \tag{10}$$

where $r_0 = 1.75$ fm denotes the effective range in $^3\mathrm{S}_1$. This is in very good agreement with our results.

4. Summary

In this Letter, we have shown that the separation of scales in the nuclear force can be successfully applied to derive a model-independent low momentum nucleon–nucleon interaction. $V_{\mathrm{low\,k}}$ is obtained by integrating out the high momentum components of various potential models, leading to a physically equivalent effective theory in the low momentum Hilbert space. The decimation filters the details of the assumed short-distance dynamics of the bare interactions, provided one requires that the low energy observables are preserved under the RG. We have argued and provided numerical evidence that the model-independence of $V_{\mathrm{low\,k}}$ is a consequence of the common long-range OPE interaction and the reproduction of the same elastic scattering phase shifts. The momentum scale $\Lambda \sim 2.1\ \mathrm{fm}^{-1}$, corresponding to laboratory energies $E_{\mathrm{lab}} \sim 350$ MeV, is precisely the scale at which the low momentum interaction becomes independent of the input models. Our Letter demonstrates that the differences in the high momentum components of the nucleon force are not constrained by fits to the low energy phase shifts and the deuteron properties.

$V_{\mathrm{low\,k}}$ does not have high momentum modes, which are related to the strong, short range-repulsion in conventional models. Therefore, the low momentum interaction is considerably softer than the bare interactions (both in a plane-wave and an oscillator basis) and does not have a repulsive core. As a consequence, when $V_{\mathrm{low\,k}}$ is used as the microscopic input in the many-body problem, the high momentum effects in the particle–particle channel do not have to be addressed by performing a Brueckner ladder resummation or short-range correlation methods. In fact, with a cutoff on relative momenta, the phase space in the particle–particle channel is comparable to the particle–hole channels, and it would seem strange to resum the particle–particle channel, while the particle–hole channels are treated perturbatively in a hole–line expansion.

The use of $V_{\text{low}\,k}$ in microscopic nuclear many-body calculations leads to model-independent results. $V_{\text{low}\,k}$ has been successfully used as shell model effective interaction in model space calculations for two valence particle nuclei such as ^{18}O and ^{134}Te [28]. The starting point of these calculations has traditionally been the Brueckner G matrix, which depends on the bare V_{NN} used as well as the particular nuclei via the Pauli blocking operator. By means of $V_{\text{low}\,k}$, the same low momentum interaction is used in different mass regions. $V_{\text{low}\,k}$ has further been incorporated into Fermi liquid theory. This connects the low momentum interaction in free space and the quasiparticle interaction in normal Fermi systems. Two constraints have been derived which relate the Fermi liquid parameters of nuclear matter to the S-wave low momentum interaction at zero relative momentum [29]. Finally, adopting the RG approach to Fermi systems proposed by Shankar, $V_{\text{low}\,k}$ is taken as the input to microscopic calculations of the quasiparticle interactions and the pairing gaps in neutron matter [30]. From the success of these applications, we believe that the model-independent $V_{\text{low}\,k}$ is a very promising starting point for microscopic nuclear many-body calculations.

Few-body calculations with $V_{\text{low}\,k}$ are in progress. Preliminary results for the ground state energies of triton and ^{3}He show that the three-body force for $V_{\text{low}\,k}$ is weaker than, e.g., the conventional three-body force constructed for the Argonne potential.

Acknowledgements

We thank Gerry Brown for his encouragement and many stimulating discussions. This work was supported by the US DOE grant DE-FG02-88ER40388, the US NSF grant PHY-0099444 and by the Ramon Areces Foundation of Spain.

References

[1] M. Lacombe, et al., Phys. Rev. C 21 (1980) 861.
[2] R. Machleidt, K. Holinde, C. Elster, Phys. Rep. 149 (1987) 1.
[3] V.G.J. Stoks, et al., Phys. Rev. C 49 (1994) 2950.
[4] R.B. Wiringa, V.G.J. Stoks, R. Schiavilla, Phys. Rev. C 51 (1995) 38.
[5] R. Machleidt, F. Sammarruca, Y. Song, Phys. Rev. C 53 (1996) 1483.
[6] R. Machleidt, Phys. Rev. C 63 (2001) 024001.
[7] C. Ordonez, L. Ray, U. van Kolck, Phys. Rev. Lett. 72 (1994) 1982.
[8] D.B. Kaplan, M.J. Savage, M.B. Wise, Nucl. Phys. B 534 (1998) 329.
[9] T.-S. Park, et al., Phys. Rev. C 58 (1998) 637.
[10] E. Epelbaum, W. Glockle, U.-G. Meissner, Nucl. Phys. A 671 (2000) 295.
[11] S.R. Beane, et al., in: M. Shifman (Ed.), At the Frontier of Particle Physics, Vol. 1, World Scientific, p. 133, nucl-th/0008064.
[12] D.R. Entem, R. Machleidt, Phys. Lett. B 524 (2001) 93.
[13] G.P. Lepage, How to Renormalize the Schrödinger Equation, Lectures given at 9th Jorge Andre Swieca Summer School: Particles and Fields, São Paulo, Brazil, February 1997, nucl-th/9706029.
[14] T.T.S. Kuo, S.Y. Lee, K.F. Ratcliff, Nucl. Phys. A 176 (1971) 65.
[15] T.T.S. Kuo, E. Osnes, Folded-Diagram Theory of the Effective Interaction in Nuclei, Atoms and Molecules, in: Lecture Notes in Physics, Vol. 364, Springer, Berlin, 1990, p. 1.
[16] S.K. Bogner, A. Schwenk, T.T.S. Kuo, G.E. Brown, nucl-th/0111042.
[17] S.Y. Lee, K. Suzuki, Phys. Lett. B 91 (1980) 173.
[18] K. Suzuki, S.Y. Lee, Prog. Theor. Phys. 64 (1980) 2091.
[19] S.K. Bogner, T.T.S. Kuo, L. Coraggio, Nucl. Phys. A 684 (2001) 432c.
[20] M.C. Birse, J.A. McGovern, K.G. Richardson, Phys. Lett. B 464 (1999) 169.
[21] S.K. Bogner, T.T.S. Kuo, Phys. Lett. B 500 (2001) 279.
[22] V.G.J. Stoks, et al., Phys. Rev. C 48 (1993) 792.
[23] S.K. Bogner, T.T.S. Kuo, A. Schwenk, Phys. Rep. 386 (2003) 1.
[24] E. Epelbaum, et al., Nucl. Phys. A 645 (1999) 413.
[25] T. Neff, Ph.D. Thesis, Darmstadt University, 2002.
[26] G.E. Brown, Unified Theory of Nuclear Models and Forces, 3rd Edition, North-Holland, Amsterdam, 1971.
[27] G.E. Brown, A.D. Jackson, The Nucleon–Nucleon Interaction, North-Holland, Amsterdam, 1976.
[28] S.K. Bogner, et al., Phys. Rev. C 65 (2001) 051301(R).
[29] A. Schwenk, G.E. Brown, B. Friman, Nucl. Phys. A 703 (2002) 745.
[30] A. Schwenk, B. Friman, G.E. Brown, Nucl. Phys. A 713 (2003) 191.

PHYSICAL REVIEW C **69**, 034329 (2004)

Family of Hermitian low-momentum nucleon interactions with phase shift equivalence

Jason D. Holt, T. T. S. Kuo, and G. E. Brown
Department of Physics, SUNY, Stony Brook, New York 11794, USA
(Received 29 September 2003; published 25 March 2004)

Using a Schmidt orthogonalization transformation, a family of Hermitian low-momentum nucleon-nucleon (*NN*) interactions is derived from the non-Hermitian Lee-Suzuki (LS) low-momentum *NN* interaction. As special cases, our transformation reproduces the Hermitian interactions of Okubo and Andreozzi. Aside from their common preservation of the deuteron binding energy, these Hermitian interactions are shown to be phase shift equivalent, all preserving the empirical phase shifts up to decimation scale Λ. Employing a solvable matrix model, the Hermitian interactions given by different orthogonalization transformations are studied; the interactions can be very different from each other particularly when there is a strong intruder state influence. However, because the parent LS low-momentum *NN* interaction is only slightly non-Hermitian, the Hermitian low-momentum nucleon interactions given by our transformations, including the Okubo and Andreozzi ones, are all rather similar to each other. Shell model matrix elements given by the LS and several Hermitian low-momentum interactions are compared.

DOI: 10.1103/PhysRevC.69.034329 PACS number(s): 21.30.Fe, 21.60.Cs

I. INTRODUCTION

A fundamental problem in nuclear physics has been the determination of the effective nucleon-nucleon (*NN*) interaction appropriate for complex nuclei. Typically, one starts from a *NN* interaction V_{NN} constrained by the deuteron properties and the empirical low-energy *NN* scattering phase shifts. Several realistic meson models [1–4] for V_{NN} have been obtained in this way, and while they all share the same one pion tail, they differ significantly in how they treat the shorter distance components. Despite this difference, these models all give approximately the same low-energy phase shifts and deuteron binding energy. This result clearly manifests the main theme of the renormalization group (RG) and effective field theory (EFT) approach, namely, physics in the infrared region is insensitive to the details of the short-distance dynamics [5–14]. It is thus possible to have infinitely many theories that differ substantially at small distances, but still give the same low-energy physics, as long as they possess the same symmetries and the "correct" long-wavelength structure. Since low-energy physics is not concerned with these high-energy details, one should just use an effective theory with the short-wavelength modes integrated out.

Following this RG-EFT idea, a low-momentum *NN* effective interaction $V_{low\text{-}k}$ was recently developed [15–20]. While similar in spirit to traditional EFT, $V_{low\text{-}k}$ is not derived via the usual RG-EFT methods; rather, it combines the standard nuclear physics approach with EFT making it a *more effective* EFT (MEEFT) [21–23], as discussed in Ref. [24]. As such, a main step in its derivation is the integrating out of the high momentum components of some realistic *NN* potential model V_{NN} such as those of Refs. [1–4]. Even though these V_{NN} models are quite different from each other, it is remarkable that the $V_{low\text{-}k}$'s derived from them are nearly identical to each other, suggesting a nearly unique $V_{low\text{-}k}$ [25]. Furthermore, shell model calculations using $V_{low\text{-}k}$ have given very encouraging results over a wide range of nuclei [16,17,19,20]. Applications of $V_{low\text{-}k}$ to quasiparticle interaction, superfluid gaps and equation of state for neutron matter have also been highly successful [18].

An important problem in deriving $V_{low\text{-}k}$ is how to obtain a low-momentum *NN* interaction which is Hermitian. The $V_{low\text{-}k}$ given by the *T*-matrix equivalence approach [15,16] is not Hermitian, and some additional transformation is needed to make it Hermitian. There are a number of methods used to obtain a Hermitian effective interaction, such as those of Okubo [26], Suzuki and Okamoto [27], and Andreozzi [28]. Which of these methods should one use? How different are the Hermitian $V_{low\text{-}k}$'s given by them? These questions seem to have not been investigated. In the present work, we shall study the $V_{low\text{-}k}$'s given by these methods as well as develop a unified method by which a family of phase-shift equivalent Hermitian low-momentum *NN* interactions can be obtained. It is, of course, important that $V_{low\text{-}k}$ preserve phase shifts, and while the non-Hermitian $V_{low\text{-}k}$ given by the *T*-matrix equivalence [15,16] can be shown to preserve phase shifts in a straightforward way, it seems to be more involved to prove the phase shift preservation by Hermitian $V_{low\text{-}k}$. Epelbaum *et al.* [8] have pointed out that phase shifts are preserved for the Okubo low-momentum interaction [8], but phase shift preservation for other Hermitian $V_{low\text{-}k}$'s seems to have not been investigated—an issue that we will also examine.

To conclude our Introduction, we shall briefly review the non-Hermitian low-momentum interaction given by the *T*-matrix equivalence approach. Then in Sec. II we present a general method, based on Schmidt orthogonalization transformation, for generating a family of Hermitian effective interactions, and show that the Hermitian interactions of Okubo, Suzuki, and Okamoto and Andreozzi all belong to this family. In Sec. III, we study our method using a simple solvable matrix model of the Hoffmann type [29]; focusing on the influence of intruder states and the difference of the Hermitian interactions given by the various methods. In Sec. IV we present a proof that phase shifts are preserved by the Hermitian interactions generated by our method. This preservation will also be checked by phase shift calculations using different Hermitian interactions. Finally in Sec. V we present our results for the $V_{low\text{-}k}$'s corresponding to the various Hermitian interactions, where we show that our approach allows us to construct a $V_{low\text{-}k}$, which preserves the deuteron wave function, in addition to the preservation of deuteron binding energy and low-energy phase shifts.

In the following let us first briefly review the $V_{low\text{-}k}$ given by T-matrix equivalence [15,16]. One starts from the half-on-shell T matrix

$$T(k',k,k^2) = V_{NN}(k',k) + P\int_0^\infty q^2 dq V_{NN}(k',q) \times \frac{1}{k^2-q^2}T(q,k,k^2). \quad (1)$$

An effective low-momentum T matrix is then defined by

$$T_{low\text{-}k}(p',p,p^2) = V_{low\text{-}k}(p',p) + P\int_0^\Lambda q^2 dq V_{low\text{-}k}(p',q) \times \frac{1}{p^2-q^2}T_{low\text{-}k}(q,p,p^2), \quad (2)$$

where the intermediate state momentum is integrated up to Λ. In the above two equations we employ the principal value boundary conditions, as indicated by the symbol P in front of the integral sign. We require that the T matrices satisfy the condition

$$T(p',p,p^2) = T_{low\text{-}k}(p',p,p^2), \quad (p',p) \leq \Lambda. \quad (3)$$

The above equations define the effective low-momentum interaction $V_{low\text{-}k}$. Using a $\hat{Q}$-box folded-diagram method [32,33], it has been shown [15–17] that the above equations are satisfied by the solution

$$V_{low\text{-}k} = \hat{Q} - \hat{Q}'\int\hat{Q} + \hat{Q}'\int\hat{Q}\int\hat{Q} - \hat{Q}'\int\hat{Q}\int\hat{Q}\int\hat{Q} + \cdots. \quad (4)$$

Here $\hat{Q}$ box denotes the irreducible vertex function whose intermediate states are all beyond Λ, and $\hat{Q}'$ is the same vertex function except that it starts with terms second order in the interaction. The low-momentum effective NN interaction of Eq. (4) can be calculated using iteration methods such as the Lee-Suzuki [30], Andreozzi [28], or Krenciglowa-Kuo [31] methods.

The above $V_{low\text{-}k}$ preserves both the deuteron binding energy and the half-on-shell T matrix of V_{NN} (which implies the preservation of the phase shifts up to $E_{lab}=2\hbar^2\Lambda^2/M$, M being the nucleon mass). This $V_{low\text{-}k}$ is not Hermitian, as indicated by Eq. (4). As we will show soon, starting from this $V_{low\text{-}k}$ a family of phase-shift equivalent Hermitian low-momentum NN interactions can be obtained.

II. FORMALISM

Before presenting our general Hermitization procedure, let us first review some basic formulations about the model space effective interaction. We start from the Schroedinger equation

$$(H_0 + V)\Psi_n = E_n\Psi_n, \quad (5)$$

where H_0 is the unperturbed Hamiltonian and V the interaction. The eigenstates of H_0 are ϕ_n with eigenvalues ϵ_n. For example, H_0 can be the kinetic energy operator and V the NN interaction V_{NN}. A model-space projection operator P is defined as $\Sigma_{i=1}^d|\phi_i\rangle\langle\phi_i|$, where d is the dimension of the model space. The projection operator complement to P is denoted as Q, and as usual, one has $P^2=P, Q^2=Q$ and $PQ=0$. In the present work, P represents all the momentum states with momentum less than the cut-off scale Λ.

A model-space effective interaction V_{eff} is introduced with the requirement that the effective Hamiltonian $P(H_0+V_{eff})P$ reproduces some of the eigenvalues and certain information about the eigenfunctions of the original Hamiltonian (H_0+V). There are a number of ways to derive V_{eff}, but, as indicated by Eq. (4), our effective interaction is obtained by the folded diagram method [32,33] and can be calculated conveniently using the Lee-Suzuki-Andreozzi [28] or Krenciglowa-Kuo [31] iteration methods. We denote this effective interaction as V_{LS}, with the corresponding model space Schroedinger equation

$$P(H_0 + V_{LS})P\chi_m = E_m\chi_m, \quad (6)$$

where $\{E_m\}$ is a subset of $\{E_n\}$ of Eq. (5) and $\chi_m = P\Psi_m$.

It is convenient to rewrite the above effective interaction in terms of the wave operator ω, namely,

$$PV_{LS}P = Pe^{-\omega}(H_0 + V)e^{\omega}P - PH_0P, \quad (7)$$

where ω possesses the usual properties

$$\omega = Q\omega P,$$

$$\chi_m = e^{-\omega}\Psi_m,$$

$$\omega\chi_m = Q\Psi_m. \quad (8)$$

While the eigenvectors Ψ_n of Eq. (5) are orthogonal to each other, it is clear that the eigenvectors χ_m of Eq. (6) are not so and the effective interaction V_{LS} is not Hermitian. We now make a Z transformation such that

$$Z\chi_m = v_m,$$

$$\langle v_m|v_{m'}\rangle = \delta_{mm'}, \quad m,m' = 1,d, \quad (9)$$

where d is the dimension of the model space. This transformation reorients the vectors χ_m such that they become orthonormal to each other. We assume that χ_m's $(m=1,d)$ are linearly independent so that Z^{-1} exists, otherwise the above transformation is not possible. Since v_m and Z exist entirely within the model space, we can write $v_m = Pv_m$ and $Z = PZP$.

Using Eq. (9), we transform Eq. (6) into

$$Z(H_0 + V_{LS})Z^{-1}v_m = E_m v_m, \quad (10)$$

which implies

$$Z(H_0 + V_{LS})Z^{-1} = \sum_{m\epsilon P} E_m|v_m\rangle\langle v_m|. \quad (11)$$

Since E_m is real [it is an eigenvalue of Eq. (5)] and the vectors v_m are orthonormal to each other, $Z(H_0+V_{LS})Z^{-1}$ must be Hermitian. The original problem is now reduced to a Hermitian model-space eigenvalue problem

$$P(H_0 + V_{herm})Pv_m = E_m v_m \quad (12)$$

with the Hermitian effective interaction

$$V_{herm} = Z(H_0 + V_{LS})Z^{-1} - PH_0P, \quad (13)$$

or equivalently

$$V_{herm} = Ze^{-\omega}(H_0 + V)e^{\omega}Z^{-1} - PH_0P. \quad (14)$$

To calculate V_{herm}, we must first have the Z transformation. Since there are certainly many ways to construct Z, this generates a family of Hermitian effective interactions, all originating from V_{LS}. For example, we can construct Z using the familiar Schmidt orthogonalization procedure, namely,

$$v_1 = Z_{11}\chi_1,$$

$$v_2 = Z_{21}\chi_1 + Z_{22}\chi_2,$$

$$v_3 = Z_{31}\chi_1 + Z_{32}\chi_2 + Z_{33}\chi_3,$$

$$v_4 = \ldots \quad (15)$$

with the matrix elements Z_{ij} determined from Eq. (9). We denote the Hermitian effective interaction using this Z transformation as V_{schm}. Clearly there are more than one such Schmidt procedures. For instance, we can use v_2 as the starting point, which gives $v_2=Z_{22}\chi_2$, $v_3=Z_{31}\chi_1+Z_{32}\chi_2$, and so forth. This freedom in how the orthogonalization is actually achieved, gives us infinitely many ways to generate a Hermitian interaction, and this is our family of Hermitian interactions produced from V_{LS}.

We now show how some well-known Hermitization transformations relate to (and in fact, are special cases of) ours. We first look at the Okubo transformation [26]. From Eq. (8) we have

$$\langle\chi_m|(\omega^\dagger\omega)|\chi_{m'}\rangle = \delta_{mm'}. \quad (16)$$

It follows that an analytic choice for the Z transformation is

$$Z = P(1 + \omega^\dagger\omega)^{1/2}P. \quad (17)$$

This leads to the Hermitian effective interaction

$$V_{okb\text{-}1} = P(1 + \omega^\dagger\omega)^{1/2}P(H_0 + V_{LS})P(1 + \omega^\dagger\omega)^{-1/2}P - PH_0P. \quad (18)$$

From Eqs. (8), (9), (16), and (17), it is easily seen that the above is equal to the Okubo Hermitian effective interaction

$$V_{okb} = P(1 + \omega^\dagger\omega)^{-1/2}(1 + \omega^\dagger)(H_0 + V)(1 + \omega)(1 + \omega^\dagger\omega)^{-1/2}P - PH_0P \quad (19)$$

giving us an alternate expression, Eq. (18), for the Okubo interaction.

There is another interesting choice for the transformation Z. As pointed out by Andreozzi [28], the positive definite operator $P(1+\omega^\dagger\omega)P$ can be decomposed into two Cholesky matrices, namely,

$$P(1 + \omega^\dagger\omega)P = PLL^TP, \quad (20)$$

where L is a lower triangle Cholesky matrix, L^T being its transpose. Since L is real and it is within the P space, we have from Eq. (16) that

$$Z = L^T \quad (21)$$

and the corresponding Hermitian effective interaction from Eq. (13) is

$$V_{cho} = PL^TP(H_0 + V_{LS})P(L^{-1})^TP - PH_0P. \quad (22)$$

This is the Hermitian effective interaction of Andreozzi [28].

The final Hermitian effective interaction we consider is that of Suzuki and Okamoto [27,34], which is of the form

$$V_{suzu} = Pe^{-G}(H_0 + V)e^GP - PH_0P \quad (23)$$

with $G=\tanh^{-1}(\omega-\omega^\dagger)$ and $G^\dagger=-G$. It has been shown that this interaction is the same as the Okubo interaction [27]. In terms of the Z transformation, it is readily seen that the operator e^{-G} in Eq. (23) is equal to $Ze^{-\omega}$ with Z given by Eq. (17). Thus, three well-known and particularly useful Hermitian effective interactions indeed belong to our family.

III. MODEL CALCULATIONS

The results from the preceding section show that a family of Hermitian effective interactions can be derived from a Schmidt-type transformation of the non-Hermitian interaction V_{LS}. We now check to see if these interactions do reproduce some of the eigenvalues of the original Hamiltonian, and how the effective interactions given by the various methods differ. In this section, we shall use a solvable matrix model to study these questions. Hoffmann *et al.* [29] have employed a matrix model to study the influence of intruder states on effective interactions. Since we are also interested to see how intruder states might effect our Hermitian potentials, we use a matrix model of this type to study V_{schm}, V_{okb}, and V_{cho}, together with their parent non-Hermitian interaction V_{LS}.

We employ a 5×5 matrix model $H=H_0+xV$, x being a strength parameter. We take $H_0=\{1,1,1,5,9\}$ and

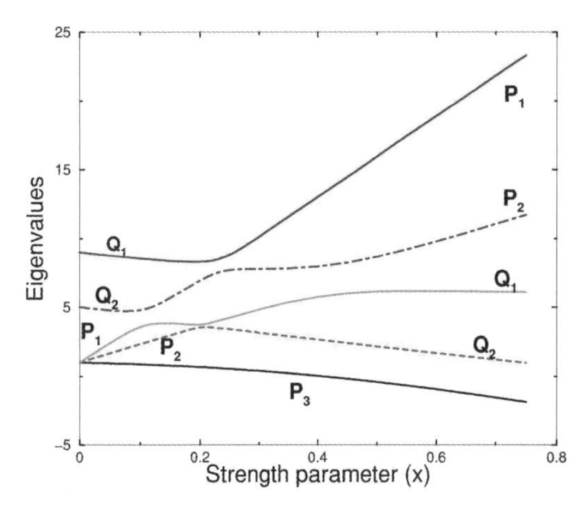

FIG. 1. (Color online) Eigenvalues of our model 5×5 interaction as a function of the strength parameter x.

$$V=\begin{pmatrix} 1 & 2 & 5 & 0 & 5 \\ 5 & 25 & 5 & 5 & 0 \\ 5 & 5 & 15 & 2 & 2 \\ 0 & 5 & 2 & -5 & 1 \\ 5 & 0 & 2 & 1 & -5 \end{pmatrix}. \quad (24)$$

Our P space is chosen as the space spanned by the three lowest eigenstates of H_0, namely, $PH_0P=\{1,1,1\}$. The rather large diagonal matrix elements are used for V so that intruder states will enter as the strength parameter x increases. How this happens is shown in Fig. 1, where the eigenvalues of H are plotted as a function of the strength parameter. In this figure the states are labeled as P_1, P_2, P_3, Q_1, and Q_2, according to the structure of their wave functions. The P states are those whose wave functions are dominated by their P-space components, i.e., $\langle\Psi|P|\Psi\rangle$. In contrast, the Q states are dominated by their Q-space components. For a small interaction (small x), the lowest three states are all P states. As the interaction strength increases, states Q_1 and Q_2 decrease in value and states P_1 and P_2 increase, and we would expect these eigenvalues to intersect. Of course, they do not actually cross, but at certain interaction strength a Q state "intrudes" into the P space, becoming lower than the rising P state. At $x=0.5$, for example, the lowest three states are P_3, Q_2, and Q_1, so Q_2 and Q_1 are intruder states in the sense that they have entered the P space when the interaction is strong. We want our model-space effective Hamiltonian $PH_{eff}P$ to reproduce the lowest three states of H. Thus at large x, we are requiring our $PH_{eff}P$ to reproduce two intruder states. In the rest of this section we study how these intruder states effect our potentials and what impact they have on the Hermitization procedure.

The effective interactions are calculated using the following procedures. First we calculate the wave operator ω using the Lee-Suzuki iteration method developed by Andreozzi [28]; we denote this method as the ALS method, and V_{LS} is then given by Eq. (8). V_{schm} is obtained from Eqs. (9), (13), and (15), and V_{cho} is calculated using Eqs. (20) and (22). For the Okubo interaction, it is convenient to calculate it using the method of Suzuki *et al.* [27,34,35]. With this method, we first find the eigenvalues and eigenfunctions defined by

$$\omega^{\dagger}\omega|\alpha\rangle = \mu_{\alpha}^{2}|\alpha\rangle. \quad (25)$$

Then the Okubo Hermitian effective interaction is given by

$$\langle\alpha|V_{okb}|\beta\rangle = (\sqrt{\mu_{\alpha}^{2}+1}\langle\alpha|V_{LS}|\beta\rangle + \sqrt{\mu_{\beta}^{2}+1}\langle\alpha|V_{LS}^{\dagger}|\beta\rangle)D(\alpha,\beta) \quad (26)$$

with

$$D(\alpha,\beta) = [\sqrt{\mu_{\alpha}^{2}+1} + \sqrt{\mu_{\beta}^{2}+1}]^{-1}. \quad (27)$$

Using these methods we can calculate V_{LS}, V_{okb}, V_{cho}, and V_{schm} for the model potential of Eq. (24). Two different strength parameters, $x=0.1$ and $x=0.55$, have been used to see how the intruder states influence the effective interactions. Our results for these two parameters are shown in Tables I and II. First let us inspect the Hermiticity of our effective interactions; clearly V_{okb}, V_{cho}, and V_{schm} are all Hermitian, irrespective of the strength parameter, as they should be. The degree of non-Hermiticity of V_{LS}, however, is highly dependent on x. In Table I, we see that V_{LS} is only slightly non-Hermitian, as the largest difference in symmetric matrix elements is only of the order of 20 %. The impact of the strength parameter on the non-Hermiticity of V_{LS} can be seen when comparing this with Table II. Here V_{LS} is strongly non-Hermitian with symmetric matrix elements differing by more than a factor of 4. Thus we see that when no intruder states are present (low strength parameter), our parent interaction is approximately Hermitian, but when the intruder states enter (high strength parameter), our parent interaction, namely, V_{LS}, loses that Hermiticity in a striking manner.

TABLE I. Comparison of 3×3 Hermitian ($okb,cho,schm$) and non-Hermitian (LS) effective interactions. The matrix model of Eq. (24) is employed with interaction strength $x=0.1$.

−0.0368	0.3657	0.3976	V_{LS}
0.3793	0.8811	0.4482	
0.4722	0.4728	1.3999	
−0.0358	0.3732	0.4368	V_{okb}
0.3732	0.8814	0.4609	
0.4368	0.4609	1.3986	
−0.0041	0.4063	0.5215	V_{cho}
0.4063	0.9020	0.5143	
0.5215	0.5143	1.3461	
−0.0251	0.3823	0.4698	V_{schm}
0.3823	0.8846	0.4695	
0.4698	0.4695	1.3846	

TABLE II. Comparison of 3×3 Hermitian $(okb,cho,schm)$ and non-Hermitian (LS) effective interactions. The matrix model of Eq. (24) is employed with interaction strength x=0.55.

−2.6246	−3.4921	−0.9479	V_{LS}
0.8528	1.1730	2.7893	
0.2651	0.7817	−3.0008	
−1.7918	−0.6049	0.0615	V_{okb}
−0.6049	0.6264	0.7647	
0.0615	0.7647	−3.2870	
−1.2079	0.4599	0.6813	V_{cho}
0.4599	−0.4459	1.3268	
0.6813	1.3268	−2.7985	
−1.9123	0.3729	0.4656	V_{schm}
0.3729	−0.5113	1.1263	
0.4656	1.1263	−2.0288	

The next point to note concerning Tables I and II is the differences between the Hermitian effective interactions given by the various methods. In Table I, we see that the Hermitization procedures produce potentials which do not differ greatly from the parent potential. This, however, is expected since the parent potential is already approximately Hermitian. For a high strength parameter, we see that the resultant Hermitized potentials $(V_{okb},V_{cho},V_{schm})$ are all indeed quite different. Thus, if we want Hermitian potentials which are similar, it is crucial that the influence of intruder states be minimal.

Finally, we examine Table III where we show the eigenenergies and wave functions for the parent and Hermitized potentials. As expected, we see that the Hermitization procedures preserve eigenenergies. Note, however, the eigenfunctions of the various interactions are very different, although they correspond to the same eigenvalues. We note also that the ground state wave function of both V_{LS} and V_{schm} are equal to the P-space projection of the ground state wave function of the full-space Hamiltonian.

TABLE III. Comparison of eigenenergies and wave functions for the Hermitian $(okb,cho,schm)$ and non-Hermitian (LS) effective interactions. The matrix model of Eq. (24) is employed with interaction strength x=0.55, its 5 eigenvalues being −0.6714, 1.9146, 6.1702, 9.2084, and 17.4281.

	Eigenenergy	Wave function
V_{LS}	−0.6714	(−0.959,0.193,0.207)
V_{okb}	−0.6714	(0.881,−0.466,−0.081)
V_{cho}	−0.6714	(−0.528,0.790,0.311)
V_{schm}	−0.6714	(−0.959,0.193,0.207)
	1.9146	(0.916,0.397,−0.051)
	1.9146	(0.472,0.853,0.223)
	1.9146	(−0.826,−0.563,0.030)
	1.9146	(0.235,0.951,0.201)
	6.1702	(−0.325,−0.179,0.929)
	6.1702	(−0.035,−0.234,0.971)
	6.1702	(0.198,−0.241,0.950)
	6.1702	(0.158,−0.242,0.957)

IV. PHASE SHIFT EQUIVALENCE

The non-Hermitian V_{LS} given by the Lee-Suzuki (or folded diagram) method is specifically constructed to preserve the half-on-shell T matrix $T(p',p,p^2)$ [15,16]; this interaction of course preserves the phase shift which is given by the fully-on-shell T matrix $T(p,p,p^2)$. It would be of interest to study if phase shifts are also generally preserved by the Hermitian interactions generated using the transformations described in Sec. II.

Let us consider two T matrices $T_1(\omega_1)=V_1+V_1g_1(\omega_1)T_1(\omega_1)$ and $T_2(\omega_2)=V_2+V_2g_2(\omega_2)T_2(\omega_2)$, with the propagators $g_1(\omega_1)=P/(\omega_1-H_0)$ and similarly for $g_2(\omega_2)$. The unperturbed state is defined by H_0, namely, $H_0|q\rangle=q^2|q\rangle$. The symbol P denotes the principal value boundary condition. These T matrices are related by the well-known two-potential formula

$$T_2^{\dagger}=T_1+T_2^{\dagger}(g_2-g_1)T_1+\Omega_2^{\dagger}(V_2-V_1)\Omega_1, \qquad (28)$$

where the wave operator Ω is defined by $T_1(\omega_1)=V_1\Omega_1(\omega_1)$ and similarly for Ω_2. Applying the above relation to the half-on-shell T matrices in momentum space, we have

$$\langle p'|T_2^{\dagger}(p'^2)|p\rangle=\langle p'|T_1(p^2)|p\rangle+\langle p'|T_2^{\dagger}(p'^2)[g_2(p'^2)-g_1(p^2)]T_1(p^2)|p\rangle+\langle\psi_2(p')|(V_2-V_1)\times|\psi_1(p)\rangle. \qquad (29)$$

Here the true and unperturbed wave functions are related by $|\psi_1(p)\rangle=\Omega_1(p^2)|p\rangle$ and similarly for ψ_2.

Using the above relation, we shall now show that the phase shifts of the full-space interaction V are preserved by the Hermitian interaction V_{herm}, for momentum $\leqslant\Lambda$. Let us denote the last term of Eq. (29) as $D(p',p)$. We use V_{herm} for V_2 and V for V_1. Recall that the eigenfunction of (H_0+V_{herm}) is v_m [see Eq. (12)] and that for $H\equiv(H_0+V)$ is Ψ_m. We define a wave operator

$$U_P=\sum_{m\in P}|v_m\rangle\langle\Psi_m|. \qquad (30)$$

Then $|v_m\rangle=U_P|\Psi_m\rangle$ and $PV_{herm}P=U_P(H_0+V)U_P^{\dagger}-PH_0P$. For our present case, $\langle\psi_2(p')|$ is $\langle v_{p'}|$ and $|\psi_1(p)\rangle$ is $|\Psi_p\rangle$. Since $\langle v_{p'}|U_P=\langle\Psi_{p'}|$, we have

$$D(p',p)=\langle\Psi_{p'}|(HU_P^{\dagger}-U_P^{\dagger}H)|\Psi_p\rangle=(p'^2-p^2)\langle v_{p'}|\Psi_p\rangle. \qquad (31)$$

Clearly $D(p,p)=0$. The second term on the right-hand side of Eq. (29) vanishes when $p'=p$. Hence

$$\langle p|T_{herm}(p^2)|p\rangle=\langle p|T(p^2)|p\rangle, \quad p\leqslant\Lambda, \qquad (32)$$

where T_{herm} is the T matrix for (H_0+V_{herm}) and T for (H_0+V). Consequently the phase shifts of V are preserved by

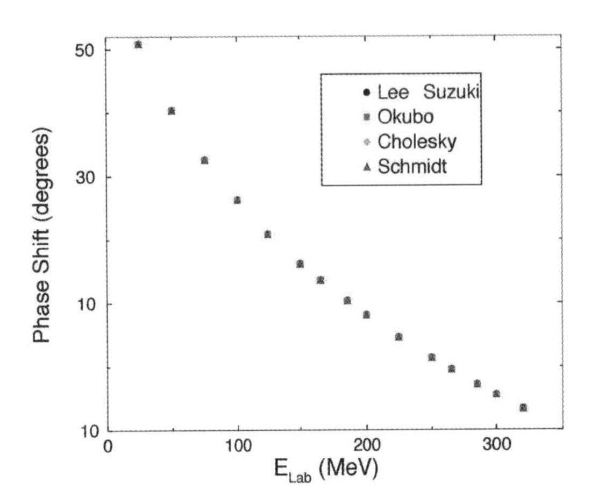

FIG. 2. (Color online) The phase shifts obtained from the non-Hermitian LS $V_{low\text{-}k}$ are compared with those obtained from the three Hermitian $V_{low\text{-}k}$ interactions in the $^1S_0$ partial wave channel.

V_{herm}. Recall that our T matrices are real, because of the principal-value boundary conditions employed.

To double check this preservation, we have calculated the phase shifts of the various Hermitian potentials (V_{okb}, V_{cho}, and V_{schm}) together with V_{LS}. As shown in Fig. 2 the $^1S_0$ phase shifts obtained from them all agree with each other very well, as is the case for other partial waves. The calculations were performed with the CD-Bonn potential and $\Lambda = 2.0\ \text{fm}^{-1}$. Since the phase shifts of V_{LS} are, by construction, the same as those of the full-space potential V, the Hermitian potentials preserve the phase shifts of V. It has been pointed out [8] that the Okubo Hermitian potential preserves the phase shifts, but we have found that there is a family of Hermitian potentials, including Okubo, which all preserve the phase shifts up to the decimation scale Λ.

V. HERMITIAN LOW-MOMENTUM INTERACTIONS

A main purpose of having a low-momentum nucleon interaction $V_{low\text{-}k}$ is to use it in nuclear many body problems such as the shell model nuclear structure calculations. As we have seen, however, there is a family of phase-shift equivalent Hermitian interactions. How different are they? Which one should one use for nuclear structure calculations? Will these Hermitian effective interactions give rise to the same physical properties? It is these questions we now seek to answer.

We have calculated the Hermitian $V_{low\text{-}k}$'s corresponding to V_{okb}, V_{cho}, and V_{schm} using several *NN* potentials. Calculations for the non-Hermitian $V_{low\text{-}k}$ corresponding to V_{LS} were also performed. In Fig. 3 we compare the results for the $^1S_0$ channel, obtained with the CD-Bonn potential and $\Lambda = 2.0\ \text{fm}^{-1}$. Clearly the Hermitian interactions are all quite similar to each other and to the parent non-Hermitian potential V_{LS}. In Fig. 4, we show a similar plot for the $^3S_1$ channel. Again, with the exception of very low momentum, the Hermitian potentials are all nearly identical to the parent V_{LS}. It

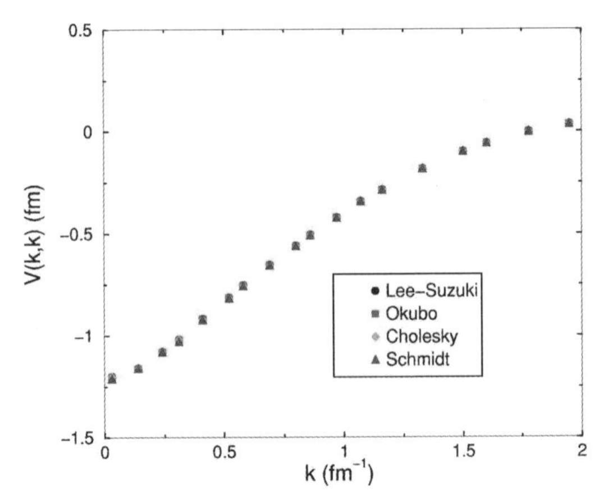

FIG. 3. (Color online) The non-Hermitian Lee-Suzuki $V_{low\text{-}k}$ interaction is compared with the three Hermitian $V_{low\text{-}k}$ interactions in the $^1S_0$ channel.

is of interest that at very low momentum, the V_{schm} matrix elements are slightly more attractive than the others. We note that the Hermitian effective interactions all preserve the deuteron binding energy (2.225 MeV). In addition, they are all phase shift equivalent up to the decimation scale Λ, as illustrated in Fig. 2.

Despite the similarities, there is, however, an additional degree of preservation that V_{schm} satisfies but the other two Hermitian interactions do not. By construction, the deuteron ground-state wave function given by V_{schm} is exactly equal to the P-space projection of the wave function of V, which is not true for V_{okb} and V_{cho}. This additional preservation is worth studying further. We refer to Figs. 5 and 6, where we plot the S- and D-state deuteron wave functions for a cutoff of $\Lambda = 2.0\ \text{fm}^{-1}$. Overall, the agreement is very good between all potentials, not just V_{LS} and V_{schm}, which is to be expected

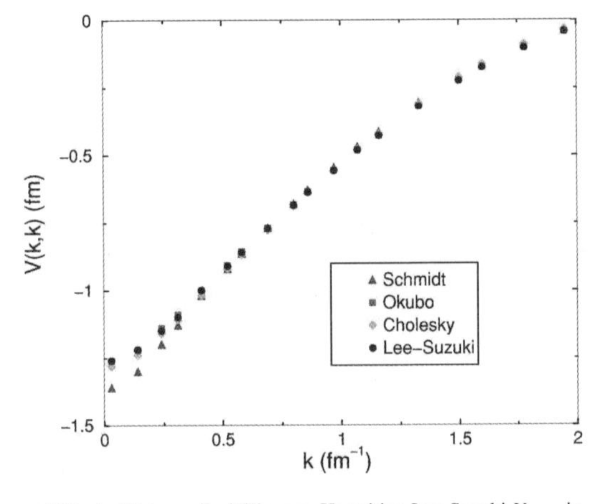

FIG. 4. (Color online) The non-Hermitian Lee-Suzuki $V_{low\text{-}k}$ interaction is compared with the three Hermitian $V_{low\text{-}k}$ interactions in the $^3S_1$ channel.

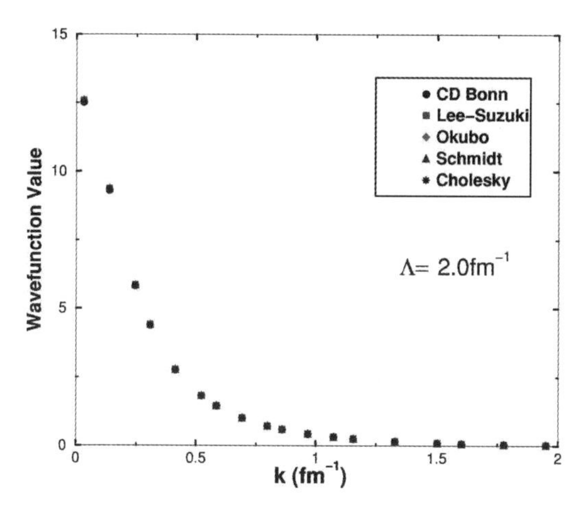

FIG. 5. (Color online) S-state deuteron wave functions for the bare and effective interactions plotted with respect to momentum, using a model space cutoff of $\Lambda = 2.0\ \mathrm{fm}^{-1}$.

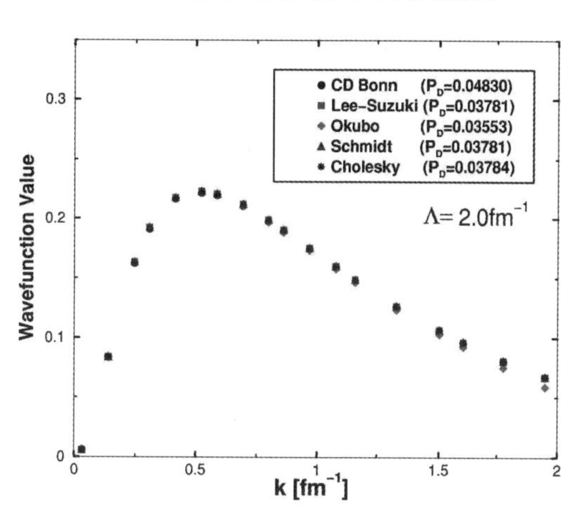

FIG. 6. (Color online) D-state deuteron wave functions for the bare and effective interactions plotted with respect to momentum, using a model space cutoff of $\Lambda = 2.0\ \mathrm{fm}^{-1}$.

considering that the interactions themselves are approximately the same. The fact that V_{schm} gives exactly the same wave function as V_{LS} can be seen from the D-state probability of the deuteron for each interaction, which we list for convenience in Fig. 6. Whereas the P_D's for the effective interactions are close, they are exact for V_{LS} and V_{schm}. This exact preservation of the ground state wave function presents us with an extra constraint that might be useful in deciding which Hermitian potential to use.

To ensure our Hermitian potentials will be useful for nuclear structure calculations, we examine in Table IV some shell model matrix elements calculated with $V_{low\text{-}k}$ corre-

TABLE IV. A comparison of shell model relative matrix element calculated using V_{LS} with those calculated using the Hermitian interactions. The oscillator length parameter b is given by $b = \sqrt{\hbar / m\omega}$, in units of fm. The matrix elements are in units of MeV.

	b(fm)	$n=0$				$n=1$			
$\langle n^1S_0 \vert V \vert n^1S_0 \rangle$		V_{LS}	V_{okb}	V_{cho}	V_{schm}	V_{LS}	V_{okb}	V_{cho}	V_{schm}
	1.4	−9.96	−9.96	−9.96	−9.96	−5.43	−5.43	−5.37	−5.37
	2.0	−4.85	−4.86	−4.86	−4.86	−4.40	−4.40	−4.41	−4.41
	2.6	−2.59	−2.59	−2.60	−2.60	−2.80	−2.80	−2.81	−2.81
$\langle n^1S_0 \vert V \vert (n+1)^1S_0 \rangle$									
	1.4	−7.06	−6.98	−6.93	−6.91	−3.16	−3.06	−2.94	−2.94
	2.0	−4.49	−4.48	−4.48	−4.48	−3.85	−3.83	−3.82	−3.82
	2.6	−2.64	−2.64	−2.65	−2.65	−2.70	−2.69	−2.70	−2.70
$\langle n^1S_0 \vert V \vert (n+2)^1S_0 \rangle$									
	1.4	−4.20	−4.04	−3.89	−3.87	−1.23	−1.05	−0.88	−0.88
	2.0	−3.83	−3.80	−3.99	−3.78	−3.18	−3.13	−3.11	−3.10
	2.6	−2.50	−2.49	−2.49	−2.49	−2.50	−2.49	−2.49	−2.49
$\langle n^3S_1 \vert V \vert n^3S_1 \rangle$									
	1.4	−11.99	−12.26	−12.05	−12.31	−8.64	−9.11	−8.62	−9.14
	2.0	−5.51	−5.50	−5.52	−5.52	−5.58	−5.69	−5.63	−5.71
	2.6	−2.90	−2.84	−2.87	−2.85	−3.30	−3.31	−3.33	−3.33
$\langle n^3S_1 \vert V \vert (n+1)^3S_1 \rangle$									
	1.4	−9.51	−10.15	−9.54	−10.65	−6.05	−6.68	−5.94	−7.16
	2.0	−5.32	−5.44	−5.38	−5.55	−5.12	−5.32	−5.17	−5.45
	2.6	−3.00	−3.01	−3.02	−3.05	−3.26	−3.31	−3.30	−3.35

sponding to V_{okb}, V_{cho}, V_{schm}, and V_{LS}; they are generally similar to each other, as would be expected owing to the similarity between the potentials themselves. As indicated by Fig. 4, the V_{schm} potential in the $^3S_1$ channel is slightly stronger than the others. As a result, some of its shell model matrix elements in this channel are also slightly stronger. We should also point out that some of the largest differences occur between the off-diagonal elements of V_{LS} and those of the Hermitian potentials, but this is precisely what would be expected since V_{LS} is non-Hermitian to begin with. The fact that these differences are generally small is a desirable result, as it implies that shell model calculations will not depend sensitively on which interaction one employs.

We have seen that the Hermitian potentials generated above are all approximately the same, and we would like to offer an explanation as to why this is so. Although $V_{low\text{-}k}$ corresponding to V_{LS} is non-Hermitian, we emphasize that it is only slightly so. In reference to our model study of Sec. II, this corresponds to the situation with a small strength parameter, and thus it would not be surprising to see the Hermitian $V_{low\text{-}k}$ potentials are so similar to the parent $V_{low\text{-}k}$. For example, this is especially transparent in the case of the Okubo interaction of Eqs. (18) and (19), as it can be written as [27,35]

$$\langle\alpha|V_{okb}|\beta\rangle = \langle\alpha|\frac{1}{2}(V_{LS}+V_{LS}^{\dagger})|\beta\rangle + \frac{\sqrt{\mu_\alpha^2+1}-\sqrt{\mu_\beta^2+1}}{\sqrt{\mu_\alpha^2+1}+\sqrt{\mu_\beta^2+1}}\langle\alpha|\frac{1}{2}(V_{LS}-V_{LS}^{\dagger})|\beta\rangle. \tag{33}$$

This tells us that V_{okb} can be well approximated by V_{LS}, if V_{LS} is only slightly non-Hermitian. In fact in this case V_{okb} is very accurately reproduced by the simple average $(V_{LS}+V_{LS}^{\dagger})/2$.

We note that the D-state probabilities given in Fig. 6 for the various effective interactions are significantly different from that of the bare potential. This is due to the fact that the effective interactions are subjected to a momentum cutoff at 2.0 fm^{-1}, while the bare CD-Bonn potential extends far beyond that. This can clearly be seen from the figure where a large portion of the wavefunction is simply cut off. While P_D is not an observable, it has theoretical relevance—it is an important characteristic of modern nucleon—nucleon potentials.

According to the MEEFT prescription, it is necessary to impose the cutoff at the limit of experimental data (in this case the limit of NN scattering experiments at 2.0 fm^{-1}), and it is at this cutoff that the effective interactions from all the bare potentials become nearly identical. Raising the cutoff would erode this approximate uniqueness and model independence. But at the above cutoff, we cannot preserve P_D. To illustrate, we refer to Fig. 7 where we plot the D-state wave functions with a cutoff of $\Lambda=3.0$ fm^{-1}. It shows that much more of the wave function is contained in the region below the cutoff, and as a result, the P_D's for the effective interactions are much closer to that of the bare potential. If we increase the cutoff to 4.0 fm^{-1}, they are almost exact. How to resolve this disparity in how to choose the cutoff warrants further study.

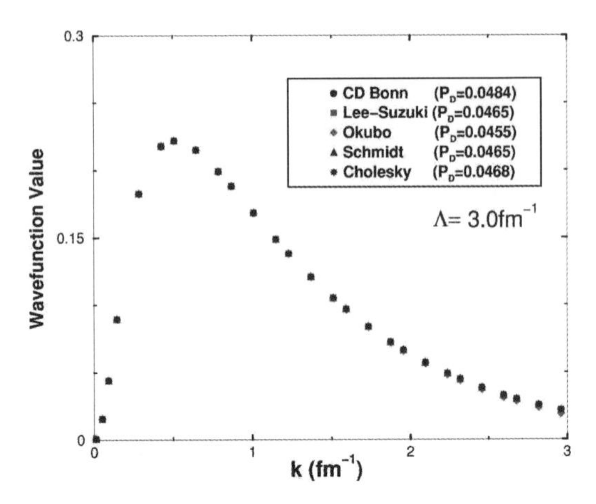

FIG. 7. (Color online) D-state deuteron wave functions for the bare and effective interactions plotted with respect to momentum, using a model space cutoff of $\Lambda=3.0$ fm^{-1}.

VI. SUMMARY AND CONCLUSION

We have studied a general method for deriving low-momentum NN interactions which are phase shift equivalent. By integrating out the high momentum components of a realistic NN potential such as the CD-Bonn potential, the Lee-Suzuki or folded-diagram method is employed to derive a parent low-momentum NN potential. This potential preserves the deuteron binding energy and phase shifts up to cutoff scale Λ. In addition it preserves the half-on-shell T-matrix up to the same scale. This Lee-Suzuki low-momentum NN interaction is not Hermitian, and further transformation is needed to obtain low-momentum interactions which are Hermitian. We have shown how to construct a family of such an interaction using the Schmidt orthogonalization procedure, and we have seen that two existing Hermitization schemes, namely, Okubo and Andreozzi, are in fact special cases of our general process. We have shown that all the Hermitian interactions so generated are phase shift equivalent, all reproducing empirical phase shifts up to scale Λ. These potentials also preserve the deuteron binding energy.

Through an analysis of this procedure using a solvable matrix model we have seen some interesting properties of Schmidt transformation method. In particular, with the entrance of intruder states, the parent potential can become highly non-Hermitian, and that the Hermitian potentials can deviate largely from each other and from the parent potential. It is fortunate that such deviations are not present for low-momentum NN potentials, for cutoff momentum $\Lambda \sim 2.0$ fm^{-1}. This is mainly because our parent Lee-Suzuki $V_{low\text{-}k}$ is only slightly non-Hermitian, and as a result, the

Hermitian low-momentum nucleon interactions generated from our orthogonalization procedure are all close to each other and close to the parent $V_{low\text{-}k}$. Shell model matrix elements of these Hermitian low-momentum nucleon interactions are found to be approximately equivalent to each other.

ACKNOWLEDGMENTS

We thank Jeremy Holt for many helpful discussions. Partial support from the U.S. Department of Energy under Contract No. DE-FG02-88ER40388 is gratefully acknowledged.

[1] R. Machleidt, Phys. Rev. C **63**, 024001 (2001).
[2] V. G. J. Stoks, R. Klomp, C. Terheggen, and J. de Schwart, Phys. Rev. C **49**, 2950 (1994).
[3] R. B. Wiringa, V. G. J. Stoks, and R. Schiavilla, Phys. Rev. C **51**, 38 (1995).
[4] M. Lacombe, B. Loiseau, J. M. Richard, R. Vinh Mau, J. Côté, P. Pirès, and R. de Tourreil, Phys. Rev. C **21**, 861 (1980).
[5] V. Bernard, N. Kaiser, and U.-G. Meissner, Int. J. Mod. Phys. E **4**, 193 (1995).
[6] P. Lepage, in *Nuclear Physics*, edited by C. A. Bertulani *et al.* (World Scientific, Singapore, 1997), p. 135.
[7] D. B. Kaplan, M. J. Savage, and M. B. Wise, Phys. Lett. B **424**, 390 (1998); Nucl. Phys. **B534**, 329 (1998); nucl-th/9802075.
[8] E. Epelbaum, W. Glöckle, A. Krüger, and Ulf-G. Meissner, Nucl. Phys. **A645**, 413 (1999).
[9] *Nuclear Physics with Effective Field Theory II*, edited by P. Bedaque *et al.* (World Scientific, Singapore, 1999).
[10] P. Bedaque and U. van Kolck, Annu. Rev. Nucl. Part. Sci. **52**, 339 (2002); nucl-th/0203055.
[11] D. R. Phillips, Czech. J. Phys. **52**, B49 (2002); nucl-th/0203040.
[12] U. van Kolck, Prog. Part. Nucl. Phys. **43**, 409 (1999).
[13] W. Haxton and C. L. Song, Phys. Rev. Lett. **84**, 5484 (2000); nucl-th/9907097.
[14] S. Beane *et al.*, in *At The Frontier of Particle Physics-Handbook of QCD*, edited by M. Shifman (World Scientific, Singapore, 2001), Vol. 1.
[15] S. Bogner, T. T. S. Kuo, and L. Coraggio, Nucl. Phys. **A684**, 432c (2001).
[16] S. Bogner, T. T. S. Kuo, L. Coraggio, A. Covello, and N. Itaco, Phys. Rev. C **65**, 051301(R) (2002).
[17] T. T. S. Kuo, S. Bogner, and L. Coraggio, Nucl. Phys. **A704**, 107c (2002).
[18] A. Schwenk, G. E. Brown, and B. Friman, Nucl. Phys. **A703**, 745 (2002).
[19] L. Coraggio, A. Covello, A. Gargano, N. Itaco, T. T. S. Kuo, D. R. Entem, and R. Machleidt, Phys. Rev. C **66**, 021303(R) (2002).
[20] L. Coraggio, A. Covello, A. Gargano, N. Itaco, and T. T. S. Kuo, Phys. Rev. C **66**, 064311 (2002).
[21] G. Brown and M. Rho, Phys. Rep. **363**, 85 (2002); nucl-th/0103102.
[22] G. E. Brown and M. Rho, nucl-th/0305089, Phys. Rep.(to be published).
[23] T. S. Park, K. Kubodera, D. P. Min, and M. Rho, Nucl. Phys. **A684**, 101 (2001).
[24] K. Kubodera, nucl-th/0308055.
[25] S. K. Bogner, T. T. S. Kuo, and A. Schwenk, Phys. Rep. **386**, 1 (2003); nucl-th/0305035.
[26] S. Okubo, Prog. Theor. Phys. **12**, 603 (1954).
[27] K. Suzuki and R. Okamoto, Prog. Theor. Phys. **70**, 439 (1983).
[28] F. Andreozzi, Phys. Rev. C **54**, 684 (1996).
[29] H. M. Hoffmann, S. Y. Lee, J. Richert, and H. A. Weidenmüller, Phys. Lett. **B45**, 421 (1973).
[30] K. Suzuki and S. Y. Lee, Prog. Theor. Phys. **64**, 2091 (1980).
[31] E. M. Krenciglowa and T. T. S. Kuo, Nucl. Phys. **A235**, 171 (1974).
[32] T. T. S. Kuo, S. Y. Lee, and K. F. Ratcliff, Nucl. Phys. **A176**, 65 (1971).
[33] T. T. S. Kuo and E. Osnes *Lecture Notes of Physics* (Springer, Berlin, 1990), Vol. 364, p. 1.
[34] K. Suzuki, R. Okamoto, P. J. Ellis, and T. T. S. Kuo, Nucl. Phys. **A567**, 576 (1994).
[35] T. T. S. Kuo, P. J. Ellis, Jifa Hao, Zibang Li, K. Suzuki, R. Okamoto, and H. Kumagai, Nucl. Phys. **A560**, 622 (1993).

ELSEVIER

Available online at www.sciencedirect.com

Nuclear Physics A 733 (2004) 153–165

www.elsevier.com/locate/npe

Counter terms for low momentum nucleon–nucleon interactions

Jason D. Holt [a], T.T.S. Kuo [a,*], G.E. Brown [a], Scott K. Bogner [b]

[a] *Department of Physics, State University of New York at Stony Brook, Stony Brook, NY 11794, USA*
[b] *Institute for Nuclear Theory, University of Washington, Seattle, WA 96195, USA*

Received 22 May 2003; received in revised form 3 November 2003; accepted 5 December 2003

Abstract

There is much current interest in treating low energy nuclear physics using the renormalization group (RG) and effective field theory (EFT). Inspired by this RG-EFT approach, we study a low-momentum nucleon–nucleon (NN) interaction, $V_{\text{low-}k}$, obtained by integrating out the fast modes down to the scale $\Lambda \sim 2\ \text{fm}^{-1}$. Since NN experiments can only determine the effective interaction in this low momentum region, our chief purpose is to find such an interaction for complex nuclei whose typical momenta lie below this scale. In this paper we find that $V_{\text{low-}k}$ can be highly satisfactorily accounted for by the counter terms corresponding to a short range effective interaction. The coefficients C_n of the power series expansion $\sum C_n q^n$ for the counter terms have been accurately determined, and results derived from several meson-exchange NN interaction models are compared. The counter terms are found to be important only for the S, P and D partial waves. Scaling behavior of the counter terms is studied. Finally we discuss the use of these methods for computing shell model matrix elements.

PACS: 21.60.Cs; 21.30.Fe; 27.80.+j

1. Introduction

Since the pioneering work of Weinberg [1], there has been much progress and interest in treating low-energy nuclear physics using the renormalization group (RG) and effective field theory (EFT) approach [2–11]. A central idea here is that physics in the infra red

* Corresponding author.
E-mail address: thomas.kuo@sunysb.edu (T.T.S. Kuo).

0375-9474/$ – see front matter © 2003 Published by Elsevier B.V.
doi:10.1016/j.nuclphysa.2003.12.004

region must be insensitive to the details of the short range (high momentum) dynamics. In low-energy nuclear physics, we are probing nuclear systems with low-energy probes of wave length λ; such probes cannot reveal the short range details at distances much smaller than λ. Furthermore, our understanding about the short range dynamics is still preliminary, and model dependent. Because of these considerations, a central step in the RG-EFT approach is to divide the fields into two categories: slow fields and fast fields, separated by a chiral symmetry breaking scale $\Lambda_\chi \sim 1$ GeV. Then by integrating out the fast fields, one obtains an effective field theory for the slow fields only. This approach has been very successful in treating low-energy nuclear systems, as discussed in the references cited above.

In order to have an effective interaction appropriate for complex nuclei in which typical nucleon momenta are $< k_F$, we separate fast and slow modes by a scale much smaller than Λ_χ, namely $\Lambda \sim 2\ \text{fm}^{-1}$. This corresponds to the pion production threshold and it represents the limit of available 2-nucleon data; experiments give a unique effective interaction only up to this scale. This cutoff has been employed recently by several authors [12–20] in a new development for the nucleon–nucleon (NN) interaction: the derivation of a low momentum NN potential, $V_{\text{low-}k}$. Despite some general similarities, we would like to stress that $V_{\text{low-}k}$ is not a traditional EFT construct, and would like to now attempt to clarify this issue. The key ingredient in our approach is the highly refined standard nuclear physics approach (SNPA) and the objective is to marry the SNPA to an EFT [21].

As an illustration of our strategy, we briefly discuss how this marriage can be effectuated. As summarized recently [22,23], a thesis now developed for some time posits that by combining the SNPA, based on potentials fit to experiments, with modern effective field theory, one can achieve more predictive power than the purist's EFT alone. This combination of SNPA and effective field theory, called EFT* by Kubodera [24], but more properly called *more effective* effective field theory (MEEFT) as noted by Kubodera, was rediscovered in [12–20], and consists of an RG-type approach which limits Λ to $\sim 2.1\ \text{fm}^{-1}$, because this is equal to the center of mass momentum up to which experiments measuring the nucleon–nucleon scattering phase shifts have been carried out.

Thus, the "more effective" in MEEFT arises from the guarantee that all experimental data will be reproduced by $V_{\text{low-}k}$, which will contain no Fourier components higher than those at which experiments have been carried out and analyzed. Therefore, $V_{\text{low-}k}$ is reliable up to $k = \Lambda$. This seems to be adequate for a description of shell model properties of complex nuclei.

In practice, the construction of $V_{\text{low-}k}$ begins with one of a number of available realistic models [25–29] for the nucleon–nucleon (NN) potential V_{NN}. While these models agree well in the low momentum (long range) region, where they are just given by the one pion exchange interaction, in the high momentum (short range) region they are rather uncertain and, in fact, differ significantly from one another. Naturally, it would be desirable to remove these uncertainties and model dependence from the high momentum components of the various modern NN potentials. Following our RG-type approach to achieve this end, it would seem to be appropriate to integrate out the high momentum modes of the various V_{NN} models. This is how the $V_{\text{low-}k}$ was derived, and in Section 2, we provide a brief outline of the derivation based on T-matrix equivalence.

The low-momentum NN potential, $V_{\text{low-}k}$, reproduces the deuteron binding energy, low-energy NN phase shifts and the low-momentum half-on-shell T-matrix. Furthermore, it is a smooth potential and can be used directly in nuclear many body calculations, avoiding the calculation of the Brueckner G-matrix [12–14,18–20]. Shell model nuclear structure calculations using $V_{\text{low-}k}$ have indeed yielded very encouraging results [13,19,20]. These calculations apply the same $V_{\text{low-}k}$ to a wide range of nuclei, including those in the sd-shell, tin region, and heavy nuclei in the lead region. At the present, no algebraic form exists for $V_{\text{low-}k}$, and it should be useful to have one. In the present work we study the feasibility of finding such an expression.

A central result of modern renormalization theory is that a general RG decimation generates an infinite series of counter terms [3–6,9] consistent with the input interaction. When we derive our low momentum interaction, the high momentum modes of the input interaction are integrated out. Does this decimation also generate a series of counter terms? If so, then what are the properties of the counter terms so generated? We study these questions in Section 3 where we carry out an accurate determination of the counter terms and show that this approach reproduces not only $V_{\text{low-}k}$, but the deuteron binding energy and low-energy phase shifts as well. We shall discuss that the counter terms represent generally a short range effective interaction and are important only for partial waves with angular momentum $l \leqslant 2$. The scaling behavior of the counter terms with respect to the decimation momentum will be studied in Section 4. Finally in Section 5, we examine the prospect for using $V_{\text{low-}k}$ and the counter term method in shell model calculations.

2. T-matrix equivalence

Since our method for deriving the low-momentum interaction $V_{\text{low-}k}$ has been described elsewhere [12–14], in the following we only outline the derivation. We obtain $V_{\text{low-}k}$ from a realistic V_{NN} model, such as the CD-Bonn model, by integrating out the high-momentum components. This integration is carried out with the requirement that the deuteron binding energy and low-energy phase shifts of V_{NN} are preserved by $V_{\text{low-}k}$. This preservation may be satisfied by the following T-matrix equivalence approach [12–14]. We start from the half-on-shell T-matrix

$$T(k', k, k^2) = V_{NN}(k', k) + \int_0^\infty q^2 dq\, V_{NN}(k', q) \frac{1}{k^2 - q^2 + i0^+} T(q, k, k^2) \tag{1}$$

noting that the intermediate state momentum q is integrated from 0 to ∞. We then define an effective low-momentum T-matrix by

$$\begin{aligned} &T_{\text{low-}k}(p', p, p^2) \\ &\quad = V_{\text{low-}k}(p', p) + \int_0^\Lambda q^2 dq\, V_{\text{low-}k}(p', q) \frac{1}{p^2 - q^2 + i0^+} T_{\text{low-}k}(q, p, p^2), \end{aligned} \tag{2}$$

where Λ denotes a momentum space cutoff and $(p', p) \leqslant \Lambda$. We choose $\Lambda \sim 2\ \text{fm}^{-1}$, essentially the momentum up to which the experiments give us information in the phase shift analysis. Note that in Eq. (2) the intermediate state momentum is integrated from 0 to Λ. We require the above T-matrices satisfying the condition

$$T(p', p, p^2) = T_{\text{low-}k}(p', p, p^2), \quad (p', p) \leqslant \Lambda. \tag{3}$$

The above equations define the effective low momentum interaction $V_{\text{low-}k}$, and are satisfied by the solution [12–14]

$$V_{\text{low-}k} = \widehat{Q} - \widehat{Q}' \int \widehat{Q} + \widehat{Q}' \int \widehat{Q} \int \widehat{Q} - \widehat{Q}' \int \widehat{Q} \int \widehat{Q} \int \widehat{Q} + \cdots \tag{4}$$

which is just the Kuo–Lee–Ratcliff folded-diagram effective interaction [30,31]. Here $\widehat{Q}$ represents the irreducible vertex function whose intermediate states are all beyond Λ; $\widehat{Q}'$ is the same as $\widehat{Q}$ except with its terms first order in the interaction removed.

For any decimation momentum Λ, the above $V_{\text{low-}k}$ can be calculated highly accurately (essentially exactly) using either the Andreozzi–Lee–Suzuki [32,33] (ALS) or the Krenciglowa–Kuo [34] iteration methods. These procedures preserve the deuteron binding energy in addition to the half-on-shell T-matrix, which implies preservation of the phase shifts. After obtaining the above $V_{\text{low-}k}$, which is not Hermitian, we further perform an Okubo transformation [35] to make it Hermitian. This is done using the method given in Refs. [36,37]. Here we first calculate the eigenvalues and eigenvectors of the operator $\omega^\dagger \omega$ where ω is the wave operator obtained with the ALS method. Then the Hermitian $V_{\text{low-}k}$ is calculated in terms of these quantities in a convenient way (see Eq. (23) of Ref. [37]).

3. Counter terms

Here we study whether the low-momentum interaction $V_{\text{low-}k}$ can be well represented by the low-momentum part of the original NN interaction supplemented by certain simple counter terms. Specifically, we consider

$$V_{\text{low-}k}(q, q') \simeq V_{\text{bare}}(q, q') + V_{\text{counter}}(q, q'), \quad (q, q') \leqslant \Lambda, \tag{5}$$

where V_{bare} is some bare NN potential V_{NN}. Since $V_{\text{low-}k}$ and V_{bare} are both known, there is no question about the existence of V_{counter} in general. Our aim is, however, to investigate if V_{counter} can be well represented by a short ranged effective interaction, such as a smeared out delta function. We shall check this conjecture by assuming a suitable momentum expansion form for V_{counter} and investigating how well it can satisfy the above equality. Since V_{bare} is generally given according to partial waves as is $V_{\text{low-}k}$, we shall proceed to determine V_{counter} separately for each partial wave. At very small radial distance, the Bessel function $j_l(qr)$ behaves like $(qr)^l$. We assume that V_{counter} is a very short ranged interaction. Hence the leading term in a momentum expansion of the partial wave matrix element $\langle ql|V_{\text{counter}}|q'l'\rangle$ is proportional to $q^l q'^{l'}$. Thus we consider the following expansion for the partial wave counter term potential, namely

$$\begin{aligned}\langle ql|V_{\text{counter}}|q'l'\rangle = q^l q'^{l'} \big[& C_0 + C_2(q^2 + q'^2) + C_4(q^2 + q'^2)^2 + C_6(q^6 + q'^6) \\ & + C_4' q^2 q'^2 + C_6' q^4 q'^2 + C_6'' q^2 q'^4 + \cdots \big].\end{aligned} \tag{6}$$

The counter term coefficients will be determined by standard χ-square fitting procedure so that the difference between $V_{\text{low-}k}$ and $(V_{\text{bare}} + V_{\text{counter}})$ is minimized.

We note that for l and l' being both S-wave, the leading term for V_{counter} is C_0 which is momentum independent and corresponds to a delta interaction. For other channels, however, the leading term for the counter potential is momentum dependent. For example, for l and l' being both P-wave, the leading term is $qq'C_0$.

If we take the projection operator P to project onto states with momentum less than Λ, namely $P = \sum_{k<\Lambda} |k\rangle\langle k|$, we can rewrite V_{bare} of Eq. (5) as $PV_{\text{bare}}P$. From now on we shall also abbreviate V_{counter} as V_{CT}. The above counter term approach for $V_{\text{low-}k}$ is similar to that used in RG-EFT [3,5], as both methods impose some cutoff, integrate out the modes above that cutoff, and then attempt to recover the information contained in those states in a counter term series. Ultimately, the main difference is the cutoff scale used and the resulting potential to which V_{CT} is added. As already noted, in the representation of $V_{\text{low-}k}$ given above, we use a momentum cutoff above which no constraining data exists, and thus our "slow" potential is a bare NN potential projected onto the low momentum states. In the EFT scheme, the "slow" potential is obtained by integrating out heavy dynamical degrees of freedom above the chiral symmetry breaking scale. Hence, counter terms are added to V_{light}, which is given by light mesons below the cutoff scale, and usually taken to be $V_{\text{one-pion}}$ with unrestricted pion momentum.

We have also tried to fit $V_{\text{low-}k}$ solely to a low order momentum expansion, and found the results to be less than satisfactory; the number of terms needed to produce an accurate representation is much higher than the order used here. Attempts at fitting $(V_{\text{one-pion}} + V_{\text{CT}})$ also generated results far less satisfactory than fitting it by $(V_{\text{bare}} + V_{\text{CT}})$. We shall discuss these points later.

As mentioned earlier, the counter term coefficients are determined using standard fitting techniques. We perform this fitting over all partial wave channels, and find consistently very good agreement. In Fig. 1 we compare some ${}^1\text{S}_0$ and ${}^3\text{S}_1$ matrix elements of $(PV_{\text{bare}}P + V_{\text{CT}})$ with those of $V_{\text{low-}k}$ for momenta below the cutoff Λ. A similar comparison for the ${}^3\text{S}_1$–${}^3\text{D}_1$ and ${}^3\text{D}_1$ channels is displayed in Fig. 2. We have also obtained very good agreement for the P-waves, as displayed in Fig. 3. It may be mentioned that here the momentum factor qq' of Eq. (6) is essential in achieving the good fit as shown. As demonstrated by these figures, the two methods yield nearly identical effective interactions, lending strong support that $V_{\text{low-}k}$ can be very accurately represented by $(PV_{\text{bare}}P + V_{\text{CT}})$. Furthermore V_{CT} is a very short ranged effective interaction.

To further check the accuracy of our counter term approach, we have calculated the phase shifts given by $V_{\text{low-}k}$ and compared them with those given by $(V_{\text{bare}} + V_{\text{CT}})$, as seen in Fig. 4, where we have found that the phase shifts are almost exactly preserved by the counter term approach. The phase shifts are plotted up to a lab energy of 325 MeV, and only at this high momentum, do they begin to differ slightly from those obtained from $V_{\text{low-}k}$. In passing, we mention that the deuteron binding energy is also accurately reproduced by our counter term approach (e.g., for the CD Bonn potential the deuteron binding energy given by the full potential versus the counter term approach are respectively 2.225 and 2.224 MeV). Thus, we can accurately reproduce $V_{\text{low-}k}$ and all its results by using $(PV_{\text{bare}}P + V_{\text{CT}})$.

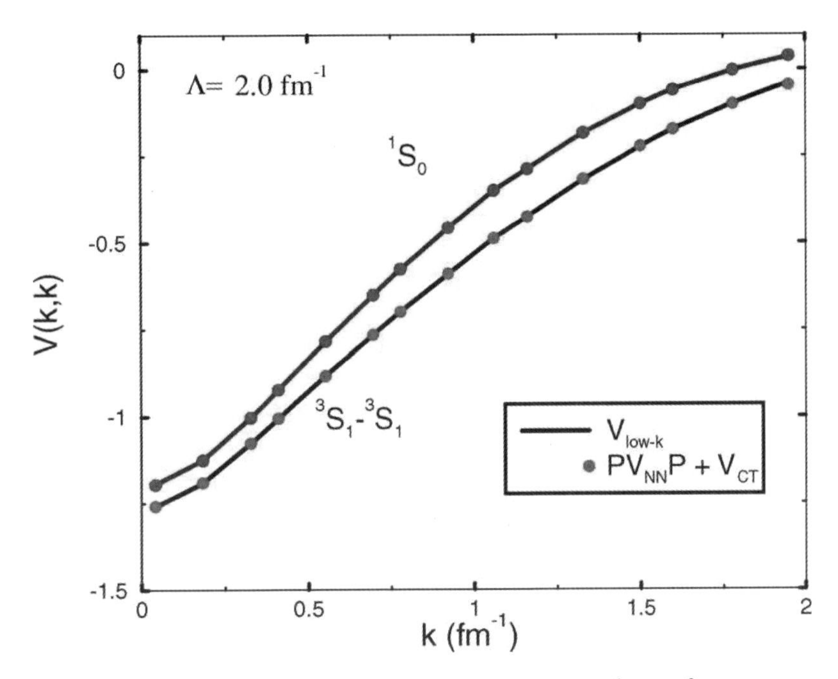

Fig. 1. Comparison of $V_{\text{low-}k}$ with $PV_{\text{NN}}P$ plus counter terms, for $^1S_0$ and $^3S_1$ channels.

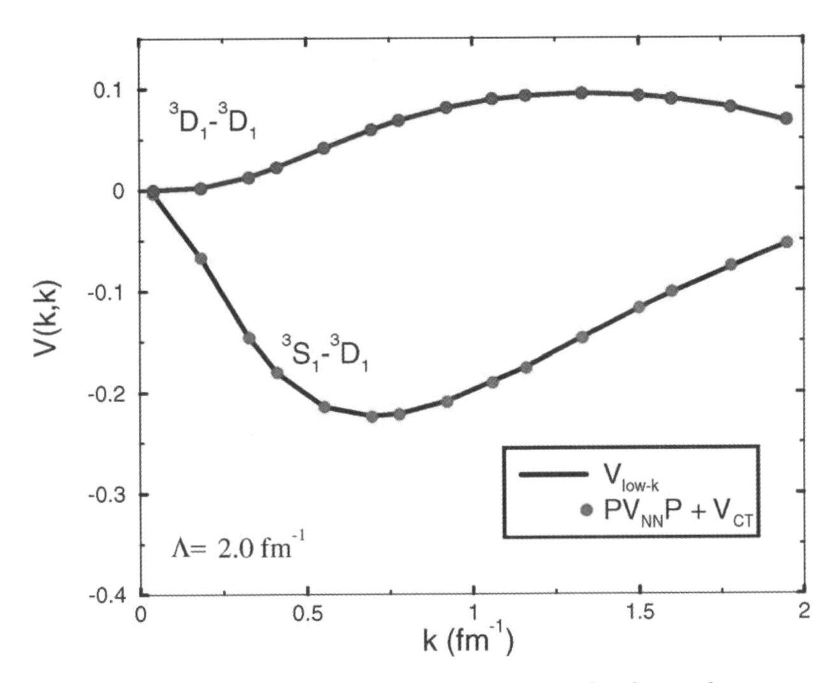

Fig. 2. Comparison of $V_{\text{low-}k}$ with $PV_{\text{NN}}P$ plus counter terms, for $^3S_1$–$^3D_1$ and $^3D_1$ channels.

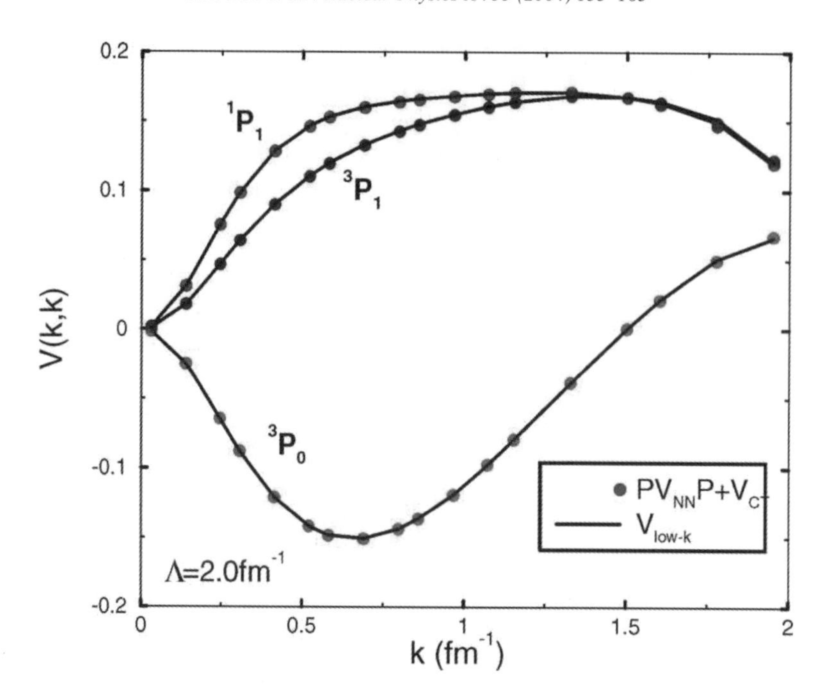

Fig. 3. Comparison of $V_{\text{low-}k}$ with $PV_{\text{NN}}P$ plus counter terms, for P-wave channels.

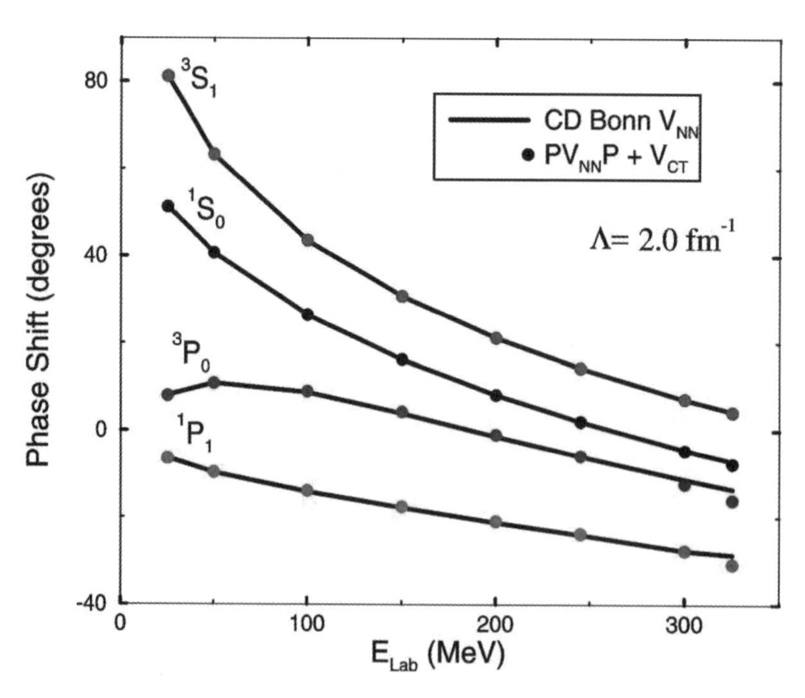

Fig. 4. Comparison of phase shifts given by V_{NN} and $PV_{\text{NN}}P$ plus counter terms.

Now let us examine the counter terms themselves. In Table 1, we list some of the counter term coefficients, using CD-Bonn as our bare potential. It is seen that the counter terms are significant only for the S, P and D partial waves. We do not list results beyond the $^3P_2$ partial waves, as the coefficients for them are all zero up to the level of our numerical accuracy (all entries in the table with magnitudes less than 10^{-4} have been set to zero). It is of interest that except for the above partial waves, $V_{\text{low-}k}$ is essentially the same as $PV_{\text{bare}}P$ alone. This behavior is clearly a reflection that V_{CT} is basically a very short ranged effective interaction. From the table, C_0 is clearly the dominant term in the expansion. Coefficients beyond C_4 are generally small and can be ignored. In the last row of the table, we list the rms deviations between $V_{\text{low-}k}$ and $PV_{\text{bare}}P + V_{\text{CT}}$; the fit is indeed very good.

Comparing counter term coefficients for different V_{bare} potentials can illustrate key differences between those potentials. Thus, in Table 2, we compare the low order counter terms obtained for the CD-Bonn [25], Nijmegen [26], Argonne [27] and Paris [29] NN potentials. The C_0 coefficients for these potentials are significantly different, indicating that one of the chief differences between these potentials is the way in which they treat

Table 1
Listing of counter terms for all partial waves obtained from the CD-Bonn potential using $\Lambda = 2\ \text{fm}^{-1}$. The units for the combined quantity $q^l q'^{l'} C_n q^n$ of Eq. (6) is fm, with momentum q in fm^{-1}. All counter terms for higher order partial waves are zero up to our numerical accuracy

Wave	C_0	C_2	C_4	C_4'	C_6	C_6'	C_6''	Δ_{rms}
$^1S_0$	−0.1580	−1.309E–2	3.561E–4	−8.469E–4	0	−1.191E–4	−1.191E–4	0.0002
$^3S_1$	−0.4651	5.884E–2	−1.824E–3	−1.163E–2	−3.243E–4	5.181E–4	5.181E–4	0.0003
$^3S_1$–$^3D_1$	2.879E–2	−3.581E–3	1.969E–3	−3.676E–3	−1.376E–4	6.300E–4	−2.449E–4	0.0025
$^3D_1$	−1.943E–3	−1.918E–4	1.709E–4	0	0	0	0	< 0.0001
$^1P_1$	−4.311E–2	−7.305E–4	7.999E–4	−2.049E–4	−2.594E–4	0	0	0.0002
$^3P_0$	−5.566E–2	4.821E–4	2.874E–4	1.098E–4	0	0	0	< 0.0001
$^3P_1$	−5.479E–2	−5.916E–4	8.636E–4	−2.065E–4	−2.721E–4	0	0	0.0002
$^3P_2$	−1.294E–2	3.075E–4	7.469E–4	−4.289E–4	−2.179E–4	0	0	0.0002

Table 2
Comparison of counter terms for $V_{\text{low-}k}$ obtained from the CD-Bonn, Argonne V18, Nijmegen, and Paris potentials using $\Lambda = 2\ \text{fm}^{-1}$. The units for the counter coefficients are the same as in Table 1

	C_0	C_2	C_4	C_4'	
$^1S_0$	−0.158	−0.0131	0.0004	−0.0008	CDB
	−0.570	0.0111	−0.0005	0.0004	V18
	−0.753	−0.0099	0.0003	0.0002	NIJ
	−1.162	−0.0187	0.0004	−0.0002	PAR
$^3S_1$	−0.465	0.0588	−0.0018	−0.0116	
	−1.081	0.0822	−0.0002	−0.0107	
	−1.147	0.0682	0.0004	−0.0100	
	−2.228	0.0251	0.0013	−0.0103	
$^3S_1$–$^3D_1$	0.0288	−0.0036	0.0020	−0.0037	
	0.0209	−0.0025	0.0018	−0.0037	
	0.0242	−0.0006	0.0040	−0.0094	
	0.0184	−0.0015	0.0019	−0.0061	

the short range repulsion. For instance, the Paris potential effectively has a very strong short-range repulsion and consequently its C_0 is of much larger magnitude than the others.

Our counter term V_{CT} has been determined by requiring a best fit between $(V_{bare} + V_{CT})$ and $V_{low\text{-}k}$. We have explored other schemes of fitting: we have tried to determine V_{CT} by requiring a best fit between $(V_{one\text{-}pion} + V_{CT})$ and $V_{low\text{-}k}$. But the results are far from satisfactory, the resulting rms deviation at best fit being too large. It may be of interest to determine V_{CT} by requiring a best fit between $V_{low\text{-}k}$ and V_{CT} alone. We have also tried this, with similar unsatisfactory results (large resulting rms deviation). A possible explanation may be the following. A main portion of $V_{low\text{-}k}$ come from $V_{two\text{-}pion}$ and other higher order processes; it appears that such contributions cannot be compensated by the V_{CT} of the simple low-order form of Eq. (6).

4. Scaling of counter terms

In Fig. 5 we display the scaling behavior of counter terms C_0 and C_2 with respect to the decimation momentum Λ. We note that for the ${}^1S_0$ channel, C_0 and C_2 display a weak Λ dependence. This is a welcome result. However, the C_0 for the ${}^3S_1$ channel varies significantly with Λ. In this channel, there is tensor force which is a mid-range interaction coming from π and ρ mesons; it has large momentum components in the intermediate momentum region of several fm^{-1}. As we lower Λ through this region, we are actually integrating out a predominant portion of the tensor force, in addition to the short range

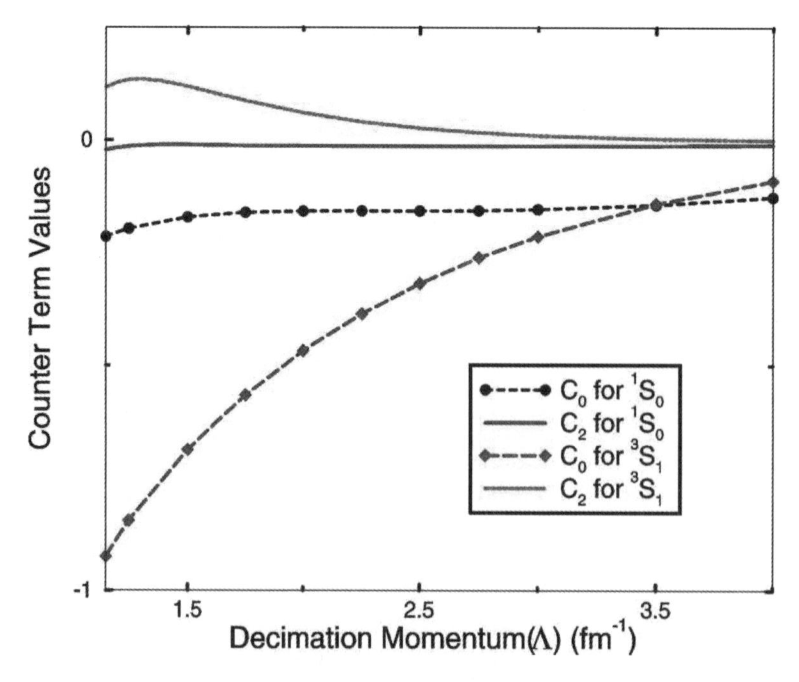

Fig. 5. Scaling behavior of S-wave leading order counter terms.

repulsion of V_{bare}. As a result, we would expect the C_0 to change to compensate for this loss of the tensor force. In contrast, the C_0 for $^1S_0$ comes mainly from integrating out only the short range repulsion, and little variation is seen. Conversely, as Λ increases, the tensor force is largely retained within the cutoff, and the two C_0's come close to each other as seen in the figure. To justify this line of thinking, we removed the tensor force to see how it would effect the $^3S_1$ C_0. We found that the scaling for this term was indeed quite different, displaying a flat behavior over the entire Λ range; C_0 only changed from 0.05 at $\Lambda = 1.2$ fm to 0.07 at $\Lambda = 3.0$ fm.

It should be observed that our Fig. 5 does not cover the small Λ region. The reason for this is that we have found that our counter term approach is not applicable to cases with $\Lambda \sim 1\ \text{fm}^{-1}$ or smaller. In this region, we have not been able to achieve a satisfactory fit between $V_{\text{low-}k}$ and $(V_{\text{bare}} + V_{\text{CT}})$. There the rms deviations are about two orders of magnitude larger than those of Table 1.

In passing, some practical differences between our counter terms V_{CT} and those used in traditional EFT may be noted. Our V_{CT} has a particular simple form corresponding to a short range effective interaction, and we have found that $V_{\text{low-}k}$ can be almost exactly reproduced by $PV_{\text{bare}}P$ plus V_{CT}. $V_{\text{low-}k}$ reproduces the empirical deuteron binding energy as well as phase shifts for all partial waves up to scale Λ. In traditional EFT, the effective potential is given as $V_{\text{one-pion}}$ plus counter terms, which represent the effect of the fast fields integrated out beyond the decimation scale. V_{bare} contains more diagrams, such as those with two-pion exchanges, than $V_{\text{one-pion}}$. Thus the actual values of these two types of counter terms can turn out to be rather different, as was found when we compared our values with those of Ref. [5]. (Note that our units for the expansion coefficients are different from theirs. In Ref. [5] the expansion is of the form $C'_n(k/\Lambda)^n$ while we use $C_n k^n$. Clearly C'_n corresponds to $C_n\Lambda^n$.) A comprehensive comparison of our counter terms with those of traditional EFT [5,39] will be very useful and of interest, and we plan to carry out such a study in the near future.

As we mentioned earlier, the counter terms are all rather small except for the S waves. This is consistent with the RG-EFT approach where the counter term potential is mainly a delta function force [3,5]. When using $\Lambda \sim 2\ \text{fm}^{-1}$, we see that the C_0 coefficients for the $^1S_0$ and $^3S_1$ channels are different. This suggests that the counter term potential is spin dependent and may be written as

$$V_{\text{CT}} \approx (f_0 + f'_0\sigma\cdot\sigma)\delta(\vec{r}), \tag{7}$$

with $f_0 = (3C_0(^3S_1) + C_0(^1S_0))/4$ and $f'_0 = (C_0(^3S_1) - C_0(^1S_0))/4$. For the coefficients of Table 1, we have $f_0 = -0.382$ fm and $f'_0 = -0.077$ fm. As indicated in Fig. 5, this spin-dependent factor f'_0 will diminish as Λ increases.

5. Comparison of shell model matrix elements

In shell model calculations, a basic input is the relative matrix elements $\langle nlSJT|V_{\text{eff}} \times |n'l'SJT\rangle$, where nl and $n'l'$ denote the oscillator wave functions for the relative motion of the two nucleons and S and T are the two-nucleon spin and isospin. The relative momenta $l(l')$ and S couple to total angular momentum J. From these matrix elements, the two-

Table 3
Shell model relative matrix elements of $V_{\text{low-}k}$ and $(PV_{\text{NN}}P + V_{\text{CT}})$ (column CT), with $\Lambda = 2.0\ \text{fm}^{-1}$ and $\hbar\omega = 19$ MeV. Matrix elements are in units of MeV

TSJ	nl	$n'l'$	CDB	CDB-CT	NIJ	V18
100	00	00	−9.053	−9.053	−9.066	−9.062
	10	00	−6.744	−6.742	−6.698	−6.609
	10	20	−3.527	−3.529	−3.363	−3.375
011	00	00	−11.00	−11.00	−10.72	−10.80
	02	00	−8.440	−8.405	−8.636	−8.627
	02	02	2.450	2.453	2.253	2.263
	10	20	−6.873	−6.850	−7.033	−6.738
	10	22	−6.969	−6.940	−7.562	−7.637
	12	22	2.281	2.278	2.070	2.079
001	01	01	3.835	3.858	3.587	3.670
	11	01	3.757	3.781	3.446	3.440
110	01	01	−2.482	−2.452	−2.407	−2.557
	11	01	−1.124	−1.099	−0.900	−1.162

particle shell model matrix elements in the laboratory frame can be calculated [38]. Thus if $V_{\text{low-}k}$ and $PV_{\text{bare}}P + V_{\text{CT}}$ are to prove useful in a broad sense, they must preserve these matrix elements.

In Table 3 we provide a comparison between these matrix elements as obtained from several input potentials. To begin, we note that the harmonic oscillator matrix elements for $V_{\text{low-}k}$ are approximately the same regardless of which of the four bare potentials we use. So, as far as providing input for shell model calculations, this is further evidence that the $V_{\text{low-}k}$ is approximately unique. Next, we come to the main result of the table, which is that the matrix elements for $V_{\text{low-}k}$ and $V_{\text{bare}} + V_{\text{CT}}$ are in very good agreement. We have used the CD Bonn for the bare potential, but this holds for the other bare potentials as well. In fact for the $^1S_0$, $^3S_1$, and $^3D_1$ channels, the elements are virtually exact out to four significant figures, and there is only minimal disagreement in the coupled channel.

6. Conclusion

While there are several models for the nucleon–nucleon potential in use today, they suffer from uncertainty and model dependence. Motivated by RG-EFT ideas, a low momentum NN interaction, $V_{\text{low-}k}$, has been constructed via integrating out the high momentum, model dependent components of these different potentials. The result appears to give an approximately unique representation of the low momentum NN potential, and the main issue addressed in this paper was whether $V_{\text{low-}k}$ could accurately be cast in a form $V_{\text{bare}} + V_{\text{CT}}$, where V_{CT} is a low order counter term series, and what the physical significance of this counter term series was. We have shown that this was indeed the case as $V_{\text{low-}k}$ is nearly identical to $V_{\text{NN}} + V_{\text{CT}}$ over all partial waves, and that $V_{\text{NN}} + V_{\text{CT}}$ reproduces both the deuteron binding energy and the NN phase shifts in the low momentum region. We have found that only the leading terms in the counter term series have much significance, indicating that the counter term potential is mainly a short range effective interaction which can be accurately represented by a simple low-order momentum expansion. Furthermore, we examined the scaling properties of the S-wave counter terms with respect to Λ and found that the tensor force is responsible for the behavioral differences exhibited between the $^3S_1$ channel and the $^1S_0$ channel.

Finally, we examined the potential for using $V_{\text{low-}k}$ and $V_{\text{NN}} + V_{\text{CT}}$ in shell model calculations. Again, it was seen that not only does $V_{\text{low-}k}$ give approximately identical input matrix elements irrespective of which bare potential was used, but that the counter term approach provided nearly the same results. Thus we have shown that the high momentum information integrated out of each bare NN potential can be accurately replaced by the counter terms, making the use of $V_{\text{low-}k}$ in a broad context much more tractable.

Acknowledgements

Many helpful discussions with Rupreht Machleidt and Achim Schwenk are gratefully acknowledged. This work was supported in part by the US DOE Grant No. DE–FG02-88ER40388.

References

[1] S. Weinberg, Phys. Lett. B 251 (1990) 288;
S. Weinberg, Nucl. Phys. B 363 (1991) 3.
[2] V. Bernard, N. Kaiser, U.-G. Meissner, Int. J. Mod. Phys. E 4 (1995) 193.
[3] P. Lepage, How to renormalize the Schroedinger equation, in: C.A. Bertulani, et al. (Eds.), Nuclear Physics, World Scientific, 1997, p. 135, nucl-th/9706029.
[4] D.B. Kaplan, M.J. Savage, M.B. Wise, Phys. Lett. B 424 (1998) 390;
D.B. Kaplan, M.J. Savage, M.B. Wise, Nucl. Phys. B 534 (1998) 329, nucl-th/9802075.
[5] E. Epelbaum, W. Glöckle, A. Krüger, U.-G. Meissner, Nucl. Phys. A 645 (1999) 413;
E. Epelbaum, W. Glöckle, A. Krüger, U.-G. Meissner, Nucl. Phys. A 671 (2000) 295.
[6] P. Bedaque, et al. (Eds.), Nuclear Physics with Effective Field Theory, vol. II, World Scientific, 1999.
[7] P. Bedaque, U. van Kolck, Annu. Rev. Nucl. Part. Sci. 52 (2002) 339, nucl-th/0203055.
[8] D.R. Phillips, Czech. J. Phys. 52 (2002) B49, nucl-th/0203040.
[9] U. van Kolck, Prog. Part. Nucl. Phys. 43 (1999) 409.
[10] W. Haxton, C.L. Song, Phys. Rev. Lett. 84 (2000) 5484, nucl-th/9907097.
[11] S. Beane, et al., in: M. Shifman (Ed.), At the Frontier of Particle Physics—Handbook of QCD, vol. 1, World Scientific, Singapore, 2001.
[12] S. Bogner, T.T.S. Kuo, L. Coraggio, Nucl. Phys. A 684 (2001) 432c, nucl-th/0204058.
[13] S. Bogner, T.T.S. Kuo, L. Coraggio, A. Covello, N. Itaco, Phys. Rev. C 65 (2002) 051301(R), nucl-th/9912056.
[14] T.T.S. Kuo, S. Bogner, L. Coraggio, Nucl. Phys. A 704 (2002) 107c.
[15] S. Bogner, A. Schwenk, T.T.S. Kuo, G.E. Brown, nucl-th/0111042.
[16] S. Bogner, T.T.S. Kuo, A. Schwenk, D.R. Entem, R. Machleidt, nucl-th/0108041.
[17] S. Bogner, T.T.S. Kuo, A. Schwenk, nucl-th/0305035.
[18] A. Schwenk, G.E. Brown, B. Friman, Nucl. Phys. A 703 (2002) 745, nucl-th/0109059.
[19] L. Coraggio, A. Covello, A. Gargano, N. Itako, T.T.S. Kuo, D.R. Entem, R. Machleidt, Phys. Rev. C 66 (2002) 021303(R), nucl-th/0206025.
[20] L. Coraggio, A. Covello, A. Gargano, N. Itako, T.T.S. Kuo, Phys. Rev. C 66 (2002) 064311.
[21] G.E. Brown, M. Rho, nucl-th/0305089.
[22] T.S. Park, K. Kubodera, D.P. Min, M. Rho, Nucl. Phys. A 684 (2001) 101.

[23] G. Brown, M. Rho, Phys. Rep. 363 (2002) 85, nucl-th/0103102.
[24] K. Kubodera, nucl-th/0308055.
[25] R. Machleidt, Phys. Rev. C 63 (2001) 024001, nucl-th/0006014.
[26] V.G.J. Stoks, R. Klomp, C. Terheggen, J. de Schwart, Phys. Rev. C 49 (1994) 2950, nucl-th/9406039.
[27] R.B. Wiringa, V.G.J. Stoks, R. Schiavilla, Phys. Rev. C 51 (1995) 38, nucl-th/9408016.
[28] D.R. Entem, R. Machleidt, H. Witala, Phys. Rev. C 65 (2002) 064005, nucl-th/0111033.
[29] M. Lacombe, et al., Phys. Rev. C 21 (1980) 861.
[30] T.T.S. Kuo, S.Y. Lee, K.F. Ratcliff, Nucl. Phys. A 176 (1971) 65.
[31] T.T.S. Kuo, E. Osnes, in: Springer Lecture Notes of Physics, vol. 364, 1990, p. 1.
[32] K. Suzuki, S.Y. Lee, Prog. Theor. Phys. 64 (1980) 2091.
[33] F. Andreozzi, Phys. Rev. C 54 (1996) 684.
[34] E.M. Krenciglowa, T.T.S. Kuo, Nucl. Phys. A 235 (1974) 171.
[35] S. Okubo, Prog. Theor. Phys. 12 (1954) 603.
[36] K. Suzuki, R. Okamoto, Prog. Theo. Phys. 70 (1983) 439.
[37] T.T.S. Kuo, P.J. Ellis, J. Hao, Zibang Li, K. Suzuki, R. Okamoto, H. Kumagai, Nucl. Phys. A 560 (1993) 622.
[38] T.T.S. Kuo, G.E. Brown, Nucl. Phys. 85 (1966) 40.
[39] D.R. Entem, R. Machleidt, nucl-th/0304018.

PHYSICAL REVIEW C **72**, 041304(R) (2005)

Low momentum shell model effective interactions with all-order core polarizations

Jason D. Holt,[1] Jeremy W. Holt,[1] T. T. S. Kuo,[1] G. E. Brown,[1] and S. K. Bogner[2]
[1]*Department of Physics, State University of New York, Stony Brook, New York 11794, USA*
[2]*Department of Physics, The Ohio State University, Columbus, Ohio 43210, USA*
(Received 5 April 2005; published 31 October 2005)

An all-order summation of core-polarization diagrams using the low-momentum nucleon-nucleon interaction $V_{\text{low}-k}$ is presented. The summation is based on the Kirson-Babu-Brown (KBB) induced interaction approach in which the vertex functions are obtained self-consistently by solving a set of nonlinear coupled equations. It is found that the solution of these equations is simplified using $V_{\text{low}-k}$, which is energy independent, and using the Green's functions in the particle-particle and particle-hole channels. We have applied this approach to the *sd*-shell effective interactions and find that the results we calculated to all orders by using the KBB summation technique are remarkably similar to those of second-order perturbation theory, the average differences being less than 10%.

DOI: 10.1103/PhysRevC.72.041304 PACS number(s): 21.60.Cs, 21.30.−x, 21.10.−k

Since the early works of Bertsch [1] and Kuo and Brown [2], the effect of core polarization (CP) in nuclear physics has received much attention. CP is particularly important in the shell-model effective interactions, where this process provides the long-range internucleon interaction mediated by excitations of the core [3]. In microscopic calculations of effective interactions, CP has played an essential role, as illustrated by the familiar situation in ^{18}O. There the spectrum calculated with the bare G matrix was too compressed compared with experiment, while the inclusion of CP had the desirable effect of both lowering the 0^+ ground state and raising the 4^+ state, leading to a much-improved agreement with experiment [1,2]. As pointed out by Zuker [4], the Kuo-Brown matrix elements, although developed quite some time ago, continue to be a highly useful shell-model effective interaction. It should be noted that the CP diagrams associated with the above interactions were all calculated to second order (in the G matrix) in perturbation theory. However, what are the effects of CP beyond second order, and how can they be calculated? In this communication we address these questions and present an all-order summation of CP diagrams for the *sd*-shell interactions.

There have been a number of important CP studies beyond second order. Third-order CP diagrams, including those with one fold, were studied in detail by Barrett and Kirson [5] for the *sd*-shell effective interactions. Hjorth-Jensen *et al.* [6] have carried out extensive investigations of the third-order CP diagrams for the tin region. A main result of these studies is that the effect of the third-order diagrams is generally comparable with that of the second order; the former cannot be ignored in comparison with the latter. As is well known, high-order CP calculations are difficult to perform, largely because the number of CP diagrams grows rapidly as one goes to higher orders in perturbation theory. The number of diagrams at third order is already quite large, though still manageable. Primarily because of this difficulty, a complete fourth-order calculation has never been carried out. It was soon realized that an order-by-order calculation of CP diagrams beyond third order is not practicable. To fully assess the effects of CP to high order, a nonperturbative method is called for.

The nonperturbative method we use is based on the elegant and rigorous induced interaction approach of Kirson [7] and Babu and Brown [8], hereafter referred to as KBB. Other successful nonperturbative summation methods have also been developed, such as the parquet summation [9] and coupled cluster expansion [10]. In the KBB formalism one obtains the vertex functions by solving a set of self-consistent equations, thereby generating CP diagrams to all orders. Using this approach, Kirson has studied ^{18}O and ^{18}F by using a G-matrix interaction, and Sprung and Jopko [11] have carried out a model study of this approach by using a separable interaction. A main conclusion of both studies is that when CP diagrams are included to all orders the effective interaction is very close to that given by the bare interaction alone. In contrast, Sjöberg [12] applied the Babu-Brown formalism to nuclear matter and found that the inclusion of CP diagrams to all orders has a significant effect on the Fermi liquid parameters in comparison with those given by the bare interaction. These conflicting results for CP studies of finite nuclei and infinite nuclear matter have served as a primary motivation for our present reexamination of the CP effect.

Our application of the KBB formalism to shell-model effective interactions is similar to that of Kirson, but our treatment is different in a number of important regards. As we discuss, the particle-core and hole-core coupling vertices used in the present work include a larger class of diagrams than has been previously studied. We show how the inclusion of these diagrams is facilitated by using the recently developed low-momentum nucleon-nucleon interaction $V_{\text{low}-k}$ [13–19] instead of the previously used G matrix. This is primarily because the G matrix [6,20] depends on both starting energy and the Pauli exclusion operator, while $V_{\text{low}-k}$ depends on neither. It is noted that the S-wave interactions calculated from the Moszkowski-Scott separation method gave essentially the same matrix elements as $V_{\text{low}-k}$ [21]. We now turn our attention to the formal aspects of our approach then discuss

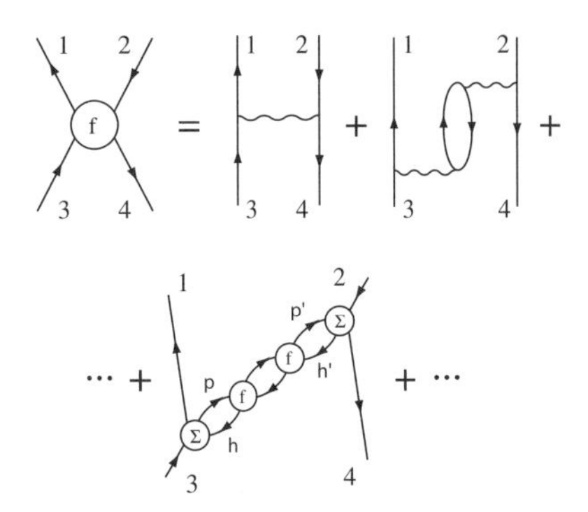

FIG. 1. Self-consistent diagrammatic expansion of the ph vertex function f, where Σ is defined in the text.

the application of these methods to the sd-shell effective interactions.

The KBB induced interaction approach provides a very appealing way for summing up planar diagrams to all orders. Its fundamental requirement is that the irreducible vertex functions be calculated self-consistently. This means that any CP term contained in a vertex function must be generated self-consistently from the same vertex function. We note that it is this requirement that plays the essential role of generating CP diagrams to all orders. To see this point, it may be convenient to first consider the particle-hole (ph) vertex function f. (We shall consider the particle-particle vertex a little later.) As shown in Fig. 1, f is generated by summation of the driving term V and CP terms, the latter being dependent on f. This then gives the self-consistent equation for f:

$$f = V + \Sigma g_{ph}\Sigma + \Sigma g_{ph} f g_{ph}\Sigma + \Sigma g_{ph} f g_{ph} f g_{ph}\Sigma + \cdots, \quad (1)$$

where g_{ph} is the free ph Green's function and Σ denotes the vertex for particle-core and hole-core coupling. The second-order CP diagram of Fig. 1 is the lowest-order term contained in $\Sigma g_{ph}\Sigma$. We note that, for simplicity, the bra and ket indices have been suppressed in the above equation as well as in the following equations. For example, in Eq. (1) the f on the left-hand side represents $\langle 12^{-1}|f|34^{-1}\rangle$, whereas the fifth and sixth Σs on the right-hand side represent $\langle 1ph^{-1}|\Sigma|3\rangle$ and $\langle 2^{-1}|\Sigma|p'h'^{-1}4^{-1}\rangle$, respectively.

The generation of high-order CP diagrams may be seen easily for the special case $\Sigma = V$. In this case Eq. (1) becomes

$$f = V + V g_{ph} V + V g_{ph} f g_{ph} V + V g_{ph} f g_{ph} f g_{ph} V + \cdots. \quad (2)$$

Since f appears on both sides of this equation, it is clear that an iterative solution for f will yield CP diagrams to all orders, including those with "bubbles inside bubbles," like those shown in diagram (a) of Fig. 2.

For nuclear many-body calculations in general, we also need the particle-particle (pp) vertex function Γ. Like f, Γ is given by a driving term plus CP terms. Furthermore, the diagrammatic representation of Γ is identical to Fig. 1 except that the hole lines 2 and 4 are replaced with corresponding particle lines. This gives the self-consistent equation for Γ:

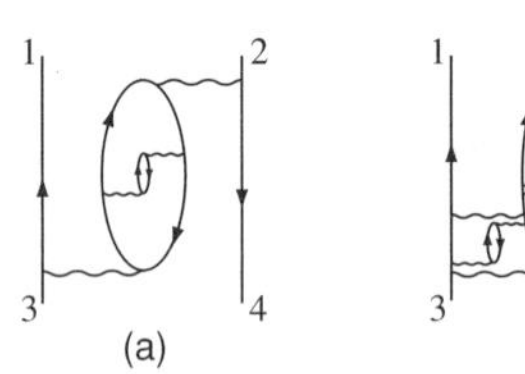

FIG. 2. Higher-order terms contributing to the vertex functions f and Γ, including (a) nested bubbles in bubbles and (b) particle-core and hole-core couplings.

$$\Gamma = V + \Sigma g_{ph}\Sigma + \Sigma g_{ph} f g_{ph}\Sigma + \Sigma g_{ph} f g_{ph} f g_{ph}\Sigma + \cdots. \quad (3)$$

To clarify our compact notation, we note that the external lines of the Σ vertices in Γ are different than those shown in Fig. 1. The upper Σ vertex, for example, now represents $\langle 2|\Sigma|ph^{-1}4\rangle$. These different Σ vertices can be related to each other, however, by means of appropriate particle-hole transformations.

Finally, the vertex functions f and Γ are coupled together by means of the coupling vertex Σ. In the present work we choose

$$\begin{aligned}\Sigma &= V + \Sigma_{ph} + \Sigma_{pp},\\ \Sigma_{ph} &= V g_{ph} V + V g_{ph} f g_{ph} V + V g_{ph} f g_{ph} f g_{ph} V + \cdots,\\ \Sigma_{pp} &= V g_{pp} V + V g_{pp}\Gamma g_{pp} V + V g_{pp}\Gamma g_{pp}\Gamma g_{pp} V + \cdots,\end{aligned} \quad (4)$$

where g_{pp} is the free pp Green's function.

The self-consistent vertex functions f and Γ are determined from Eqs. (1), (3), and (4). These are similar to the equations used by Kirson [7], except that our Σ includes both Σ_{ph} and Σ_{pp}, while the equivalent term in Kirson's calculations includes only Σ_{ph} [7,22]. To see the role of the Σ vertices, let us consider diagram (b) of Fig. 2. Here the lower particle-core vertex, which contains repeated particle-particle interactions, belongs to Σ_{pp}, while the upper one, which contains repeated particle-hole interactions, belongs to Σ_{ph}. It is, of course, necessary to include Σ_{pp} in order to have such CP diagrams in the all-order sum. Our equations are equivalent to those of Kirson when Σ_{pp} is set to zero, and in this case Γ does not enter the calculation of f.

Solving the above equations for f and Γ may seem complicated, but we have found their solution can be simplified significantly through use of the true ph and pp Green's functions:

$$\begin{aligned}G_{ph} &= g_{ph} + g_{ph} f G_{ph},\\ G_{pp} &= g_{pp} + g_{pp}\Gamma G_{pp}.\end{aligned} \quad (5)$$

When these Green's functions are used to partially sum and regroup our series, the self-consistent Eqs. (1), (3), and (4) assume a much simpler form

$$\begin{aligned} f &= V + \Sigma G_{ph} \Sigma, \\ \Gamma &= V + \Sigma G_{ph} \Sigma, \\ \Sigma &= V + V G_{ph} V + V G_{pp} V. \end{aligned} \tag{6}$$

The above simplifications also aid our numerical efforts, and, using the following iterative method, we find that our coupled equations can be solved rather efficiently. For the nth iteration, we start from $f^{(n)}$ and $\Gamma^{(n)}$ to first calculate $G_{ph}^{(n)}$ and $G_{pp}^{(n)}$ followed by $\Sigma^{(n)}$, as seen from Eqs. (5) and (6). We then obtain the vertex functions for the subsequent iteration by taking $f^{(n+1)} = V + \Sigma^{(n)} G_{ph}^{(n)} \Sigma^{(n)}$ and $\Gamma^{(n+1)} = V + \Sigma^{(n)} G_{ph}^{(n)} \Sigma^{(n)}$. The entire iterative process begins from the initial $f^{(0)} = V + V g_{ph} V$ and $\Gamma^{(0)} = V + V g_{ph} V$ and typically converges after just a few iterations.

In the present work, we have included folded diagrams to all orders. As detailed in [23], we use this method to reduce the full-space nuclear many-body problem $H\Psi_n = E_n \Psi_n$ to a model-space problem $H_{\text{eff}} \chi_m = E_m \chi_m$, where $H = H_0 + V$, $H_{\text{eff}} = H_0 + V_{\text{eff}}$, and V denotes the bare NN interaction. The effective interaction V_{eff} is given by the folded-diagram expansion

$$V_{\text{eff}} = \hat{Q} - \hat{Q}' \int \hat{Q} + \hat{Q}' \int \hat{Q} \int \hat{Q} - \cdots. \tag{7}$$

We consider the effective interactions for valence nucleons, and in this case $\hat{Q}$ is the irreducible pp vertex function that we shall calculate by using the KBB equations. In Ref. [7], the effect of higher-order CP diagrams to the nonfolded $\hat{Q}$ term was extensively studied. In the present work, we first calculate $\hat{Q}$ including CP diagrams to all orders. Then we sum the above folded-diagram series for V_{eff} to all orders by using the Lee-Suzuki iteration method as discussed in Ref. [6]. In this way, folded CP diagrams are included to all orders.

For the present calculation we have chosen to use the low-momentum nucleon-nucleon interaction, $V_{\text{low}-k}$. Since the vertex functions f and Γ both depend on the starting energy, there would be off-energy-shell effects present in many CP diagrams if the G-matrix interaction were chosen. This would make the calculation very complicated. $V_{\text{low}-k}$, on the other hand, is energy independent, so no such difficulties are encountered. Because detailed treatments of $V_{\text{low}-k}$ have been given elsewhere [13–19], here we provide only a brief description. We define $V_{\text{low}-k}$ through the T-matrix equivalence $T(p', p, p^2) = T_{\text{low}-k}(p', p, p^2); (p', p) \leqslant \Lambda$, where T is given by the full-space equation $T = V_{NN} + V_{NN} g T$ and $T_{\text{low}-k}$ is given by the model-space (momenta $\leqslant \Lambda$) equation $T_{\text{low}-k} = V_{\text{low}-k} + V_{\text{low}-k} g T_{\text{low}-k}$. Here V_{NN} represents some realistic NN potential and Λ is the decimation momentum beyond which the high-momentum components of V_{NN} are integrated out. $V_{\text{low}-k}$ preserves both the deuteron binding energy and the low-energy scattering phase shifts of V_{NN}. Since empirical nucleon scattering phase shifts are available up to only the pion production threshold ($E_{\text{lab}} \sim 350$ MeV), beyond this momentum the realistic NN potentials cannot be uniquely determined. Accordingly, we choose $\Lambda \approx 2.0\ \text{fm}^{-1}$, thereby retaining only the information from a given potential that is constrained by experiment. In fact, for this Λ, the $V_{\text{low}-k}$ derived from various NN potentials [24–27] are all nearly identical [17]. Except where noted otherwise, in our calculations we employ the $V_{\text{low}-k}$ derived from the change-dependent (CD) Bonn potential [24].

As an initial study, we carried out a restricted all-order CP calculation for the sd-shell effective interactions. In particular, we summed only the TDA diagrams for the Green's functions G_{pp} and G_{ph}, leaving a study of RPA diagrams to a future publication. We used two choices for the shell-model space: One with four shells (10 orbits from $0s_{1/2}$ to $1p_{1/2}$) and the other with five shells (15 orbits from $0s_{1/2}$ to $3s_{1/2}$), both with oscillator constant $\hbar\omega = 14$ MeV. Only core excitations within this space are included. Vary, Sauer, and Wong [28] have pointed out that for CP diagrams one needs to include intermediate states of high excitation energies (up to $\sim 10\hbar\omega$) in order for the second-order CP term to converge. In their work a G matrix derived from the Reid soft-core potential was used, and our use of $V_{\text{low}-k}$ may yield different results as it has greatly reduced high-momentum components. We found that the difference between our five-shell and four-shell calculations was minimal; the results differing by about 2% or less. This finding is supported by recent studies that show desirable convergence properties of $V_{\text{low}-k}$ [29]. We plan to study this convergence problem for $V_{\text{low}-k}$ in the near future.

With these restrictions, we calculated V_{eff} from Eqs. (5)–(7). A large number of angular-momentum recouplings are involved in calculating the CP diagrams. In this regard, we followed closely the diagram rules in [30]. Previous second-order calculations [6,13] included in the $\hat{Q}$ box the first-order pp diagram, the second-order pp and hh ladder diagrams, and the second-order CP diagram. Our all-order calculation includes these same diagrams except the second-order CP diagram is replaced with the all-order CP diagrams from KBB. In Fig. 3 we compare the sd-shell V_{eff} matrix elements we

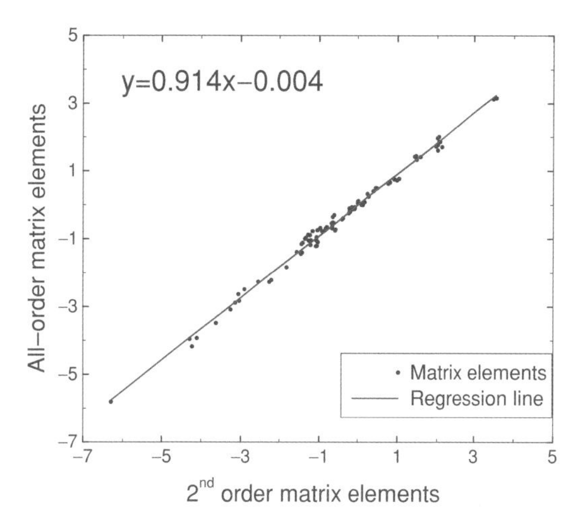

FIG. 3. (Color online) A comparison of the second-order CP matrix elements with those of the all-order KBB calculation.

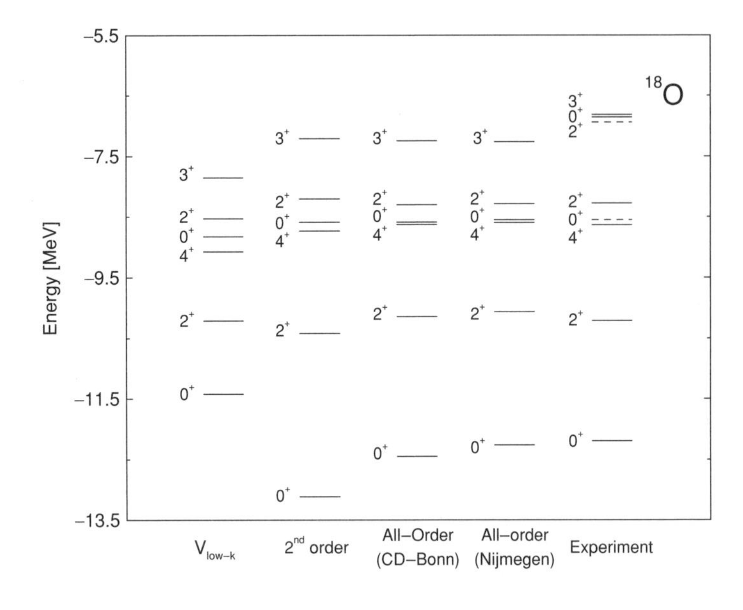

FIG. 4. Spectra for the ^{18}O system calculated to different orders in perturbation theory. Dashed lines for the experimental levels [31] indicate levels with large intruder state mixing [32,33]. All calculations were performed with the experimental single-particle energies of ^{17}O.

calculated from the second-order and all-order $\hat{Q}$ boxes just described, both by using the five-shell space mentioned above. A least-squares fit was applied to the data, and it is apparent that the effect of including CP to all orders in our calculation is a mild suppression of the second-order contributions. This conclusion is further born out in the calculation of the ^{18}O and ^{18}F spectra, the results of which are shown in Figs. 4 and 5. Here we observe a weak suppression of the second-order effects in ^{18}O but a moderate suppression in ^{18}F. In the same figures we also observe that the spectra for different $V_{\text{low}-k}$ derived from the CD Bonn and Nijmegen bare potentials are nearly identical.

In summary, we have presented a method based on the KBB induced interaction formalism for efficiently summing CP diagrams to all orders in perturbation theory. This summation is carried out by way of the KBB self-consistent equations whose solution is significantly simplified by the use of the true *pp* and *ph* Green's functions, and by the use of the energy-independent $V_{\text{low}-k}$. Although our calculation was restricted in several important aspects, we find that our final renormalized interaction

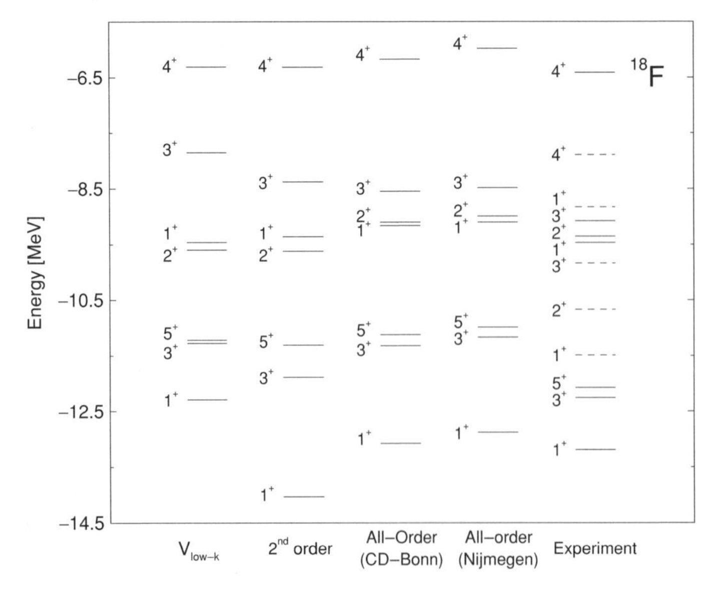

FIG. 5. Spectra for the ^{18}F system calculated to different orders in perturbation theory. See the caption to Fig. 4 for details.

is remarkably close to that of second-order perturbation theory. This is of practical importance and a welcoming result, for it allows one to use the results from a second-order calculation to approximate the contributions resulting from a large class of higher-order diagrams. In the future we intend both to expand our treatment by including additional diagrammatic contributions (RPA) and to generalize our method to study the all-order CP effects for effective operators such as magnetic moments.

We thank M. Kirson for helpful discussions. Partial support from the U.S. Department of Energy under contract DE-FG02-88ER40388 is gratefully acknowledged.

[1] G. F. Bertsch, Nucl. Phys. **74**, 234 (1965).
[2] T. T. S. Kuo and G. E. Brown, Nucl. Phys. **85**, 40 (1966).
[3] G. E. Brown, *Unified Theory of Nuclear Models and Forces* (North-Holland, Amsterdam, 1971).
[4] A. P. Zuker, Phys. Rev. Lett. **90**, 042502 (2003).
[5] B. R. Barrett and M. W. Kirson, Nucl. Phys. **A148**, 145 (1970).
[6] M. Hjorth-Jensen, T. T. S. Kuo, and E. Osnes, Phys. Rep. **261**, 126 (1995), and references therein.
[7] M. W. Kirson, Ann. Phys. (NY) **66**, 624 (1971); **68**, 556 (1971); **82**, 345 (1974).
[8] S. Babu and G. E. Brown, Ann. Phys. (NY) **78**, 1 (1973).
[9] A. Jackson, A. Lande, and R. A. Smith, Phys. Rep. **86**, 55 (1982); A. Lande and R. A. Smith, Phys. Rev. A **45**, 913 (1992) .
[10] D. J. Dean and M. Hjorth-Jensen, Phys. Rev. C **69**, 054320 (2004).
[11] D. W. L. Sprung and A. M. Jopko, Can. J. Phys. **50**, 2768 (1972).
[12] O. Sjöberg, Ann. Phys. (NY) **78**, 39 (1973).
[13] S. Bogner, T. T. S. Kuo, L. Coraggio, A. Covello, and N. Itaco, Phys. Rev. C **65**, 051301(R) (2002).
[14] T. T. S. Kuo, S. K. Bogner, and L. Coraggio, Nucl. Phys. **A704**, 107c (2002).
[15] L. Coraggio *et al.*, Phys. Rev. C **66**, 021303(R) (2002).
[16] L. Coraggio, A. Covello, A. Gargano, N. Itaco, and T. T. S. Kuo, Phys. Rev. C **66**, 064311 (2002).
[17] S. K. Bogner, T. T. S. Kuo, and A. Schwenk, Phys. Rep. **386**, 1 (2003).
[18] J. D. Holt, T. T. S. Kuo, G. E. Brown, and S. K. Bogner, Nucl. Phys. **A733**, 153 (2004).
[19] J. D. Holt, T. T. S. Kuo, and G. E. Brown, Phys. Rev. C **69**, 034329 (2004).
[20] E. M. Krenciglowa, C. L. Kung, T. T. S. Kuo, and E. Osnes, Ann. Phys. (NY) **101**, 154 (1976).
[21] J. W. Holt and G. E. Brown, nucl-th/0408047.
[22] P. J. Ellis and E. Osnes, Rev. Mod. Phys. **49**, 777 (1977).
[23] T. T. S. Kuo and E. Osnes, in *Lecture Notes in Physics* (Springer-Verlag, New York, 1990), Vol. 364.
[24] R. Machleidt, Phys. Rev. C **63**, 024001 (2001).
[25] R. B. Wiringa, V. G. J. Stoks, and R. Schiavilla, Phys. Rev. C **51**, 38 (1995).
[26] V. G. J. Stoks, R. A. M. Klomp, C. P. F. Terheggen, and J. J. de Swart, Phys. Rev. C **49**, 2950 (1994).
[27] D. R. Entem, R. Machleidt, and H. Witala, Phys. Rev. C **65**, 064005 (2002).
[28] J. P. Vary, P. U. Sauer, and C. W. Wong, Phys. Rev. C **7**, 1776 (1973).
[29] S. K. Bogner, A. Schwenk, R. J. Furnstahl, and A. Nogga, nucl-th/0504043.
[30] T. T. S. Kuo, J. Shurpin, K. C. Tam, E. Osnes, and P. J. Ellis, Ann. Phys. (NY) **132**, 237 (1981).
[31] D. R. Tilley, H. R. Weller, C. M. Cheves, and R. M. Chasteler, Nucl. Phys. **A595**, 1 (1995).
[32] B. H. Wildenthal, Prog. Part. Nucl. Phys. **11**, 5 (1984).
[33] B. A. Brown and B. H. Wildenthal, Annu. Rev. Nucl. Part. Sci. **38**, 29 (1988).

PHYSICAL REVIEW C **73**, 011001(R) (2006)

Low-momentum hyperon-nucleon interactions

B.-J. Schaefer,[1,*] M. Wagner,[1] J. Wambach,[1,2] T. T. S. Kuo,[3] and G. E. Brown[3]
[1]*Institut für Kernphysik, TU Darmstadt, D-64289 Darmstadt, Germany*
[2]*Gesellschaft für Schwerionenforschung GSI, D-64291 Darmstadt, Germany*
[3]*Department of Physics and Astronomy, State University of New York, Stony Brook, New York, 11794-3800, USA*
(Received 22 June 2005; published 27 January 2006)

We present a first exploratory study for hyperon-nucleon interactions using renormalization group techniques. The effective two-body low-momentum potential $V_{\text{low }k}$ is obtained by integrating out the high-momentum components from realistic Nijmegen *YN* potentials. A *T*-matrix equivalence approach is employed, so that the low-energy phase shifts are reproduced by $V_{\text{low }k}$ up to a momentum scale $\Lambda \sim 500$ MeV. Although the various bare Nijmegen models differ somewhat from each other, the corresponding $V_{\text{low }k}$ interactions show convergence in some channels, suggesting a possible model-independent *YN* interaction at low momenta.

DOI: 10.1103/PhysRevC.73.011001 PACS number(s): 13.75.Ev, 21.30.−x

Starting from modern nucleon-nucleon interactions and performing a renormalization group (RG) decimation it has become possible to derive a "universal" low-momentum effective interaction $V_{\text{low }k}$ [1]. The basic idea is to integrate out the short-distance physics encoded in hard-core interactions that are not well constrained by the available phase-shift data. The resulting effective interactions form the starting point for ab initio nuclear structure calculations in few-body systems [2], shell-model studies [3,4], and mean-field treatments via density-functional methods [5]. They also serve as input for the derivation of Landau Fermi-liquid interactions and provide predictions of pairing gaps in nuclei and homogeneous neutron matter [6].

In this article we generalize the $V_{\text{low }k}$ approach to the hyperon-nucleon (*YN*) sector. The ultimate goal is to provide effective potentials of similar quality as in the *NN* case that could serve as the starting point for realistic calculations of the structure of hypernuclei and homogeneous hyperonic matter. At present such a program is hampered by the lack of a comparable database and the collapse to a model-independent low-momentum potential is far from obvious. In a first exploratory study we wish to address this point by considering the low-momentum decimation of various potentials by the Nijmegen group. We focus on hyperons with strangeness $S = -1$ for which $I = 1/2$ and $I = 3/2$ isospin states are available. For $I = 1/2$, several hyperon-nucleon channels occur that require new technical developments for the coupled RG flow equations.

The effective, low-momentum potential $V_{\text{low }k}$ for elastic two-body scattering is obtained by integrating out high-momentum components of a realistic bare potential V interaction. This is achieved by imposing a cutoff Λ on all loop integrals in the half-on-shell (HOS) *T*-matrix equation and replacing the bare potential V with the effective $V_{\text{low }k}$ potential. Because the physical low-energy quantities must not depend on the cutoff, the HOS *T*-matrix should be preserved for relative momenta $k', k < \Lambda$. This results in a modified Lippmann-Schwinger equation with a cutoff-dependent effective potential $V_{\text{low }k}$

$$T(k', k; k^2) = V_{\text{low }k}(k', k) + \frac{2}{\pi}\mathcal{P}\int_0^\Lambda q^2 dq \frac{V_{\text{low }k}(k', q)T(q, k; k^2)}{k^2 - q^2}.$$

By demanding $dT(k', k; k^2)/d\Lambda = 0$, an exact RG flow equation for $V_{\text{low }k}$ can be obtained [7]

$$\frac{d}{d\Lambda}V_{\text{low }k}(k', k) = \frac{2}{\pi}\frac{V_{\text{low }k}(k', \Lambda)T(\Lambda, k; \Lambda^2)}{1 - k^2/\Lambda^2}. \qquad (1)$$

Integrating this flow equation with a given initial bare potential at a large cutoff (small distance) one obtains the physically equivalent effective theory ($V_{\text{low }k}$) at a smaller cutoff Λ (larger distance).

Instead of solving the RG Eq. (1) directly as a differential equation with, e.g., standard Runge-Kutta methods, we use the Andreozzi-Lee-Suzuki (ALS) iteration method, which is based on a similarity transformation [8,9]. With folded diagram techniques it has been shown [10] that this iteration method indeed yields a solution of Eq. (1).

By construction, the resulting $V_{\text{low }k}$ for the *NN* interaction reproduces the empirical deuteron binding energy and scattering phase shifts up to $E_{\text{lab}} = 2\hbar^2\Lambda^2/M$. Most importantly, it is found that for $\Lambda < 2$ fm^{-1} ($E_{\text{lab}} < 330$ MeV) $V_{\text{low }k}$ is independent of the particular V_{NN} model, i.e., all matrix elements of the different high-precision potentials collapse to a single universal low-momentum effective potential [7]. This can be largely attributed to the long-range one-pion exchange (OPE), which is common to all realistic potentials and dominates the low-momentum scattering. The main effect of the RG evolution is a constant shift of the bare matrix elements that removes the ambiguities in the short-range part of the potential.

The energy-independent $V_{\text{low }k}$ is non-Hermitian. This can readily be seen from the RG Eq. (1) because the momenta are treated asymmetrically. With a second similarity transformation the non-Hermiticity of $V_{\text{low }k}$ can be eliminated. Phase shifts are preserved by this second transformation and there are a number of such phase-shift equivalent transformations [11] such as the well-known Okubo one [12]. In the present work we have used the Okubo transformation

*E-mail: bernd-jochen.schaefer@physik.tu-darmstadt.de

to obtain the Hermitian $V_{\text{low }k}$. For the *NN* interaction, the diagonal matrix elements are almost unchanged by the second transformation.

As realistic *YN* interactions we use the soft-core potentials by the Nijmegen group [13]. They are based on one-boson-exchange models (OBE) of the *NN* potential and use $SU(3)_F$ symmetry to infer the coupling vertices in the presence of a hyperon. Because the flavor symmetry is broken by the finite quark masses the pertinent coupling strengths have to be adjusted to data. Six different fits are available referred to NSC97a–NSC97f in the following. Each potential comes in two basis representations (isospin and physical particle basis). In this article we work on the isospin basis that was also originally used for the potential construction by the Nijmegen group. Therefore all isomultiplets are degenerate. The corresponding isospin-averaged masses are given by $m_N = 938.9$ MeV, $m_\Lambda = 1115.7$ MeV, and $m_\Sigma = 1193.1$ MeV. In the $I = 3/2$ channel no Λ hyperon is involved.

The different Nijmegen fits describe the known *YN* cross-section data equally well ($\chi^2/N \sim 0.55$) but exhibit differences on a more detailed level. Because of the few available data points (only 35 altogether), the phase shifts and some scattering lengths exhibit large variations for different fits [14]. This is in marked contrast to the *NN* case, where the wealth of the database allows for high-precision potentials. These differ only in their short-range properties that are not constrained by the available data. The collapse of the low-momentum potential $V_{\text{low }k}$ for the *NN* interaction to a model-independent effective potential, observed after the RG decimation [1] is basically driven by the precision of the measured phase shifts. Because of the scarceness and uncertainties in the *YN* data we cannot expect this model independence for the low-momentum *YN* $V_{\text{low }k}$ potential.

In addition, for the *YN* interaction the choice of the cutoff Λ is not so obvious. In this work a cutoff around $\Lambda \sim 500$ MeV is chosen which corresponds to 2.5 fm^{-1}.

We first consider the $I = 3/2$ case which corresponds to the $\Sigma N \to \Sigma N$ channel. The ALS iteration is exactly the same as for the *NN* interaction albeit with the appropriate substitution of the hyperon mass. For the ALS iteration we have used a cutoff $\Lambda \equiv \Lambda_P = 500$ MeV in the model space (*P* space) with 64 grid points and a cutoff $\Lambda_Q = 10$ GeV for the complementary *Q* space, also with 64 grid points. To verify the insensitivity of the results on these quantities we have varied the grid points in an interval (32,70) and the cutoff Λ_Q between the range $\Lambda_Q \pm 2$ GeV. The standard ALS method converges rapidly. Details will be presented in Ref. [15].

The results for the $^1S_0$ partial wave are shown in Fig. 1. In the lower panel the diagonal matrix elements for three different bare potentials (dashed lines) and the corresponding RG evolved $V_{\text{low }k}$ potentials are displayed versus the relative momentum *k*. The RG decimation yields a soft-core $V_{\text{low }k}$ potential that is more attractive (dotted curves). As already mentioned, the $V_{\text{low }k}$'s are basically non-Hermitian after the RG decimation. By means of an Okubo transformation we obtain Hermitian $V_{\text{low }k}$ potentials. As in the *NN* case [1] the differences are negligible.

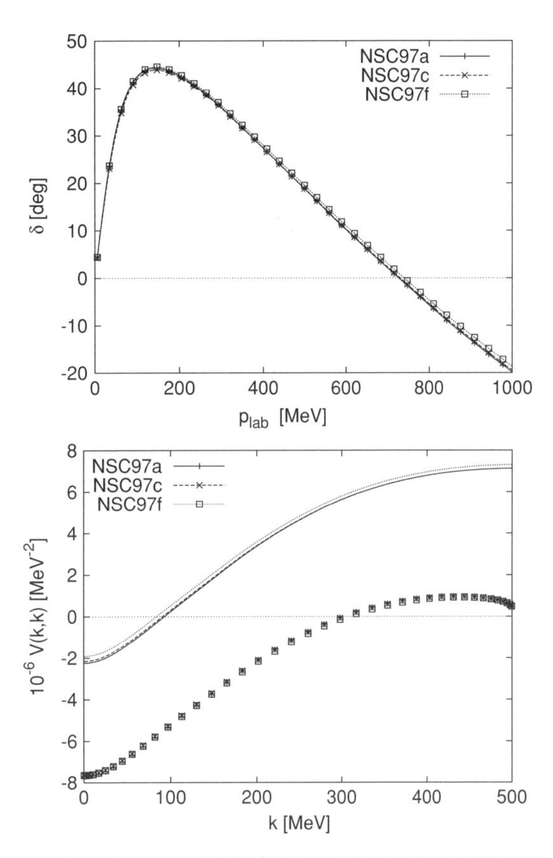

FIG. 1. (Upper panel) The $^1S_0$ phase shifts for the six different potentials as a function of the momentum in the lab frame p_{LAB}. (Lower panel) Three diagonal bare potentials V_{bare} (NSC97a,-c,-f; dashed lines) and three (Hermitian) $V_{\text{low }k}$ matrix elements (dotted) for $\Sigma N \to \Sigma N$ ($I = 3/2, {}^1S_0$) versus the relative momentum *k*.

By construction, the on-shell *T*-matrix is phase-shift equivalent that must result in identical phase shifts for a given bare potential fit and momenta below the cutoff Λ. This is demonstrated in the upper panel of Fig. 1. This comparison also serves as a test for our numerics. For the $^1S_0$ partial wave, the bare potentials do not differ strongly and thus yield almost the same $V_{\text{low }k}$ for all fits considered.

As discussed below, this changes for higher partial waves where the different bare potential fits deviate significantly and therefore no universal $V_{\text{low }k}$ is found.

As in the *NN* case, the *YN* interaction contains tensor components and hence partial waves can mix. The generalization of the ALS iteration in the presence of tensor forces is straightforward. For $S = 1$ one has to enlarge the *T*-matrix to a (2×2) block structure in the standard way, corresponding to the orbital angular momentum combinations $L = J \pm 1$. Also for this case we have verified that $V_{\text{low }k}$ is phase-shift equivalent to the bare potentials, as is shown in the upper

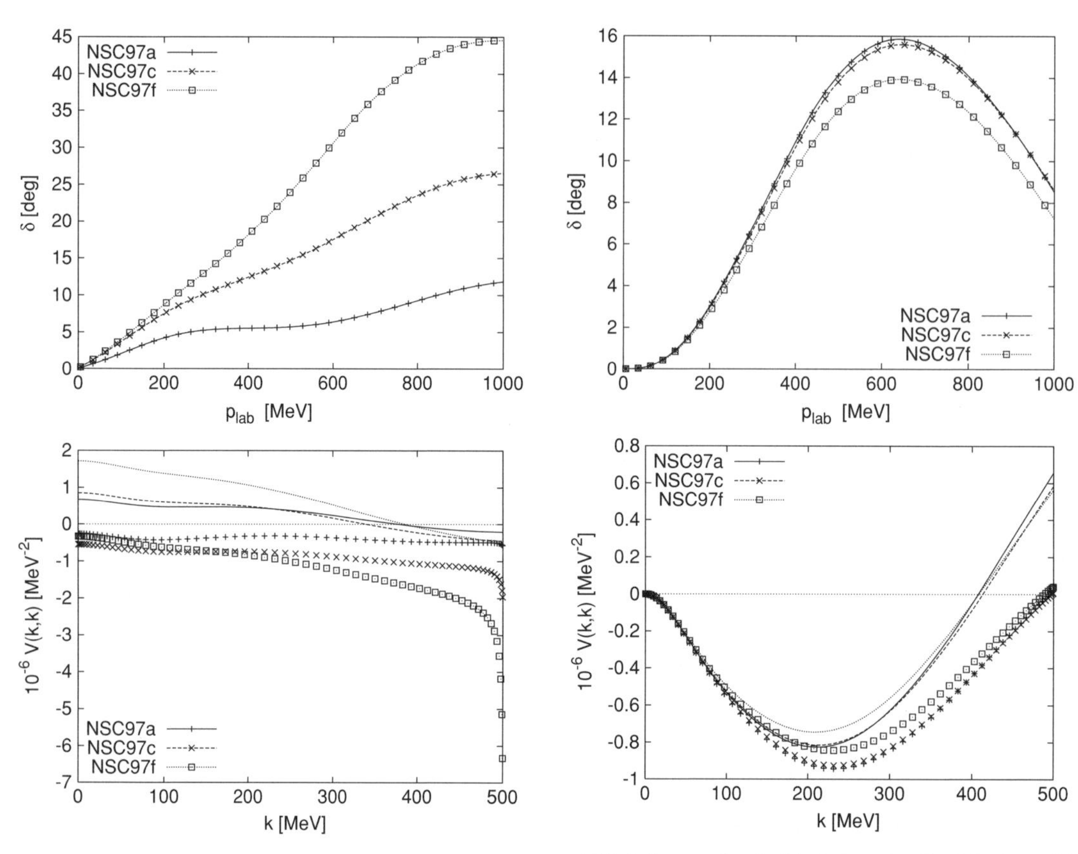

FIG. 2. The same as described in the legend to Fig. 1 for the partial wave ${}^3S_1$. (Upper panel) Nuclear bar phase shift.

FIG. 3. Same as described in the legend to Fig. 1 for the partial wave ${}^1P_1$ and $I = 3/2$. (Upper panel) Nuclear bar phase shift.

panel of Fig. 2 for the ${}^3S_1$ partial wave. The lower panel displays the corresponding $V_{\text{low}\,k}$ potentials together with the bare ones. They again become more attractive for all Nijmegen fits. Because of strong differences in the phase shifts we do not find a collapse of $V_{\text{low}\,k}$ to one potential, especially at larger momenta where deviations are most pronounced.

The RG flow Eq. (1) implies a pole at the cutoff boundary $k = \Lambda$. In the vicinity of this pole the slope for the $V_{\text{low}\,k}$ diverges that can be clearly seen in the lower panel of Fig. 2.

To complete the analysis for the $I = 3/2$ channel we show in Fig. 3 the ${}^1P_1$ partial wave that serves as an example for the spin singlet-spin triplet transition. Such transitions that are induced by the antisymmetric spin-orbit force are negligible in the *NN* interaction because of the small mass differences. However, for the *YN* interaction this is not the case anymore and these transitions can be significant. Because of the smaller deviations in the phase shift, the corresponding $V_{\text{low}\,k}$ interactions collapse to single potential (especially for small momenta).

The treatment of of the $I = 1/2$ channel is much more complicated. For this isospin four coupled channels available correspond to the transitions: $(\Lambda N, \Sigma N) \to (\Lambda N, \Sigma N)$. This is a completely new situation for the $V_{\text{low}\,k}$ approach. Because now channels with different masses couple, new phenomena are to be expected concerning, e.g., the convergence behavior of the ALS iteration.

In flavor space the Lippmann-Schwinger equation becomes a coupled 2×2 matrix equation where the diagonal matrix elements describe respectively the $\Lambda N \to \Lambda N$ and $\Sigma N \to \Sigma N$ channels, whereas the off-diagonal elements describe the $\Lambda N \to \Sigma N$ and $\Sigma N \to \Lambda N$ transitions. Using a notation where we only list the hyperons $(Y, Y' = \Lambda, \Sigma)$

$$T^{Y'Y}\big(k', k; E_k^Y\big) = V^{Y'Y}(k', k) + \sum_{Z=\Lambda,\Sigma} \frac{2}{\pi} \mathcal{P} \int_0^\infty dq q^2 \frac{V^{Y'Z}(k', q) T^{ZY}\big(q, k; E_k^Y\big)}{E_k^Y - H_0^Z} \tag{2}$$

with the free Hamiltonian $H_0^Y(q) = q^2/2\mu_Y + m_Y + m_N$, the energy $E_k^Y = k^2/2\mu_Y + m_Y + m_N$ and the reduced mass $\mu_Y = m_Y m_N / m_Y + m_N$ we have (with the inclusion of tensor forces) in general four coupled equations to solve. For

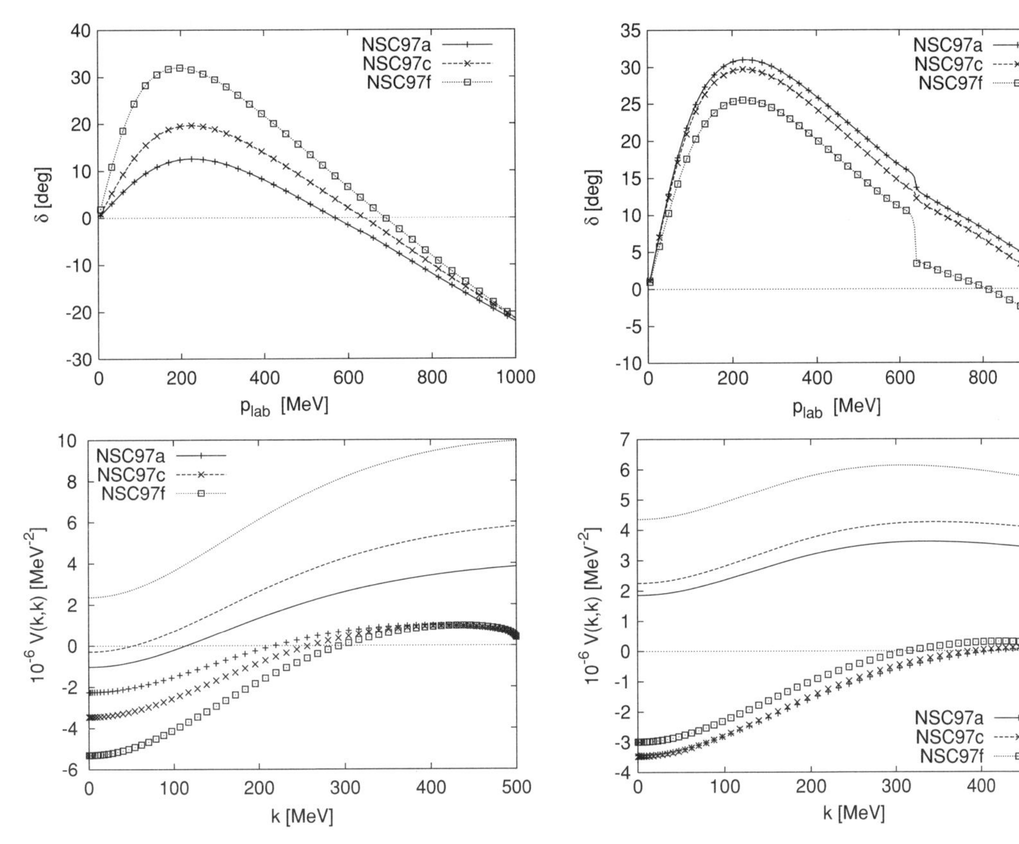

FIG. 4. ${}^1S_0$ partial wave for the $I = 1/2$ $\Lambda N \to \Lambda N$ channel. (Upper panel) Phase shifts. (Lower panel) Bare potentials and $V_{\text{low }k}$ potentials. The labeling in both panels is the same as in Fig. 1.

FIG. 5. Same as Fig. 4 but for the partial wave ${}^3S_1$. (Upper panel) Nuclear bar phase shifts.

some channels, the mass difference $m_\Sigma - m_\Lambda$ enters in the denominator of Eq. (2), which induces, e.g., a threshold behavior for the Σ hyperon. The mass differences that are not present in the *NN* interaction enter also in the ALS iteration. It is found that the standard ALS iteration procedure does not converge to the proper consecutive set of eigenvalues. As a consequence, a wrong sorting of the eigenvalues in the different P and Q spaces emerges. Using a modified ALS iteration or introducing an energy cutoff solves this problem and convergence to the correct eigenvalues can be found [15].

As an example we show in Fig. 4 the ${}^1S_0$ partial wave for the $\Lambda N \to \Lambda N$ channel. All $V_{\text{low }k}$ potentials are again more attractive and a more narrow grouping as compared to the bare potentials can be observed. One also observes that the shift of the $V_{\text{low }k}$ potential is largest for the bare NSC97f potential and smallest for the NSC97a potential in contrast to all other partial waves. The $V_{\text{low }k}$s are shifted in such a way that the potentials collapse for relative momenta near the cutoff reflecting the corresponding trends in the phase shifts.

In Fig. 5 the $\Lambda N \to \Lambda N$ channel with tensor coupling is shown for the ${}^3S_1$ partial wave. Here again the RG decimation pushes the bare potentials down basically by a (large) constant to attractive $V_{\text{low }k}$s. Because this channel includes the ΣN transition, the Σ threshold is visible in the phase shifts for lab momenta above 600 MeV. The jump in the phase shift at the threshold depends strongly on the model.

To complete this analysis results for the spin singlet-spin triplet transition in the $\Lambda N \to \Lambda N$ channel for the ${}^1P_1$ partial wave are presented in Fig. 6. For small momenta no RG decimation takes place. For this partial wave the bare interaction is repulsive and so is $V_{\text{low }k}$.

Recently, it has been shown that the $V_{\text{low }k}$ approach that is based on RG techniques provides a novel and powerful tool to obtain phase-shift equivalent low-momentum *NN* interactions. After the RG decimation a model-independent low-momentum potential $V_{\text{low }k}$ for different high-precision *NN* interactions was found. This model independence of the $V_{\text{low }k}$ matrix elements is an important property that is basically driven by the

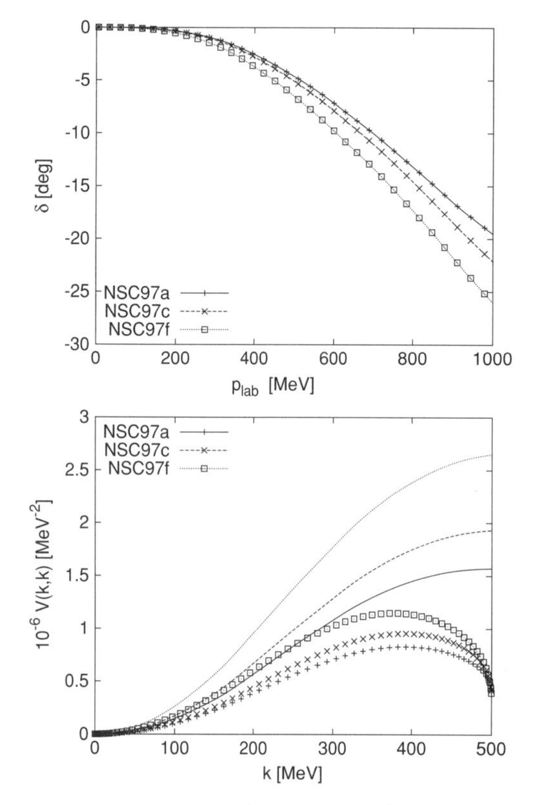

FIG. 6. Same as Fig. 4 but for the partial wave $^1P_1$. (Upper panel) Nuclear bar phase shift.

phase-shift equivalence and the common one-pion exchange of the input models.

In the present work the model dependence of the $V_{\text{low}\,k}$ for the *YN* interaction is investigated. Because of coupled-channel effects in flavor space that are not present in the *NN* case, the RG evolution is technically more complicated but can be treated. Because of the few experimental data currently available, the model fits by the Nijmegen group do not allow for a model-independent low-momentum *YN* interaction. It is therefore important to calculate $V_{\text{low}\,k}$ for other *YN* models such as the revised Jülich potential [16]. Such calculations are in preparation [15].

Although no model independence of the *YN* low-momentum interaction is obtained at present, a convergence of the different $V_{\text{low}\,k}$s is seen generally for all the Nijmegen potentials. Especially, for partial waves that do not deviate strongly for different bare potentials the model independence of $V_{\text{low}\,k}$ is pronounced. All $V_{\text{low}\,k}$ potentials are much softer than the bare ones. Softer interactions lead to stronger binding that should be of relevance in microscopic hypernuclei calculations.

By construction, all low-energy two-body observables are cutoff independent. Bogner *et al.* argue that any (new) induced cutoff dependence is because of higher-body forces [2]. For the *NN* interaction they conclude that such contributions are rather small. For the *YN* interaction this is still an open issue and should be tested in light hypernuclei.

ACKNOWLEDGMENTS

We thank A. Schwenk for helpful discussions. One of the authors (B.J.S.) also thanks S. K. Bogner for numerous enlightening discussions. He also expresses his gratitude to G. E. Brown and T. T. S. Kuo for the invitation to Stony Brook, where this work was initiated. M. W. is supported by BMBF grant 06DA116. Partial support to T. T. S. K. and G. E. B. from the U.S. Department of Energy under contract DE-FG02-88ER/40388 is gratefully acknowledged.

[1] S. K. Bogner, T. T. S. Kuo, and A. Schwenk, Phys. Rep. **386**, 1 (2003).
[2] A. Nogga, S. K. Bogner, and A. Schwenk, Phys. Rev. C **70**, 061002(R) (2004), nucl-th/0405016.
[3] L. Coraggio and N. Itaco, Phys. Lett. **B616**, 43 (2005).
[4] A. Schwenk and A. P. Zuker (2005), nucl-th/0501038.
[5] S. K. Bogner, A. Schwenk, R. J. Furnstahl, and A. Nogga, Nucl. Phys. **A763**, 59 (2005), nucl-th/0504043.
[6] A. Schwenk, B. Friman, and G. E. Brown, Nucl. Phys. **A713**, 191 (2003).
[7] S. K. Bogner, A. Schwenk, T. T. S. Kuo, and G. E. Brown (2001), nucl-th/0111042.
[8] F. Andreozzi, Phys. Rev. C **54**, 684 (1996).
[9] K. Suzuki and S. Y. Lee, Prog. Theor. Phys. **64**, 2091 (1980).
[10] S. K. Bogner, T. T. S. Kuo, L. Coraggio, A. Covello, and N. Itaco, Phys. Rev. C **65**, 051301(R) (2002).
[11] J. D. Holt, T. T. S. Kuo, and G. E. Brown, Phys. Rev. C **69**, 034329 (2004).
[12] S. Okubo, Prog. Theor. Phys. **12**, 603 (1954).
[13] T. A. Rijken, V. G. J. Stoks, and Y. Yamamoto, Phys. Rev. C **59**, 21 (1999).
[14] *NN-Online*, URL http://nn-online.org.
[15] B.-J. Schaefer, M. Wagner, J. Wambach, T. T. S. Kuo, and G. E. Brown, to be published.
[16] J. Haidenbauer and U.-G. Meissner, Phys. Rev. C **72**, 044005 (2005), nucl-th/0506019.

Hans Bethe and the Nuclear Many-Body Problem

Jeremy Holt and Gerald E. Brown

1. The Atomic and Nuclear Shell Models

Before the Second World War, the inner workings of the nucleus were a mystery. Fermi and collaborators in Rome had bombarded nuclei by neutrons, and the result was a large number of resonance states, evidenced by sharp peaks in the cross section; i.e., in the off-coming neutrons. These peaks were the size of electron volts (eV) in width. (The characteristic energy of a single molecule flying around in the air at room temperature is the order of 1/40 of an electron volt. Thus, one electron volt is the energy of a small assemblage of these particles.) On the other hand, the difference in energy between low-lying states in light and medium nuclei is the order of MeV, one million electron volts.

Niels Bohr's point[1] was that if the widths of the nuclear levels (compound states) were more than a million times smaller than the typical excitation energies of nuclei, then this meant that the neutron did not just fall into the nucleus and come out again, but that it collided with the many particles in the nucleus, sharing its energy. According to the energy-time uncertainty relation, the width ΔE of such a resonance is related to the lifetime of the state by $\tau = \hbar/\Delta E \sim \hbar/(1\,\text{eV}) \simeq 10^{-15}$ sec. In fact, the time $t = \hbar/(10\,\text{MeV}) \simeq 10^{-22}$ sec is the characteristic time for a nucleon to circle around once in the nucleus, the dimension of which is Fermis (1 Fermi $= 10^{-13}$ cm). Thus, a thick "porridge" of all of the nucleons in the nucleus was formed, and only after a relatively long time would this mixture come back to the state in which one single nucleon again possessed all of the extra energy, enough for it to escape. The compound states in which the incoming energy was shared by all other particles in the porridge looked dauntingly complicated.

At the time, atomic physics was considered to be very different from nuclear physics, the atomic many-body problem being that of explaining the makeup of atoms. Niels Bohr[2] had shown, starting with the hydrogen atom, that each negatively charged electron ran around the much smaller nucleus in one of many "stationary states." Once in such a state, it was somehow "protected" from spiraling into the positively charged nucleus. (This "protection" was understood only later with the discovery of wave mechanics by Schrödinger and Heisenberg.) The allowed states of the hydrogen atom were obtained in the following way. The Coulomb attraction between the electron and proton provides the centripetal force, yielding

$$\frac{e^2}{r^2} = \frac{mv^2}{r}, \tag{1}$$

where $-e$ is the electron charge, $+e$ is the proton charge, m is the electron mass, v is its velocity, and r is the radius of the orbit. Niels Bohr carried out the quantization, the meaning of which will be clear later, using what is called the classical action, but a more transparent (and equivalent) way is to use the particle-wave duality picture of de Broglie (Prince L. V. de Broglie received the Nobel prize in 1929 for his discovery of the wave nature of electrons) in which a particle with mass m and velocity v is assigned a wavelength

$$\lambda = \frac{h}{mv}, \tag{2}$$

where h is Planck's constant. If the wave is to be stationary, it must fit an integral number of times around the circumference of the orbit, leading to

$$n\frac{h}{mv} = 2\pi r. \tag{3}$$

Eliminating v from Eqs. (1) and (3), we find the radius of the orbit to be

$$r = \frac{n^2\hbar^2}{me^2}, \quad \text{where} \quad \hbar = \frac{h}{2\pi}. \tag{4}$$

Using Eq. (1), the kinetic energy of the electron is

$$T = \frac{1}{2}mv^2 = \frac{e^2}{2r}. \tag{5}$$

The potential energy is

$$V = -\frac{e^2}{r} = -2T\,, \tag{6}$$

which follows easily from Eq. (5). Equation (6) is known as a "virial relation," an equation that relates the kinetic and potential energies to one another. In the case of a potential depending on r as $1/r$, the kinetic energy is always equal to $-1/2$ of the potential energy. From this relation we find that the total energy is given by

$$E = T + V = -T\,. \tag{7}$$

Finally, one finds by combining Eqs. (7), (5), and (4) that

$$E_n = -\frac{1}{2}\frac{me^4}{n^2\hbar^2} = -\frac{1}{n^2}Ry\,, \tag{8}$$

where Ry is the Rydberg unit for energy

$$Ry \simeq 13.6 \text{ electron volts.} \tag{9}$$

Thus, we find only a discrete set of allowed energies for an electron bound in a hydrogen atom, and we label the corresponding states by their values of n. Since $n = 1$ for the innermost bound orbit in hydrogen, called an s-state for reasons we discuss later, 13.6 electron volts is the ionization energy, the energy necessary to remove the electron.

Electrons are fermions, such that only one particle can occupy a given quantum state at a time. This is called the "exclusion principle." (Wolfgang Pauli received the Nobel prize in 1945 for the discovery of the exclusion principle, also called the "Pauli principle.") Once an electron has been put into a state, it excludes other electrons from occupying it. However, electrons have an additional property called spin, which has the value 1/2 (in units of $\hbar$), and this spin can be quantized along an arbitrary axis to be either up or down. Thus two electrons can occupy the $1s$ state. Putting two electrons around a nucleus consisting of two neutrons and two protons makes the helium atom. Helium is the lightest element of the noble gases. (They are called "noble" because they interact very little with other chemical elements.) Since the $1s$ shell is filled with two electrons, the helium atom is compact and does not have an empty $1s$ orbital which would like to grab a passing electron, unlike hydrogen which is very chemically active due to a vacancy in its $1s$ shell.

To go further in the periodic table we have to put electrons into the $n = 2$ orbit. A new addition is that an electron in the $n = 2$ state can have an orbital angular momentum of $l = 0$ or 1, corresponding to the states $2s$ and $2p$, respectively. (The spectroscopic ordering is $s\,p\,d\,f\,g\,h\,i$ for $l = 0 - 6$, a notation that followed from the classification of atomic spectra well before the Bohr atom was formulated.) The angular momentum $l = 1$ of the $2p$ state can be projected on an arbitrary axis to give components $m = 1, 0$, or -1. So altogether, including spin, six particles can be put into the $2p$ state and two into the $2s$ state. Thus, adding eight electrons in the $2s$- and $2p$-states, the next member of the noble gases, neon, is obtained. It is particularly compact and the electrons are well bound, because both the $n = 1$ and $n = 2$ shells are filled.

Consider an element in which the $n = 2$ shell is not filled, oxygen, which has eight electrons, two in the $n = 1$ shell and six in the $n = 2$ shell. Oxygen in the bloodstream or in the cellular mitochondria is always on the alert to fill in the two empty orbits. We call such a "grabbing" behavior "oxidation," even though the grabbing of electrons from other chemical elements is done not only by oxygen. The molecules that damage living cells by stealing their electrons are called "free radicals," and it is believed that left alone, they are a major cause of cancer and other illnesses. This is the origin of the term "oxidative stress." We pay immense amounts of money for vitamins and other "antioxidants" in order to combat free radicals by filling in the empty states.

We need to add one further piece to the picture of the atomic shell model, the so-called j–j coupling, which really gave the key success of the nuclear shell model, as we explain later. The j we talk about is the total angular momentum, composed of adding the orbital angular momentum l and the spin angular momentum s. The latter can take on projections of $+1/2$ or $-1/2$ along an arbitrary axis, so j can be either $l + 1/2$ or $l - 1/2$. The possible projections of j are $m = j, j - 1, \ldots, -j$. Thus, if we reconsider the p-shell in an atom, which has $l = 1$, the projections are reclassified to be those of $j = 3/2$ and $j = 1/2$. The former has projections $3/2$, $1/2$, $-1/2$, and $-3/2$, the projections differing by integers, whereas the latter has $+1/2$ and $-1/2$. Altogether there are six states, the same number that we found earlier, through projections of $m_l = -1, 0, 1$ and of $m_s = +1/2, -1/2$. The classification in terms of j is important because there is a spin-orbit coupling; i.e., an interaction between spin and orbital motion which depends on the relative angle between the spin and orbital angular momentum. This interaction has the effect of increasing the energy of a state for which the spin is in the same direction as the orbital angular momentum, as in the $2p_{3/2}$ state being higher in energy than the $2p_{1/2}$ state. In this notation, the subscript refers to the total angular momentum. The filling of the shells in the j–j classification scheme is the same as in the l scheme in that the same number of electrons fill a shell in either scheme, the only difference being the subshell labels.

The above is the atomic shell model, so called because electrons are filled in shells. The electric interaction is weak compared with the nuclear one, down by a factor of $\alpha = e^2/\hbar c = 1/137$ from the nuclear interaction (called the strong interaction). The atomic shell model is also determined straightforwardly because the nucleus is very small (roughly 10,000 times smaller in radius than the atom), and it chiefly acts as a center about which the electrons revolve. In atoms the number of negatively charged electrons is equal to the number of positively charged protons, the neutrons being of neutral charge. In light nuclei there are equal numbers of neutrons and protons, but the number of neutrons relative to the protons grows as the mass number A increases, because it is relatively costly in energy to concentrate the repulsion of the protons in the nucleus. On the whole, electric forces play only a minor role in the forces between nucleons inside the nucleus, except for determining the ratio of protons to neutrons. By the time we get to ^{208}Pb with 208 nucleons, 126 neutrons and 82 protons, we come to a critical situation for the nuclear shell model, which requires the spin-orbit force, as we discuss later.

After a decade or two of nuclear "porridge," imagine people's surprise when the nuclear shell model was introduced in the late 1940's[3,4] and it worked; i.e., it explained a lot of known nuclear characteristics, especially the "magic numbers." That is, as shells were filled in a prescribed order, those nuclei with complete shells turned out to be substantially more bound than those in which the shells were not filled. In particular, a particle added to a closed shell had an abnormally low binding energy.

What determines the center in the nuclear shell model? In the case of the atomic shell model, the much heavier and much smaller nucleus gave a center about which electrons could be put into shells. Nuclei were known to be tightly bound. There must be a center, and most simply the center would be exactly in the middle of the charge distribution. This charge distribution can be determined by considering two nuclei that differ from one another by the exchange of a proton and a neutron. In this case the two nuclei will have different binding energies due to the extra Coulomb energy associated with the additional proton. This difference could be measured by the energy of the radioactive decay of the one nucleus into the other and estimated roughly as Ze^2/R (where R is the radius of the nucleus and Z is the number of protons). This gave $R \sim 1.5A^{1/3} \times 10^{-13}$ cm, where A is the mass number. (Later electron scattering experiments, acting like a very high resolution electron scattering microscope, gave the detailed shape of the charge distribution and basically replaced the 1.5 by 1.2.)

The most common force is zero force, that is, matter staying at rest in equilibrium. The force, according to Newton's law, is the (negative) derivative of the potential. If the potential at short distances is some constant times the square of r; i.e., $V = Cr^2$, then the force $F = -dV/dr = -2Cr$ is zero at $r = 0$. Furthermore, the negative sign indicates that any movement away from $r = 0$ is met by an attractive force directed back toward the center, leading to stable equilibrium. Such a potential is called a harmonic oscillator. It occurs quite commonly in nature, since most matter is more or less in equilibrium.

The energy of a particle in such a potential can be expressed classically as

$$E = \frac{p^2}{2M} + \frac{1}{2}M\omega^2 r^2 , \tag{10}$$

where p is the particle's momentum, M is its mass, and ω describes the strength of the restoring force. Increasing ω will bind the nucleons closer together, which allows ω to be related to the nuclear radius R by

$$\hbar\omega \sim \frac{\hbar^2}{MR^2} . \tag{11}$$

Using harmonic oscillator wave functions with ω determined so that

$$R = 1.2A^{1/3} \times 10^{-13}\text{cm} \tag{12}$$

gives a remarkably good fit to the shape of the charge distribution (in which only the protons are included) of the nucleus. The possible energy levels in such a potential are shown in Fig. 1 where we have labeled them in the same way as in the atomic shell model. Note that the distances between neighboring levels are always the same, $\hbar\omega$.[a] We redraw the potential in Fig. 2, incorporating the j–j coupling and a spin-orbit interaction.

[a]The situation is analogous to that with light quanta, which have energy $\hbar\omega$ each, so that an integral number of quanta is emitted for any given frequency $\hbar\omega$.

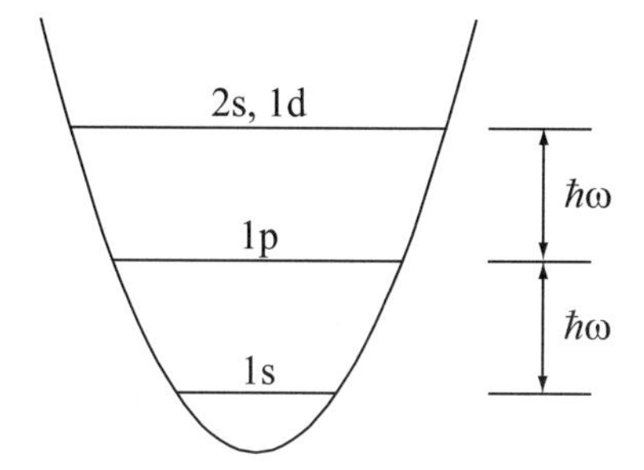

Fig. 1 Harmonic oscillator potential with possible single-particle states.

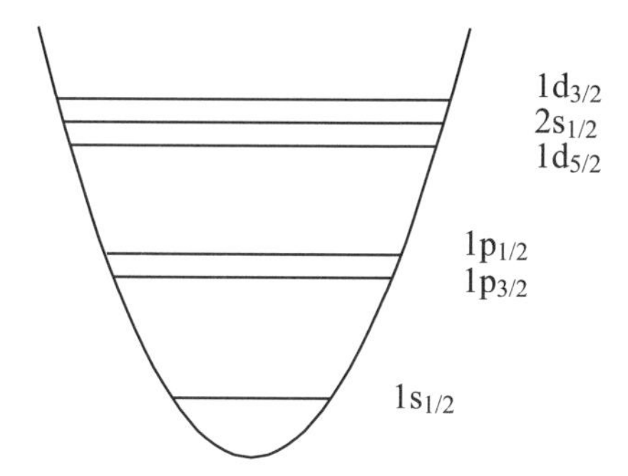

Fig. 2 Harmonic oscillator potential with possible single particle states in j–j coupling.

Note that the spin-orbit splitting, that is, the splitting between two states with the same l, is somewhat smaller than the distance between shells,[b] so that the classification of levels according to l as shown in Fig. 1 gives a good zero-order description.

At first sight the harmonic oscillator potential appears unreasonable because the force drawing the particle back to the center increases as the particle moves farther away, but nucleons can certainly escape from nuclei when given enough energy. However, the last nucleon in a nucleus is typically bound by about 8 MeV, and this has the effect that its wave function drops off rapidly outside of the potential; i.e., the probability of finding it outside the potential is generally small. In fact, the harmonic oscillator potential gives remarkably good wave functions, which can only be improved upon by very detailed calculations.

[b]In heavier nuclei such as ^{208}Pb the angular momenta become so large that the spin-orbit splitting is not small compared to the shell spacing, as we shall see.

In the paper of the "Bethe Bible" coauthored with R. F. Bacher,[5] nuclear masses had been measured accurately enough in the vicinity of ^{16}O so that "It may thus be said safely that the completion of the neutron-proton shell at ^{16}O is established beyond doubt from the data about nuclear masses." The telltale signal of the closure of a shell is that the binding energy of the next particle added to the closed shell nucleus is anomalously small. Bethe and Bacher (p. 173) give the shell model levels in the harmonic oscillator potential, the infinite square well, and the finite square well. These figures would work fine for the closed shell nuclei ^{16}O and ^{40}Ca. In Fig. 1, ^{16}O would result from filling the $1s$ and $1p$ shells, ^{40}Ca from filling additionally the $2s$ and $1d$ shells. However, other nuclei that were known to be tightly bound, such as ^{208}Pb, could not be explained by these simple potentials.

The key to the Goeppert Mayer-Jensen success was the spin-orbit splitting, as it turned out. There were the "magic numbers," the large binding energies of ^{16}O, ^{40}Ca, and ^{208}Pb. Especially lead, with 82 protons and 126 neutrons, is very tightly bound. Now it turns out (see Fig. 3) that with the strong spin-orbit splitting the $1h_{11/2}$ level for protons and the $1i_{13/2}$ level for neutrons lie in both cases well below the next highest levels. The notation here is different from that used in atomic physics. The "1" denotes that this is the first time that an h ($l = 5$) level would be filled in adding protons to the nucleus and for the neutron levels similarly, i denoting $l = 6$.

So how could a model in which nucleons move around in a common potential without hitting each other be reconciled with the previous "nuclear porridge" of Niels Bohr? One explanation is that the thorough mixture of particles in the porridge arose because the neutron was dropped into the nucleus with an energy of around 8 MeV above the ground state energy in the cases of "porridge." In 1958 Landau formulated his theory of Fermi liquids, showing that as a particle (fermion) was added with an energy just above the highest occupied state (just at the top of the Fermi sea), the added particle would travel forever without exciting the other particles. In more physical terms, the mean free path (between collisions) of the particle is proportional to the inverse of the square of the difference of its momentum from that of the Fermi surface; i.e.,

$$\lambda \sim \frac{C}{(k - k_{\rm F})^2}, \tag{13}$$

where C is a constant that depends on the interaction. Thus, as $k \to k_{\rm F}$, $\lambda \to \infty$ and the particle never scatters. Here $k_{\rm F}$, the Fermi momentum, is that of the last filled orbit.

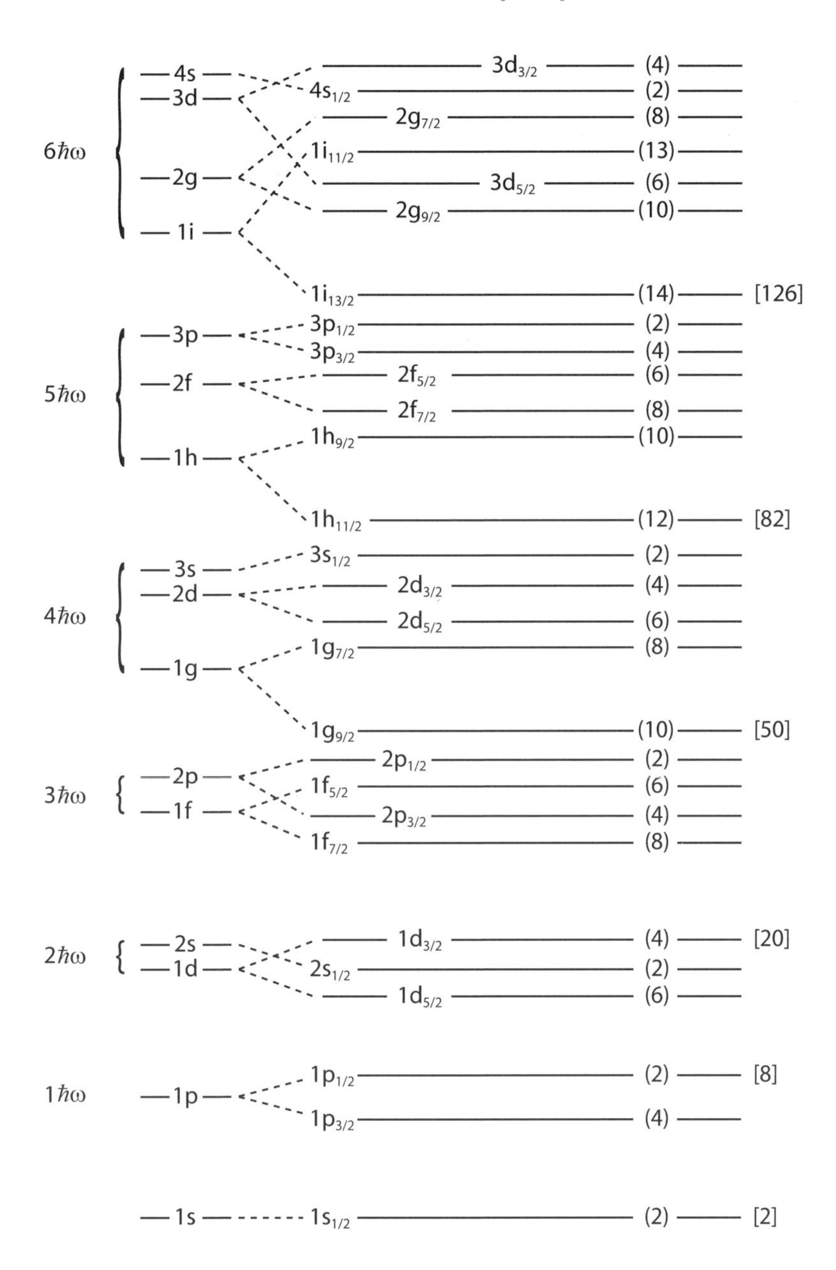

Fig. 3 Sequence of single particle states in the nuclear shell model.

Once the surprise had passed that one could assign a definite shell model state to each nucleon and that these particles moved rather freely, colliding relatively seldomly, the obvious question was how the self-consistent potential the particles moved in could be constructed from the interactions between the particles. The main technical problem was that these forces were very strong. Indeed, Jastrow[6] characterized the short range force between two nucleons as a vertical hard core of infinite height and radius of 0.6×10^{-13} cm, about one-third the average distance between two nucleons in nuclei. (Later, the theoretical radius of the core shrank to 0.4×10^{-13} cm.) The core was much later found to be a rough characterization of the short-range repulsion from vector meson exchange.

The whole postwar development of quantum electrodynamics by Feynman and Schwinger was in perturbation theory, with expansion in the small parameter $\alpha = e^2/\hbar c = 1/137$. Of course, infinities were encountered, but since they shouldn't be there, they were set to zero. The concept of a hard core potential of finite range, the radius a reasonable fraction of the average distance between particles, was new. Lippmann and Schwinger[7] had already developed a formalism that could deal with such an interaction. In perturbation theory; i.e., expansion of an interaction which is weak relative to other quantities, the correction to the energy resulting from the interaction is obtained by integrating the perturbing potential between the wave functions; i.e.,

$$\Delta E \simeq \int \psi_0^\dagger(x,y,z)\, V(x,y,z)\, \psi_0(x,y,z)\, dx\, dy\, dz\,. \tag{14}$$

Here $\psi_0(x,y,z)$ is the solution of the Schrödinger wave equation for the zero-order problem, that with $V(x,y,z) = 0$, and $\psi_0^\dagger$ is the complex conjugate of ψ_0. The wavefunction gives the probability of finding a particle located in a small region $dx\, dy\, dz$ about the point (x,y,z):

$$P(x,y,z) = \psi_0^\dagger(x,y,z)\, \psi_0(x,y,z)\, dx\, dy\, dz\,. \tag{15}$$

The integral in Eq. (14) is carried out over the entire region in which $V(x,y,z)$ is nonzero.

We see the difficulty that arises if V is infinite, as in the hard core potential; namely, the product of an infinite V and finite ψ is infinite. Equation (14) is just the first-order correction to the energy. More generally, perturbation theory yields a systematic expansion for the energy shift in terms of integrals involving higher powers of the interaction $(V^2, V^3, \cdots)$. As long as $\psi(x,y,z)$ is nonzero, all of these terms are infinite.

Watson[8,9] realized that a repulsion of any strength, even an infinitely high hard core, could be handled by the formalism of Lippmann and Schwinger, which was invented in order to handle two-body scattering. Quantum mechanical scattering is described by the T-matrix:

$$T = V + V\frac{1}{E - H_0}V + V\frac{1}{E - H_0}V\frac{1}{E - H_0}V + \cdots, \tag{16}$$

where V is the two-body potential, E is the unperturbed energy, and H_0 is the unperturbed Hamiltonian, containing only the kinetic energy. This infinite number of terms could be summed to give the result

$$T = V + V\frac{1}{E - H_0}T. \tag{17}$$

One can see that this is true by rewriting Eq. (16) as

$$T = V + V\frac{1}{E - H_0}\left(V + V\frac{1}{E - H_0}V + V\frac{1}{E - H_0}V\frac{1}{E - H_0}V + \cdots\right), \tag{18}$$

where the term in parentheses is clearly just T. Watson realized that the T-matrix made sense also for an extremely strong repulsive interaction in a system of many nucleons. An incoming particle could be scattered off each of the nucleons, one by one, and the scattering amplitudes could be added up, the struck nucleon being left in the same state as it was initially. The sum of the amplitudes could be squared to give the total amplitude for the scattering off the nucleus.

Keith Brueckner, who was a colleague of Watson's at the University of Indiana at the time, saw the usefulness of this technique for the nuclear many-body problem. He was the first to recognize that the strong short-range interactions, such as the infinite hard core, would scatter two nucleons to momenta well above those filled in the Fermi sea. Thus, the exclusion principle would have little effect and could be treated as a relatively small correction.

Basically, for the many-body problem, the T-matrix is called the G-matrix, and the latter obeys the equation

$$G = V + V\frac{Q}{e}G. \tag{19}$$

Whereas in the Lippmann-Schwinger formula Eq. (17) the energy denominator e was taken to be $E - H_0$, there was considerable debate about what to put in for e in Eq. (19). We will see later that it is most conveniently chosen to be $E - H_0$, as in Eq. (17).[c] In Eq. (19), the operator Q excludes from

[c]This conclusion will be reached only after many developments, as we outline in Sec. 3.

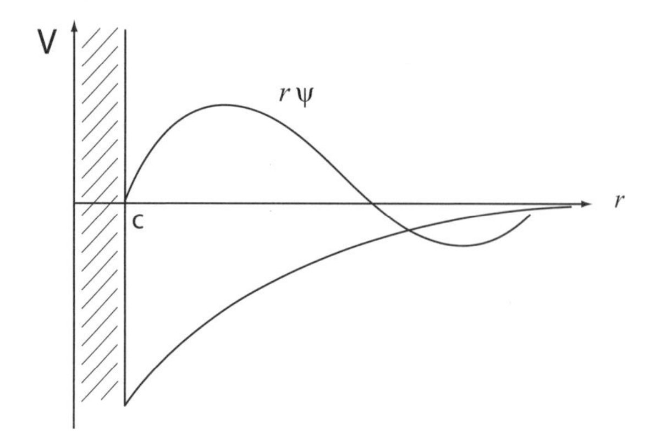

Fig. 4 The wave function ψ for the relative motion of two particles interacting via a potential with an infinite hard core of radius c and an attractive outside potential. It is convenient to deal with $r\psi$ rather than ψ.

the intermediate states not only all of the occupied states below the Fermi momentum k_{F}, because of the Pauli principle, but also all states beyond a maximum momentum k_{max}. The states below k_{max} define what is called the "model space," which can generally be chosen at the convenience of the investigator. The basic idea is that the solution Eq. (19) of G is to be used as an effective interaction in the space spanned by the Fermi momentum k_{F} and k_{max}. This effective interaction G is then to be diagonalized within the model space.

In the above discussion leading to Eq. (19) we have written the nucleon-nucleon interaction as V. In fact, the interaction is complicated, involving various combinations of spins and angular momenta of the two interacting nucleons. In the time of Brueckner and the origin of his theory, it took a lot of time and energy just to keep these combinations straight. This large amount of bookkeeping is handled today fairly easily with electronic computers. The major problem, however, was how to handle the strong short-range repulsion, and this problem was discussed in terms of the relatively simple, what we call "central," interaction shown in Fig. 4. In fact, the G of Eq. (19) is still calculated from that equation today in the most successful effective nuclear forces. The k_{max} is taken to be the maximum momentum at which experiments have been analyzed, $k_{\mathrm{max}} = 2.1\ \mathrm{fm}^{-1} = \Lambda$, where Λ is now interpreted as a cutoff. Since experiments of momenta higher than Λ have not been carried out and analyzed, at least not in such a way as to bear directly on the determination of the potential, one approach is to leave them out completely. The only important change since the 1950's is that the $V(r)$, the first term on the right-hand side of Eq. (19), is now rewritten in terms of a sum over momenta, by what is called a Fourier transform, and this sum is truncated at Λ, the higher momenta being discarded. The resulting effective interaction, which replaces G, is now called $V_{\text{low}-k}$.[10,11] We expand on this discussion at the end of this article.

The G-matrix of Brueckner was viewed by nuclear physicists as a complicated object. However, it is clear what the effect of a repulsion at core radius c that rises to infinite height will be — it will stop the two interacting particles from going into the region of $r < c$. In non-relativistic quantum mechanics, this means that their wave function of relative motion must be zero. In other words, the wave function, whose square gives the probability of finding the particle in a given region, must be zero inside the hard core. Also, the wave function must be continuous outside, so that it must start from zero at $r = c$. Therefore, we know that the wave function must look something like that shown in Fig. 4. In any case, given the boundary condition $\psi = 0$ at $r = c$, and the potential energy V, the wave equation can be solved for $r > c$. It is not clear at this point how $V(r)$ is to be determined and what the quantity ψ is. We put off further discussion of this until Sec. 3, in which we develop Hans Bethe's "Reference Spectrum."

One of the most important, if not the most important, influences on Hans Bethe in his efforts to give a basis for the nuclear shell model was the work of Feshbach, Porter, and Weisskopf.[12] These authors showed that although the resonances formed by neutrons scattered by nuclei were indeed very narrow, their strength function followed the envelope of a single-particle potential; i.e., of the strength function for a single neutron in a potential $V + iW$. The strength function for the compound nucleus resonances is defined as $\bar{\Gamma}_n/D$, where Γ_n is the width of the resonance for elastic neutron scattering (scattering without energy loss) and D is the average spacing between resonances. This function gives the strength of absorption, averaged over many of the resonances. Parameters of the one-body potential are given in the caption to Fig. 5. The parameter a is the surface thickness of the one-body potential.

The peaks in the neutron strength function occur at those mass numbers where the radius of the single-particle potential is just big enough to bind another single-particle state with zero angular momentum. Although the single-particle resonance is split up into the many narrow states discussed by Bohr, a vestige of the single-particle shell model resonance still remains.

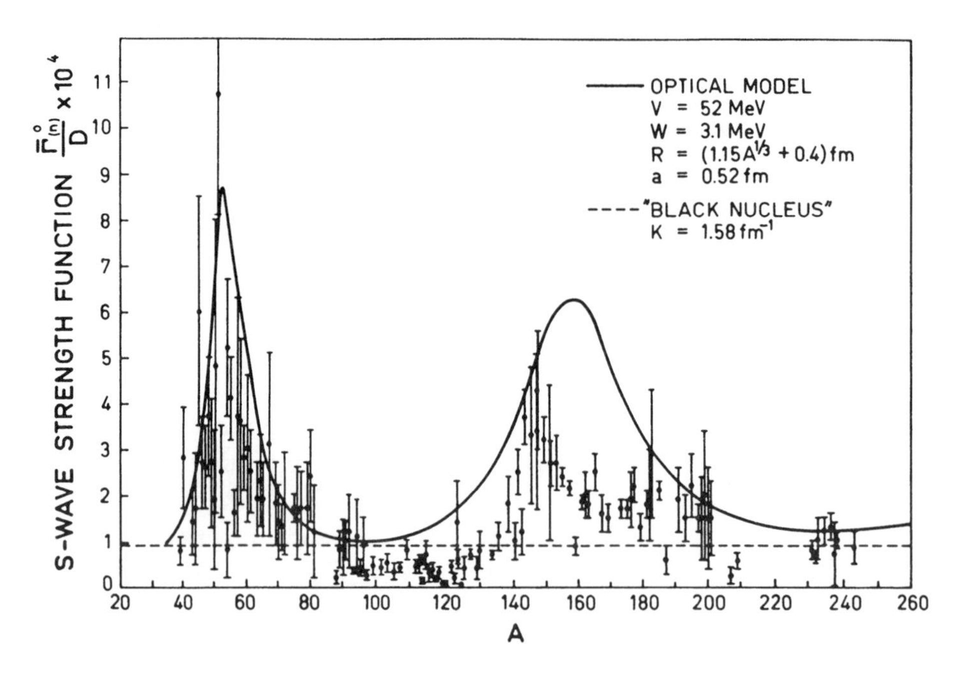

Fig. 5 The S-wave neutron strength function as a function of mass number. This figure is taken from A. Bohr and B. R. Mottelson, *Nuclear Structure*, Vol. 1, p. 230, W. A. Benjamin, New York, 1969.

By contrast, an earlier literal interpretation of Bohr's model was worked out by Feshbach, Peaslee, and Weisskopf[13] in which the neutron is simply absorbed as it enters the nucleus. This curve is the dashed line called "Black Nucleus" in Fig. 5. It has no structure and clearly does not describe the variations in the averaged neutron strength function. The Feshbach, Porter, and Weisskopf paper was extremely important in showing that there was an underlying single-particle shell model structure in the individually extremely complicated neutron resonances.

In fact, the weighting function $\rho(E-E')$ used by Feshbach, Porter, and Weisskopf[12] to calculate the average of the scattering amplitude $S(E)$

$$\langle S(E)\rangle_{Av} = \int_{-\infty}^{\infty} \rho(E-E')F(E')dE' , \tag{20}$$

where $F(E)$ is an arbitrary (rapidly varying) function of energy E, was a square one that had end effects which needed to be thrown away. A much more elegant procedure was suggested by Jim Langer (see Brown[14]), which involved using the weighting function

$$\rho(E-E') = \frac{I}{\pi}\frac{1}{(E-E')^2+I^2}\,. \tag{21}$$

With this weighting function, the average scattering amplitude was

$$\left\{\sum_n \frac{\Gamma(n)}{W_n - E}\right\}_{Av} = \sum_n \frac{\Gamma(n)}{W_n - E - iI}\,, \tag{22}$$

where W_n are the energies of the compound states with eV widths. Since the W_n are complex numbers all lying in the lower half of the complex plane, the evaluation of the integral is carried out by contour integration, closing the contour about the upper half plane.

Now if the I is chosen to be about equal to the widths of the single-neutron states in the optical model $V + iW$; i.e., of the order of $W = 3.1$ MeV, then the imaginary part of this average can be obtained as

$$\mathrm{Im}\left(\left\{\sum_n \frac{\Gamma(n)}{W_n - E}\right\}_{Av}\right) = \pi\frac{\bar{\Gamma}(n)}{D} = \mathrm{Im}\sum_m\left(\frac{\Gamma_m}{\hat{E}_m - E}\right), \tag{23}$$

where Γ_m and $\hat{E}_m$ are the widths and (complex) energies of the single neutron states in the complex potential $V+iW$. This shows how the averaged strength function is reproduced by the single-particle levels.

In the summer of 1958 Hans Bethe invited me (G.E.B.) to Cornell, giving me an honorarium from a fund that the AVCO Company, for which he consulted on the physics of the nose cones of rockets upon reentry into the atmosphere, had given him for that purpose. I was unsure of the convergence of the procedure by which I had obtained the above results. Hans pointed out that the width of the single-neutron state $\mathrm{Im}\,(\Gamma_m)$ would be substantially larger than the widths of the two-particle, one-hole states that the single-particle state would decay into, because in the latter the energy would be divided into the three excitations, two particles and one hole, and the widths went quadratically with available energy as can be obtained from Eq. (13); ergo the two-particle, one-hole widths would be down by a factor of $\sim 3/9 = 1/3$ from the single-particle width, but nonetheless would acquire the same imaginary part I, which is the order of the single particle width in the averaging. Thus, my procedure would be convergent. So I wrote my *Reviews of Modern Physics* article,[14] which I believe was quite elegant, beginning with Jim Langer's idea about averaging and ending with Hans Bethe's argument about convergence. I saw then clearly the advantage of having good collaborators, but had to wait two more decades until I could lure Hans back into astrophysics where we would really collaborate tightly.

2. Hans Bethe in Cambridge, England

We pick up Hans Bethe in 1955 at the time of his sabbatical in Cambridge, England. The family, wife Rose, and children, Monica and Henry, were with him.

There was little doubt that Keith Brueckner had a promising approach to attack the nuclear many-body problem; i.e., to describe the interactions between nucleons in such a way that they could be collected into a general self-consistent potential. That potential would have its conceptual basis in the Hartree-Fock potential and would turn out to be the shell model potential, Hans realized.

Douglas Hartree was a fellow at St. John's College when he invented the self-consistent Hartree fields for atoms in 1928. He put Z electrons into wave functions around a nucleus of Z protons, A–Z neutrons, the latter having no effect because they had no charge. The heavy compact nucleus was taken to be a point charge at the origin of the coordinate system, because the nucleon mass is nearly 2000 times greater than the electron and the size of the nucleus is about 10,000 times smaller in radius than that of the atom. The nuclear charge Ze is, of course, screened by the electron charge as electrons gather about it. The two innermost electrons are very accurately in $1s$ orbits (called K-electrons in the historical nomenclature). Thus, the other $Z-2$ electrons see a screened charge of $(Z-2)e$, and so it could go, but Hartree used instead the so-called Thomas-Fermi method to get a beginning approximation to the screened electric field.

Given this screened field as a function of distance r, measured from the nucleus located at $r = 0$, Hartree then sat down with his mechanical computer punching buttons as his Monromatic or similar machine rolled back and forth, the latter as he hit a return key. (G.E.B. — These machines tore at my eyes, giving me headaches, so I returned to analytical work in my thesis in 1950. Hans used a slide rule, whipping back and forth faster than the Monromatic could travel, achieving three-figure accuracy.) When Hartree had completed the solution of the Schrödinger equation for each of the original Z electrons, he took its wave function and squared it. This gave him the probability, $\rho_i(x_i, y_i, z_i)$, of finding electron i at the position given by the coordinates x_i, y_i, z_i. Then, summing over i, with $x_i = x$, $y_i = y$, and $z_i = z$

$$\sum_i \rho_i(x, y, z) = \rho_e^{(1)}(x, y, z) \tag{24}$$

gave him the total electron density at position (x, y, z). Then he began over again with a new potential

$$V^{(1)} = \frac{Ze}{r} - e\rho_e^{(1)}(x, y, z)\,, \tag{25}$$

where superscript (1) denotes that this is a first approximation to the self-consistent potential. To reach approximation (2) he repeated the process, calculating the Z electronic wave functions by solving the Schrödinger equation with the potential $V^{(1)}$. This gave him the next Hartree potential $V^{(2)}$. He kept going until the potential no longer changed upon iteration; i.e., until $V^{(n+1)} \simeq V^{(n)}$, the $\simeq$ meaning that they were approximately equal, to the accuracy Hartree desired. Such a potential is called "self-consistent" because it yields an electron density that reproduces the same potential.

Of course, this was a tedious job, taking months for each atom (now only seconds with electronic computers). Some of the papers are coauthored, D. R. Hartree and W. Hartree. The latter Hartree was his father, who wanted to continue working after he retired from employment in a bank. In one case, Hartree made a mistake in transforming his units Z and e to more convenient dimensionless units, and he performed calculations with these slightly incorrect units for some months. Nowadays, young investigators might nonetheless try to publish the results as referring to a fractional Z, hoping that fractionally charged particles would attach themselves to nuclei, but Hartree threw the papers in the wastebasket and started over.

Douglas Hartree was professor in Cambridge in 1955 when Hans Bethe went there to spend his sabbatical. The Hartree method was improved upon by the Russian professor Fock, who added the so-called exchange interaction which enforced the Pauli exclusion principle, guaranteeing that two electrons could not occupy the same state. We shall simply enforce this principle by hand in the following discussion, our purpose here being to explain what "self-consistent" means in the many-body context. Physicists generally believe self-consistency to be a good attribute of a theory, but Hartree did not have to base his work only on beliefs. Given his wave functions, a myriad of transitions between atomic levels could be calculated, and their energies and probabilities could be compared with experiment. Douglas Hartree became a professor at Cambridge. This indicates the regard in which his work was held.

The shell model for electrons reached success in the Hartree-Fock self-consistent field approach. This was very much in Hans Bethe's mind, when he set out to formulate Brueckner theory so that he could obtain a self-consistent potential for nuclear physics. He begins his 1956 paper[15] on the "Nuclear Many-Body Problem" with "Nearly everybody in nuclear physics has marveled at the success of the shell model. We shall use the expression 'shell model' in its most general sense, namely as a scheme in which each nucleon is given its individual quantum state, and the nucleus as a whole is described by a 'configuration,' i.e., by a set of quantum numbers for the individual nucleons."

He goes on to note that even though Niels Bohr had shown that low-energy neutrons disappear into a "porridge" for a very long time before re-emerging (although Hans was not influenced by Bohr's paper, because he hadn't read it; this may have given him an advantage as we shall see), Feshbach, Porter, and Weisskopf had shown that the envelope of these states followed that of the single particle state calculated in the nuclear shell model, as we noted in the last section.

Bethe confirms "while the success of the model [in nuclear physics] has thus been beyond question for many years, a theoretical basis for it has been lacking. Indeed, it is well established that the forces between two nucleons are of short range, and of very great strength, and possess exchange character and probably repulsive cores. It has been very difficult to see how such forces could lead to any over-all potential and thus to well-defined states for the individual nucleons."

He goes on to say that Brueckner has developed a powerful mathematical method for calculating the nuclear energy levels using a self-consistent field method, even though the forces are of short range.

"In spite of its apparent great accomplishments, the theory of Brueckner *et al.* has not been readily accepted by nuclear physicists. This is in large measure the result of the very formal nature of the central proof of the theory. In addition, the definitions of the various concepts used in the theory are not always clear. Two important concepts in the theory are the wave functions of the individual particles, and the potential V_c 'diagonal' in these states. The paper by Brueckner and Levinson[16] defines rather clearly how the potential is to be obtained from the wave functions, but not how the wave functions can be constructed from the potential V_c. Apparently, BL assume tacitly that the nucleon wave functions are plane waves, but in this case, the method is only applicable to an infinite nucleus. For a finite nucleus, no prescription is given for obtaining the wave functions."

Hans then goes on to define his objective, which will turn out to develop into his main activity for the next decade or more: "It is the purpose of the present paper to show that the theory of Brueckner gives indeed the foundation of the shell model."

Hans was rightly very complimentary to Brueckner, who had "tamed" the extremely strong short-ranged interactions between two nucleons, often taken to be infinite in repulsion at short distances at the time. On the other hand, Brueckner's immense flurry of activity, changing and improving on previous papers, made it difficult to follow his work. Also, the Watson input scattering theory seemed to give endless products of scattering operators, each appearing to be ugly mathematically ("Taming" thus was the great accomplishment of Watson and Brueckner.) And in the end, the real goal was to provide a quantitative basis for the nuclear shell model, based on the nucleon-nucleon interaction, which was being reconstructed from nucleon-nucleon scattering experiments at the time.

The master at organization and communication took over, as he had done in formulating the "Bethe Bible" during the 1930's.

One can read from the Hans Bethe archives at Cornell that Hans first made 100 pages of calculations reproducing the many results of Brueckner and collaborators, before he began numbering his own pages as he worked out the nuclear many-body problem. He carried out the calculations chiefly analytically, often using mathematical functions, especially spherical Bessel functions, which he had learned to use while with Sommerfeld. When necessary, he got out his slide rule to make numerical calculations.

During his sabbatical year, Hans gave lectures in Cambridge on the nuclear many-body problem. Two visiting American graduate students asked most of the questions, the British students being reticent and rather shy. But Professor Nevill Mott asked Hans to take over the direction of two graduate students, Jeffrey Goldstone and David Thouless, the former now professor at M.I.T. and the latter professor at the University of Washington in Seattle. We shall return to them later.

As noted earlier, the nuclear shell model has to find its own center "self-consistently." Since it is spherically symmetrical, this normally causes no problem. Simplest is to assume the answer: begin with a deep square well or harmonic oscillator potential and fill it with single particle eigenstates as a zero-order approximation as Bethe and Bacher[5] did in their 1936 paper of the Bethe Bible. Indeed, now that the sizes and shapes of nuclei have been measured by high resolution electron microscopes (high energy electron scattering), one can reconstruct these one-particle potential wells so that filling them with particles reproduces these sizes and shapes. The wave functions in such wells are often used as assumed solutions to the self-consistent potential that would be obtained by solving the nuclear many-body problem.

We'd like to give the flavor of the work at the stage of the Brueckner theory, although the work using the "Reference-Spectrum" by Bethe, Brandow, and Petschek,[17] which we will discuss later, will be much more convenient for understanding the nuclear shell model. The question we consider is the magnitude of the three-body cluster terms; i.e., the contribution to the energy from the interaction of three particles. (Even though the elementary interaction may be only a two-particle one, an effective three-body interaction arises inside the nucleus, as we shall discuss.) One particular three-body cluster will be shown in Fig. 10. The three-body term, called a three-body cluster in the Brueckner expansion, is

$$\Delta E_3 = \sum_{ijk} \left\langle \Phi_0 \left| I_{ij} \frac{Q}{e_{ji}} I_{jk} \frac{Q}{e_{ki}} I_{ki} \right| \Phi_0 \right\rangle . \tag{26}$$

Here the three nucleons i, j, and k are successively excited. There are, of course, higher order terms in which more nucleons are successively excited and de-excited. In this cluster term, working from right to left, first there is an interaction I_{ki} between particles i and k. In fact, this pair of particles is allowed to interact any number of times, the number being summed into the G-matrix G_{ik}. We shall discuss the G-matrix in great detail later. Then the operator Q excludes all intermediate states that are occupied by other particles — the particle k can only go into an unoccupied state before interacting with particle j. The denominator e_{ki} is the difference in energy between the specific state k that particle i goes to and the state that it came from. The particle k can make virtual transitions, transitions that don't conserve energy, because of the Heisenberg uncertainty principle

$$\Delta E \Delta t \gtrsim \hbar . \tag{27}$$

Thus, if $\Delta E \Delta t > \hbar$, then the particle can stay in a given state only at time

$$t = \hbar / e_{ki} \tag{28}$$

so the larger the e_{ki} the shorter the time that the particle in a given state can contribute to the energy ΔE_3. (Of course, the derivation of ΔE_3 is carried out in the standard operations of quantum mechanics. We bring in the uncertainty principle only to give some qualitative understanding of the result.) Once particle k has interacted with particle j, particle j goes on to interact with the original particle i, since the nucleus must be left in the same ground state Φ_0 that it began in, if the three-body cluster is to contribute to its energy.

Bethe's calculation gave

$$\Delta E_3/A = -0.66 \text{ MeV} \tag{29}$$

corrected a bit later in the paper to -0.12 MeV once only the fraction of spin-charge states allowed by selection rules are included. This is to be compared with Brueckner's -0.007 MeV. Of course, these are considerably different, but this is not the main point. The main point is that both estimates are small compared with the empirical nuclear binding energy.[d]

Thus, it was clear that the binding energy came almost completely from the two-body term G_{ik}, and that the future effort should go into evaluating this quantity, which satisfies the equation

$$G = V + V\frac{Q}{e}G\,. \tag{30}$$

(We will have different G's for the different charge-spin states.)

Although written in a deceptively simple way, this equation is ugly, involving operators in both the coordinates x, y, z, and their derivatives. However, G can be expressed as a two-body operator (see Sec. 3). It does not involve a sum over the other particles in the nucleus, so the interactions can be evaluated one pair at a time.

Thus, the first paper of Bethe on the nuclear many-body problem collected the work of Brueckner and collaborators into an orderly formalism in which the evaluation of the two-body operators G would form the basis for calculating the shell model potential $V(r)$. Bethe went on with Jeffrey Goldstone to investigate the evaluation of G for the extreme infinite-height hard core potential, and he gave David Thouless the problem that, given the empirically known shell model potential $V_{SM}(r)$, what properties of G would reproduce it.

As we noted in the last section, an infinite sum over the two-body interaction is needed in order to completely exclude the two-nucleon wave functions from the region inside the (vertical) hard core. As noted earlier,

[d]At least at that time there appeared to be a good convergence in the so-called cluster expansion. But see Sec. 3!

Jastrow[6] had proposed a vertical hard core, rising to ∞, initially of radius $R_c = 0.6 \times 10^{-13}$ cm. This is about 1/3 of the distance between nucleons, so that removing this amount of space from their possible occupancy obviously increases their energy. (From the Heisenberg principle $\Delta p_x \Delta x \gtrsim \hbar$, which is commonly used to show that if the particles are confined to a smaller amount of volume, the Δx is decreased, so the Δp_x, which is of the same general size as p_x, is increased.) Thus, as particles were pushed closer together with increasing density, their energy would be greater. Therefore, the repulsive core was thought to be a great help in saturating matter made up of nucleons. One of the authors (G.E.B.) heard a seminar by Jastrow at Yale in 1949, and Gregory Breit, who was the leading theorist there, thought well of the idea. Indeed, we shall see later that Breit played an important role in providing a physical mechanism for the hard core as we shall discuss in the next section. This mechanism actually removed the "sharp edges" which gave the vertical hard core relatively extreme properties. So we shall see later that the central hard core in the interaction between two nucleons isn't vertical, rather it's of the form

$$\text{hard core} = \frac{C}{r} \exp\left(-\frac{m_\omega c r}{\hbar}\right), \tag{31}$$

where $\hbar/m_\omega c$ is 0.25×10^{-13} cm, and m_ω is the mass of the ω-meson. The constant C is large compared with unity

$$C \gg 1 \tag{32}$$

so that the height of the core is many times the Fermi energy; i.e., the energy measured from the bottom of the shell model potential well up to the last filled level. Thus, although extreme, treating the hard core potential gives a caricature problem. Furthermore, solving this problem in "Effect of a repulsive core in the theory of complex nuclei," by H. A. Bethe and J. Goldstone[18] gave an excellent training to Jeffrey Goldstone, who went on to even greater accomplishments. (The Goldstone boson is named after him; it is perhaps the most essential particle in QCD.)

We will find in the next section that the Moszkowski-Scott separation method is a more convenient way to treat the hard core. And this method is more practical because it includes the external attractive potential at the same time. However, even though the wave function of relative motion of the two interacting particles cannot penetrate the hard core, it nonetheless has an effect on their wave function in the external region. This effect is conveniently included by changing the reduced mass of the interacting nucleons from $M^\star$ to $0.85M^\star$ for a hard core radius $r_c = 0.5 \times 10^{-13}$ cm, and to $\sim 0.9M^\star$ for the presently accepted $r_c = 0.4 \times 10^{-13}$ cm.

3. The Reference-Spectrum Method for Nuclear Matter

We take the authors' prerogative to jump in history to 1963, to the paper of H. A. Bethe, B. H. Brandow, and A. G. Petschek[17] with the title of that of this section, because this work gives a convenient way of discussing essentially all of the physical effects found by Bethe and collaborators, and also the paper by S. A. Moszkowski and B. L. Scott[19] the latter paper being very important for a simple understanding of the G-matrix. In any case, the reference-spectrum included in a straightforward way all of the many-body effects experienced by the two interacting particles in Brueckner theory, and enabled the resulting G-matrix to be written as a function of r alone, although separate functions $G_l(r)$ had to be obtained for each angular momentum l.

Let us prepare the ground for the reference-spectrum, including also the Moszkowski-Scott method, by a qualitative discussion of the physics involved in the nucleon-nucleon interaction inside the nucleus, say at some reasonable fraction of nuclear matter density. We introduce the latter so we can talk about plane waves locally, as an approximation. For simplicity we take the short-range repulsion to come from a hard core, of radius 0.4×10^{-13} cm.

(i) In the region just outside the hard core, the influence of neighboring nucleons on the two interacting ones is negligible, because the latter have been kicked up to high momentum states by the strong hard-core interaction and that of the attractive potential, which is substantially stronger than the local Fermi energy of the nucleons around the two interacting ones. Thus, one can begin integrating out the Schrödinger equation for $r\psi(r)$, starting from $r\psi(r) = 0$ at $r = r_c$. The particles cannot penetrate the infinitely repulsive hard core (see Fig. 6), so their wave function begins from zero there. In fact, the Schrödinger equation is one of relative motion; i.e., a one-body equation in which the mass is the reduced mass of the two particles, $M_N/2$ for equal mass nucleons. As found by Bethe and Goldstone, the M_N will be changed to $M_N^\star$ in the many-body medium, but the mass is modified not only by the hard core but also by the attractive part of the two-body potential, which we have not yet discussed.

(ii) We consider the spin-zero $S = 0$ and spin-one $S = 1$ states; i.e., for angular momentum $l = 0$. These are the most important states. We compare

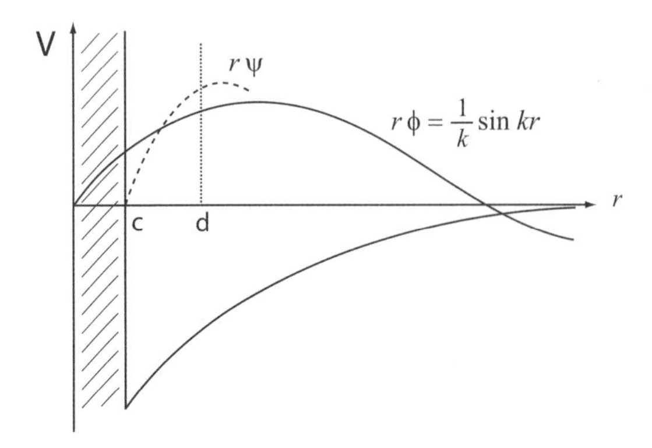

Fig. 6 The unperturbed S-wave function $r\phi = \frac{1}{k}\sin kr$ and the wave function $r\psi$ obtained by integrating the Schrödinger equation in the presence of the potential out from $r = c$.

the spin-one state in the presence of the potential with the unperturbed one; i.e., the one in the absence of a potential.

Before proceeding further with Hans Bethe's work, let us characterize the nice idea of Moszkowski and Scott in the simplest possible way.

Choose the separation point d such that

$$\left.\frac{d\psi(r)/dr}{\psi(r)}\right|_{r=d} = \left.\frac{d\phi(r)/dr}{\phi(r)}\right|_{r=d} \tag{33}$$

Technically, this is called the equality of logarithmic derivatives. As shown in Fig. 6, such a point will be there, because $r\psi$, although it starts from zero at $r = c$, has a greater curvature than $r\phi$. Now if only the potential inside of d were present; i.e., if $V(r) = 0$ for $r > d$, then Eq. (33) in quantum mechanics is just the condition that the inner potential would produce zero scattering, the inner attractive potential for $r < d$ just canceling the repulsion from the hard core. Of course, if Eq. (33) is true for a particular k, say $k \lesssim k_{\rm F}$, then it will not be exactly true for other k's. On the other hand, since the momenta in the short-distance wave function are high compared with $k_{\rm F}$, due to the infinite hard core, and the very deep interior part of the attractive potential that is needed to compensate for it, Eq. (33) is nearly satisfied for all momenta up to $k_{\rm F}$ if the equality is true for one of the momenta.

The philosophy here is very much as in Bethe's work "Theory of the Effective Range in Nuclear Scattering."[20] This work is based on the fact that the inner potential (we could define it as $V(r)$ for $r < d$) is deep in comparison with the energy that the nucleon comes in with, so that the scattering depends only weakly on this (asymptotic) energy. In fact, for a potential which is Yukawa in nature

$$V = \infty\,, \qquad \text{for } r < c = 0.4 \times 10^{-13} \text{ cm} \tag{34}$$

$$V = -V_0 e^{-(r-c)/R} \quad \text{for } r > c\,, \tag{35}$$

then $V_0 = 380$ MeV and $R = 0.45 \times 10^{-13}$ cm in order to fit the low energy neutron-proton scattering.[21] The value of 380 MeV is large compared with the Fermi energy of ~ 40 MeV. In the collision of two nucleons, the equality of logarithmic derivatives, Eq. (33), would mean that the inner part of the potential interaction up to d, which we call V_s, would give zero scattering. All the scattering would be given by the long-range part which we call V_l.

(iii) Now we know that the wave function $\psi(r)$ must "heal" to $\phi(r)$ as $r \to \infty$, because of the Pauli principle. There is no other place for the particle to go, because for k below k_{F} all other states are occupied. Delightfully simple is to approximate the healing by taking ψ equal to ϕ for $r > d$.

The conclusion of the above is that

$$G(r) \simeq G_s + v_l(r) + v_l \frac{Q}{e} v_l + \cdots\,, \tag{36}$$

where

$$v_l = V(r) \quad \text{for } r \gtrsim d\,. \tag{37}$$

We shall discuss G_s, the G-matrix which would come from the short-range part of the potential for $r < d$ later. It will turn out to be small.

Now we have swept a large number of problems under the rug, and we don't apologize for it because Eq. (36) gives a remarkably accurate answer. However, first we note that there must be substantial attraction in the channel considered in order for the short-range repulsion to be canceled. So the separation method won't work for cases where there is little attraction. Secondly, a number of many-body effects have been discarded, and these had to be considered by Bethe, Brandow, and Petschek, one by one.

We do not want to fight old battles over again, but simply note that Bethe, Brandow, and Petschek found that they could take into account the many-body effects including the Pauli principle by choosing a single-particle spectrum, which is approximated by

$$\epsilon_m = \frac{k_m^2}{2m_{RS}^\star} + A\,. \tag{38}$$

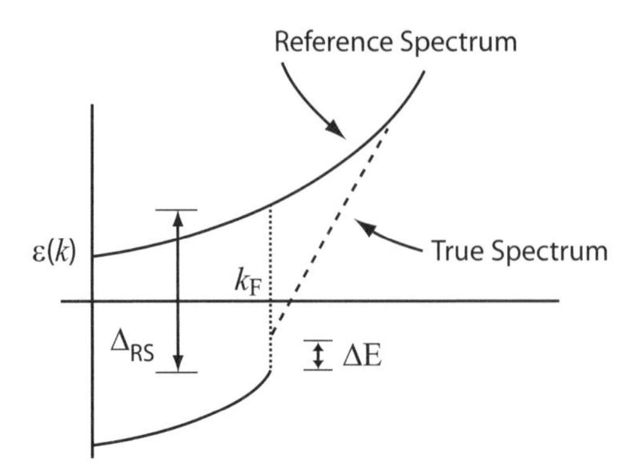

Fig. 7 Energy spectrum in the Brueckner theory (inclusive of self-energy insertion). The true spectrum has a small gap ΔE at $k_{\rm F}$ and is known, from the arguments of Bethe *et al.*, to join the reference-spectrum at $k \approx 3$ fm^{-1}. The dotted line is an interpolation between these two points.

The hole energies, i.e., the energies of the initially bound particles, can be treated fairly roughly since they are much smaller than the energies of the particles which have quite high momenta (see Fig. 7). The parameters $m^\star_{RS}$ and A were arrived at by a self-consistency process. Given an input $m^\star_{RS}$ and A, the same $m^\star_{RS}$ and A must emerge from the calculation. Thus, the many-body effects can be included by changing the single-particle energy through the coefficient of the kinetic energy and by adding the constant A. In this way the many-body problem is reduced to the Lippmann-Schwinger two-body problem with changed parameters, $m \to m^\star_{RS}$ and the addition of A. As shown in Fig. 7, the effect is to introduce a gap Δ_{RS} between particle and hole states.[e]

A small gap, shown as ΔE, occurs naturally between particle and hole states, resulting from the unsymmetrical way they are treated in Brueckner theory, the particle-particle scattering being summed to all orders. The main part of the reference-spectrum gap Δ_{RS} is introduced so as to numerically reproduce the effects of the Pauli Principle. That this can be done is not surprising, since the effect of either is to make the wave function heal more rapidly to the noninteracting one.

Before we move onwards from the reference-spectrum, we want to show the main origin of the reduction of m to $m^\star_{RS} \sim 0.8m$. Such a reduction obviously increases the energies of the particles in intermediate states $\epsilon(k)$

[e]We call the states initially in the Fermi sea "hole states" because holes are formed when the two-body interaction transfers them to the particle states which lie above the Fermi energy, leaving holes behind in intermediate states.

which depends inversely on $m^\star_{RS}$. A lowering of $m^\star$ from m was already found in the Bethe-Goldstone solution of the scattering from a hard core alone with $m^\star = 0.85m$ for a hard core radius of $r_c = 0.5 \times 10^{-13}$ cm and somewhat less for $r_c = 0.4 \times 10^{-13}$ cm, a more reasonable value. The $m^\star_{RS}$ are not much smaller than this.

Now we have to introduce the concept of off-energy-shell self energies. Let's begin by defining the on-shell self-energy which is just the energy of a particle in the shell model or optical model potential. It would be given by the process shown in Fig. 8.

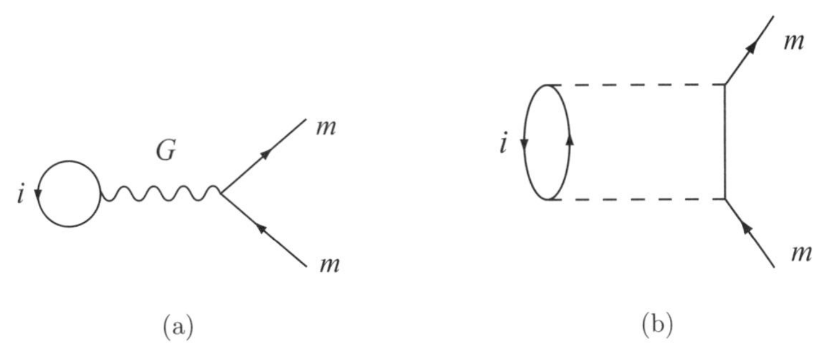

Fig. 8 (a) The on-shell self energy involves the interaction of a particle in state m with the filled states in the Fermi sea i. The wiggly line G implies a sum over all orders in a ladder of V, the second-order term in dashed lines being shown in (b). Together with the kinetic energy, this gives the self energy.

The on-shell $U(k_m)$ should just give the potential energy of a nucleon of momentum k in the optical model potential $U + iW$ used by Feshbach, Porter, and Weisskopf, so $U(k_m) + k_m^2/2m$ gives the single-particle energy.

However, the energies $U(k_m)$ to be used in Brueckner theory are not on-shell energies. This is because another particle must be excited when the one being considered is excited, as shown in Fig. 9. Thus, considering the second

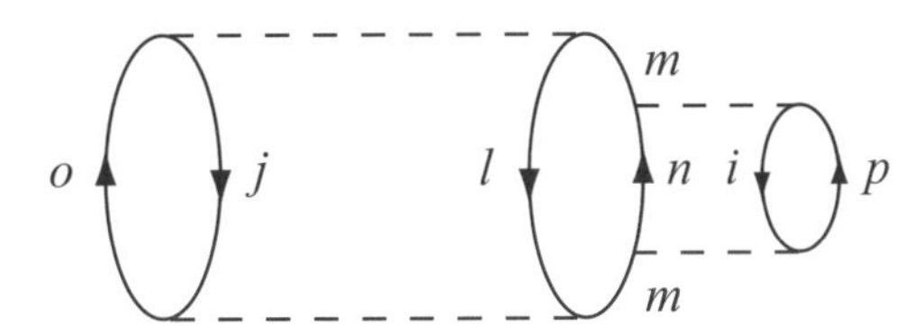

Fig. 9 Second-order contribution to the self energy of a particle in state m when present in a process contributing to the ground-state energy. The states m, n, o, p refer to particles and i, j, l to holes.

order in V interaction between the particle being considered and those in the nucleus one has

$$\tilde{\epsilon}_m^{(2)} = - \sum_{p,n>k_{\rm F};i<k_{\rm F}} \frac{|\langle np|V|im\rangle|^2}{\epsilon_n + \epsilon_p + \epsilon_o - \epsilon_j - \epsilon_i - \epsilon_l} \tag{39}$$

i.e., although the particle in state o and the hole in state j do not enter actively into the self energy of p, one must include their energies in the energy denominator. Defining

$$\tilde{\epsilon} = \epsilon_n + \epsilon_p + \epsilon_o - \epsilon_j - \epsilon_i - \epsilon_l\,, \tag{40}$$

$$e = \epsilon_n + \epsilon_p - \epsilon_i - \epsilon_l \tag{41}$$

one sees that the additional energy in the denominator in Eq. (39) is

$$\Delta e = \tilde{\epsilon} - \epsilon = \epsilon_o - \epsilon_j\,. \tag{42}$$

This Δe represents the energy that must be "borrowed" in order to excite the particle from state j to o, even though this particle does not directly participate in the interaction on state p which gives its self energy. The point is that because j must also be excited, as well as the particles on the right in the diagram of Fig. 9, the total $\tilde{e}$ is that much greater than the e necessary to give the on-shell energy. Thus, the time that this interaction can go on for is decreased, since, from the uncertainty principle the time that energy can be borrowed for goes as $\Delta t = \hbar/\Delta E$. As noted earlier, the wave function $\psi(r)$ lies outside the hard core, so that integrals over the G-matrix are only over the attractive interaction. These are cut down by the additional energy denominator that comes in by the interaction being off shell. In the higher momentum region this can cut the self energy down by 15–20 MeV. We shall see later that when this is handled properly, only $\sim 1/3$ of this survives.

The strategy that T. T. S. Kuo and one of the authors (G.E.B.) have employed over many years, beginning with Kuo and Brown[22] to calculate the G-matrix in nuclei has been to use the separation method for the G in the $l = 0$ states (S-states), which have a large enough attraction so that a separation distance d can be defined, but to use the Bethe, Brandow, and Petschek reference-spectrum in angular momentum states which do not have this strong attraction.

The reference-spectrum was as far as one could get using only two-body clusters; i.e., summing up the two-body interaction. In order to make further progress, three and four-body clusters had to be considered; i.e., processes in which one pair of the initially interacting particles was left excited while another pair interacted, and only then returned to its initial state as in the calculations of the lowest order three-body cluster ΔE_3, Eq. (26).

Hans Bethe investigated[23] the four-body clusters, following the work[24] on three-nucleon clusters in nuclear matter by his post-doc R. (Dougy) Rajaraman, a paper just following the Bethe, Brandow, and Petschek reference-spectrum paper. Rajaraman suggested on the basis of including the other three-body clusters than that shown in Fig. 8, that the off-shell effect should be decreased by a factor of $\sim 1/2$. (We shall see from Bethe's work below that it should actually be more like $\sim 1/3$.) Bethe first made a fairly rough calculation of the four-body clusters, but good enough to show when summed to all orders in G, it was tremendous, giving

$$\Delta E_4 \simeq -35 \text{ MeV/particle} \tag{43}$$

to the binding energy.

Visiting CERN in the summer of 1964, Bethe learned of Faddeev's solution of the three-body problem.[25] This formalism for three-body scattering is very similar to that for two-body scattering in Eq. (17) in that one can define a T-matrix

$$T\Phi = V\psi, \tag{44}$$

where ψ and Φ are the three-particle correlated and uncorrelated wave functions, and T satisfies

$$T = V + V\frac{1}{e}T \tag{45}$$

where e now represents the energy denominator of all three particles. The T can be split into

$$T = T^{(1)} + T^{(2)} + T^{(3)} \tag{46}$$

but now one defines

$$T^{(1)} = V_{23} + V_{23}\frac{1}{e}T\,, \tag{47}$$

etc. In other words, $T^{(1)}$ denotes that part of T in which particle "1" did not take part in the last interaction. Hans found that he could write expressions for the three-body problem using the G-matrix as effective interaction; e.g.

$$T^{(1)}\Phi = G_{23}\psi^{(1)}\,, \tag{48}$$

where $\psi^{(1)}$ is defined by

$$\psi^{(1)} = \Phi + \frac{1}{e}(T^{(2)} + T^{(3)})\Phi\,. \tag{49}$$

The ground state energy is given by

$$E_0 = \langle \Phi\,|T|\,\Phi\rangle, \tag{50}$$

where Φ is the unperturbed (free-particle) wave function. In short, one could use the Brueckner G-matrices as effective interactions in the Faddeev formalism, which then summed the three-body cluster terms to all orders in the solution of these equations.

Hans came to Copenhagen in late summer of 1964, with the intention of solving the Faddeev equations for the three-body system using the reference-spectrum approximation Eq. (38) for the effective two-body interaction. David Thouless, Hans' Ph.D. student in Cambridge, was visiting Nordita (Nordic Institute of Theoretical Atomic Physics) at that time. I (G.E.B.) got David together with Hans the morning after Hans arrived.[f] Hans wrote down the three coupled equations on the blackboard and began to solve them by some methods Dougy Rajaraman had used for summing four-body clusters. David took a look at the three coupled equations with Hans' G-matrix reference-spectrum approximation for the two-body interaction and said "These are just three coupled linear equations. Why don't you solve them analytically?" By late morning Hans had the solution, given in his 1965 paper. The result, which can be read from this paper, is that the off-shell correction to the three-body cluster of Fig. 10 should be cut down by a factor of 1/3. Hans' simple conclusion was "when three nucleons are close

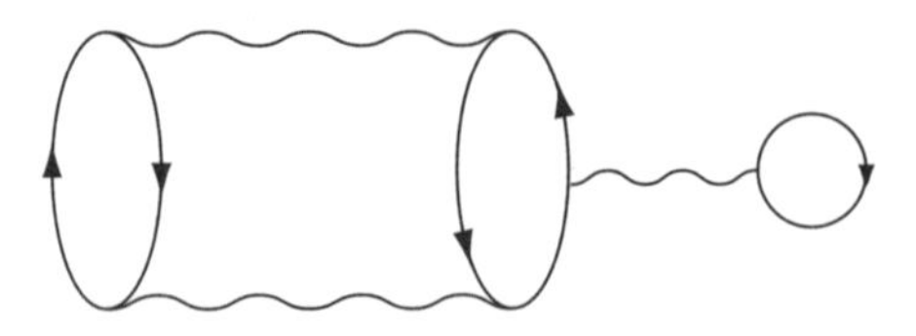

Fig. 10 The contribution of the three-body cluster to the off-shell self energy of a particle in state p. The wiggly line here represents the G-matrix. One particular contribution to this is shown in Fig. 9.

[f]Our most fruitful discussions invariably came the morning after he arrived in Copenhagen.

together, an elementary treatment would give us three repulsive pair interactions. In reality, we cannot do more than exclude the wave function from the repulsive region, hence we get only one core interaction rather than 3."

In the summer of 1967, I (G.E.B.) gave the summary talk of the International Nuclear Physics meeting held in Tokyo, Japan. This talk is in the proceedings. I designed my comments as letters to the speakers, as did Herzog in Saul Bellow's novel by that name.

Dear Professor Bethe,

First of all, your note is too short to be intelligible. But by valiant efforts, and a high degree of optimism in putting together corrections of the right sign, you manage to get within 3 MeV/particle of the binding energy of nuclear-matter. It is nice that there are still some discrepancies, because we must have some occupation for theorists, in calculating three-body forces and other effects.

Most significantly, you confirm that it is a good approximation to use plane-wave intermediate states in calculation of the G-matrix. This simplifies life in finite nuclei immensely.

I cannot agree with you that there is no difference between hard- and soft-core potentials. I remember your talk at Paris, where you showed that the so-called dispersion term (the contribution of G_s — G.E.B.), *which is a manifestation of off-energy-shell effects, differs by* ~ 3 *or* ~ 4 *MeV for hard- and soft-core potentials, and that this should be the only difference. I remain, therefore, a strong advocate of soft-core potentials, and am confident that careful calculation will show them to be significantly better.*

Let me remind you that we (Kuo and Brown) always left out this dispersion correction in our matrix elements for finite nuclei, in the hopes of softer ones.

......

Yours, etc.

In summary, the off-energy-shell effects at the time of the reference-spectrum through the dispersion correction G_s contributed about +6 MeV to the binding energy, a sizable fraction of the ~ 15 MeV total binding energy. However, Bethe's solution of the three-body problem via Faddeev cut this by 1/3. The reference-spectrum paper was still using a vertical hard core, whereas the introduction of the Yukawa-type repulsion by Breit[26] and independently by Gupta,[27] although still involving Fourier (momentum) components of $p \sim m_\omega c$ with m_ω the 782 MeV/c^2 mass of the ω-meson, $\sim 2\frac{1}{2}$ times greater than $k_{\rm F}$, cut the dispersion correction down somewhat more, so we were talking in 1967 about a remaining ~ 1 MeV compared with the -16 MeV binding energy per particle. At that stage we agreed to neglect interactions in the particle intermediate energy states, as is done in the Schwinger-Dyson Interaction Representation. However, in the latter, any interactions that particles have in intermediate states are expressed in terms of higher-order corrections, whereas we just decided ("decreed") in 1967 that these were negligible. An extensive discussion of all of the corrections to the dispersion term from three-body clusters and other effects is given in Michael Kirson's Cornell 1966 thesis, written under Hans' direction. In detail, Kirson was not quite as optimistic about dropping all off-shell effects as we have been above, but he does find them to be at most a few MeV.

Once it was understood that the short-range repulsion came from the vector meson exchange potentials, rather than a vertical hard core, the short-range G-matrix G_s in Eq. (36) could be evaluated and it was found to be substantially smaller than that for the vertical hard core. Because of the smoothness in the potential, the off-shell effects were smaller, sufficiently so that G_s could be neglected. Thus, the short-range repulsion was completely tamed and one had to deal with only the well-behaved power expansion

$$G = v_l + v_l \frac{Q}{e} v_l + \cdots \tag{51}$$

In fact, in the $^1S_0$ states the first term gave almost all of the attraction, whereas the iteration of the tensor force in the second term basically gave the amount that the effective $^3S_1$ potential exceeded the $^1S_0$ one in magnitude. (The tensor force contributed to only the triplet states because it required $S = 1$.)

Bethe's theory of the effective range in nuclear scattering[20] showed how the scattering length and effective range, the first two terms in the expansion of the scattering amplitude with energy, could be obtained from any potential. The scattering length and effective range could be directly obtained from the experimental measurements of the scattering.

Thus, one knew that any acceptable potential must reproduce these two constants, and furthermore, from our previous discussion, contain a strong short-range repulsion. To satisfy these criteria, Kallio and Kolltveit[28] therefore took singlet and triplet potentials

$$V_i(r) = \begin{cases} \infty & \text{for } r \le 0.4 \text{ fm} \\ -A_i e^{-\alpha_i (r-0.4)} & \text{for } r > 0.4 \text{ fm} \end{cases} \quad \text{for } i = s, t\,, \tag{52}$$

where

$$\begin{aligned} A_s &= 330.8 \text{ MeV}, \quad \alpha_s = 2.4021 \text{ fm}^{-1} \\ A_t &= 475.0 \text{ MeV}, \quad \alpha_t = 2.5214 \text{ fm}^{-1} \end{aligned} \tag{53}$$

By using this potential, Kallio and Kolltveit obtained good fits to the spectra of light nuclei.

Surprising developments, summarized in the *Physics Report* "Model-Independent Low Momentum Nucleon Interaction from Phase Shift Equivalence,"[10] showed how one could define an effective interaction between nucleons by starting from a renormalization group approach in which momenta larger than a cutoff Λ are integrated out of the effective G-matrix interaction. In fact, the data which went into the phase shifts obtained by various groups came from experiments with center of mass energies less than 350 MeV, which corresponds to a particle momentum of 2.1 fm^{-1}. Thus, setting Λ equal to 2.1 fm^{-1} gave an effective interaction $V_{\text{low}-k}$ which included all experimental measurements.

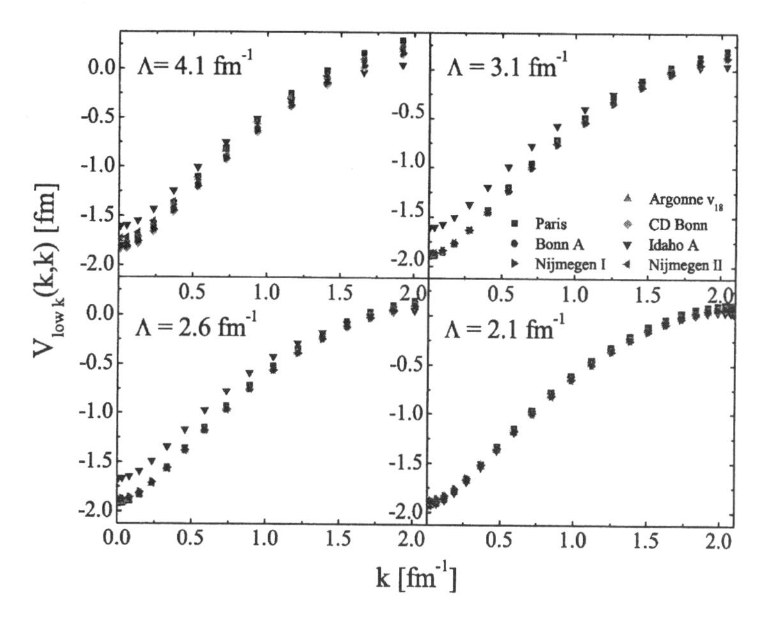

Fig. 11 Diagonal matrix elements of $V_{\text{low}-k}$ for different high-precision potentials in the $^1S_0$ partial wave with various cutoffs Λ.

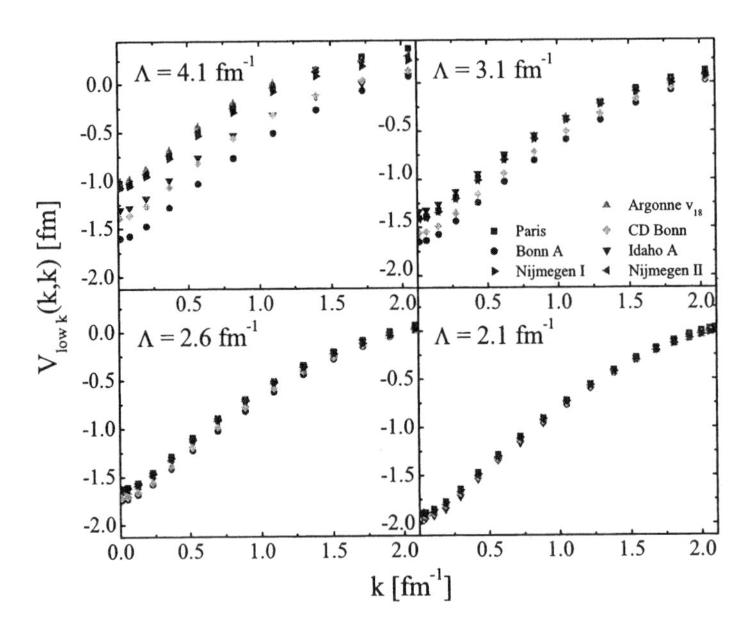

Fig. 12 Diagonal matrix elements of $V_{\text{low}-k}$ for different high-precision potentials in the $^3S_1$ partial wave with various cutoffs Λ.

We show in Figs. 11 and 12 (Figs. 11 and 12 in Ref. 10) the collapse of the various potentials obtained from different groups as Λ is lowered from 4.1 fm^{-1} to 2.1 fm^{-1}. The reason for this collapse is that as the data included in the phase shift determinations is lowered from a large energy range to that range which includes only measured data; the various groups were all using the same data in order to determine their scattering amplitudes and therefore obtained the same results. Because of this uniqueness, $V_{\text{low}-k}$ is much used in nuclear structure physics these days.

The above renormalization group calculations were all carried out in momentum space. What is the relation of $V_{\text{low}-k}$ to the effective interaction of Eq. (51)? Although in the Mozskowski-Scott method the separation distance d should be obtained separately for each channel, for comparison with $V_{\text{low}-k}$ (which has the same momentum cutoff in all channels) it is most convenient to determine it for the $^1S_0$ channel ($d = 1.025$) and then use the same d in the $^3S_1$ channel. We show a comparison between $V_{\text{low}-k}$ and the G of Eq. (51) in Figs. 13 and 14.

In Fig. 15 we show the diagonal matrix elements of the Kallio-Kolltveit potential, essentially the G of Eq. (51). Note that although the separation of short and long distance was made in configuration space, there are essentially no Fourier components above $k \sim 2$ fm^{-1} in momentum space.

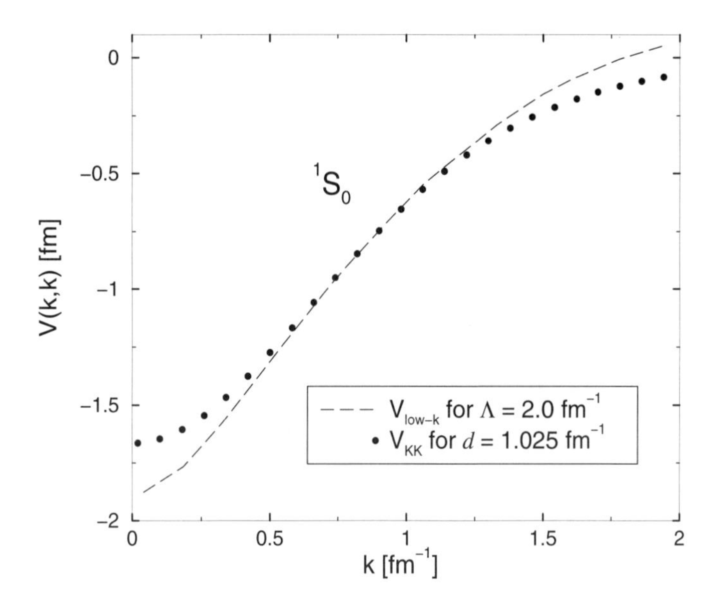

Fig. 13 The $^1S_0$ diagonal matrix elements of $V_{\mathrm{low}-k}$ and the Kallio-Kolltveit potential for a configuration space cutoff of 1.025 fm.

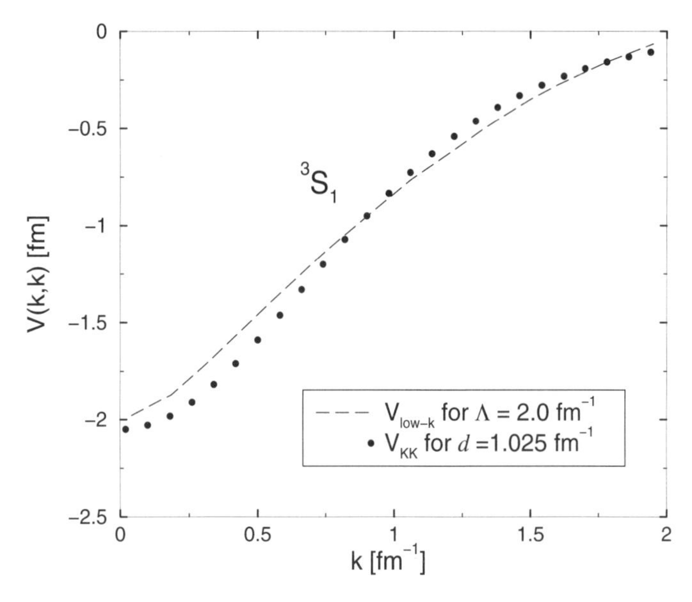

Fig. 14 The $^3S_1$ diagonal matrix elements of $V_{\mathrm{low}-k}$ and the Kallio-Kolltveit potential for a configuration space cutoff of 1.025 fm.

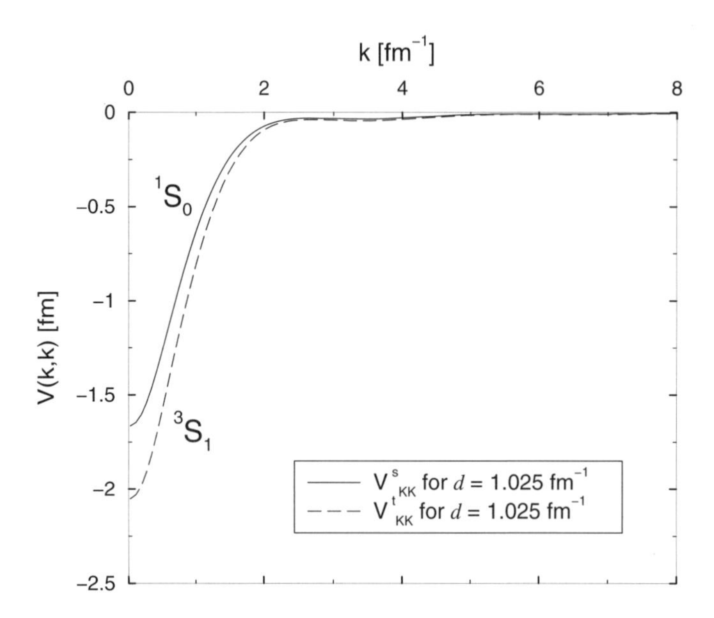

Fig. 15 Diagonal matrix elements of the long-distance Kallio-Kolltveit potential v_l, including momenta above the $V_{\mathrm{low}-k}$ cutoff of 2.1 fm^{-1}.

It is, therefore, clear that the G of Eq. (51) that Bethe ended with is essentially $V_{\mathrm{low}-k}$, the only difference being that he made the separation of scales in configuration space whereas in $V_{\mathrm{low}-k}$ it is made in momentum space. Thus, we believe that Hans Bethe arrived at the right answer in "The Nuclear Many-Body Problem," but only much later did research workers use it to fit spectra.

References

1. N. Bohr, *Nature* **137**, 344 (1936).
2. N. Bohr, *Phil. Mag.* **26**, 476 (1913).
3. M. G. Mayer, *Phys. Rev.* **75**, 1969 (1949).
4. O. Haxel, J. H. D. Jensen and H. E. Suess, *Z. Phys.* **128**, 295 (1950).
5. H. A. Bethe and R. F. Bacher, *Rev. Mod. Phys.* **8**, 82 (1936).
6. R. Jastrow, *Phys. Rev.* **81**, 165 (1951).
7. B. A. Lippmann and J. Schwinger, *Phys. Rev.* **79**, 469 (1950).
8. K. M. Watson, *Phys. Rev.* **89**, 575 (1953).
9. N. C. Francis and K. M. Watson, *Phys. Rev.* **92**, 291 (1953).
10. S. K. Bogner, T. T. S. Kuo and A. Schwenk, *Phys. Rep.* **386**, 1 (2003).
11. G. E. Brown and M. Rho, *Phys. Rep.* **396**, 1 (2004).
12. H. Feshbach, C. E. Porter and V. F. Weisskopf, *Phys. Rev.* **96**, 448 (1954).

13. H. Feshbach, D. C. Peaslee and V. F. Weisskopf, *Phys. Rev.* **71**, 145 (1947).
14. G. E. Brown, *Rev. Mod. Phys.* **31**, 893 (1959).
15. H. A. Bethe, *Phys. Rev.* **103**, 1353 (1956).
16. K. A. Brueckner and C. A. Levinson, *Phys. Rev.* **97**, 1344 (1955).
17. H. A. Bethe, B. H. Brandow and A. G. Petschek, *Phys. Rev.* **129**, 225 (1963).
18. H. A. Bethe and J. Goldstone, *Proc. Roy. Soc. A* **238**, 551 (1957).
19. S. A. Moszkowski and B. L. Scott, *Ann. Phys.* **14**, 107 (1961).
20. H. A. Bethe, *Phys. Rev.* **76**, 38 (1949).
21. G. E. Brown and A. D. Jackson, *The Nucleon-Nucleon Interaction*, North-Holland, Amsterdam, 1976.
22. T. T. S. Kuo and G. E. Brown, *Nucl. Phys.* **85**, 40 (1966).
23. H. A. Bethe, *Phys. Rev.* **138**, B804 (1965).
24. R. Rajaraman, *Phys. Rev.* **129**, 265 (1963).
25. L. D. Faddeev, *Sov. Phys. JETP* **12**, 1014 (1961); *Sov. Phys. Doklady* **6**, 384 (1962).
26. G. Breit, *Phys. Rev.* **120**, 287 (1960).
27. S. N. Gupta, *Phys. Rev.* **117**, 1146 (1960).
28. A. Kallio and K. Kolltveit, *Nucl. Phys.* **53**, 87 (1964).

Progress in Particle and Nuclear Physics 62 (2009) 135–182

Contents lists available at ScienceDirect

Progress in Particle and Nuclear Physics

journal homepage: www.elsevier.com/locate/ppnp

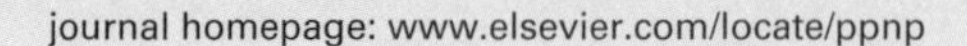

Review

Shell-model calculations and realistic effective interactions

L. Coraggio [a], A. Covello [a,b,*], A. Gargano [a], N. Itaco [a,b], T.T.S. Kuo [c]

[a] *Istituto Nazionale di Fisica Nucleare, Complesso Universitario di Monte S. Angelo, Via Cintia, I-80126 Napoli, Italy*

[b] *Dipartimento di Scienze Fisiche, Università di Napoli Federico II, Complesso Universitario di Monte S. Angelo, Via Cintia, I-80126 Napoli, Italy*

[c] *Department of Physics, SUNY, Stony Brook, NY 11794, USA*

ARTICLE INFO

Keywords:
Shell model
Realistic effective interactions
Nuclear forces
Low-momentum nucleon–nucleon potential
Nuclei around ^{132}Sn

ABSTRACT

A review is presented of the development and current status of nuclear shell-model calculations, in which the two-body effective interaction between valence nucleons is derived from the free nucleon–nucleon potential. The significant progress made in this field within the last decade is emphasized, in particular as regards the so-called $V_{\text{low-k}}$ approach to the renormalization of the bare nucleon–nucleon interaction. In the last part of the review, we first give a survey of realistic shell-model calculations from early to present days. Then, we report recent results for neutron-rich nuclei near doubly magic ^{132}Sn, and for the whole even-mass $N = 82$ isotonic chain. These illustrate how shell-model effective interactions derived from modern nucleon–nucleon potentials are able to provide an accurate description of nuclear structure properties.

Contents

* Corresponding author.
E-mail address: covello@na.infn.it (A. Covello).

0146-6410/$ – see front matter © 2008 Elsevier B.V. All rights reserved.
doi:10.1016/j.ppnp.2008.06.001

1. Introduction

The shell model is the basic framework for nuclear structure calculations in terms of nucleons. This model, which entered into nuclear physics more than fifty years ago [1,2], is based on the assumption that, as a first approximation, each nucleon inside the nucleus moves independently from the others in a spherically symmetric potential including a strong spin-orbit term. Within this approximation the nucleus is considered as an inert core, made up by shells filled up with neutrons and protons paired to angular momentum $J = 0$, plus a certain number of external nucleons, the "valence" nucleons. As is well known, this extreme single-particle shell model, supplemented by empirical coupling rules, proved very soon to be able to account for various nuclear properties [3], like the angular momentum and parity of the ground-states of odd-mass nuclei. It was clear from the beginning [4], however, that for a description of nuclei with two or more valence nucleons, the "residual" two-body interaction between valence nucleons had to be taken explicitly into account, the term residual meaning that part of the interaction which is not absorbed into the central potential. This removes the degeneracy of the states belonging to the same configuration, and produces a mixing of different configurations. A fascinating account of the early stages of the nuclear shell model is given in the comprehensive review by Talmi [5].

In any shell-model calculation, one has to start by defining a "model space", namely by specifying a set of active single-particle (SP) orbits. It is in this truncated Hilbert space that the Hamiltonian matrix has to be set up and diagonalized. A basic input, as mentioned above, is the residual interaction between valence nucleons. This is in reality a "model-space effective interaction", which differs from the interaction between free nucleons in various respects. In fact, besides being residual in the sense mentioned above, it must account for the configurations excluded from the model space.

It goes without saying that a fundamental goal of nuclear physics is to understand the properties of nuclei starting from the forces between nucleons. Nowadays, the A nucleons in a nucleus are understood as non-relativistic particles interacting via a Hamiltonian consisting of two-body, three-body, and higher-body potentials, with the nucleon–nucleon (NN) term being the dominant one.

In this context, it is worth emphasizing that there are two main lines of attack to attain this ambitious goal. The first one is comprised of the so-called *ab initio* calculations where nuclear properties, such as binding and excitation energies, are calculated directly from the first principles of quantum mechanics, using an appropriate computational scheme. To this category belong the Green's function Montecarlo Method (GFMC) [6,7], the no-core shell-model (NCSM)) [8,9], and the coupled-cluster methods (CCM)) [10,11]. The GFMC calculations, on which we shall briefly comment in Section 2.3, are at present limited to nuclei with $A \leq 12$. This limit may be overcome by using limited Hilbert spaces, and introducing effective interactions, which is done in the NCSM and CCM. Actually, coupled-cluster calculations employing modern NN potentials have been recently performed for ^{16}O and its immediate neighbors [12–14].

The main feature of the NCSM is the use of a large, but finite number of harmonic oscillator basis states to diagonalize an effective Hamiltonian containing realistic two- and three- nucleon interactions [15]. In this way, no closed core is assumed, and all nucleons in the nucleus are treated as active particles. Very recently, this approach has been applied to the study of $A = 10 - 13$ nuclei [16]. Clearly, all *ab initio* calculations need huge amounts of computational resources and therefore, as mentioned above, are currently limited to light nuclei.

The second line of attack is just the theme of this review paper, namely the use of the traditional shell-model with two-nucleon effective interactions derived from the bare NN potential. In this case, only the valence nucleons are treated as active particles. However, as we shall discuss in detail in Section 3, core polarization effects are taken into account perturbatively in the derivation of the effective interaction. Of course, this approach allows one to perform calculations for medium- and heavy-mass nuclei which are far beyond the reach of ab initio calculations. Many-body forces beyond the NN potential may also play a role in the shell-model effective Hamiltonian. However, this is still an open problem, and it is the main scope of this review to assess the progress made in the derivation of a two-body effective interaction from the free NN potential.

While efforts in this direction started some forty years ago [17,18], for a long time there was widespread skepticism about the practical value of what had become known as "realistic shell model" calculations. This was mainly related to the highly complicated nature of the nucleon–nucleon force, in particular to the presence of a very strong repulsion at short distances, which made solving the nuclear many-body problem especially difficult. As a consequence, a major problem, and correspondingly a main source of uncertainty, in shell-model calculations has long been the two-body effective interaction between the valence nucleons. An early survey of the various approaches to this problem can be found in the Cargèse lectures by Elliott [19], where a classification of the various categories of nuclear structure calculations was made by the number of free parameters in the two-body interaction being used in the calculation.

Since the early 1950s through the mid 1990s, hundreds of shell-model calculations have been carried out, many of them being very successful in describing a variety of nuclear structure phenomena. In the vast majority of these calculations, either empirical effective interactions containing adjustable parameters have been used, or the two body-matrix elements themselves have been treated as free parameters. The latter approach, which was pioneered by Talmi [20], is generally limited to relatively small spaces owing to the large number of parameters to be least-squares fitted to experimental data.

Early calculations of this kind were performed for the p-shell nuclei by Cohen and Kurath [21], who determined fifteen matrix elements of the two-body interaction and two single-particle energies from 35 experimental energies of nuclei from $A = 8$ up to $A = 16$. The results of these calculations turned out to be in very satisfactory agreement with experiment.

A more recent and very successful application of this approach is that by Brown and Wildenthal [22–24], where the $A = 17 - 39$ nuclei were studied in the complete sd-shell space. In the final version of this study, which spanned about 10 years, 66 parameters were determined by a least squares fit to 447 binding and excitation energies. A measure of the quality of the results is given by the rms deviation, whose value turned out to be 185 keV. In this connection, it is worth mentioning that in the recent work of Brown and Richter [25] the determination of a new effective interaction for the sd-shell has been pursued by the inclusion of an updated set of experimental data. As regards the empirical effective interactions, they may be schematically divided into two categories. The first is based on the use of simple potentials, such as a Gaussian or Yukawa central force with various exchange operators consistent with those present in the interaction between free nucleons. These contain several parameters, such as the strengths and ranges of singlet and triplet interactions, which are usually determined by a fit to the spectroscopic properties under study. In the second category one may put the so-called "schematic interactions". These contain few free parameters, typically one or two, at the price of being an oversimplified representation of the real potential. Into this category fall the well known pairing [26–28], pairing plus quadrupole [29], and surface delta [30–32] interactions. The successful spin and isospin dependent Migdal interaction [33] also belongs to this category. While these simple interactions are able to reproduce some specific nuclear properties [see, e.g., [34,35]], they are clearly inadequate for detailed quantitative studies.

We have given above a brief sketch of the various approaches to the determination of the shell-model effective interaction that have dominated the field for more than four decades. This we have done to place in its proper perspective the great progress made over the last decade by the more fundamental approach employing realistic effective interactions derived from modern NN potentials. We refer to Ref. [5] for a detailed review of shell-model calculations, based on the above empirical approaches.

As we shall discuss in detail in the following sections, from the late 1970s on there has been substantial progress toward a microscopic approach to nuclear structure calculations starting from the free NN potential V_{NN}. This has concerned both the two basic ingredients which come into play in this approach, namely the NN potential and the many-body methods for deriving the model space effective interaction, V_{eff}. As regards the first point, NN potentials have been constructed, which reproduce quite accurately the NN scattering data. As regards the derivation of V_{eff}, the first problem one is confronted with is that all realistic NN potentials have a strong repulsive core which prevents their direct use in nuclear structure calculations. As is well known, this difficulty can be overcome by resorting to the Brueckner G-matrix method (see Section 4.1), which was originally designed for nuclear matter calculations.

While in earlier calculations one had to make some approximations in calculating the G matrix for finite nuclei, the developments of improved techniques, as for instance the Tsai-Kuo method [36,37] (see Section 4.1), allowed one to calculate it in a practically exact way. Another major improvement consisted of the inclusion in the perturbative expansion for V_{eff} of folded diagrams, whose important role had been recognized by many authors, within the framework of the so-called $\hat{Q}$-box formulation ([38], see Section 3).

These improvements brought about a revival of interest in shell-model calculations with realistic effective interactions. This started in the early 1990s, and continued to increase during the following years. Given the skepticism mentioned above, the main aim of this new generation of realistic calculations was to give an answer to the key question of whether they could provide an accurate description of nuclear structure properties. By the end of the 1990s the answer to this question turned out to be in the affirmative. In fact, a substantial body of results for nuclei in various mass regions (see, for instance [39]) proved the ability of shell-model calculations employing two-body matrix elements derived from modern NN potentials to provide a description of nuclear structure properties at least as accurate as that provided by traditional, empirical interactions. It should be noted that in this approach the single-nucleon energies are generally taken from experiment (see, e.g., Section 5.3.3), so that the calculation contains essentially no free parameters. Based on these results, in the last few years the use of realistic effective interactions has been rapidly gaining ground in nuclear structure theory.

As will be discussed in detail in Section 5, the G matrix has been routinely used in practically all realistic shell-model calculations through 2000. However, the G matrix is model-space dependent as well as energy dependent; these dependences make its actual calculation rather involved (see Section 4.1). In this connection, it may be recalled that an early criticism of the G-matrix method dates back to the 1960s [40] which led to the development of a method for deriving directly from the phase shifts a set of matrix elements of V_{NN} in oscillator wave functions [40,19,41]. This resulted in the well-known Sussex interaction which has been used in several nuclear structure calculations.

Recently, a new approach to overcome the difficulty posed by the strong short-range repulsion contained in the free NN potential has been proposed [42–44]. The basic idea underlying this approach is inspired by the recent applications of effective field theory and renormalization group to low-energy nuclear systems. Starting from a realistic *NN* potential, a low-momentum *NN* potential, $V_{\rm low-k}$, is constructed that preserves the physics of the original potential V_{NN} up to a certain cutoff momentum Λ. In particular, the scattering phase shifts and deuteron binding energy calculated from V_{NN} are reproduced by $V_{\rm low-k}$. This is achieved by integrating out, in the sense of the renormalization group, the high-momentum components of V_{NN}. The resulting $V_{\rm low-k}$ is a smooth potential that can be used directly in nuclear structure calculations without first calculating the *G* matrix. The practical value of the $V_{\rm low-k}$ approach has been assessed by several calculations, which have shown that it provides an advantageous alternative to the *G*-matrix one (see Section 5.3.1).

The purpose of the present paper is to give a review of the basic formalism and the current status of realistic shell-model calculations, and a self-contained survey of the major developments in the history of the field, as regards both the *NN* potential and the many-body approach to the derivation of the effective interaction. During the last four decades, there have been several reviews focused on either of these two subjects, and we shall have cause to refer to most of them in the following sections. Our review is similar in spirit to the one by Hjorth-Jensen et al. [45], in the sense that it aims at giving an overall view of the various aspects of realistic shell-model calculations, including recent selected results. The novelty of the present paper is that it covers the developments of the last decade which have brought these calculations into the mainstream of nuclear structure [see, for instance, [5]]. As mentioned above, from the mid 1990s on, there has been a growing success in explaining experimental data by means of two-body effective interactions derived from the free *NN* potential, which has evidenced the practical value of realistic shell-model calculations. It is worth emphasizing that a major step in this direction has been the introduction of the low-momentum potential $V_{\rm low-k}$, which greatly simplifies the microscopic derivation of the shell-model effective interaction. On these grounds, we may consider that a first important phase in the microscopic approach to shell model, started more than 40 years ago, has been completed. It is just this consideration at the origin of the present review.

Four more recent reviews [46,11,47,48] reporting on progress in shell-model studies are in some ways complementary to ours, in that they discuss aspects which we have considered to be beyond the scope of the present review. These regard, for instance, a phenomenologically oriented survey of shell-model applications [46] or large-scale shell-model calculations [11,47,48].

We start in Section 2 with a review of the *NN* interaction, trying to give an idea of the long-standing, painstaking work that lies behind the development of the modern high-precision potentials. In Section 3 we discuss the derivation of the shell-model effective interaction within the framework of degenerate perturbation theory. The crucial role of folded diagrams is emphasized. Section 4 is devoted to the handling of the short-range repulsion contained in the free *NN* potential. We first discuss in Section 4.1 the traditional Brueckner *G*-matrix method and then introduce in Section 4.2 the new approach based on the construction of a low-momentum *NN* potential. In Section 5 we first give a survey of realistic shell-model calculations performed over the last four decades (Sections 5.1 and 5.2) and then present some results of recent calculations. More precisely, in Section 5.3.1 a comparison is made between the *G*-matrix and $V_{\rm low-k}$ approaches while in Section 5.3.2 results obtained with different *NN* potentials are presented. In Section 5.3.3 we report selected results of calculations for nuclei neighboring doubly magic ^{132}Sn and compare them with experiment. Finally, in Section 5.3.4 we discuss the role of the many-body contributions to the effective interaction by investigating the results of a study of the even $N = 82$ isotones. The last section, Section 6, contains a brief summary and concluding remarks.

2. Nucleon–nucleon interaction

2.1. Historical overview

The nucleon–nucleon interaction has been extensively studied since the discovery of the neutron, and in the course of time there have been a number of Conferences [49–51] and review papers [52–54] marking the advances in the understanding of its nature. Here, we shall start by giving a brief historical account and a survey of the main aspects relevant to nuclear structure, the former serving the purpose to look back and recall how hard it has been making progress in this field.

As is well known, the theory of nuclear forces started with the meson exchange idea introduced by Yukawa [55]. Following the discovery of the pion, in the 1950s much effort was made to describe the *NN* interaction in terms of pion-exchange models. However, while by the end of the 1950s the one-pion exchange (OPE) had been experimentally established as the long-range part of V_{NN}, the calculations of the two-pion exchange were plagued by serious ambiguities. This led to several pion-theoretical potentials differing quite widely in the two-pion exchange effects. This unpleasant situation is well reflected in various review papers of the period of the 1950s, for instance the article by Phillips [56]; a comprehensive list of references can be found in Ref. [52].

While the theoretical effort mentioned above was not very successful, substantial progress in experimental studies of the properties of the *NN* interaction was made during the course of the 1950s. In particular, from the examination of the *pp* scattering data at 340 MeV in the laboratory system Jastrow [57] inferred the existence of a strong short-range repulsion, which he represented by a hard sphere for convenience in calculation. As we shall discuss in detail later, this feature, which prevents the direct use of V_{NN} in nuclear structure calculations, has been at the origin of the Brueckner *G*-matrix method (Section 4.1) and of the recent $V_{\rm low-k}$ approach (Section 4.2).

At this point it must be recalled that as early as 1941 an investigation of the possible types of nonrelativistic *NN* interaction at most linear in the relative momentum $\boldsymbol{p}$ of the two nucleons and limited by invariance conditions was carried out by

Eisenbud and Wigner [58]. It turned out that the general form of V_{NN} consists of central, spin–spin, tensor and spin-orbit terms. Some twenty years later, the most general V_{NN} when all powers of $\boldsymbol{p}$ are allowed was given by Okubo [59], which added a quadratic spin-orbit term. When sufficiently reliable phase-shift analyses of *NN* scattering data became available (see for instance Ref. [60]), these studies were a key guide for the construction of phenomenological *NN* potentials. In the early stages of this approach, the inclusion of all the four types of interaction resulting from the study of Eisenbud and Wigner (1941), with the assumption of charge independence, led to the Gammel–Thaler potential [61], which may be considered the first quantitative *NN* potential. In this potential, following the suggestion of Jastrow (1951), a strong short-range repulsion represented by a hard core (infinite repulsion) at about 0.4 fm was used. As we shall see later, it took a decade before soft-core potentials were considered.

In the early 1960s two vastly improved phenomenological potentials appeared, both going beyond the Eisenbud-Wigner form with addition of a quadratic spin-orbit term. These were developed by the Yale group [62] and by Hamada and Johnston (HJ) [63]. Both potentials have infinite repulsive cores and approach the one-pion-exchange-potential at large distances. Historically, the HJ potential occupies a special place in the field of microscopic nuclear structure. In fact, it was used in the mid 1960s in the work of Kuo and Brown [18], which was the first successful attempt to derive the shell-model effective interaction from the free *NN* potential. We therefore find it appropriate to summarize here its main features. This may also allow a comparison with the today's high-quality phenomenological potentials, as for instance Argonne V_{18} (see Section 2.2). The HJ potential has the form

$$V = V_{\mathrm{C}}(r) + V_{\mathrm{T}}(r)S_{12} + V_{LS}(r)\boldsymbol{L}\cdot\boldsymbol{S} + V_{LL}(r)L_{12}, \tag{1}$$

where C, T, *LS* and *LL* denote respectively central, tensor, spin-orbit and quadratic spin-orbit terms. The operator S_{12} is the ordinary tensor operator and the quadratic spin-orbit operator is defined by

$$L_{12} = [\delta_{lJ} + (\boldsymbol{\sigma}_1\cdot\boldsymbol{\sigma}_2)]\boldsymbol{L}^2 - (\boldsymbol{L}\cdot\boldsymbol{S})^2. \tag{2}$$

The V_i (i = C, T, *LS* and *LL*) are spin-parity dependent, and hard cores, with a common radius of 0.485 fm, are present in all states. With about 30 parameters the HJ potential model reproduced in a quantitative way the *pp* and *np* data below 315 MeV.

As mentioned above, the era of soft-core potentials started in the late 1960s with the work of Reid [64] and Bressel et al. [65]. The original Reid soft-core potential Reid68 has been updated some 25 years later [66] producing a high-quality potential denoted as Reid93 (see Section 2.2).

Let us now come back to the meson-theory based potentials. The discovery of heavy mesons in the early 1960s revived the field. This resulted in the development of various one-boson-exchange (OBE) potentials and in a renewed confidence in the theoretical approach to the study of the *NN* interaction. The optimistic view of the field brought about by the advances made during the 1960s is reflected in the Summary [49] of the 1967 International Conference on the Nucleon-Nucleon Interaction held at the University of Florida in Gainesville. A concise and clear account of the early OBE potentials (OBEP), including a list of relevant references, can be found in the review of the meson theory of nuclear forces by Machleidt [52].

During the 1960s, a sustained effort was made to try to understand the properties of complex nuclei in terms of the fundamental *NN* interaction. This brought in focus the problem of how to handle the serious difficulty resulting from the strong short-range repulsion contained in the free *NN* potential. We shall discuss this point in detail in Section 4. Here, it should be mentioned that the idea of overcoming the above difficulty by constructing a smooth, yet realistic, *NN* potential that could be used directly in nuclear structure calculations was actively explored in the mid 1960s. This led to the development of a non-local, separable potential fitting two-nucleon scattering data with reasonable accuracy [67,68]. This potential, known as Tabakin potential, was used by the MIT group in several calculations of the structure of finite nuclei within the framework of the Hartree–Fock method [69–72]. An early account of the results of nuclear structure calculations using realistic *NN* interactions was given at the above mentioned Gainesville Conference by Moszkowski [73].

As regards the experimental study of the *NN* scattering, this was also actively pursued in the 1960s (see [49]), leading to the much improved phase-shift analysis of McGregor et al. [74], which included 2066 *pp* and *np* data up to 450 MeV. This set the stage for the theoretical efforts of the 1970s, which were addressed to the construction of a quantitative *NN* potential (namely, able to reproduce with good accuracy all the known *NN* scattering data) within the framework of the meson theory. In this context, the main goal was to go beyond the OBE model by taking into account multi-meson exchange, in particular the 2π-exchange contribution. These efforts were essentially based on two different approaches: dispersion relations and field theory.

The work along these two lines, which went on for more than one decade, resulted eventually in the Paris potential [75–78,50] and in the so called "Bonn full model" [79], the latter including also contributions beyond 2π. In the sector of the OBE model a significant progress was made through the work of the Nijmegen group [80]. This was based on Regge-pole theory, and led to a quite sophisticated OBEP which is known as the Nijmegen78 potential. The Nijmegen, Paris, and Bonn potentials fitted the world *NN* data below 300 MeV available in 1992 with a χ^2/datum = 5.12, 3.71, and 1.90, respectively [53].

To have a firsthand idea of the status of the theory of the *NN* interaction around 1990, we refer to Ref. [51], while a detailed discussion of the above three potentials can be found in [53]. Here we would like to emphasize that they mark the beginning of a new era in the field of nuclear forces, and may be considered as the first generation of *NN* realistic potentials. In particular, as will be discussed in Section 5.1, the Paris and Bonn potentials have played an important role in the revival of interest in nuclear structure calculations starting from the bare *NN* interaction. We shall therefore give here a brief outline of the main characteristics of these two potentials, as well as of the energy-independent OBE parametrization of the Bonn full model, which has been generally employed in nuclear structure applications.

In addition to the 2π-exchange contribution, the Paris potential contains the OPE and ω-meson exchange. This gives the long-range and medium-range part of the *NN* interaction, while the short-range part is of purely phenomenological nature. In its final version [78] the Paris potential is parametrized in an analytical form, consisting of a regularized discrete superposition of Yukawa-type terms. This introduces a large number of free parameters, about 60 [53], that are determined by fitting the *NN* scattering data.

As already mentioned, the Bonn full model is a field-theoretical meson-exchange model for the *NN* interaction. In addition to the 2π-exchange contribution, this model contains single π, ω, and δ exchanges and $\pi\rho$ contributions. It has been shown [79] that the latter are essential for a quantitative description of the phase shifts in the lower partial waves, while additional 3π and 4π contributions are not very important. The Bonn full model has in all 12 parameters which are the coupling constants and cutoff masses of the meson-nucleon vertices involved. This model is an energy-dependent potential, which makes it inconvenient for application in nuclear structure calculations. Therefore, an energy independent one-boson parametrization of this potential has been developed within the framework of the relativistic three-dimensional Blanckenbecler-Sugar (BbS) reduction of the Bethe-Salpeter equation [79,52]. This OBEP includes exchanges of two pseudoscalar (π and η), two scalar (σ and δ), and two vector (ρ and ω) mesons. As in the Bonn full model, there are only twelve parameters which have to be determined through a fit of the *NN* scattering data.

At this point, it must be pointed out that there are three variants of the above relativistic OBE potential, denoted by Bonn A, Bonn B and Bonn C. The parameters of these potentials and the predictions of Bonn B for the two-nucleon system are given in [52]. The latter are very similar to the ones by the Bonn full model. The main difference between the three potentials is the strength of the tensor force as reflected in the predicted *D*-state probability of the deuteron P_D. With $P_D = 4.4\%$ Bonn A has the weakest tensor force. Bonn B and Bonn C predict 5% and 5.6%, respectively. Note that for the Paris potential $P_D = 5.8\%$. We shall have cause to come back to this important point later.

We should now mention that there also exist three other variants of the OBE parametrization of the Bonn full model. These are formulated within the framework of the Thompson equation [52,81], and use the pseudovector coupling for π and η, while the potential defined within the BbS equation uses pseudoscalar coupling. It may be mentioned that the results obtained with the Thompson choice differ little from those obtained with the BbS reduction. A detailed discussion on this point is given in [81].

As we shall see later, the potential with the weaker tensor force, namely Bonn A, has turned out to give the best results in nuclear structure calculations. Unless otherwise stated, in the following we shall denote by Bonn A, B, and C the three variants of the energy-independent approximation to the Bonn full model defined within the BbS equation. However, to avoid any confusion when consulting the literature on this subject, the reader may take a look at Tables A.1 and A.2 in [52].

2.2. High-precision NN potentials

From the early 1990s on there has been much progress in the field of nuclear forces. In the first place, the *NN* phase shift analysis was greatly improved by the Nijmegen group [82–85]. They performed a multienergy partial-wave analysis of all *NN* scattering data below 350 MeV laboratory energy, after rejection of a rather large number of data (about 900 and 300 for the *np* and *pp* data, respectively) on the basis of statistical criteria. In this way, the final database consisted of 1787 *pp* and 2514 *np* data. The *pp*, *np* and combined $pp + np$ analysis all yielded a χ^2/datum ≈ 1, significantly lower than any previous multienergy partial-wave analysis. This analysis has paved the way to a new generation of high-quality *NN* potentials which, similar to the analysis, fit the *NN* data with the almost perfect χ^2/datum ≈ 1. These are the potentials constructed in the mid 1990s by the Nijmegen group, NijmI, NijmII and Reid93 [66], the Argonne V_{18} potential [86], and the CD-Bonn potential [87,88].

The two potentials NijmI and NijmII are based on the original Nijm78 potential [80] discussed in the previous section. They are termed Reid-like potentials since, as is the case for the Reid68 potential [64], each partial wave is parametrized independently. At very short distances these potentials are regularized by exponential form factors. The Reid93 potential is an updated version of the Reid68 potential, where the singularities have been removed by including a dipole form factor. While the NijmII and the Reid93 are totally local potentials, the NijmI contains momentum-dependent terms which in configuration space give rise to nonlocalities in the central force component. Except for the OPE tail, these potentials are purely phenomenological with a total of 41, 47 and 50 parameters for NijmI, NijmII and Reid93, respectively. They all fit the *NN* scattering data with an excellent χ^2/datum $= 1.03$ [66]. As regards the *D*-state probability of the deuteron, this is practically the same for the three potentials, namely P_D in % $= 5.66$ for NijmI, 5.64 for NijmII, and 5.70 for Reid93. It is worth mentioning that in the work by Stoks et al. [66] an improved version of the Nijm78 potential, dubbed Nijm93, was also presented, which with 15 parameters produced a χ^2/datum of 1.87.

The CD-Bonn potential [88] is a charge-dependent OBE potential. It includes the π, ρ, and ω mesons plus two effective scalar-isoscalar σ bosons, the parameters of which are partial-wave dependent. As is the case for the early OBE Bonn potentials, CD-Bonn is a nonlocal potential. It predicts a deuteron *D*-state probability substantially lower than that yielded by the potentials of the Nijmegen family, namely $P_D = 4.85\%$. This may be traced to the nonlocalities contained in the tensor force [88]. While the CD-Bonn potential reproduces important predictions by the Bonn full model, the additional fit freedom obtained by adjusting the parameters of the σ_1 and σ_2 bosons in each partial wave produces a χ^2/datum of 1.02 for the 4301 data of the Nijmegen database, the total number of free parameters being 43. In this connection, it may be mentioned that the Nijmegen database has been updated [88] by adding the *pp* and *np* data between January 1993 and December 1999. This 1999 database contains 2932 *pp* data and 3058 *np* data, namely 5990 data in total. The χ^2/datum for the CD-Bonn potential in regard to the latter database remains 1.02.

The Argonne V_{18} model [86], so named for its operator content, is a purely phenomenological (except for the OPE tail) nonrelativistic *NN* potential with a local operator structure. It is an updated version of the Argonne V_{14} potential [89], which was constructed in the early 1980s, with the addition of three charge-dependent and one charge-asymmetric operators. In operator form the V_{18} potential is written as a sum of 18 terms,

$$V_{ij} = \sum_{p=1,18} V_p(r_{ij}) O_{ij}^p. \tag{3}$$

To give an idea of the degree of sophistication reached by modern phenomenological potentials, it may be instructive to write here explicitly the operator structure of the V_{18} potential [86]. The first 14 charge independent operators are given by:

$$\begin{aligned} O_{ij}^{p=1,14} &= 1, (\boldsymbol{\tau}_i \cdot \boldsymbol{\tau}_j), (\boldsymbol{\sigma}_i \cdot \boldsymbol{\sigma}_j), (\boldsymbol{\sigma}_i \cdot \boldsymbol{\sigma}_j)(\boldsymbol{\tau}_i \cdot \boldsymbol{\tau}_j), S_{ij}, S_{ij}(\boldsymbol{\tau}_i \cdot \boldsymbol{\tau}_j), \boldsymbol{L} \cdot \boldsymbol{S}, \boldsymbol{L} \cdot \boldsymbol{S}(\boldsymbol{\tau}_i \cdot \boldsymbol{\tau}_j), \\ & L^2, L^2(\boldsymbol{\tau}_i \cdot \boldsymbol{\tau}_j), L^2(\boldsymbol{\sigma}_i \cdot \boldsymbol{\sigma}_j), L^2(\boldsymbol{\sigma}_i \cdot \boldsymbol{\sigma}_j)(\boldsymbol{\tau}_i \cdot \boldsymbol{\tau}_j), (\boldsymbol{L} \cdot \boldsymbol{S})^2, (\boldsymbol{L} \cdot \boldsymbol{S})^2(\boldsymbol{\tau}_i \cdot \boldsymbol{\tau}_j). \end{aligned} \tag{4}$$

The four additional operators breaking charge independence are given by

$$O_{ij}^{p=15,18} = T_{ij}, (\boldsymbol{\sigma}_i \cdot \boldsymbol{\sigma}_j) T_{ij}, S_{ij} T_{ij}, (\tau_{zi} + \tau_{zj}), \tag{5}$$

where $T_{ij} = 3\tau_{zi}\tau_{zj} - \boldsymbol{\tau}_i \cdot \boldsymbol{\tau}_j$ is the isotensor operator analogous to the S_{ij} operator. As is the case for the NijmI and NijmII potentials, at very short distances the V_{18} potential is regularized by exponential form factors. With 40 adjustable parameters, this potential gives a χ^2/datum of 1.09 for the 4301 data of the Nijmegen database. As regards the deuteron *D*-state probability, this is $P_D = 5.76\%$, very close to that predicted by the potentials of the Nijmegen family.

All the high-precision *NN* potentials described above have a large number of free parameters, say about 45, which is the price one has to pay to achieve a very accurate fit of the world *NN* data. This makes it clear that, to date, high-quality potentials with an excellent χ^2/datum ≈ 1 can only be obtained within the framework of a substantially phenomenological approach. Since these potentials fit almost equally well the *NN* data up to the inelastic threshold, their on-shell properties are essentially identical, namely they are phase-shift equivalent. In addition, they all predict almost identical deuteron observables (quadrupole moment and *D*/*S*-state ratio) [54]. While they have also in common the inclusion of the OPE contribution, their off-shell behavior may be quite different. In fact, the short-range (high-momentum) components of these potentials are indeed quite different, as we shall discuss later in Section 4.2. This raises a central question of how much nuclear structure results may depend on the *NN* potential one starts with. We shall consider this important point in Section 5.3.2.

The brief review of the *NN* interaction given above has been mainly aimed at highlighting the progress made in this field over a period of about 50 years. As already pointed out in the Introduction, and as we shall discuss in detail in Sections 5.1 and 5.2, this has been instrumental in paving the way to a more fundamental approach to nuclear structure calculations than the traditional, empirical one. It is clear, however, that from a first-principle point of view a substantial theoretical progress in the field of the *NN* interaction is still in demand. It seems fair to say that this is not likely to be achieved along the lines of the traditional meson theory. Indeed, in the past few years, efforts in this direction have been made within the framework of the chiral effective theory. The literature on this subject, which is still actively pursued, is by now very extensive and there are several comprehensive reviews [90–92], to which we refer the reader. Therefore, in the next section we shall only give a brief survey focusing attention on chiral potentials which have been recently employed in nuclear structure calculations.

2.3. *Chiral potentials*

The approach to the *NN* interaction based upon chiral effective field theory was started by Weinberg [93,94] some fifteen years ago, and since then it has been developed by several authors. The basic idea [93] is to derive the *NN* potential, starting from the most general Lagrangian for low-energy pions and nucleons consistent with the symmetries of quantum chromodynamics (QCD), in particular the spontaneously broken chiral symmetry. All other particle types are "integrated out", their effects being contained in the coefficients of the series of terms in the pion-nucleon Lagrangian. The chiral Lagrangian provides a perturbative framework for the derivation of the nucleon–nucleon potential. In fact, it was shown by Weinberg [94] that a systematic expansion of the nuclear potential exists in powers of the small parameter Q/Λ_χ, where Q denotes a generic low-momentum and $\Lambda_\chi \approx 1$ GeV is the chiral symmetry breaking scale. This perturbative low-energy theory is called chiral perturbation theory (χPT). The contribution of any diagram to the perturbation expansion is characterized by the power ν of the momentum Q, and the expansion is organized by counting powers of Q. This procedure [94] is referred to as power counting.

Soon after the pioneering work by Weinberg, where only the lowest order *NN* potential was obtained, Ordóñez et al. [95] extended the effective chiral potential to order $(Q/\Lambda_\chi)^3$ [next-to-next-to-leading order (NNLO), $\nu = 3$] showing that this accounted, at least qualitatively, for the most relevant features of the nuclear potential. Later on, this approach was further pursued by Ordóñez, Ray and van Kolck [96,97], who derived at NNLO a *NN* potential both in momentum and coordinate space. With 26 free parameters this potential model gave a satisfactory description of the Nijmegen phase shifts up to about 100 MeV [97]. These initial achievements prompted extensive efforts to understand the *NN* force within the framework of chiral effective field theory.

A clean test of chiral symmetry in the two-nucleon system was provided by the work of Kaiser, Brockmann and Weise [98] and Kaiser, Gerstendörfer and Weise [99]. Restricting themselves to the peripheral nucleon–nucleon interaction, these authors obtained at NNLO, without adjustable parameters, an accurate description of the empirical phase shifts in the partial waves with $L \geq 3$ up to 350 MeV and up to about (50–80) MeV for the D-waves.

Based on a modified Weinberg power counting, Epelbaum, Glöckle and Meissner constructed a chiral *NN* potential at NNLO consisting of one- and two-pion exchange diagrams and contact interactions (which represent the short-range force) [100,101]. The nine parameters related to the contact interactions were determined by a fit to the *np* *S*- and *P*-waves and the mixing parameter ϵ_1 for $E_{\rm Lab} < 100$ MeV. This potential gives a χ^2/datum for the *NN* data of the 1999 database below 290 MeV laboratory energy of more than 20 [102].

In their program to develop a *NN* potential based upon chiral effective theory, Entem and Machleidt set themselves the task to achieve an accuracy for the reproduction of the *NN* data comparable to that of the high-precision potentials constructed in the 1990s, which have been discussed in Section 2.2. The first outcome of this program was a NNLO potential, called Idaho potential [103]. This model includes one- and two-pion exchange contributions up to chiral order three, and contact terms up to order four. For the latter, partial wave dependent cutoff parameters are used, which introduces more parameters bringing the total number up to 46. This potential gives a χ^2/datum for the reproduction of the 1999 *np* database up to $E_{\rm Lab} = 210$ MeV of 0.98 [104].

The next step taken by Entem and Machleidt was the investigation of the chiral 2π-exchange contributions to the *NN* interaction at fourth order, which was based on the work by Kaiser [105,106], who gave analytical results for these contributions in a form suitable for implementation in a next-to-next-to-next-to-leading (N^3LO, fourth order) calculation. This eventually resulted in the first chiral *NN* potential at N^3LO [102]. This model includes 24 contact terms (24 parameters) which contribute to the partial waves with $L \leq 2$. With 29 parameters in all, it gives a χ^2/datum for the reproduction of the 1999 *np* and *pp* data below 290 MeV of 1.10 and 1.50, respectively. The deuteron *D*-state probability is $P_D = 4.51\%$.

Very recently a *NN* potential at N^3LO has been constructed by Epelbaum, Glöckle and Meissner (2005) which differs in various ways from that of Entem and Machleidt, as discussed in detail in [107]. It consists of one-, two- and three-pion exchanges, and a set of 24 contact interactions. The total number of free parameters is 26. These have been determined by a combined fit to some *nn* and *pp* phase shifts from the Nijmegen analysis, together with the *nn* scattering length. The description of the phase shifts and deuteron properties at N^3LO turns out to be improved compared to that previously obtained by the same authors at NLO and NNLO [108]. As regards the deuteron *D*-state probability, this N^3LO potential gives $P_D = 2.73\%$–3.63%, a value which is significantly smaller than that predicted by any other modern *NN* potential.

In regard to potentials at N^3LO, it is worth mentioning that it has been shown [109,110] that the effects of three-pion exchange, which starts to contribute at this order, are very small and therefore of no practical relevance. Accordingly, they have been neglected in both the above studies.

The foregoing discussion has all been focused on the two-nucleon force. The role of three-nucleon interactions in light nuclei has been, and is currently, actively investigated within the framework of *ab initio* approaches, such as the GFMC and the NCSM. Let us only remark here that in recent years the Green's function Monte Carlo method has proved to be a valuable tool for calculations of properties of light nuclei using realistic two-nucleon and three-nucleon potentials [6,111]. In particular, the combination of the Argonne V_{18} potential and Illinois-2 three-nucleon potential has yielded good results for energies of nuclei up to ^{12}C [112]. For a review of the GFMC method and applications up to $A = 8$ we refer the interested reader to the paper by Pieper and Wiringa [7].

In this context, it should be pointed out that an important advantage of the chiral perturbation theory is that at NNLO and higher orders it generates three-nucleon forces. This has prompted applications of the complete chiral interaction at NNLO to the three- and four-nucleon systems [113]. These applications are currently being extended to light nuclei with $A > 4$ [114].

However, as regards the derivation of a realistic shell-model effective interaction, the 3*N* forces have not been taken into account up to now. As mentioned in the Introduction, in this review we shall give a brief discussion of the three-body effects, as inferred from the study of many valence-nucleon systems.

3. Shell-model effective interaction

3.1. Generalities

As mentioned in the Introduction, a basic input to nuclear shell-model calculations is the model-space effective interaction. It is worth recalling that this interaction differs from the interaction between two free nucleons in several respects. In the first place, a large part of the *NN* interaction is absorbed into the mean field, which is due to the average interaction between the nucleons. In the second place, the *NN* interaction in the nuclear medium is affected by the presence of the other nucleons; one has certainly to take into account the Pauli exclusion principle, which forbids two interacting nucleons to scatter into states occupied by other nucleons. Finally, the effective interaction has to account for effects of the configurations excluded from the model space. Ideally, the eigenvalues of the shell-model Hamiltonian in the model space should be a subset of the eigenvalues of the full nuclear Hamiltonian in the entire Hilbert space.

In a microscopic approach this shell-model Hamiltonian may be constructed starting from a realistic NN potential by means of many-body perturbation techniques. This approach has long been a central topic of nuclear theory. The following subsections are devoted to a detailed discussion of it.

First, let us introduce the general formalism which is needed in the effective interaction theory. We would like to solve the Schrödinger equation for the A-nucleon system:

$$H|\Psi_\nu\rangle = E_\nu|\Psi_\nu\rangle, \tag{6}$$

where

$$H = H_0 + H_1, \tag{7}$$

and

$$H_0 = \sum_{i=1}^{A}(t_i + U_i), \tag{8}$$

$$H_1 = \sum_{i<j=1}^{A} V_{ij}^{NN} - \sum_{i=1}^{A} U_i. \tag{9}$$

An auxiliary one-body potential U_i has been introduced, in order to break up the nuclear Hamiltonian as the sum of a one-body term H_0, which describes the independent motion of the nucleons, and the interaction H_1.

In the shell model, the nucleus is represented as an inert core, plus n valence nucleons moving in a limited number of SP orbits above the closed core and interacting through a model-space effective interaction. The valence or model space is defined in terms of the eigenvectors of H_0

$$|\Phi_i\rangle = [a_1^\dagger a_2^\dagger \dots a_n^\dagger]_i|c\rangle, \quad i = 1, \dots, d, \tag{10}$$

where $|c\rangle$ represents the inert core and the subscripts 1, 2,..., n denote the SP valence states. The index i stands for all the other quantum numbers needed to specify the state.

The aim of the effective interaction theory is to reduce the eigenvalue problem of Eq. (6) to a model-space eigenvalue problem

$$PH_{\text{eff}}P|\Psi_\alpha\rangle = E_\alpha P|\Psi_\alpha\rangle, \tag{11}$$

where the operator P,

$$P = \sum_{i=1}^{d} |\Phi_i\rangle\langle\Phi_i|, \tag{12}$$

projects from the complete Hilbert space onto the model space. The operator $Q = 1 - P$ is its complement. The projection operators P and Q satisfy the properties

$$P^2 = P, \qquad Q^2 = Q, \qquad PQ = QP = 0.$$

In the following, the concept of the effective interaction is introduced by a very general and simple method [115,116]. Let us define the operators

$$PHP = H_{PP}, \qquad PHQ = H_{PQ},$$
$$QHP = H_{QP}, \qquad QHQ = H_{QQ}.$$

Then the Schrödinger equation (6) can be written as

$$H_{PP}P|\Psi_\nu\rangle + H_{PQ}Q|\Psi_\nu\rangle = E_\nu P|\Psi_\nu\rangle \tag{13}$$

$$H_{QP}P|\Psi_\nu\rangle + H_{QQ}Q|\Psi_\nu\rangle = E_\nu Q|\Psi_\nu\rangle. \tag{14}$$

From the latter equation we obtain

$$Q|\Psi_\nu\rangle = \frac{1}{E_\nu - H_{QQ}} H_{QP}P|\Psi_\nu\rangle, \tag{15}$$

and substituting the r.h.s. of this equation into Eq. (13) we have

$$\left(H_{PP} + H_{PQ}\frac{1}{E_\nu - H_{QQ}}H_{QP}\right)P|\Psi_\nu\rangle = E_\nu P|\Psi_\nu\rangle. \tag{16}$$

If the l.h.s. operator, which acts only within the model space, is denoted as

$$H_{\rm eff}(E_\nu) = H_{PP} + H_{PQ}\frac{1}{E_\nu - H_{QQ}}H_{QP}, \tag{17}$$

Eq. (16) reads

$$H_{\rm eff}(E_\nu)P|\Psi_\nu\rangle = E_\nu P|\Psi_\nu\rangle, \tag{18}$$

which is of the form of Eq. (11). Moreover, since the operators P and Q commute with H_0, we can write Eq. (17) as

$$H_{\rm eff}(E_\nu) = PH_0P + V_{\rm eff}(E_\nu), \tag{19}$$

with

$$V_{\rm eff}(E_\nu) = PH_1P + PH_1Q\frac{1}{E_\nu - H_{QQ}}QH_1P. \tag{20}$$

This equation defines the effective interaction as derived by Feshbach in nuclear reaction studies [116].

Now, on expanding $(E_\nu - H_{QQ})^{-1}$ we can write

$$V_{\rm eff}(E_\nu) = PH_1P + PH_1\frac{Q}{E_\nu - QH_0Q}H_1P + PH_1\frac{Q}{E_\nu - QH_0Q}H_1\frac{Q}{E_\nu - QH_0Q}H_1P + \cdots, \tag{21}$$

which is equivalent to the Bloch-Horowitz form of the effective interaction [115]:

$$V_{\rm eff}(E_\nu) = PH_1P + PH_1\frac{Q}{E_\nu - QH_0Q}V_{\rm eff}(E_\nu). \tag{22}$$

Eqs. (20) and (22) are the desired result. In fact, they represent effective interactions which, used in a truncated model space, give a subset of the true eigenvalues. Bloch and Horowitz [115] have studied the analytic properties of the eigenvalue problem in terms of the effective interaction of Eq. (22). It should be noted, however, that the above effective interactions depend on the eigenvalue E_ν. This energy dependence is a serious drawback, since one has different Hamiltonians for different eigenvalues.

Some forty years ago, the theoretical basis for an energy-independent effective Hamiltonian was set down by Brandow in the frame of a time-independent perturbative method [117]. Starting from the degenerate version of the Brillouin-Wigner perturbation theory, the energy terms were expanded out of the energy denominators. Then a rearrangement of the series was performed, leading to a completely linked-cluster expansion. The energy dependence was eliminated by introducing a special type of diagram, the so-called folded diagrams.

A linked-cluster expansion for the shell-model effective interaction was also derived in Refs. [118–121] within the framework of the time-dependent perturbation theory. In the following subsection we shall describe in some detail the time-dependent perturbative approach by Kuo, Lee and Ratcliff [121]. We have tried to give a brief, self-contained presentation of this subject which as matter of fact is rather complex and multi-faceted. To this end, we have discussed the various elements entering this approach without going into the details of the proofs. Furthermore, we have found it useful to first introduce in Sections 3.2.1 and 3.2.2 the concept of folded diagrams and the decomposition theorem, respectively, which are two basic tools for the derivation of the effective interaction, as is shown in Section 3.2.3. A complete review of this approach can be found in [38], to which we refer the reader for details.

3.2. Degenerate time-dependent perturbation theory: Folded-diagram approach

3.2.1. Folded diagrams

In this section, we focus on the case of two-valence nucleons and therefore the nucleus is a doubly closed core, plus two valence nucleons. We denote as active states those SP levels above the core, which are made accessible to the two valence nucleons. The higher-energy SP levels and the filled ones in the core are called passive states. In such a frame the basis vector $|\Phi_i\rangle$ is

$$|\Phi_i\rangle = [a_1^\dagger a_2^\dagger]_i|c\rangle. \tag{23}$$

In the complex time limit, the time-development operator in the interaction representation is given by

$$U(0,-\infty) = \lim_{\epsilon\to 0^+}\lim_{t\to-\infty(1-{\rm i}\epsilon)} U(0,t) = \lim_{\epsilon\to 0^+}\lim_{t\to-\infty(1-{\rm i}\epsilon)} {\rm e}^{{\rm i}Ht}{\rm e}^{-{\rm i}H_0t},$$

which can be expanded as

$$U(0,-\infty) = \lim_{\epsilon\to 0^+}\lim_{t'\to-\infty(1-{\rm i}\epsilon)} \sum_{n=0}^{+\infty}(-{\rm i})^n\int_{t'}^{0}{\rm d}t_1\int_{t'}^{t_1}{\rm d}t_2\cdots\int_{t'}^{t_{n-1}}{\rm d}t_n H_1(t_1)H_1(t_2)\cdots H_1(t_n), \tag{24}$$

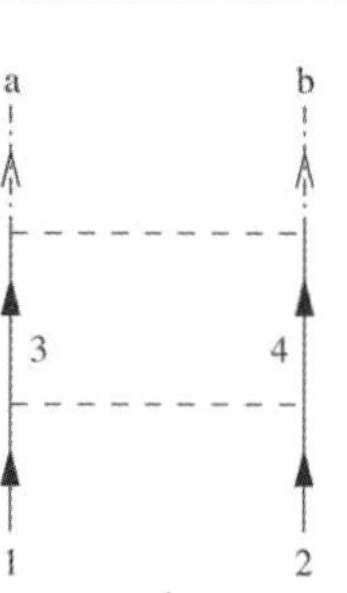

Fig. 1. A second-order time-ordered Goldstone diagram.

where

$$H_1(t) = e^{iH_0t}H_1e^{-iH_0t}.$$

Let us now act on $|\Phi_i\rangle$ with $U(0, -\infty)$:

$$U(0, -\infty)|\Phi_i\rangle = U(0, -\infty)[a_1^\dagger a_2^\dagger]_i|c\rangle. \tag{25}$$

The action of the time-development operator on the unperturbed wave function $|\Phi_i\rangle$ may be represented by an infinite collection of diagrams. The type of diagram we consider here is referred to as time-ordered Goldstone diagrams (for a description of Goldstone diagrams see for instance Ref. [117])

As an example, we show in Fig. 1 a second-order time-ordered Goldstone diagram, which is one of the diagrams appearing in (25). The dashed vertex lines denote V_{NN}-interactions (for the sake of simplicity we take $H_1 = V_{NN}$), a and b represent two passive particle states while 1, 2, 3, and 4 are valence states. From now on in the present section, the passive particle states will be represented by letters and dashed-dotted lines.

The diagram A of Fig. 1 gives a contribution

$$A = a_a^\dagger a_b^\dagger|c\rangle \times \frac{1}{4}V_{ab34}V_{3412}I(A), \tag{26}$$

where $V_{\alpha\beta\gamma\delta}$ are non-antisymmetrized matrix elements of V_{NN}. $I(A)$ is the time integral

$$I(A) = \lim_{\epsilon\to 0^+}\lim_{t'\to-\infty(1-i\epsilon)}(-i)^2\int_{t'}^{0}dt_1\int_{t'}^{t_1}dt_2 e^{-i(\epsilon_3+\epsilon_4-\epsilon_a-\epsilon_b)t_1}e^{-i(\epsilon_1+\epsilon_2-\epsilon_3-\epsilon_4)t_2}, \tag{27}$$

where the ϵ_α's are SP energies. A folded diagram arises upon factorization of diagram A, as shown in Fig. 2. From this figure, we see that diagram B represents a factorization of diagram A into the product of two independent diagrams. The time sequence for diagram A is $0 \geq t_1 \geq t_2 \geq t'$, while in diagram B it is $0 \geq t_1 \geq t'$ and $0 \geq t_2 \geq t'$, with no constraint on the relative ordering of t_1 and t_2.

Therefore, diagrams A and B are not equal, unless subtracting from B the time-incorrect contribution represented by the folded diagram C. It is worth noting that lines 3 and 4 in C are not hole lines, but folded active particle lines. From now on the folded lines will be denoted by drawing a little circle. Explicitly, diagram C is given by

$$C = a_a^\dagger a_b^\dagger|c\rangle \times \frac{1}{4}V_{ab34}V_{3412}I(C), \tag{28}$$

where

$$I(C) = \lim_{\epsilon\to 0^+}\lim_{t'\to-\infty(1-i\epsilon)}(-i)^2\int_{t'}^{0}dt_1\int_{t_1}^{0}dt_2 e^{-i(\epsilon_3+\epsilon_4-\epsilon_a-\epsilon_b)t_1}e^{-i(\epsilon_1+\epsilon_2-\epsilon_3-\epsilon_4)t_2}. \tag{29}$$

Note that the rules to evaluate the folded diagrams are identical to those for standard Goldstone diagrams, except counting as hole lines the folded active lines in the energy denominator.

We now introduce the concept of generalized folded diagram and derive a convenient method to compute it in a degenerate model space. Let us consider, for example, the diagrams shown in Fig. 3. All the three diagrams A, B, and C have identical integrands and constant factors, the only difference being in the integration limits. The integration limits of diagram A correspond to the time ordering $0 \geq t_1 \geq t_2 \geq t_3 \geq t_4 \geq t'$. As pointed out before, the factorization of A into two independent diagrams (diagram B) violates the above time ordering. To correct for the time ordering in B, one has to subtract the folded diagram C, whose time constraints are $0 \geq t_1 \geq t_2 \geq t'$, $0 \geq t_3 \geq t_4 \geq t'$, and $t_3 \geq t_2$. Five different

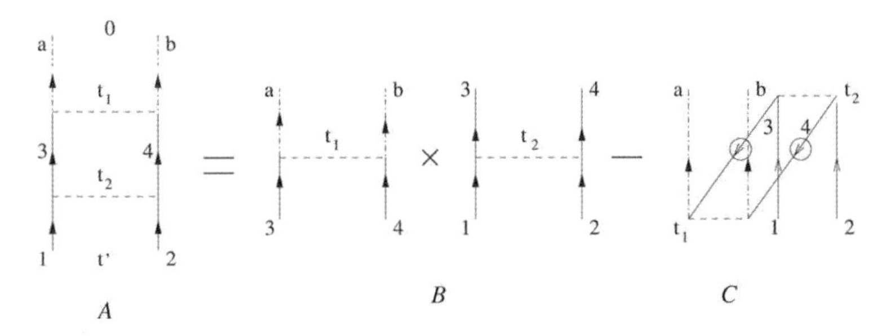

Fig. 2. A diagrammatic identity which defines the folded diagram.

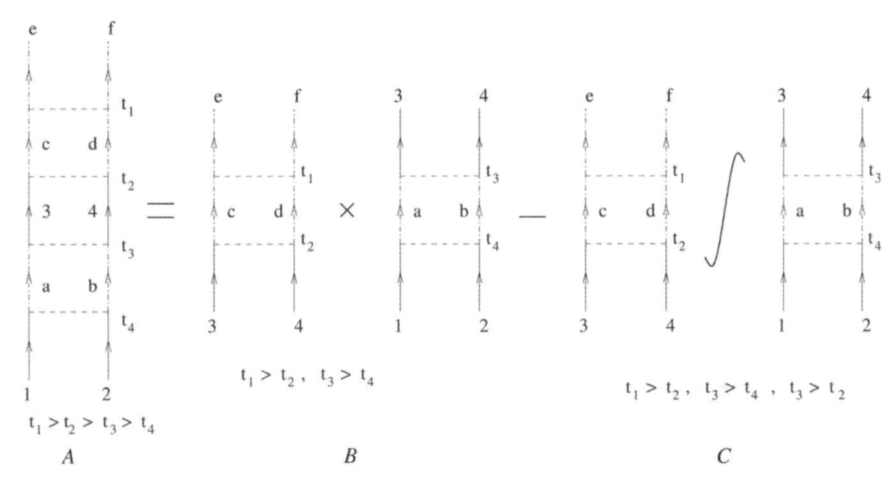

Fig. 3. A diagrammatic identity which illustrates the generalized folded diagram.

time sequences satisfy the above three constraints, thus C consists of five ordinary folded diagrams (see Fig. 4) and is called generalized folded diagram. From now on, the integral sign will denote the generalized folding operation.

An advantageous method to evaluate generalized folded diagrams in a degenerate model space is as follows. Let us consider diagram C of Fig. 3. As pointed out before, A, B, and C have identical integrands and constant factors, so that we may write the time integral $I(C)$ as ([38], pp. 16–18)

$$I(C) = I(B) - I(A) = \frac{1}{\epsilon_1 + \epsilon_2 - \epsilon_3 - \epsilon_4} \left[\frac{1}{(\epsilon_3 + \epsilon_4 - \epsilon_e - \epsilon_f)(\epsilon_3 + \epsilon_4 - \epsilon_c - \epsilon_d)} - \frac{1}{(\epsilon_1 + \epsilon_2 - \epsilon_e - \epsilon_f)(\epsilon_1 + \epsilon_2 - \epsilon_c - \epsilon_d)} \right] \frac{1}{\epsilon_1 + \epsilon_2 - \epsilon_a - \epsilon_b}. \tag{30}$$

In a degenerate model space, the first factor is infinite, while the second is zero. However, $I(C)$ can be determined by a limiting procedure. If we write $\epsilon_1 + \epsilon_2 = \epsilon_3 + \epsilon_4 + \Delta$, where $\Delta \to 0$, we can put Eq. (30) into the form

$$\begin{aligned} I(C) = {} & \lim_{\Delta \to 0} \frac{1}{\Delta} \left[\frac{1}{(\epsilon_3 + \epsilon_4 - \epsilon_e - \epsilon_f)(\epsilon_3 + \epsilon_4 - \epsilon_c - \epsilon_d)} - \frac{1}{(\epsilon_3 + \epsilon_4 - \epsilon_e - \epsilon_f + \Delta)(\epsilon_3 + \epsilon_4 - \epsilon_c - \epsilon_d + \Delta)} \right] \\ & \times \frac{1}{(\epsilon_3 + \epsilon_4 - \epsilon_a - \epsilon_b + \Delta)} \\ = {} & \left[\frac{1}{(\epsilon_3 + \epsilon_4 - \epsilon_e - \epsilon_f)^2 (\epsilon_3 + \epsilon_4 - \epsilon_c - \epsilon_d)} + \frac{1}{(\epsilon_3 + \epsilon_4 - \epsilon_e - \epsilon_f)(\epsilon_3 + \epsilon_4 - \epsilon_c - \epsilon_d)^2} \right] \\ & \times \frac{1}{(\epsilon_3 + \epsilon_4 - \epsilon_a - \epsilon_b)}. \end{aligned} \tag{31}$$

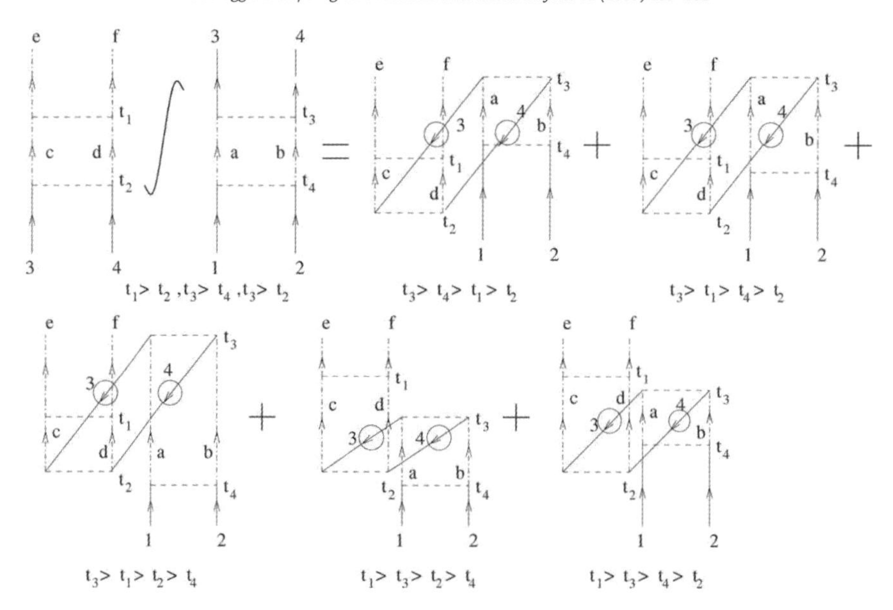

Fig. 4. Generalized folded diagram expressed as sum of ordinary folded diagrams.

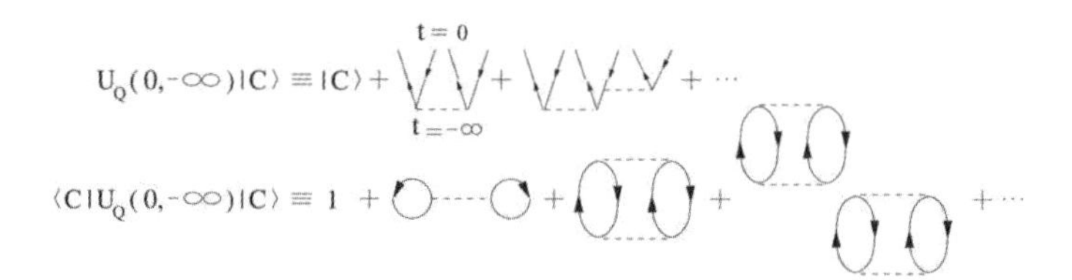

Fig. 5. Diagrammatic representation of $U_Q(0,-\infty)|c\rangle$ and $\langle c|U(0,-\infty)|c\rangle$.

Thus, the energy denominator of the generalized folded diagram C can be expressed as the derivative of the energy denominator of the l.h. part of the diagram with respect to the energy variable ω, calculated at $\epsilon_1+\epsilon_2$:

$$I(C) = -\frac{1}{\epsilon_1+\epsilon_2-\epsilon_a-\epsilon_b}\frac{\mathrm{d}}{\mathrm{d}\omega}\left(\frac{1}{\omega-\epsilon_c-\epsilon_d}\frac{1}{\omega-\epsilon_e-\epsilon_f}\right)_{\omega=\epsilon_1+\epsilon_2}. \tag{32}$$

The last equation will prove to be very useful to evaluate the folded diagrams.

3.2.2. The decomposition theorem

Let us consider again the wave function $U(0,-\infty)|\Phi_i\rangle$. We can rewrite it as

$$U(0,-\infty)|\Phi_i\rangle = U_L(0,-\infty)|\Phi_i\rangle \times U(0,-\infty)|c\rangle, \tag{33}$$

where the subscript L indicates that all the H_1 vertices in $U_L(0,-\infty)|\Phi_i\rangle$ are valence linked, i.e. are linked directly or indirectly to at least one of the valence lines. We now factorize each of the two terms on the r.h.s. of Eq. (33) in order to write $U(0,-\infty)|\Phi_i\rangle$ in a form useful for the derivation of the model-space secular equation.

We first consider $U(0,-\infty)|c\rangle$, which can rewritten as

$$U(0,-\infty)|c\rangle = U_Q(0,-\infty)|c\rangle \times \langle c|U(0,-\infty)|c\rangle, \tag{34}$$

where $U_Q(0,-\infty)|c\rangle$ denotes the collection of diagrams in which every vertex is connected to the time $t=0$ boundary. In fact, $U_Q(0,-\infty)|c\rangle$ is proportional to the true ground-state wave function of the closed-shell system, while $\langle c|U(0,-\infty)|c\rangle$ represents all the vacuum fluctuation diagrams. These two terms are illustrated in the first and second line of Fig. 5, respectively.

A similar factorization of $U_L(0,-\infty)|\Phi_i\rangle$ can be performed, which can be expressed in terms of the so-called $\hat{Q}$-boxes. The $\hat{Q}$-box, which should not be confused with the projection operator Q introduced in Section 3.1, is defined as the sum of all

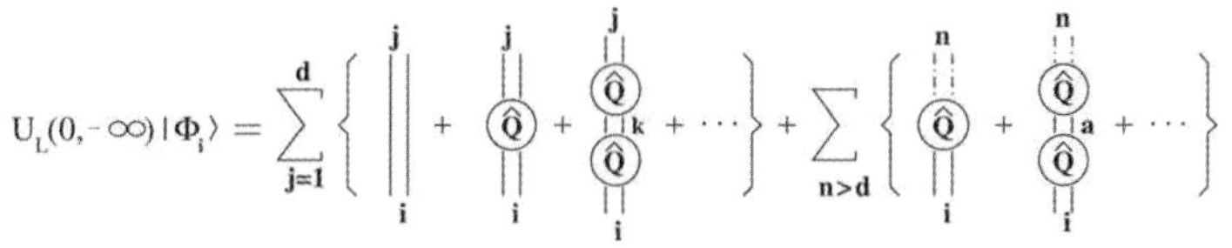

Fig. 6. Diagrammatic representation of $U_L(0, -\infty)|\Phi_i\rangle$.

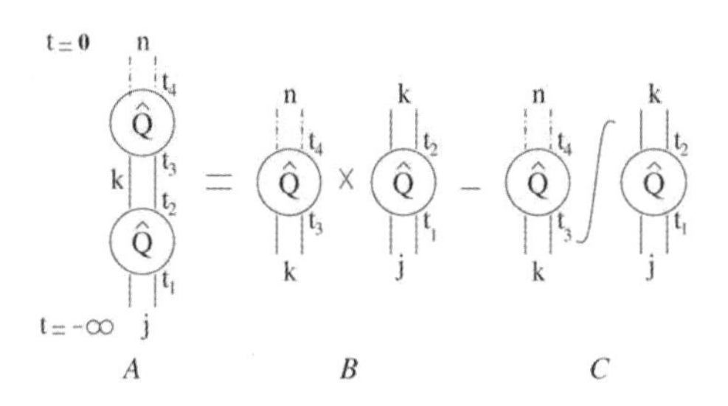

Fig. 7. Factorization of a 2-$\hat{Q}$-box sequence.

diagrams that have at least one H_1-vertex, are valence linked and irreducible (i.e., with at least one passive line between two successive vertices). Clearly, $U_L(0, -\infty)|\Phi_i\rangle$ must terminate either in an active or passive state at $t = 0$, thus we can write

$$U_L(0, -\infty)|\Phi_i\rangle = |\chi_i^P\rangle + |\chi_i^Q\rangle, \tag{35}$$

as shown in Fig. 6. Note that in this figure the intermediate indices k, a, ... represent summations over all P-space states.

It is also possible to factorize out of $|\chi_i^Q\rangle$ a term belonging to $|\chi_i^P\rangle$, by means of folded-diagram factorization. In Fig. 7 we show, as an example, how a 2-$\hat{Q}$-box sequence can be factorized.

It is worth noting the similarity between Figs. 7 and 3. As a matter of fact, when factorizing diagram A time incorrect contributions arise in diagram B, that are compensated by subtracting from it the generalized folded diagram C.

Using the generalized folding procedure, we are able to factorize out of each term in $|\chi_i^Q\rangle$ a diagram belonging to $|\chi_i^P\rangle$ (see Fig. 8). Applying this factorization to all the terms in $|\chi_i^Q\rangle$, collecting columnwise the diagrams on the r.h.s. in Fig. 8 and adding them up, we may represent $|\chi_i^Q\rangle$ as shown in Fig. 9.

The collection of diagrams in the upper parenthesis of Fig. 9 is simply $\langle\Phi_j|U_L(0, -\infty)|\Phi_i\rangle$, which, according to Eq. (35), is related to $|\chi_i^P\rangle$ by

$$|\chi_i^P\rangle = \sum_{j=1}^{d} |\Phi_j\rangle\langle\Phi_j|U_L(0, -\infty)|\Phi_i\rangle. \tag{36}$$

Therefore, taking into account Figs. 6 and 9, and Eq. (36) we can express $U_L(0, -\infty)|\Phi_i\rangle$ as

$$U_L(0, -\infty)|\Phi_i\rangle = \sum_{j=1}^{d} U_{QL}(0, -\infty)|\Phi_j\rangle\langle\Phi_j|U_L(0, -\infty)|\Phi_i\rangle, \tag{37}$$

where we have represented diagrammatically $U_{QL}(0, -\infty)|\Phi_j\rangle$ in Fig. 10.

Eqs. (33), (34) and (37) define the decomposition theorem, which we rewrite as follows:

$$U(0, -\infty)|\Phi_i\rangle = \sum_{j=1}^{d} U_Q(0, -\infty)|\Phi_j\rangle\langle\Phi_j|U(0, -\infty)|\Phi_i\rangle, \tag{38}$$

where

$$\langle\Phi_j|U(0, -\infty)|\Phi_i\rangle = \langle\Phi_j|U_L(0, -\infty)|\Phi_i\rangle \times \langle c|U(0, -\infty)|c\rangle, \tag{39}$$

and

$$U_Q(0, -\infty)|\Phi_j\rangle = U_{QL}(0, -\infty)|\Phi_j\rangle \times U_Q(0, -\infty)|c\rangle. \tag{40}$$

The decomposition theorem, as given by Eq. (38), states that the action of $U(0, -\infty)$ on $|\Phi_i\rangle$ can be represented as the sum of the wave functions $U_Q(0, -\infty)|\Phi_j\rangle$ weighted with the matrix elements $\langle\Phi_j|U(0, -\infty)|\Phi_i\rangle$. Eq. (38) will play a crucial role in the next Section 3.2.3, where we shall give an expression for the shell-model effective Hamiltonian.

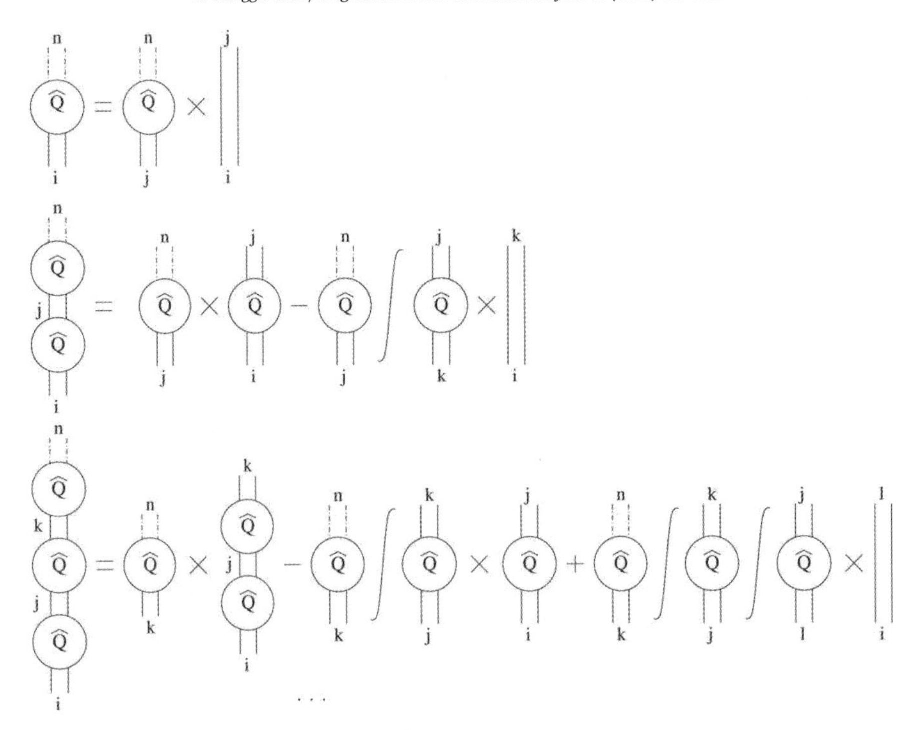

Fig. 8. Factorization of $|\chi_i^Q\rangle$ using the generalized folding procedure.

Fig. 9. $|\chi_i^Q\rangle$ wave function.

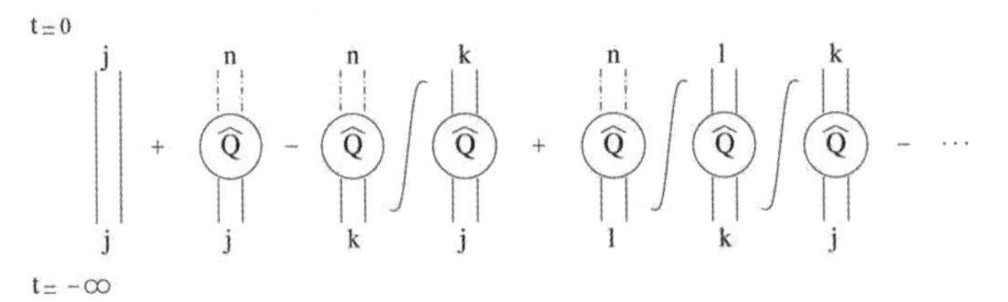

Fig. 10. Diagrammatic representation of $U_{QL}(0,-\infty)|\Phi_j\rangle$.

3.2.3. The model-space secular equation

For the sake of clarity, let us recall the model-space secular Eq. (11)

$$PH_{\mathrm{eff}}P|\Psi_\alpha\rangle = E_\alpha P|\Psi_\alpha\rangle,$$

where $\alpha = 1, \ldots, d$, $|\Psi_\alpha\rangle$ and E_α are the true eigenvectors and eigenvalues of the full Hamiltonian H.

From now on, we shall use, for the convenience of the proof, the Schrödinger representation. However, it is worth to point out that the results obtained hold equally well in the interaction picture.

First of all, we establish a one-to-one correspondence between some model-space parent states $|\rho_\alpha\rangle$ and d true eigenfunctions $|\Psi_\alpha\rangle$. Let us start with a trial parent state

$$|\rho_1\rangle = \frac{1}{\sqrt{d}}\sum_{i=1}^{d}|\Phi_i\rangle, \tag{41}$$

and act with the time development operator $U(0,-\infty)$ on it. More precisely, we construct the wave function

$$\frac{U(0,-\infty)|\rho_1\rangle}{\langle\rho_1|U(0,-\infty)|\rho_1\rangle} = \lim_{\epsilon\to 0^+}\lim_{t'\to-\infty(1-\mathrm{i}\epsilon)}\frac{\mathrm{e}^{\mathrm{i}Ht'}|\rho_1\rangle}{\langle\rho_1|\mathrm{e}^{\mathrm{i}Ht'}|\rho_1\rangle}. \tag{42}$$

By inserting a complete set of eigenstates of H between the time evolution operator and $|\rho_1\rangle$, we obtain

$$\frac{U(0,-\infty)|\rho_1\rangle}{\langle\rho_1|U(0,-\infty)|\rho_1\rangle} = \lim_{\epsilon\to 0^+}\lim_{t'\to-\infty}\frac{\sum_\lambda \mathrm{e}^{\mathrm{i}E_\lambda t'}\mathrm{e}^{E_\lambda\epsilon t'}|\Psi_\lambda\rangle\langle\Psi_\lambda|\rho_1\rangle}{\sum_\beta \mathrm{e}^{\mathrm{i}E_\beta t'}\mathrm{e}^{E_\beta\epsilon t'}\langle\rho_1|\Psi_\beta\rangle\langle\Psi_\beta|\rho_1\rangle} = \frac{|\Psi_1\rangle}{\langle\rho_1|\Psi_1\rangle} \equiv |\tilde{\Psi}_1\rangle. \tag{43}$$

Here, $|\Psi_1\rangle$ is the lowest eigenstate of H for which $\langle\Psi_1|\rho_1\rangle \neq 0$, this stems from the fact that the real exponential damping factor in the above equation suppresses all the other non-vanishing terms.

This procedure can be easily continued, thus obtaining a set of wave functions

$$|\tilde{\Psi}_\alpha\rangle = \frac{U(0,-\infty)|\rho_\alpha\rangle}{\langle\rho_\alpha|U(0,-\infty)|\rho_\alpha\rangle}, \tag{44}$$

where

$$\langle\rho_\alpha|\rho_\alpha\rangle = 1, \tag{45}$$

$$\langle\rho_\alpha|\Psi_\alpha\rangle \neq 0, \tag{46}$$

$$\langle\rho_\alpha|\Psi_1\rangle = \langle\rho_\alpha|\Psi_2\rangle = \cdots = \langle\rho_\alpha|\Psi_{\alpha-1}\rangle = 0. \tag{47}$$

The above correspondence Eqs. (44)–(47) holds if the parent states $|\rho_\alpha\rangle$ are linearly independent. Under this assumption, we can write

$$|\rho_\alpha\rangle = \sum_{i=1}^{d} C_i^\alpha|\Phi_i\rangle, \tag{48}$$

with

$$\sum_{i=1}^{d} C_i^\alpha C_i^\beta = \delta_{\alpha\beta}. \tag{49}$$

By construction, $|\tilde{\Psi}_\alpha\rangle$ is an eigenstate of H, so, using Eq. (44), we can write

$$H\frac{U(0,-\infty)|\rho_\alpha\rangle}{\langle\rho_\alpha|U(0,-\infty)|\rho_\alpha\rangle} = E_\alpha\frac{U(0,-\infty)|\rho_\alpha\rangle}{\langle\rho_\alpha|U(0,-\infty)|\rho_\alpha\rangle}. \tag{50}$$

Now, making use of Eq. (48) and applying the decomposition theorem as expressed by Eq. (38), the above equation becomes

$$H\frac{\sum_{ij} U_Q(0,-\infty)|\Phi_j\rangle\langle\Phi_j|U(0,-\infty)|\Phi_i\rangle C_i^\alpha}{\sum_{km}\langle\Phi_k|U(0,-\infty)|\Phi_m\rangle C_k^\alpha C_m^\alpha} = E_\alpha\frac{\sum_{ij} U_Q(0,-\infty)|\Phi_j\rangle\langle\Phi_j|U(0,-\infty)|\Phi_i\rangle C_i^\alpha}{\sum_{km}\langle\Phi_k|U(0,-\infty)|\Phi_m\rangle C_k^\alpha C_m^\alpha}. \tag{51}$$

In order to simplify the above expression, we define the coefficients b_j^α

$$b_j^\alpha = \frac{\sum_i\langle\Phi_j|U(0,-\infty)|\Phi_i\rangle C_i^\alpha}{\sum_{km}\langle\Phi_k|U(0,-\infty)|\Phi_m\rangle C_k^\alpha C_m^\alpha} = \frac{\sum_i\langle\Phi_j|U_L(0,-\infty)|\Phi_i\rangle C_i^\alpha}{\sum_{km}\langle\Phi_k|U_L(0,-\infty)|\Phi_m\rangle C_k^\alpha C_m^\alpha}, \tag{52}$$

where the r.h.s. of the above equation has been obtained by use of Eq. (39), which cancels out the vacuum fluctuations diagrams $\langle c|U(0,-\infty)|c\rangle$ of Fig. 5. Multiplying Eq. (51) by $\langle\Phi_k|$, it becomes

$$\sum_{j=1}^{d}\langle\Phi_k|HU_Q(0,-\infty)|\Phi_j\rangle b_j^\alpha = E_\alpha b_k^\alpha, \tag{53}$$

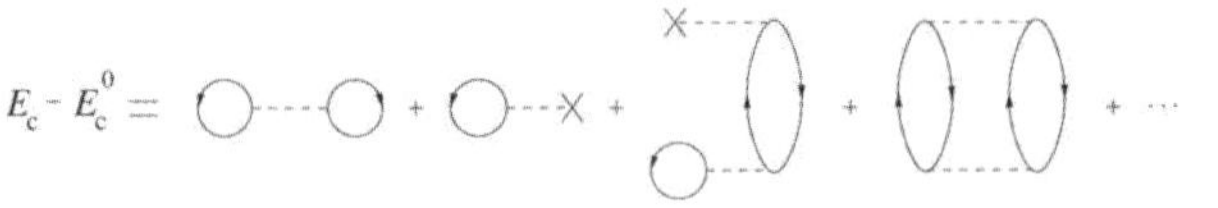

Fig. 11. Goldstone linked-diagram expansion of $E_c - E_c^0$. The cross insertions represent the H_0 operator.

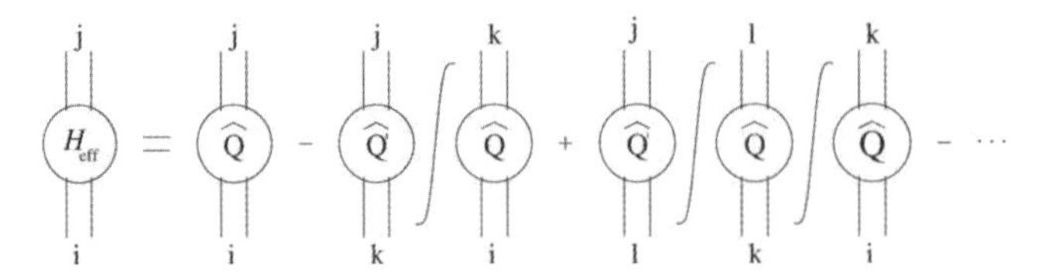

Fig. 12. Diagrammatic representation of $H_{\rm eff}$ matrix elements.

where use has been made of the relation $\langle\Phi_k|U_Q(0,-\infty)|\Phi_j\rangle = \delta_{kj}$ (see Ref. [121]). The above equation is the model-space secular equation we needed, where $H_{\rm eff}$ is given by $HU_Q(0,-\infty)$ and b_j^α represents the projection of $|\tilde{\Psi}_\alpha\rangle$ onto the model-space wave function $|\Phi_j\rangle$.

In Eq. (53) we can write the Hamiltonian H as $H = H_0 + H_1$. First, let us consider the contribution from H_0. Since $|\Phi_i\rangle$ is an eigenstate of H_0 and $\langle\Phi_i|U_Q(0,-\infty)|\Phi_j\rangle = \delta_{ij}$, we obtain

$$\langle\Phi_i|H_0U_Q(0,-\infty)|\Phi_j\rangle = \langle\Phi_i|H_0|\Phi_j\rangle = \delta_{ij}(E_c^0 + E_v^0), \tag{54}$$

where E_c^0 is the unperturbed core energy, and E_v^0 is the unperturbed energy of the two valence nucleons with respect to E_c^0.

As for the matrix element $\langle\Phi_i|H_1U_Q(0,-\infty)|\Phi_j\rangle$, we see from inspection of Eq. (40) that it contains a collection of diagrams in which H_1 is not linked to any valence line at $t = 0$. These diagrams are obtained acting with H_1 on $U_Q(0,-\infty)|c\rangle$ and their contribution to the l.h.s. of Eq. (53) is $\delta_{ij}\langle c|H_1U_Q(0,-\infty)|c\rangle = \delta_{ij}(E_c - E_c^0)$, E_c being the true ground-state energy of the closed-shell system. The diagram expansion of $(E_c - E_c^0)$ is given in Fig. 11 as illustrated in [122].

The other terms of $\langle\Phi_i|H_1U_Q(0,-\infty)|\Phi_j\rangle$ are all linked to the external active lines. By denoting, for simplicity, the collection of these terms as

$$\langle\Phi_i|[H_1U_Q(0,-\infty)]_L|\Phi_j\rangle, \tag{55}$$

the secular Eq. (53) can be rewritten in the following form:

$$E_v^0 + \sum_{j=1}^{d}\langle\Phi_i|[H_1U_Q(0,-\infty)]_L|\Phi_j\rangle b_j^\alpha = (E_\alpha - E_c)b_i^\alpha. \tag{56}$$

We now define

$$H_{\rm eff}^1 = [H_1U_Q(0,-\infty)]_L, \tag{57}$$

and show in Fig. 12 a diagrammatic representation of its matrix elements, which has been obtained starting from the definition of $U_{QL}(0,-\infty)|\Phi_j\rangle$ given in Fig. 10. It should be noted that we have two kinds of $\hat{Q}$-box, $\hat{Q}$ and $\hat{Q}'$. $\hat{Q}$ and $\hat{Q}'$ are a collection of irreducible, valence-linked diagrams with at least one and two H_1-vertices, respectively. The fact that the lowest order term in $\hat{Q}'$ is of second order in H_1 is just because of the presence of H_1 in the matrix elements of Eq. (55).

Formally, $H_{\rm eff}^1$ can be written in operator form as

$$H_{\rm eff}^1 = \hat{Q} - \hat{Q}'\int\hat{Q} + \hat{Q}'\int\hat{Q}\int\hat{Q} - \hat{Q}'\int\hat{Q}\int\hat{Q}\int\hat{Q} + \cdots, \tag{58}$$

where the integral sign represents a generalized folding operation. It is worth noting that, by definition, the $\hat{Q}$-box contains diagrams at any order in H_1. Actually, when performing realistic shell-model calculations it is customary to include diagrams up to a finite order in H_1. A complete list of all the $\hat{Q}$-box diagrams up to third order can be found in Ref. [45].

The Schrödinger Eq. (6) is finally reduced to the model-space eigenvalue problem of Eq. (56), whose eigenvalues are the energies of the A-nucleus relative to the core ground-state energy E_c. As mentioned above, the latter can be calculated by way of the Goldstone expansion [122], expressed as the sum of diagrams shown in Fig. 11. It is also worth noting that the operator $H_{\rm eff}^1$, as defined by Eq. (58), contains both one- and two-body contributions since in our derivation of the of Eq. (56) we have considered nuclei with two-valence nucleons. All the one-body contributions, the so-called $\hat{S}$-box [123], once summed to the eigenvalues of H_0, E_v^0, give the SP term of the effective shell-model Hamiltonian. The eigenvalues of this term represent the energies of the nucleus with one-valence nucleon relative to the core. This identification justifies the commonly used subtraction procedure [123] where only the two-body terms of $H^1_{\rm eff}$ (the effective two-body interaction $V_{\rm eff}$) are retained, while the single-particle energies are taken from experiment.

In this context, it is worth to mentioning that in some recent papers [124,125] the SP energies and the two-body interaction employed in realistic shell-model calculations are derived consistently in the framework of the linked-cluster expansion. In particular, in Ref. [124], where the light p-shell nuclei have been studied using the CD-Bonn potential renormalized through the $V_{\rm low-k}$ procedure (see Section 4.2), a Hartree–Fock basis is derived which is used to calculate the binding energy of ^{4}He and the effective shell-model Hamiltonian composed of one- and two-body terms.

In concluding this brief discussion of Eq. (56), it is worth pointing out that for systems with more than two valence nucleons $H^1_{\rm eff}$ contains 1-, 2-, 3-, ..., n-body components, even if the original Hamiltonian of Eq. (7) contains only a two-body force [126]. The role of these effective many-body forces as well as that of a genuine 3-body potential in the shell model is still an open problem (see [127] and references therein) and is outside the scope of this review. However, we shall come back to this point in Section 5.3.4.

In Section 3.2.1 we have shown how, in a degenerate model space, a generalized folding diagram can be evaluated. In [128], it has been shown that a term like $-\hat{Q}' \int \hat{Q}$ may be written as

$$-\hat{Q}' \int \hat{Q} = \frac{{\rm d}\hat{Q}'(\omega)}{{\rm d}\omega}\hat{Q}(\omega), \tag{59}$$

where ω is equal to the energy of the incoming particles.

The above result may be extended to obtain a convenient prescription to calculate $H^1_{\rm eff}$ as given by Eq. (58). We can write

$$H^1_{\rm eff} = \sum_{i=0}^{\infty} F_i, \tag{60}$$

where

$$\begin{aligned} F_0 &= \hat{Q}, \\ F_1 &= \hat{Q}_1\hat{Q}, \\ F_2 &= \hat{Q}_2\hat{Q}\hat{Q} + \hat{Q}_1\hat{Q}_1\hat{Q}, \\ &\dots \end{aligned} \tag{61}$$

and

$$\hat{Q}_m = \frac{1}{m!}\frac{d^m\hat{Q}(\omega)}{d\omega^m}\bigg|_{\omega=\omega_0}, \tag{62}$$

ω_0 being the energy of the incoming particles at $t = -\infty$.

Note that in Eq. (61) we have made use of the fact that, by definition,

$$\frac{{\rm d}\hat{Q}'(\omega)}{{\rm d}\omega} = \frac{{\rm d}\hat{Q}(\omega)}{{\rm d}\omega}. \tag{63}$$

The number of terms in F_i grows dramatically with i. Two iteration methods to partially sum up the folded diagram series have been introduced in Refs. [37,129,130]. These methods are known as the Krenciglowa–Kuo (KK) and the Lee–Suzuki (LS) procedure, respectively. In [130], it has been shown that, when converging, the KK partial summation converges to those states with the largest model space overlap, while the LS one converges to the lowest states in energy.

The LS iteration procedure was proposed within the framework of an approach to the construction of the effective interaction known as Lee–Suzuki method, which is based on the similarity transformation theory. In the following subsection we shall briefly present this method to illustrate the LS iterative technique used to sum up the folded-diagram series (60).

3.3. *The Lee–Suzuki method*

Let us start with the Schrödinger equation for the A-nucleon system as given in Eq. (6) and consider the similarity transformation

$$\mathcal{H} = X^{-1}HX, \tag{64}$$

where X is a transformation operator defined in the whole Hilbert space.

If we require that

$$Q\mathcal{H}P = 0, \tag{65}$$

then it can be easily proved that the P-space effective Hamiltonian satisfying Eq. (11) is just $P\mathcal{H}P$. Eq. (65) is the so-called decoupling equation, whose solution leads to the determination of $H_{\rm eff}$.

There is, of course, more than one choice for the transformation operator X. We take

$$X = e^{\Omega}, \tag{66}$$

where the wave operator Ω satisfies the conditions:

$$\Omega = Q\Omega P, \tag{67}$$

$$P\Omega P = Q\Omega Q = P\Omega Q = 0. \tag{68}$$

Taking into account Eq. (67), we can write $X = 1 + \Omega$ and consequently

$$H_{\text{eff}} = P\mathcal{H}P = PHP + PH_1Q\Omega, \tag{69}$$

with H_1 defined in Eq. (9), while the decoupling Eq. (65) becomes

$$QH_1P + QHQ\Omega - \Omega PHP - \Omega PH_1Q\Omega = 0. \tag{70}$$

We now introduce the P-space effective interaction R by subtracting the unperturbed energy PH_0P from H_{eff}:

$$R = PH_1P + PH_1Q\Omega. \tag{71}$$

In a degenerate model space $PH_0P = \omega_0 P$, we can consider a linearized iterative equation for the solution of the decoupling Eq. (70)

$$[E_0 - (QHQ - \Omega_{n-1}PH_1Q)]\Omega_n = QH_1P - \Omega_{n-1}PH_1P. \tag{72}$$

We now write the $\hat{Q}$-box introduced in Section 3.2.2 in operatorial form as given in [130]

$$\hat{Q} = PH_1P + PH_1Q\frac{1}{E_0 - QHQ}QH_1P, \tag{73}$$

and define R_n as the n-th order iterative effective interaction

$$R_n = PH_1P + PH_1Q\Omega_n. \tag{74}$$

Then, if we start with $\Omega_0 = 0$ in Eq. (72), R_n can be written in terms of the $\hat{Q}$-box and its derivatives:

$$R_1 = \hat{Q},$$

$$R_2 = [1 - \hat{Q}_1]^{-1}\hat{Q},$$

$$R_3 = [1 - \hat{Q}_1 - \hat{Q}_2R_2]^{-1}\hat{Q},$$

$$\cdots$$

$$R_n = \left[1 - \hat{Q}_1 - \sum_{m=2}^{n-1}\hat{Q}_m \prod_{k=n-m+1}^{n-1} R_k\right]^{-1}\hat{Q} \tag{75}$$

where $\hat{Q}_n$ is defined in Eq. (62).

The solution R_n so obtained corresponds to a certain resummation of the folded diagrams to infinite order. In fact, if we consider, for example, R_2 and expand it in power series of $\hat{Q}_1$, we obtain

$$R_2 = [1 - \hat{Q}_1]^{-1}\hat{Q} = 1 + \hat{Q}_1\hat{Q} + \hat{Q}_1\hat{Q}_1\hat{Q} + \cdots \tag{76}$$

It is clear from the above expression that R_2 contains terms corresponding to an infinite number of folds.

It can be shown that the Lee–Suzuki method yields converged results after a small number of iterations [123], making this procedure very advantageous to sum up the folded-diagram series.

4. Handling the short-range repulsion of the *NN* potential

As already pointed out, a most important goal of nuclear shell-model theory is to derive the effective interaction between valence nucleons directly from the free *NN* potential. In Section 3, we have shown how this effective interaction may be calculated microscopically within the framework of a many-body theory. As is well known, however, V_{NN} is not suitable for this kind of approach. In fact, owing to the contribution from the repulsive core, the matrix elements of V_{NN} are generally very large, and an order-by-order perturbative calculation of the effective interaction in terms of V_{NN} is clearly not meaningful. A resummation method has to be employed in order to take care of the strong short-range repulsion contained in V_{NN}.

This point is more evident, if we consider the extreme model of a hard-core potential, as is the case of the potential models developed in the early 1960s. This situation is illustrated in Fig. 13. For these potentials, the perturbation expansion of the effective interaction in terms of V_{NN} is meaningless, since each term of the series involving matrix elements of the *NN* potential between unperturbed two-body states is infinite. This is because the unperturbed wave function, in contrast to the true wave function, gives a non-zero probability of finding a particle located inside the hard-core distance.

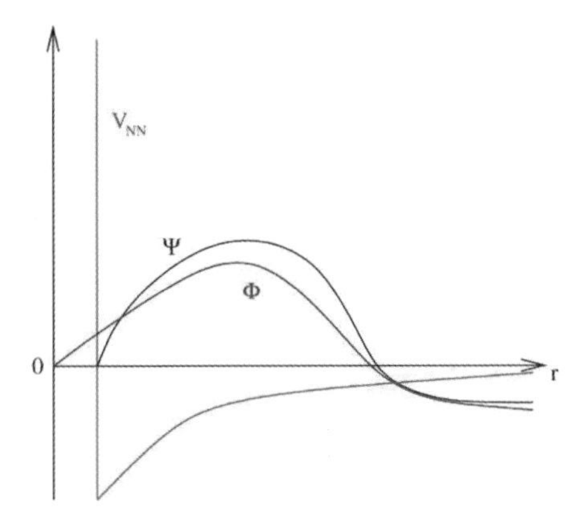

Fig. 13. Radial dependence of the relative wave function Ψ of two nucleons interacting via a hard-core potential V_{NN}. Φ refers to the uncorrelated wave function.

The traditional way out of this problem is the so-called Brueckner reaction matrix G, which is based on the idea of treating exactly the interaction between a given pair of nucleons [131]. The G matrix, defined as a sum of all ladder-type interactions (see Section 4.1.2), is used to replace the NN interaction vertices once a rearrangement of the effective interaction perturbative series has been performed.

Recently, a new method to renormalize the NN interaction has been proposed [42,43]. A low-momentum model space defined up to a cutoff momentum Λ is introduced and an effective potential $V_{\text{low-k}}$ is derived from V_{NN}. This $V_{\text{low-k}}$ satisfies a decoupling condition between the low- and high-momentum spaces. Moreover, it is a smooth potential which preserves exactly the on-shell properties of the original potential and it is thus suitable to advantageously replace V_{NN} in realistic many-body calculations.

Sections 4.1 and 4.2 are devoted to the description of the reaction G matrix and $V_{\text{low-k}}$ potential, respectively.

Before doing this, however, it may be worth recalling that a method to avoid the G-matrix treatment to eliminate effects of the repulsive core in the NN potential was proposed in the late 1960s [132,40]. As already mentioned in the Introduction, this consists in using the experimental NN phase shifts to deduce matrix elements of the NN potential in a basis of relative harmonic oscillator states. These matrix elements, which have become known as the Sussex matrix elements (SME), have been used in several nuclear structure calculations, but the agreement with experiment has been generally only semi-quantitative. A comparison between the results obtained by Sinatkas et al. [133] using the SME and those obtained with a realistic effective interaction derived from the Bonn A potential is made for the $N = 50$ isotones in [134].

4.1. The Brueckner G-matrix approach

4.1.1. Historical introduction

The concept of G matrix originates from the theory of multiple scattering of Watson [135,136]. In this approach, the elastic scattering of a fast particle by a nucleus was described by way of a transformed potential obtained in terms of the Lippmann–Schwinger matrix [137] for two-body scattering. The procedure of Watson for constructing such an "equivalent two-body potential" was generalized to the study of nuclear many-body systems by Brueckner and co-workers [131,138]. They introduced a reaction matrix for the scattering of two nucleons while they are moving in the nuclear medium. This matrix, which is known as the Brueckner reaction matrix, includes all two-particle correlations via summing all ladder-type interactions, and made it possible to perform Hartree–Fock self-consistent calculations for nuclear matter.

Only a few years later Goldstone [122] proved a new perturbation formula for the ground-state energy of nuclear matter, which gave the formal basis of the Brueckner theory. The Goldstone linked-diagram theory applies to systems with non-degenerate ground state, which is the case of nuclear matter as well as of closed-shell finite nuclei.

The Brueckner theory was seen to be the key to solve the paradox on which the attention of many nuclear physicists was focused during the early 1950s. In fact, it allowed one to reconcile a description of the nucleus in terms of an overall potential and the peculiar features of the two-body nuclear force. This was very well evidenced in the paper by Bethe [139], whose main purpose was indeed to establish that the Brueckner theory provided the theoretical foundation for the shell model.

After their first works, Brueckner and co-workers published a series of papers on the same subject [140–144], where further analyses and developments of the method, as well as numerical calculations, were given. Important advances as regards nuclear matter were provided by the work of Bethe and Goldstone [145] and Bethe et al. [146]. For several years up to the 1960s, nuclear physicists were indeed very active in this field, as may be seen from the review papers by Day [147], Rajaraman and Bethe [148], and Baranger [149], where comprehensive lists of references can be found. For a recent review of the developments in this area made Bethe and coworkers we refer to [150].

As regards open-shell nuclei, we have shown in the previous section that the model-space effective interaction may be obtained by way of a linked-diagram expansion containing both folded and non-folded Goldstone diagrams. An order-by-order perturbative calculation of such diagrams in terms of V_{NN} is not appropriate, and one has to resort again to the reaction matrix G.

The renormalization of V_{NN} through the Brueckner theory has been the standard procedure to derive realistic effective interactions since the pioneering works of the early 1960s. A survey of shell-model calculations employing the G matrix is given in Sections 5.1 and 5.2.

In Section 4.1.2 we shall define the reaction matrix G, and discuss the main problems related to its definition. Then, in Section 4.1.3, we shall address the problem of how the reaction matrix may be calculated, focusing attention on the G_T matrix for which plane waves are used as intermediate states. It is, in fact, this matrix which is most commonly used nowadays in realistic shell-model calculations. For simplicity, almost everywhere in this section we shall denote the NN potential by V.

4.1.2. Essentials of the theory

In the literature, the G matrix is typically introduced by way of the Goldstone expansion for the calculation of the ground-state energy in nuclear matter and closed shell nuclei. As a first step, the ground-state energy is written as a linked-cluster perturbation series. Then, all diagrams differing one from another only in the number of V interactions between two particles lines are summed. This corresponds to define a well-behaved two-body operator, the reaction matrix G, that replaces the potential V in the series. A very clear and simple presentation of the G matrix along this line is provided in the paper by Day [147].

Here, we shall not discuss the details of the G-matrix theory, but simply give the elements needed to make clear its definition in connection with the derivation of the model-space effective interaction for open-shell nuclei. Therefore, in the following we refer, as in the previous section, to an A-nucleon system with the Hamiltonian given by Eqs. (7)–(9) and represented as a doubly closed core plus valence nucleons moving in a limited number of SP orbits above the core.

The two-body operator G is defined by the integral equation

$$G(\omega) = V + V \frac{Q_{2p}}{\omega - H_0} G(\omega), \tag{77}$$

where ω is an energy variable known as "starting energy" and Q_{2p} an operator which projects onto particle–particle states, namely states composed of two SP levels above the doubly closed core. As we shall discuss later, this operator may be chosen in different ways depending on the specific context in which it is used. Here it is worth noting that its presence in Eq. (77) reminds us that the G matrix, differently from the Lippmann–Schwinger T matrix, is defined in the nuclear medium. It may be also noted that the starting energy ω is not a free parameter. Rather, its value is determined by the physical problem being studied. For the T matrix, ω denotes the scattering energy for two particles in free space, but depends on the nuclear medium for the G matrix. In fact, we shall see that for a G matrix to be used within the linked-diagram expansion of the effective interaction ω depends on the diagram where G appears. Only in some cases it represents the energy of the two-particle incoming state, which was instead the meaning of the energy variable ω used in Section 3.

Let us start by writing the Q_{2p} operator as

$$Q_{2p} = \sum_{ij} C(ij) |\phi_{ij}\rangle\langle\phi_{ij}|, \tag{78}$$

where the state $|\phi_{ij}\rangle$, which is the antisymmetrized product of the two SP states $|i\rangle$ and $|j\rangle$, is an eigenstate of H_0 with energy $\epsilon_i + \epsilon_j$. The constant $C(ij)$ is 1 if the state $|\phi_{ij}\rangle$ pertains to the space defined by Q_{2p}, 0 otherwise. We now introduce the operator P_{2p} which projects on the complementary space

$$P_{2p} = 1 - Q_{2p}. \tag{79}$$

Taking matrix elements of G between states of the P_{2p}-space we have

$$\langle\phi_{kl}|G(\omega)|\phi_{nm}\rangle = \langle\phi_{kl}|V|\phi_{nm}\rangle + \sum_{ij} \langle\phi_{kl}|V|\phi_{ij}\rangle \frac{C(ij)}{\omega - \epsilon_i - \epsilon_j} \langle\phi_{ij}|G(\omega)|\phi_{nm}\rangle, \tag{80}$$

which in a series expansion form becomes

$$\begin{aligned}\langle\phi_{kl}|G(\omega)|\phi_{nm}\rangle &= \langle\phi_{kl}|V|\phi_{nm}\rangle + \sum_{ij} \langle\phi_{kl}|V|\phi_{ij}\rangle \frac{C(ij)}{\omega - \epsilon_i - \epsilon_j} \langle\phi_{ij}|V|\phi_{nm}\rangle \\ &\quad + \sum_{iji'j'} \langle\phi_{kl}|V|\phi_{ij}\rangle \frac{C(ij)}{\omega - \epsilon_i - \epsilon_j} \langle\phi_{ij}|V|\phi_{i'j'}\rangle \frac{C(i'j')}{\omega - \epsilon_{i'} - \epsilon_{j'}} \langle\phi_{i'j'}|V|\phi_{nm}\rangle + \cdots .\end{aligned} \tag{81}$$

The above expansion can be represented diagrammatically by the series shown in Fig. 14. The diagrams on the r.h.s. of this figure are known as "ladder" diagrams, and each diagram corresponds to a situation in which a pair of particles interacts

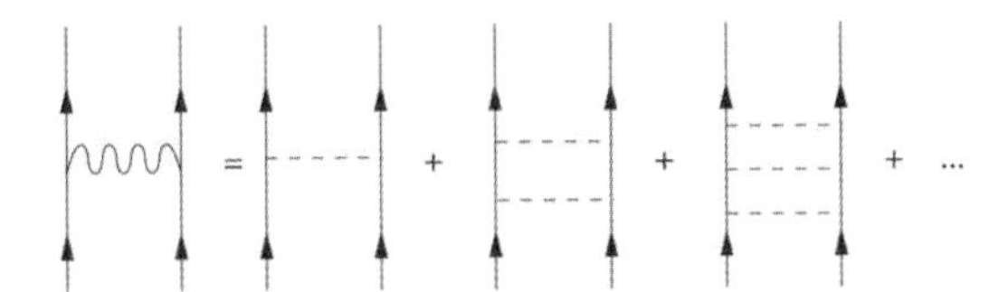

Fig. 14. Diagrammatic representation defining the matrix G.

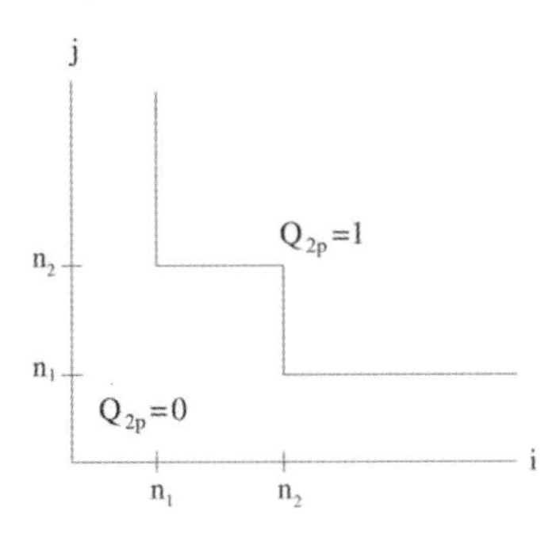

Fig. 15. A graph of the projection operator Q_{2p} appropriate for open-shell nuclei.

a certain number of times, with the restriction that the intermediate states involved in the scattering must be those defined through the Q_{2p} operator. In other words, two particles initially in a state of the P_{2p}-space undergo a sequence of scatterings into states of the Q_{2p}-space and then, after several such scatterings, go back to a state of the original space.

For the sake of completeness, we introduce the correlated wave function $|\psi_{nm}(\omega)\rangle$ [145],

$$|\psi_{nm}(\omega)\rangle = |\phi_{nm}\rangle + \frac{Q_{2p}}{\omega - H_0} V |\psi_{nm}(\omega)\rangle, \tag{82}$$

which once iterated becomes

$$|\psi_{nm}(\omega)\rangle = |\phi_{nm}\rangle + \frac{Q_{2p}}{\omega - H_0} G(\omega) |\phi_{nm}\rangle, \tag{83}$$

where use has been made of the integral Eq. (77). Eq. (83) allows one to write the action of G on an unperturbed state as

$$G(\omega)|\phi_{nm}\rangle = V|\psi_{nm}(\omega)\rangle, \tag{84}$$

which makes evident that the G operator may be considered an effective potential. The correlated wave function (82) and its properties are extensively discussed in Refs. [149,151,152].

At this point, we go further in our discussion clarifying the meaning of the starting energy ω and illustrating some possible choices of the projection operator Q_{2p}.

Let us begin with Q_{2p}. We define the Q_{2p} operator by specifying its boundaries labeled by the three numbers (n_1, n_2, n_3), each representing a SP level, the levels being numbered starting from the bottom of the potential well. Explicitly, using Eq. (78) the operator Q_{2p} is written as

$$C(ij) = \begin{cases} 1 & \text{if } i,j > n_2 \\ 1 & \text{if } n_1 < i \leq n_2 \text{ and } j > n_2 \text{ or vice-versa} \\ 0 & \text{otherwise.} \end{cases}$$

The graph of Fig. 15 makes more clear its definition. Note that the index n_3 denotes the number of levels in the full space. In principle, it should be infinite; in practice, as we shall see in Section 4.1.3, it is chosen to be a large, but finite number. As regards the indices n_1 and n_2, the only mandatory requirement is that none of them should be below the last occupied orbit of the doubly closed core. It is customary to take for n_1 the number the SP levels below the Fermi surface while n_2 may be chosen starting from the last SP valence level and going up. How this choice is performed is better illustrated by an example.

Let us consider the nucleus ^{18}O which in the shell-model framework is described as consisting of the doubly closed ^{16}O and two valence neutrons which are allowed to occupy the three levels of the *sd* shell. The model-space effective interaction for ^{18}O may be derived by using the linked-diagram expansion of Section 3, with the G matrix replacing the NN potential in all the irreducible, valence-linked diagrams composing the $\hat{Q}$-box. In so doing, one has to be careful to exclude from the $\hat{Q}$-box those diagrams containing a ladder sequence already included in the G matrix. We may take the matrix G with a Q_{2p} operator as specified, for instance, by $(3, 6, \infty)$. In this case, one of two SP levels composing the intermediate state in the calculation of the G matrix has to be beyond the *sd* shell, while the other one may be also an *sd* level. Then, when calculating

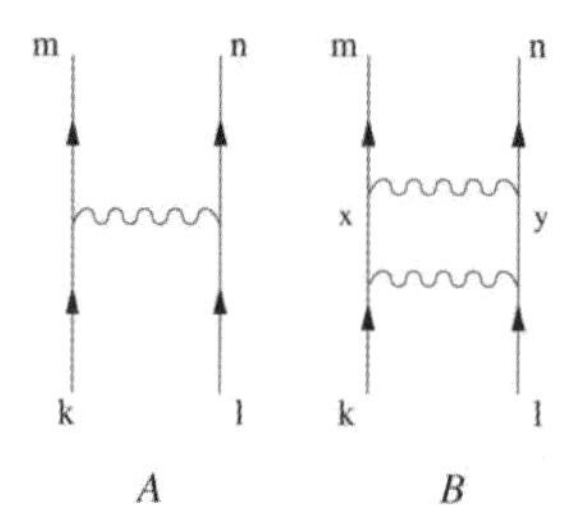

Fig. 16. Second-order ladder diagram contained in the $\hat{Q}$-box.

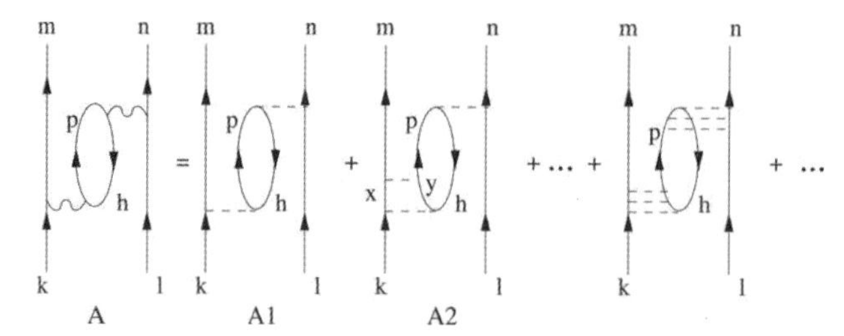

Fig. 17. Illustration of the summation procedure to replace V vertices with G vertices.

the $\hat{Q}$-box the included diagrams strictly depend on the considered G matrix. As an example, we have reported in Fig. 16 a first- and second-order diagram of the $\hat{Q}$-box, the wavy lines denoting the G-interactions. As discussed in Section 3.2.2, the incoming and outcoming lines of the A and B diagrams are levels of the *sd* shell, while the the intermediate state of diagram B should have at least one passive line. This means that none of the two SP states, x or y, could be a $0s$ or a $0p$ level and at least one of them must be beyond the *sd* shell. Therefore, if we take the G matrix with the $(3, 6, \infty)$ Q_{2p} operator, all possible B diagrams of Fig. 16 are already contained in the A diagram. On the other hand, if we increase the n_2 using a G matrix with the $(3, 10, \infty)$ Q_{2p} operator, the B diagrams should explicitly appear in the calculation of the $\hat{Q}$-box providing that $x, y \leq 10$ and x and/or $y \geq 7$. A detailed discussion on the points illustrated here is in Ref. [153].

We now come to the energy variable ω which may be seen as the "starting energy" at which G is computed. We know that any G-vertex in a $\hat{Q}$-box diagram is employed as the substitute for the ladder series of Fig. 14, and therefore the corresponding starting energy depends on the diagram as a whole, and in particular on the location of the G-vertex in the diagram. For instance, the G matrices in the three vertices of diagrams A and B of Fig. 16 have to be calculated at $\omega = \epsilon_k + \epsilon_l$, corresponding to the energy of the two-particle incoming state. As an other example, we have shown in Fig. 17 the diagram A. To illustrate how ω is evaluated, we have written it as the sum of the ladder sequence A1, A2, ..., An, ... The lowest vertex of diagram A corresponds to $\langle \phi_{mp} | G(\omega) | \phi_{kh} \rangle$ and its starting energy can be determined by looking at the two lowest vertices in diagram A2, whose contribution may be written as

$$\frac{\langle \phi_{mp} | V | \phi_{xy} \rangle \langle \phi_{xy} | V | \phi_{kh} \rangle}{\epsilon_k + \epsilon_h - \epsilon_x - \epsilon_y}. \tag{85}$$

Note that the energy denominator of the G matrix is given by $\omega - H_0$, H_0 acting on the intermediate two-particle state $|\phi_{xy}\rangle$. Therefore, to obtain the denominator of Eq. (85) as $\omega - \epsilon_x - \epsilon_y$, one has to choose $\omega = \epsilon_k + \epsilon_h$. Similarly, for the upper G-vertex of diagram A $\omega = \epsilon_k + \epsilon_l + \epsilon_h - \epsilon_m$. In both the previous examples ω is different from the energy of the incoming state, which results to be $\epsilon_k + \epsilon_l$.

In concluding this subsection, it is important to consider the dependence of the G matrix on the SP potential U, which explicitly appears in the denominator through H_0. The advantages of different choices have been extensively investigated essentially in connection with the Goldstone expansion and its convergence properties (see the review paper by Baranger [149]).

It is worth recalling that, as a general rule, in a perturbation approach to nuclear structure calculations, one starts by summing and subtracting the SP potential U to the Hamiltonian, so as to have as perturbation $H_1 = V - U$. As a consequence, the diagrams of the $\hat{Q}$-box for the shell-model effective interaction contain $-U$ as well as V vertices. Diagrams where $-U$ vertices are attached to particle lines between successive V interactions may be taken into account directly through the reaction matrix. This means that each diagram of the expansion shown in Fig. 14 is replaced with a new diagram, in which any number of $-U$ insertions has been introduced in the particle lines. A typical diagram is shown in Fig. 18. In this way, a new reaction matrix may be introduced, where the $-U$ interactions for the intermediate particle states are summed up to all orders. More precisely, by changing the propagator of the particle lines from H_0 to $H_0 - Q_{2p}UQ_{2p}$ one obtains the reaction matrix with plane waves as intermediate states, which is usually called G_T. Note that the term $Q_{2p}UQ_{2p}$ instead of

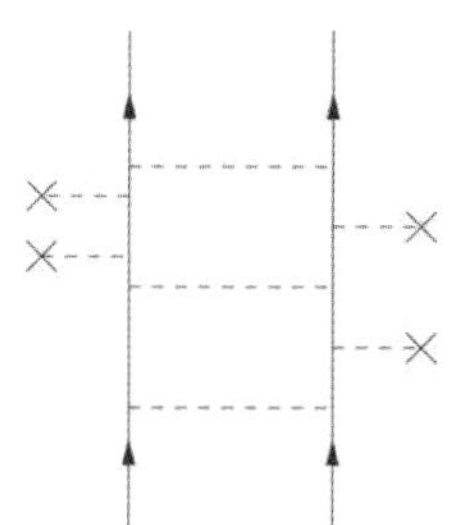

Fig. 18. Typical U insertions to the G-matrix intermediate states.

U is subtracted from H_0 since, as pointed out in [149], particle and hole states are defined with respect to the SP potential U. In other words, the Hamiltonian $H_0 - Q_{2p}UQ_{2p}$, which commutes with Q_{2p}, preserves the particle and hole states as defined by U. The reaction matrix G_T satisfies the integral equation

$$G_T(\omega) = V + VQ_{2p}\frac{1}{\omega - Q_{2p}tQ_{2p}}Q_{2p}G_T(\omega), \tag{86}$$

where one has to recall that t represents the kinetic energy.

An extensive study of the properties and features of G_T in the framework of the shell model can be found in [153], where arguments in favor of the use of G_T are given. We find it interesting to briefly recall these arguments here.

First of all, the G_T matrix of Eq. (86) minimizes the dependence on the SP potential. In fact, it depends only on Q_{2p} while the reaction matrix G of Eq. (77) depends also on the spectrum of the SP states of the Q_{2p} space.

The second argument is essentially based on the paper by Bethe et al. [146], where it was shown that all $-U$ and Brueckner–Hartree–Fock self-energy insertions cancel each other when the self-energy is calculated on energy shell. This result may be plausibly extended [153] to low-energy intermediate particle states not far off the energy shell, but such a cancelation is very unlikely for the high-energy states. On the other hand, in nuclear matter calculations [152], it has been shown that the sum of self-energy-insertion diagrams and other three-body clusters is almost negligible, in other words they tend to cancel each other. If we assume that this nuclear-matter result still holds for finite nuclei, we reasonably expect that only the $-U$ insertions survive for high-energy intermediate states, so leading to the reaction matrix G_T. The use of the G matrix of Eq. (77) is instead suggested for low-lying states assuming that $-U$ and Brueckner–Hartree–Fock self-energy insertions cancel approximately. In other words the reaction matrix calculation may be performed following a two-step procedure, which is discussed in detail in [153]. Here we only give a sketch of such procedure.

Let us start by dividing the intermediate states of the Q_{2p} space into high-energy and low-energy states, the Q_{2p} operator being written as

$$Q_{2p} = Q_{2p}^h + Q_{2p}^l, \tag{87}$$

with orthogonalized plane waves used for Q_{2p}^h and harmonic oscillator wave functions for Q_{2p}^l. The fist step consists in the calculation of the G_T matrix of Eq. (86) with the projection operator taken as Q_{2p}^h. Then using this reaction matrix instead of the NN potential, the G matrix of Eq. (77) is calculated with a projection operator restricted to Q_{2p}^l. Actually, the second step is carried out by means of a pertubative expansion, which implies that the final reaction matrix, which we call $\tilde{G}_T$, is written as

$$\tilde{G}_T(\omega) = G_T(\omega) + G_T(\omega)\frac{Q_{2p}^l}{\omega - H_0}G_T(\omega) + G_T(\omega)\frac{Q_{2p}^l}{\omega - H_0}G_T(\omega)\frac{Q_{2p}^l}{\omega - H_0}G_T(\omega) + \cdots \tag{88}$$

Hence $\tilde{G}_T$ is obtained by summing to G_T the second -, third-, $\cdots$ order two-particle ladder diagrams, with $G_T(\omega)$ vertices and intermediate states restricted to the low-lying states defined by the Q_{2p}^l operator. These ladder terms may be directly included in the evaluation of the $\hat{Q}$-box, once the G_T of Eq. (86) with Q_{2p}^h as projection operator has been calculated. Note that in so doing no $-U$ or Brueckner-self-energy insertions are explicitly included in the $\hat{Q}$-box diagrams since, on the basis of the previous arguments, one may assume that they cancel each other.

It is worth pointing out that no firm criteria exist on how to perform the partition of Eq. (87). It is customary to take n_2 in the definition of the Q_{2p}^h operator so that it encompasses the shell above the model space [153]. For example, if we consider again ^{18}O we specify Q_{2p} by $(3, 10, \infty)$ to define the G_T matrix, then Q_{2p}^l while corresponds to the $1p0f$-shell and the ladder terms with two particle in this shell are taken into account directly when the $\hat{Q}$-box is evaluated.

4.1.3. Calculation of the reaction matrix

Starting from the introduction of the G matrix, several approximate techniques have been developed for the solution of Eq. (77). A review of them, together with a critical discussion, can be found in [147,149,45,154]. Among the early methods, we would just like to mention the separation method of Moszkowski and Scott [155] and the reference spectrum method of Bethe et al. [146], which have both been extensively applied to both nuclear matter and finite nuclei. For many years, they were the primary methods for G-matrix calculation and were first employed by Kuo and Brown [18] to calculate from the Hamada–Johnston potential [63] the G matrix elements to be used in shell-model calculations for ^{18}O and ^{18}F. It is interesting to note that during the 1970s significant advances were made in the calculation of the reaction matrix, as regards the G matrix of Eq. (77) with harmonic oscillator intermediate states, as well as the G_T matrix of Eq. (86) having plane-wave intermediate states. In fact, accurate solutions of Eq. (77) were proposed by Barrett et al. [156] and of Eq. (86) by Tsai and Kuo [36], both of them providing an exact treatment of the projection operator Q_{2p}.

Here we shall only focus on the technique to compute the reaction matrix of Eq. (86), which we have used in our shell-model calculations. More precisely, we shall describe the method proposed by Tsai and Kuo [36], which was further investigated and applied by Krenciglowa et al. [153]. This method, based on a matrix inversion in the momentum space, allows, as mentioned above, an exact treatment of the projection Q_{2p} operator except for the only approximation of using a finite value instead of infinity for n_3 defining Q_{2p}.

Let us start by noting that a solution of Eq. (86) is

$$G_T(\omega) = V + VQ_{2p}\frac{1}{Q_{2p}(e-V)Q_{2p}}Q_{2p}V, \tag{89}$$

as it can be easily shown using the operator identity

$$\frac{1}{A-B} = \frac{1}{A} + \frac{1}{A}B\frac{1}{A-B}, \tag{90}$$

with $e = \omega - t$, $A = Q_{2p}eQ_{2p}$, and $B = Q_{2p}VQ_{2p}$.

The propagator of Eq. (89) was shown [36] to take the form

$$Q_{2p}\frac{1}{Q_{2p}(e-V)Q_{2p}}Q_{2p} = \frac{1}{e-V} - \frac{1}{e-V}P_{2p}\frac{1}{P_{2p}\frac{1}{e-V}P_{2p}}P_{2p}\frac{1}{e-V}, \tag{91}$$

where P_{2p} is defined in Eq. (79). The identity (91) allows us to rewrite Eq. (89) as

$$G_T(\omega) = G_{TF}(\omega) + \Delta G(\omega), \tag{92}$$

with

$$\Delta G(\omega) = -V\frac{1}{e-V}P_{2p}\frac{1}{P_{2p}\frac{1}{e-V}P_{2p}}P_{2p}\frac{1}{e-V}V \tag{93}$$

and

$$G_{TF}(\omega) = V + V\frac{1}{e-V}V = V + V\frac{1}{e}G_{TF}(\omega), \tag{94}$$

where we have used the identity (90) with $Q_{2p} = 1$ to obtain relation (94).

An equivalent and very convenient expression of $\Delta G(\omega)$, which has also the advantage to be applicable when V is a hard-core potential, is

$$\Delta G(\omega) = -G_{TF}(\omega)\frac{1}{e}P_{2p}\frac{1}{P_{2p}[(1/e)+(1/e)G_{TF}(1/e)]P_{2p}}P_{2p}\frac{1}{e}G_{TF}(\omega). \tag{95}$$

The set of Eqs. (92), (94) and (95) provides a very convenient way to calculate the G_T matrix, once a finite-n_3 approximation is made, namely a truncated P_{2p}-space is taken. Let us define

$$\begin{aligned} G_R(\omega) &= G_{TF}(1/e), \\ G_L(\omega) &= (1/e)G_{TF}, \\ G_{LR}(\omega) &= (1/e) + (1/e)G_{TF}(1/e), \end{aligned} \tag{96}$$

then Eq. (92) in a matrix formalism can be written as

$$\langle\phi_{kl}|G_T(\omega)|\phi_{lm}\rangle = \langle\phi_{kl}|G_{TF}(\omega)|\phi_{lm}\rangle - \sum_{ijkl\in P_{2p}}\langle\phi_{kl}|G_L(\omega)|\phi_{ij}\rangle \times \langle\phi_{ij}|G_{LR}^{-1}(\omega)|\phi_{kl}\rangle\langle kl|G_R(\omega)|\phi_{lm}\rangle. \tag{97}$$

This equation makes clear the necessity of a finite-n_3 approximation. In fact, the G_{LR}^{-1} matrix appearing in this equation is defined in the P_{2p} space and represents the inverse of the $P_{2p}G_{LR}P_{2p}$ matrix, which has to be a finite matrix.

The basic ingredient for the calculation of the G_T matrix through Eq. (92) is the matrix G_{TF} which is identical to the G_T defined by Eq. (86), except that it is free from the projection operator Q_{2p}. A very convenient way to calculate $G_{TF}(\omega)$ is to employ the momentum–space matrix inversion method [157]. Once the G_{TF} matrix elements in the momentum space are known, the matrices G_L, G_R, and G_{LR} can be easily evaluated in the same space. As a final step, these matrices are all transformed in the harmonic oscillator basis and the G_T matrix is obtained by means of Eq. (97) through simple matrix operations.

Here we have only given a sketch of the procedure to solve Eqs. (92), (94) and (95). A detailed description of the whole procedure can be found in [153], where the validity of an n_3-finite approximation is discussed. It is worth mentioning that this point is also discussed in Ref. [45].

4.2. The $V_{\text{low-k}}$ approach

In this subsection we shall describe in detail the derivation and main features of the low-momentum *NN* interaction $V_{\text{low-k}}$, which has been introduced in Refs. [42–44]. The idea underlying the construction of $V_{\text{low-k}}$ from a realistic model for V_{NN} is based on the renormalization group (RG) and the effective field theory (EFT) approaches [158,92,90,91]. As mentioned in Section 2, while the high-quality *NN* potentials all reproduce the empirical deuteron properties and low-energy phase shifts very accurately, they differ significantly from one another in high momentum behavior. A cutoff momentum Λ that separates fast and slow modes is then introduced and from the original V_{NN} an effective potential $V_{\text{low-k}}$, satisfying a decoupling condition between the low- and high-momentum spaces, is derived by integrating out the high-momentum components. The main result is that $V_{\text{low-k}}$ is free from the model-dependent high-momentum modes. In consequence, it is a smooth potential which preserves exactly the onshell properties of the original V_{NN}, and is suitable to be used directly in nuclear structure calculations. In Section 5.3.1, we will show that the $V_{\text{low-k}}$ approach provides a real alternative to the Brueckner *G*-matrix renormalization procedure.

4.2.1. Derivation of the low-momentum NN potential $V_{\text{low-k}}$

In carrying out the above high-momentum integration, or decimation, an important requirement is that the low-energy physics of V_{NN} is exactly preserved by $V_{\text{low-k}}$. For the two-nucleon problem, there is one bound state, namely the deuteron. Thus one must require that the deuteron properties given by V_{NN} are preserved by $V_{\text{low-k}}$. In the nuclear effective interaction theory, there are several well-developed model-space reduction methods. One of them is the Kuo–Lee–Ratcliff (KLR) folded diagram method [121,38], which was originally formulated for discrete (bound state) problems. A detailed discussion of this method has been given earlier in Section 3.

For the two-nucleon problem, we want the effective interaction $V_{\text{low-k}}$ to also preserve low-energy scattering phase shifts, in addition to the deuteron binding energy. Thus we need an effective interaction for scattering (unbound) problems. A convenient framework for this purpose is the *T*-matrix equivalence approach, as described below.

We use a continuum model space specified by

$$P = \int d\mathbf{p} |\mathbf{p}\rangle\langle\mathbf{p}|, \quad p \leq \Lambda \tag{98}$$

where $\mathbf{p}$ is the two-nucleon relative momentum and Λ the cutoff momentum, which is also known as the decimation momentum. Its typical value is about 2 fm^{-1} as we shall discuss later. Our purpose is to look for an effective interaction $PV_{\text{eff}}P$, with P defined above, which preserves certain properties of the full-space interaction V_{NN} for both bound and unbound states. This effective interaction is referred to as $V_{\text{low-k}}$.

We start from the full-space half-on-shell *T*-matrix (written in a single partial wave channel)

$$T(k', k, k^2) = V_{NN}(k', k) + \mathcal{P} \int_0^\infty q^2 dq V_{NN}(k', q) \frac{1}{k^2 - q^2} T(q, k, k^2). \tag{99}$$

This is the *T*-matrix for the two-nucleon problem with the Hamiltonian

$$H = H_0 + V_{NN} = t + V_{NN}, \tag{100}$$

t being the relative kinetic energy. We then define a *P*-space low-momentum *T*-matrix by

$$T_{\text{low-k}}(p', p, p^2) = V_{\text{low-k}}(p', p) + \mathcal{P} \int_0^\Lambda q^2 dq V_{\text{low-k}}(p', q) \frac{1}{p^2 - q^2} T_{\text{low-k}}(q, p, p^2), \tag{101}$$

where $(p', p) \leq \Lambda$ and the integration interval is from 0 to Λ. In the above, the symbol $\mathcal{P}$ in front of the integration sign denotes principal value integration. We require the equivalence condition

$$T(p', p, p^2) = T_{\text{low-k}}(p', p, p^2); \quad (p', p) \leq \Lambda. \tag{102}$$

The above equations define the effective low-momentum interaction; it is required to preserve the low-momentum ($\leq \Lambda$) half-on-shell *T*-matrix. Since phase shifts are given by the full-on-shell *T*-matrix $T(p, p, p^2)$, low-energy phase shifts given by the above $V_{\text{low-k}}$ are clearly the same as those of V_{NN}.

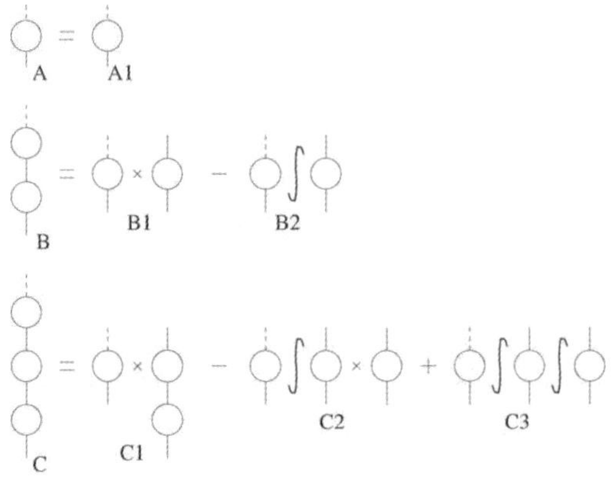

A A1

B B1 B2

C C1 C2 C3

Fig. 19. Folded-diagram factorization of the half-on-shell T-matrix.

In the following, let us show that a solution of the above equations may be found by way of the KLR folded-diagram method [121,38] described in Section 3 as a means to construct the shell-model effective interaction. Within this approach the $V_{\rm low-k}$ may be written in the operator form as

$$V_{\rm low-k} = \hat{Q} - \hat{Q}' \int \hat{Q} + \hat{Q}' \int \hat{Q} \int \hat{Q} - \hat{Q}' \int \hat{Q} \int \hat{Q} \int \hat{Q} + \cdots, \tag{103}$$

which contains a $\hat{Q}$-box whose explicit definition is given below.

In the time-dependent formulation, the T-matrix of Eq. (99) can be written as $\langle k'|VU(0,-\infty)|k\rangle$, $U(0,-\infty)$ being the time-evolution operator. In this way a diagrammatic analysis of the T-matrix can be made. A general term of T may be written as

$$\langle k'| \left[V + V\frac{1}{e(k^2)}V + V\frac{1}{e(k^2)}V\frac{1}{e(k^2)}V + \cdots \right] |k\rangle, \tag{104}$$

where $e(k^2) \equiv (k^2 - H_0)$.

Note that the intermediate states (represented by 1 in the numerator) cover the entire space. In other words, we have $1 = P + Q$ where P denotes the model-space projection operator and Q its complement. Expanding it out in terms of P and Q, a typical term of T is of the form

$$V\frac{Q}{e}V\frac{Q}{e}V\frac{P}{e}V\frac{Q}{e}V\frac{P}{e}V. \tag{105}$$

Let us now define the $\hat{Q}$-box as

$$\langle k'|\hat{Q}(k^2)|k\rangle = \langle k'| \left[V + V\frac{Q}{e(k^2)}V + V\frac{Q}{e(k^2)}V\frac{Q}{e(k^2)}V + \cdots \right] |k\rangle, \tag{106}$$

where all intermediate states belong to Q. One readily sees that the P-space portion of the T-matrix can be regrouped in terms of a $\hat{Q}$-box series, more explicitly as

$$\langle p'|T|p\rangle = \langle p'| \left[\hat{Q} + \hat{Q}\frac{P}{e(p^2)}\hat{Q} + \hat{Q}\frac{P}{e(p^2)}\hat{Q}\frac{P}{e(p^2)}\hat{Q} + \cdots \right] |p\rangle. \tag{107}$$

Note that all the $\hat{Q}$-boxes have the same energy variable, namely p^2. This regrouping is shown in Fig. 19, where each $\hat{Q}$-box is denoted by a circle, and the solid line represents the propagator $\frac{P}{e}$. The diagrams A, B and C are respectively the one- and two- and three-$\hat{Q}$-box terms of T, and clearly $T = A + B + C + \cdots$. Note that the dashed vertical line is not a propagator; it is just a "ghost" line to indicate the external indices.

Now a folded-diagram factorization for the T-matrix can be performed, following the same procedure as in Section 3 [121, 38]. As is discussed in Section 3.2.1, diagram B of Fig. 19 may be factorized into the product of two parts (see B1), where the time integrations of the two parts are independent from each other, each integrating from $-\infty$ to 0. In this way, we have introduced a time-incorrect contribution which must be corrected. In other words, B is not equal to B1, rather it is equal to B1 minus the folded-diagram correction B2. Schematically, the above factorization is written as

$$\hat{Q}\frac{P}{e}\hat{Q} = \hat{Q} \times \hat{Q} - \hat{Q}' \int \hat{Q}, \tag{108}$$

where the last term is the once-folded two $\hat{Q}$-box term. The meaning of the correction term B2 can also be visualized in terms of the differences in the $\hat{Q}$-box energy variables. With such energies explicitly written out, the above equation represents

$$\langle p'|\hat{Q}(p^2)\frac{P}{e(p^2)}\hat{Q}(p^2)|p\rangle = \sum_{p''}\langle p'|\hat{Q}(p''^2)|p''\rangle\langle p''|\frac{P}{e(p^2)}\hat{Q}(p^2)|p\rangle - \langle p'|\hat{Q}'\int\hat{Q}|p\rangle. \quad (109)$$

Note that the energy variable for the first $\hat{Q}$-box on the right hand side of this equation is p''^2, instead of p^2. Before factorization, the energy variable for this $\hat{Q}$-box is p^2. We recall that the integral sign represents a generalized folding operation and the leading $\hat{Q}$-box of any folded term must be at least of second order in V_{NN}.

In the same way, we factorize the three-$\hat{Q}$-box term C as shown in the third line of Fig. 19:

$$\hat{Q}\frac{P}{e}\hat{Q}\frac{P}{e}\hat{Q} = \hat{Q}\times\hat{Q}\frac{P}{e}\hat{Q} - \left[\hat{Q}'\int\hat{Q}\right]\times\hat{Q} + \hat{Q}'\int\hat{Q}\int\hat{Q}. \quad (110)$$

Higher-order $\hat{Q}$-box terms are also factorized following the same folded-diagram procedure. Let us now collect the terms in the figure in a "slanted" way. The sum of terms A1, B2, C3... is just the low-momentum effective interaction of Eq. (103). The sum B1, C2, D3... is $V_{\text{low-k}}\frac{P}{e}\hat{Q}$. Similarly the sum C1 + D2 + E3 + $\cdots$ is just $V_{\text{low-k}}\frac{P}{e}\hat{Q}\frac{P}{e}\hat{Q}$. It is worth to mention that diagrams D1, D2, ..., E1, E2, ... are not shown in Fig. 19. Continuing this procedure, it can be seen that the P-space portion of T-matrix is $V_{\text{low-k}} + V_{\text{low-k}}\frac{P}{e}T$. Namely, the $V_{\text{low-k}}$ defined by Eq. (103) is a solution of Eqs. (99)–(102).

An advantageous method to calculate the series in Eq. (103) is the Lee–Suzuki one, as described in Section 3.3. Following this algebraic approach the $V_{\text{low-k}}$ is expressed as

$$V_{\text{low-k}} = PV_{NN}P + PV_{NN}Q\omega, \quad (111)$$

where the operator ω is given by the solution of the decoupling Eq. (70). Note that in this latter equation the letter Ω was used instead of ω to avoid confusion with the energy variable ω defined in the same section. Here, since there is no problem of ambiguity, the same letter of the original paper by Lee and Suzuki [129] has been used and, for the sake of clarity, Eq. (70) is re-written for the Hamiltonian (100)

$$QV_{NN}P + QHQ\omega - \omega PHP - \omega PV_{NN}Q\omega = 0. \quad (112)$$

The iterative technique presented in Section 3.3 to solve the decoupling equation is formulated for degenerate model spaces. Therefore, it is not useful in this case since $PH_0P = PtP$ is obviously non-degenerate. We can then resort to the iterative technique for non-degenerate model spaces, proposed by Andreozzi [159] to solve the decoupling equation [see Eq. (112)] of the Lee–Suzuki method. This procedure, which will be referred to the Andreozzi–Lee–Suzuki (ALS) procedure, is quite convenient for obtaining the low-momentum NN interaction.

We give now a sketch of this procedure. Let us define the operators:

$$p(\omega) = PHP + PHQ\omega, \quad (113)$$

$$q(\omega) = QHQ - \omega PHQ, \quad (114)$$

in terms of which the basic equations of the ALS iterative procedure read

$$\begin{aligned} x_0 &= -(QHQ)^{-1}QHP, \\ x_1 &= q(x_0)^{-1}x_0p(x_0), \\ &\dots \\ x_n &= q(x_0 + x_1 + \cdots + x_{n-1})^{-1}x_{n-1}p(x_0 + x_1 + \cdots + x_{n-1}). \end{aligned} \quad (115)$$

Once the iterative procedure has converged, $x_n \to 0$, [159], the operator ω is given by

$$\omega_n = \sum_{i=0}^{n} x_i, \quad (116)$$

and the $V_{\text{low-k}}$ is obtained by Eq. (111).

In applying the ALS method, we have employed a momentum-space discretization procedure by introducing an adequate number of Gaussian mesh points [153].

4.2.2. *Phase-shift equivalent Hermitian $V_{\text{low-k}}$'s*

It should be pointed out that the low-momentum NN interaction given by the above folded-diagram expansion is not Hermitian. This is not convenient for applications, and one would like to have an interaction which is Hermitian. We shall describe how a family of Hermitian $V_{\text{low-k}}$'s can be derived, if we relax the half-on-shell T-matrix preservation. This subject has been studied in [160], where a general framework for constructing hermitian effective interactions is introduced, and it is shown how this generates a family of hermitian phase-shift equivalent low-momentum NN interactions.

In the following, we outline the essential steps for the derivation of a Hermitian $V_{\text{low-k}}$. We denote the folded-diagram low-momentum NN interaction obtained from the ALS procedure as V_{LS}. The model-space secular equation for V_{LS} is

$$P(H_0 + V_{LS})P|\chi_m\rangle = E_m|\chi_m\rangle, \tag{117}$$

where $\{E_m\}$ is a subset of the eigenvalues $\{E_n\}$ of the full-space Schrödinger equation, $(H_0 + V)|\Psi_n\rangle = E_n|\Psi_n\rangle$, and $|\chi_m\rangle$ is the P-space projection of the full-space wave function $|\Psi_m\rangle$, namely $|\chi_m\rangle = P|\Psi_m\rangle$. The above effective interaction may be rewritten in terms of the wave operator ω [see Eqs. (66)–(68)]

$$PV_{LS}P = Pe^{-\omega}(H_0 + V)e^{\omega}P - PH_0P. \tag{118}$$

While the full-space eigenvectors $|\Psi_n\rangle$ are orthogonal to each other, the model-space eigenvectors $|\chi_m\rangle$'s are clearly not so and consequently the effective interaction V_{LS} is not Hermitian. We now make a Z transformation such that

$$Z|\chi_m\rangle = |v_m\rangle;$$
$$\langle v_m|v_{m'}\rangle = \delta_{mm'}; \quad m, m' = 1, d, \tag{119}$$

where d is the dimension of the model space. This transformation reorients the vectors $|\chi_m\rangle$'s such that they become orthonormal. We assume that the $|\chi_m\rangle$'s are linearly independent so that Z^{-1} exists, otherwise the above transformation is not possible. Since $|v_m\rangle$ and Z exist entirely within the model space, we can write $|v_m\rangle = P|v_m\rangle$ and $Z = PZP$.

Using Eq. (119), we transform Eq. (117) into

$$Z(H_0 + V_{LS})Z^{-1}|v_m\rangle = E_m|v_m\rangle, \tag{120}$$

which implies

$$Z(H_0 + V_{LS})Z^{-1} = \sum_{m\epsilon P} E_m|v_m\rangle\langle v_m|. \tag{121}$$

Since E_m is real (it is an eigenvalue of $(H_0 + V)$ which is Hermitian) and the vectors $|v_m\rangle$ are orthonormal, $Z(H_0 + V_{LS})Z^{-1}$ is clearly Hermitian. The non-Hermitian secular Eq. (117) is now transformed into a Hermitian model-space eigenvalue problem

$$P(H_0 + V_{herm})P|v_m\rangle = E_m|v_m\rangle, \tag{122}$$

with the Hermitian effective interaction given by

$$V_{herm} = Z(H_0 + V_{LS})Z^{-1} - PH_0P, \tag{123}$$

or equivalently

$$V_{herm} = Ze^{-\omega}(H_0 + V)e^{\omega}Z^{-1} - PH_0P. \tag{124}$$

To calculate V_{herm}, we must first have the Z transformation. Since there are certainly many ways to construct Z, this generates a family of Hermitian effective interactions, all originating from V_{LS}. For example, we can construct Z using the familiar Schmidt orthogonalization procedure. We denote the hermitian effective interaction obtained using this Z transformation as V_{schm}.

We can also construct Z analytically in terms of the wave operator. From the properties of the wave operator ω, we have

$$\langle\chi_m|(1 + \omega^+\omega)|\chi_{m'}\rangle = \delta_{mm'}. \tag{125}$$

It follows that an analytic choice for the Z transformation is

$$Z = P(1 + \omega^+\omega)^{1/2}P. \tag{126}$$

This leads to the Hermitian effective interaction

$$V_{\text{okb}} = P(1 + \omega^+\omega)^{1/2}P(H_0 + V_{LS})P(1 + \omega^+\omega)^{-1/2}P - PH_0P, \tag{127}$$

or equivalently

$$V_{\text{okb}} = P(1 + \omega^+\omega)^{-1/2}(1 + \omega^+)(H_0 + V)(1 + \omega)(1 + \omega^+\omega)^{-1/2}P - PH_0P. \tag{128}$$

This is just the Okubo Hermitian effective interaction [161]. The Hermitian effective interaction of Suzuki and Okamoto [162, 10] is equivalent to the above.

There is another interesting choice for the transformation Z. As pointed out in [159], the positive definite operator $P(1 + \omega^+\omega)P$ admits the Cholesky decomposition, namely

$$P(1 + \omega^+\omega)P = PLL^{\mathrm{T}}P, \tag{129}$$

where L is a lower triangular matrix, L^{T} being its transpose. Since L is real and within the P-space, we have

$$Z = L^{\mathrm{T}}, \tag{130}$$

and the corresponding Hermitian effective interaction from Eq. (123) is

$$V_{\mathrm{chol}} = PL^{\mathrm{T}}P(H_0 + V_{LS})P(L^{-1})^{\mathrm{T}}P - PH_0P. \tag{131}$$

We have found that this Hermitization procedure is numerically more convenient than the other methods. In the practical applications presented in this review, we have employed this procedure. From now on, we shall denote the Hermitian low-momentum NN interaction as $V_{\mathrm{low-k}}$.

Using a solvable matrix model, it has been shown that the above Hermitian effective interactions V_{schm}, V_{okb}, and V_{chol} can be in general quite different [160] from each other and from V_{LS}, especially when V_{LS} is largely non-Hermitian. For the V_{NN} case, it is fortunate that the $V_{\mathrm{low-k}}$ corresponding to V_{LS} is only slightly non-Hermitian. As a result, the Hermitian low-momentum NN interactions V_{schm}, V_{okb}, and V_{chol} are all quite similar to each other and to the one corresponding to V_{LS}, as discussed in [160].

We have just shown that a family of Hermitian effective interactions $V_{\mathrm{low-k}}$'s can be generated starting from the Lee–Suzuki transformation. In the following, we shall prove that all such interactions preserve the full-on-shell T-matrix $T(p, p, p^2)$, $p \leq \Lambda$, and consequently they are all phase-shift equivalent.

Let us consider two T-matrices

$$T_1(p^2) = V_1 + V_1 g(p^2) T_1(p), \tag{132}$$

and

$$T_2(p'^2) = V_2 + V_2 g(p'^2) T_2(p'), \tag{133}$$

which are defined starting from two different potentials, V_1 and V_2, and/or two different propagators, $g(p^2) = \mathcal{P}(p_1^2 - H_0)^{-1}$ and $g(p'^2) = \mathcal{P}(p_2'^2 - H_0)^{-1}$. These T-matrices are related by the well known two-potential formula [146]

$$T_2^{\dagger}(p'^2) = T_1(p^2) + T_2^{\dagger}(p'^2)[g(p'^2) - g(p^2)]T_1(p^2) + X_2(p'^2)[V_2 - V_1]X_1(p^2), \tag{134}$$

where the operator $X = \mathrm{e}^{\omega}$ [see Eq. (66)] is defined so as to have $T = VX$.

Applying the above relation to the half-on-shell T-matrix elements in momentum space, we have

$$\langle p'|T_2^{\dagger}(p'^2)|p\rangle = \langle p'|T_1(p^2)|p\rangle + \langle p'|T_2^{\dagger}(p'^2)[g(p'^2) - g(p^2)]T_1(p^2)|p\rangle + \langle \psi_2(p')|(V_2 - V_1)|\psi_1(p)\rangle. \tag{135}$$

Here the true and unperturbed wave functions are related by $|\psi_1(p)\rangle = X_1(p^2)|p\rangle$ and similarly for ψ_2. Using the above relation, we shall now show that the phase shifts of the full-space interaction V_{NN} are preserved by the Hermitian interaction $V_{\mathrm{low-k}}$ for momentum $\leq \Lambda$. Let us denote the last term of Eq. (135) as $D(p', p)$. We use $V_{\mathrm{low-k}}$ for V_2 and V_{NN} for V_1. Recall that the eigenfunction of $(H_0 + V_{\mathrm{low-k}})$ is $|v_m\rangle$ [see Eq. (117)] and that for $H \equiv (H_0 + V_{NN})$ is $|\Psi_m\rangle$. We define a wave operator

$$U_P = \sum_{m \in P} |v_m\rangle\langle\Psi_m|. \tag{136}$$

Then $|v_m\rangle = U_P|\Psi_m\rangle$ and $PV_{\mathrm{low-k}}P = U_P(H_0 + V_{NN})U_P^{\dagger} - PH_0P$. For our present case, $\langle\psi_2(p')|$ is $\langle v_{p'}|$ and $|\psi_1(p)\rangle$ is $|\Psi_p\rangle$. Since $\langle v_{p'}|U_P = \langle\Psi_{p'}|$, we have

$$D(p', p) = \langle\Psi_{p'}|(HU_P^{\dagger} - U_P^{\dagger}H)|\Psi_p\rangle = (p'^2 - p^2)\langle v_{p'}|\Psi_p\rangle. \tag{137}$$

Clearly $D(p, p) = 0$. The second term on the right hand side of Eq. (135) vanishes when $p' = p$. Hence

$$\langle p|T_{\mathrm{low-k}}(p^2)|p\rangle = \langle p|T(p^2)|p\rangle, \quad p \leq \Lambda, \tag{138}$$

where $T_{\mathrm{low-k}}$ is the T-matrix for $(PH_0P + V_{\mathrm{low-k}})$ and T for $(H_0 + V_{NN})$. Thus we have shown that the phase shifts of V_{NN} are preserved by $V_{\mathrm{low-k}}$. Recall that our T-matrices are real, because of the principal-value boundary conditions employed.

4.2.3. Properties of $V_{\mathrm{low-k}}$

We have shown above that low-energy observables given by the full-momentum potential V_{NN} are preserved by the low-momentum potential $V_{\mathrm{low-k}}$. This preservation is an important point, and it is worth checking numerically that the deuteron binding energy and low-energy phase shifts given by V_{NN} are indeed reproduced by $V_{\mathrm{low-k}}$. In this subsection we would like to report some results in this regard, together with some discussion about the differences between the various modern potentials.

Before presenting the results, let us first address an important question concerning the choice of the momentum cutoff Λ in the derivation of $V_{\mathrm{low-k}}$. Phase shifts are given by the full-on-shell T-matrix $T(p, p, p^2)$, hence, for a chosen Λ, $V_{\mathrm{low-k}}$ can reproduce phase shifts up to $E_{\mathrm{lab}} = 2\hbar^2\Lambda^2/M$, M being the nucleon mass. Realistic NN potentials are constructed to fit

Table 1
Calculated binding energies BE_d of the deuteron (MeV) for different values of the cutoff Λ (fm^{-1})

Λ	0.5	1.0	1.5	2.0	2.5	3.0
	2.2246	2.2246	2.2246	2.2246	2.2246	2.2246

The $V_{\text{low-k}}$ has been calculated from the CD-Bonn potential for which $BE_d = 2.224575$ MeV [88].

Table 2
np $^1S_0$ phase shifts (deg) as predicted by the CD-Bonn potential and its $V_{\text{low-k}}$ ($\Lambda = 2.0\ \text{fm}^{-1}$)

E_{lab}	CD-Bonn	$V_{\text{low-k}}$	Expt.
1	62.1	62.1	62.1
10	60.0	60.0	60.0
25	50.9	50.9	50.9
50	40.5	40.5	40.5
100	26.4	26.4	26.8
150	16.3	16.3	16.9
200	8.3	8.3	8.9
250	1.6	1.6	2.0
300	−4.3	−4.3	−4.5

Table 3
Deuteron quadrupole moments (fm^2) as predicted by the Argonne V_{18} potential and its $V_{\text{low-k}}$ for different values of Λ (fm^{-1})

Argonne V_{18}	Λ	$V_{\text{low-k}}$
0.270	1.8	0.268
	2.0	0.269
	2.2	0.270
	2.4	0.270
	2.6	0.270
	2.8	0.270

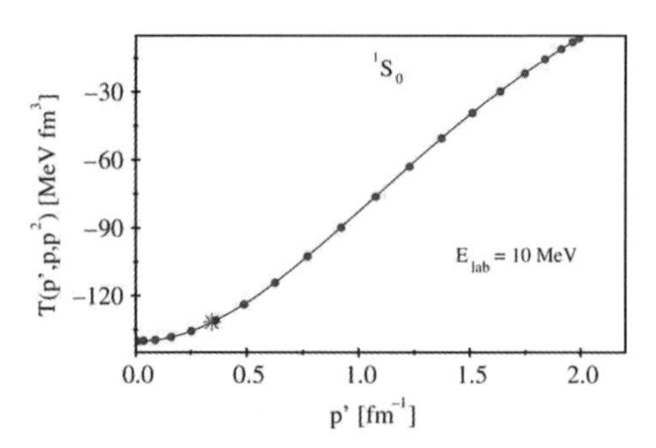

Fig. 20. Comparison of the half-on-shell $T(p', p, p^2)$ matrix elements (continuous line) with those of $T_{\text{low-k}}(p', p, p^2)$ for $p = 0.35\ \text{fm}^{-1}$ ($E_{\text{lab}} = 10$ MeV). The asterisk indicates the on-shell matrix element. The cutoff momentum is $\Lambda = 2\ \text{fm}^{-1}$.

empirical phase shifts up to $E_{\text{lab}} \approx 350$ MeV [85], which is the pion production threshold. It is reasonable then to require $V_{\text{low-k}}$ to reproduce phase shifts also up to this energy. Thus one should use Λ in the vicinity of $2\ \text{fm}^{-1}$.

We first check the deuteron binding energy BE_d given by $V_{\text{low-k}}$. For a range of Λ, such as $0.5\ \text{fm}^{-1} \leq \Lambda \leq 3\ \text{fm}^{-1}$, BE_d given by $V_{\text{low-k}}$ agrees very accurately (to 4 places after the decimal point) with that given by V_{NN}. In Table 1, we present the results for the $V_{\text{low-k}}$ derived from the CD-Bonn NN potential [88].
We have also checked the phase shifts and the half-on-shell $T(p', p, p^2)$ matrix elements with $(p', p) \leq \Lambda$.

In Table 2 we report the neutron–proton $^1S_0$ phase shifts calculated both with the CD-Bonn potential and its $V_{\text{low-k}}$ (with a cutoff momentum $\Lambda = 2\ \text{fm}^{-1}$). It is worth noting that the degree of accuracy shown in Table 2 allows $V_{\text{low-k}}$ to reproduce the χ^2/datum of the original potential up to $E_{\text{lab}} = 2\hbar^2\Lambda^2/M$.

In Table 3, for sake of completeness, we also present the deuteron quadrupole moments for different values of the cutoff momentum Λ, calculated with the corresponding $V_{\text{low-k}}$'s, derived from the Argonne V_{18} potential. These values are compared with the one predicted by the original potential.

In Figs. 20 and 21 we compare np $^1S_0$ $T(p', p, p^2)$ matrix elements given by the CD-Bonn potential and its $V_{\text{low-k}}$ with a cutoff momentum $\Lambda = 2.0\ \text{fm}^{-1}$ and $p = 0.35, 1.34\ \text{fm}^{-1}$. From these figures it can be seen that, even if the half-on-shell T-matrix is not exactly preserved by the ALS transformation, it is reproduced to a good degree of accuracy.

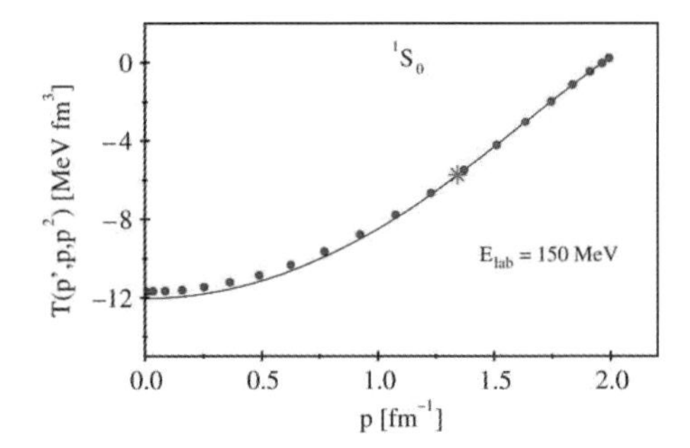

Fig. 21. As in Fig. 20, but for $p = 1.34$ fm^{-1} ($E_{lab} = 150$ MeV).

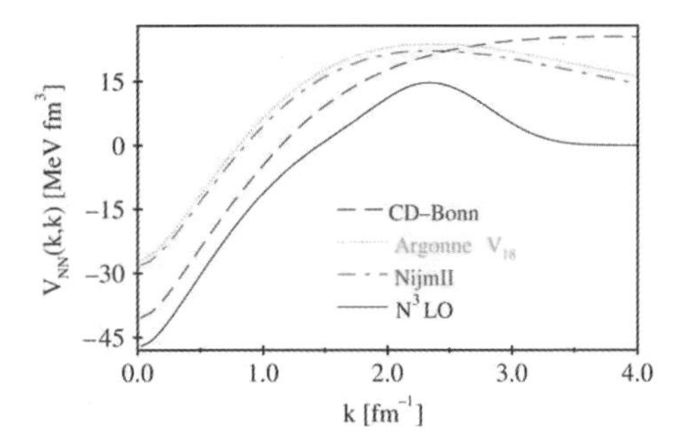

Fig. 22. Matrix elements in the $^1S_0$ channel for different NN potentials.

Table 4
Calculated P_D's with different NN potentials and with the corresponding V_{low-k}'s

	NijmII	Argonne V_{18}	CD-Bonn	N^{3}LO
Full potential	5.63	5.76	4.85	4.51
V_{low-k}	4.32	4.37	4.04	4.32

As mentioned earlier, modern NN potentials all possess a strong short-range repulsion. As a result, the k-space matrix elements $\langle k'|V_{NN}|k\rangle$ are large for large momenta, as illustrated in Fig. 22, where we compare the diagonal matrix elements of the phase-shift equivalent potentials NijmII, Argonne V_{18}, CD-Bonn, and N^{3}LO.

It is of interest to examine the behavior of V_{low-k}. In Fig. 23, we display some momentum–space matrix elements of V_{low-k} derived from the above V_{NN}'s. Here, we see that V_{low-k} matrix elements vary smoothly and become rather small near the cutoff momentum $\Lambda = 2.1$ fm^{-1}. Clearly, V_{low-k} is a smooth potential, no longer having the strong short-range repulsion contained in V_{NN}.

From Fig. 22 it is evident that the matrix elements of V_{NN} for the various potentials considered are all very different, even if the latter are onshell equivalent. It is interesting to note that the V_{low-k}'s derived from them are very close to each other, as shown in Fig. 23. This suggests that the various V_{low-k}'s are almost identical.

In Ref. [163] this V_{low-k} property has been evidenced through the study of the ground-state properties of some doubly-magic nuclei starting from the above mentioned NN potentials. As a matter of fact, it has been found that the calculated binding energies per nucleon and the rms charge radii are scarcely sensitive to the choice of the potential from which the V_{low-k} is derived. This insensitivity may be traced to the fact that when renormalizing the short-range repulsion of the various potentials, the differences existing between their offshell properties are attenuated. In this regard, let us consider the offshell tensor force strength. This is related to the D-state probability of the deuteron P_D, which implies that, when comparing NN potentials, offshell differences are seen in P_D differences (see for instance Ref. [53]). In Table 4 the predicted P_D's for each of the potentials under consideration are reported and compared with those calculated with the corresponding V_{low-k}'s ($\Lambda = 2.1$ fm^{-1}). It can be seen that while the P_D's given by the full potentials are substantially different, ranging from 4.5 to 5.8 %, they become very similar after renormalization. This is an indication that the starting "onshell equivalent" potentials have been made almost "offshell equivalent" by the renormalization procedure.

It is worth pointing out that when the short-range repulsion of different V_{NN}'s is renormalized by means of the G-matrix approach the differences between these potentials are strongly quenched too. This is illustrated in Fig. 24. There we report the correlation plot between Nijmegen II and CD-Bonn matrix elements in the HO basis with the oscillator parameter $\hbar\omega = 14$ MeV as well the correlation plot between the corresponding G_T matrix elements with a Pauli operator tailored

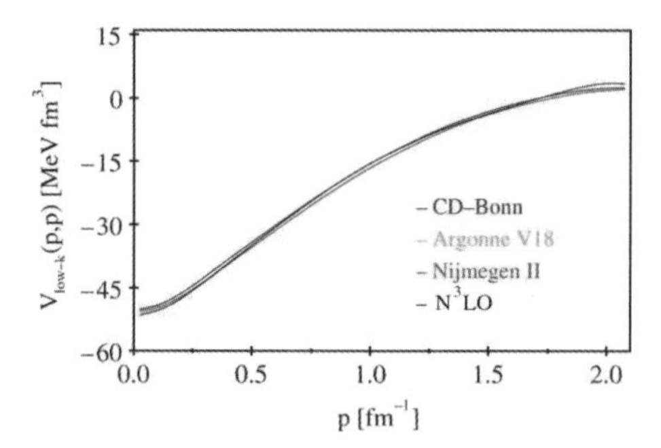

Fig. 23. As in Fig. 22, but for the respective $V_{\text{low-k}}$'s with a cutoff momentum $\Lambda = 2.1\ \text{fm}^{-1}$.

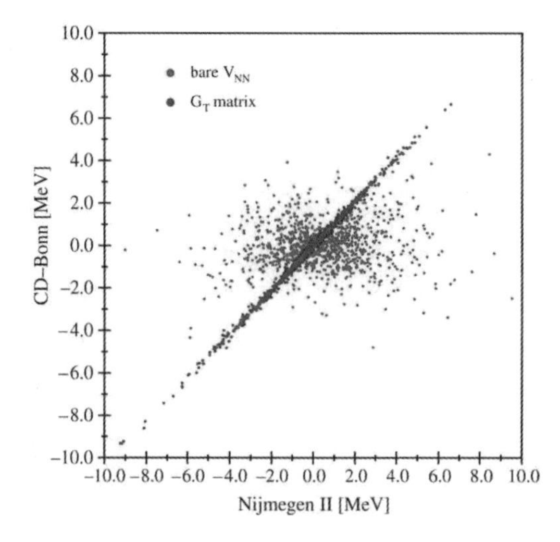

Fig. 24. Correlation plots between Nijmegen II and CD-Bonn matrix elements and between the corresponding G_T matrix elements (see text for details). The red and blue dots correspond to bare V_{NN} and G_T-matrix elements, respectively.

for the *sd* shell (see Section 4.1.2) and a starting energy of −5 MeV. We see that while the bare matrix elements are quite different, this is not the case for the G_T matrix ones.

Before closing this section, a brief discussion of $V_{\text{low-k}}$ in the framework of the RG-EFT approach may be in order. A main point of the RG-EFT approach is that low-energy physics is not sensitive to fields beyond a cutoff scale Λ. Thus for treating low-energy physics, one just integrates out the fields beyond Λ, thereby obtaining a low-energy effective field theory. In RG-EFT, this integration out, or decimation, generates an infinite series of counter terms [158] which is a simple power series in momentum.

The question as to whether the decimation of high-momentum modes in the derivation of $V_{\text{low-k}}$ generates a series of counter terms has been recently investigated in Refs. [164,165]. In these studies it is assumed that $V_{\text{low-k}}$ can be expressed as

$$V_{\text{low-k}}(q, q') \simeq V_{NN}(q, q') + V_{\text{counter}}(q, q');\ (q, q') \leq \Lambda, \tag{139}$$

where the counter potential is given as a power series

$$V_{\text{counter}}(q, q') = C_0 + C_2 q^2 + C_2' q'^2 + C_4(q^4 + q'^4) + C_4' q^2 q'^2 + C_6(q^6 + q'^6) + C_6' q^4 q'^2 + C_6'' q^2 q'^4 + \cdots \tag{140}$$

The counter term coefficients are determined using standard fitting techniques, so that the right hand side of Eq. (139) provides a best fit to the left-hand side of the same equation. In Ref. [164] this fitting was performed over all partial wave channels and a consistently good agreement was obtained.

An examination of the counter terms was made in [164,165], where the coefficients of some of them are also listed. A main result is that the counter terms are all rather small except for C_0 and C_2 of the S waves. This is consistent with the RG-EFT approach where the counter term potential is given as a delta function plus its derivatives [158].

5. Realistic shell-model calculations

5.1. The early period

As discussed in the Introduction, while the first attempts to derive the shell-model effective interaction from the free *NN* potential go back to some forty years ago, the practical value of this approach has emerged only during the last decade or so. On this basis, we may call the "early period" that extending over the previous three decades. It is the aim of this section to give a brief survey of the main efforts and developments made in the field during this long period of time.

The first attempt to perform a realistic shell-model calculation may be considered the one by Dawson, Talmi and Walecka [17]. They chose ^{18}O as testing ground and approximately solved a Bethe–Goldstone equation for the wave function of the two valence neutrons using the free *NN* potential of Brueckner–Gammel–Thaler [142] which contains a hard core. This initial approach, in which the effective interaction was simply taken to be the Brueckner reaction matrix, was followed by several authors [166–169]. In this context, the merit of the various methods used to obtain the nuclear reaction matrix elements was a matter of discussion [170–173].

A quantitative estimate of core polarization effects was first given by Bertsch [174], who found, for the Kallio–Kolltveit force [168], that they can produce a correction to the interaction of two valence nucleons in ^{18}O and ^{42}Sc as large as 30% of the *G*-matrix interaction. Soon afterwards, a turning point in the development of the field was marked by the work of Kuo and Brown [18], who derived an *sd*-shell effective interaction from the Hamada–Johnston potential [63]. After having carefully evaluated the reaction matrix elements for this potential, they took into account the corrections arising from the one-particle–one-hole ($1p$–$1h$) excitation of the ^{16}O core (the so-called "bubble diagram" G_{1p1h}). Comparison of the calculated and experimental spectra of ^{18}O and ^{19}F provided clear evidence of the crucial role played by the core polarization.

This remarkable achievement prompted the derivation of effective interactions for nuclei in other mass regions, such as the calcium [175], nickel [176,177] and lead region [178–180]. These effective interactions were all deduced from the Hamada–Johnston potential by use of the *G*-matrix theory and including core-polarization effects. In this context, the individual effects of the various second-order diagrams were studied [181,175] and the conclusion was reached that the inclusion of the core polarization diagram, G_{1p1h}, is generally sufficient to obtain a satisfactory agreement between calculated and experimental spectra [182].

The substantial soundness of the effective interactions derived along the lines of the Kuo–Brown approach is evidenced by the fact that, with empirical modifications, they have been used in a number of successful shell-model calculations to date. This has brought about various "semi-realistic" versions of the original Kuo–Brown interaction. In this connection, it is worth pointing out that these empirical modifications essentially concern the monopole parts of the effective interaction, as described in detail in Ref. [183], where the KB1, KB2, and KB3 interactions were introduced and applied to the study of *fp*-shell nuclei. This kind of approach, which is still favored by several authors, especially in the context of large scale shell model calculations, is outside the scope of the present article. We only mention here that, recently, three-body monopole corrections to realistic forces have been considered as a simple way to avoid a full treatment of three-body forces [184]. For a detailed discussion of the various effective interactions and the results of several applications, the reader is referred to the review papers by Brown [46], Otsuka et al. [48], Talmi [5], and Caurier et al. [47], where a full list of references can also be found.

It is interesting to note that, in spite of the very encouraging results produced by the first generation of realistic shell-model calculations, the general attitude of workers in the field was rather pessimistic. This is well reflected in the final statements made by Kirson in his Summary talk [185] at the 1975 Tucson International Conference:

> "To summarize my summary, I would say that there are definitely major obstacles in the way of doing really convincing calculations. We have a theory, we are aware of the weakness of our computations, and we should certainly invest the effort needed to plug the more obvious holes. But we should recognize the lack of precision inherent in our inability to go to high orders in perturbation theory and adopt more qualitative methods of extracting information and gaining understanding. New ideas, new viewpoints are much needed. We should also try to withstand the temptations of seductive ^{18}O, and pay some attention to other systems."

The aim of the Tucson Conference [186] was to define the current status of effective interaction and operator theory in the mid 1975s. The main issues of this Conference were expounded in the subsequent excellent review by Ellis and Osnes [126]. As it emerges from the first three talks [187–189] presented at the Tucson Conference and as mentioned in Section 3, in the years around 1970 the theoretical bases for an energy-independent derivation of the effective interaction were laid down within the framework of the time-independent [117] as well as time-dependent [120,121] perturbation theory. At the same time, the problems inherent in the practical application of the theory became a matter of discussion. In this context, considerable effort was made to study the convergence properties of the perturbative approach to the derivation of the effective interaction. As representative examples for these studies, we may mention here the works of Refs. [190–193]. In the first of these papers all significant terms through third order and a few selected fourth-order terms in the *G* matrix were calculated, and the convergence order-by-order of the perturbation series was investigated, while in the second one, the focus was on the convergence rate of the sum over intermediate-particle states in the second-order core polarization contribution to the effective shell-model interaction. In the third work the authors raised the problem of the divergence of the perturbation series due to the occurrence of the so-called "intruder states".

A most important turn in the field of realistic shell-model calculations resulted from the explicit inclusion [194,123] of the folded diagrams in the derivation of the effective interaction within the framework of the Kuo, Lee, and Ratcliff theory [121]. In Refs. [194] and [123] starting from a realistic G matrix, effective interactions for *sd*-shell nuclei were derived by means of the folded-diagram expansion, and employed in calculations for $A = 18$ to 24. Note that in Ref. [123] the very convenient method proposed in [130] (see Section 3.3) was used to sum up the folded diagram series. It turned out that the contribution of the folded diagrams to the effective interaction is quite large, typically about 30%. In the earlier study [194] the G matrix was obtained starting from the Reid soft core potential while in the later study [123] both this potential and the Paris potential were used. It may be worth noting that this latter paper was the first one where the Paris potential was employed in a microscopic calculation of finite open-shell nuclei.

As pointed out in Section 2.1, the advent of advanced meson-theoretic *NN* potentials, such as the Paris and Bonn ones, was instrumental in reviving the interest in realistic effective interactions during the 1980s. In this context, a main issue was the sensitivity of nuclear structure calculations to the strength of the tensor-force component contained in the *NN* potential, for which a practical measure is the predicted *D*-state probability of the deuteron P_D (see Section 2.1).

Earlier work on this problem was done by Sommermann et al. [195], who reexamined the intermediate state convergence problem for the core polarization diagram in the expansion of an effective shell-model interaction using the Bonn–Jülich (MDFPΔ2) potential [196,197] in comparison to the Reid soft-core interaction [64]. It is a remarkable achievement of this work to have shown that the so-called Vary–Sauer–Wong effect, namely the slow rate of convergence found for the Reid potential [191,198], was of negligible importance for the Bonn–Jülich potential, as a consequence of the much weaker tensor force contained in the latter.

Some ten years later, using a G-matrix folded-diagram method, the *sd*-shell effective interaction was derived from various potentials, and applied to calculate the spectra of some light *sd*-shell nuclei [199]. In particular, the three variants of the OBE parametrization of the Bonn full model, Bonn A, Bonn B, and Bonn C, differing in the strength of the tensor force (see Section 2.1) as well as the Paris potential, were considered. It turned out that the best agreement with experiment was achieved with the weakest tensor force potential (Bonn A).

The development of more realistic *NN* potentials, such as the Bonn and Paris potentials, led also in the early 1990s to new studies of the role of the third- and higher-order perturbative contributions to the effective interaction, which had been investigated some twenty years before in the work of Ref. [190]. Within the framework of the folded diagram theory, the inclusion of the third-order diagrams in G in the calculation of the $\hat{Q}$-box was studied, and the order-by-order convergence of the effective interaction in terms of the mass number A was examined [200–203]. The main findings of the last study of this series [203] were that the convergence of the perturbation expansion seems to be rather insensitive to the mass number, and that the effects of third-order contributions in the $T = 1$ channel are almost negligible. We refer to Ref. [45] for a review of the status of the theory of realistic effective interactions in the mid 1990s.

We have given above a brief review of the great deal of work done in the field of realistic shell-model calculations, during what we have called, somewhat arbitrarily, the "early period". Before finishing this subsection, a word is in order concerning this choice of presentation, which, in our opinion, reflects how hard it has been to make this kind of calculation a reliable tool for quantitative nuclear structure studies.

We draw the reader's attention to the fact that, after the calculations performed by Kuo and collaborators from the mid 1960's up to the mid 1970's [175–180], practically no other calculations were done for about two decades having as the only aim a detailed description of nuclear structure properties. This lack of confidence in the predictive power of realistic effective interactions is well evidenced by the fact that in almost all the papers on this subject published from the mid 1970's up to about the early 1990's, the focus has been on nuclei of the *sd*-shell, considered as a necessary testing ground for the effective interaction before embarking on a more ambitious program of applications. In this connection, the warning by Kirson (see above citation) against the temptations of seductive ^{18}O acquires a particular significance.

5.2. Modern calculations

As discussed in the previous subsection, during the 1980s substantial progress was made in the practical application of shell model with realistic effective interactions. This has resulted in a new generation of realistic shell-model calculations oriented to the study of the spectroscopic properties of nuclei in various mass regions.

The first attempts in this direction focused on the Sn isotopes [204,45,205], in particular on the neutron deficient ones. In the work of Ref. [204] the Bonn A potential was used while in that of Ref. [45] results obtained with the stronger tensor-force versions of this potential (Bonn B and Bonn C) were also reported. In both these works all two-body diagrams through third order in the G matrix were included in the $\hat{Q}$-box. In the study of Andreozzi et al. [205], both the Paris and Bonn A potentials were employed and the $\hat{Q}$-box was taken to be composed of diagrams through second order in G. A comparative discussion of the above studies can be found in [205]. The results of these initial "modern" calculations, in particular those obtained with the Bonn A potential, turned out to be in remarkably good agreement with the experimental data.

These achievements encouraged further work along these lines. In the few following years, realistic shell-model calculations were carried out for a number of nuclei around doubly magic ^{100}Sn [134], ^{132}Sn [206–209], and ^{208}Pb [210–212]. Note that in all these calculations, except the one of Ref. [209] where the CD-Bonn 1996 potential [87] was employed, the effective interaction was derived from the Bonn A potential. We refer to Covello et al. [39] for a survey of the status of realistic shell-model calculations at the end of the 1990s. Suffice it to say here that the rms deviation for the spectra of 15 medium- and heavy-mass nuclei reported in this paper is in 7 cases less than 100 keV, only in one case reaching 165 keV.

The success achieved by these calculations gave a clear-cut answer to the crucial, long-standing problem of how accurate a description of nuclear structure properties can be provided by realistic effective interactions, opening finally the way to a more fundamental approach to the nuclear shell model than the traditional, empirical one.

As is well known, much attention is currently being focused on nuclei in the regions of shell closures off stability, as they allow the exploration of possible shell-structure modifications when approaching the proton and neutron drip lines. The experimental study of these nuclei is very difficult, but in recent years new data have become available for some of them, which provide a challenging testing ground for realistic effective interactions. Starting in the early 2000s, several "exotic" nuclei around doubly magic ^{100}Sn and ^{132}Sn have been studied employing realistic effective interactions derived from the high-precision CD-Bonn *NN* potential [88]. In the first paper of this series [213] shell-model calculations were performed for Sn isotopes beyond $N = 82$, in particular the heaviest Sn isotope known to date, ^{134}Sn. In this work, similarly to what had been done in their previous papers, the authors employed a *G*-matrix folded-diagram formalism. A very good agreement was found between the calculated results and the available experimental data.

A most important turn in the derivation of the effective interaction from the free *NN* potential has resulted from the development of the $V_{\text{low-k}}$ approach (see Section 4.2). This approach to renormalize the short-range repulsion of V_{NN} has been recently used in several calculations. Attention has been focused on exotic nuclei neighboring ^{100}Sn [214], and ^{132}Sn [215–217,163,218,219]. Actually, most of this work is concerned with nuclei in the latter region, for which there are new interesting experimental data. The agreement between theory and experiment has turned out to be very good in all cases. This supports confidence in the predictions of these realistic calculations, which may therefore stimulate, and be helpful to, future experiments. To illustrate this important point, we shall review in Section 5.3.3 some selected results for ^{132}Sn neighbors with neutrons beyond the 82 shell.

While most of the realistic shell-model calculations carried out in last few years have employed effective interactions derived from the CD-Bonn *NN* potential, some attempts to put to the test chiral potentials have also been made. Representative examples for these studies are the works of Refs. [220–223], all of them making use of the $V_{\text{low-k}}$ renormalization method. Actually, only in [220], where the NNLO Idaho potential [104] was used, shell model calculations for various two valence-particle nuclei were performed. In the other three works [221–223] ground-state properties of doubly magic nuclei ^{4}He, ^{16}O, and ^{40}Ca were calculated within the framework of the Goldstone expansion [122] with the N^{3}LO potential [102]. The results obtained in [220] turned out to be comparable to those produced by effective interactions derived from the CD-Bonn potential. Clearly, this is related to the important question of how much nuclear structure results depend on the *NN* potential one starts with. We shall discuss this point in Section 5.3.2.

5.3. Results

5.3.1. Comparison between G-matrix and $V_{\text{low-k}}$ approaches

As evidenced by the discussion given in Sections 5.1 and 5.2, for about forty years the Brueckner *G* matrix represented practically the only tool to renormalize the repulsive core contained in the *NN* potential. After its introduction, the $V_{\text{low-k}}$ potential has been readily seen as an alternative to the *G* matrix. In fact, $V_{\text{low-k}}$ is a smooth free-space potential, and its matrix elements between unperturbed two-particle states can be directly used as input to the $\hat{Q}$-box folded-diagram procedure of Section 3. Therefore, a main issue has been to assess the merit of the $V_{\text{low-k}}$ approach in practical applications.

First work along this line was done by Kuo et al. [224], who carried out shell-model calculations for ^{18}O using two-body matrix elements of effective interactions derived from the Bonn A and Paris potentials. Other studies [225,43,226] were then performed which, starting from the CD-Bonn potential, extended the comparison between the *G* matrix and $V_{\text{low-k}}$ approaches to nuclei heavier than ^{18}O. In particular, the calculations of Ref. [225] concern ^{132}Sn neighbors, while results for ^{18}O, ^{134}Te, and ^{135}I are presented in [43]. A comparison between the *G* matrix and $V_{\text{low-k}}$ spectra for the heavy-mass nucleus ^{210}Po can be found in [226]. In all these calculations the cutoff parameter Λ is chosen in the vicinity of 2 fm^{-1}.

It has been a remarkable finding of all these studies that the $V_{\text{low-k}}$ results are as good, or even slightly better than, the *G*-matrix ones. The rather close similarity between the results of the two approaches finds an exception for the Paris potential [224]. We shall comment on this point in Section 5.3.2. It is also important to point out that the $V_{\text{low-k}}$ approach appears to work equally well in different mass regions. A detailed description of the above mentioned calculations can be found in the cited references. Here, for the sake of completeness, we report two examples concerning ^{18}O and ^{134}Te, which are both taken from [43], where more details about the calculations can be found.

The experimental and calculated spectra of ^{18}O including the five lowest-lying states, and those of ^{134}Te with the eight lowest-lying states are reported in Figs. 25 and 26, respectively. The calculated spectra have been obtained from the CD-Bonn potential through both the $V_{\text{low-k}}$ and *G*-matrix folded-diagram approaches. As for the $V_{\text{low-k}}$ calculations, two values of the cutoff momentum have been used, $\Lambda = 2.0$ and 2.2 fm^{-1}. For ^{18}O we have adopted a model space with the two valence neutrons in the $(0d_{5/2}, 0d_{3/2}, 1s_{1/2})$ shell, while for ^{134}Te the two valence protons are allowed to occupy the levels of the $(0g_{7/2}, 1d_{5/2}, 1d_{3/2}, 2s_{1/2}, 0h_{11/2})$ shell. In both cases the SP energies are taken from experiment, in particular from the spectra of ^{17}O and ^{133}Sb for the *sd* and *sdgh* shells, respectively. The energy of the $2s_{1/2}$ proton level, which is still missing, is from the study by Andreozzi et al. [206]. Note that in Figs. 25 and 26, the energies are relative to the ground state of the

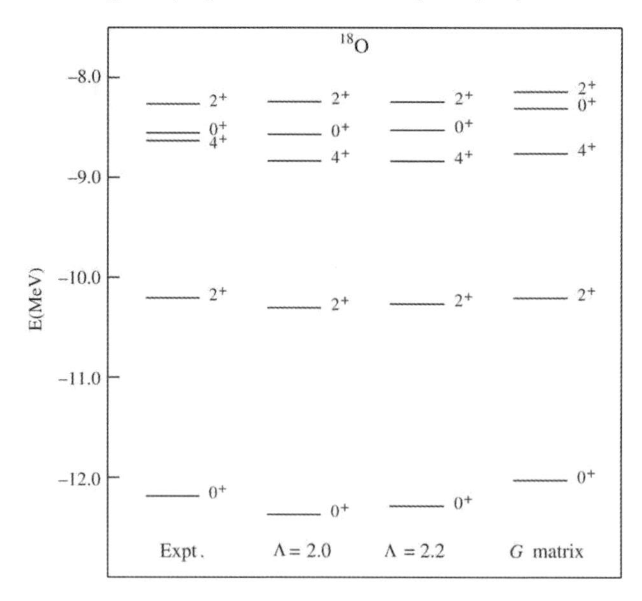

Fig. 25. Low-lying states in ^{18}O.

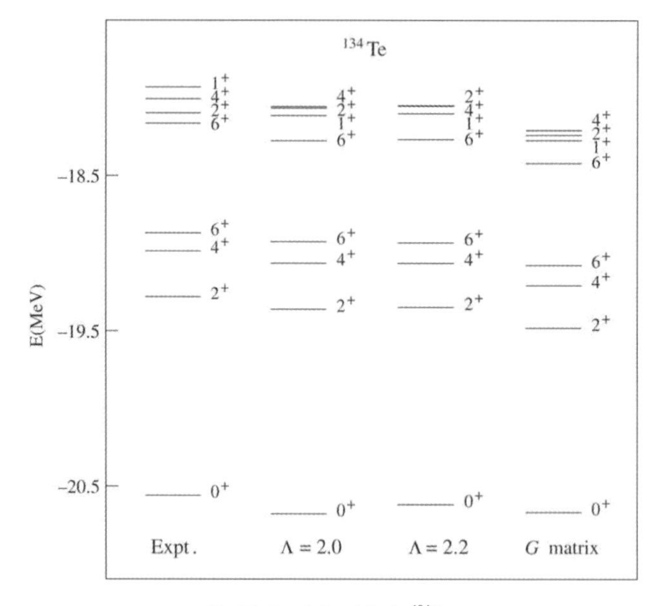

Fig. 26. Low-lying states in ^{134}Te.

neighboring doubly closed nucleus ^{16}O and ^{132}Sn, respectively, with the mass-excess values for ^{17}O and ^{133}Sb needed for the absolute scaling of the SP energies coming from Refs. [227,228]. As for ^{134}Te, the Coulomb contribution was taken into account by adding to the CD-Bonn effective interaction the matrix element of the Coulomb force between two $g_{\frac{7}{2}}$ protons with $J = 0$.

As pointed out above, we see from Figs. 25 and 26 that the $V_{\text{low-k}}$ and G-matrix results are rather close to each other, and in quite good agreement with experiment. It is worth noting that the $V_{\text{low-k}}$ results for the two values of Λ do not differ significantly. We shall come back to this point later.

In this context, we may also mention the work of Refs. [229,230], where harmonic-oscillator matrix elements of the $V_{\text{low-k}}$ potential and G matrix are compared in 4 major shells. It was found that both the $T = 0$ and $T = 1$ matrix elements are very similar.

On the above grounds, the conclusion is reached that the $V_{\text{low}-k}$ approach is a reliable way to renormalize the bare *NN* potential before employing it in the derivation of the shell-model effective interaction.

Another relevant point to be stressed is that the $V_{\text{low}-k}$ is far easier to be used than the *G* matrix. In fact, $V_{\text{low}-k}$ is a soft *NN* potential and as such does not depend either on the starting energy or on the model space, as is instead the case of the *G* matrix, which is defined in the nuclear medium. This has the desirable consequence that the $V_{\text{low}-k}$ matrix element to be used in an interaction vertex of a given $\hat{Q}$ box diagram is completely determined by the incoming and outcoming lines attached to the vertex under consideration. Hence, it does not depend on any other part of the diagram. Also, it is worth noting that the same $V_{\text{low}-k}$ can be used in different mass regions.

Finally, when considering the features of the $V_{\text{low}-k}$ approach, one has also to keep in mind that, as discussed in Section 4.2.3, different *NN* potentials lead to low-momentum potentials, which are remarkably similar to each other. In the same section, we have showed that the differences between the different potentials are also quenched within the *G* matrix approach. This will be further discussed in Section 5.3.2.

We conclude by mentioning that the $V_{\text{low}-k}$ approach has also proved to be an advantageous alternative to the *G* matrix in the calculation of the ground-state properties of doubly closed nuclei. As is well known, the traditional approach to this problem is the Brueckner–Hartree–Fock (BHF) theory. In the study of Ref. [221] it has been showed that the $V_{\text{low}-k}$ is suitable for being used directly in the Goldstone expansion. Good results for ^{16}O and ^{40}Ca were obtained by solving the Hartree–Fock (HF) equations for $V_{\text{low}-k}$ and then using the resulting self-consistent basis to compute higher-order Goldstone diagrams. In this connection, it should be mentioned also the work of Ref. [231], where use was made of the $V_{\text{low}-k}$ potential in a HF and BHF approach to evaluate the binding energy and the saturation density of nuclear matter and finite nuclei. Their conclusion was that for nuclear matter at small and medium densities as well as for finite nuclei HF and BHF results with $V_{\text{low}-k}$ are nearly identical.

5.3.2. Calculations with different NN potentials

As discussed in Section 5, a main problem relevant to realistic shell-model calculations has been to assess the accuracy of the effective interaction derived from the *NN* potential. However, it is also an important issue what information comes out on the *NN* potential from this kind of calculation. As we have seen in Section 2, from the early 1950s on several different potentials have been constructed. The potentials of the last generation all fit equally well the *NN* phase shifts, but, except for the OPE tail, they are quite different from each other. Actually, these potentials are well-constrained by the deuteron properties and the *NN* scattering data for $E_{\text{lab}} < 350$ MeV, and therefore their offshell behavior is subject to ambiguities. This is reflected in the different values predicted for the *D*-state probability of the deuteron P_D (see Section 2), as well as in the different binding energies calculated for $A = 3, 4$ systems [232]. In fact, for $A > 2$ the interaction may be off the energy shell, since two nucleons may have different energies before and after they interact.

In this context, it is of great interest to perform nuclear structure calculations to see to which extent the differences between *NN* potentials persist once the short-range repulsion is renormalized through the *G*-matrix or the $V_{\text{low}-k}$ approach. This problem has long been investigated by studying properties of both nuclear matter and doubly closed nuclei. A list of references on this subject can be found in the work of Ref. [222], which we have briefly discussed in Sections 4.2.3 and 5.2. As for open-shell nuclei, some attempts have been made within the framework of the shell model.

In the early work by Sommermann et al. [195], the comparison between Bonn–Jülich and Reid results aimed at investigating the intermediate-state convergence problem, as discussed in Section 5.1. The paper by Shurpin et al. [123] had instead as a main purpose to compare the spectra of several *sd*-shell nuclei calculated with two different potentials, Paris and Reid. The results showed that the Paris potential was as good as, if not better than, the Reid potential. The idea of comparing shell-model calculations with different *NN* potentials was revived at the end of the 1980s. In the work of Refs. [233,199,234] *sd*-shell nuclei were studied using as input several different *NN* potentials, all constructed before the 1990s. This work was based on the use of the reaction matrix *G* to derive effective interactions through the $\hat{Q}$-box folded-diagram method, including up to second-order diagrams. The outcome of these studies was that the effective interaction matrix elements become more attractive when a *NN* potential with a weaker tensor-force strength is used. Jiang et al. [199] noted that the nonlocality of the potential gives a further contribution in this direction. All these calculations indicated in general that the best agreement with experiment was produced by the Bonn potential.

A confirmation of the important role played by weaker tensor force potentials came out from the paper of Ref. [205], where Paris and Bonn A results for neutron deficient Sn isotopes are compared. A better agreement with experiment was obtained with the Bonn A potential. The merit of potentials with weak tensor force was also evidenced in [45], where realistic shell-model calculations were performed for nuclei in the mass regions of ^{16}O, ^{40}Ca, and ^{100}Sn using the three variants (A, B, and C) of the Bonn potential.

A comparative study of shell-model results obtained from different *NN* potentials, including the high-precision CD-Bonn potential, was made for the first time in [235]. In this paper, the nucleus ^{134}Te was used as a testing ground to investigate the effects of the CD-Bonn potential with respect to those of the three potentials Bonn A, Paris, and Nijm93. As in all previous calculations, it was found that a weak tensor force leads to a better description of nuclear structure properties. Furthermore, it turned out that the calculated spectra, except that obtained from the Paris potential, are only slightly different from each other.

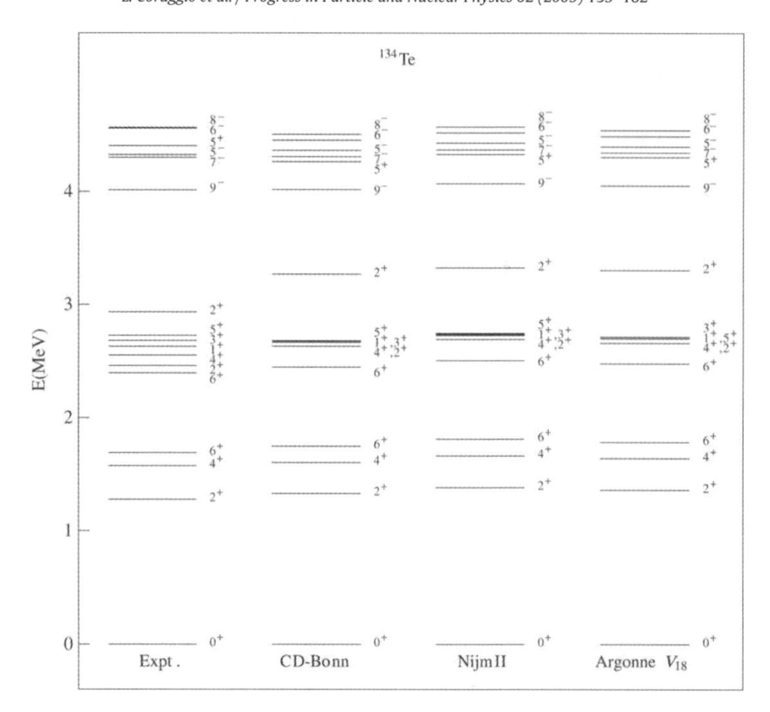

Fig. 27. Spectrum of ^{134}Te. Predictions by various *NN* potentials are compared with experiment.

All the above calculations made use of the reaction matrix G and their results provide evidence that the differences existing between various potentials are quenched by the G-matrix renormalization procedure (see Fig. 24). One may wonder what happens when the $V_{\rm low-k}$ potential is used. The first answer to this question may be found in [224], where the $V_{\rm low-k}$ approach is used to calculate the Paris and Bonn A spectra for ^{18}O. It is interesting to note that these two spectra are much closer to each other than those obtained with the G matrix approach [199]. This reflects the fact that the $V_{\rm low-k}$ procedure is more effective in reducing the differences between the two potentials than the G-matrix approach. The matrix elements in the $^3S_1$ channel for both Paris and Bonn A potentials are reported in Fig. 1 of [224], where it is shown that they become similar after the $V_{\rm low-k}$ renormalization, the latter being much larger for the strongly repulsive Paris potential.

In Section 4.2.3 (see Figs. 22 and 23), we have reported a similar comparison in the $^1S_0$ channel for the chiral potential N^{3}LO, and the three high-precision potentials CD-Bonn, NijmII, and Argonne V_{18}. The differences between the matrix elements of these potentials practically disappear when the $V_{\rm low-k}$ matrix elements are considered. On these grounds, one expects that shell-model effective interactions derived from phase-shift equivalent *NN* potentials through the $V_{\rm low-k}$ approach should lead to very similar results.

This we have verified by calculating the spectra of ^{134}Te starting from the CD-Bonn, NijmII, and Argonne V_{18} potentials. For all the three $V_{\rm low-k}$'s derived from these *NN* potentials, use has been made of the same cutoff momentum $\Lambda = 2.2$ fm^{-1}. Other details on how these calculations have been performed may be found in Section 5.3.3, where we present results for nuclei beyond ^{132}Sn, which have been obtained using the CD-Bonn potential.

In Fig. 27, the three calculated spectra are shown and compared with the experimental one. Note that, with respect to Fig. 26, the energies are relative to the ground state and higher lying levels are included. We see that the calculated spectra are very similar, the differences between the predicted energies not exceeding 80 keV. It is also seen that the agreement with experiment is very good for all the three potentials. In fact, the rms deviation is 115, 128, and 143 keV for CD-Bonn, Argonne V_{18}, and NijmII, respectively.

Finally, it is of interest to make a few comments regarding the value of the cutoff momentum Λ. In Section 4.2.3 the plausibility of using a value in the vicinity of 2 fm^{-1} was discussed. It was also shown in Section 5.3.1 that shell-model results do not change significantly when varying Λ from 2 to 2.2 fm^{-1}. A preliminary study of the dependence of shell-model results on Λ can be found in the work of Ref. [236], where results for ^{134}Te are reported using the CD-Bonn, Argonne V_{18}, and NijmII potentials with Λ varying over a rather large range (1.5–2.5 fm^{-1}). It has been found that the best agreement with experiment for each of these three potentials corresponds to a somewhat different value of Λ. However, changes around this value do not significantly modify the quality of agreement for all the three potentials. As an example, the rms deviation for NijmII remains below 150 keV for any value of Λ between 1.8 and 2.2 fm^{-1}. This subject certainly deserves further study.

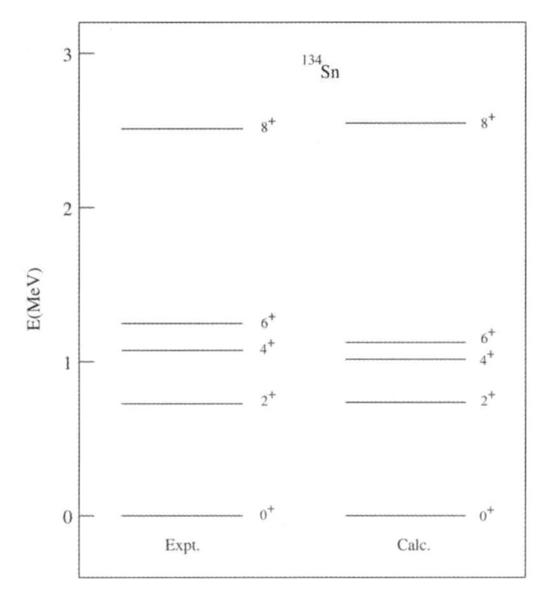

Fig. 28. Experimental and calculated spectra of ^{134}Sn.

5.3.3. Neutron rich nuclei beyond ^{132}Sn: Comparing theory and experiment

The study of exotic nuclei is a subject of a current experimental and theoretical interest. Experimental information for nuclei in the vicinity of ^{78}Ni, ^{100}Sn, ^{132}Sn, which have been long inaccessible to spectroscopic studies, is now available thanks to new advanced facilities and techniques. The advent of radioactive ion beams is opening new perspectives in this kind of study. Nuclei in the regions of shell closures are a source of direct information on shell structure. New data offer therefore the opportunity to test the shell model and look for a possible evolution of shell structure when going toward proton or neutron drip lines.

This is stimulating shell-model studies in these regions. Several realistic shell-model calculations have been recently performed for nuclei around ^{100}Sn and ^{132}Sn, as mentioned in Section 5.2. We report here some selected results on ^{132}Sn neighbors with $N > 82$. More precisely, we consider the three nuclei ^{134}Sn, ^{134}Sb, and ^{135}Sb, which have been very recently studied in [219,218,163], to which we refer for a more detailed discussion.

The experimental data which have been now become available for these neutron-rich nuclei may suggest a modification in the shell structure. They are, in fact, somewhat different from what one might expect by extrapolating the existing results for $N < 82$, and as a possible explanation a change in the single-proton level scheme has been suggested. The latter, caused by a more diffuse nuclear surface, may be seen as a precursor of major effects, which should show up at larger neutron excess.

We shall see now that the properties of these three nuclei are all well accounted for by a shell-model calculation with a two-body effective interaction derived from a modern *NN* potential. In so doing, we evidence the merit of realistic shell-model calculations, showing, at the same time, that there is no need to invoke new effects to describe few-valence-particle nuclei just above ^{132}Sn. In this connection, it is worth mentioning that different shell-model calculations have been performed for the above nuclei [237–242]. None of these studies, however, has succeeded in accounting simultaneously for the peculiar features of all the three nuclei.

In our calculations, we assume that ^{132}Sn is a closed core and let the valence neutrons occupy the six levels $0h_{9/2}$, $1f_{7/2}$, $1f_{5/2}$, $2p_{3/2}$, $2p_{1/2}$, and $0i_{13/2}$ of the 82–126 shell, while for the proton the model space includes the five levels $0g_{7/2}$, $1d_{5/2}$, $1d_{3/2}$, $2s_{1/2}$, and $0h_{11/2}$ of the 50–82 shell. The proton and neutron single-particle energies have been taken from the experimental spectra of ^{133}Sb and ^{133}Sn, respectively, the only exceptions being those of the $s_{1/2}$ proton and $i_{13/2}$ neutron levels which are still missing. The energy of the former, as mentioned in Section 5.3.1, is from [206], while that of the latter from [213]. The two-body matrix elements of the effective interaction have been derived from the CD-Bonn *NN* potential. This potential is renormalized by constructing the $V_{\text{low-k}}$ with a cutoff momentum $\Lambda = 2.2\ \text{fm}^{-1}$. The effective interaction has been then derived within the framework of the $\hat{Q}$-box plus folded diagram method including diagrams up to second order in $V_{\text{low-k}}$. The computation of these diagrams is performed within the harmonic-oscillator basis, using intermediate states composed of all possible hole states and particle states restricted to the five shells above the Fermi surface. This guarantees stability of the results when increasing the number of intermediate particle states. The oscillator parameter used is $\hbar\omega = 7.88$ MeV and for protons the Coulomb force has been explicitly added to the $V_{\text{low-k}}$ potential.

The experimental and calculated spectra of ^{134}Sn, ^{134}Sb, and ^{135}Sb are compared in Figs. 28–30, respectively. Note that both the calculated and experimental spectra of ^{134}Sb include all levels lying below 1 MeV excitation energy, while for ^{135}Sb

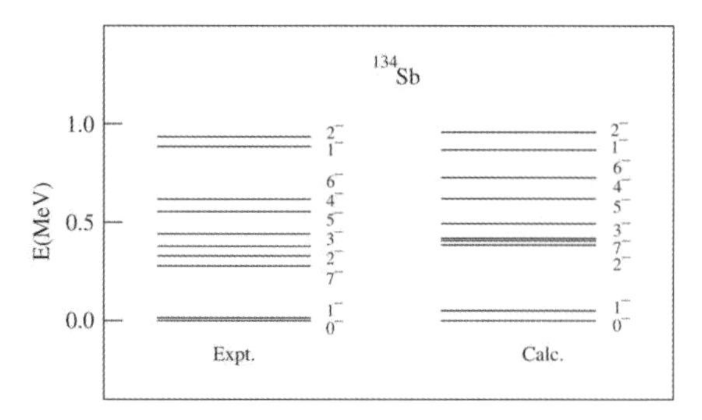

Fig. 29. Experimental and calculated spectra of ^{134}Sb.

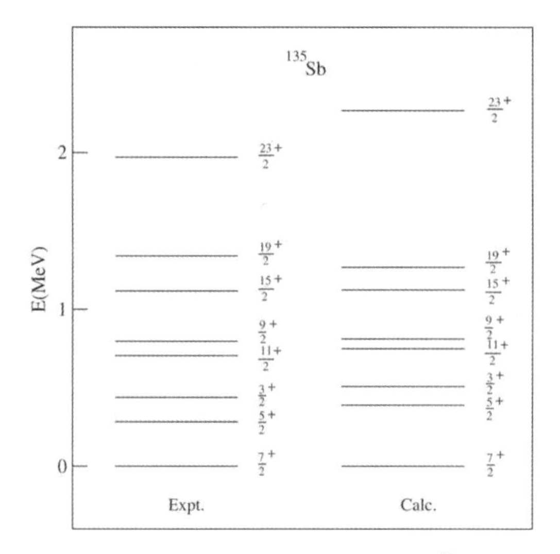

Fig. 30. Experimental and calculated spectra of ^{135}Sb.

all the yrast states have been reported. As for ^{134}Sn, only the levels reported in Fig. 28 have been observed, while in the calculated spectrum five states, predicted in between the 6^+ and 8^+ states, are omitted.

The two isotones with $A = 134$ are the most appropriate systems to study the effects of the effective interaction. In fact, ^{134}Sn is a direct source of information on the neutron–neutron channel while ^{134}Sb on the proton–neutron one. The nucleus ^{135}Sb represents a further interesting test, since it allows one to investigate at once the role of the effective interaction in both channels.

From the above figures, we see that the experimental levels in all the three nuclei are very well reproduced by the theory, the discrepancies being of the order of few tens of keV for most of the states. Note that the very low-energy positions of both the first-excited 2^+ and 1^- states in ^{134}Sn and ^{134}Sb, respectively, are well accounted for, as is also the case for the low excitation energy of the $5/2^+$ state in ^{135}Sb. We refer to the above mentioned papers for more details on each of the three calculations and for a discussion of the structure of the states in ^{134}Sb and ^{135}Sb. As for the latter, it is shown in [163] that it is the admixed nature of the $5/2^+$ state that explains its anomalously low position and the strongly hindered $M1$ transition to the ground state. The role of the effective interaction as well as of the $5/2^+$ single-proton energy in determining this situation has been examined, with particular attention to the effects induced by the proton–neutron interaction. The relevant role of the latter was also evidenced in ^{134}Sb [218]. For this nucleus, a direct relation can be established between the diagonal matrix elements of the proton–neutron effective interaction and the calculated energies reported in Fig. 29, since the corresponding wave functions are characterized by very little configuration mixing. In particular, the eight lowest-lying states can be identified with the eight members of the $\pi g_{7/2}\nu f_{7/2}$ multiplet while the two higher states, with $J^\pi = 1^-$ and 2^-, are the only members of $\pi d_{5/2}\nu f_{7/2}$ multiplet having an experimental counterpart.

In this connection, it may be interesting to summarize here the analysis of the proton–neutron matrix elements reported in the work of Ref. [218]. Focusing on the $\pi g_{7/2}\nu f_{7/2}$ configuration, we examine the various terms which contribute to the effective interaction, namely the $V_{\text{low-k}}$ and all the second-order two-body diagrams composing the $\hat{Q}$ box. It is worth

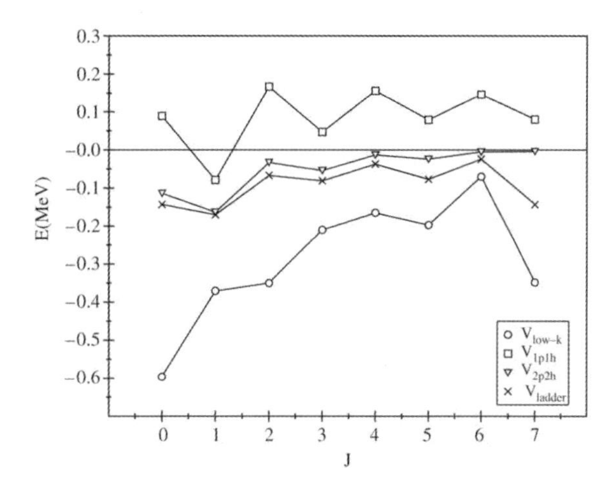

Fig. 31. Diagonal matrix elements of the $V_{\text{low}-\text{k}}$ and contributions from the two-body second-order diagrams for the $\pi g_{7/2}\nu f_{7/2}$ configuration.

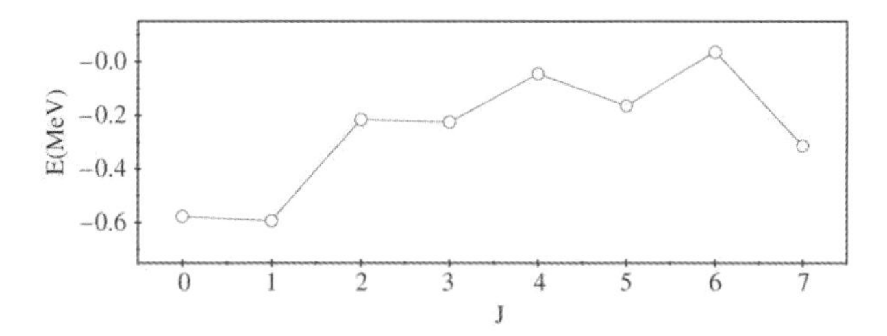

Fig. 32. Diagonal matrix elements of the two-body effective interaction for the $\pi g_{7/2}\nu f_{7/2}$ configuration.

pointing out that the effective interaction V_{eff} is obtained by summing to the $\hat{Q}$-box the folded-diagram series [see Eq. (58)], but the contribution of the latter provides a common attenuation for all matrix elements, thus not affecting the following discussion.

The diagonal matrix elements of $V_{\text{low}-\text{k}}$ for the $\pi g_{7/2}\nu f_{7/2}$ configuration are reported in Fig. 31 as a function of J together with the contributions arising from the $1p$–$1h$, $2p$–$2h$, and ladder excitations. In Fig. 32 we present for the same configuration the diagonal matrix elements of the effective interaction which we have obtained starting from this $V_{\text{low}-\text{k}}$. Note that the behavior of V_{eff} shown in Fig. 32 is quite similar to that of the $\pi g_{7/2}\nu f_{7/2}$ multiplet in ^{134}Sb, whose members, as mentioned above, are almost pure. This behavior, however, is quite different from that of $V_{\text{low}-\text{k}}$, especially as regards the $0^- - 1^-$ spacing. This means that the second-order two-body contributions accounting for configurations left out of the chosen model space are of key importance to renormalize the $V_{\text{low}-\text{k}}$ and get the correct behavior. In particular, a crucial role is played by the core-polarization effects arising from the $1p$–$1h$ excitations.

At this point one may wonder why our effective interaction is able to explain the properties of these neutron-rich nuclei above ^{132}Sn while the realistic shell-model calculations of Refs. [237–239] do not. In the latter calculations, the effective interaction has been derived within the G-matrix folded-diagram approach including in the $\hat{Q}$-box diagrams up to third order in G, and intermediate states with at most $2\hbar\omega$ excitation energy [203]. With this interaction, a satisfactory agreement with experiment was obtained only for ^{134}Sn when using experimental SP energies. A downshift of the $\pi d_{5/2}$ level with respect to the $\pi g_{7/2}$ one by 300 keV turned out to be necessary to obtain a good description of ^{135}Sb [237,239], but did not help for ^{134}Sb [238]. We have verified that the differences between the results presented in this section and those of Refs. [237–239] can be traced mainly to the different dimension of the intermediate state space. In fact, we have found that including intermediate states only up to $2\hbar\omega$ excitation energy in our derivation of the effective interaction, the results of the two calculations become very similar.

In summary, we have shown that modern realistic shell-model calculations are able to describe with quantitative accuracy the spectroscopic properties of ^{134}Sn, ^{134}Sb, and ^{135}Sb, in particular, their peculiar features. This leads to the conclusion that a consistent description of ^{132}Sn neighbors with neutrons beyond the $N = 82$ shell is possible. In this connection, it is worth mentioning two more recent works [243,244], where shell-model results for other nuclei around doubly magic ^{132}Sn are reported, including predictions for ^{136}Sn.

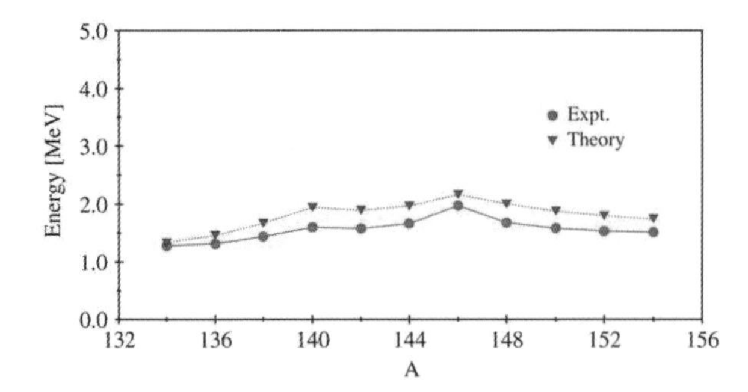

Fig. 33. Excitation energies of the $J = 2^+$ yrast states for $N = 82$ isotones as a function of mass number A.

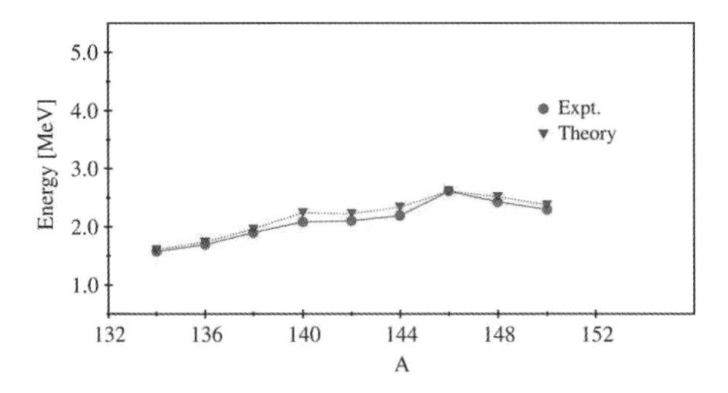

Fig. 34. Same as in Fig. 33 but for $J = 4^+$.

5.3.4. The role of many-body forces

In this subsection, we shall discuss the role that the 3*N* forces, both real and effective, play in shell-model calculations based on the results we obtain for the even-mass $N = 82$ isotonic chain. This study has been performed with the same input for all nuclei of the isotonic chain, namely the SP energies and the matrix elements of the two-body effective interaction employed in the calculations discussed in Section 5.3.3. Here, the shell-model calculations have been carried out by using the Oslo code [245].

Let us first consider the two valence-proton system ^{134}Te. As mentioned in Section 2.3, a genuine 3*N* potential, which may affect the one- and two-body components of the shell-model effective interaction H^1_{eff} of Eq. (58), has never been considered explicitly. The contributions of the 3*N* potential to the one-body term of H^1_{eff} may be assumed as taken into account when, as usual, the SP energies are taken from the experimental data. As regards the contribution of the 3*N* potential to the two-body component of H^1_{eff}, it may be reduced with an appropriate choice of the cutoff momentum Λ of the $V_{\text{low-k}}$. In [229], it was suggested to choose Λ around 2 fm^{-1}, since this value seems to minimize the role of the 3*N* potential when calculating the ^{3}H and ^{4}He binding energies. In this context, it is worth recalling that here we have used experimental SP energies and an effective interaction derived from a $V_{\text{low-k}}$ with $\Lambda = 2.2$ fm^{-1}.

For the ground-state energy of ^{134}Te, relative to the doubly magic ^{132}Sn, we obtain a value of −20.796 MeV to be compared with the experimental one −20.584 MeV. As regards the energy spectrum, the rms deviation is 0.115 MeV (see Fig. 27). These results seem to validate for many-nucleons systems the hypothesis made in [229] about the choice of Λ.

Let us now come to nuclei with more than two valence nucleons. We know that in this case many-body terms appear in the shell-model effective interaction, even if the original potential is only a two-body one. The effects of these terms have not yet been quantitatively evaluated [11,38,246,247]. However, it is generally believed that effective many-body forces have little influence on the excitation energies, while the binding energies may be significantly affected [127]. In this connection, it may be mentioned that Talmi [248] has shown that the experimental binding energies of the tin isotopes can be reproduced quite accurately employing an empirical two- and three-body interaction.

In order to study the effective many-body forces, we have calculated the excitation energies of the lowest-lying yrast states in the even $N = 82$ isotones. In Figs. 33–35, the calculated energy of the $J^\pi = 2^+, 4^+$, and 6^+ states, respectively, are reported as a function of the mass number A and compared with the experimental data. It is seen that the agreement between theory and experiment is of the same quality all over the isotonic chain, thus confirming that effective many-body forces play a negligible role in the description of the relative energy spectra.

In Fig. 36, we report the ground-state energy per valence nucleon. It appears that the experimental and theoretical behaviors diverge when increasing the number of protons. As in the case of tin isotopes [249], the binding energy is

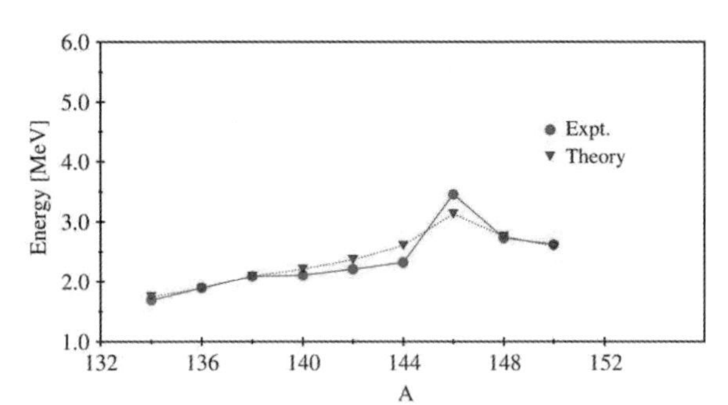

Fig. 35. Same as in Fig. 33 but for $J = 6^+$.

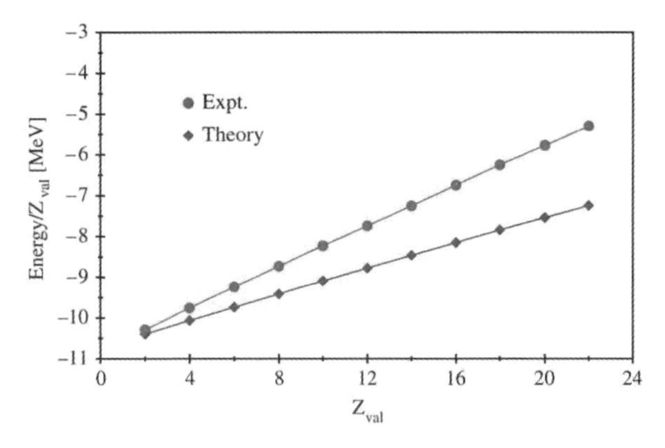

Fig. 36. Ground-state energies per valence proton of $N = 82$ isotones as a function of the valence-proton number Z.

increasingly overestimated. This deficiency may be healed by adding a small repulsive monopole contribution, whose origin may be traced to effective many-body forces [248].

6. Summary and conclusions

In this paper, we have tried to give a self-contained review of the present status of shell-model calculations employing two-body effective interactions derived from the free nucleon–nucleon potential. A main feature of these kinds of calculations which are commonly referred to as realistic shell-model calculations, is that no adjustable parameter appears in the determination of the effective interaction. This removes the uncertainty inherent in the traditional use of empirical interactions.

As discussed in the Introduction, only in the last few years has this approach, which started in the 1960s, acquired a reputation for being able to provide an accurate description of nuclear structure properties. This means that to place the current status of the field in its proper perspective, it is helpful to look back and survey the difficulties faced, and the progress achieved, over about four decades. The present review, while aimed at giving an insight into modern realistic shell-model calculations, is also concerned with the above aspects in regard to both the *NN* potential and the many-body methods for deriving the effective interaction.

As regards the first issue, in Section 2 we have discussed in some detail the main advances in the understanding of the *NN* interaction from the early stages of development until the current efforts based on chiral effective field theory. This we have done to show how the advances in this field have been intimately related to the development of realistic nuclear structure calculations. The present stage of development is certainly enough to provide a sound starting point for a quantitative description of nuclear structure properties. It remains a major task for the near future to assess the role of the three-body forces in many-nucleon systems. A brief discussion on this matter is given in Section 5.3.4 based on the results for the even $N = 82$ isotones.

In Section 3 we have described how the shell-model effective interaction can be derived within the framework of degenerate time-dependent perturbation theory.

The problem of the short-range repulsion contained in the *NN* potential has been discussed in Section 4, where, after an outline of the traditional Brueckner *G*-matrix method, we have described a new approach to the renormalization of V_{NN}, which consists in constructing a low-momentum *NN* potential, $V_{\text{low-k}}$, that preserves the physics of V_{NN} up to a cutoff momentum Λ. Since V_{low-k} is already a smooth potential, it can be used directly as input for the derivation of the shell-model effective interaction instead of the usual G matrix. While its merit in this respect has been assessed by several calculations (Section 5.3.1) the *NN* potential V_{low-k} is currently attracting much attention. On the one hand, it is being applied in various contexts, for instance in the study of few-body problems [250,114]. On the other hand, its properties are being studied in detail, in particular as regards the dependence on the original potential model [251,222,236]. Concerning shell-model calculations, results obtained with different *NN* potentials using the V_{low-k} approach have been presented in Section 5.3.2.

A main motivation for this review has been the practical success achieved by realistic shell-model calculations in the last few years, after a long period of hope and dismay. In Section 5. we have first surveyed the main efforts in this field, going from early times to present days. We have then presented, by way of illustration, some selected results of recent calculations for nuclei around doubly magic ^{132}Sn. These neutron-rich nuclei, which lie well away from the valley of stability, offer the opportunity for a stringent test of the matrix elements of the effective interaction. The very good agreement with the available experimental data shown in Section 5.3.3 supports confidence in the predictive power of realistic shell-model calculations in the regions of shell closures off stability, which are of great current interest.

As already pointed out above, in this review our focus has been on shell-model calculations employing genuine free *NN* potentials. As a consequence of this choice, even a brief summary of the many successful shell-model calculations employing modified versions of free *NN* potentials would have been beyond the scope of the present article. Here, we have also not touched upon another important aspect of shell-model work, namely the development of new methods, such as the Monte Carlo shell model (MCSM), as well as of high-quality codes, such as ANTOINE [252], which has greatly extended the feasibility of large-scale calculations. However, both these issues are discussed in detail in the recent article by Caurier et al. [47] while for a specialized review of the MCSM we refer to Ref. [48].

To conclude, we may say that the stage for realistic shell-model calculations is by now well set, which is a major step towards the understanding of the properties of complex nuclei in terms of the forces among nucleons. In this context, the V_{low-k} approach provides a new effective tool to handle the short-range repulsion of the free *NN* potential.

Acknowledgments

Helpful discussions with Gerry Brown, Scott Bogner, Jason Holt and Achim Schwenk are gratefully acknowledged. This work was supported in part by the Italian Ministero dell'Istruzione, dell'Università e della Ricerca (MIUR), and by the U.S. DOE Grant No. DE-FG02-88ER40388.

References

[1] M.G. Mayer, Phys. Rev. 75 (1949) 1969.
[2] O. Haxel, J.H.D. Jensen, H.E. Suess, Phys. Rev. 75 (1949) 1766.
[3] M.G. Mayer, J.H.D. Jensen, Elementary Theory of Nuclear Shell Structure, John Wiley, New York, 1955.
[4] M.G. Mayer, Phys. Rev. 78 (1950) 22.
[5] I. Talmi, Adv. Nucl. Phys. 27 (2003) 1.
[6] S.C. Pieper, et al., Phys. Rev. C 64 (2001) 014001.
[7] S.C. Pieper, R.B. Wiringa, Annu. Rev. Nucl. Part. Sci. 51 (2001) 53.
[8] P. Navratil, B.R. Barrett, Phys. Rev. C 54 (1996) 2986.
[9] P. Navratil, J.P. Vary, B.R. Barrett, Phys. Rev. Lett. 84 (2000) 5728.
[10] K. Suzuki, et al., Nucl. Phys. A 567 (1994) 576.
[11] D.J. Dean, et al., Prog. Part. Nucl. Phys. 53 (2004) 419.
[12] K. Kowalski, et al., Phys. Rev. Lett. 92 (2004) 132501.
[13] M. Wloch, et al., Phys. Rev. Lett. 94 (2005) 212501.
[14] J.R. Gour, et al., Phys. Rev. C 74 (2006) 024310.
[15] A. Nogga, et al., Phys. Rev. C 73 (2006) 064002.
[16] P. Navratil, et al., Phys. Rev. Lett. 99 (2007) 042501.
[17] J.F. Dawson, I. Talmi, J.D. Walecka, Ann. Phys. (NY) 18 (1962) 339.
[18] T.T.S. Kuo, G.E. Brown, Nucl. Phys. 85 (1966) 40.
[19] J.P. Elliott, in: M. Jean (Ed.), Cargèse Lectures in Physics, vol. 3, Gordon and Breach, New York, 1969, p. 337.
[20] I. Talmi, Rev. Modern Phys. 34 (1963) 704.
[21] S. Cohen, D. Kurath, Nucl. Phys. 73 (1965) 1.
[22] B.H. Wildenthal, in: A. Bohr, R.A. Broglia (Eds.), Proceedings of the International School of Physics "Enrico Fermi", Course XLIX, Varenna, 1976, Nort-Holland, 1977, p. 383.
[23] B.H. Wildenthal, Prog. Part. Nucl. Phys. 11 (1984) 5.
[24] B.A. Brown, B.H. Wildenthal, Ann. Rev. Nucl. Part. Sci. 38 (1988) 29.
[25] B.A. Brown, W.A. Richter, Phys. Rev. C 74 (2006) 0343150.
[26] S.T. Belayev, Mat.-Fys. Medd. K. Dan. Vidensk. Selsk. 31 (1959).
[27] L.S. Kisslinger, R.A. Sorensen, Mat.-Fys. Medd. K. Dan. Vidensk. Selsk. 32 (1960).
[28] A.K. Kerman, R.D. Lawson, M.H. MacFarlane, Phys. Rev. 124 (1961) 162.
[29] L.S. Kisslinger, R.A. Sorensen, Rev. Modern Phys. 35 (1963) 853.
[30] L.M. Green, S.A. Moszkowski, Phys. Rev. B 139 (1965) 790.
[31] R. Arvieu, S.A. Moszkowski, Phys. Rev. 145 (1966) 830.
[32] A. Plastino, R. Arvieu, S.A. Moszkowski, Phys. Rev. 145 (1966) 837.
[33] A.B. Migdal, Theory of Fermi Systems and Applications to Atomic Nuclei, Interscience, New York, 1967.
[34] F. Andreozzi, et al., Phys. Rev. C 41 (1990) 250.
[35] F. Andreozzi, et al., Phys. Rev. C 45 (1992) 2008.
[36] S.F. Tsai, T.T.S. Kuo, Phys. Lett. B 39 (1972) 427.
[37] E.M. Krenciglowa, T.T.S. Kuo, Nucl. Phys. A 235 (1974) 171.

[38] T.T.S. Kuo, E. Osnes, Lecture Notes in Physics, vol. 364, Springer-Verlag, Berlin, 1990.
[39] A. Covello, et al., Acta Phys. Polon B 32 (2001) 871.
[40] J.P. Elliott, et al., Nucl. Phys. A 121 (1968) 241.
[41] A.D. Jackson, J.P. Elliott, Nucl. Phys. A 125 (1969) 276.
[42] S. Bogner, T.T.S. Kuo, L. Coraggio, Nucl. Phys. A 684 (2001) 432c..
[43] S. Bogner, et al., Phys. Rev. C 65 (2002) 051301(R).
[44] S. Bogner, T.T.S. Kuo, A. Schwenk, Phys. Rep. 386 (2003) 1.
[45] M. Hjorth-Jensen, T.T.S. Kuo, E. Osnes, Phys. Rep. 261 (1995) 125.
[46] B.A. Brown, Prog. Part. Nucl. Phys. 47 (2001) 517.
[47] E. Caurier, et al., Rev. Modern Phys. 77 (2005) 427.
[48] T. Otsuka, et al., Prog. Part. Nucl. Phys. 47 (2001) 319.
[49] A.E.S. Green, M.H. MacGregor, R. Wilson, Rev. Modern Phys. 39 (1967) 495.
[50] R. Vinh-Mau, in: P. Blasi, R.A. Ricci (Eds.), Proceedings of International Conference on Nuclear Physics, Florence, 1983, Tipografia Compositori, Bologna, 1983, p. 61.
[51] R. Vinh-Mau, in: M.S. Hussein, et al. (Eds.), Proceedings of the 1989 International Nuclear Conference, São Paulo, World Scientific, Singapore, 1990, p. 189.
[52] R. Machleidt, Adv. Nucl. Phys. 19 (1989) 189.
[53] R. Machleidt, G.Q. Li, Phys. Rep. 242 (1994) 5.
[54] R. Machleidt, I. Slaus, J. Phys. G 27 (2001) R69.
[55] Y. Yukawa, Proc. Phys. Math. Soc. Japan 17 (1935) 48.
[56] R.J.N. Phillips, Rep. Prog. Phys. 22 (1959) 562.
[57] R. Jastrow, Phys. Rev. 81 (1951) 165.
[58] L. Eisenbud, E.P. Wigner, Proc. Natl. Acad. Sci., Wash. 27 (1941) 281.
[59] S. Okubo, R.E. Marshak, Ann. Phys., NY 4 (1958) 166.
[60] H.P. Stapp, T.J. Ypsilantis, N. Metropolis, Phys. Rev. 105 (1957) 302.
[61] J.L. Gammel, R.M. Thaler, Phys. Rev. 107 (1957) 291.
[62] K.E. Lassila, et al., Phys. Rev. 126 (1962) 881.
[63] T. Hamada, I.D. Johnston, Nucl. Phys. 34 (1962) 382.
[64] R.V. Reid, Ann. Phys., NY 50 (1968) 411.
[65] C.N. Bressel, A.K. Kerman, B. Rouben, Nucl. Phys. A 124 (1969) 624.
[66] V.G.J. Stoks, et al., Phys. Rev. C 49 (1994) 2950.
[67] F. Tabakin, Ann. Phys., NY 30 (1964) 51.
[68] F. Tabakin, Phys. Rev. 174 (1968) 1208.
[69] A.K. Kerman, J.P. Svenne, F.M.H. Villars, Phys. Rev. 147 (1966) 710.
[70] W.H. Bassichis, A.K. Kerman, J.P. Svenne, Phys. Rev. 160 (1967) 746.
[71] A.K. Kerman, M.K. Pal, Phys. Rev. 162 (1967) 970.
[72] A.K. Kerman, in: M. Jean (Ed.), Proceedings of the Cargése Lectures in Physics, vol. 3, Gordon and Breach, New York, 1969, p. 395.
[73] S.A. Moszkowski, Rev. Modern Phys. 39 (1967) 657.
[74] M.H. McGregor, R.A. Arndt, R.A. Wright, Phys. Rev. 182 (1969) 1714.
[75] W.N. Cottingham, et al., Phys. Rev. D 8 (1973) 800.
[76] R. Vinh-Mau, et al., Phys. Lett. B 44 (1973) 1.
[77] M. Lacombe, et al., Phys. Rev. D 12 (1975) 1495.
[78] M. Lacombe, et al., Phys. Rev. C 21 (1980) 861.
[79] R. Machleidt, K. Holinde, C. Elster, Phys. Rep. J. Phys. G 149 (1987) 1.
[80] M.M. Nagels, T.A. Rijken, J.J.D. Swart, Phys. Rev. D 17 (1978) 768.
[81] R. Brockmann, R. Machleidt, Phys. Rev. C 42 (1990) 1965.
[82] J.R. Bergervoet, et al., Phys. Rev. C 41 (1990) 1435.
[83] R.A.M. Klomp, V.G.J. Stoks, J.J. de Swart, Phys. Rev. C 42 (1991) R1258.
[84] R.A.M. Klomp, V.G.J. Stoks, J.J. de Swart, Phys. Rev. C 45 (1992) 2023.
[85] V.G.J. Stoks, et al., Phys. Rev. C 48 (1993) 792.
[86] R.B. Wiringa, V.G.J. Stoks, R. Schiavilla, Phys. Rev. C 51 (1995) 38.
[87] R. Machleidt, F. Sammarruca, Y. Song, Phys. Rev. C 53 (1996) R1483.
[88] R. Machleidt, Phys. Rev. C 63 (2001) 024001.
[89] R.B. Wiringa, R.A. Smith, T.L. Ainsworth, Phys. Rev. C 29 (1984) 1207.
[90] U. van Kolck, Prog. Part. Nucl. Phys. 43 (1999) 337.
[91] S.R. Beane, et al., in: M. Shifman (Ed.), At the Frontiers of Particle Physics, in: Handbook of QCD, vol. 1, World Scientific, Singapore, 2001, p. 133.
[92] P.F. Bedaque, U. van Kolck, Annu. Rev. Nucl. Part. Sci. 52 (2002) 339.
[93] S. Weinberg, Phys. Lett. B 251 (1990) 288.
[94] S. Weinberg, Nucl. Phys. B 363 (1991) 3.
[95] C. Ordóñez, U. van Kolck, Phys. Lett. B 291 (1992) 459.
[96] C. Ordóñez, L. Ray, U. van Kolck, Phys. Rev. Lett. 72 (1994) 1982.
[97] C. Ordóñez, L. Ray, U. van Kolck, Phys. Rev. C 53 (1996) 2086.
[98] N. Kaiser, R. Brockmann, W. Weise, Nucl. Phys. A 625 (1997) 758.
[99] N. Kaiser, S. Gerstendörfer, W. Weise, Nucl. Phys. A 637 (1998) 395.
[100] E. Epelbaoum, W. Glöckle, U.G. Meissner, Nucl. Phys. A 637 (1998) 107.
[101] E. Epelbaum, W. Glöckle, U.G. Meissner, Nucl. Phys. A 671 (2000) 295.
[102] D.R. Entem, R. Machleidt, Phys. Rev. C 68 (2003) 041001(R).
[103] D.R. Entem, R. Machleidt, in: A. Covello (Ed.), Challenges of Nuclear Structure, Proceedings of the 7th International Spring Seminar on Nuclear Physics, World Scientific, Singapore, 2002, p. 113.
[104] D.R. Entem, R. Machleidt, H. Witala, Phys. Rev. C 65 (2002) 064005.
[105] N. Kaiser, Phys. Rev. C 64 (2001) 057001.
[106] N. Kaiser, Phys. Rev. C 65 (2002) 017001.
[107] E. Epelbaum, W. Glöckle, U.G. Meissner, Nucl. Phys. A 747 (2005) 362.
[108] E. Epelbaum, W. Glöckle, U.G. Meissner, Eur. Phys. J. A 19 (2004) 401.
[109] N. Kaiser, Phys. Rev. C 61 (1999) 014003.
[110] N. Kaiser, Phys. Rev. C 62 (2000) 024001.
[111] S.C. Pieper, K. Varga, R.B. Wiringa, Phys. Rev. C 66 (2002) 044310.
[112] S.C. Pieper, Nucl. Phys. A 751 (2005) 516c.
[113] E. Epelbaum, et al., Phys. Rev. C 66 (2002) 064001.
[114] A. Nogga, et al., Nucl. Phys. A 737 (2004) 236.
[115] C. Bloch, J. Horowitz, Nucl. Phys. 8 (1958) 91.

[116] H. Feshbach, Ann. Phys. 19 (1962) 287.
[117] B.H. Brandow, Rev. Modern Phys. 39 (1967) 771.
[118] T. Morita, Prog. Theor. Phys. 29 (1963) 351.
[119] G. Oberlechner, F. Owono-N'-Guema, J. Richert, Nuovo Cimento B 68 (1970) 23.
[120] M. Johnson, M. Baranger, Ann. Phys. 62 (1971) 172.
[121] T.T.S. Kuo, S.Y. Lee, K.F. Ratcliff, Nucl. Phys. A 176 (1971) 65.
[122] J. Goldstone, Proc. R. Soc. Lond. Ser. A 293 (1957) 267.
[123] J. Shurpin, T.T.S. Kuo, D. Strottman, Nucl. Phys. A 408 (1983) 310.
[124] L. Coraggio, N. Itaco, Phys. Lett. B 616 (2005) 43.
[125] L. Coraggio, et al., Phys. Rev. C 75 (2007) 024311.
[126] P.J. Ellis, E. Osnes, Rev. Modern Phys. 49 (1977) 777.
[127] P.J. Ellis, et al., Phys. Rev. C 71 (2005) 034401.
[128] T.T.S. Kuo, E.M. Krenciglowa, Nucl. Phys. A 342 (1980) 454.
[129] S.Y. Lee, K. Suzuki, Phys. Lett. B 91 (1980) 173.
[130] K. Suzuki, S.Y. Lee, Prog. Theor. Phys. 64 (1980) 2091.
[131] K.A. Brueckner, C.A. Levinson, H.M. Mahmoud, Phys. Rev. 95 (1954) 217.
[132] J.P. Elliott, H.A. Mavromatis, E.A. Sanderson, Phys. Lett. B 24 (1967) 358.
[133] J. Sinatkas, et al., J. Phys. G 18 (1992) 1377.
[134] L. Coraggio, et al., J. Phys. G 26 (2000) 1697.
[135] K.M. Watson, Phys. Rev. 89 (1953) 575.
[136] N.C. Francis, K.M. Watson, Phys. Rev. 92 (1953) 291.
[137] B.A. Lippmann, J. Schwinger, Phys. Rev. 79 (1950) 469.
[138] K.A. Brueckner, C.A. Levinson, Phys. Rev. 97 (1955) 1344.
[139] H.A. Bethe, Phys. Rev. 103 (1956) 1353.
[140] K.A. Brueckner, Phys. Rev. 97 (1955) 1353.
[141] K.A. Brueckner, Phys. Rev. 100 (1955) 36.
[142] K.A. Brueckner, J.L. Gammel, Phys. Rev. 109 (1958) 1023.
[143] K.A. Brueckner, J.L. Gammel, H. Weitzner, Phys. Rev. 110 (1958) 431.
[144] K.A. Brueckner, A.M. Lockett, M. Rotenberg, Phys. Rev. 121 (1961) 255.
[145] H.A. Bethe, J. Goldstone, Proc. R. Soc. Lond. Ser. A 238 (1957) 551.
[146] H.A. Bethe, B.H. Brandow, A.G. Petscheck, Phys. Rev. 129 (1963) 225.
[147] B.D. Day, Rev. Modern Phys. 39 (1967) 719.
[148] R. Rajaraman, H.A. Bethe, Rev. Modern Phys. 39 (1967) 745.
[149] M. Baranger, in: M. Jean (Ed.), Rendiconti della Scuola Internazionale di Fisica "Enrico Fermi", vol. XL, Academic Press, New York and London, 1969, p. 511.
[150] G.E. Brown, J. Holt, in: G.E. Brown, C.H. Lee (Eds.), Hans bethe: The Nuclear Many Body Problem, Hans Bethe and His Physics, World Scientific, Singapore, 2006.
[151] A. de Shalit, H. Feshbach, Theoretical Nuclear Physics. Volume I: Nuclear Structure, John Wiley and Sons, New York, 1974.
[152] H.A. Bethe, Ann. Rev. Nucl. Sci. 21 (1971) 93.
[153] E.M. Krenciglowa, C.L. Kung, T.T.S. Kuo, Ann. Phys. 101 (1976) 154.
[154] B.R. Barrett, Czech. J. Phys. 49 (1999) 1.
[155] S.A. Moszkowski, B.L. Scott, Ann. Phys., NY 11 (1960) 65.
[156] B.R. Barrett, R.G.L. Hewitt, R.J. McCarthy, Phys. Rev. C 3 (1971) 1137.
[157] G.E. Brown, A.D. Jackson, T.T.S. Kuo, Nucl. Phys. A 133 (1969) 481.
[158] G.P. Lepage, in: C.A. Bertulani, et al. (Eds.), Nuclear Physics: Proceedings of the VIII Jorge Andre' Swieca Summer School, World Scientific, Singapore, 1997, p. 135.
[159] F. Andreozzi, Phys. Rev. C 54 (1996) 684.
[160] J.D. Holt, T.T.S. Kuo, G.E. Brown, Phys. Rev. C 69 (2004) 034329.
[161] S. Okubo, Prog. Theor. Phys. 12 (1954) 603.
[162] K. Suzuki, R. Okamoto, Prog. Theor. Phys. 70 (1983) 439.
[163] L. Coraggio, et al., Phys. Rev. C 72 (2005) 057302.
[164] J.D. Holt, et al., Nucl. Phys. A 733 (2004) 153.
[165] T.T.S. Kuo, in: A. Covello (Ed.), Key Topics in Nuclear Structure: Proceedings of the 8th International Spring Seminar on Nuclear Physics, Paestum, 2004, World Scientific, Singapore, 2005, p. 105.
[166] A. Kahana, E. Tomusiak, Nucl. Phys. 71 (1965) 402.
[167] A. Kahana, H.C. Lee, K. Scott, Phys. Rev. 185 (1969) 1378.
[168] A. Kallio, K. Kolltveit, Nucl. Phys. 53 (1964) 87.
[169] T. Engeland, A. Kallio, Nucl. Phys. 59 (1964) 211.
[170] R.K. Badhuri, E.L. Tomusiak, Proc. Phys. Soc. (London) 86 (1965) 451.
[171] A. Kallio, Phys. Lett. 18 (1965) 51.
[172] T.T.S. Kuo, G.E. Brown, Phys. Lett. 18 (1965) 54.
[173] C.W. Wong, Nucl. Phys. A 91 (1967) 399.
[174] G.F. Bertsch, Nucl. Phys. 74 (1965) 234.
[175] T.T.S. Kuo, G.E. Brown, Nucl. Phys. A 114 (1968) 241.
[176] R.D. Lawson, M.H. MacFarlane, T.T.S. Kuo, Phys. Lett. 22 (1966) 168.
[177] T.T.S. Kuo, Nucl. Phys. A 90 (1967) 199.
[178] T.T.S. Kuo, Nucl. Phys. A 122 (1968) 325.
[179] G.H. Herling, T.T.S. Kuo, Nucl. Phys. A 181 (1972) 113.
[180] J.B. McGrory, T.T.S. Kuo, Nucl. Phys. A 247 (1975) 283.
[181] G.E. Brown, T.T.S. Kuo, Nucl. Phys. A 92 (1967) 481.
[182] T.T.S. Kuo, Ann. Rev. Nucl. Part. Sci. 24 (1974) 101.
[183] A. Poves, A.P. Zuker, Phys. Rep. 70 (1981) 235.
[184] A.P. Zuker, Phys. Rev. Lett. 90 (2003) 042502.
[185] M.W. Kirson, in: B.R. Barrett (Ed.), Lecture Notes in Physics, vol. 40, Springer-Verlag, Berlin, 1975, p. 330.
[186] B.R. Barrett (Ed.), Proceedings of the International Topical Conference on Effective Interactions and Operators, in: Lecture Notes in Physics, vol. 40, Springer-Verlag, Berlin, 1975.
[187] B.H. Brandow, in: B.R. Barrett (Ed.), Lecture Notes in Physics, vol. 40, Springer-Verlag, Berlin, 1975, p. 1.
[188] M.B. Johnson, in: B.R. Barrett (Ed.), Lecture Notes in Physics, vol. 40, Springer-Verlag, Berlin, 1975, p. 25.
[189] K.F. Ratcliff, in: B.R. Barrett (Ed.), Lecture Notes in Physics, vol. 40, Springer-Verlag, Berlin, 1975, p. 42.
[190] B.R. Barrett, M.W. Kirson, Nucl. Phys. A 148 (1970) 145.
[191] J.P. Vary, P.U. Sauer, C.W. Wong, Phys. Rev. C 7 (1973) 1776.

[192] T.H. Schucan, H.A. Weidenmüller, Ann. Phys., NY 73 (1972) 108.
[193] T.H. Schucan, H.A. Weidenmüller, Ann. Phys., NY 76 (1973) 483.
[194] J. Shurpin, et al., Phys. Lett. B 69 (1977) 395.
[195] H.M. Sommermann, et al., Phys. Rev. C 23 (1981) 1765.
[196] K. Holinde, et al., Phys. Rev. C 15 (1977) 1432.
[197] K. Holinde, et al., Phys. Rev. C 18 (1978) 870.
[198] C.L. Kung, T.T.S. Kuo, K.F. Ratcliff, Phys. Rev. C 19 (1979) 1063.
[199] M.F. Jiang, et al., Phys. Rev. C 46 (1992) 910.
[200] M. Hjorth-Jensen, E. Osnes, H. Müther, Ann. Physics 213 (1992) 102.
[201] M. Hjorth-Jensen, et al., Nucl. Phys. A 541 (1992) 105.
[202] M. Hjorth-Jensen, E. Osnes, T.T.S. Kuo, Nucl. Phys. A 540 (1992) 145.
[203] M. Hjorth-Jensen, et al., J. Phys. G 22 (1996) 321.
[204] T. Engeland, et al., Phys. Scr. T 56 (1995) 58.
[205] F. Andreozzi, et al., Phys. Rev. C 54 (1996) 1636.
[206] F. Andreozzi, et al., Phys. Rev. C 56 (1997) R16.
[207] A. Covello, et al., Prog. Part. Nucl. Phys. 38 (1997) 165.
[208] A. Holt, et al., Nucl. Phys. A 618 (1997) 107.
[209] A. Holt, et al., Nucl. Phys. A 634 (1998) 41.
[210] L. Coraggio, et al., Phys. Rev. C 58 (1998) 3346.
[211] F. Andreozzi, et al., Phys. Rev. C 59 (1999) 746.
[212] L. Coraggio, et al., Phys. Rev. C 60 (1999) 064306.
[213] L. Coraggio, et al., Phys. Rev. C 65 (2002) 051306.
[214] L. Coraggio, et al., Phys. Rev. C 70 (2004) 034310.
[215] L. Coraggio, et al., Phys. Rev. C 66 (2002) 064311.
[216] J. Genevey, et al., Phys. Rev. C 67 (2003) 051306.
[217] A. Scherillo, et al., Phys. Rev. C 70 (2004) 054318.
[218] L. Coraggio, et al., Phys. Rev. C 73 (2006) 031302(R).
[219] A. Covello, et al., BgNSTrans. 10 (2) (2005) 10.
[220] L. Coraggio, et al., Phys. Rev. C 66 (2002) 021303(R).
[221] L. Coraggio, et al., Phys. Rev. C 68 (2003) 034320.
[222] L. Coraggio, et al., Phys. Rev. C 71 (2005) 014307.
[223] L. Coraggio, et al., Phys. Rev. C 73 (2006) 014304.
[224] T.T.S. Kuo, S. Bogner, L. Coraggio, Nucl. Phys. A 704 (2002) 107c.
[225] A. Covello, et al., in: A. Covello (Ed.), Challenges in Nuclear Sructure: Proceedings of the 7th International Spring Seminar on Nuclear Physics, Maiori, 2001, World Scientific, Singapore, 2002, p. 139.
[226] A. Covello, in: A. Molinari, et al. (Eds.), Proceedings of the International School of Physics "E. Fermi", Course CLIII, IOS Press, Amsterdam, 2003, p. 79.
[227] G. Audi, A.H. Wapstra, Nucl. Phys. A 565 (1993) 1.
[228] B. Fogelberg, et al., Phys. Rev. Lett. 82 (1999) 1823.
[229] A. Schwenk, J. Phys. G 31 (2005) S1273.
[230] A. Schwenk, A.P. Zuker, Phys. Rev. C 74 (2006) 0161302.
[231] J. Kuckei, et al., Nucl. Phys. A 723 (2003) 32.
[232] A. Nogga, H. Kamada, W. Glöckle, Phys. Rev. Lett. 85 (2000) 944.
[233] M.F. Jiang, et al., Phys. Rev. C 40 (1989) R1857.
[234] E. Maglione, L.S. Ferreira, Phys. Lett. B 262 (1991) 179.
[235] A. Covello, et al., Proceedings of the Nuclear Structure 98 Conference, Gatlinburg, Tennessee, 1998, in: C. Baktash (Ed.), AIP Conf. Proc., vol. 481, AIP, New York, 1999, p. 56.
[236] A. Covello, et al., J. Phys. Conf. Ser. 20 (2005) 137.
[237] J. Shergur, et al., Phys. Rev. C 65 (2002) 034313.
[238] J. Shergur, et al., Phys. Rev. C 71 (2005) 064321.
[239] J. Shergur, et al., Phys. Rev. C 72 (2005) 024305.
[240] W.T. Chou, E.K. Warburton, Phys. Rev. C 45 (1992) 1720.
[241] S. Sarkar, M.S. Sarkar, Phys. Rev. C 64 (2001) 014312.
[242] A. Korgul, et al., Eur. Phys. J. A 15 (2002) 181.
[243] A. Covello, et al., Eur. Phys. J. ST 150 (2007) 93.
[244] A. Covello, et al., Prog. Part. Nucl. Phys. (2007) doi:10.1016/j.ppnp.2007.01.001.
[245] T. Engeland, The Oslo shell-model code, unpublished, 1991–2006.
[246] H. Müther, A. Polls, T.T.S. Kuo, Nucl. Phys. A 435 (1985) 548.
[247] A. Polls, et al., Nucl. Phys. A 401 (1983) 125.
[248] I. Talmi, Contemporary Concepts in Physics, vol. 7, Harwood Academic Publishers, Chur, Switzerland, 1993.
[249] T. Engeland, et al., in: A. Covello (Ed.), Highlights of Modern Nuclear Structure: Proceedings of the 6th International Spring Seminar on Nuclear Physics, S. Agata sui due Golfi, 1998, World Scientific, Singapore, 1999, p. 117.
[250] S. Fujii, et al., Phys. Rev. C 70 (2004) 024003.
[251] J.D. Holt, et al., Phys. Rev. C 70 (2004) 061002.
[252] E. Caurier, F. Nowacki, Acta Phys. Polon B 30 (1999) 705.

CHAPTER IV
Brown–Rho Scaling in Nuclei, Nuclear Matter and Neutron Stars

Hadrons can experience profound changes in their properties through interactions with a background medium. Even within the dense center of normal nuclei, the masses of the mesons carrying the nuclear force may be altered, thereby modifying the nuclear force from that in free space. In this chapter we discuss the implications of Brown–Rho scaling, which predicts how hadronic masses scale in a dense medium.

[G.E. Brown and M. Rho (1991)]: Brown–Rho scaling was one of the first attempts in nuclear physics to connect medium-dependent meson masses to the partial restoration of chiral symmetry as the temperature or density increases. In this first of these papers it was shown that by including scale invariance in chiral effective Lagrangians through a scalar glueball field χ, one could define an in-medium pion decay constant f_π^* and obtain scaling relations between this quantity and the masses of light hadrons (except for the pion). The scaling law predicted that these light hadron masses decrease proportionally with the in-medium pion decay constant and that at nuclear matter density these in-medium hadron masses should be approximately 80% of their vacuum values. How then can one explain the successes of traditional nuclear physics which employed free-space NN interactions without medium-modified meson masses? Due to the large cancellations that occur between different meson exchange forces, if most of the mesons masses are scaled approximately the same, then the net effect will still be small.

[J.W. Holt, G.E. Brown, J.D. Holt and T.T.S. Kuo (2007)]: Since hadronic masses decrease with the nuclear density in Brown–Rho scaling, one of the simplest systems in which to study these effects is symmetric nuclear matter at constant density n. In Landau's theory of normal Fermi liquids, nuclear matter is described in terms of quasiparticles interacting at the Fermi surface. Although certain ground state properties of the interacting system are beyond the scope of Fermi liquid theory, one can obtain information about the low-energy excitations about equilibrium. In this study it was found that low-momentum interactions, iterated in the Babu–Brown induced interaction formalism and with meson masses decreased according to Brown–Rho scaling, improved the symmetry energy, compression modulus, and anomalous orbital gyromagnetic moment obtained in Fermi liquid theory.

[J.W. Holt, G.E. Brown, T.T.S. Kuo, J.D. Holt and R. Machleidt (2008)]: Describing the anomalously long β-decay lifetime of ^{14}C has long been a challenge in nuclear structure physics. The spin and isospin quantum numbers of the ^{14}C and ^{14}N ground states satisfy the selection rules for an allowed Gamow–Teller transition, yet the known lifetime of 5730 years is nearly six orders of magnitude longer than would be expected from typical β-decay lifetimes in light nuclei. In this study we employed Brown–Rho scaling for the vector meson masses and form factor cutoffs, while the "σ" meson was treated explicitly as correlated 2π exchange. It was found that the ground state to ground state Gamow–Teller strength is quite sensitive to effects from Brown–Rho scaling, namely the decreasing of the tensor force and suppression of S-wave attraction. At approximately 80% nuclear matter density (typical of the average density experienced by the valence p-shell nucleons in ^{14}C), the Gamow–Teller matrix element vanishes. In contrast, the Gamow–Teller strengths from excited states of ^{14}C to the ^{14}N ground state are only mildly sensitive to these medium effects.

[L.-W. Siu, J.W. Holt, T.T.S. Kuo and G.E. Brown (2009)]: It is ironic that for all of the different free-space nucleon–nucleon potentials constructed to date, it remains a long-standing problem in nuclear theory to reproduce simultaneously just two properties of idealized nuclear matter: its saturation binding energy per nucleon E/A and saturation density ρ_0. Numerous methods for calculating the many-body ground state energy using realistic NN potential models have consistently found saturation energies and densities that were 'not' in simultaneous agreement with the experimental values of $E/A = -16$ MeV and $\rho_0 = 0.16$ fm^{-3}. In this paper symmetric nuclear matter was calculated using a ring-diagram method where the $pp-hh$ ring diagrams with the V_{low-k} interaction were summed to all orders. But the resulting saturation energy and density were $E/A \simeq -20$ MeV and $\rho_0 \simeq 0.5$ fm^{-3}, still rather far from experiments. In the above calculations, the free-space V_{NN} interactions were employed. A main result of this paper was that the inclusion of the medium dependent corrections to V_{NN} from Brown–Rho scaling gave E/A and ρ_0 both in very good agreement with the expermental results. It is likely that the inclusion of Brown–Rho scaling or other similar density dependent corrections is indispensable for describing nuclear saturation. It is of interest that the saturation effect from Brown–Rho scaling was found in this paper to be remarkably close to that generated by the well-known empirical Skyrme three-nucleon force.

[H. Dong, T.T.S. Kuo and R. Machleidt (2009)]: Given the large nuclear densities predicted to occur within neutron stars, the density dependence of the nuclear interaction should be necessary for understanding neutron star structure and evolution. In this work microscopic calculations for neutron stars were carried out using a β-stable equation of state where the $pphh$ ring diagrams were summed to all orders with low-momentum NN interactions. The addition of Brown–Rho (BR) scaling to the free-space NN interaction was found to be essential: The maximum neutron star mass, radius and moment of inertia were respectively about ($1.2M_\odot$, 7.3km, $24M_\odot$km^2) when using the free space V_{NN}, all being too small.

However, the inclusion of the effects from BR scaling enhanced them to ($1.8M_\odot$, 8.9km, $60M_\odot\text{km}^2$) respectively, all in good agreement with empirical values. BR scaling has been important for light nuclei such as ^{14}C. This paper demonstrated that it is also important for the 'largest' nucleus — the neutron star.

VOLUME 66, NUMBER 21 PHYSICAL REVIEW LETTERS 27 MAY 1991

Scaling Effective Lagrangians in a Dense Medium

G. E. Brown
Department of Physics, State University of New York at Stony Brook, Stony Brook, New York 11794

Mannque Rho
Service de Physique Théorique, Centre d'Etudes Nucléaires de Saclay, 91191 Gif-sur-Yvette, France
(Received 8 January 1991)

By using effective chiral Lagrangians with a suitable incorporation of the scaling property of QCD, we establish the approximate *in-medium* scaling law, $m_\sigma^*/m_\sigma \approx m_N^*/m_N \approx m_\rho^*/m_\rho \approx m_\omega^*/m_\omega \approx f_\pi^*/f_\pi$. This has a highly nontrivial implication for nuclear processes at and above nuclear-matter density. Some concrete cases are cited in this paper.

PACS numbers: 21.65.+f, 11.40.Fy

One of the most exciting new directions in nuclear physics is to study how nuclear phenomena change as the environment changes. Thus relativistic heavy-ion experiments are to address the state of nuclear matter at high temperature and/or density and high-energy high-duty-cycle electron machines are to probe the properties of individual hadrons in close encounter with other strongly interacting matter. These are the processes that reflect the change of the strong-interaction vacuum as density and temperature are "dialed." Given a fundamental theory of strong interactions, i.e., QCD, one should, in principle, be able to calculate all the observables unambiguously as the environment is modified. It is possible that this feat will be eventually accomplished but it is highly unlikely that it will come soon enough to make contact with either on-going experiments or experiments to come in the near future. In this Letter, we propose, focusing on the density effect, to approach this problem from an effective-Lagrangian point of view, starting from low-energy effective theories based on spontaneously broken chiral symmetry that have been phenomenologically successful in describing low-energy and low-density hadronic interactions.

Briefly our strategy is follows. We start with a known structure of effective Lagrangians at low energy and zero density (i.e., free space), dictated by symmetries and other constraints of QCD (e.g., chiral symmetry with its current algebra and anomaly, trace anomaly, etc.). We are interested in how this theory evolves as density (or temperature) is increased. Embedding a hadron in dense matter is equivalent to changing the vacuum, thereby modifying quark and gluon condensates in QCD variables. Our first key assumption is that as the condensates change, the symmetries of the Lagrangian remain more or less *intact*, while the relevant scale is changed in a prescribed way that we will explain below. This means that we will have, as density increases, the same Lagrangian but with the masses and coupling constants of the theory *modified according to the symmetry constraints of QCD*. Some of the steps we take may appear to be *ad hoc* and drastic but we will argue that there is strong evidence from experiments that our scheme is *supported* by nature.

To illustrate our point, we start with the original Skyrme Lagrangian[1] consisting of the current-algebra term characterized by a dimensional constant f_π and the quartic stabilizing term characterized by a dimensionless constant ϵ. (More precise definitions of these quantities will be given later.) It is convenient to work with physical quantities by taking f_π to be the pion decay constant (experimentally ≈ 93 MeV) and the axial-vector coupling constant g_A in place of ϵ^2.[2,3] If we accept that baryons arise as solitons from the Skyrme Lagrangian, which seems to be fairly well established by now, we can use simple scaling arguments[3,4] to show that, modulo overall constants, the size and mass of the baryon go as

$$\langle r^2\rangle \sim g_A/f_\pi^2, \quad M \sim g_A^{1/2} f_\pi. \tag{1}$$

We now ask what happens to these quantities when the Skyrmion is embedded in a dense medium. As in Refs. 3 and 4, we will argue that as long as there is no phase transition that changes symmetries of the "vacuum" (or more precisely the ground state), the leading modification in the theory is in the basic constants of the Lagrangian, i.e.,

$$\langle r^2(\rho)\rangle^* \sim g_A^*(\rho)/f_\pi^{*2}(\rho), \quad M^*(\rho) \sim [g_A^*(\rho)]^{1/2} f_\pi^*(\rho). \tag{2}$$

Here we indicate the density-dependent quantities by asterisks. (From now on we will omit the density dependence in quantities with asterisks, unless explicitly required.) We identify (2) as the quasiparticle size and mass, respectively, for single baryons in the medium. Further (residual) interactions, say, between quasiparticles (or quasi-Skyrmions), can be introduced in a way analogous to the zero-density case using a given effective Lagrangian as specified more precisely below.

Now what about mesons? Chiral effective theories contain, in the large-N_c limit, Goldstone bosons, ordi-

VOLUME 66, NUMBER 21 PHYSICAL REVIEW LETTERS 27 MAY 1991

nary mesons, and a $U_A(1)$ boson called η', in addition to the baryons. The η' decouples from the rest of the world in the large-N_c limit and does not concern us for our purpose. In considering the properties of mesons in a dense medium, it is more convenient to introduce explicit degrees of freedom associated with other mesons than the Goldstone bosons. The most important of them all are the strong vector mesons (e.g., $\rho, \omega, \ldots$) and the glueballs. (In considering baryons, it is legitimate to "integrate out" these other meson degrees of freedom, as, e.g., in the Skyrme model.) A convenient framework to incorporate the strong vector mesons is through the hidden-gauge-symmetry strategy advocated by Bando, Kugo, and Yamawaki.[5] The glueballs that are needed for our purpose will be introduced through QCD anomalies.

Given a chiral Lagrangian which contains pseudoscalar (Goldstone) and vector (hidden-gauge) bosons, what is the effect of changing vacuum structure by density? We will argue that similarly to the baryon case, the leading effect is in the effective masses of the mesons and possibly their coupling constants.

The principal in-medium scaling law[6] conjectured in Ref. 7 which we would like to justify is

$$m_\sigma^*/m_\sigma \approx m_N^*/m_N \approx m_\rho^*/m_\rho \approx m_\omega^*/m_\omega\,, \tag{3}$$

where the masses without asterisks stand for free-space values. To do this, we follow the approach of Campbell, Ellis, and Olive[8] and introduce scale invariance into the effective Lagrangian. Their approach can be summarized as follows. The standard chiral Lagrangian lacks the trace anomaly which is an essential ingredient of QCD. (The axial anomaly, another ingredient of QCD, can be incorporated easily via the η' field but we are not concerned with it here.) In order to understand what this is, we consider the scale transformation

$$x \rightarrow {}^\lambda x = \lambda^{-1} x, \quad \lambda \geq 0\,, \tag{4}$$

under which an arbitrary field ϕ transforms with the canonical mass dimension (called "scale dimension") d_ϕ,

$$\phi(x) \rightarrow \lambda^{d_\phi}\phi(\lambda x)\,. \tag{5}$$

The conventional scale dimensions are 1 for scalar and vector fields and $\frac{3}{2}$ for fermion fields. An action with a Lagrangian density with scale dimension 4 is clearly invariant under scale transformation. This means that the classical QCD action in the chiral limit (i.e., with zero quark mass) is scale invariant. The quark mass term has scale dimension 3, and thus breaks scale invariance. Quantum mechanics introduces a profound modification to this structure. A well-known property of QCD is that the pure Yang-Mills action is not scale invariant at the quantum level due to dimensional transmutation. This can be stated in terms of the divergence of the dilatation current D_μ whose nonzero value signals the breaking of scale invariance or equivalently the nonvanishing trace of the energy-momentum tensor. Including the effect of quark masses, the trace anomaly is given (apart from an anomalous dimension contribution) by[9]

$$\partial^\mu D_\mu = \theta_\mu^\mu = \sum_q m_q \bar{q} q - \frac{\beta(g)}{g} \mathrm{Tr} G_{\mu\nu} G^{\mu\nu}\,, \tag{6}$$

where $\beta(g)$ is the usual beta function of QCD. The quantum nature of the trace anomaly is evident from the dependence on the strong fine-structure constant $g^2/4\pi$. *Our second main assumption is this: In order to be consistent with the scale property of QCD, effective Lagrangians must reproduce faithfully the trace anomaly (6) in terms of effective fields.* This assumption leads to the universal scaling that we argue holds well in the Nambu-Goldstone phase of chiral symmetry.

Now the chiral field $U = e^{i\pi/f_\pi}$ with $\pi = \boldsymbol{\tau}\cdot\boldsymbol{\pi}(x)$ has scale dimension zero.[10] A derivative brings in a scale dimension 1 and hence the current-algebra term $(f_\pi^2/4) \times \mathrm{Tr}(\partial_\mu U \partial^\mu U^\dagger)$ has scale dimension 2, the mass term $c\,\mathrm{Tr}(MU + \mathrm{H.c.})$, with c a constant and M the quark mass matrix, has scale dimension 0, and the Skyrme quartic term has scale dimension 4. Clearly this effective Lagrangian in its original form does not satisfy the trace condition (6). In order to restore consistency with the trace condition, it is convenient to define the scalar "glueball" field χ,

$$\langle 0 | (\mathrm{Tr} G_{\mu\nu} G^{\mu\nu})^{1/4} | 0 \rangle \propto \langle 0 | \chi | 0 \rangle \equiv \chi_0\,, \tag{7}$$

where $|0\rangle$ stands for the vacuum at zero density. (The vacuum at nonzero density or, more properly, the ground state will be denoted by $|0^*\rangle$.) As defined, the χ field has scale dimension 1. In terms of this field, the second term of Eq. (6), i.e., the trace anomaly, can be reproduced by a potential term of the form

$$V(\chi) \propto \chi^4 \ln(\chi/e^{1/4}\chi_0)\,. \tag{8}$$

With the help of the χ field, the Skyrme Lagrangian can be rewritten so as to be consistent with the QCD scale property,

$$\begin{aligned}\mathcal{L} = {} & \frac{f_\pi^2}{4}\left(\frac{\chi}{\chi_0}\right)^2 \mathrm{Tr}(\partial_\mu U \partial^\mu U^\dagger) + \frac{\epsilon^2}{4}\mathrm{Tr}[U^\dagger \partial_\mu U, U^\dagger \partial_\nu U]^2 \\ & + \tfrac{1}{2}\partial_\mu \chi \partial^\mu \chi \\ & + c\left(\frac{\chi}{\chi_0}\right)^3 \mathrm{Tr}(MU + \mathrm{H.c.}) + V(\chi)\,,\end{aligned} \tag{9}$$

where a scale-invariant kinetic-energy term for the glueball field is introduced. As written, the scale breaking resides entirely in the last line of Eq. (9), all the rest being scale invariant. It should be particularly noted that the Skyrme quartic term is scale invariant by itself. This is a key point for later discussions.[11]

One immediate consequence of the above discussion is that the quark condensate in a dense medium scales as

$$\langle 0^* | \bar{q} q | 0^* \rangle / \langle 0 | \bar{q} q | 0 \rangle = (\chi_*/\chi_0)^3, \quad \chi_* \equiv \langle 0^* | \chi | 0^* \rangle\,. \tag{10}$$

VOLUME 66, NUMBER 21 PHYSICAL REVIEW LETTERS 27 MAY 1991

This suggests defining an *effective* pion decay constant as

$$f_\pi^* = f_\pi \chi_*/\chi_0 , \tag{11}$$

so that

$$\langle 0^*|\bar{q}q|0^*\rangle/\langle 0|\bar{q}q|0\rangle = (f_\pi^*/f_\pi)^3 \tag{12}$$

which is the relation deduced in Ref. 7. Let us assume that as density increases, the potential develops a minimum at $\chi = \chi_*$. This suggests expanding the χ field as

$$\chi = \chi_* + \chi' , \tag{13}$$

with χ' a fluctuating field. In the vacuum characterized by χ_*, the Lagrangian for the chiral field can be written as

$$\mathcal{L}_U = \frac{f_\pi^{*2}}{4}\mathrm{Tr}(\partial_\mu U \partial^\mu U^\dagger) + \frac{\epsilon^2}{4}\mathrm{Tr}[U^\dagger \partial_\mu U, U^\dagger \partial_\nu U]^2 + c\left(\frac{f_\pi^*}{f_\pi}\right)^3 \mathrm{Tr}(MU + \mathrm{H.c.}) + \cdots , \tag{14}$$

with

$$U(x) = \exp(i\pi^*/f_\pi^*), \quad \pi^* \equiv \pi\chi_*/\chi_0 . \tag{15}$$

The ellipsis stands for other fields including the χ' field with which we are not directly concerned here. We have thus effectively assigned a scale dimension 1 to the f_π^* and to the pion field π^*. There is no explicit σ field of the linear σ model (i.e., the $\bar{q}q$ scalar in the Nambu-Jona-Lasinio model). However, a strong mixing is expected between the glueball field χ' and two pions in the scalar channel so as to give the effective σ with mass $m_\sigma \sim 560$ MeV needed in nuclear physics. Thus in nature we should identify the "observed" lowest-mass scalar to be a mixture of a quark-containing scalar and the glueball scalar. This immediately suggests the scaling

$$m_\chi^*/m_\chi \approx f_\pi^*/f_\pi \to m_\sigma^*/m_\sigma \approx f_\pi^*/f_\pi , \tag{16}$$

with the σ field now understood to be the effective *nuclear physics* scalar.

The Lagrangian (14), together with (1) and (2), implies that the nucleon mass scales as

$$m_N^*/m_N \approx (g_A^*/g_A)^{1/2} f_\pi^*/f_\pi . \tag{17}$$

Since, modulo loop corrections, g_A is scale invariant (because the coefficient ϵ^2 of the Skyrme quartic term to which g_A is related is scale invariant), we may set $g_A^*/g_A \approx 1$. Alternatively we may invoke the phenomenological observation that the in-medium constant g_A^* saturates rapidly in light nuclei at 1 and stays constant as density (or mass number) increases.[12] In some problems[13] the density dependence of g_A^* is important but not at densities higher than nuclear matter ρ_0. The rapid variation in g_A between $\rho = 0$ and $\rho = \rho_0$ is a loop effect as suggested in Ref. 4, indicating a possible new (lower) scale *induced* in nuclei. Accounting for such effects is an important subject for classical nuclear physics but not relevant at the mean-field order we are concerned with. We thus establish that

$$m_N^*/m_N \approx f_\pi^*/f_\pi . \tag{18}$$

This result was also obtained using *in-medium* QCD sum rules.[14]

Before proceeding to complete the scaling relation (3), we pause to emphasize that Eq. (18) is extremely practical. It tells us that knowledge of the effective mass of the nucleon m_N^* supplies the order parameter f_π^* of the broken-symmetry regime at finite density. Essentially every nuclear physicist knows what m_N^* is at (at least) nuclear matter density although there is no consensus on its exact value. Quite conservatively (from the standpoint of the difference of m_N^* from m_N),

$$m_N^*(\rho_0)/m_N \approx 0.8 . \tag{19}$$

This, coupled with Eqs. (18) and (12), has the remarkable consequence that

$$\langle 0^*|\bar{q}q|0^*\rangle/\langle 0|\bar{q}q|0\rangle \approx \tfrac{1}{2} \text{ at } \rho = \rho_0 ; \tag{20}$$

that is, the quark condensate has dropped by $\sim 50\%$ already in the middle of nuclei.

Let us now consider the vector-meson masses. In a generalized model with vector mesons (i.e., the hidden-gauge-symmetric theory[5]), one has the Kawarabayashi-Suzuki-Riazuddin-Fayyazuddin relation

$$m_V^2 = 2g^2 f_\pi^2 , \tag{21}$$

where g is the hidden-gauge coupling. We will ignore $O(1/N_c)$ correction and set $m_V = m_\rho = m_\omega$. In terms of the original Skyrme model, we can identify the gauge coupling g^2 with $(8\epsilon^2)^{-1}$ and hence find,[15] thanks to the scale invariance of the Skyrme quartic term,

$$m_V^*/m_V \approx f_\pi^*/f_\pi . \tag{22}$$

This completes the relation (3). [A striking omission from Eq. (3) is the scaling for Goldstone bosons, in particular for the pion which figures importantly in nuclear physics. The Lagrangian (14) implies that the pion mass m_π^* scales as $(f_\pi^*)^{1/2}$ which differs from Eq. (3). The story turns out to be much more intricate in the case of Goldstone bosons. As discussed in Ref. 16, there is an intricate effect associated with the explicit chiral-symmetry-breaking term which turns out to restore approximately, at least at the high temperature relevant in high-energy heavy-ions collisions, the universal scaling to the pion.]

We stress that our scaling relation is a *mean-field* relation that emerges at the tree level of the effective Lagrangian (14), the key point of our discussion being that (14) with scaled masses be taken as the effective Lagrangian to calculate physical observables according to the chiral perturbation scheme as, e.g., discussed in Ref. 17.

We see a lot of consequences of these dropping masses

in nature, some of which are discussed in Refs. 13, 14, and 18–21. In fact, quite detailed recent work by Hosaka and Toki[22] who investigated effective matrix elements of the in-medium two-body interaction in the s,d shell shows that a uniform decrease in various masses yields G-matrix elements which are consistent with empirical matrix elements. Perhaps more significantly, they are precursors to QCD phase transitions: From Eqs. (1) and (2), combined with the preceding arguments, we see that the nucleon radius increases rapidly beyond $\rho=\rho_0$,[23] i.e.,

$$\langle r^2\rangle^*/\langle r^2\rangle \approx f_\pi^2/f_\pi^{*2}\,. \quad (23)$$

This rapid increase in radius with higher density may be interpreted as deconfinement as suggested in Ref. 8. Implications of the scaling presented in this paper on QCD phase transitions in heavy-ion collisions were recently discussed by Brown, Bethe, and Pizzochero.[24]

We are living in a "swelled" world with all the masses dropping as density increases, how have the many calculations in conventional nuclear physics held up for so long in such a drastically different scenario? The answer to this puzzle is that to the extent that there is a common scale, it can be scaled out, leaving relatively small effects. To put it more precisely, to the extent that m_i^*/m_i equals a common $\lambda(\rho)$ for any i, we have to leading order (in the mean-field sense defined above)

$$H(m_i^*,r)\approx\lambda(\rho)H(m_i,x)\,, \quad (24)$$

with $m_i^*r=m_ix$ defining x. Since $\lambda(\rho)\gtrsim 0.8$ in nuclei, the energy is shifted little, say, less than 20%. But the distance swells by $\lambda(\rho)^{-1}$, i.e., $r=(m_i/m_i^*)x$. This "swelling" must of course show up in certain processes. Indeed one such case is the spin-orbit interaction in nuclei: It has been pointed out[21,25] that the spin-orbit interaction in nuclei most sensitively displays the density-dependent masses, giving us a factor $(m/m^*)^3$, as is obvious from its dimensionality in r.

We would like to thank John Ellis for communication of his work which showed us how to tie in our scaling arguments with the trace anomaly. This work was partially supported by the U.S. Department of Energy under Grant No. DE-FG02-88 ER 4038.

---

[1] T. H. R. Skyrme, Nucl. Phys. **31**, 556 (1962).

[2] G. E. Brown, A. D. Jackson, M. Rho, and V. Vento, Phys. Lett. **140B**, 285 (1984).

[3] I. Zahed and G. E. Brown, Phys. Rep. **142**, 1 (1986).

[4] M. Rho, Phys. Rev. Lett. **54**, 767 (1988).

[5] M. Bando, T. Kugo, and K. Yamawaki, Phys. Rep. **164**, 217 (1988).

[6] The Nambu–Jona-Lasinio model provides a microscopic description of the "dropping" masses as first shown by V. Bernard, U.-G. Meissner, and I. Zahed, Phys. Rev. Lett. **59**, 966 (1987). It predicts, in a medium, $m_N^*/m_\sigma^*=3/2$ *for all density* and hence $m_\sigma^*/m_\sigma=m_N^*/m_N$. However, since the cutoff mass used in the model is comparable to vector-meson masses, it is difficult to trust its prediction of the in-medium behavior of the vector mesons.

[7] G. E. Brown, Nucl. Phys. **A488**, 659c (1988); Z. Phys. C **38**, 291 (1988).

[8] B. A. Campbell, J. Ellis, and K. A. Olive, Nucl. Phys. **B345**, 325 (1990).

[9] J. Collins, A. Duncan, and S. Joglekar, Phys. Rev. D **16**, 438 (1977); N. K. Nielsen, Nucl. Phys. **B120**, 212 (1977).

[10] J. Ellis, Nucl. Phys. **B22**, 478 (1976).

[11] The scale invariance of this term explains why the Skyrme Lagrangian gives, in the high-density limit, the same asymptotic property as noninteracting quarks.

[12] B. Buck and F. Perez, Phys. Rev. Lett. **50**, 1975 (1983).

[13] G. E. Brown and M. Rho, Phys. Lett. B **222**, 324 (1989).

[14] G. E. Brown, in Proceedings of Akito Arima Symposium, May 1990 [Nucl. Phys. A (to be published)].

[15] This together with Eq. (21) gives $[(g^*/m_V^*)/(g/m_V)]^2 \approx (f_\pi^*/f_\pi)^{-2}\approx(m_N^*/m_N)^{-2}$ as suggested by G. E. Brown and M. Rho [Phys. Lett. B **237**, 3 (1990)] to argue for the disappearance of tensor forces in nuclear matter.

[16] G. E. Brown, V. Koch, and M. Rho, State University of New York, Stony Brook, report, 1990 (to be published).

[17] S. Weinberg, Phys. Lett. B **251**, 288 (1990); M. Rho, Phys. Rev. Lett. **66**, 1275 (1991).

[18] Brown and Rho (Ref. 15).

[19] G. E. Brown, C. B. Dover, P. B. Siegel, and W. Weise, Phys. Rev. Lett. **60**, 2723 (1988).

[20] T. Hatsuda, H. Hogaasen, and M. Prakash, Phys. Rev. C **42**, 2212 (1990).

[21] G. E. Brown, A. Sethi, and N. M. Hintz (to be published).

[22] A. Hosaka and H. Toki, University of Pennsylvania Report No. 0449T, 1990 (to be published).

[23] Up to $\rho=\rho_0$, the nucleon radius is roughly constant because of the interplay between the factor g_A^*/g_A which decreases and f_π/f_π^* which increases. This effect is important for y scaling in nuclei as shown in Ref. 13.

[24] G. E. Brown, H. A. Bethe, and P. Pizzochero, "The Hadron to Quark/Gluon Transition in Relativistic Heavy Ion Collisions," Caltech report, 1991 (to be published).

[25] G. E. Brown, H. Müther, and M. Prakash, Nucl. Phys. **A506**, 565 (1990).

ELSEVIER

Available online at www.sciencedirect.com
ScienceDirect

Nuclear Physics A 785 (2007) 322–338

Nuclear matter with Brown–Rho-scaled Fermi liquid interactions

Jeremy W. Holt *, G.E. Brown, Jason D. Holt, T.T.S. Kuo

Department of Physics, SUNY, Stony Brook, NY 11794, USA

Received 18 October 2006; accepted 18 December 2006

Available online 29 December 2006

Abstract

We present a description of symmetric nuclear matter within the framework of Landau Fermi liquid theory. The low momentum nucleon–nucleon interaction $V_{\text{low-}k}$ is used to calculate the effective interaction between quasiparticles on the Fermi surface, from which we extract the quasiparticle effective mass, the nuclear compression modulus, the symmetry energy, and the anomalous orbital gyromagnetic ratio. The exchange of density, spin, and isospin collective excitations is included through the Babu–Brown induced interaction, and it is found that in the absence of three-body forces the self-consistent solution to the Babu–Brown equations is in poor agreement with the empirical values for the nuclear observables. This is improved by lowering the nucleon and meson masses according to Brown–Rho scaling, essentially by including a scalar tadpole contribution to the meson and nucleon masses, as well as by scaling g_A. We suggest that modifying the masses of the exchanged mesons is equivalent to introducing a short-range three-body force, and the net result is that the Brown–Rho double decimation [G.E. Brown, M. Rho, Phys. Rep. 396 (2004) 1] is accomplished all at once.

1. Introduction

Landau's theory of normal Fermi liquids [2–4] describes strongly interacting many-body systems in terms of weakly interacting quasiparticles. Provided that the quasiparticles lie sufficiently close to the Fermi surface, they will be long-lived and constitute appropriate degrees of freedom for the system. The central aim of the theory is to determine the quasiparticle interaction, either

* Corresponding author.
E-mail address: jeholt@grad.physics.sunysb.edu (J.W. Holt).

0375-9474/$ – see front matter
doi:10.1016/j.nuclphysa.2006.12.099

phenomenologically or microscopically, with which it is possible to describe the low-energy, long-wavelength excitations of the system. This, in turn, is sufficient for the description of many bulk equilibrium properties of the interacting Fermi system. The initial application of Fermi liquid theory to nuclear physics was the phenomenological description of finite nuclei and nuclear matter by Migdal [5,6], and later a microscopic approach to Fermi liquid theory based on the Brueckner–Bethe–Goldstone reaction matrix theory was developed by Bäckman [7] and others [8,9] to describe nuclear matter. Although the latter approach was quantitatively successful, it was observed [9] that Brueckner–Bethe–Goldstone theory is less reliable in the vicinity of the Fermi surface due to the use of angle-averaged Pauli operators and the unsymmetrical treatment of particle and hole self energies, which leads to an unphysical energy gap at the Fermi surface.

With the recent development of a nearly universal low-momentum nucleon–nucleon (NN) interaction $V_{\text{low-}k}$ [10] derived from renormalization group methods, the application of Fermi liquid theory to nuclear matter has received renewed attention [11–13]. The strong short-distance repulsion incorporated into all high-precision NN potential models is integrated out in low momentum interactions, rendering them suitable for perturbation theory calculations. Although limited Brueckner–Hartree–Fock studies [14] indicate that saturation is not achieved with $V_{\text{low-}k}$ at a fixed momentum cutoff Λ, it has recently been shown [15] that by supplementing $V_{\text{low-}k}$ with the leading-order chiral three-nucleon force, nuclear matter does saturate, thereby justifying the use of $V_{\text{low-}k}$ in studies of nuclear matter.

Although many properties of the interacting ground state are beyond the scope of Fermi liquid theory, the quasiparticle interaction is directly related to several nuclear observables, including the compression modulus, symmetry energy, and anomalous orbital gyromagnetic ratio. As originally shown by Landau, the quasiparticle interaction is obtained from a certain limit of the four-point vertex function in the particle–hole channel. It is well known that using realistic NN interactions in the lowest order approximation to the quasiparticle interaction is insufficient to stabilize nuclear matter, as evidenced by a negative value of the compression modulus. This general phenomenon is observed in our calculations with $V_{\text{low-}k}$ as well. However, stability is achieved by treating the exchange of density, spin, and isospin collective excitations to all orders in perturbation theory. The inclusion of these virtual collective modes in the quasiparticle interaction is carried out through the induced interaction formalism of Babu and Brown [16], which was originally developed for the description of liquid ^{3}He and later applied to nuclear matter by Sjöberg [17,18]. Subsequent work [19,20] has confirmed the importance of the induced interaction in building up correlations around a single quasiparticle, thereby increasing the compression modulus.

Our study is motivated in part by the work of Schwenk et al. [11], who were able to predict the spin-dependent parameters of the quasiparticle interaction from the experimentally extracted spin-independent parameters. Crucial to these calculations was a novel set of sum rules, derived from the induced interaction formalism, based on a similar treatment by Bedell and Ainsworth [21] to liquid ^{3}He. In this paper we present a fully self-consistent solution to the Babu–Brown induced interaction equations for symmetric nuclear matter. Our iterative solution turns out to be qualitatively similar to the results of [11], but we find that at nuclear matter density the compression modulus and symmetry energy are smaller than the experimentally observed values while the anomalous orbital gyromagnetic ratio is too large, suggesting the possibility that important phenomena have been neglected.

We propose to extend this study by including hadronic modifications associated with the partial restoration of chiral symmetry at nuclear matter density, as suggested in [22]. In this scenario, referred to as Brown–Rho scaling, the dynamically generated hadronic masses drop in the approach to chiral restoration, and at nuclear matter density it is expected that the masses of the light hadrons (other than the masses of the pseudoscalar mesons, which are protected by their Goldstone nature) decrease by approximately 20%. The success of one-boson-exchange and chiral EFT potentials in describing the nucleon–nucleon interaction suggests that a modification of meson masses in medium ought to have verifiable consequences in low energy nuclear physics. Although there is much current theoretical and experimental effort devoted to the program of assessing these medium modifications, the consequences for low-energy nuclear physics have yet to be fully explored.

Applying the mass scaling suggested in [22] to our calculations of nuclear matter, we obtain a set of Fermi liquid coefficients in better agreement with both experiment and the nontrivial sum rules derived in [11]. Explicit three-body forces, though essential for a complete description of nuclear matter, have been neglected in this study. However, we argue that modifying the vector meson masses is equivalent to including a specific short-ranged three-body force. We conclude with a discussion of the consequences of Brown–Rho scaling on the tensor force, which is diminished by the increasing strength of ρ-meson exchange.

2. Fermi liquid theory

In this section we present a short description of Fermi liquid theory and its application to nuclear physics with emphasis on the microscopic foundation of the theory. The main assumption underlying Landau's description of many-body Fermi systems is that there is a one-to-one correspondence between states of the ideal system and states of the interacting system. As one gradually turns on the interaction, the noninteracting particles become "dressed" through interactions with the many-body medium and evolve into weakly interacting quasiparticles. The interacting system is in many ways similar to an ideal system in that the classification of energy states remains unchanged and there is a well-defined Fermi surface, but the quasiparticles acquire an effective mass m^* and finite lifetimes $\tau \sim (k - k_F)^{-2}$. The energy of the interacting system is a complicated functional of the quasiparticle distribution function, and in general the exact dependence is inaccessible. But one can extract important information about bulk properties of the system by considering small changes in the distribution function. Expanding to second order, one finds

$$\delta E = \sum_{\mathbf{k}_1} \epsilon^{(0)}_{\mathbf{k}_1} \delta n(\mathbf{k}_1) + \frac{1}{2\Omega} \sum_{\mathbf{k}_1,\mathbf{k}_2} f(\mathbf{k}_1, \mathbf{k}_2) \delta n(\mathbf{k}_1) \delta n(\mathbf{k}_2) + \mathcal{O}(\delta n^3). \tag{1}$$

In this equation Ω is the volume of the system, $\epsilon^{(0)}_{\mathbf{k}_1}$ is the energy added to the system by introducing a single quasiparticle with momentum $\mathbf{k}_1$ (note that for $|\mathbf{k}_1| \equiv k_1 = k_F$, $\epsilon^{(0)}_{\mathbf{k}_1}$ is just the chemical potential), and $f(\mathbf{k}_1, \mathbf{k}_2)$ describes the interaction between two quasiparticles.

Since the quasiparticle interaction $f(\mathbf{k}_1, \mathbf{k}_2)$ is the fundamental quantity of interest in Fermi liquid theory, we will carefully discuss its properties and its relationship to nuclear observables. Assuming the interaction to be purely exchange, it can be written as

$$f(\mathbf{k}_1, \mathbf{k}_2) = \frac{1}{N_0} \big[F(\mathbf{k}_1, \mathbf{k}_2) + F'(\mathbf{k}_1, \mathbf{k}_2) \tau_1 \cdot \tau_2 + G(\mathbf{k}_1, \mathbf{k}_2) \sigma_1 \cdot \sigma_2 + G'(\mathbf{k}_1, \mathbf{k}_2) \tau_1 \cdot \tau_2 \sigma_1 \cdot \sigma_2 \big], \tag{2}$$

where we have factored out the density of states per unit volume at the Fermi surface, $N_0 = \frac{2m^* k_F}{\hbar^2 \pi^2}$, which leaves dimensionless Fermi liquid parameters denoted by F, G, F', G'. The spin–

orbit interaction is neglected because it vanishes in the long wavelength limit in which we will be interested. Also, we have not included tensor operators (which would greatly complicate our calculation) because the tensor force contributes almost completely in second order, as shown in the original paper by Kuo and Brown [23], as an effective central interaction in the $^3S_1$ state. In [11] the tensor Fermi liquid parameters for symmetric nuclear matter were calculated from $V_{\text{low-}k}$ in which the dominant second-order contributions from one-pion exchange were included. Since quasiparticles are well-defined only near the Fermi surface, we assume that $k_1 = k_F = k_2$. In this case the dimensionless Fermi liquid parameters F, F', G, G' depend on only the angle between $\mathbf{k}_1$ and $\mathbf{k}_2$, which we call θ. Then it is convenient to perform a Legendre polynomial expansion as follows

$$F(\mathbf{k},\mathbf{k}') = \sum_l F_l P_l(\cos\theta), \qquad G(\mathbf{k},\mathbf{k}') = \sum_l G_l P_l(\cos\theta), \quad \text{etc.} \tag{3}$$

The Fermi liquid parameters $F_l, G_l, \ldots$ decrease rapidly for larger l, and so there are only a small number of parameters that can either be fit to experiment or calculated microscopically.

In the original application of the theory to liquid ^{3}He and nuclear systems, the quasiparticle interaction was obtained phenomenologically by fitting the dimensionless Fermi liquid parameters to relevant data. For nuclear matter several important relationships exist between nuclear observables and the Fermi liquid parameters. Galilean invariance can be used [2] to connect the Landau parameter F_1 to the quasiparticle effective mass

$$\frac{m^*}{m} = 1 + \frac{F_1}{3}. \tag{4}$$

Adding a small number of neutrons and removing the same number of protons from the system will increase and decrease, respectively, the density of protons and neutrons in the system (and therefore the Fermi energies of the two species). The change in the energy, described by the symmetry energy β, can be related [6] to the Landau parameter F_0'

$$\beta = \frac{\hbar^2 k_F^2}{6m^*}(1 + F_0'). \tag{5}$$

In a similar way, the equal increase or decrease of the proton and neutron densities leads to a relationship between the scalar–isoscalar Landau parameter F_0 and the compression modulus $\mathcal{K}$

$$\mathcal{K} = \frac{3\hbar^2 k_F^2}{m^*}(1 + F_0). \tag{6}$$

Finally, it can be shown [6] that an odd nucleon added just above the Fermi sea induces a polarization of the medium leading to an anomalous contribution to the orbital gyromagnetic ratio of the form

$$g_l^p = [1 - \delta g_l]\mu_N, \qquad g_l^n = [\delta g_l]\mu_N, \tag{7}$$

where δg_l is given by

$$\delta g_l = \frac{1}{6}\frac{F_1' - F_1}{1 + F_1/3}. \tag{8}$$

Clearly there are certain values of the Landau parameters that are physically unreasonable. For instance, if $F_1 < -3$ or $F_0 < -1$, the effective mass or compression modulus would be negative. Quite generally it can be shown [24] that the Landau parameters must satisfy stability conditions

$$X_l > -(2l + 1), \tag{9}$$

where X represents F, G, F', G'.

A rigorous foundation for the assumptions underlying Landau's theory can be obtained through formal many-body techniques [25,26]. It is not our goal to reproduce the original arguments [4], but rather to give a clear motivation for the diagrammatic expansion leading to the quasiparticle interaction. Starting from the usual definition of the four-point Green's function in momentum space

$$\begin{aligned} &G_{\alpha\beta,\gamma\delta}(k_1,k_2;k_3,k_4) \\ &= (2\pi)^8\delta^{(4)}(k_1+k_2-k_3-k_4)\Big[G_{\alpha\gamma}(k_1)G_{\beta\delta}(k_2)\delta^{(4)}(k_1-k_3)-G_{\alpha\delta}(k_1)G_{\beta\gamma}(k_2) \\ &\times\delta^{(4)}(k_2-k_3)+\frac{i}{(2\pi)^4}G(k_1)G(k_2)G(k_3)G(k_4)\Gamma_{\alpha\beta,\gamma\delta}(k_1,k_2;k_3,k_4)\Big], \end{aligned} \tag{10}$$

where $G(k_1)$ is the Fourier transform of $G(xt,x't')$ and $k_1,\ldots,k_4$ represent four-vectors (e.g. $k_1=(\mathbf{k}_1,\omega_1)$), it can be shown that the quasiparticle interaction is related to a certain limit of the four-point vertex function $\Gamma_{\alpha\beta,\gamma\delta}(k_1,k_2;k_3,k_4)$. From energy–momentum conservation $(k_1+k_2=k_3+k_4)$ we can write $k_3-k_1=K=k_2-k_4$ and therefore define $\Gamma(k_1,k_2;K)=\Gamma(k_1,k_2;k_3,k_4)$. The important point is that since we are considering only low-energy long-wavelength excitations, the particle–hole energy–momentum K should be small. We can write a Bethe–Salpeter equation for the fully reducible vertex function Γ in terms of the ph irreducible vertex function $\tilde{\Gamma}$ in the direct channel with momentum transfer K:

$$\begin{aligned} &\Gamma_{\alpha\beta,\gamma\delta}(k_1,k_2;K) \\ &= \tilde{\Gamma}_{\alpha\beta,\gamma\delta}(k_1,k_2;K)-i\sum_{\epsilon,\eta}\int\frac{d^4q}{(2\pi)^4}\tilde{\Gamma}_{\alpha\epsilon,\gamma\eta}(k_1,q;K)G(q)G(q+K)\Gamma_{\eta\beta,\epsilon\delta}(q,k_2;K) \end{aligned} \tag{11}$$

shown diagrammatically in Fig. 1. The product of propagators may have singularities in the limit that $K\to 0$, in which case the poles can be replaced by δ-functions inside the integral:

$$G(q)G(q+K)=\frac{2i\pi z^2\hat{\mathbf{q}}\cdot\mathbf{K}}{\omega-v_F\hat{\mathbf{q}}\cdot\mathbf{K}}\delta(\epsilon-\mu)\delta(q-k_F)+\phi(\mathbf{q}), \tag{12}$$

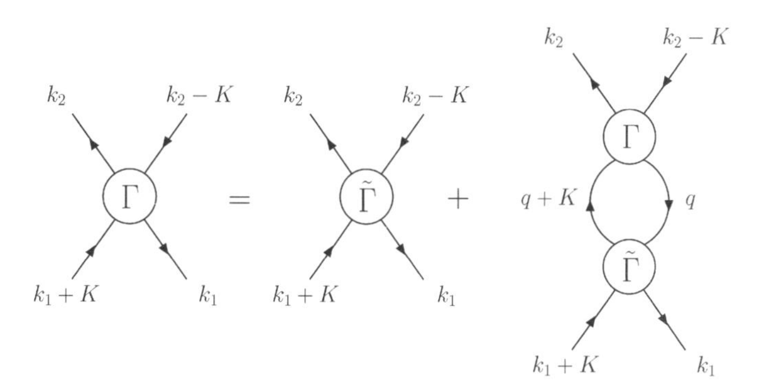

Fig. 1. The Bethe–Salpeter equation for the fully irreducible vertex function Γ in terms of the ph irreducible vertex function $\tilde{\Gamma}$.

where z is the renormalization at the quasiparticle pole and $\phi(\mathbf{q})$ accounts for the multipair background. The limit $K = (\omega, \mathbf{K}) \to 0$ depends on the relative ordering of the two limits $\mathbf{K} \to 0$ and $\omega \to 0$. Defining

$$\begin{aligned} \Gamma^{\omega}(k_1, k_2) &= \lim_{\omega\to 0}\lim_{\mathbf{K}\to 0} \Gamma(k_1, k_2; K) \quad \text{and} \\ \Gamma^{K}(k_1, k_2) &= \lim_{\mathbf{K}\to 0}\lim_{\omega\to 0} \Gamma(k_1, k_2; K), \end{aligned} \tag{13}$$

from Eq. (12) we see that the product of propagators is regular for Γ^{ω}. Thus, to calculate Γ^{ω} we must first calculate the ph irreducible diagrams belonging to $\tilde{\Gamma}$ and then iterate via the Bethe–Salpeter equation with the intermediate multipair background ϕ. The δ-function singularities in Γ^{K} can be used to perform the integrals over q_0 and $|\mathbf{q}|$, and through algebraic manipulation it is possible to combine Γ^{ω} and Γ^{K} into a single integral equation

$$\begin{aligned} &\Gamma^{K}_{\alpha\beta,\gamma\delta}(k_1, k_2) \\ &\quad = \Gamma^{\omega}_{\alpha\beta,\gamma\delta}(k_1, k_2) - \frac{1}{16\pi} N_0 z^2 \sum_{\epsilon,\eta} \int d\Omega_q\, \Gamma^{\omega}_{\alpha\epsilon,\gamma\eta}(k_1, q)\Gamma^{K}_{\eta\beta,\epsilon\delta}(q, k_2). \end{aligned} \tag{14}$$

Physically, Γ^{ω} represents the exchange of virtual excitations between quasiparticles, and Γ^{K} represents the forward scattering of quasiparticles at the Fermi surface. By relating these vertex functions to the equations describing zero sound, Landau [4] was able to make the identifications

$$f(k_1, k_2) = z^2\Gamma^{\omega}(k_1, k_2) \quad \text{and} \quad a(k_1, k_2) = z^2\Gamma^{k}(k_1, k_2), \tag{15}$$

where $f(k_1, k_2)$ is just the quasiparticle interaction introduced earlier and $a(k_1, k_2)$ is the physical scattering amplitude.

3. Induced interaction

In principle one could exactly calculate the quasiparticle interaction by summing up all ph irreducible diagrams contributing to the ph vertex function in the limit $k/\omega \to 0$. Since this is not practicable in general, one must limit the calculation to a certain subset of diagrams. We could proceed by calculating the relevant diagrams order by order, but this would miss an essential point, which we now elaborate. From Eqs. (14) and (15), we see that the physical scattering amplitude $a(k_1, k_2)$ iterates the quasiparticle interaction to all orders through an integral equation shown schematically in Fig. 2. If only a finite set of diagrams are included in the quasiparticle interaction, then the scattering amplitude will not be antisymmetric. For instance, if we include only the bare particle–hole antisymmetrized vertex shown in Fig. 3(a), then diagram (b) will be contained in the equation for the scattering amplitude but its exchange diagram, labeled (c), will not. Quantitatively, the fact that the scattering amplitude is antisymmetric requires that it vanish in singlet-odd and triplet-odd states as the Landau angle θ approaches 0. This leads to two constraints [4,27] on the Fermi liquid parameters in the form of sum rules:

$$\sum_l \left(\frac{F_l}{1 + F_l/(2l+1)} + 3\frac{G'_l}{1 + G'_l/(2l+1)} \right) = 0, \tag{16}$$

$$\sum_l \left(\frac{2}{3}\frac{F_l}{1 + F_l/(2l+1)} + \frac{F'_l}{1 + F'_l/(2l+1)} + \frac{G_l}{1 + G_l/(2l+1)} \right) = 0. \tag{17}$$

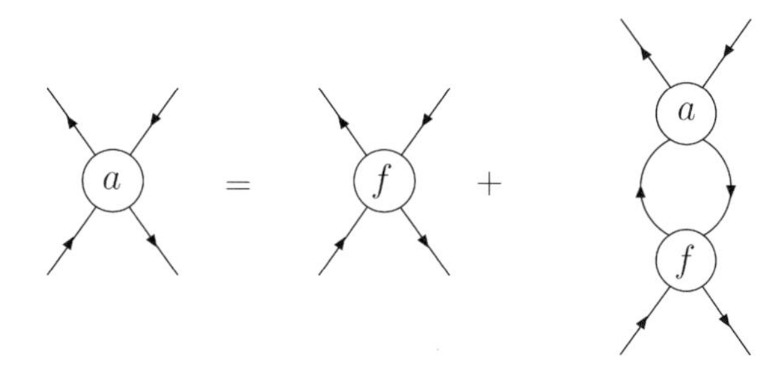

Fig. 2. The diagrammatic relationship between the physical scattering amplitude a and the quasiparticle interaction f.

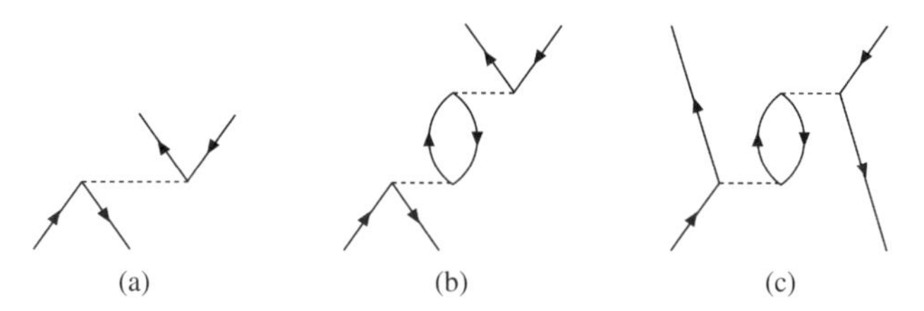

Fig. 3. Diagrams contributing to the quasiparticle interaction f and the scattering amplitude a. Diagrams (a) and (c) contribute to f, whereas all three contribute to a.

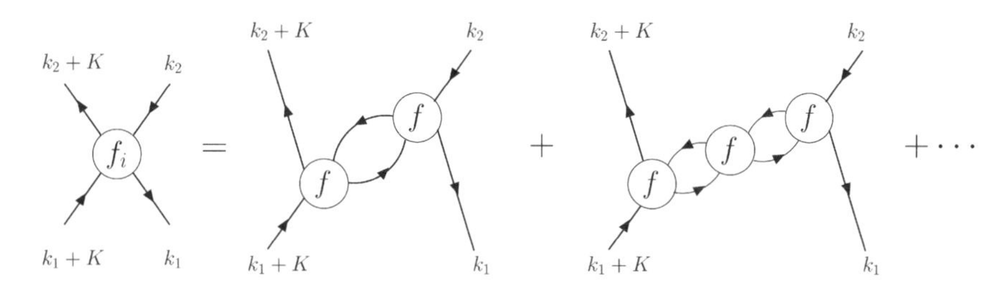

Fig. 4. The diagrammatic form of the induced interaction. In the limit that $k_1 = k_2$ it can be shown that the external lines exactly couple to particle–hole excitations through the f function.

Clearly, the sum rules must be satisfied for the "correct" set of Fermi liquid parameters describing nuclear matter. To account for this infinite set of exchange diagrams, Babu and Brown [16] proposed separating the quasiparticle interaction into a *driving term* and an *induced term*:

$$f(k,k') = f_d(k,k') + f_i(k,k'), \tag{18}$$

where the induced interaction is defined to contain those diagrams that would be the exchange terms necessary to preserve the antisymmetry of $a(k_1,k_2)$. Then the induced interaction is given by a diagrammatic expansion shown in Fig. 4. Physically, the induced interaction represents that part of the quasiparticle interaction that results from the exchange of virtual collective modes, which can be classified as density, spin, or isospin excitations. In the limit that $\mathbf{k}_1 \to \mathbf{k}_2$ it can be rigorously proved [16] that the coupling of quasiparticles to these collective excitations is precisely through the quasiparticle interaction itself, thereby justifying the diagrammatic expression in Fig. 4.

The relationship between the induced interaction and the full quasiparticle interaction was derived by Babu and Brown [16] for liquid ^{3}He and applied to nuclear matter by Sjöberg [17]. To lowest order in the Fermi liquid parameters, the induced interaction is given by

$$4F_i = \left[\frac{F_0^2}{1+F_0\alpha_0} + \frac{3F_0'^2}{1+F_0'\alpha_0} + \frac{3G_0^2}{1+G_0\alpha_0} + \frac{9G_0'^2}{1+G_0'\alpha_0}\right]\alpha_0,$$
$$4G_i = \left[\frac{F_0^2}{1+F_0\alpha_0} + \frac{3F_0'^2}{1+F_0'\alpha_0} - \frac{G_0^2}{1+G_0\alpha_0} - \frac{3G_0'^2}{1+G_0'\alpha_0}\right]\alpha_0,$$
$$4F_i' = \left[\frac{F_0^2}{1+F_0\alpha_0} - \frac{F_0'^2}{1+F_0'\alpha_0} + \frac{3G_0^2}{1+G_0\alpha_0} - \frac{3G_0'^2}{1+G_0'\alpha_0}\right]\alpha_0,$$
$$4G_i' = \left[\frac{F_0^2}{1+F_0\alpha_0} - \frac{F_0'^2}{1+F_0'\alpha_0} - \frac{G_0^2}{1+G_0\alpha_0} + \frac{G_0'^2}{1+G_0'\alpha_0}\right]\alpha_0, \tag{19}$$

where

$$\alpha_0 = \alpha_0(\mathbf{q}, 0) = \frac{1}{2} + \frac{1}{2}\left(\frac{q}{4k_F} - \frac{k_F}{q}\right)\ln\frac{k_F - q/2}{k_F + q/2} \tag{20}$$

is the Lindhard function, which is related to the density–density correlation function $\chi_{\rho\rho}$ by

$$\chi_{\rho\rho}(\mathbf{q}, \omega) = \frac{-\alpha_0(\mathbf{q}, \omega)}{1 + F_0\alpha_0(\mathbf{q}, \omega)}, \tag{21}$$

and $\mathbf{q} = \mathbf{k}_1 - \mathbf{k}_2$. The interpretation of Eq. (19) is as follows. The Landau parameters in the numerator describe the coupling of quasiparticles to particular collective modes. For instance, the F_0 represents the coupling to density excitations, G_0 the coupling to spin excitations, etc., and the denominators enter from the summation of bubbles to all orders. Including the $l = 1$ Fermi liquid parameters, the induced interaction is given by

$$4F_i = \left[\frac{F_0^2\alpha_0}{1+F_0\alpha_0} + \left(1 - \frac{q^2}{4k_F^2}\right)\frac{F_1^2\alpha_1}{1+F_1\alpha_1} + \frac{3G_0^2\alpha_0}{1+G_0\alpha_0} + \left(1 - \frac{q^2}{4k_F^2}\right)\frac{3G_1^2\alpha_1}{1+G_1\alpha_1}\right.$$
$$\left. + \frac{3F_0'^2\alpha_0}{1+F_0'\alpha_0} + \left(1 - \frac{q^2}{4k_F^2}\right)\frac{3F_1'^2\alpha_1}{1+F_1'\alpha_1} + \frac{9G_0'^2\alpha_0}{1+G_0'\alpha_0} + \left(1 - \frac{q^2}{4k_F^2}\right)\frac{9G_1'^2\alpha_1}{1+G_1'\alpha_1}\right], \tag{22}$$

where α_1 defined by

$$\alpha_1(\mathbf{q}, 0) = \frac{1}{2}\left[\frac{3}{8} - \frac{k_F^2}{2q^2} + \left(\frac{k_F^3}{2q^3} + \frac{k_F}{4q} - \frac{3q}{32k_F}\right)\ln\left(\frac{k_F + q/2}{k_F - q/2}\right)\right] \tag{23}$$

is related to the current–current correlation function, and analogous expressions hold for the spin- and isospin-dependent parts of the induced interaction. These equations were first obtained in [20], carried far enough to include velocity-dependent effects in terms of an effective mass, in the approximation of quadratic spectrum.

Having characterized the induced part of the quasiparticle interaction, let us now elaborate on the driving term. By definition, this component of the interaction consists of those diagrams that cannot be separated into two diagrams by cutting one particle line and one hole line. Some of the low order terms contributing to the driving term are shown in Fig. 5, where the interaction vertices are assumed to be antisymmetrized. Some higher-order terms, such as diagram (*d*) in Fig. 5,

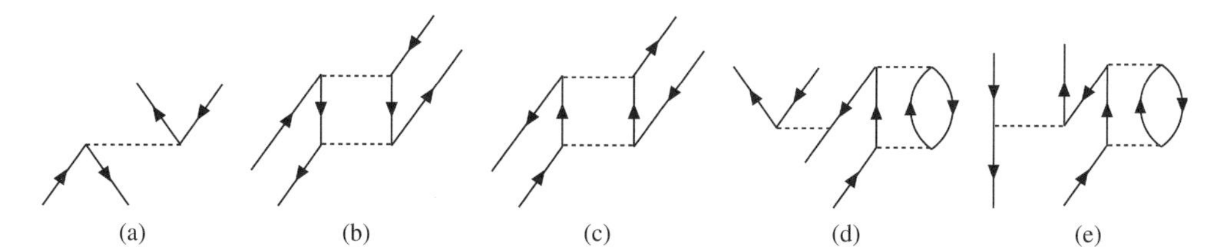

Fig. 5. A selection of diagrams contributing to the driving term in the quasiparticle interaction. Diagrams (d) and (e) are included implicitly through the renormalization at the quasiparticle pole.

are included implicitly through the quasiparticle renormalization z and need not be calculated explicitly, as described in detail in [28]. In order to preserve the Pauli principle sum rules (16) and (17) the driving term must be antisymmetrized. Thus, including Fig. 5(d) requires that (e) also be included in order for the scattering amplitude to be antisymmetric.

4. Calculations and results

According to the discussion in the previous section, the starting point of a microscopic derivation of the quasiparticle interaction is a calculation of the antisymmetrized driving term to some specified order in the bare potential. Nearly all previous calculations have used the G-matrix, since it is well known that the unrenormalized high-precision NN potentials are unsuitable for perturbation theory calculations due to the presence of a strong short-distance repulsion. The resummation of particle-particle ladder diagrams in the G-matrix softens the potential but introduces several undesirable features from the perspective of Fermi liquid theory. Most important is the unphysical gap in the single particle energy spectrum at the Fermi surface due to the fact that hole lines receive self-energy corrections but particle lines do not. In the past it was suggested [9,18] that introducing a model space, within which particles and holes are treated symmetrically, could overcome this difficulty.

An alternative method for taming the repulsive core is to integrate out the high momentum components of the interaction in such a way that the low energy dynamics are preserved [10,29]. This is accomplished by rewriting the half-on-shell T-matrix

$$T\big(p', p, p^2\big) = V_{NN}(p', p) + \frac{2}{\pi}\mathcal{P}\int_0^\infty \frac{V_{NN}(p', q)T(q, p, p^2)}{p^2 - q^2} q^2\, dq \tag{24}$$

with an explicit momentum cutoff Λ, which yields the low momentum T-matrix defined by

$$T_{\text{low-}k}\big(p', p, p^2\big) = V_{\text{low-}k}(p', p) + \frac{2}{\pi}\mathcal{P}\int_0^\Lambda \frac{V_{\text{low-}k}(p', q)T_{\text{low-}k}(q, p, p^2)}{p^2 - q^2} q^2\, dq. \tag{25}$$

Enforcing the requirement that $T_{\text{low-}k}(p', p, p^2) = T(p', p, p^2)$ for $p', p < \Lambda$ preserves the low energy physics encoded in the scattering phase shifts. Remarkably, under this construction all high-precision NN potentials flow to a nearly universal low momentum interaction $V_{\text{low-}k}$ as the momentum cutoff Λ is lowered to 2.1 fm^{-1}. In fact, $k = 2.1$ fm^{-1} is precisely the CM momentum beyond which the experimental phase shift analysis has not been incorporated in the high-precision NN interactions.

For an initial approximation to the driving term, we include the first-order antisymmetrized matrix element shown diagrammatically in Fig. 5(a) as well as the higher order diagrams, such as

(d) and (e), that are included implicitly through the renormalization strength at the quasiparticle pole. The quasiparticles are confined to a thin model space P near the Fermi surface

$$P = \lim_{\delta \to 0} \sum_{k_F < k < k_F + \delta} |\vec{k}\rangle\langle\vec{k}|, \tag{26}$$

and the first-order contribution is given by

$$\langle \vec{k}_1 \vec{k}_2 ST \big| V \big| (\vec{k}_3 \vec{k}_4 - \vec{k}_4 \vec{k}_3) ST \rangle = \langle k, \theta ST | V | k, \theta ST \rangle, \tag{27}$$

where $k_1 = k_2 = k_3 = k_4 = k_F$, θ is the angle between the two momenta, and the relative momentum $k = k_F \sin(\theta/2)$. Given the $V_{\text{low-}k}$ matrix elements in the basis $|klSTJ\rangle$, we project onto the central components and change from a spherical wave basis to a plane wave basis. Then the dimensionful driving term is given by

$$\langle kST | V_d | kST \rangle = z^2 \frac{4\pi}{2S+1} \sum_{J,l} (2J+1)\big(1 - (-1)^{l+S+T}\big) \langle klSJT | V_{\text{low-}k} | klSJT \rangle. \tag{28}$$

Inserting the form of the quasiparticle interaction in Eq. (2) into the left-hand side of Eq. (28), we obtain the Fermi liquid parameters in terms of $V_{ST}(k) = \langle kST | V | kST \rangle$. The result is

$$\begin{aligned}
f &= \frac{1}{16} V_{00} + \frac{3}{16} V_{01} + \frac{3}{16} V_{10} + \frac{9}{16} V_{11}, \\
g &= -\frac{1}{16} V_{00} - \frac{3}{16} V_{01} + \frac{1}{16} V_{10} + \frac{3}{16} V_{11}, \\
f' &= -\frac{1}{16} V_{00} + \frac{1}{16} V_{01} - \frac{3}{16} V_{10} + \frac{3}{16} V_{11}, \\
g' &= \frac{1}{16} V_{00} - \frac{1}{16} V_{01} - \frac{1}{16} V_{10} + \frac{1}{16} V_{11},
\end{aligned} \tag{29}$$

where the momentum dependence has been suppressed for simplicity. From Eq. (4) it can be shown that

$$\frac{m^*}{m} = \frac{1}{1 - \mu f_1/3}, \tag{30}$$

where $\mu = 2mk_F/\pi^2\hbar^2 = \frac{m}{m^*} N_0$, from which we construct the dimensionless Fermi liquid parameters. In all of our calculations we include partial waves up to $J = 6$. In Table 1 we show the Landau parameters of the driving term derived from three different low momentum interactions obtained from the Nijmegen I & II potentials [30] and the CD-Bonn potential [31] for a momentum cutoff of $\Lambda = 2.1\ \text{fm}^{-1}$ and a Fermi momentum of $k_F = 1.36\ \text{fm}^{-1}$. From the available theoretical analyses of nucleon momentum distributions [32], we take the quasiparticle renormalization strength to be $z = 0.7$ for nuclear matter.

The induced interaction is obtained by iterating equations (18) and (22) until a self-consistent solution is reached. The density–density and current–current correlation functions in (22) introduce a momentum dependence in the induced interaction, and the Fermi liquid parameters for the induced interaction are obtained by projecting onto the Legendre polynomials

$$F_{i,l} = \frac{2l+1}{2} \int_{-1}^{1} F_i(\theta) P_l(\cos\theta)\, d(\cos\theta), \quad \text{etc.} \tag{31}$$

Table 1
The Fermi liquid parameters of the NN interaction $V_{\text{low-}k}$ derived from the Nijmegen potentials and CD-Bonn potential for a cutoff of $\Lambda = 2.1\ \text{fm}^{-1}$ and Fermi momentum $1.36\ \text{fm}^{-1}$

	Nijmegen I				Nijmegen II				CD-Bonn			
l	F_l	G_l	F'_l	G'_l	F_l	G_l	F'_l	G'_l	F_l	G_l	F'_l	G'_l
0	−1.230	0.130	0.392	0.619	−1.475	0.248	0.549	0.583	−1.199	0.135	0.350	0.603
1	−0.506	0.241	0.252	0.118	−0.445	0.161	0.172	0.225	−0.498	0.240	0.259	0.118
2	−0.201	0.120	0.101	0.021	−0.213	0.127	0.106	0.020	−0.200	0.122	0.101	0.022
3	−0.110	0.054	0.051	0.009	−0.120	0.060	0.056	0.007	−0.111	0.055	0.051	0.010

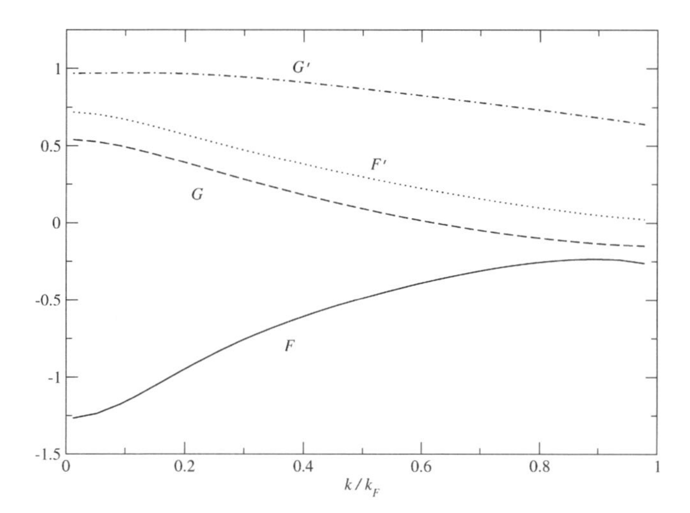

Fig. 6. The self-consistent solution for the full quasiparticle interaction as a function of $k = \frac{1}{2}|\mathbf{k}_1 - \mathbf{k}_2|$ derived from the low momentum CD-Bonn potential.

For the first iteration we use the Landau parameters obtained from the bare low momentum interaction as an estimate for the full quasiparticle interaction in Eq. (22). However, since F_0 does not satisfy the stability criteria (9) for either the Nijmegen or CD-Bonn potentials, in the first iteration we replace it in both cases with an arbitrary value that does. The convergence of the iteration scheme is generally rapid and relatively insensitive to the set of initial parameters chosen for the low-momentum Nijmegen I and CD-Bonn potentials. In contrast, the low momentum Nijmegen II potential exhibits poor convergence properties, though a solution to the coupled equations can still be found. For a completely consistent solution at each iteration we recalculate the driving term with the new effective mass.

The final self-consistent result for the quasiparticle interaction is shown in Fig. 6 for the CD-Bonn potential, and the Fermi liquid parameters for the driving term, induced interaction, and full quasiparticle interaction are shown in Table 2. For comparison we list the Fermi liquid parameters obtained in [11] where the spin-independent Landau parameters were taken from experiment and used to calculate the spin-dependent parameters with a set of nontrivial sum rules:

$$F_0 = -0.27, \quad G_0 = 0.15 \pm 0.3, \quad F'_0 = 0.71, \quad G'_0 = 1.0 \pm 0.2,$$
$$F_1 = -0.85, \quad G_1 = 0.45 \pm 0.3, \quad F'_1 = 0.14, \quad G'_1 = 0.0 \pm 0.2.$$

Table 2
The self-consistent solution of the Babu–Brown equations for the low momentum CD-Bonn potential. The full Fermi liquid parameters are obtained by projecting the quasiparticle interaction in Fig. 6 onto the Legendre polynomials

l	Full				Driving				Induced			
	F	G	F'	G'	F_d	G_d	F'_d	G'_d	F_i	G_i	F'_i	G'_i
0	−0.476	0.025	0.221	0.784	−1.276	0.144	0.373	0.642	0.801	−0.119	−0.152	0.142
1	−0.335	0.263	0.273	0.171	−0.530	0.256	0.275	0.125	0.195	0.007	−0.002	0.048
2	−0.238	0.139	0.117	0.020	−0.212	0.130	0.107	0.024	−0.026	0.009	0.010	−0.003
3	−0.101	0.055	0.050	0.014	−0.119	0.059	0.054	0.011	0.018	−0.004	−0.004	0.003

Table 3
Nuclear observables obtained from the self-consistent solution of the Babu–Brown equations and deviations δS_1 and δS_2 from the Pauli principle sum rules

	Nijmegen I	Nijmegen II	CD-Bonn
m^*/m	0.887	0.930	0.888
$\mathcal{K}$ [MeV]	136	102	136
β [MeV]	18.1	20.5	17.6
δg_l [μ_N]	0.682	0.452	0.685
δS_1	0.20	0.16	0.27
δS_2	−0.04	−0.02	−0.04

Although the experimental values for the spin-independent parameters are appreciably different from the self-consistent solution we have obtained, our values for the spin-dependent parameters fall within the errors predicted from the sum rules. However, the main effect of the induced interaction is to cut down the strong attraction in the spin-independent, isospin-independent part of the quasiparticle interaction. In fact, the repulsion in this channel coming from the induced interaction is large enough for the resulting F_0 to satisfy the stability condition in (9). The effective mass, compression modulus, and symmetry energy are shown in Table 3 together with the deviations δS_1 and δS_2 from the sum rules (16) and (17). We list the results for the three different bare potentials with a momentum cutoff of $\Lambda = 2.1\ \text{fm}^{-1}$. In calculating the contributions to (16) and (17) we have included Landau parameters for $l \leqslant 3$. The compression modulus for nuclear matter is extrapolated from the data on giant monopole resonances in heavy nuclei, with the expected value being 200–300 MeV [33,34]. The symmetry energy is determined by fitting the data on nuclear masses to various versions of the semi-empirical mass formula [35], and currently the accepted value is $\beta = 25$–35 MeV [34,36]. Both the compression modulus and the symmetry energy shown in Table 3 are significantly smaller than the experimental values. On the other hand, the anomalous orbital gyromagnetic ratio, determined from giant dipole resonances in heavy nuclei, is too large compared with the experimental value of $\delta g_l^p = 0.23 \pm 0.03$ [37].

As suggested in the introduction, we propose to remedy these discrepancies by considering the effects of Brown–Rho scaling on hadronic masses. The proposed scaling law for light hadrons—other than the pseudoscalar mesons, whose masses are protected by chiral invariance—is [22, 38]

$$\frac{m_V^*}{m_V} = \frac{m_\sigma^*}{m_\sigma} = \sqrt{\frac{g_A}{g_A^*}}\frac{m_N^*}{m_N} = 1 - C\frac{n}{n_0}, \tag{32}$$

where the subscript V denotes either the ρ or ω vector meson, σ refers to the scalar meson, g_A is the axial vector coupling, and n/n_0 is the ratio of the medium density to nuclear matter

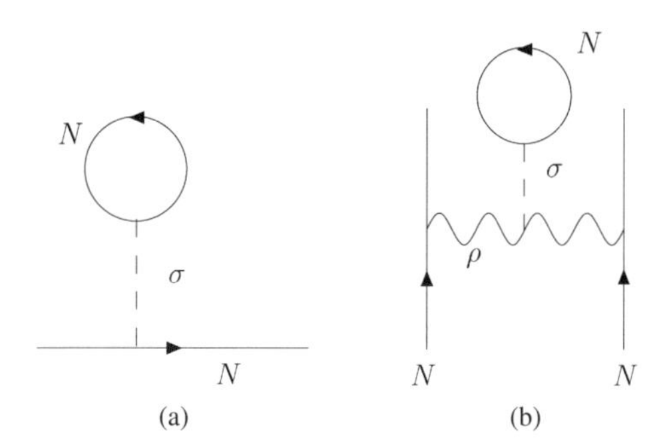

Fig. 7. Walecka mean field contribution of the scalar tadpole to the nucleon mass (a) and its extrapolation to constituent quarks in vector mesons (b).

density. This scaling can be thought of as extending Walecka mean field theory, in which the scalar tadpole contribution to the nucleon self-energy lowers the effective mass, to the level of constituent quarks. Attaching a scalar tadpole on the nucleon line, as shown in Fig. 7(a), lowers the mass according to (32), and a scalar tadpole connected to the vector mesons gives an effective three-body force as shown in Fig. 7(b). Including the in-medium scaling of the axial-vector coupling, which should approach $g_A^* = 1$ at chiral restoration, the net result is a lowering of the in-medium m_V^* by $\sim 2/3$ as much as m_N^*. Recent experimental results [39,40] are consistent with the scaling law (32) for $C = 0.14$ and 0.092, respectively. The Brown–Rho "parametric scaling" has $C = 0.2$. However, the dense loop term ΔM [41] gives a shift of the ρ-meson pole upwards. So far no one has been able to calculate it at finite density.

A number of previous studies [42–44] were successful in describing nuclear matter by starting from a chiral Lagrangian with nuclear, scalar, and vector degrees of freedom in which the hadronic masses were scaled with density according to (32). In particular, the compression modulus and anomalous orbital gyromagnetic ratio were found to be in excellent agreement with experiment, which suggests that a similar approach may prove fruitful in our present analysis. An alternative approach, complementary to the chiral Lagrangian method, is to include medium modifications directly into a one-boson-exchange potential. Such a calculation was carried out in [45] to study the saturation of nuclear matter. In their work it was suggested that the σ particle should be constructed microscopically as a pair of correlated pions interacting largely through crossed-channel ρ exchange. Medium modifications to the σ mass then arise naturally from the density-dependence of the ρ mass. The final conclusion established in [45] is that at low densities the σ scales according to (32) but that toward nuclear matter density the scaling is slowed to such an extent that saturation can be achieved.

We proceed along the lines of [45] and introduce medium modifications directly into a one-boson-exchange potential. The most refined NN potentials in this category are the Nijmegen I, Nijmegen II, and CD-Bonn potentials. The Nijmegen potentials include contributions from the exchange of ρ, ω, ϕ, σ, f_0, and a_0 mesons, as well as the pseudoscalar particles which do not receive medium modifications in Brown–Rho scaling. The CD-Bonn potential includes two vector particles (the ρ and ω) and two scalars (σ_1 and σ_2). For both potentials we scale the vector meson masses by 15% and the scalar meson masses by 7%. In this way we roughly account for the decreased scaling of the scalar particle mass observed in [45]. In the full many-body calculation we also scale the nucleon mass by 15% and with an additional $\sqrt{g_A^*/g_A} \simeq 1/\sqrt{1.25}$ at nuclear matter density. It is essential to also scale the form factor cutoffs Λ_f of the vector mesons in the boson-exchange potentials.

Table 4
Nuclear observables obtained from the self-consistent solution to the Babu–Brown equations incorporating Brown–Rho scaling. Four different bare potentials—the CD-Bonn potential (V_{CDB}), Nijmegen I (V_{NI}), Nijmegen II (V_{NII}), and Nijmegen93 (N93) potentials—were used to construct low momentum interactions for a cutoff of $\Lambda = 2.1\ \mathrm{fm}^{-1}$. In Eq. (32) the parameter $C = 0.15$

	V_{NI}	V_{NII}	V_{N93}	V_{CDB}
m^*/m	0.721	0.763	0.696	0.682
$\mathcal{K}$ [MeV]	218	142	190	495
β [MeV]	20.4	25.5	23.7	19.2
δg_l	0.246	0.181	0.283	0.267

Table 5
Fermi liquid coefficients for the self-consistent solution to the Babu–Brown equations using Brown–Rho scaled nucleon and meson masses in the four low momentum CD-Bonn and Nijmegen potentials listed in Table 4. The tabulated values display the average and spread from the four different potentials and not the actual uncertainties associated with the Fermi liquid parameters

l	F_l	G_l	F'_l	G'_l
0	-0.20 ± 0.39	0.04 ± 0.11	0.24 ± 0.16	0.53 ± 0.09
1	-0.86 ± 0.10	0.19 ± 0.06	0.18 ± 0.05	0.17 ± 0.12
2	-0.21 ± 0.01	0.12 ± 0.01	0.10 ± 0.02	0.01 ± 0.02
3	-0.09 ± 0.01	0.05 ± 0.01	0.05 ± 0.01	0.01 ± 0.01

In Table 4 we show the effective mass, compression modulus, symmetry energy, and anomalous orbital gyromagnetic ratio for the Nijmegen I & II and CD-Bonn potentials with the in-medium modifications. We also show for comparison the results from the Nijmegen93 one-boson-exchange potential, which has only 15 free parameters and is not fine-tuned separately in each partial wave. We observe that the iterative solution is in better agreement with all nuclear observables. The anomalously large compression modulus in the CD-Bonn potential results almost completely from the presence of a large ω coupling constant $g^2_{\omega NN}/4\pi = 20.0$. With the same $g^2_{\omega NN}/4\pi$ and Bonn-B potential, Rapp et al. [45] obtained $\mathcal{K} = 356$ MeV. The compression modulus is very sensitive to this parameter, as we have checked that dropping this coupling by 20% cuts the compression modulus in half but alters the other nuclear observables by less than 5%. The naive quark model predicts a ratio of $g^2_{\omega NN}/g^2_{\rho NN} = 9$ between the ω and ρ coupling constants, which is largely violated in the CD-Bonn potential $g^2_{\omega NN}/g^2_{\rho NN} = 24$ though roughly satisfied in the Nijmegen potentials $g^2_{\omega NN}/g^2_{\rho NN} = 11$, perhaps resulting in better agreement with experiment.

Thus, by extension of the Walecka mean field on nucleons to those on constituent quarks, we obtain the Fermi liquid parameters for the theory that is now essentially Brown–Rho scaled, as shown in Table 5. One should note that these results are only for infinite nuclear matter and especially the three-body term will act in many different diagrams in the finite systems. However, our arguments suggest that the three-body terms intrinsic to Brown–Rho scaling will be useful in stabilizing light nuclei.

5. Discussion of the tensor force with dropping ρ-mass in saturation of nuclear matter

The tensor force contributes chiefly in second order perturbation theory as an effective central force in the $I = 1$ channel. As the density increases, some of the intermediate states are blocked

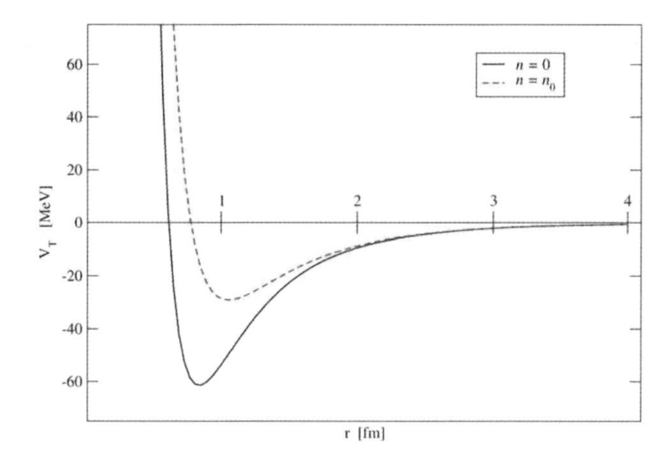

Fig. 8. Reduction in the strength of the tensor force due to a scaled ρ-meson mass. Contributions from both π-meson and ρ-meson exchange are included in both curves. We have used the Brown–Rho parametric scaling, so that at nuclear matter density $m_\rho^* = 0.8 m_\rho$.

by the Pauli principle. In the two-body system the tensor force contributes to the $^3S_1$ state, but not to the $^1S_0$ state, and gives most of the attractive interaction difference between the $^3S_1$ and $^1S_0$ states, effectively binding the deuteron. However, the intermediate state energies relevant for the second-order tensor force are > 225 MeV (see Fig. 69 of [46] which is for ^{40}Ca. For nuclear matter the intermediate state momenta would be higher), well above the Fermi energy of nuclear matter, and most intermediate momenta are above the $V_{\text{low-}k}$ upper model space limit of 420 MeV/c, so the tensor force is largely integrated out.

However, since the beginning of Brown–Rho scaling it has been understood that the tensor force is rapidly cut down with increasing density. That is because the pion mass does not change with density, being protected by chiral invariance, but the ρ-meson mass, which is dynamically generated, decreases by 20% (parametric scaling) in going from a density of $n = 0$ to nuclear matter density $n = n_0$. Since the ρ-meson exchange contributes with opposite sign from that of the pion, this cuts down the tensor force substantially. In Fig. 8 we show the total tensor force from π and ρ exchange at zero density and nuclear matter density n_0. Since it enters in the square, this means a factor of several drop in the tensor contribution to the binding energy, as shown in Fig. 9.

We believe that the work of Ref. [39] shows unambiguously that the mass of the ω-meson is $\sim 14\%$ lower at nuclear matter density than in free space. It is remarkable that nuclear structure calculations have been carried out for many years without density-dependent masses but with results usually in quantitative agreement with experiment. In [1] Brown and Rho showed that in cases where the exchange of the π-meson is not important, such as in Dirac phenomenology, there is a scale invariance such that if the masses of all relevant mesons are changed by the same amount, the results for the physical phenomena are very little changed.

Since the pion exchange gives the longest range part of the nucleon–nucleon interaction, it is amazing that there are not clearcut examples in nuclear spectroscopy such as level orderings that are altered by the ρ-meson exchange playing counterpoint to the π-meson exchange, as we find in this paper for nuclear saturation. The in-medium decrease in the ρ-mass increases the effect of ρ-exchange, which enters so as to cut down the overall tensor force, the ρ and π exchange entering with opposite sign.

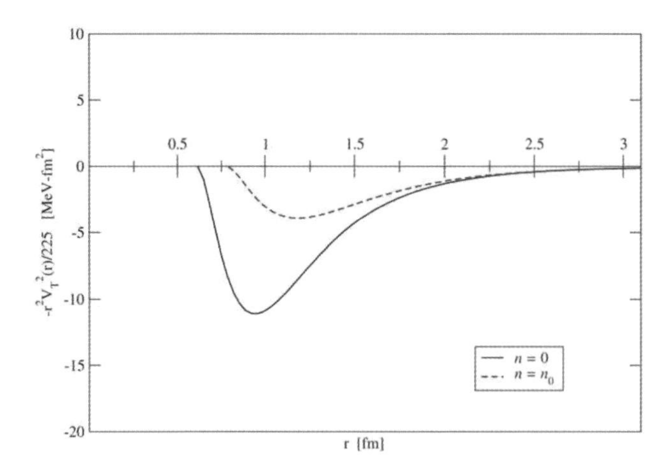

Fig. 9. Reduction of the tensor force in second order perturbation theory due to a scaled ρ-meson mass. The intermediate state energy is approximated as 225 MeV. Contributions from both π-meson and ρ-meson exchange are included in both curves. At nuclear matter density, n_0, we have used the parametric scaling $m_\rho^* = 0.8 m_\rho$.

Finally, nearly forty years since the Kuo–Brown nucleon–nucleon forces were first published, it was shown [47] that the summation of core polarization diagrams to all orders is well-approximated by a single bubble. However, in light of the double decimation of [1] being carried out here in one step, these forces should be modified to include the medium dependence of the masses. Phenomenologically this can be done by introducing three-body terms, as we did here, but from our treatment of the second-order tensor force it is clear that this should be done at constituent quark level.

6. Conclusion

We believe that by discussing the nuclear many-body problem within the context of Fermi liquid theory with the interaction $V_{\text{low-}k}$ following the work of Schwenk et al. [11] we have a format for understanding connections between the physical properties of the many-body system and the nuclear potentials. We carried out an iterative solution of the Babu–Brown equations, which include both density–density and current–current correlation functions, calculating input potentials via a momentum space decimation to $V_{\text{low-}k}$. By including Brown–Rho scaling through scalar tadpoles, as suggested by Walecka theory, our iterative solution provides the empirical Fermi liquid quantities. Our nucleon effective mass is on the low side of those usually employed, as is common in Walecka mean field theory.

Acknowledgements

We thank Achim Schwenk for helpful discussions. This work was partially supported by the US Department of Energy under Grant No. DE-FG02-88ER40388.

References

[1] G.E. Brown, M. Rho, Phys. Rep. 396 (2004) 1.
[2] L.D. Landau, Sov. Phys. JETP 3 (1957) 920.

[3] L.D. Landau, Sov. Phys. JETP 5 (1957) 101.
[4] L.D. Landau, Sov. Phys. JETP 8 (1959) 70.
[5] A.B. Migdal, A.I. Larkin, Sov. Phys. JETP 18 (1964) 717.
[6] A.B. Migdal, Theory of Finite Fermi Systems and Applications to Atomic Nuclei, Interscience, New York, 1967.
[7] S.O. Bäckman, Nucl. Phys. A 120 (1968) 593;
S.O. Bäckman, Nucl. Phys. A 130 (1969) 481.
[8] G.E. Brown, Rev. Mod. Phys. 43 (1971) 1.
[9] R.S. Poggioli, A.D. Jackson, Nucl. Phys. A 165 (1971) 582.
[10] S.K. Bogner, T.T.S. Kuo, A. Schwenk, Phys. Rep. 386 (2003) 1.
[11] A. Schwenk, G.E. Brown, B. Friman, Nucl. Phys. A 703 (2002) 745.
[12] A. Schwenk, B. Friman, G.E. Brown, Nucl. Phys. A 713 (2003) 191.
[13] A. Schwenk, B. Friman, Phys. Rev. Lett. 92 (2004) 082501.
[14] J. Kuckei, F. Montani, H. Muther, A. Sedrakian, Nucl. Phys. A 723 (2003) 32.
[15] S.K. Bogner, A. Schwenk, R.J. Furnstahl, A. Nogga, Nucl. Phys. A 763 (2005) 59.
[16] S. Babu, G.E. Brown, Ann. Phys. 78 (1973) 1.
[17] O. Sjöberg, Ann. Phys. 78 (1973) 39.
[18] O. Sjöberg, Nucl. Phys. A 209 (1973) 363.
[19] W.H. Dickhoff, A. Faessler, H. Müther, S.S. Wu, Nucl. Phys. A 405 (1983) 534.
[20] S.O. Bäckman, G.E. Brown, J.A. Niskanen, Phys. Rep. 124 (1985) 1.
[21] K.S. Bedell, T.L. Ainsworth, Phys. Lett. A 102 (1984) 49.
[22] G.E. Brown, M. Rho, Phys. Rev. Lett. 66 (1991) 2720.
[23] T.T.S. Kuo, G.E. Brown, Nucl. Phys. 85 (1966) 40.
[24] I.Ya. Pomeranchuk, Zh. Eksp. Teor. Fiz. 35 (1958) 524.
[25] A.A. Abrikosov, L.P. Gorkov, I.E. Dzyaloshinki, Methods of Quantum Field Theory in Statistical Physics, Dover, New York, 1963.
[26] F. Nozieres, Theory of Interacting Fermi Systems, Addison–Wesley, Reading, MA, 1964.
[27] B.L. Friman, A.K. Dar, Phys. Lett. B 85 (1979) 1.
[28] G.E. Brown, Many-Body Problems, North-Holland, Amsterdam, 1972.
[29] S.K. Bogner, T.T.S. Kuo, A. Schwenk, D.R. Entem, R. Machleidt, Phys. Lett. B 576 (2003) 265.
[30] R.B. Wiringa, V.G.J. Stoks, R. Schiavilla, Phys. Rev. C 51 (1995) 38.
[31] R. Machleidt, Phys. Rev. C 63 (2001) 024001.
[32] V. Pandharipande, I. Sick, P. de Witt Huberts, Rev. Mod. Phys. 69 (1997) 981.
[33] D.H. Youngblood, H.L. Clark, Y.-W. Lui, Phys. Rev. Lett. 82 (1999) 691.
[34] A.W. Steiner, M. Prakash, J.M. Lattimer, P.J. Ellis, Phys. Rep. 411 (2005) 325.
[35] P. Möller, J.R. Nix, W.D. Myers, W.J. Swiatecki, At. Data Nucl. Data Tables 59 (1995) 185.
[36] P. Danielewicz, Nucl. Phys. A 727 (2003) 233.
[37] R. Nolte, A. Baumann, K.W. Rose, M. Schumacher, Phys. Lett. B 173 (1986) 388.
[38] G.E. Brown, M. Rho, Phys. Rep. 269 (1996) 333.
[39] D. Trnka, et al., Phys. Rev. Lett. 94 (2005) 192303.
[40] M. Naruki, et al., Phys. Rev. Lett. 96 (2006) 092301.
[41] M. Harada, Y. Kim, M. Rho, Phys. Rev. D 66 (2002) 016003.
[42] B. Friman, M. Rho, Nucl. Phys. A 606 (1996) 303.
[43] C. Song, G.E. Brown, D.-P. Min, M. Rho, Phys. Rev. C 56 (1997) 2244.
[44] C. Song, Phys. Rep. 347 (2001) 289.
[45] R. Rapp, R. Machleidt, J.W. Durso, G.E. Brown, Phys. Rev. Lett. 82 (1999) 1827.
[46] G.E. Brown, Unified Theory of Nuclear Models and Forces, North-Holland, Amsterdam, 1967.
[47] J.D. Holt, J.W. Holt, T.T.S. Kuo, G.E. Brown, S.K. Bogner, Phys. Rev. C 72 (2005) 041304(R).

PRL 100, 062501 (2008) PHYSICAL REVIEW LETTERS week ending 15 FEBRUARY 2008

Shell Model Description of the ^{14}C Dating β Decay with Brown-Rho-Scaled NN Interactions

J. W. Holt,[1] G. E. Brown,[1] T. T. S. Kuo,[1] J. D. Holt,[2] and R. Machleidt[3]

[1]*Department of Physics, SUNY, Stony Brook, New York 11794, USA*
[2]*TRIUMF, 4004 Wesbrook Mall, Vancouver, British Columbia, Canada, V6T 2A3*
[3]*Department of Physics, University of Idaho, Moscow, Idaho 83844, USA*

(Received 21 September 2007; published 15 February 2008)

We present shell model calculations for the beta decay of ^{14}C to the ^{14}N ground state, treating the states of the $A = 14$ multiplet as two $0p$ holes in an ^{16}O core. We employ low-momentum nucleon-nucleon (NN) interactions derived from the realistic Bonn-B potential and find that the Gamow-Teller (GT) matrix element is too large to describe the known lifetime. By using a modified version of this potential that incorporates the effects of Brown-Rho scaling medium modifications, we find that the GT matrix element vanishes for a nuclear density around 85% that of nuclear matter. We find that the splitting between the $(J^\pi, T) = (1^+, 0)$ and $(J^\pi, T) = (0^+, 1)$ states in ^{14}N is improved using the medium-modified Bonn-B potential and that the transition strengths from excited states of ^{14}C to the ^{14}N ground state are compatible with recent experiments.

DOI: 10.1103/PhysRevLett.100.062501 PACS numbers: 21.30.Fe, 21.60.Cs, 23.40.−s, 27.20.+n

The beta decay of ^{14}C to the ^{14}N ground state has long been recognized as a unique problem in nuclear structure. Its connection to the radiocarbon dating method, which has had a significant impact across many areas of science, makes the decay of broad interest even beyond nuclear physics. But *a priori* one would not expect the beta decay of ^{14}C to be a good transition for radiocarbon dating over archaeological times, because the quantum numbers of the initial state $(J^\pi, T) = (0^+, 1)$ and final state $(J^\pi, T) = (1^+, 0)$ satisfy the selection rules for an allowed Gamow-Teller transition. The expected half-life would therefore be on the order of hours, far from the unusually long value of 5730 years [1] observed in nature. The corresponding nuclear transition matrix element is very small ($\simeq 2 \times 10^{-3}$) and is expected to result from an accidental cancellation among the different components contributing to the transition amplitude. This decay has therefore been used to investigate phenomena not normally considered in studies of allowed transitions, such as meson exchange currents [2,3], relativistic effects [4], and configuration mixing [5,6]. Of broader importance, however, is that this decay provides a very sensitive test for the in-medium nuclear interaction and in particular for the current efforts to extend the microscopic description of the nuclear force beyond that of a static two-body potential fit to the experimental data on two-nucleon systems. One such approach is to include hadronic medium modifications, in which the masses of mesons and nucleons are altered at finite density due to the partial restoration of chiral symmetry [7–9] or many-body interactions with either intermediate nucleon-antinucleon excitations [10] or resonance-hole excitations [11]. These effects are traditionally incorporated in models of the three-nucleon force, which have been well-tested in *ab initio* nuclear structure calculations of light nuclei [12,13].

In this Letter we suggest that a large part of the observed ^{14}C beta decay suppression arises from in-medium modifications to the nuclear interaction. We study the problem from the perspective of Brown-Rho scaling (BRS) [14,15], which was the first model to make a comprehensive prediction for the masses of hadrons at finite density. In BRS the masses of nucleons and most light mesons (except the pion whose mass is protected by its Goldstone boson nature) decrease at finite density as the ratio of the in-medium to free-space pion decay constant:

$$\sqrt{\frac{g_A}{g_A^*}}\frac{m_N^*}{m_N} = \frac{m_\sigma^*}{m_\sigma} = \frac{m_\rho^*}{m_\rho} = \frac{m_\omega^*}{m_\omega} = \frac{f_\pi^*}{f_\pi} = \Phi(n), \quad (1)$$

where g_A is the axial-vector coupling constant, Φ is a function of the nuclear density n with $\Phi(n_0) \simeq 0.8$ at nuclear matter density, and the star indicates in-medium values of the given quantities. Since all realistic models of the NN interaction are based on meson exchange and fit to only free-space data, Eq. (1) prescribes how to construct a density-dependent nuclear interaction that accounts for hadronic medium modifications. This program has been carried out in several previous studies of symmetric nuclear matter [16,17], where it was found that one could well describe saturation and several bulk equilibrium properties of nuclear matter using such Brown-Rho-scaled NN interactions.

The case of the ^{14}C beta decay provides a nearly ideal situation in nuclear structure physics for testing the hypothesis of Brown-Rho scaling. Just below a double shell closure, the valence nucleons of ^{14}C inhabit a region with a large nuclear density. But more important is the sensitivity of this GT matrix element to the nuclear tensor force, which as articulated by Zamick and collaborators [18,19] is one of the few instances in nuclear structure where the role of the tensor force is clearly revealed. In fact, with a

0031-9007/08/100(6)/062501(4)

residual interaction consisting of only central and spin-orbit forces it is not possible to achieve a vanishing matrix element in a pure p^{-2} configuration [20]. Jancovici and Talmi [21] showed that by including a strong tensor force one could construct an interaction which reproduces the lifetime of ^{14}C as well as the magnetic moment and electric quadrupole moment of ^{14}N, although agreement with the known spectroscopic data was unsatisfactory.

The most important contributions to the tensor force come from π and ρ meson exchange, which act opposite to each other:

$$\begin{aligned} V_\rho^T(r) &= \frac{f_{N\rho}^2}{4\pi} m_\rho \tau_1 \tau_2 \left(-S_{12} \left[\frac{1}{(m_\rho r)^3} + \frac{1}{(m_\rho r)^2} + \frac{1}{3 m_\rho r} \right] e^{-m_\rho r} \right), \\ V_\pi^T(r) &= \frac{f_{N\pi}^2}{4\pi} m_\pi \tau_1 \tau_2 \left(S_{12} \left[\frac{1}{(m_\pi r)^3} + \frac{1}{(m_\pi r)^2} + \frac{1}{3 m_\pi r} \right] e^{-m_\pi r} \right). \end{aligned} \tag{2}$$

Since the ρ meson mass is expected to decrease substantially at nuclear matter density while the π mass remains relatively constant, an unambiguous prediction of BRS is the decreasing of the tensor force at finite density, which should be clearly seen in the GT matrix element. In fact, recent shell model calculations [22] performed in a larger model space consisting of $p^{-2} + 2\hbar\omega$ excitations have shown that the beta decay suppression requires the in-medium tensor force to be weaker and the in-medium spin-orbit force to be stronger in comparison to a typical G-matrix calculation starting with a realistic NN interaction. We show in Fig. 1 the radial part of the tensor interaction $V^T(r) = V_\pi^T(r) + V_\rho^T(r)$ at zero density and nuclear matter density assuming that $m_\rho^*(n_0)/m_\rho = 0.80$.

Experiments to determine the properties of hadrons in medium have been performed for all of the light mesons important in nuclear structure physics. Studies of deeply bound pionic atoms [23] find only a small increase in the π^- mass at nuclear matter density and a related decrease in the π^+ mass. Experimental information on the scalar and vector particles comes from mass distribution measurements of in-medium decay processes. Recent photoproduction experiments [24] of correlated pions in the $T = J = 0$ channel (σ meson) have found that the distribution is shifted to lower masses in medium. The vector mesons have been the most widely studied. Whereas the situation is clear with the ω meson, the mass of which drops by $\sim 14\%$ at nuclear matter density [25], with the ρ meson it is still unclear [26,27]. We believe that our present study tests the decrease in ρ mass more simply.

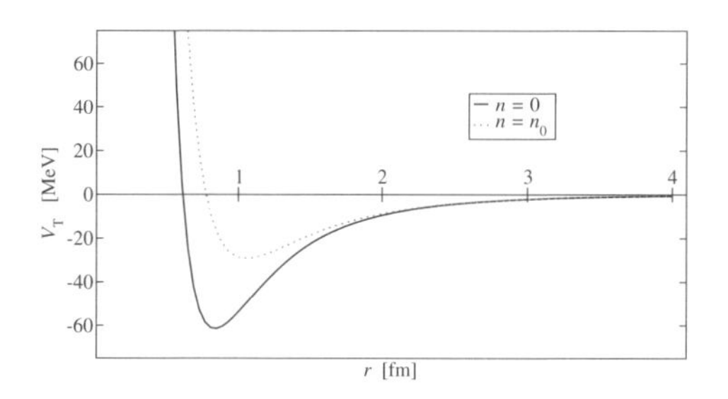

FIG. 1 (color online). The radial part of the nuclear tensor force given in Eq. (2) from π and ρ meson exchange at zero density and nuclear matter density under the assumption of BRS.

Today there are a number of high precision NN interactions based solely on one-boson exchange. In the present work we use the Bonn-B potential [28] which includes the exchange of the π, η, σ, a_0, ρ, and ω mesons. In [16] the consequences of BRS on the free-space NN interaction were incorporated into the Bonn-B potential and shown to reproduce the saturation properties of nuclear matter in a Dirac-Brueckner-Hartree-Fock calculation. The masses of the pseudoscalar mesons were unchanged, and the vector meson masses as well as the corresponding form factor cutoffs were decreased according to

$$\frac{m_\rho^*}{m_\rho} = \frac{m_\omega^*}{m_\omega} = \frac{\Lambda^*}{\Lambda} = 1 - 0.15 \frac{n}{n_0}. \tag{3}$$

The medium-modified (MM) Bonn-B potential is unique in its microscopic treatment of the scalar σ particle as correlated 2π exchange. Finite density effects arise through medium modifications to the exchanged ρ mesons in the pionic s-wave interaction as well as through the dressing of the in-medium pion propagator with Δ-hole excitations. These modifications to the vector meson masses and pion propagator would traditionally be included in the chiral three-nucleon contact interaction and the three-nucleon force due to intermediate Δ states, respectively.

Using realistic NN interactions in many-body perturbation theory is problematic due to the strong short distance repulsion in relative S states. The modern solution is to integrate out the high momentum components of the interaction in such a way that the low energy physics is preserved. The details for constructing such a low momentum interaction, $V_{\text{low-k}}$, are described in [29,30]. We define $V_{\text{low-k}}$ through the T-matrix equivalence $T(p', p, p^2) = T_{\text{low-k}}(p', p, p^2)$ for $(p', p) \leq \Lambda$, where T is given by the full-space equation $T = V_{NN} + V_{NN} g T$ and $T_{\text{low-k}}$ by the model-space (momenta $\leq \Lambda$) equation $T_{\text{low-k}} = V_{\text{low-k}} + V_{\text{low-k}} g T_{\text{low-k}}$. Here V_{NN} represents the Bonn-B NN potential and Λ is the decimation momentum beyond which the high momentum components of V_{NN} are integrated out. Since pion production starts around $E_{\text{lab}} \simeq 300$ MeV, the concept of a real NN potential is not valid beyond that energy. Consequently, we choose $\Lambda \approx 2.0\ \text{fm}^{-1}$, thereby retaining only the information from a given potential that is constrained by experiment. In fact for this Λ, the $V_{\text{low-k}}$ derived from various NN potentials are all nearly identical [30].

TABLE I. The coefficients of the LS-coupled wave functions defined in Eq. (5) and the associated GT matrix element as a function of the nuclear density n.

n/n_0	x	y	a	b	c	M_{GT}
0	0.844	0.537	0.359	0.168	0.918	−0.615
0.25	0.825	0.564	0.286	0.196	0.938	−0.422
0.5	0.801	0.599	0.215	0.224	0.951	−0.233
0.75	0.771	0.637	0.154	0.250	0.956	−0.065
1.0	0.737	0.675	0.103	0.273	0.956	0.074

We use the folded diagram formalism to reduce the full-space nuclear many-body problem $H\Psi_n = E_n\Psi_n$ to a model space problem $H_{eff}\chi_m = E_m\chi_m$ as detailed in [31]. Here $H = H_0 + V$, $H_{eff} = H_0 + V_{eff}$, $E_n = E_n(A = 14) - E_0(A = 16, \text{core})$, and V denotes the bare NN interaction. The effective interaction V_{eff} is derived following closely the folded diagram method detailed in [32]. A main difference is that in the present work the irreducible vertex function ($\hat{Q}$ box) is calculated from the low-momentum interaction V_{low-k}, while in [32] from the Brueckner reaction matrix (G matrix). In the $\hat{Q}$ box we include hole-hole irreducible diagrams of first and second order in V_{low-k}. Previous studies [33–35] have found that V_{low-k} is suitable for perturbative calculations; in all of these references satisfactory converged results were obtained including terms only up to second order in V_{low-k}.

Our calculation was carried out in jj coupling, where in the basis $\{p_{3/2}^{-2}, p_{3/2}^{-1}p_{1/2}^{-1}, p_{1/2}^{-2}\}$ one must diagonalize

$$[V_{eff}^{ij}] + \begin{bmatrix} 0 & 0 & 0 \\ 0 & \epsilon & 0 \\ 0 & 0 & 2\epsilon \end{bmatrix} \qquad (4)$$

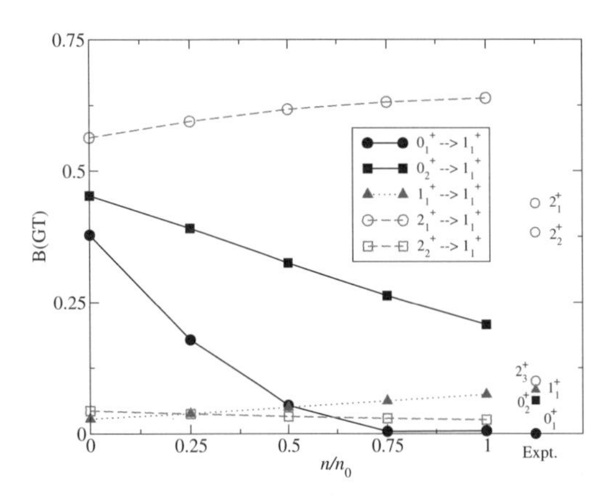

FIG. 2 (color online). The $B(\text{GT})$ values for transitions from the states of ^{14}C to the ^{14}N ground state as a function of the nuclear density and the experimental values from [36]. Note that there are three experimental low-lying 2^+ states compared to two theoretical 2^+ states in the p^{-2} configuration.

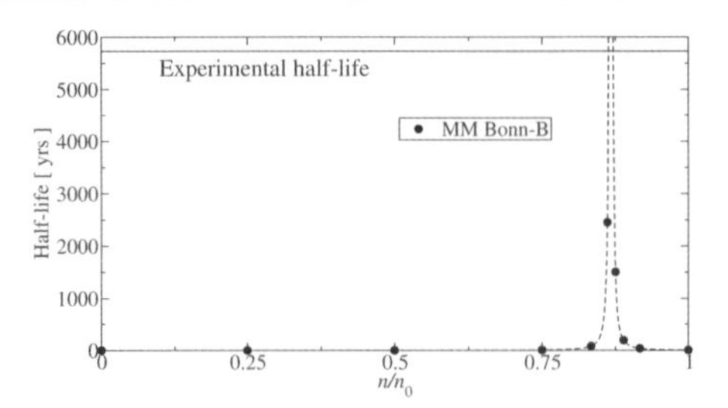

FIG. 3. The half-life of ^{14}C, as a function of the nuclear density, calculated from the MM Bonn-B potential.

to obtain the ground state of ^{14}N (and a similar 2×2 matrix for ^{14}C). We used $\epsilon = E(p_{1/2}^{-1}) - E(p_{3/2}^{-1}) = 6.3$ MeV, which is the experimental excitation energy of the first $\frac{3}{2}^-$ state in ^{15}N. One can transform the wave functions to LS coupling, where the ^{14}C and ^{14}N ground states are

$$\psi_i = x|^1S_0\rangle + y|^3P_0\rangle$$
$$\psi_f = a|^3S_1\rangle + b|^1P_1\rangle + c|^3D_1\rangle \qquad (5)$$

and the Gamow-Teller matrix element M_{GT} is given by [21]

$$\sum_k \langle\psi_f\|\sigma(k)\tau_+(k)\|\psi_i\rangle = -\sqrt{6}(xa - yb/\sqrt{3}). \qquad (6)$$

Since x and y are expected to have the same sign [20], the GT matrix element can vanish only if a and b have the same sign, which requires that the $\langle ^3S_1|V_{eff}|^3D_1\rangle$ matrix element furnished by the tensor force be large enough [21]. In Table I we show the ground state wave functions of ^{14}C and ^{14}N, as well as the GT matrix element, calculated with the MM Bonn-B interaction.

In Fig. 2 we plot the resulting $B(\text{GT}) = \frac{1}{2J_i+1}|M_{GT}|^2$ values for transitions between the low-lying states of ^{14}C and the ^{14}N ground state for the in-medium Bonn-B NN interaction taken at several different densities. Recent experiments [36] have determined the GT strengths from the ^{14}N ground state to excited states of ^{14}C and ^{14}O using the charge exchange reactions ^{14}N$(d, ^2\text{He})^{14}$C and ^{14}N$(^3\text{He}, t)^{14}$O, and our theoretical calculations are in

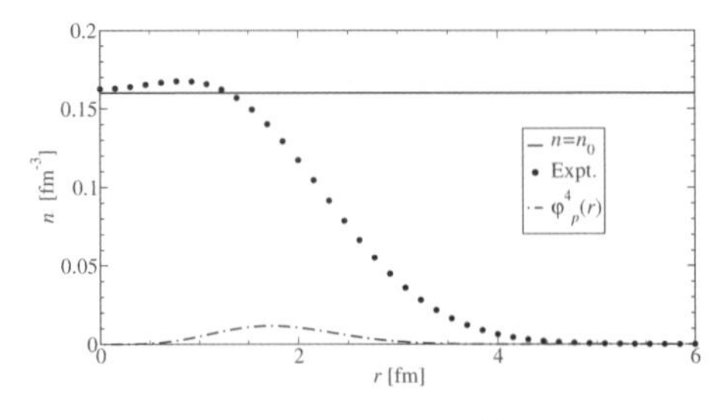

FIG. 4. Twice the charge distribution of ^{14}N taken from [37,38] and the fourth power of the p-shell wave functions.

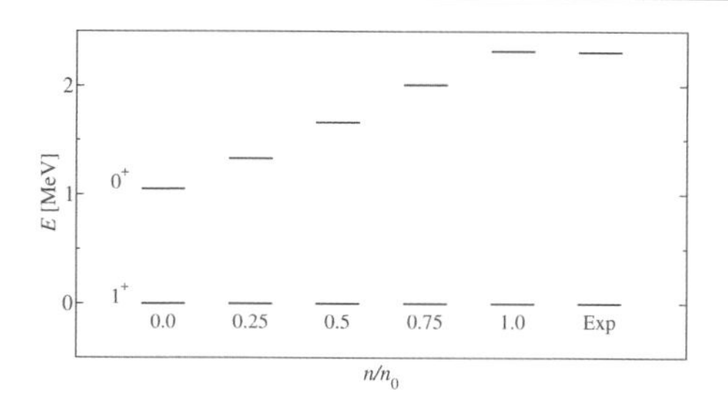

FIG. 5. The splitting between the 1_1^+ and 0_1^+ levels in ^{14}N for different values of the nuclear density. Also included is the experimental value.

good overall agreement. The most prominent effect we find is a robust inhibition of the ground state to ground state transition for densities in the range of $0.75–1.0n_0$. In contrast, the other transition strengths are more mildly influenced by the density dependence in BRS. In Fig. 3 we show the resulting half-life of ^{14}C calculated from the MM Bonn-B potential.

We emphasize that the nuclear density experienced by p-shell nucleons is actually close to that of nuclear matter; in Fig. 4 we compare twice the charge distribution of ^{14}N obtained from electron scattering experiments [37,38] with the radial part of the $0p$ wave functions, indicating clearly that the nuclear density for $0p$ nucleons is $\sim 0.8n_0$. The first excited 0^+ state of ^{14}N together with the ground states of ^{14}O and ^{14}C form an isospin triplet. We have calculated the splitting in energy between this state and the ground state of ^{14}N for a range of nuclear densities. Our results are presented in Fig. 5, where the experimental value is 2.31 MeV.

In summary, we have shown that by incorporating hadronic medium modifications into the Bonn-B potential the decay of ^{14}C is strongly suppressed at densities close to that experienced by valence nucleons in ^{14}C. In a more traditional approach such medium modifications would be built in through three-nucleon forces, and we suggest that calculations with free-space two-nucleon interactions supplemented with three-nucleon forces should also inhibit the GT transition.

We thank Igal Talmi for helpful correspondences. This work was partially supported by the U.S. Department of Energy under Grant No. DE-FG02-88ER40388 and the U.S. National Science Foundation under Grant No. PHY-0099444. TRIUMF receives federal funding via a contribution agreement through the National Research Council of Canada, and support from the Natural Sciences and Engineering Research Council of Canada is gratefully acknowledged.

[1] F. Ajzenberg-Selove, J. H. Kelley, and C. D. Nesaraja, Nucl. Phys. A **523**, 1 (1991).
[2] B. Goulard, B. Lorazo, H. Primakoff, and J. D. Vergados, Phys. Rev. C **16**, 1999 (1977).
[3] R. L. Huffman, J. Dubach, R. S. Hicks, and M. A. Plum, Phys. Rev. C **35**, 1 (1987).
[4] Y. Jin, L. E. Wright, C. Bennhold, and D. S. Onley, Phys. Rev. C **38**, 923 (1988).
[5] A. García and B. A. Brown, Phys. Rev. C **52**, 3416 (1995).
[6] I. S. Towner and J. C. Hardy, Phys. Rev. C **72**, 055501 (2005).
[7] V. Bernard, U.-G. Meissner, and I. Zahed, Phys. Rev. Lett. **59**, 966 (1987).
[8] T. Hatsuda and S. H. Lee, Phys. Rev. C **46**, R34 (1992).
[9] S. Klimt, M. Lutz, and W. Weise, Phys. Lett. B **249**, 386 (1990).
[10] G. E. Brown, W. Weise, G. Baym, and J. Speth, Comments Nucl. Part. Phys. **17**, 39 (1987).
[11] F. Klingl, N. Kaiser, and W. Weise, Nucl. Phys. A **624**, 527 (1997).
[12] P. Navrátil, J. P. Vary, and B. R. Barrett, Phys. Rev. Lett. **84**, 5728 (2000).
[13] S. C. Pieper and R. B. Wiringa, Annu. Rev. Nucl. Part. Sci. **51**, 53 (2001).
[14] G. E. Brown and M. Rho, Phys. Rev. Lett. **66**, 2720 (1991).
[15] G. E. Brown and M. Rho, Phys. Rep. **396**, 1 (2004).
[16] R. Rapp, R. Machleidt, J. W. Durso, and G. E. Brown, Phys. Rev. Lett. **82**, 1827 (1999).
[17] J. W. Holt, G. E. Brown, J. D. Holt, and T. T. S. Kuo, Nucl. Phys. A **785**, 322 (2007).
[18] L. Zamick, D. C. Zheng, and M. S. Fayache, Phys. Rev. C **51**, 1253 (1995).
[19] M. S. Fayache, L. Zamick, and B. Castel, Phys. Rep. **290**, 201 (1997).
[20] D. R. Inglis, Rev. Mod. Phys. **25**, 390 (1953).
[21] B. Jancovici and I. Talmi, Phys. Rev. **95**, 289 (1954).
[22] M. S. Fayache, L. Zamick, and H. Müther, Phys. Rev. C **60**, 067305 (1999).
[23] H. Geissel *et al.*, Phys. Rev. Lett. **88**, 122301 (2002).
[24] J. G. Messchendorp *et al.*, Phys. Rev. Lett. **89**, 222302 (2002).
[25] D. Trnka *et al.*, Phys. Rev. Lett. **94**, 192303 (2005).
[26] M. Naruki *et al.*, Phys. Rev. Lett. **96**, 092301 (2006).
[27] R. Nasseripour *et al.*, Phys. Rev. Lett. **99**, 262302 (2007).
[28] R. Machleidt, Adv. Nucl. Phys. **19**, 189 (1989).
[29] S. K. Bogner *et al.*, Phys. Rev. C **65**, 051301(R) (2002).
[30] S. K. Bogner, T. T. S. Kuo, and A. Schwenk, Phys. Rep. **386**, 1 (2003).
[31] T. T. S. Kuo and E. Osnes, *Lecture Notes in Physics* (Springer-Verlag, New York, 1990), Vol. 364.
[32] M. Hjorth-Jensen, T. T. S. Kuo, and E. Osnes, Phys. Rep. **261**, 125 (1995) and references therein.
[33] S. K. Bogner, A. Schwenk, R. J. Furnstahl, and A. Nogga, Nucl. Phys. A **763**, 59 (2005).
[34] J. D. Holt *et al.*, Phys. Rev. C **72**, 041304(R) (2005).
[35] L. Coraggio *et al.*, Phys. Rev. C **75**, 057303 (2007).
[36] A. Negret *et al.*, Phys. Rev. Lett. **97**, 062502 (2006).
[37] L. A. Schaller, L. Schellenberg, A. Ruetschi, and H. Schneuwly, Nucl. Phys. A **343**, 333 (1980).
[38] W. Schütz, Z. Phys. A **273**, 69 (1975).

PHYSICAL REVIEW C **79**, 054004 (2009)

Low-momentum NN interactions and all-order summation of ring diagrams of symmetric nuclear matter

L.-W. Siu, J. W. Holt, T. T. S. Kuo,* and G. E. Brown
Department of Physics and Astronomy, Stony Brook University, New York 11794-3800, USA
(Received 26 November 2008; revised manuscript received 13 April 2009; published 13 May 2009)

We study the equation of state for symmetric nuclear matter using a ring-diagram approach in which the particle-particle hole-hole (pphh) ring diagrams within a momentum model space of decimation scale Λ are summed to all orders. The calculation is carried out using the renormalized low-momentum nucleon-nucleon (NN) interaction $V_{\text{low-}k}$, which is obtained from a bare NN potential by integrating out the high-momentum components beyond Λ. The bare NN potentials of CD-Bonn, Nijmegen, and Idaho have been employed. The choice of Λ and its influence on the single particle spectrum are discussed. Ring-diagram correlations at intermediate momenta ($k \simeq 2$ fm^{-1}) are found to be particularly important for nuclear saturation, suggesting the necessity of using a sufficiently large decimation scale so that the above momentum region is not integrated out. Using $V_{\text{low-}k}$ with $\Lambda \sim 3$ fm^{-1}, we perform a ring-diagram computation with the above potentials, which all yield saturation energies E/A and Fermi momenta $k_F^{(0)}$ considerably larger than the empirical values. On the other hand, similar computations with the medium-dependent Brown-Rho scaled NN potentials give satisfactory results of $E/A \simeq -15$ MeV and $k_F^{(0)} \simeq 1.4$ fm^{-1}. The effect of this medium dependence is well reproduced by an empirical three-body force of the Skyrme type.

DOI: 10.1103/PhysRevC.79.054004 PACS number(s): 21.30.−x, 21.65.Mn, 21.45.Ff

I. INTRODUCTION

Obtaining the energy per nucleon (E/A) as a function of the Fermi momentum (k_F) for symmetric nuclear matter is one of the most important problems in nuclear physics. Empirically, nuclear matter saturates at $E/A \simeq -16$ MeV and $k_F \simeq 1.36$ fm^{-1}. A great amount of effort has been put into computing the above quantities starting from a microscopic many-body theory. For many years, the Brueckner-Hartree-Fock (BHF) theory [1–3] was the primary framework for nuclear matter calculations. However, BHF represents only the first-order approximation in the general hole-line expansion [4]. Conclusive studies [5–7] have shown that the hole-line expansion converges at the third order (or the second order with a continuous single-particle spectrum) and that such results are in good agreement with variational calculations [8] of the binding energy per nucleon. Nonetheless, all such calculations have shown that it is very difficult to obtain *both* the empirical saturation energy and the saturation Fermi momentum simultaneously. In fact, such calculations using various models of the nucleon-nucleon interaction result in a series of saturation points that actually lie along a band, often referred to as the Coester band [9], which deviates significantly from the empirical saturation point. For this reason it is now widely believed that free-space two-nucleon interactions alone are insufficient to describe the properties of nuclear systems close to saturation density and that accurate results can only be achieved by introducing higher-order effects, e.g., three-nucleon forces [10] or relativistic effects [11].

In the present work, we carry out calculations of the nuclear binding energy for symmetric nuclear matter using a framework based on a combination of the recently developed low-momentum NN interaction $V_{\text{low-}k}$ [12–17] and the ring-diagram method for nuclear matter of Song, Yang, and Kuo [18], which is a model-space approach where the particle-particle hole-hole (pphh) ring diagrams for the potential energy of nuclear matter are summed to all orders. In previous studies a model space of size $\Lambda \sim 3$ fm^{-1} was used to obtain improved results compared with those from the BHF method. Such an improvement can be attributed to the following desirable features in the ring-diagram approach. First, the ground-state energy shift ΔE_0 in the BHF approach is given by just the lowest-order reaction matrix (G matrix) diagram (corresponding to Fig. 1(b) with the dashed vertex representing G). It does not include diagrams corresponding to the particle-hole excitations of the Fermi sea. Such excitations represent the effect of long-range correlations. In contrast, the pphh ring diagrams, such as those in Figs. 1(c) and 1(d), are included to all orders in the ring-diagram approach. Second, the single-particle (s.p.) spectrum used in the ring-diagram approach is different from that in early BHF calculations, where one typically employed a self-consistent s.p. spectrum for momenta $k \leqslant k_F$ and a free-particle spectrum otherwise. Thus the s.p. spectrum had a large artificial discontinuity at k_F. The s.p. spectrum used in the ring-diagram approach is a continuous one. The importance of using a continuous s.p. spectrum in nuclear matter theory has been discussed and emphasized in Refs. [6] and [7]. Within the above ring-diagram framework, previous calculations [19] using G-matrix effective interactions and $\Lambda \sim 3$ fm^{-1} have yielded saturated nuclear matter that is slightly overbound ($E/A \simeq -18$ MeV) and that saturates at too high a density ($k_F \simeq 1.6$ fm^{-1}) compared to empirical data. These results are consistent, within theoretical errors, with calculations based on the third-order hole-line expansion and variational methods (see Refs. [5,6,8]).

*thomas.kuo@stonybrook.edu

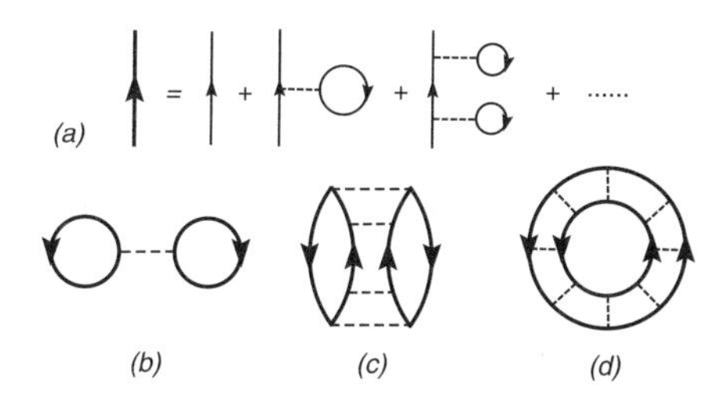

FIG. 1. Diagrams included in the pphh ring-diagram summation for the ground-state energy shift of symmetric nuclear matter. Included are (a) self-energy insertions on the single-particle propagator and (b)–(d) pphh correlations.

In the past, the above ring-diagram approach [18,19] employed the G-matrix interaction, which is energy dependent, meaning that the whole calculation must be done in a "self-consistent" way. The calculation would be greatly simplified if this energy dependence, and thus the self-consistency procedure, were removed. Such an improvement has occurred in the past several years with the development of a low-momentum NN interaction, $V_{\text{low-}k}$, constructed from renormalization group techniques [12–17]. As discussed in these references, the $V_{\text{low-}k}$ interaction has a number of desirable properties, such as being nearly unique as well as being a smooth potential suitable for perturbative many-body calculations [20]. Furthermore, $V_{\text{low-}k}$ is energy independent, making it a convenient choice for the interaction used in ring-diagram calculations of nuclear matter.

The $V_{\text{low-}k}$ interaction has been extensively used in nuclear shell-model calculations for nuclei with a few valence nucleons outside a closed shell. As reviewed recently by Coraggio *et al.* [21], the results obtained from such shell-model calculations are in very good agreement with experiments. However, applications of the $V_{\text{low-}k}$ interaction to nuclear matter have been relatively few [20,22–24]. A main purpose of the present work is to study the suitability of describing symmetric nuclear matter using $V_{\text{low-}k}$. A concern about such applications is that the use of $V_{\text{low-}k}$ alone may not provide satisfactory nuclear saturation. As illustrated in Ref. [22], Hartree-Fock (HF) calculations of nuclear matter using $V_{\text{low-}k}$ with a cutoff momentum of $\Lambda \sim 2.0\ \text{fm}^{-1}$ do not yield nuclear saturation—the calculated E/A decreases monotonically with k_F up to the decimation scale Λ.

In this work, we carry out a ring-diagram calculation of symmetric nuclear matter with $V_{\text{low-}k}$. We show in detail that satisfactory results for the saturation energy and saturation Fermi momentum can be obtained when one takes into account the following two factors: a suitable choice of the cutoff momentum and the in-medium modification of meson masses. As we discuss, ring-diagram correlations at intermediate momenta ($k \sim 2.0\ \text{fm}^{-1}$) have strong medium dependence and are important for nuclear saturation. To include their effects one needs to use a sufficiently large decimation scale Λ so that the above momentum range is not integrated out. We have carried out ring-diagram calculations for symmetric nuclear matter using $\Lambda \sim 3\ \text{fm}^{-1}$ with several modern high-precision NN potentials, and the results yield nuclear saturation. However, E/A and k_F at saturation are both considerably larger in magnitude than the corresponding empirical values. Great improvement can be obtained when one takes into account medium modifications to the exchanged mesons. Clearly mesons in a nuclear medium and those in free space are different: the former are "dressed" while the latter are "bare." Brown and Rho have suggested that the dependence of meson masses on nuclear density can be described by a simple equation known as Brown-Rho scaling [25,26]:

$$\sqrt{\frac{g_A}{g_A^*}}\frac{m_N^*}{m_N} = \frac{m_\sigma^*}{m_\sigma} = \frac{m_\rho^*}{m_\rho} = \frac{m_\omega^*}{m_\omega} = \frac{f_\pi^*}{f_\pi} = \Phi(n), \tag{1}$$

where g_A is the axial coupling constant, Φ is a function of the nuclear density n, and the star indicates in-medium values of the given quantities. At saturation density, $\Phi(n_0) \simeq 0.8$. In a high-density medium such as nuclear matter, these medium modifications of meson masses are significant and can render V_{NN} quite different from that in free space. Thus, in contrast to shell-model calculations for nuclei with only a few valence particles, for nuclear matter calculations it may be necessary to use a V_{NN} with medium modifications *built in*. In the present work, we carry out such a ring-diagram summation using a Brown-Rho scaled NN interaction.

The Skyrme [27] interaction is one of the most successful effective nuclear potentials. An important component of this interaction is a zero-range three-body force, which is equivalent to a density-dependent two-body force. Note that the importance of three-body interactions in achieving nuclear saturation with low-momentum interactions has been extensively discussed in the literature (see Ref. [20] and references quoted therein). In the last part of our work, we study whether the density dependence from Brown-Rho scaling can be well represented by that from an empirical density-dependent force of the Skyrme type.

The organization of this article is as follows. In Secs. II and III we outline our model-space pphh ring-diagram calculation for the nuclear binding energy and the concept of Brown-Rho scaling, respectively. In Sec. IV we present our computational results. A brief conclusion can be found in Sec. V.

II. SUMMATION OF pphh RING DIAGRAMS

In this section we describe how to calculate the properties of symmetric matter using the low-momentum ring-diagram method. We employ a momentum model space where all nucleons have momenta $k \leqslant \Lambda$. By integrating out the $k > \Lambda$ components, the low-momentum interaction $V_{\text{low-}k}$ is constructed for summing the pphh ring diagrams within the model space.

The ground-state energy shift $\Delta E_0 = E_0 - E_0^{\text{free}}$ for nuclear matter is defined as the difference between the true ground-state energy E_0 and the corresponding quantity for the noninteracting system E_0^{free}. In the present work, we consider ΔE_0 as given by the all-order sum of the pphh ring diagrams as shown in Figs. 1(b)–1(d).

We shall calculate the all-order sum, denoted as ΔE_0^{pp}, of such diagrams. Each vertex in a ring diagram is the renormalized effective interaction $V_{\text{low-}k}$ corresponding to the model space $k \leqslant \Lambda$. It is obtained from the following T-matrix equivalence method [12–17]. Let us start with the T-matrix equation

$$T(k', k, k^2) = V(k', k) + \mathcal{P}\int_0^{\infty} q^2 dq \frac{V(k', q)T(q, k, k^2)}{k^2 - q^2}, \quad (2)$$

where V is a bare NN potential. In the present work we use the CD-Bonn [28], Nijmegen-I [29], and Idaho (chiral) [30] NN potentials. Notice that in the above equation the intermediate-state momentum q is integrated from 0 to ∞. We then define an effective low-momentum T-matrix by

$$T_{\text{low-}k}(p', p, p^2) = V_{\text{low-}k}(p', p) + \mathcal{P}\int_0^{\Lambda} q^2 dq \times \frac{V_{\text{low-}k}(p', q)T_{\text{low-}k}(q, p, p^2)}{p^2 - q^2}, \quad (3)$$

where the intermediate-state momentum is integrated from 0 to Λ, the momentum space cutoff. The low-momentum interaction $V_{\text{low-}k}$ is then obtained from the above equations by requiring the T-matrix equivalence condition to hold, namely,

$$T(p', p, p^2) = T_{\text{low-}k}(p', p, p^2); \quad (p', p) \leqslant \Lambda. \quad (4)$$

The iteration method of Lee-Suzuki-Andreozzi [17,31,32] has been used in obtaining the above $V_{\text{low-}k}$.

With $V_{\text{low-}k}$, our ring-diagram calculations are relatively simple, compared to the G-matrix calculations of Ref. [18]. Within the model space, we use the Hartree-Fock s.p. spectrum calculated with the $V_{\text{low-}k}$ interaction, and outside the model space we use the free particle spectrum. In other words,

$$\epsilon_k = \begin{cases} \hbar^2k^2/2m + \sum_{h<k_F} \langle kh|V_{\text{low-}k}|kh\rangle; & k \leqslant \Lambda \\ \hbar^2k^2/2m; & k > \Lambda. \end{cases} \quad (5)$$

The above s.p. spectrum is medium (k_F) dependent.

Our next step is to solve the model-space RPA equation

$$\sum_{ef}[(\epsilon_i + \epsilon_j)\delta_{ij,ef} + \lambda(\bar{n}_i\bar{n}_j - n_in_j)\langle ij|V_{\text{low-}k}|ef\rangle]Y_n(ef, \lambda) = \omega_n Y_n(ij, \lambda); (i, j, e, f) \leqslant \Lambda, \quad (6)$$

where $n_a = 1$ for $a \leqslant k_F$ and $n_a = 0$ for $a > k_F$; also $\bar{n}_a = (1 - n_a)$. The strength parameter λ is introduced for calculational convenience and varies between 0 and 1. Note that the above equation is within the model space as indicated by $(i, j, e, f) \leqslant \Lambda$. The transition amplitudes Y of the above equation can be classified into two types, one dominated by hole-hole components and the other by particle-particle components. We use only the former, denoted by Y_m, for the calculation of the all-order sum of the pphh ring diagrams. This sum is given by [18,24,33]

$$\Delta E_0^{\text{pp}} = \int_0^1 d\lambda \sum_m \sum_{ijkl<\Lambda} Y_m(ij, \lambda) \times Y_m^*(kl, \lambda)\langle ij|V_{\text{low-}k}|kl\rangle, \quad (7)$$

where the normalization condition for Y_m is $\langle Y_m|\frac{1}{Q}|Y_m\rangle = -1$ and $Q(i, j) = (\bar{n}_i\bar{n}_j - n_in_j)$. In the above, Σ_m means we sum over only those solutions of the RPA equation (6) that are dominated by hole-hole components as indicated by the normalization condition.

The all-order sum of the pphh ring diagrams as indicated by Figs. 1(b)–1(d) is given by the above ΔE_0^{pp}. Because we use the HF s.p. spectrum, each propagator of the diagrams contains the HF insertions to all orders as indicated by Fig. 1(a). Clearly our ring diagrams are medium dependent; their s.p. propagators have all-order HF insertions that are medium dependent, as is the occupation factor $(\bar{n}_i\bar{n}_j - n_in_j)$ of the RPA equation.

III. BROWN-RHO SCALING AND IN-MEDIUM NN INTERACTIONS

Nucleon-nucleon interactions are mediated by meson exchange, and clearly the in-medium modification of meson masses is important for NN interactions. These modifications could arise from the partial restoration of chiral symmetry at finite density/temperature or from traditional many-body effects. Particularly important are the vector mesons, for which there is now evidence from both theory [34–36] and experiment [37,38] that the masses may decrease by approximately 10–15% at normal nuclear matter density and zero temperature. This in-medium decrease of meson masses is often referred to as Brown-Rho scaling [25,26]. For densities below that of nuclear matter, it is suggested [34] that the masses decrease linearly with the density n:

$$\frac{m_V^*}{m_V} = 1 - C\frac{n}{n_0}, \quad (8)$$

where m_V^* is the vector meson mass in-medium, n_0 is nuclear matter saturation density, and C is a constant of value ~0.10–0.15.

We study the consequences for nuclear many-body calculations by replacing the NN interaction in free space with a density-dependent interaction with medium-modified meson exchange. A simple way to obtain such potentials is by modifying the meson masses and relevant parameters of the one-boson-exchange NN potentials (e.g., the Bonn and Nijmegen interactions). The saturation of nuclear matter is an appropriate phenomenon for studying the effects of dropping masses [23,39], because the density of nuclear matter is constant and large enough to significantly affect the nuclear interaction through the modified meson masses.

One unambiguous prediction of Brown-Rho scaling in dense nuclear matter is the decreasing of the tensor force component of the nuclear interaction. The two most important contributions to the tensor force come from π and ρ meson exchange, which act opposite to each other:

$$V_\rho^T(r) = -\frac{f_\rho^2}{4\pi}m_\rho\tau_1 \cdot \tau_2 S_{12} f_3(m_\rho r), \quad (9)$$

$$V_\pi^T(r) = \frac{f_\pi^2}{4\pi}m_\pi\tau_1 \cdot \tau_2 S_{12} f_3(m_\pi r), \quad (10)$$

$$f_3(mr) = \left(\frac{1}{(mr)^3} + \frac{1}{(mr)^2} + \frac{1}{3mr}\right)e^{-mr}. \quad (11)$$

In Brown-Rho scaling the ρ meson is expected to decrease in mass at finite density while the pion mass remains nearly unchanged due to chiral invariance. Therefore, the overall strength of the tensor force at finite density will be significantly smaller than that in free space. As we shall discuss later, this decrease in the tensor force plays an important role for nuclear saturation.

The Skyrme effective interaction has been widely used in nuclear physics and has been very successful in describing the properties of finite nuclei as well as nuclear matter [27]. This interaction has both two-body and three-body terms, having the form

$$V_{\rm skyrme} = \sum_{i<j} V(i,j) + \sum_{i<j<k} V(i,j,k). \quad (12)$$

Here $V(i,j)$ is a momentum ($\vec{k}$)-dependent zero-range interaction, containing two types of terms: one with no momentum dependence and the other depending quadratically on $\vec{k}$. $V(i,j)$ corresponds to a low-momentum expansion of an underlying NN interaction. Its three-body term is a zero-range interaction,

$$V(i,j,k) = t_3\delta(\vec{r}_i - \vec{r}_j)\delta(\vec{r}_j - \vec{r}_k), \quad (13)$$

which is equivalent to a density-dependent two-body interaction of the form

$$V_\rho(1,2) = \tfrac{1}{6} t_3 \delta(\vec{r}_1 - \vec{r}_2)\rho(\vec{r}_{\rm av}), \quad (14)$$

with $\vec{r}_{\rm av} = \frac{1}{2}(\vec{r}_1 + \vec{r}_2)$.

The general structure of $V_{\rm skyrme}$ is rather similar to the effective interactions based on effective field theories (EFT) [20], with $V(i,j)$ corresponding to $V_{\text{low-}k}$ and $V(i,j,k)$ to the EFT three-body force. The Skyrme three-body force, however, is much simpler than that in EFT. We compare in the next section the density-dependent effect generated by the medium modified NN interaction with that from an empirical three-body force of the Skyrme type.

IV. RESULTS AND DISCUSSIONS

In this section, we report computational results for the binding energy of symmetric nuclear matter calculated with an all-order summation of low-momentum pphh ring diagrams. The method is already outlined and discussed in the above sections. As mentioned above, we employ a model-space approach. Starting from various bare NN interactions, we first construct the low-momentum interactions $V_{\text{low-}k}$ with a particular choice of the cutoff momentum Λ. The low-momentum ($<\Lambda$) pphh ring diagrams are then summed to all orders as given by Eq. (7) to give the binding energy.

A. Single-particle spectrum and nuclear binding energy

First, we look carefully into the role of Λ in our ring-diagram calculation. Let us start with the s.p. energy ϵ_k. Obtaining ϵ_k is the first step in our ring-diagram calculation. Within our model-space approach, ϵ_k is given by the Hartree-Fock spectrum for $k \leqslant \Lambda$, while for $k > \Lambda$, ϵ_k is taken as the free spectrum [see Eq. (5)]. As emphasized before, the s.p. spectrum obtained in this way will in general have a

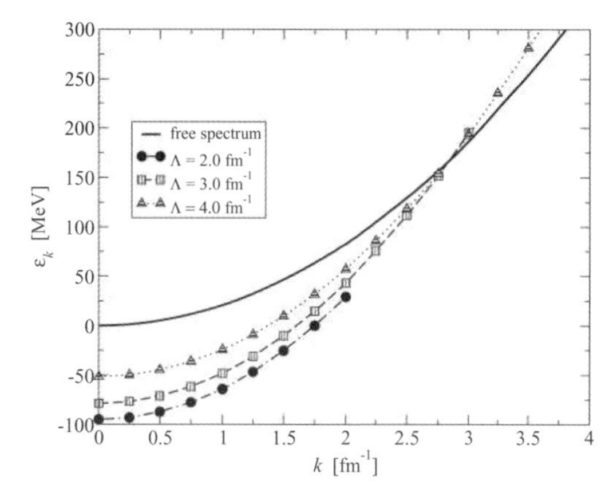

FIG. 2. Dependence of the model-space s.p. spectrum on the decimation scale Λ for symmetric nuclear matter at the empirical saturation density. The CD-Bonn potential is used in the construction of $V_{\text{low-}k}$.

discontinuity at Λ. Such a discontinuity is a direct consequence of having a finite model space. It is of much interest to study the s.p. spectrum as Λ is varied. In Fig. 2, we plot the spectrum for different values of Λ ranging from 2 to 4 fm^{-1}. We observed that with $\Lambda = 2.0$ fm^{-1}, the discontinuity at Λ is relatively large; there is a gap of about 50 MeV between the s.p. spectrum just inside Λ and that outside. However, this discontinuity decreases if Λ is increased to around 3 fm^{-1}. At this point, the s.p. spectrum is most "satisfactory" in the sense of being almost continuous. A further increase in Λ will result in an "unreasonable" situation where the s.p. spectrum just inside Λ becomes significantly higher than that outside. This is clearly shown in the data of $\Lambda = 4.0$ fm^{-1}. The above results suggest that to have a nearly continuous s.p. spectrum, which is physically desirable, it is necessary to use $\Lambda \sim 3$ fm^{-1}.

Next, we look into the effect of Λ on the nuclear binding energy. Once the s.p. energies are obtained, the all-order ring-diagram summation can be carried out [see Eqs. (6) and (7)]. Let us first discuss the computational results based on the CD-Bonn potential. Results from various Λ ranging from 2 to 3.2 fm^{-1} are shown in Fig. 3. Let us focus on (i) the overall saturation phenomena and (ii) the numerical values of the binding energy and the saturation momentum.

(i) We observe that the nuclear binding energy exhibits saturation only when Λ is $\sim$3 fm^{-1} and beyond. This reflects the importance of ring diagrams in the intermediate-momentum region ($k \sim 2$ fm^{-1}). To illustrate, let us compare the results for the cases of $\Lambda = 2$ and 3 fm^{-1}. As indicated by Eqs. (2)–(4), $V_{\text{low-}k}$ includes only the $k > \Lambda$ pp ladder interactions between a pair of "free" nucleons; there is no medium correction included. Thus the above two cases treat correlations in the momentum region between 2 and 3 fm^{-1} differently: the former includes for this momentum region only pp ladder interactions with medium effect neglected, while the latter includes both pp and hh correlations with medium effect, such as that from the Pauli blocking, included. Our results

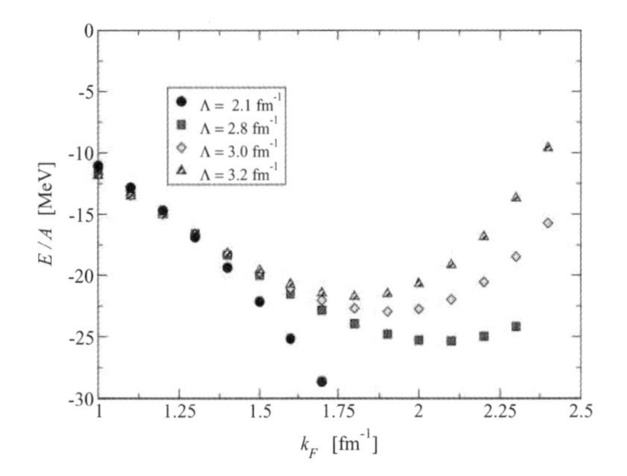

FIG. 3. Results for the energy per nucleon (E/A) of symmetric nuclear matter obtained by summing up the pphh ring diagrams to all orders. Low-momentum NN interactions, constructed from the CD-Bonn potential, with various cutoffs Λ are used in the ring-diagram summation.

indicate that the medium effect in the above momentum region is vital for saturation.

For nuclear matter binding energy calculations, there is no first-order contribution from the tensor force (V_T); its leading contribution is second order of the form $\langle {}^3S_1|V_T \frac{Q}{e} V_T|{}^3S_1\rangle$, where Q stands for the Pauli blocking operator and e the energy denominator. Thus the contribution from the tensor force depends largely on the availability of the intermediate states; this contribution is large for low k_F but is suppressed for high k_F. To illustrate this point, we plot the potential energy of nuclear matter from the ${}^1S_0$ and ${}^3S_1 - {}^3D_1$ channels separately in Fig. 4. The behavior of the potential energy in these two channels differ in a significant way. The ${}^1S_0$ channel is practically independent of the choice of Λ, as displayed in the upper panel of the figure. This indicates that for this channel the effects from medium corrections and hh correlations are not important. Also the PE/A from this channel does not exhibit saturation at a reasonable k_F. In the lower panel of the figure, we display the PE/A for the ${}^3S_1 - {}^3D_1$ channel where the tensor force is important. As seen, PE/A does not exhibit saturation when using $\Lambda = 2$ fm^{-1}. On the contrary, the result using $\Lambda = 3$ fm^{-1} shows a clear saturation behavior. This is mainly because in the former case the Pauli blocking effect is ignored for the momentum region 2–3 fm^{-1} while it is included for the latter. To have saturation, we should not integrate out the momentum components in the NN interaction that are crucial for saturation. Considering also the effect of Λ on the s.p. spectrum, we believe that $\Lambda = 3.0$ fm^{-1} is a suitable choice for our ring-diagram nuclear matter calculation. Notice that a model space $\sim$3 fm^{-1} has been used in other similar ring-summation calculations using G-matrix effective interactions [18,19].

(ii) We have performed a similar ring summation with the Nijmegen I and Idaho potentials. Results with $\Lambda = 3.0$ fm^{-1} are compared with those from CD-Bonn as shown in Fig. 5. The saturation energies for these three potentials are located between -19 and -23 MeV, while the saturation momentum ranges from 1.75 to 1.85 fm^{-1}. These quantities are considerably larger than the empirical values of -16 MeV and 1.4 fm^{-1}, respectively. We believe that improvements can be obtained if one takes into account the medium dependence of the NN interaction. Namely, instead of using a $V_{\text{low-}k}$ constructed from a bare NN interaction, one should employ a $V_{\text{low-}k}$ constructed from a "scaled" NN interaction according to the nuclear density. Below we report how we incorporate such effects into our ring-diagram summation.

B. Nuclear binding energy with Brown-Rho scaling

The concept of Brown-Rho scaling has already been discussed in Sec. III. The medium effects on the NN interaction resulting from the in-medium modification of meson masses

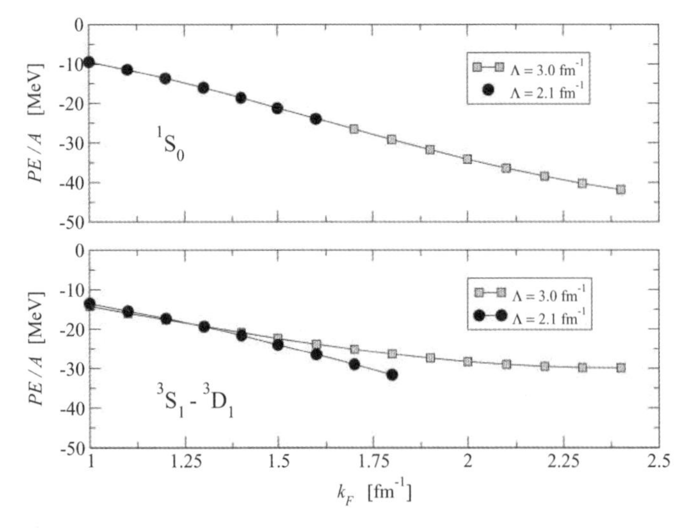

FIG. 4. Potential energy per nucleon (PE/A) in the ${}^1S_0$ and ${}^3S_1 - {}^3D_1$ channels of symmetric nuclear matter from summing up pphh ring diagrams to all orders. The CD-Bonn potential is used in the construction of $V_{\text{low-}k}$.

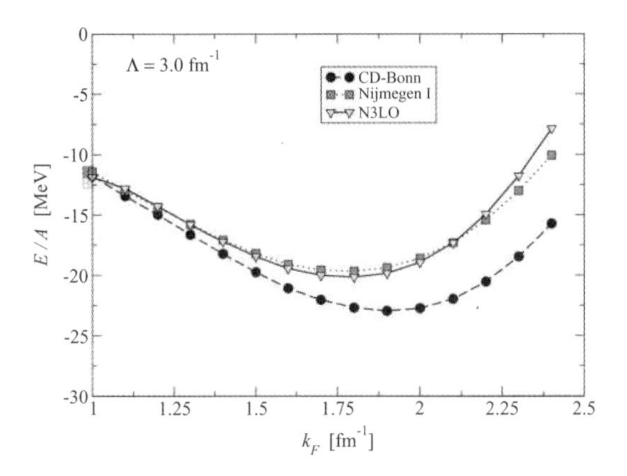

FIG. 5. The binding energy of symmetric nuclear matter from the low-momentum ring-diagram summation using various NN potentials. A momentum-space cutoff of $\Lambda = 3.0$ fm^{-1} is used.

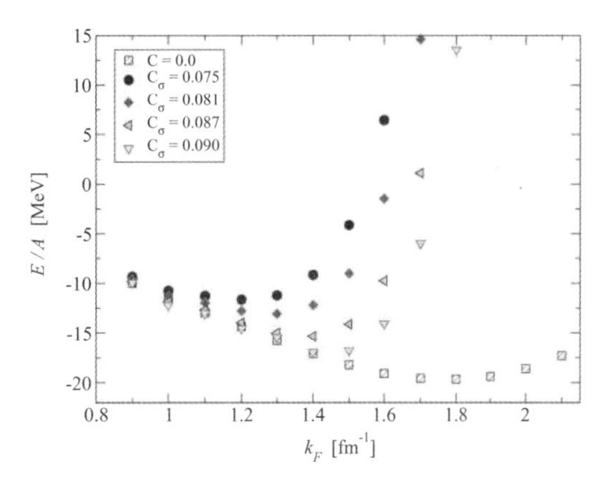

FIG. 6. (Color online) The binding energy of symmetric nuclear matter from the Brown-Rho scaled low-momentum Nijmegen II interaction using the ring-diagram summation with $\Lambda = 3.0$ fm^{-1}. Calculations for different choices of the σ meson scaling constant C_σ are shown.

have a profound effect on nuclear binding. To incorporate this in our ring-diagram calculation we work with the Nijmegen potential, which is one of the pure one-boson-exchange NN potentials. The bare Nijmegen is first Brown-Rho scaled [see Eq. (8)] with the dropping mass ratio C chosen to be 0.15. Vector meson masses in a nuclear medium have been widely studied both theoretically and experimentally, but the σ meson mass is not well constrained. Previous calculations [39] of nuclear matter saturation within the Dirac-Brueckner-Hartree-Fock formalism showed that there is too much attraction when the σ meson is scaled according to Eq. (8). However, a microscopic treatment [39] of σ meson exchange in terms of correlated 2π exchange showed that the medium effects on the σ are much weaker than those in Eq. (8). Therefore, in our ring-diagram summation using the Brown-Rho scaled Nijmegen II interaction, we employ a range of scaling parameters C_σ between 0.075 and 0.09. Our calculations are shown in Fig. 6. With Brown-Rho scaling, the numerical values for both the saturation energy and saturation momentum are greatly improved. Whereas the unscaled potential gives a binding energy $\mathrm{BE}/A \simeq 20$ MeV and $k_F^0 \simeq 1.8$ fm^{-1}, the scaled potential gives $\mathrm{BE}/A \simeq 14$–17 MeV and $k_F^0 \simeq 1.30$–1.45 fm^{-1} for a σ meson scaling constant $C_\sigma \sim 0.08$–0.09, in very good agreement with the empirical values.We conclude, first, that the medium dependence of nuclear interactions is crucial for a satisfactory description of nuclear saturation and, second, that within the framework of one-boson-exchange NN interaction models one can obtain an adequate description of nuclear matter saturation by including Brown-Rho scaled meson masses.

C. Nuclear binding energy with three-body force of the Skyrme type

As discussed earlier in Sec. III, the widely used Skyrme interaction contains a three-body term that is equivalent to a density-dependent two-body interaction. It is of much interest to study whether our result with Brown-Rho scaled Nijmegen potential can be reproduced with the unscaled Nijmegen plus an effective three-body interaction of the Skyrme type that is characterized by a strength parameter, t_3 [see Eq. (14)]. In Fig. 7 we compare the results using $t_3 = 1250$ with our previous calculations using the Brown-Rho scaled Nijmegen II potential with a σ meson scaling constant of $C_\sigma = 0.087$. In all calculations $\Lambda = 3.0$ is used. We note that satisfactory results for the saturation energy and Fermi momentum are obtained using either Brown-Rho scaling or a 3NF of the Skyrme type. However, the nuclear incompressibility is considerably larger in the case of Brown-Rho scaling.

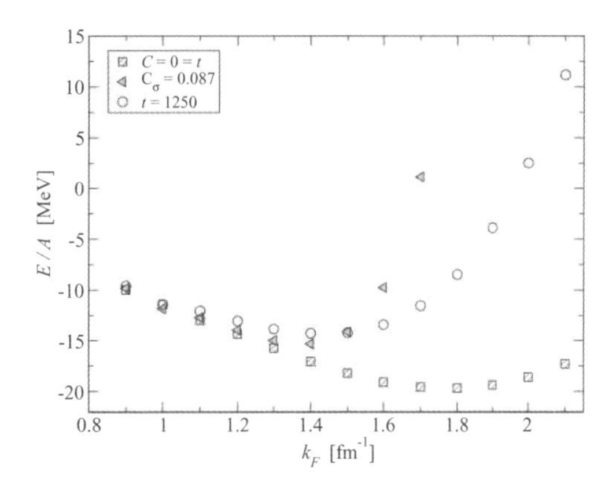

FIG. 7. The binding energy of symmetric nuclear matter from the low-momentum ring-diagram summation with $\Lambda = 3.0$ fm^{-1}. Three different interactions are used: (1) the medium-independent Nijmegen II interaction, the Brown-Rho scaled interaction with $C_\sigma = 0.087$, and finally the Nijmegen II interaction supplemented with a 3NF of the Skyrme type with $t_3 = 1250$.

V. CONCLUSION

We have studied the equation of state for symmetric nuclear matter using the low-momentum nucleon-nucleon (NN) interaction $V_{\text{low-}k}$. Particle-particle hole-hole (pphh) ring diagrams within a momentum model space $k < \Lambda$ were summed to all orders. The significant role of the intermediate-momentum range ($\sim$2.0 fm^{-1}) for nuclear saturation was discussed. We concluded that, in the ring-diagram summation, having a sufficiently large model space is important to capture the saturation effect from the intermediate-momentum components. Various bare NN potentials including CD-Bonn, Nijmegen, and Idaho have been employed, resulting in nuclear saturation with $\Lambda = 3.0$ fm^{-1}. However, the resulting binding energy and saturation momentum are still much larger than the empirical values. Improvement can be obtained when we take into account the medium modification of NN interaction. We first constructed $V_{\text{low-}k}$ from a medium-dependent Brown-Rho scaled NN potential and then implemented this into the ring-diagram summation. Satisfactory results of $E/A \simeq -15$ MeV and $k_F^{(0)} \simeq 1.4$ fm^{-1} could then be obtained. We showed that these saturation properties are well reproduced by the first ring-diagram approach with the addition of an empirical three-body force of the Skyrme type.

In the future, it is of much interest to carry out a BCS calculation on nuclear matter with $V_{\text{low-}k}$, particularly for the $^3S_1 - {}^3D_1$ channel where earlier calculations using bare NN interactions revealed a gap of 10 MeV around normal nuclear matter densities [40,41]. Recently, $V_{\text{low-}k}$ has been applied to obtain the equation of state of neutron matter [24,42] and the $^1S_0$ pairing gap [42,43]. A similar calculation on nuclear matter that incorporates the tensor correlations is obviously important and we plan to investigate it in the future.

ACKNOWLEDGMENTS

We thank R. Machleidt for many helpful discussions. This work was supported in part by US Department of Energy under Grant DF-FG02-88ER40388.

[1] H. A. Bethe, Annu. Rev. Nucl. Sci. **21**, 93 (1971).
[2] R. Machleidt, Adv. Nucl. Phys. **19**, 189 (1989).
[3] J. W. Holt and G. E. Brown, in *Hans Bethe and His Physics*, edited by G. E. Brown and C.-H. Lee (World Scientific, Singapore, 2006).
[4] B. D. Day, Rev. Mod. Phys. **50**, 495 (1978).
[5] B. D. Day, Phys. Rev. C **24**, 1203 (1981).
[6] H. Q. Song, M. Baldo, G. Giansiracusa, and U. Lombardo, Phys. Lett. **B411**, 237 (1997).
[7] H. Q. Song, M. Baldo, G. Giansiracusa, and U. Lombardo, Phys. Rev. Lett. **81**, 1584 (1998).
[8] B. D. Day and R. B. Wiringa, Phys. Rev. C **32**, 1057 (1985).
[9] F. Coester, S. Cohen, B. D. Day, and C. M. Vincent, Phys. Rev. C **1**, 769 (1970).
[10] R. B. Wiringa, V. Fiks, and A. Fabrocini, Phys. Rev. C **38**, 1010 (1988).
[11] R. Brockmann and R. Machleidt, Phys. Rev. C **42**, 1965 (1990).
[12] S. K. Bogner, T. T. S. Kuo, and L. Coraggio, Nucl. Phys. **A684**, 432 (2001).
[13] S. K. Bogner, T. T. S. Kuo, L. Coraggio, A. Covello, and N. Itaco, Phys. Rev. C **65**, 051301(R) (2002).
[14] L. Coraggio, A. Covello, A. Gargano, N. Itaco, T. T. S. Kuo, D. R. Entem, and R. Machleidt, Phys. Rev. C **66**, 021303(R) (2002).
[15] A. Schwenk, G. E. Brown, and B. Friman, Nucl. Phys. **A703**, 745 (2002).
[16] S. K. Bogner, T. T. S. Kuo, and A. Schwenk, Phys. Rep. **386**, 1 (2003).
[17] J. D. Holt, T. T. S. Kuo, and G. E. Brown, Phys. Rev. C **69**, 034329 (2004).
[18] H. Q. Song, S. D. Yang, and T. T. S. Kuo, Nucl. Phys. **A462**, 491 (1987).
[19] M. F. Jiang, T. T. S. Kuo, and H. Müther, Phys. Rev. C **38**, 2408 (1988).
[20] S. K. Bogner, A. Schwenk, R. J. Furnstahl, and A. Nogga, Nucl. Phys. **A763**, 59 (2005).
[21] L. Coraggio, A. Covello, A. Gargano, N. Itako, and T. T. S. Kuo, Prog. Part. Nucl. Phys. **62**, 135 (2009).
[22] J. Kuckei, F. Montani, H. Müther, and A. Sedrakian, Nucl. Phys. **A723**, 32 (2003).
[23] J. W. Holt, G. E. Brown, J. D. Holt, and T. T. S. Kuo, Nucl. Phys. **A785**, 322 (2007).
[24] L.-W. Siu, T. T. S. Kuo, and R. Machleidt, Phys. Rev. C **77**, 034001 (2008).
[25] G. E. Brown and M. Rho, Phys. Rev. Lett. **66**, 2720 (1991).
[26] G. E. Brown and M. Rho, Phys. Rep. **396**, 1 (2004).
[27] P. Ring and P. Schuck, *The Nuclear Many-Body Problem* (Springer-Verlag, New York, 1980), and references therein.
[28] R. Machleidt, Phys. Rev. C **63**, 024001 (2001).
[29] V. G. J. Stoks, R. A. M. Klomp, C. P. F. Terheggen, and J. J. de Swart, Phys. Rev. C **49**, 2950 (1994).
[30] D. R. Entem and R. Machleidt, Phys. Rev. C **68**, 041001(R) (2003).
[31] K. Suzuki and S. Y. Lee, Prog. Theor. Phys. **64**, 2091 (1980).
[32] F. Andreozzi, Phys. Rev. C **54**, 684 (1996).
[33] T. T. S. Kuo and Y. Tzeng, Int. J. Mod. Phys. E **3**, No. 2, 523 (1994).
[34] T. Hatsuda and S. H. Lee, Phys. Rev. C **46**, R34 (1992).
[35] M. Harada and K. Yamawaki, Phys. Rep. **381**, 1 (2003).
[36] F. Klingl, N. Kaiser, and W. Weise, Nucl. Phys. **A624**, 527 (1997).
[37] D. Trnka *et al.*, Phys. Rev. Lett. **94**, 192303 (2005).
[38] M. Naruki *et al.*, Phys. Rev. Lett. **96**, 092301 (2006).
[39] R. Rapp, R. Machleidt, J. W. Durso, and G. E. Brown, Phys. Rev. Lett. **82**, 1827 (1999).
[40] H. Müther and W. H. Dickhoff, Phys. Rev. C **72**, 054313 (2005).
[41] M. Baldo, I. Bombaci, and U. Lombardo, Phys. Lett. **B283**, 8 (1992).
[42] A. Schwenk, B. Friman, and G. E. Brown, Nucl. Phys. **A713**, 191 (2003).
[43] K. Hebeler, A. Schwenk, and B. Friman, Phys. Lett. **B648**, 176 (2007).

PHYSICAL REVIEW C **80**, 065803 (2009)

Neutron stars, β-stable ring-diagram equation of state, and Brown-Rho scaling

Huan Dong and T. T. S. Kuo*
Department of Physics and Astronomy, Stony Brook University, Stony Brook, New York 11794-3800, USA

R. Machleidt
Department of Physics, University of Idaho, Moscow, Idaho 83844, USA

(Received 1 September 2009; revised manuscript received 10 November 2009; published 14 December 2009)

Neutron star properties, such as mass, radius, and moment of inertia, are calculated by solving the Tolman-Oppenheimer-Volkov (TOV) equations using the ring-diagram equation of state (EOS) obtained from realistic low-momentum NN interactions $V_{\text{low-k}}$. Several NN potentials (CDBonn, Nijmegen, Argonne V18, and BonnA) have been employed to calculate the ring-diagram EOS where the particle-particle hole-hole ring diagrams are summed to all orders. The proton fractions for different radial regions of a β-stable neutron star are determined from the chemical potential conditions $\mu_n - \mu_p = \mu_e = \mu_\mu$. The neutron star masses, radii, and moments of inertia given by the aforementioned potentials all tend to be too small compared with the accepted values. Our results are largely improved with the inclusion of a Skyrme-type three-body force based on Brown-Rho scalings where the in-medium meson masses, particularly those of ω, ρ, and σ, are slightly decreased compared with their in-vacuum values. Representative results using such medium-corrected interactions are maximum neutron-star mass $M \sim 1.8M_\odot$ with radius $R \sim 9$ km and moment of inertia $\sim 60 M_\odot\,\text{km}^2$, values given by the four NN potentials being nearly the same. The effects of nuclei-crust EOSs on the properties of neutron stars are discussed.

DOI: 10.1103/PhysRevC.80.065803 PACS number(s): 21.65.Mn, 26.60.−c, 21.30.Fe

I. INTRODUCTION

Neutron stars are very interesting physical systems and their properties, such as masses and radii, can be derived from the equation of state (EOS) of the nuclear medium contained in them. In carrying out such derivation, there is, however, a well-known difficulty, namely, the EOS is not fully known. Determination of the EOS for neutron stars is an important yet challenging undertaking. As reviewed in Refs. [1–5], this topic has been extensively studied and much progress has been made. Generally speaking, there are two complementary approaches to determine the EOS. One is to deduce it from heavy-ion collision experiments, and crucial information about the EOS has already been obtained [1,6–8]. Another approach is to calculate the EOS microscopically from a many-body theory. (See, e.g., Refs. [2,9,10] and references quoted therein.) As is well known, there are a number of difficulties in this approach. Before discussing them, let us first briefly outline the derivation of neutron star properties from its EOS. One starts from the Tolman-Oppenheimer-Volkov (TOV) equations

$$\frac{dp(r)}{dr} = -\frac{GM(r)\epsilon(r)}{c^2r^2}\frac{\left[1+\frac{p(r)}{\epsilon(r)}\right]\left[1+\frac{4\pi r^3 p(r)}{M(r)c^2}\right]}{\left[1-\frac{2GM(r)}{rc^2}\right]},$$
$$\frac{dM(r)}{dr} = 4\pi r^2\epsilon(r), \tag{1}$$

where $p(r)$ is the pressure at radius r and $M(r)$ is the gravitational mass inside r. G is the gravitational constant and $\epsilon(r)$ is the energy density inclusive of the rest mass density. The solutions of these equations are obtained by integrating them out from the neutron star center till its edge where p is zero. (Excellent pedagogical reviews on neutron stars and TOV equations can be found in, e.g., Refs. [5,11].) In solving the aforementioned equations, an indispensable ingredient is clearly the nuclear matter EOS for the energy density $\epsilon(n)$, n being the medium density. As the density at the neutron star center is typically very high (several times higher than normal nuclear saturation density of $n_0 \simeq 0.16\,\text{fm}^{-3}$), we need to have the aforementioned EOS over a wide range of densities, from very low to very high.

In the present work we shall calculate the nuclear EOS directly from a fundamental nucleon-nucleon (NN) interaction V_{NN} and then use it to calculate neutron star properties by way of the TOV equations. There have been neutron star calculations using a number of EOSs, most of which are empirically determined, and the mass-radius trajectories given by them are widely different from each other (see, e.g., Fig. 2 of Ref. [1]). To determine the EOS with less uncertainty would certainly be desirable. There are a number of different NN potential models such as the CDBonn [12], Nijmegen [13], Argonne V18 [14], and BonnA [15] potentials. These potentials all possess strong short-range repulsions and to use them in many-body calculations one needs first take care of their short-range correlations by way of some renormalization methods. We shall use in the present work the recently developed renormalization group method which converts V_{NN} into an effective low-momentum NN interaction $V_{\text{low-k}}$ [16–21]. An advantage of this interaction is its near uniqueness, in the sense that the $V_{\text{low-k}}$'s derived from different realistic NN interactions are nearly the same. Also $V_{\text{low-k}}$ is a smooth potential suitable for being directly used in many-body calculations. This $V_{\text{low-k}}$ will then be used to calculate the nuclear EOS using a recently developed low-momentum ring-diagram approach [22], where the particle-particle hole-hole (pphh) ring diagrams of the EOS are summed to all orders. The aforementioned procedures are discussed in more detail later on.

*thomas.kuo@stonybrook.edu

We also study the effects of Brown-Rho (BR) scaling [23–26] on neutron star properties. As discussed in Ref. [22], low-momentum ring-diagram calculations using two-body NN interactions alone are not able to reproduce the empirical properties for symmetric nuclear matter; the calculated energy per particle (E_0/A) and saturation density (n_0) are both too high compared with the empirical values of $E_0/A \simeq -16$ MeV and $n_0 \simeq 0.16\,\mathrm{fm}^{-3}$. A main idea of the BR scaling is that the masses of in-medium mesons are generally suppressed, because of their interactions with the background medium, compared with mesons in free space. As a result, the NN interaction in the nuclear medium can be significantly different from that in free space, particularly at high density. Effects from such medium modifications have been found to be very helpful in reproducing the empirical properties of symmetric nuclear matter [22]. Dirac-Brueckner-Hartree-Fock (DBHF) nuclear matter calculations have been conducted with and without BR scaling [27–29]. In addition, BR scaling has played an essential role in explaining the extremely long lifetime of ^{14}C β decay [30]. As mentioned earlier, the central density of neutron stars is typically rather high, $\sim 8n_0$ or higher, n_0 being the saturation density of normal nuclear matter. At such high density, the effect of BR scaling should be especially significant. Neutron stars may provide an important test for BR scaling.

In the following, we first describe in Sec. II the derivation of the low-momentum NN interaction $V_{\text{low-k}}$, on which our ring-diagram EOS is based. Our method for the all order summation of the ring diagrams is also addressed. Previously such summation has been carried out for neutron matter [31] and for symmetric nuclear matter [22]. In the present work we consider β-stable nuclear matter composed of neutrons, protons, electrons, and muons. Thus we need to calculate ring diagrams for asymmetric nuclear matter whose neutron and proton fractions are different. An improved treatment for the angle-averaged proton-neutron Pauli exclusion operator is discussed. In Sec. III we outline the BR scaling for in-medium NN interactions. We discuss that the effect of BR scaling can be satisfactorily simulated by an empirical three-body force of the Skyrme type. In Sec. IV, we present and discuss our results of neutron star calculations based on ring-diagram pure-neutron EOSs; ring-diagram β-stable EOSs consisting of neutrons, protons, electrons, and muons; and the well-known EOS of Baym, Pethick, and Sutherland (BPS) [32] for the nuclei crust of neutron stars.

II. RING-DIAGRAM EOS FOR ASYMMETRIC NUCLEAR MATTER

In our calculation of neutron star properties, we employ a nuclear EOS derived microscopically from realistic NN potentials V_{NN}. Such microscopic calculations would provide a test if it is possible to derive neutron star properties starting from an underlying NN interaction. In this section, we describe the methods for this derivation. A first step in this regard is to derive an effective low-momentum interaction $V_{\text{low-k}}$ by way of a renormalization procedure, the details of which have been described in Refs. [16–21]. Here we just briefly outline its main steps. It is generally believed that the low-energy properties of physical systems can be satisfactorily described by an effective theory confined within a low-energy (or low-momentum) model space [20]. In addition, the high-momentum (short-range) parts of various V_{NN} models are model dependent and rather uncertain [20].

Motivated by the aforementioned two considerations, the following low-momentum renormalization (or model space) approach has been introduced. Namely, one employes a low-momentum model space where all particles have momentum less than a cutoff scale Λ. The corresponding renormalized effective NN interaction is $V_{\text{low-k}}$, which is obtained by integrating out the $k > \Lambda$ momentum components of V_{NN}. This "integrating-out" procedure is carried out by way of a T-matrix equivalence approach. We start from the full-space T-matrix equation

$$T(k', k, k^2) = V_{NN}(k', k) + \mathcal{P}\int_0^\infty q^2 dq \frac{V_{NN}(k', q)T(q, k, k^2)}{k^2 - q^2}, \tag{2}$$

where $\mathcal{P}$ denotes the principal-value integration. Notice that in the above equation the intermediate state momentum q is integrated from 0 to ∞. We then define an effective low-momentum T matrix by

$$T_{\text{low-k}}(p', p, p^2) = V_{\text{low-k}}(p', p) + \mathcal{P}\int_0^\Lambda q^2 dq \times \frac{V_{\text{low-k}}(p', q)T_{\text{low-k}}(q, p, p^2)}{p^2 - q^2}, \tag{3}$$

where the intermediate state momentum is integrated from 0 to Λ, the momentum space cutoff. The low-momentum interaction $V_{\text{low-k}}$ is then obtained from the previous equations by requiring the T-matrix equivalence condition to hold, namely,

$$T(p', p, p^2) = T_{\text{low-k}}(p', p, p^2);\ (p', p) \leqslant \Lambda. \tag{4}$$

Note that T and $T_{\text{low-k}}$ are both half on energy shell, and they are equivalent within Λ. The low-energy ($<\Lambda^2$) phase shifts of V_{NN} are preserved by $V_{\text{low-k}}$ and so is the deuteron binding energy. As we shall discuss later, for neutron star calculations we need to employ a cutoff scale of $\Lambda \sim 3\,\mathrm{fm}^{-1}$.

For many years, the Brueckner-Hartree-Fock (BHF) [15,33–36] and the DBHF [37] methods were the primary framework for nuclear matter calculations. (DBHF is a relativistic generalization of BHF). In both BHF and DBHF the G-matrix interaction is employed. This G-matrix interaction is energy dependent. This energy dependence adds complications to calculations. In contrast, the above $V_{\text{low-k}}$ is energy independent, which facilitates the calculation of nuclear EOS.

We use $V_{\text{low-k}}$ to calculate the EOS of asymmetric nuclear matter of total density n and asymmetric parameter α,

$$n = n_n + n_p;\quad \alpha = \frac{n_n - n_p}{n_n + n_p}, \tag{5}$$

where n_n and n_p denote, respectively, the neutron and proton density and they are related to the respective Fermi momentum by $k_{Fn}^3/(3\pi^2)$ and $k_{Fp}^3/(3\pi^2)$. The proton fraction is $\chi = (1 - \alpha)/2$.

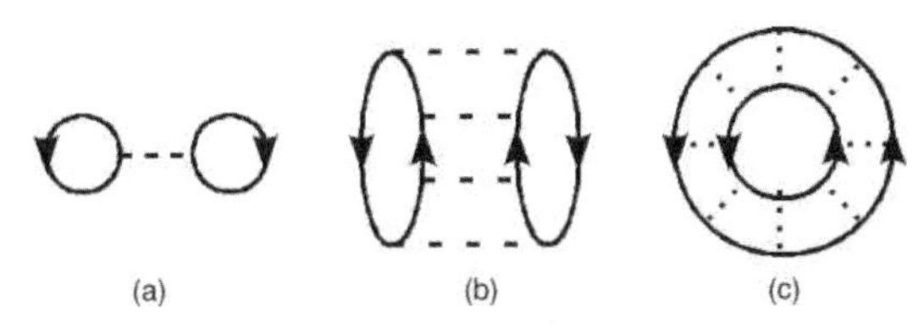

FIG. 1. Diagrams included in the all-order pphh ring-diagram summation for the ground-state energy shift of nuclear matter. Each dashed line represents a $V_{\text{low-k}}$ vertex.

In our ring-diagram EOS calculation, the ground-state energy shift ΔE is given by the all-order sum of the pphh ring diagrams as illustrated in Fig. 1, where (a), (b), and (c) are, respectively, 1st-, 4th-, and 8th-order such diagrams. (ΔE_0 is defined as $(E_0 - E_0^{\text{free}})$, where E_0 is the true ground-state energy and E_0^{free} is that for the noninteracting system.) Following Ref. [22], we have

$$\Delta E_0(n,\alpha) = \int_0^1 d\lambda \sum_m \sum_{ijkl<\Lambda} Y_m(ij,\lambda) \times Y_m^*(kl,\lambda)\langle ij|V_{\text{low-k}}|kl\rangle, \quad (6)$$

where the transition amplitudes are given by the RPA equation

$$\sum_{kl}[(\epsilon_i+\epsilon_j)\delta_{ij,kl} + \lambda(\bar{f}_i\bar{f}_j - f_if_j)\langle ij|V_{\text{low-k}}|kl\rangle]Y_n(kl,\lambda) = \omega_n Y_n(ij,\lambda); \quad (i,j,k,l) \leqslant \Lambda. \quad (7)$$

In the previous equations, the single particle (s.p.) indices $(i,j,\ldots k,l)$ denote both protons and neutrons. The s.p. energies ϵ are the Hartree-Fock energies given by

$$\epsilon_k = \hbar^2k^2/2m + \sum_{h<k_F(h)} \langle kh|V_{\text{low-k}}|kh\rangle, \quad (8)$$

where $k_F(h) = k_{Fn}$ if h is a neutron and $= k_{Fp}$ if it is a proton. Clearly ϵ_k depends on the nuclear matter density n and the asymmetric parameter α. The occupation factors f_i and f_j of Eq. (7) are given by $f_a = 1$ for $k \leqslant k_F(a)$ and $f_a = 0$ for $k > k_F(a)$; also $\bar{f}_a = (1 - f_a)$. Again $k_F(a) = k_{Fn}$ if a is a neutron and $= k_{Fp}$ if it is a proton. The factor $(\bar{f}_i\bar{f}_j - f_if_j)$ is clearly also dependent on n and α. Note that the normalization condition for Y_m in Eq. (6) is $\langle Y_m|\frac{1}{Q}|Y_m\rangle = -1$ and $Q(\vec{k}_i,\vec{k}_j) = (\bar{f}_i\bar{f}_j - f_if_j)$ [38]. In addition, Σ_m in Eq. (6) means we sum over only those solutions of the RPA equation [Eq. (7)] that are dominated by hole-hole components as indicated by the normalization condition. Note that there is a strength parameter λ in the Eqs. (6) and (7), and it is integrated from 0 to 1.

The amplitudes Y in Eq. (6) actually represent the overlap matrix elements

$$Y_m^*(kl,\lambda) = \langle\Psi_m(\lambda, A-2)|a_la_k|\Psi_0(\lambda, A)\rangle, \quad (9)$$

where $\Psi_0(\lambda, A)$ denotes the true ground state of nuclear matter, which has A particles, while $\Psi_m(\lambda, A-2)$ denotes the mth true eigenstate of the $(A-2)$ system. If there is no ground-state correlation (i.e., Ψ_0 is a closed Fermi sea), we have $Y_m^*(kl,\lambda) = f_kf_l$ and Eq. (6) reduces to the HF result. Clearly our EOS includes the effect of the ground-state correlation generated by the all-order sum of pphh ring diagrams.

A computational aspect may be mentioned. We solve ring-diagram equations on the relative and center-of-mass (RCM) coordinates $\vec{k}$ and $\vec{K}$ [39]. [$\vec{k} = (\vec{k}_i - \vec{k}_j)/2$ and $\vec{K} = \vec{k}_i + \vec{k}_j$.] In so doing, the treatment of the Pauli operator $Q(\vec{k}_i,\vec{k}_j) \equiv (\bar{f}_i\bar{f}_j - f_if_j)$ of Eq. (7) plays an important role. This operator is defined in the laboratory frame, with its value being either 1 or -1 (for pphh ring diagrams). In our calculation, however, we need Q in the RCM coordinates. Angle-average approximations are commonly used in nuclear matter calculations, and with them we can obtain angle-averaged $\bar{Q}(k,K)$. For symmetric nuclear matter, detailed expressions for $\bar{Q}$ have been given, see Eqs. (4.9) and (4.9a) of Ref. [39]. The results for this case, where the Fermi surfaces for neutron and proton are equal, are already fairly complicated. Derivation of the angle-averaged Pauli operator for asymmetric nuclear matter has been worked out in Ref. [40]. Their derivation and results are both considerably more complicated than the symmetric case. It would be desirable if they can be simplified. We have found that this can be attained by way of a scale transformation. Namely, we introduce new s.p. neutron and proton momentum coordinates $\vec{k}'_n$ and $\vec{k}'_p$ defined by

$$\vec{k}'_n = \vec{k}_n\sqrt{k_{Fp}/k_{Fn}}; \quad \vec{k}'_p = \vec{k}_p\sqrt{k_{Fn}/k_{Fp}}. \quad (10)$$

On this new frame, the Fermi surfaces for neutron and proton become equivalent, both being $(k_{Fn}k_{Fp})^{1/2}$; asymmetric nuclear matter can then be treated effectively as a symmetric one as far as the Pauli operator is concerned. We have found that this transformation largely simplifies the derivation and calculation of the angle-averaged asymmetric Pauli operator $\bar{Q}$ and is employed in the present work.

III. IN-MEDIUM NN INTERACTIONS BASED ON BROWN-RHO SCALING

A main purpose of the present work is to study whether neutron star properties can be satisfactorily described by EOS microscopically derived from NN interactions. Before proceeding, it is important to also check if such EOS can satisfactorily describe the properties of normal nuclear matter. The EOS and NN interaction one uses for neutron star should also be applicable to normal nuclear matter. As discussed in Ref. [22], many-body calculations for normal nuclear matter using two-body NN interactions alone are generally not capable of reproducing empirical nuclear matter saturation properties. To remedy this shortcoming one needs to consider three-body forces and/or NN interactions with in-medium modifications [22,27,41].

A central result of the BR scaling is that the masses of mesons in nuclear medium are suppressed (dropped) compared to those in free space [23–26]. Nucleon-nucleon interactions are mediated by meson exchange, and clearly in-medium modifications of meson masses can significantly alter the NN interaction. These modifications could arise from the partial restoration of chiral symmetry at finite density/temperature or from traditional many-body effects. Particularly important are the vector mesons, for which there is now evidence from both theory [42–44] and experiment [45,46] that the masses may decrease by approximately 10–15% at normal nuclear matter density and zero temperature. For densities below that of nuclear matter, a linear approximation for the in-medium mass decrease has been suggested [42], namely,

$$\frac{m_V^*}{m_V} = 1 - C\frac{n}{n_0}, \tag{11}$$

where m_V^* is the vector meson mass in-medium, n is the local nuclear matter density, and n_0 is the nuclear matter saturation density. C is a constant of value $\sim$0.10–0.15. BR scaling has been found to be very important for nuclear matter saturation properties in the ring-diagram calculation of symmetric nuclear matter of Ref. [22].

It is of interest that the effect of BR scaling in nuclear matter can be well represented by an empirical Skyrme three-body force [22]. The Skyrme force has been a widely used effective interaction in nuclear physics and it has been very successful in describing the properties of both finite nuclei and nuclear matter [47]. It has both two-body and three-body terms, namely,

$$V_{\text{Skyrme}} = \sum_{i<j} V(i,j) + \sum_{i<j<k} V_{3b}(i,j,k). \tag{12}$$

Here $V(i,j)$ is a momentum-dependent zero-range interaction. Its three-body term is also a zero-range interaction,

$$V_{3b}(i,j,k) = t_3\delta(\vec{r}_i - \vec{r}_j)\delta(\vec{r}_j - \vec{r}_k), \tag{13}$$

which is usually expressed as a density-dependent two-body interaction of the form

$$V_n(1,2) = \tfrac{1}{6}(1 + x_3P_\sigma)t_3\delta(\vec{r}_1 - \vec{r}_2)n(\vec{r}_{\text{av}}), \tag{14}$$

where P_σ is the spin-exchange operator and $\vec{r}_{\text{av}} = \frac{1}{2}(\vec{r}_1 + \vec{r}_2)$. t_3 and x_3 are parameters determined by fitting certain experimental data. The general structure of V_{Skyrme} is rather similar to the effective interactions based on effective field theories (EFT) [41], with $V(i,j)$ corresponding to $V_{\text{low-k}}$ and $V(i,j,k)$ to the EFT three-body force. The Skyrme three-body force, however, is much simpler than that in EFT.

IV. RESULTS AND DISCUSSIONS

A. Symmetric nuclear matter and BR scaling

When an EOS is used to calculate neutron star properties, it is important and perhaps necessary to first test if the EOS can satisfactorily describe the properties of symmetric nuclear matter such as its energy per particle E_0/A and saturation density n_0. In principle, only those EOSs that have done well in this test are suitable for being used in neutron star calculation. In this subsection, we calculate properties of symmetric nuclear matter using the low-momentum ring-diagram EOS that we use in our neutron star calculations, to test if it can meet the aforementioned test. As described in Sec. II, we first calculate the $V_{\text{low-k}}$ interaction for a chosen decimation scale Λ. Then we calculate the ground-state energy per particle E_0/A using Eq. (6) ($E_0 = E_0^{\text{free}} + \Delta E_0$.) with the pphh ring diagrams summed to all orders.

In the aforementioned calculation, the choice of Λ plays an important role. As discussed in Ref. [22], the tensor force is important for nuclear saturation and therefore one should use a sufficiently large Λ so that the tensor force is not integrated out during the derivation of $V_{\text{low-k}}$. Because the main momentum components of the tensor force has $k \sim 2\,\text{fm}^{-1}$, one needs to use $\Lambda \sim 3\,\text{fm}^{-1}$ or larger. There is another consideration concerning the choice of Λ. The density of neutron star interior is very high, several times larger than n_0. To accommodate such high density, it is necessary to use sufficiently large Λ, suggesting a choice of Λ larger than $\sim$3 fm^{-1}. As discussed in Ref. [20], a nice feature of $V_{\text{low-k}}$ is its near uniqueness: The $V_{\text{low-k}}$'s derived from various different realistic NN potentials are practically identical to each other for $\Lambda < \sim$2.1 fm^{-1}, while for larger Λ's the resulting $V_{\text{low-k}}$'s begin to have noticeable differences but are still similar to each other for Λ up to about 3.5 fm^{-1}. This and the aforementioned considerations have led us to choose Λ between $\sim$3 and $\sim$3.5 fm^{-1} for our present study. The dependence of our results on the choice of Λ is discussed later on.

We have carried out $V_{\text{low-k}}$ ring-diagram calculations for symmetric nuclear matter using several NN potentials (CDBonn [12], Nijmegen [13], Argonne V18 [14], and BonnA [15]) with several values of Λ ranging from 3.0 to 3.5 fm^{-1}. In Fig. 2 we present some representative results using CDBonn and BonnA potentials, the results for other potentials and Λ's being very similar. As shown, the results for small densities are nearly independent of Λ within the range considered. But for larger densities, the results have significant variations with Λ and potentials, this being possibly a reflection of the different short-range repulsions contained in the potential models. A common feature of our results is, as displayed in the figure, that the calculated E_0/A and saturation Fermi momentum k_F^0 are all both too high compared with the empirical values ($E_0 \sim -16$ MeV and $k_F^0 \sim 1.35\,\text{fm}^{-1}$ or $n_0 \sim 0.16\,\text{fm}^{-3}$).

As discussed in Sec. III, the aforementioned situation can be largely improved by way of using a V_{NN} with BR scaling. In Ref. [27] authors have carried DBHF calculations for symmetric matter using a BR-scaled BonnB NN potential, and they have obtained results in good agreement with the empirical values, largely improved over those from the unscaled potential. In Ref. [22], ring-diagram EOS calculations for symmetric nuclear matter have been performed using the

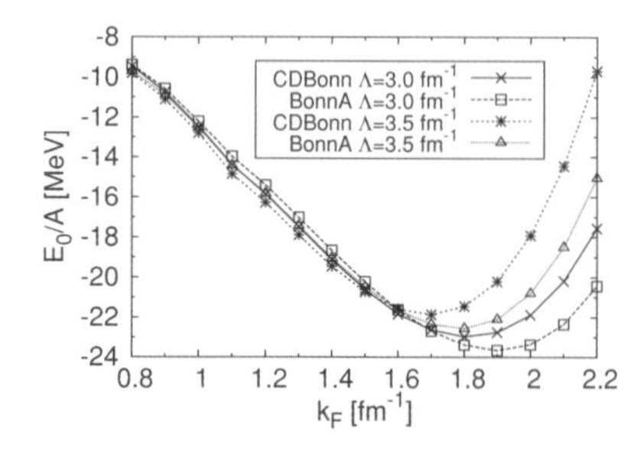

FIG. 2. Ring-diagram EOSs for symmetric nuclear matter with $V_{\text{low-k}}$ derived from CDBonn and BonnA potentials and $\Lambda = 3$ and 3.5 fm^{-1}.

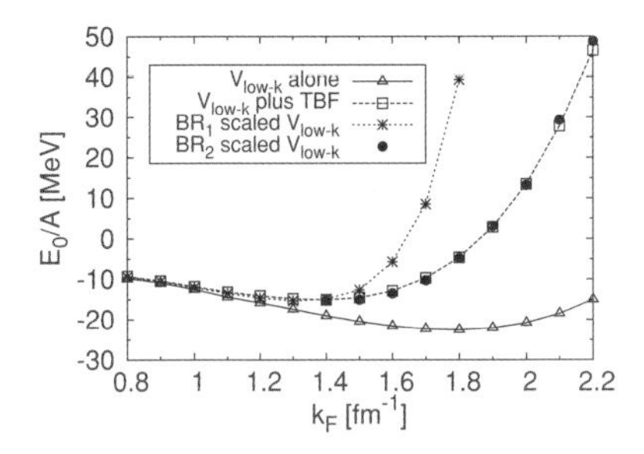

FIG. 3. Ring-diagram EOSs for symmetric nuclear matter given by $V_{\text{low-k}}$ alone, $V_{\text{low-k}}$ with linear (BR_1) and nonlinear (BR_2) scalings, and $V_{\text{low-k}}$ plus the three-body force (TBF) V_{3b} of Eq. (13). $\Lambda = 3.5\,\text{fm}^{-1}$ used for all cases. See text for other explanations.

Nijmegen potential without and with BR scaling, the latter giving highly improved results for nuclear matter saturation. It should be useful if the above effect on nuclear saturation from BR scaling also holds for other NN potentials. To study this, we use in the present work a different potential, the BonnA potential [15], for investigating the effect of BR scaling on ring-diagram calculations for symmetric nuclear matter. In Fig. 3, results of such ring-diagram calculations for symmetric nuclear matter with and without BR scaling are presented. For the scaled calculation, the mesons (ρ, ω, σ) of the BonnA potential are slightly scaled according to Eq. (11) with the choice of $C_\rho = 0.113$, $C_\omega = 0.128$, and $C_\sigma = 0.102$. These values are chosen so that the calculated E_0/A and k_F^0 are in satisfactory agreement with the empirical values. The EOS given by the aforementioned BR-scaled potential is shown by the top curve of Fig. 3 (labeled as "BR_1"), and it has $E_0/A = -15.3\,\text{MeV}$ and $k_F^0 = 1.33\,\text{fm}^{-1}$, in good agreement with the empirical values. In addition, it has compression modulus $\kappa = 225\,\text{MeV}$. The result using $V_{\text{low-k}}$ alone is also shown in Fig. 3 (bottom curve). Clearly BR scaling is also important and helpful for the BonnA potential in reproducing empirical nuclear matter saturation properties.

We shall now discuss if the above effect of BR scaling can be simulated by an empirical three-body force of the Skyrme type. It is generally agreed that the use of two-body force alone cannot satisfactorily describe nuclear saturation; certain three-body forces are needed to ensure nuclear saturation [41]. There are basic similarities between three-body force and BR scaling. To see this, let us consider a meson exchanged between two interacting nucleons. When this meson interacts with a third spectator nucleon, this process contributes to BR scaling or equivalently it generates the three-body interaction. In Ref. [22], it was already found that the ring-diagram results of BR-scaled $V_{\text{low-k}}$ derived from the Nijmegen potential can be well reproduced by the same calculation except for the use of the interaction given by the sum of the $V_{\text{low-k}}$ plus the empirical three-body force (TBF) V_{3b} of Eq. (13). (Note that V_{3b} is calculated using Eq. (14) with n being the local nuclear matter density.) Here we repeat this calculation using a different potential, namely, the BonnA potential. The strength parameter t_3 is adjusted so that the low-density ($<\sim n_0$) EOS given by the ($V_{\text{low-k}} + V_{3b}$) calculation are in good agreement with that from the BR-scaled $V_{\text{low-k}}$. (We fix the parameter x_3 of Eq. (14) as zero, corresponding to treating the $^1S_0$ and $^3S_1$ channels on the same footing.) Results for such a calculation, with t_3 chosen as $2000\,\text{MeV fm}^6$, are presented as the middle curve of Fig. 3 (labeled as "$V_{\text{low-k}}$ plus TBF"). As shown, for $k_F \leqslant \sim 1.4\,\text{fm}^{-1}$ they agree very well with the results from the BR-scaled $V_{\text{low-k}}$. The aforementioned ($V_{\text{low-k}} + V_{3b}$) calculation gives $E_0/A = -14.7\,\text{MeV}$ and $k_F^0 = 1.40\,\text{fm}^{-1}$ in satisfactory agreement with the BR-scaled results given earlier. Its compression modulus is $\kappa = 140\,\text{MeV}$.

It should be noticed, however, for $k_F > \sim 1.4\,\text{fm}^{-1}$ the curve for "BR_1-scaled $V_{\text{low-k}}$" rises much more rapidly (more repulsive) than the "$V_{\text{low-k}}$ plus TBF" one. The compression modulus given by them is also quite different, 225 versus 140 MeV. These differences may be related to the linear BR scaling adopted in Eq. (11). This scaling is to be used for density less than $\sim n_0$. For density significantly larger than n_0, such as in the interior of neutron stars, this linear scaling is clearly not suitable.

To our knowledge, how to scale the mesons at high densities is still an open question [23–26,42]. In the present work, we have considered two schemes for extending the BR scaling to higher densities: One is the above Skyrme-type extrapolation; the other is an empirical modification where in the high-density region a nonlinear scaling is assumed, namely, $m^*/m = [1 - C(n/n_0)^B]$, with B chosen empirically. The exponent B is 1 in the linear BR scaling of Eq. (11). As seen in Fig. 3, the linear BR-scaled EOS agrees well with the "$V_{\text{low-k}}$ plus TBF" EOS only in the low-density ($<\sim n_0$) region, but not so for densities beyond. Can a different choice of B give better agreement for the high-density region? As seen in Fig. 3, to obtain such better agreements we need to use a scaling with weaker density dependence than that of BR_1. Thus we have considered $B < 1$ and have found that the EOSs with B near $1/3$ have much improved agreements with the Skyrme EOS in the high-density region. To illustrate, we have repeated the "BR_1" EOS of Fig. 3 with only one change, namely, changing B from 1 to 0.3. (The scaling parameters C are not changed, for convenience of comparison.) The new results, labeled "BR_2," are also presented in Fig. 3. As seen, "BR_2" and "$V_{\text{low-k}}$ plus TBF" are nearly identical in a wide range of densities beyond $\sim n_0$. This is an interesting result, indicating that below $\sim n_0$ the "$V_{\text{low-k}}$ plus TBF" EOS corresponds to the linear BR_1-scaled EOS, whereas beyond $\sim n_0$ it corresponds to the nonlinear BR_2 one. The BR_1 and BR_2 EOSs have a small discontinuity (in slope) at n_0, and the aforementioned EOS with TBF is practically a continuous EOS with good fitting to both. As we discuss later, the above three-body force is also important and desirable for neutron star calculations involving much higher densities. Possible microscopic connections between the Skyrme three-body force and BR scalings are being further studied, and we hope to report our results soon in a separate publication.

The ring-diagram nuclear matter EOSs using the "$V_{\text{low-k}}$ plus TBF" interaction are in fact rather insensitive to the choice Λ. As discussed earlier, a suitable range for Λ is from ~ 3 to ~ 3.5 fm. So in carrying out the previous calculations,

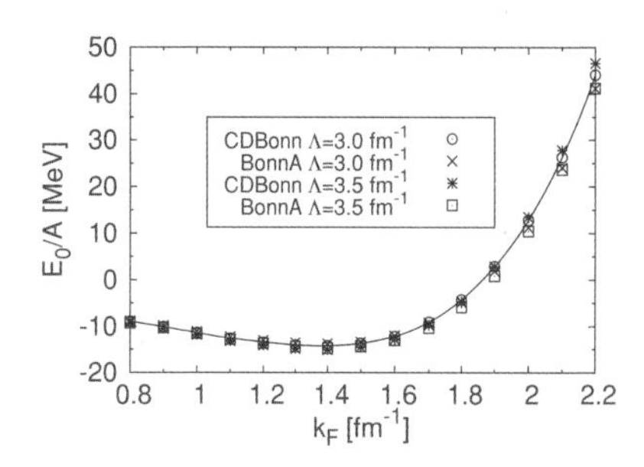

FIG. 4. Ring-diagram EOS for symmetric nuclear matter with the interaction being the sum of $V_{\text{low-k}}$ and the three-body force of Eq. (13). Four sets of results are shown for CDBonn and BonnA potentials with $\Lambda = 3$ and $3.5\,\text{fm}^{-1}$. A common three-body force of $t_3 = 2000$ MeV fm^6 is employed.

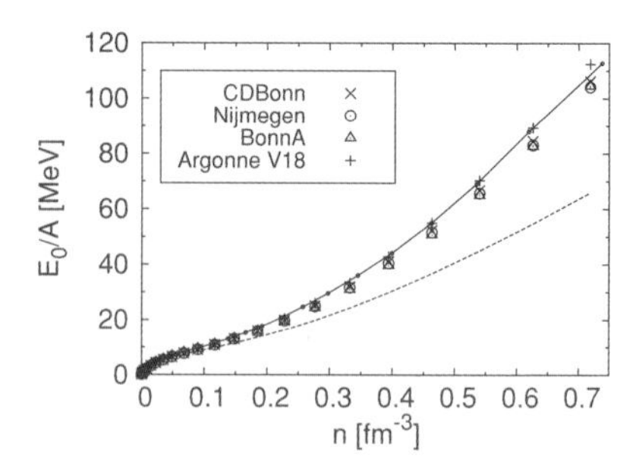

FIG. 5. Ring-diagram neutron matter EOS obtained from four realistic NN potentials. The interaction "$V_{\text{low-k}}$ plus TBF" is used. The solid line with small open circles represents the results from the variational many-body calculation of Friedman-Pandharipande [48]. The dotted line denotes the EOS using CDBonn-$V_{\text{low-k}}$ only (TBF suppressed).

one first chooses a Λ within the aforementioned range. Then t_3 is determined by the requirement that the low-density ($<\sim n_0$) ring-diagram EOS given by BR$_1$-scaled $V_{\text{low-k}}$ is reproduced by that from ($V_{\text{low-k}} + V_{3b}$). In Fig. 4 we present some sample results for $\Lambda = 3$ and $3.5\,\text{fm}^{-1}$ with CDBonn and BonnA potentials, all using $t_3 = 2000$ MeV fm^6. Note that this t_3 value is for $\Lambda = 3.5\,\text{fm}^{-1}$ and the BonnA potential; for convenience in comparison it is here used also for the other three cases. It is encouraging to see that within the aforementioned Λ range our results are remarkably stable with regard to the choice of both Λ and t_3. The four curves of Fig. 4 are nearly overlapping, and their $(E_0/A, k_F^0, \kappa)$ values are all close to (-15 MeV, $1.40\,\text{fm}^{-1}$, 150 MeV). We have repeated the aforementioned calculations for the Nijmegen and Argonne V18 potentials and have obtained highly similar results. As we discuss in the next section, the inclusion of V_{3b} is also important in giving a satisfactory neutron-matter ring-diagram EOS. Calculations of neutron star properties using the aforementioned ($V_{\text{low-k}} + V_{3b}$) interaction are also presented there. Unless otherwise specified, we use from now on $\Lambda = 3.5\,\text{fm}^{-1}$ for the decimation scale and $t_3 = 2000$ MeV fm^6 for the three-body force V_{3b}.

B. Neutron star with neutrons only

As a preliminary test of our ring-diagram EOS, in this section we consider neutron stars as composed of pure-neutron matter only. This simplified structure is convenient for us to describe our methods of calculation. In addition, this also enable us to check how well the properties of neutron stars can be described under the pure-neutron-matter assumption. Realistic neutron stars have of course more complicated compositions; they have nuclei crust and their interior composed of neutrons as well as other elementary particles [1,9]. We study the effects of using β-stable and nuclei-crust EOSs in our neutron star calculations in the next section. In the present work we consider neutron stars at zero temperature.

Using the methods outlined in Sec. II, we first calculate the ground-state energy per particle E_0/A for neutron matter. Then the energy density ε, inclusive of the rest-mass energy, is obtained as

$$\varepsilon(n) = n\left(\frac{E_0}{A} + m_n c^2\right), \tag{15}$$

where c is the speed of light and m_n the nucleon mass. By differentiating E_0/A with density, we obtain the pressure-density relation

$$p(n) = n^2 \frac{d(E_0/A)}{dn}. \tag{16}$$

From the previous two results, the EOS $\varepsilon(p)$ is obtained. It is the EOS $\varepsilon(p)$ that is used in the solution of the TOV equations.

To accommodate the high densities in the interior of neutron stars, we have chosen $\Lambda = 3.5\,\text{fm}^{-1}$ for our present neutron star calculation. Our ring-diagram EOS for neutron matter is then calculated using the interaction ($V_{\text{low-k}} + V_{3b}$) with the parameter $t_3 = 2000$ MeV fm^6. Note this value was determined for symmetric nuclear matter, as discussed in Sec. IV A. Is this t_3 also appropriate for the neutron matter EOS? We address this question here. In Fig. 5 we present results from the previous neutron matter EOS calculations for four interactions (CDBonn, Nijmegen, Argonne V18, BonnA). It is seen that the EOSs given by them are quite close to each other, giving a nearly unique neutron-matter EOS. Friedman and Pandharipande (FP) [48] have carried out variational many-body calculations for neutron matter EOS using the two- and three-nucleon interactions V_{14} and TNI, respectively; their EOS results also shown in Fig. 5. Brown [49] has carried out extensive studies of neutron matter EOS and has found that the FP EOS can be reproduced by the EOS given by certain empirical Skyrme effective interactions (with both two- and three-body parts). As seen in Fig 5, our results agree with the FP EOS impressively well. For comparison, we present in Fig. 5 also the CDBonn EOS without the inclusion of V_{3b} (i.e., $t_3 = 0$). It is represented by the dotted-line and is much lower than the FP EOS, particularly at high densities. For $n < \sim n_0/2$ the effect of V_{3b} is rather small, and in this density range one may calculate the EOS using $V_{\text{low-k}}$ alone. Clearly the inclusion of V_{3b} with $t_3 = 2000$ MeV fm^6 is essential for attaining the

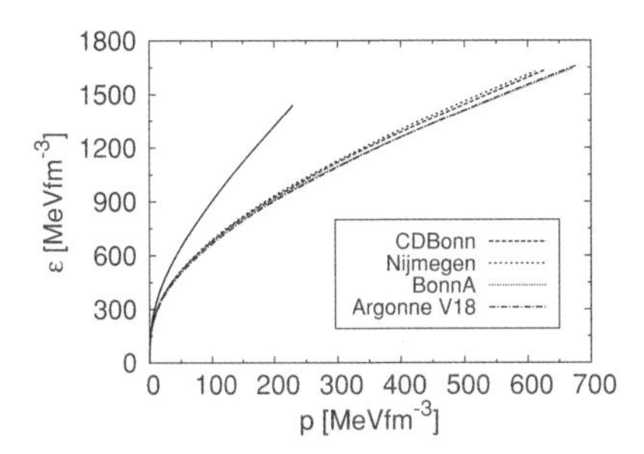

FIG. 6. Neutron matter $\varepsilon(p)$ obtained from four realistic NN potentials. The upper-left thin line denotes the $\varepsilon(p)$ from CDBonn-$V_{\text{low-k}}$ only (TBF suppressed).

above good agreement between our EOSs and the FP one. It is of interest that the t_3 value determined for symmetric nuclear matter turns out to be also appropriate for neutron matter.

In Fig. 6, our results for the EOS $\varepsilon(p)$ are presented, where the EOSs given by various potentials are remarkably close to each other. The inclusion of V_{3b} is found to be also important here. As also shown in Fig. 6, the EOS given by $V_{\text{low-k}}$ alone (without V_{3b}) lies considerably higher than those with V_{3b}. It is of interest that, for a given pressure, the inclusion of V_{3b} has a large effect in reducing the energy density. We have chosen to use $\Lambda = 3.5\,\text{fm}^{-1}$, and this limits the highest pressure $p_{\Lambda,}$ which can be provided by our ring-diagram EOS calculation. As shown in the figure, the highest pressure there is about $650\,\text{MeV/fm}^3$. But in neutron star calculation we need EOS at higher pressure such as $1000\,\text{MeV/fm}^3$ (or $\sim 4 \times 10^{-4} M_\odot\, c^2/\text{km}^3$). The EOS at such high pressure is indeed uncertain, and some model EOS has to be employed. In the present work we adopt a polytrope approach; namely, we fit a section of the calculated EOS near the maximum-pressure end by a polytrope $\varepsilon(p) = \alpha p^\gamma$ and use this polytrope to determine the energy density for pressure beyond p_Λ. [In our fitting the section is chosen as ($\sim$0.8 to 1)p_Λ.] The polytrope EOS has been widely and successfully used in neutron star calculations [11,50]. In fact we have found that our calculated EOS, especially the section near its high-pressure end, can be very accurately fitted by a polytrope. In Table I we list the polytropes obtained from the aforementioned fitting for four NN interactions. It is seen that the four polytropes are close to each other. The exponent γ plays an important role in determining the neutron-star maximum mass.

TABLE I. Fitted polytrope αp^γ for high-pressure region. See text for other explanations.

Potentials	α[a]	γ
CDBonn	69.69 ± 1.01	0.4876 ± 0.0022
Nijmegen	69.99 ± 1.01	0.4885 ± 0.0021
BonnA	72.30 ± 1.01	0.4779 ± 0.0021
Argonne V18	67.71 ± 1.01	0.4887 ± 0.0021

[a]Unit of α is $(\text{MeV/fm}^3)^{1-\gamma}$.

TABLE II. Neutron stars with different center pressures.

P_c ($M_\odot\, c^2/\text{km}^3$)	M ($M_\odot$)	R (km)	I ($M_\odot$ km^2)
8.07×10^{-7}	0.101	11.58	3.78
7.18×10^{-6}	0.347	10.12	13.51
5.38×10^{-5}	1.037	10.10	50.02
2.33×10^{-4}	1.597	10.00	70.69

In obtaining the neutron star properties, we numerically solve the TOV equations [Eq. (1)] by successive integrations. In so doing, we need to have the pressure P_c at the center of the neutron star to begin the integration. As we shall see soon, different P_c's will give, for example, different masses for neutron stars. We also need the EOS $\varepsilon(p)$ for a wide range of pressure. As discussed earlier, we shall use the ring-diagram EOS for pressure less than p_Λ and the fitted polytrope EOS for larger pressure. In Table II, we list some typical results for the neutron star mass M and its corresponding radius R and static moment of inertia I. (The calculation of I is discussed later.) They were obtained with four different center pressures P_c and as seen these properties of the neutron star vary significantly with P_c.

We present some of our calculated results for the mass-radius trajectories of neutron stars in Fig. 7. They were obtained using the CDBonn $V_{\text{low-k}}$ ($\Lambda = 3.5\,\text{fm}^{-1}$) with and without the three-body force V_{3b} ($t_3 = 2000$ MeV fm^6) discussed earlier. As seen, the inclusion of V_{3b} significantly increases both the maximum neutron star mass M and its corresponding radius R; the former increased from $\sim 1.2 M_\odot$ to $\sim 1.8 M_\odot$ and the latter from $\sim$7 to $\sim$9 km. The aforementioned results are understandable, because V_{3b} makes the EOS stiffer and consequently enhances both M and R. Note that our results are within the causality limit. We have repeated the above calculations using the Nijmegen, Argonne, and BonnA potentials, with results quite similar to the CDBonn ones. In Fig. 8, we present the density profiles corresponding to the maximum-mass neutron stars of Fig. 7. It is clearly seen that the inclusion of the three-body force TBF has an important

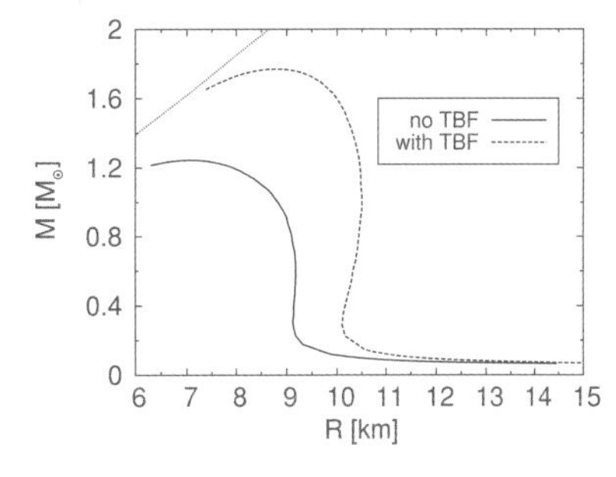

FIG. 7. Mass-radius trajectories of pure neutron stars from ring-diagram EOSs given by the CD-Bonn $V_{\text{low-k}}$ interaction with and without the three-body force (TBF) V_{3b}. Only stars to the right of maximum mass are stable against gravitational collapse. The causality limit is indicated by the straight line in the upper left corner.

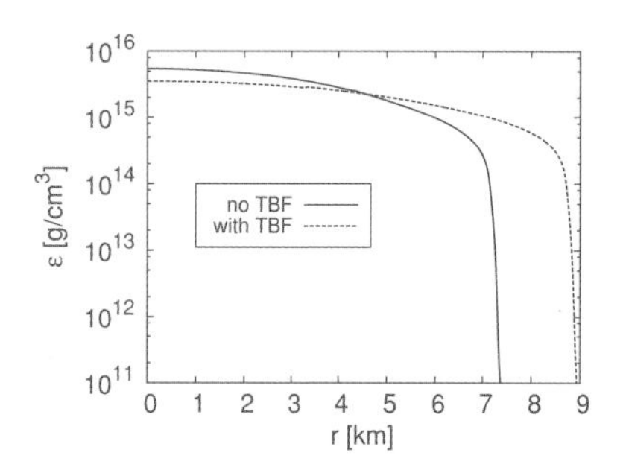

FIG. 8. Density profiles of maximum-mass neutron stars of Fig. 7.

effect on the neutron star's density distribution, reducing the central density and enhancing the outer one.

We have also performed calculations using the BR_1-scaled $V_{\text{low-k}}$ interaction (BonnA and $\Lambda = 3.5\,\text{fm}^{-1}$) without V_{3b}. The resulting maximum mass and its radius given are, respectively, $\sim 3.2 M_\odot$ and ~ 12 km, both being considerably larger than the values of Fig. 7. This is also reasonable, because, as was shown in Fig. 3, the BR_1-scaled EOS is much stiffer than the "$V_{\text{low-k}}$ plus TBF" one. It may be mentioned that, if the neutron matter EOS given by the BR-scaled interaction is plotted in Fig. 5, it would be very much higher, especially in the high-density region, than the FP EOS shown there. However, the "$V_{\text{low-k}}$ plus TBF" ones are very close to the FP one as shown earlier. We feel that the aforementioned comparison is a further indication that the linear BR_1 scaling of Eq. (11) is not suitable for high density. It is suitable only for density up to about $\sim n_0$.

Moment of inertia is an important property of neutron stars [51,52]. Here we would like to calculate this quantity using our $V_{\text{low-k}}$ ring-diagram formalism. Recall that we have used the TOV equations (1) to calculate neutron star mass and radius, and in so doing we also obtain the density distribution inside neutron stars. From this distribution, the moment of inertia I of neutron stars is readily calculated. It may be noted that the TOV equations are for spherical and static (nonrotating) neutron stars, and the I so obtained is the static one for spherical neutron stars. The moments of inertia for rotating stars are more complicated to calculate, but for low rotational frequencies (less than $\sim$300 Hz) they are rather close to the static ones [52]. In Fig. 8, we present our results for two calculations, the interactions used being the same as those in Fig. 7. It is seen that the the inclusion of our three-body force V_{3b} (TBF) largely enhances the moment of inertia of a maximum-mass neutron star.

The measurement of a neutron-star moment of inertia is still rather uncertain, and the best determined value so far is that of the Crab pulsar ($97 \pm 38 M_\odot\,\text{km}^2$) [53]. For $M \geqslant 1.0 M_\odot$, Lattimer and Schutz [51] have determined an empirical formula relating the moment of inertia I of neutron stars to their mass M and radius R, namely,

$$I \approx (0.237 \pm 0.008) M R^2 \times \left[1 + 4.2 \frac{M}{M_\odot}\frac{\text{km}}{R} + 90\left(\frac{M}{M_\odot}\frac{\text{km}}{R}\right)^4\right]. \quad (17)$$

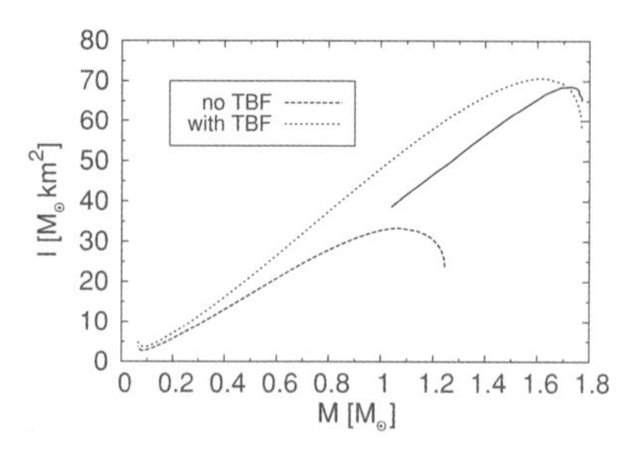

FIG. 9. Pure neutron stars' moments of inertia calculated from CDBonn potential with and without three-body force (TBF). Results from the empirical formula (17) for $M \geqslant 1.0 M_\odot$ with TBF are denoted by the solid line.

To check if our calculated (M, R, I) are consistent with this empirical relation, we have computed I using our calculated M and R values (with TBF as in the top curve of Fig. 9) as inputs to Eq. (17). Results of this computation are also shown in Fig. 9. As shown, they are in good general agreement with the empirical formula. Especially our moment of inertia at maximum mass agrees remarkably well with the corresponding empirical value. We have also repeated the aforementioned computation with other potentials and obtained similar results.

C. Effects from β-stable and nuclei-crust EOSs

In the preceding section, we considered neutron stars as composed of neutrons only, and we have obtained rather satisfactory results. Would the quality of them be significantly changed when we use a more realistic composition? As a small-step improvement, in this section we first carry out calculations using the ring-diagram β-stable EOS composed of neutrons, protons, electrons, and muons only. The results of them are briefly compared with those obtained with neutrons only. Calculations using a combination of the BPS EOS [32] inside the nuclei crust and our β-stable EOS for the interior are also carried out. The crust of the neutron star is composed of two parts, the outer crust and the inner crust [32,54–57]. The choice of the density regions defining these crusts and how to match the EOSs at the boundaries between different regions are discussed.

Let us first discuss our β-stable EOS, where the composition fractions of its constituents are determined by the chemical equilibrium equations

$$\mu_n = \mu_p + \mu_e, \quad (18)$$

$$\mu_e = \mu_\mu, \quad (19)$$

together with the charge and mass conservation conditions

$$n_p = n_e + n_\mu, \quad (20)$$

$$n = n_n + n_p + n_e + n_\mu. \quad (21)$$

Here, μ_n, μ_p, μ_e, and μ_μ are the chemical potentials for neutron, proton, electron, and muon, respectively, and their densities are, respectively, n_n, n_p, n_e and n_μ. The total

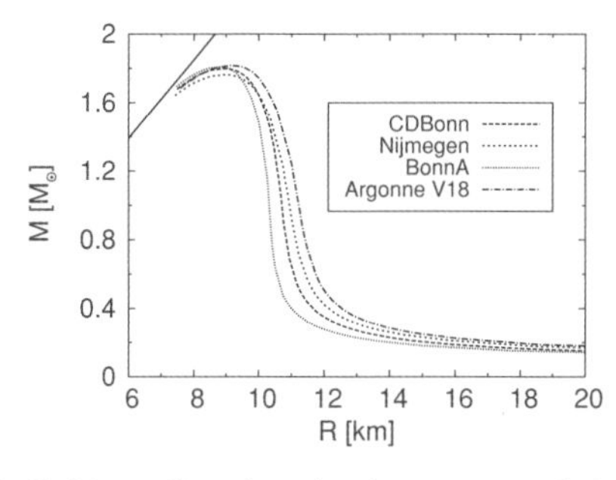

FIG. 12. Mass-radius trajectories of neutron stars calculated with a combination of β-stable ring-diagram EOS for the core and the nuclei-crust EOSs for the crusts. Ring-diagram EOSs given by four NN potentials, all with the V_{3b} three-body force; $\Lambda = 3.5\,\text{fm}^{-1}$ and $t_3 = 2000$ MeV fm^6 are used. $n_t = 0.04\,\text{fm}^{-3}$ is used for the inner-crust boundary. See the legend of Fig. 7 for other explanations.

EOS will be employed. The EOS in the inner-crust region is somewhat uncertain, and so is the transition density separating the inner crust and core. We use in our calculations $n_t = 0.05$ and $0.04\,\text{fm}^{-3}$ [54,55] to illustrate the effect of the nuclei crust. Following Refs. [54,57], we use in the inner-crust region a polytropic EOS, namely, $p = a + b\varepsilon^{4/3}$, with the constants a and b determined by requiring a continuous matching of the three EOSs at n_{out} and n_t.

In Fig. 11, our results for the mass-radius trajectories using the previous three EOSs with $n_t = 0.04$ and $0.05\,\text{fm}^{-3}$, labeled β-crust$_1$ and β-crust$_2$ respectfully, are compared with the trajectory given by the β-stable alone (namely, $n_t = 0$). As seen, the effect from the nuclei-crust EOS on the maximum neutron-star mass and its radius is rather small, merely increasing the maximum mass by $\sim 0.02 M_\odot$ and its radius ~ 0.1 km as compared with the β-alone ones. However, its effect is important in the low-mass large-radius region, significantly enhancing the neutron-star mass there. That the maximum mass is not significantly changed by the inclusion nuclei-crust EOS is consistent with Fig. 8, which indicates the mass of maximum-mass neutron stars being confined predominantly in the core region. It may be mentioned that our ring-diagram EOS is microscopically calculated from realistic NN interactions, whereas the crust EOSs are not. So there are disparities between them. It would be useful and of much interest if the crust EOSs can also be derived from realistic

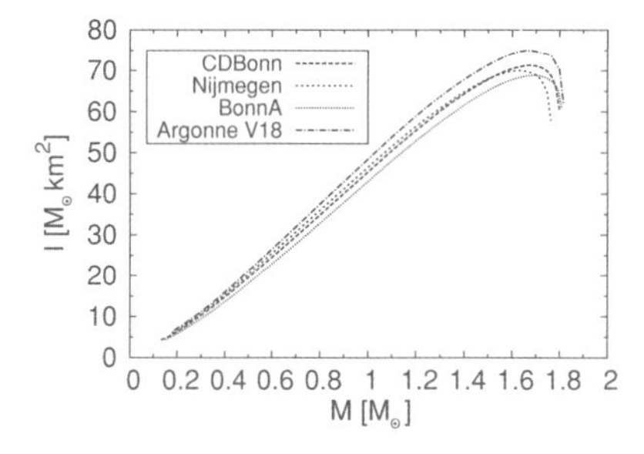

FIG. 13. Moments of inertia of neutron stars of Fig. 12.

TABLE III. Maximum mass and the corresponding radius and moment of inertia of β-stable neutron stars with nuclei-crust boundary $n_t = 0.04$ fm^{-3}. The three-body force V_{3b} is included for the results in the first four rows, but is not in the last.

Potentials	M ($M_\odot$)	R (km)	I ($M_\odot$ km^2)
CDBonn	1.80	8.94	60.51
Nijmegen	1.76	8.92	57.84
BonnA	1.81	8.86	61.09
Argonne V18	1.82	9.10	62.10
CDBonn ($V_{3b} = 0$)	1.24	7.26	24.30

NN interactions using similar microscopic methods. Further studies in this direction are needed.

In. Fig. 12 we present our mass-radius results using the above three EOSs with $n_t = 0.04\,\text{fm}^{-3}$. Four NN potentials are employed, and they give similar trajectories, especially in the high- and low-mass regions. A corresponding comparison for the moment of inertia is presented in Fig. 13; again the results from the four potentials are similar. In Table III, our results for the maximum neutron mass and its radius and moment of inertia using the aforementioned combined EOSs are presented, and as seen the results for the maximum-mass neutron star given by the four potentials are indeed close to each other. It is also seen that the effect of the three-body force is quite important for M, R, and I, as illustrated by the CDBonn case.

V. SUMMARY AND CONCLUSION

We have performed neutron-star calculations based on three types of EOSs: the pure-neutron ring-diagram, the β-stable (n, p, e, μ) ring-diagram, and the BPS nuclei-crust EOS. The ring-diagram EOSs, where the pphh ring diagrams are summed to all orders, are microscopically derived using the low-momentum interaction $V_{\text{low-k}}$ obtained from four realistic NN potentials (CDbonn, Nijmegen, Argonne V18, BonnA). We require that the EOS used for neutron stars should give satisfactory saturation properties for symmetric nuclear matter, but this requirement is not met by our calculations using the aforementioned potentials as they are. Satisfactory nuclear matter saturation properties can be attained by using the aforementioned potentials with the commonly used linear BR scaling (BR$_1$) where the masses of in-medium mesons are slightly suppressed compared with their masses in vacuum. But this linear scaling is not suitable for neutron stars; the maximum mass of a neutron star given by our BR$_1$ ring-diagram calculation is $\sim 3.2 M_\odot$, which is not satisfactory. BR$_1$ is suitable only for low densities; it needs some extension so that it can be applied to the high densities inside the neutron star. We have used an extrapolation method for this extension, namely, we add an empirical Skyrme-type three-body force V_{3b} to $V_{\text{low-k}}$. We have found that the EOS given by this extrapolation agrees well with the EOS obtained from linear BR$_1$ scaling for low densities, but for high densities it agrees well with that from a nonlinear BR$_2$ scaling. The EOS using the aforementioned extrapolation gives satisfactory saturation properties for symmetric nuclear matter, and for neutron matter it agrees well with the FP EOS for neutron matter.

The effects from V_{3b} have been found to be both important and desirable. Compared with the results given by the unscaled $V_{\text{low-k}}$, it increases the maximum mass of the neutron star and its radius and moment of inertia by $\sim$40%, $\sim$20%, and $\sim$150%, respectively. The proton fractions are found to be generally small ($<7\%$), making our neutron star results using the pure-neutron EOS and those using the β-stable EOS being nearly the same. We have estimated the effect from the nuclei-crust EOSs by using a combination of three EOSs: the BPS EOS for the outer crust, a fitted polytropic EOS for the inner crust, and our β-stable ring-diagram EOS for the core region. The effect from the nuclei-crust EOSs on the maximum neutron-star mass and its radius is found to be rather small, as compared with those given by the calculation where the β-stable EOS is used throughout. However, its effect is important in the low-mass large-radius region, significantly enhancing the neutron-star mass there. Using the aforementioned combined three EOSs, our results for neutron star's maximum mass and its radius and moment of inertia are, respectively, $\sim 1.8 M_{\odot}$, $\sim$9 km, and $\sim 60 M_{\odot}\,\text{km}^2$, all in good agreement with accepted values.

How to extend the BR scaling to high densities is still an open question. Although we have obtained satisfactory results by using a nonlinear scaling for the high-density region, or equivalently a Skyrme-type three-body force, for the extension, it would still be certainly useful and interesting to explore other ways for doing so. Further studies in this direction would be very helpful in determining the medium-dependent modifications to the NN potentials in the high-density region.

ACKNOWLEDGMENTS

We thank G. E. Brown, Edward Shuryak, Izmail Zahed, and Shu Lin for many helpful discussions. This work is supported in part by the US Department of Energy under Grant DF-FG02-88ER40388 and by the National Science Foundation under Grant PHY-0099444.

[1] J. M. Lattimer and M. Prakash, Phys. Rep. **442**, 109 (2007).
[2] M. Prakash, J. M. Lattimer, J. A. Pons, A. W. Steiner, and S. Reddy, Lect. Notes Phys. **578**, 364 (2001).
[3] A. Sedrakian, Prog. Part. Nucl. Phys. **58**, 168 (2007).
[4] F. Weber, *Pulsars as Astrophysical Laboratories for Nuclear and Particle Physics* (IOP publishing, Bristol, Great Britain, 1999).
[5] S. L. Shapiro and S. A. Teukolsky, *Black Holes, White Dwarfs and Neutron Stars: The Physics of Compact Objects* (Wiley, New York, 1983).
[6] P. Danielewicz, R. Lacey, and W. A. Lynch, Science **298**, 1592 (2002).
[7] B.-A. Li, L.-W. Chen, and C. M. Ko, Phys. Rep. **464**, 113 (2008).
[8] M. B. Tsang, Y. Zhang, P. Danielewicz, M. Famiano, Z. Li, W. G. Lynch, and A. W. Steiner, Phys. Rev. Lett. **102**, 122701 (2009).
[9] H. Heiselberg and M. Hjorth-Jensen, Phys. Rep. **328**, 237 (2000).
[10] F. Sammarruca and P. Liu, arXiv:0906.0320 [nucl-th].
[11] R. R. Silbar and S. Reddy, Am. J. Phys. **72**, 892 (2004).
[12] R. Machleidt, Phys. Rev. C **63**, 024001 (2001).
[13] V. G. J. Stoks, R. A. M. Klomp, C. P. F. Terheggen, and J. J. de Swart, Phys. Rev. C **49**, 2950 (1994).
[14] R. B. Wiringa, V. G. J. Stoks, and R. Schiavilla, Phys. Rev. C **51**, 38 (1995).
[15] R. Machleidt, Adv. Nucl. Phys. **19**, 189 (1989).
[16] S. K. Bogner, T. T. S. Kuo, and L. Coraggio, Nucl. Phys. **A684**, 432 (2001).
[17] S. K. Bogner, T. T. S. Kuo, L. Coraggio, A. Covello, and N. Itaco, Phys. Rev. C **65**, 051301(R) (2002).
[18] L. Coraggio, A. Covello, A. Gargano, N. Itaco, T. T. S. Kuo, D. R. Entem, and R. Machleidt, Phys. Rev. C **66**, 021303(R) (2002).
[19] A. Schwenk, G. E. Brown, and B. Friman, Nucl. Phys. **A703**, 745 (2002).
[20] S. K. Bogner, T. T. S. Kuo, and A. Schwenk, Phys. Rep. **386**, 1 (2003).
[21] J. D. Holt, T. T. S. Kuo, and G. E. Brown, Phys. Rev. C **69**, 034329 (2004).
[22] L. W. Siu, J. W. Holt, T. T. S. Kuo, and G. E. Brown, Phys. Rev. C **79**, 054004 (2009).
[23] G. E. Brown and M. Rho, Phys. Rev. Lett. **66**, 2720 (1991).
[24] G. E. Brown and M. Rho, Phys. Rep. **269**, 333 (1996).
[25] G. E. Brown and M. Rho, Phys. Rep. **363**, 85 (2002).
[26] G. E. Brown and M. Rho, Phys. Rep. **396**, 1 (2004).
[27] R. Rapp, R. Machleidt, J. W. Durso, and G. E. Brown, Phys. Rev. Lett. **82**, 1827 (1999).
[28] D. Alonso and F. Sammarruca, Phys. Rev. C **68**, 054305 (2003).
[29] G. E. Brown, W. Weise, G. Baym, and J. Speth, Comments Nucl. Part. Phys. **17**, 39 (1987).
[30] J. W. Holt, G. E. Brown, T. T. S. Kuo, J. D. Holt, and R. Machleidt, Phys. Rev. Lett. **100**, 062501 (2008).
[31] L. W. Siu, T. T. S. Kuo, and R. Machleidt, Phys. Rev. C **77**, 034001 (2008).
[32] G. Baym, C. Pethick, and P. Sutherland, Astrophys. J. **170**, 299 (1971).
[33] H. A. Bethe, Annu. Rev. Nucl. Sci. **21**, 93 (1971).
[34] J. W. Holt and G. E. Brown, "Hans Bethe and the Nuclear Many-Body Problem" in *Hans Bethe and His Physics*, edited by G. E. Brown and C.-H. Lee (World Scientific, Singapore, 2006).
[35] H. Q. Song, M. Baldo, G. Giansiracusa, and U. Lombardo, Phys. Lett. **B411**, 237 (1997).
[36] H. Q. Song, M. Baldo, G. Giansiracusa, and U. Lombardo, Phys. Rev. Lett. **81**, 1584 (1998).
[37] R. Brockmann and R. Machleidt, Phys. Rev. C **42**, 1965 (1990).
[38] T. T. S. Kuo and Y. Tzeng, Int. J. Mod. Phys. E **3**, 523 (1994).
[39] H. Q. Song, S. D. Yang, and T. T. S. Kuo, Nucl. Phys. **A462**, 491 (1987).
[40] H. Q. Song, Z. X. Wang, and T. T. S. Kuo, Phys. Rev. C **46**, 1788 (1992).
[41] S. K. Bogner, A. Schwenk, R. J. Furnstahl, and A. Nogga, Nucl. Phys. **A763**, 59 (2005).
[42] T. Hatsuda and S. H. Lee, Phys. Rev. C **46**, R34 (1992).
[43] M. Harada and K. Yamawaki, Phys. Rep. **381**, 1 (2003).
[44] F. Klingl, N. Kaiser, and W. Weise, Nucl. Phys. **A624**, 527 (1997).
[45] D. Trnka *et al.*, Phys. Rev. Lett. **94**, 192303 (2005).
[46] M. Naruki *et al.*, Phys. Rev. Lett. **96**, 092301 (2006).

[47] P. Ring and P. Schuck, *The Nuclear Many-Body Problem* (Springer-Verlag, New York, 1980).
[48] B. Friedman and V. R. Pandharipande, Nucl. Phys. **A361**, 502 (1981).
[49] B. A. Brown, Phys. Rev. Lett. **85**, 5296 (2000).
[50] J. Cooperstein, Phys. Rev. C **37**, 786 (1988).
[51] J. M. Lattimer and B. F. Schutz, Astrophys. J. **629**, 979 (2005).
[52] A. Worley, P. G. Krastev, and B.-A. Li, Astrophys. J. **685**, 390 (2008).
[53] M. Bejger and P. Haensel, Astron. Astrophys. **405**, 747 (2003).
[54] J. Xu, L.-W. Chen, B.-A. Li, and H.-R. Ma, Phys. Rev. C **79**, 035802 (2009).
[55] A. W. Steiner, Phys. Rev. C **77**, 035805 (2008).
[56] S. B. Ruster, M. Hempel, and J. Schaffner-Bielich, Phys. Rev. C **73**, 035804 (2006).
[57] J. Carriere, C. J. Horowitz, and J. Piekarewicz, Astrophys. J. **593**, 463 (2003).
[58] A. Szmagliński, W. Wójcik, and M. Kutschera, Acta Phys. Pol. B **37**, 277 (2006).